Commercial Chicken Production Manual

Commercial Chicken Production Manual

Fourth Edition

MACK O. NORTH

Poultry Management Consultant
Oceanside, California

DONALD D. BELL

Poultry Specialist
University of California
Riverside, California

An avi Book
Published by Van Nostrand Reinhold
New York

In *Commercial Chicken Production Manual,* the names of many medicinal products appear, often with the trade names, which may vary from country to country. But nothing contained herein is to be construed as an endorsement for a named product, either by trade or chemical name, nor is criticism of similar products implied when not mentioned.

Products such as pesticides, rodenticides, disinfectants, drugs, antibiotics, and vaccines are usually licensed for use. These licenses are specific for each country or state, require verification of effectiveness and safety, and must be accompanied by detailed labels explaining the application, dosage, species, safety precautions, and limitations associated with the material. The user is responsible for adhering to these instructions and to use such products only as advised.

This book cannot include all the products licensed for sale, nor can it give the many benefits or limitations associated with their use. Always check with local agricultural authorities regarding the legal use of any of these medicaments.

Often, materials that are effective in one region may fail in others. Consult with local university advisors before using any such products.

An AVI Book
(AVI is an imprint
  of Van Nostrand Reinhold)

Copyright © 1990
  by Van Nostrand Reinhold
Library of Congress
  Catalog Card Number 89-37826
ISBN 0-442-31882-2

Van Nostrand Reinhold
115 Fifth Avenue
New York, New York 10003

Van Nostrand Reinhold
  International Company Limited
11 New Fetter Lane
London EC4P 4EE, England

Van Nostrand Reinhold
480 La Trobe Street
Melbourne, Victoria 3000,
  Australia

Nelson Canada
1120 Birchmont Road
Scarborough, Ontario
Canada M1K 5G4

Printed in the United States of America

16 15 14 13 12 11 10 9 8 7 6 5 4 3 2 1

Library of Congress Cataloging-in-Publication Data
North, Mack O.
    Commercial chicken production manual / Mack
O. North and Donald D. Bell. — 4th ed.
        p.   cm.
    Includes bibliographical references.
    ISBN 0-442-31881-2
    1. Chickens—Handbooks, manuals, etc. I. Bell,
Donald D., 1933–   II. Title.
SF487.N77 1990
636.5—dc20   89-37826
                                                    CIP

Contents

Preface

Not since the *Commercial Chicken Production Manual* was originally published in 1972, has there been a change in authorship. The first three editions were written by Mack O. North, but with this fourth edition, Professor Donald D. Bell is a coauthor. With his lifelong dedication to the science of chicken production in the fields of research, computer application, writing, teaching, and lecturing, he brings a wealth of experience to this new edition. As State Poultry Specialist of the University of California located at the Riverside campus, he is internationally known and has been the principal or keynote speaker at more than 150 worldwide events. His research has been comprehensive, practical, and thorough, and quoted time after time.

This entire fourth edition of *Commercial Chicken Production Manual* has been revised as a joint venture of the two authors. A great amount of new material has been added, obsolete material has been deleted, and photographs have been included. Management procedures, production standards, and cost data have been upgraded to reflect genetic, nutritional, economic, and productive improvements in the industry.

Basically, this book is a guide to the best methods of raising chickens commercially. Page after page of explicit directions are given in 40 chapters that cover every aspect of chicken production. The very popular organization and detail in presenting material has been maintained to again make the contents of the book concise, factual, and easy to read.

This new edition contains more than 300 tables, figures, and diagrams, many full-page, and more than 15,000 facts and recommendations to aid one in attaining a high degree of perfection in chicken raising.

Because this manual is used in almost every country in the world, figures are given in both the English and metric systems. Originally the book was prepared as a service manual and guide, but it is now being used by many agricultural colleges as a text for their poultry courses. A Spanish translation of the third edition is also available.

The global acceptance of this book as a complete reference for commercial chicken production is greatly appreciated and has warranted the writing of this fourth edition to update the material.

List of Abbreviations

alternating current	ac	hydrogen ion	
ampere	A	concentration	pH
ante meridiem	A.M.	inch	in.
average	avg	infrared	IR
board foot	bd. ft.	joule	J
body weight	BW	kilo	k
British thermal unit	Btu	kilogram	kg
bushel	bu	kilometer	km
calorie, large	C	kilowatt	kW
calorie, small	c	kilowatthour	kWh
candella	cd	linear foot	ln ft
cent	¢	liter	L
centimeter	cm	lumen	lm
cycle per minute	c/min	lux	lx
day	da	metabolizable energy	ME
decibel	dB	meter	m
degree Celsius	°C	mile	mi
degree Fahrenheit	°F	mile per hour	mi/h
deoxyribonucleic		milligram	mg
acid	DNA	milliliter	ml
direct current	dc	millimeter	mm
dollar	$	millimeter mercury	mmHg
dozen	doz	minute	min
each	ea	month	mo
east	E.	north	N.
foot	ft	ounce (avoirdupois)	oz
foot-candle	fc	ounce each	oz/ea
foot per minute	ft/min	ounce per dozen	oz/doz
gallon	gal	parts per million	ppm
gallon per minute	gal/min	percent	%
gram	g	pint	pt
gram each	g/ea	plaque-forming units	pfu
horse power	hp	post meridiem	P.M.
hour	h	pound per square	
hundred weight	cwt	inch	PSI

probable error	pe	total sulfur amino	
protein	pro	acids	TSAA
quart	qt	United States	U.S.
relative humidity	rh	United States of	
revolutions per		America	USA
minute	r/min	U.S. gallon	U.S. gal
revolutions per		U.S. Pharmacopeia	U.S.P.
second	r/s	volt	V
ribonucleic acid	RNA	watt	W
second	s	watthour	Wh
south	S.	week	wk
tablespoonful	tbsp	weight	wt
teaspoonful	tsp	west	W.
therm	thm	yard	yd
ton	ton	year	yr

Commercial Chicken Production Manual

1
Modern Breeds of Chickens

During the past two centuries more than 300 pure breeds and varieties of chickens have been developed. However, few have survived commercialism in the poultry industry to be used by modern chicken breeders. Many of the early breeds are kept for exhibition purposes only; some have been lost forever; others are maintained by government breeding stations so they will be available to specialized breeders if necessary. These *gene pools* are important because they maintain certain genetic characteristics found in these rare breeds.

1-A. VARIETIES USED FOR MODERN BREEDING

In the early days of the commercial poultry industry, most of the chicks sold represented pure breeds or varieties. Breeding practices at that time were confined to improving the economic potential of these pure lines. Gradually, however, two or more breeds were crossed to improve their productivity. Eventually, particularly in the case of those birds bred for the production of meat, new synthetic lines were developed. Although many pure breeds were incorporated in their production, these new synthetics did not represent any former breed or variety. They were new and different; many more are being developed regularly.

Most of the breeds and varieties of chickens used in today's breeding programs, or used to develop synthetic lines, are included in the following sections.

Single Comb White Leghorn

The Single Comb White Leghorn is one of several varieties of Leghorns, but the only one used for commercial egg production. All Leghorns have yellow skin and lay eggs with white shells. Although only one variety is used, there are many strains in existence. Some may be feather-sexed when 1 day old.

Single Comb Rhode Island Red

The Single Comb Rhode Island Red has a long block-like body, single comb, and lays a brown egg. It has yellow skin, and the feathers are red with some black in the tail, hackle, neck, and wings. Several years ago many strains of this variety were in existence, most of which were excellent egg producers.

In recent years the Rhode Island Red has been used almost exclusively to produce *sex linkage* in the offspring chicks. When a male of this breed (genetically gold and nonbarred) is mated with a female that is genetically silver or barred, it is possible to determine the sex of the resultant day-old chicks by differences in the color of the chick down. Today, a good many of the commercial brown-egg layers are the result of crossing special strains of Rhode Island Reds and Barred Plymouth Rocks. The offspring are excellent producers of large brown eggs.

New Hampshire

The New Hampshire has light-red color, yellow skin, single comb, and produces a light-brown egg. At first the New Hampshire was known for its high egg production, but later it became recognized as a bird with good meat qualities. For several years it was the leading breed for the production of broiler chicks. Later, New Hampshire females were crossed with males of another meat-type variety to produce crossbred broiler chicks.

Today, the pure New Hampshire is merchandised by only a few breeders of meat-type birds. Its popularity has waned because of the broiler offspring's mediocre growth compared with other varieties and dark pinfeathers and hair that make it difficult to produce cleanly picked carcasses in modern processing plants. However, it is sold in many foreign countries for meat purposes.

The New Hampshire has been used in developing many of the synthetic lines of meat-type chickens and is still used for this purpose. Its ability to produce a large number of eggs that hatch well has made it a valuable asset to many breeding combinations.

White Plymouth Rock

The White Plymouth Rock has yellow skin and single comb. Although a pure variety was used by early broiler parent breeders, it now makes up the background for many synthetic lines. The white feathers are beneficial to broiler production and commercial processing plants, which do a better job of picking chickens with white feathers than those with colored feathers.

Most original strains of White Plymouth Rocks were genetically *slow feathering*, a disadvantage to a quality broiler chicken. Now however, most strains are *fast feathering.*

Cornish

Cornish chickens have pea combs, lay a brown egg, and have yellow skin. They have a body type very different from most other breeds. The legs are short, the body is broad, and the breast is very wide and muscular.

The Cornish features are desirable from a meat standpoint, but the birds lay only a few small eggs with poor hatchability. In order to utilize the strain's meat qualities, Cornish males are crossed with females from such breeds as Barred Plymouth Rock, New Hampshire, White Plymouth Rock, and synthetic lines.

Barred Plymouth Rock

The Barred Plymouth Rock has feathers with bars of white and black running crosswise, giving the bird a gray appearance. It has a single comb, yellow skin, and lays a brown egg.

As the demand for commercial eggs increased, consumers in many countries showed a preference for eggs with white shells rather than brown, and the Barred Plymouth Rock dropped in popularity. Today, the breed is mainly used as the female side of the mating with a Rhode Island Red male to produce *autosexing* in the offspring chicks used as commercial brown-egg layers. The autosexing feature has helped to make the cross popular.

Light Sussex

The Sussex is predominantly a British meat-type breed with several varieties, of which the Light Sussex is the most popular. It has white skin, lays a brown egg, and is a good meat producer. In England and some European countries, broiler chickens with white skin are preferred to those with yellow skin.

1-B. PRESENT-DAY EGG PRODUCTION LINES

Egg production lines are those used to produce egg-type pullets for the production of commercial eggs with either a white shell or a brown shell. The birds are relatively small in size, lay a large number of eggs with sound shells, live well, and produce eggs economically.

White-Egg Lines

Today, practically all commercial white-egg lines of chickens are Single Comb White Leghorns. In the early days the lines were pure; that is, they were not involved with strain crossing. Today, however, most breeders cross birds of two or more lines to produce the commercial pullet.

Single line. The breeder usually uses a *closed flock*, continually selecting the better birds in each generation and breeding from them. Only a small percentage of the better birds are used in the matings. Normally, the pullets are kept in egg production for a year in order to measure several factors responsible for economic production of quality eggs. Selection of the best birds is made at the end of the first year of egg production. Many features will enter into the selection, such as

body weight	egg weight
growth rate	egg production
growing livability	eggshell quality
pullet quality	interior quality of eggs
age at sexual maturity	adult livability

Hybrid vigor. Birds within certain breeds and varieties of chickens are more homozygous for some of the above features than for others. When mated together, the heterozygosity of the offspring is increased; new dominant genes are brought together, and the offspring are superior to the parent lines. This so-called *hybrid vigor,* or *heterosis,* implies a physical well-being as the cause of the improvement in the offspring, but actually it is the increased heterozygosity. Recessive genes—those that produce poorer results—are masked by the better dominant genes (see Chapters 21 and 22).

Male line and female line. It is obvious that in making any cross between two egg lines, a male from one line must be mated with a female from the other line. The opposite sex in each case is of no value—the sex is terminal—and all are destroyed at 1 day old because it is uneconomical to grow them. This is true not only with the parent lines but with the production of the commercial pullet; the cockerel chicks are destroyed.

Strain cross. Rather than select for superiority of all good traits within a single strain, many breeders resort to a technique of selecting for only a few in a line, then crossing two or more lines to produce the commercial pullet.

Two-line cross. Crossing two or more lines increases heterosis in the offspring, defined as a marked improvement in vigor or capacity for increased productivity. But to get as much improvement as possible from the cross, one parent line is bred to excel in only certain qualities; the other parent excels in others. A simplified example of a two-line cross would be as follows:

Male line	*Female line*
Bred for	Bred for
livability	egg production
large body size	shell quality
egg size	interior egg quality

Although there may be several other factors involved with each line, the above listing represents the major ones. When the two lines are crossed, the resulting pullets would be used for the production of commercial eggs; these pullets would have

good livability	good egg production
relatively large body size	good shell quality
good egg size	good interior egg quality

Three-line cross. Three lines are developed, each with different qualities. Two lines are crossed, and the offspring from these two are crossed with the third line. Although additional lines generally add to the cost of producing the commercial pullets, the advantages may outweigh the additional expense.

Four-line cross. Four lines are developed. Two of the four lines are crossed; then the remaining two are crossed. The male offspring from one of the above crosses are mated with the female offspring of the other cross to produce the commercial pullet.

Strains used for crossing must nick. Two lines of chickens, which when mated together complement each other, are said to *nick*. The poultry breeder will develop many lines of egg-laying strains, and will mate many of them together. Some of the crosses will give improved results in the offspring, some will not. A few of those that do nick will be used for the production of commercial pullets. In this way it is also possible to develop commercial birds that excel in only one, or at least a few, particular traits. For example, the breeder might develop a commercial pullet that lays exceptionally large eggs. Another pullet, the result of a different cross, would live unusually well. In each of these cases, the nickability is especially involved with one particular trait.

Inbred crosses. Some breeders resort to heavy inbreeding within certain lines by mating brothers and sisters—or other closely related individuals—for several generations; then, two of the inbred lines are crossed to produce a commercial pullet. This increases homozygosity within the inbred lines that in turn betters uniformity. Although inbreeding decreases performance, it is more than restored when inbred lines are crossed.

Brown-Egg Lines

While it is known that shell color has no effect on the nutritive value of eggs, shell color is a consumer preference in certain localities. In the United States and Germany white shells are preferred, while brown shells are preferred in France, the United Kingdom, and the Far East.

Several breeders have developed special lines and crosses for the production of commercial pullets that lay eggs with brown shells. In some instances, two breeds or varieties are used to make the cross. Not only do the offspring lay brown eggs but the chicks may be *sexed* at 1 day of age by differentiation in down color. One example of such a parental cross is

| *Parent male line* | *Parent female line* |
| Rhode Island Red | Barred Plymouth Rock |

When a Rhode Island Red male is mated with a Barred Plymouth Rock female the offspring male chicks are black with a white head spot; the offspring female chicks are black with no white head spot. Consequently, this cross makes it possible to sex the chicks at hatching time according to down color, sometimes known as *color sexing*. Special egg-producing lines of Rhode Island Reds and Barred Plymouth Rocks that nick are used to make the cross.

In some instances, other varieties are mated together to produce offspring that lay eggs with brown shells. Most of these crosses make it possible to color-sex the offspring through the use of genes for silver and gold (see Chapter 21-F). Synthetic lines have also been developed, and when crossed many of these produce differences in the color of the sexes at hatching time. An example is

| *Parent male line* | *Parent female line* |
| Rhode Island Red | Silver (_____) |

Male offspring from this cross are white (silver) or mostly white, while female offspring are gold or buff.

Body size. Birds producing brown-shelled eggs are 30 to 50% larger than those producing eggs with white shells. This relatively large size increases the feed cost to produce eggs because a larger bird consumes more feed than a smaller one. Usually it costs more to produce a dozen eggs with brown shells than a dozen with white shells.

Egg production. Most lines of birds producing commercial eggs with brown shells lay as well as those producing eggs with white shells. In most instances the brown-egg lines lay eggs that are larger than those produced by white-egg lines, but the shells are usually thinner.

Identifying Crosses

Poultry breeders follow a simplified method of identifying crosses they use to produce the commercial egg-type pullet as follows:

Cross symbol	Identification
P S	Pure strain
B X	Breed cross
C B	Crossbred
S X	Strain cross
IN X	Incross
IB X	Inbred cross
XL-Link	Sex-link
Syn.	Synthetic

1-C. PRESENT-DAY MEAT PRODUCTION LINES

Certain varieties and lines of chickens have been bred with emphasis on the production of meat rather than eggs. They are capable of producing economical gains in weight when raised as broilers or roasters.

Generally, it is impossible to breed a single line of chickens that will produce both eggs and meat in abundance; the breeding program must go one way or the other. When strains are selected for high meat production, their ability to lay a large number of eggs decreases.

Female Meat Lines

In the past, meat-type breeders specialized in developing the necessary line for either the male or female parent of the mating to produce commercial broiler chicks. Today, however, most, but not all, meat-line breeders develop both the male and female sides of the mating.

Since the females of the mating lay the eggs and are responsible for their hatchability, the female lines are bred to produce a fairly large number of eggs that hatch well, yet the birds are large and genetically have good growth-promoting characteristics.

Male Meat Lines

Originally some breeders concentrated only on a meat-type male to mate with parent meat-type females produced by another breeder. But nowadays most breeders develop both the male and female side of the mating.

Male meat lines have exceptionally heavy fleshing; they are large, grow rapidly, and have good feed conversion. To get these traits within a meat strain egg production and hatchability have been sacrificed.

Today, such male lines are predominantly synthetics incorporating genes necessary for meat production, conformity, and ease of processing with little emphasis on egg production and hatchability.

Cornish used for meat lines. Probably all meat lines on the male side incorporate some blood of the Dark Cornish or Light Cornish. Such varieties give the synthetic line a broad breast, short legs, and a plump carcass. But because these two Cornish varieties have colored feathers the synthetic line must be bred to have white feathers.

White-feathered male meat lines. Not only do the birds from these male meat lines have white feathers but when males are mated with colored females the offspring have white or nearly white feathers. This is a decided advantage at processing time because it is easier to pick white chickens than colored. Genetically, the synthetic male lines are *dominant white* for plumage color (see Chapter 22-B).

Yellow and white skin color. Consumers in most countries have a preference for broilers and roasters with yellow skin and practically all current male and female broiler breeding strains have yellow skin, but in England and some European countries white skin is preferred. The most practical way to produce such broilers is to mate a white-skin male with a yellow-skin female. The Light Sussex with white skin is the predominant male used; the females with yellow skin are from the regular female lines already in existence. The offspring from such a mating have white skin because white skin is dominant to yellow skin (see Chapter 21-C).

Special Lines for Meat Production

Sex-linked meat lines. Certain feather colors and patterns and speed of feather growth can be linked with the sex of the bird. When gold (buff or red) males are mated with certain silver (white) females, the offspring female chicks are gold or buff and the male chicks are silver (white). Similarly, if fast-feathering males are mated with slow-feathering females, the characteristics are reversed in the offspring and the differentiation can be observed in the wings of the newly hatched chicks. Either of the matings makes it possible to determine the sex of the chicks at 1 day of age. The procedure is used to produce what are known as *sex-linked chicks* and makes it possible to visually segregate males and females at hatching time. There are many such matings used today (see Chapters 21-F and 22-K).

Lines for roaster production. Roasters are larger than broilers and require special lines of birds that will grow rapidly to the heavier weights. Genetic breeders have developed special strains or crosses that produce chicks with this desired trait.

Squab broilers. Squab broilers are usually sold at 2.0 to 2.5 lb (0.9 to1.1 kg), straight-run, live weight. Although a misnomer, the popular name of *Rock Cornish Game Hen* is often used to merchandise the processed squab broiler. They are sold whole-body and never cut up. Special strains incorporating a high percentage of Cornish blood to give a wide breast in the young offspring have been developed.

1-D. THE MINI-BREEDS

During the past few years several strains of *mini-chickens* have appeared on the market. These birds are not to be confused with bantam chickens that originated many years ago. Dwarfism in mini-breeds is due to the sex-linked recessive gene, *dw.* However, the gene expresses itself differently in meat and egg lines (see Chapter 22-F).

Dwarfism in Mini-Leghorns

Dwarfism in modern Mini-Leghorn females when compared with normal pullets shows many differences, for example

1. The birds are 5 to 10% lighter at 8 weeks and 25% lighter at 25 to 30 weeks of age. The reduced body size is due to lower thyroid and growth hormone activity.
2. Shank length is about 20% shorter.
3. Red blood cell count is higher.
4. Feed consumption of the laying bird is less by 10 to 20%.
5. Egg production may be slightly lower.
6. Egg weight is about 10% lighter.
7. Feed required to produce one dozen eggs is about 5 to 10% less.
8. Laying house livability is approximately the same.

Important: Mini-Leghorns are smaller and require less floor space in the poultry house or cage, allowing more birds to be housed in a given area. This is an economic advantage, of course. But because of the small size of the birds, the carcass value at the end of the laying year is very small; sometimes the *spent hens* are of no value.

Considering the advantages and disadvantages, the Mini-Leghorn has not been able to take the place of Leghorns of conventional size; profitability is not always satisfactory. Results are highly variable when measured on return per dollar invested; the mini-breeds often show a lower return than normal birds.

Dwarfism in the Mini-Meat Strains

The *dw* gene has also been incorporated into several strains of meat birds primarily to produce broiler breeder females (female parent of commercial broiler chicks) capable of laying eggs at less cost. When compared with normal females the mini-meat-type females show several differences, namely

1. They require 35% less floor space.
2. The laying pullets consume 10 to 15% less feed.
3. Layers produce 5% more hatching eggs.
4. Eggs are about 5% smaller.

However, the broilers grown from the eggs produced by a mating of mini-parents are decidedly smaller, and this is a serious disadvantage in any strain being used to produce meat regardless of how well the parents do in the above categories.

Important: From the standpoint of the broiler mini-mother, she may be used efficiently to produce hatching eggs. To overcome the disadvantage of some of the lower growth rate in her offspring, she is mated with a normal meat-type male. This cross produces female broilers of normal size, and male broilers with the *dw* gene in the heterozygous state that are 3 to 4% lower in weight than normal male broilers (see Chapter 22-F).

1-E. THE PACKAGE DEAL

Today, most of the meat-line breeders, but not all, produce both the male and female lines. In such cases, the *primary breeder* sells both the cockerel and pullet parent-line day-old chicks as a *package deal* in which the customer is supplied 12 to 15 cockerel chicks with every 100 female chicks.

Primary breeders of egg lines, producing eggs with either white or brown shells almost without exception, have bred all the lines needed for the production of the commercial egg-type pullet. This has been necessary because of the intricate methods of making the matings to produce the parent males and females used in their breeding programs. Only certain combinations will nick. Therefore, male and female egg-type breeder parents always come from the same breeder, and are shipped to the customer as a package deal.

1-F. NATIONAL POULTRY IMPROVEMENT PLAN

This plan, specific for the United States, began in 1935 as a voluntary program administered by agreement between the states, the U.S. Department of Agriculture, and the poultryman.

There are two phases:

1. To improve the production and market quality of chickens.
2. To reduce losses from certain diseases commonly associated with hatchery and breeder flock dissemination, particularly
   a. pullorum disease
   b. fowl typhoid
   c. *Mycoplasma gallisepticum* infection
   d. *Mycoplasma synoviae* infection

Associated with the disease control section of the plan is a program of blood testing the breeders to determine if they are carriers of any of these four diseases.

Those interested in the many details of the plan should secure a copy of "National Poultry Improvement Plan" from the U.S. Department of Agriculture, APHIS-VS, National Poultry Improvement Plan; Room 828 FB, Hyattesville, MD 20782.

2

Structure
of the Chicken

The chicken is a warm-blooded vertebrate, and during the course of evolution it owes its origin to reptiles. Although there are many similarities between the two, there are also vast differences.

Reptiles are *poikilotherms*. That is, they are cold blooded, meaning their body temperature is not regulated and is usually that of their ambient temperature. Chickens are *homeotherms*. They are warm blooded, meaning their deep body temperature is relatively high and usually almost constant. They are also *endotherms*. They have the ability to generate deep body heat to increase their body temperature.

Both reptiles and chickens lay eggs, and they are incubated outside their bodies, but the female reptile buries its eggs in the sand or soil, and the surrounding temperature is adequate for the growth of the egg-contained embryo. During natural embryonic development, eggs of the chicken are covered by the hen and they retain her body temperature during the entire incubating process.

Most birds can fly; reptiles cannot.

2-A. SURFACE OF THE CHICKEN

The chicken is covered with feathers, skin, and scales, the latter being a derivative of reptiles (see Figures 2-1 and 2-2).

Feathers

Birds are almost completely covered with feathers, which makes them different from other vertebrates. During the evolutionary process of the

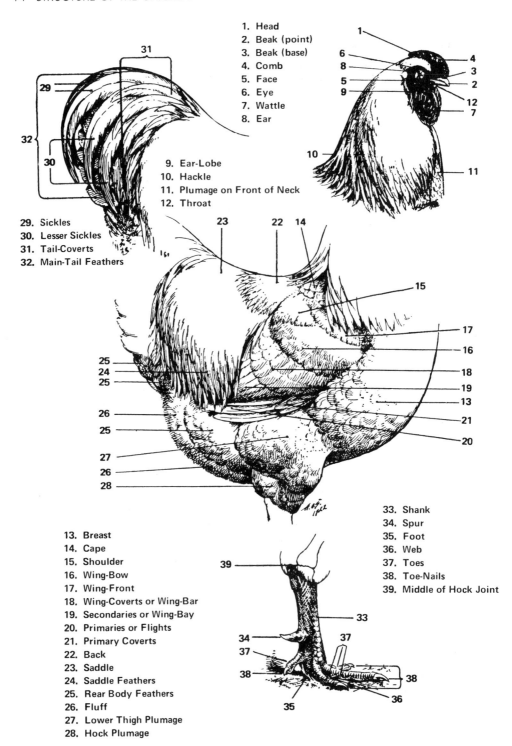

1. Head
2. Beak (point)
3. Beak (base)
4. Comb
5. Face
6. Eye
7. Wattle
8. Ear

9. Ear-Lobe
10. Hackle
11. Plumage on Front of Neck
12. Throat

29. Sickles
30. Lesser Sickles
31. Tail-Coverts
32. Main-Tail Feathers

13. Breast
14. Cape
15. Shoulder
16. Wing-Bow
17. Wing-Front
18. Wing-Coverts or Wing-Bar
19. Secondaries or Wing-Bay
20. Primaries or Flights
21. Primary Coverts
22. Back
23. Saddle
24. Saddle Feathers
25. Rear Body Feathers
26. Fluff
27. Lower Thigh Plumage
28. Hock Plumage

33. Shank
34. Spur
35. Foot
36. Web
37. Toes
38. Toe-Nails
39. Middle of Hock Joint

Figure 2-1. Nomenclature of the male.
(Courtesy of American Poultry Assoc.)

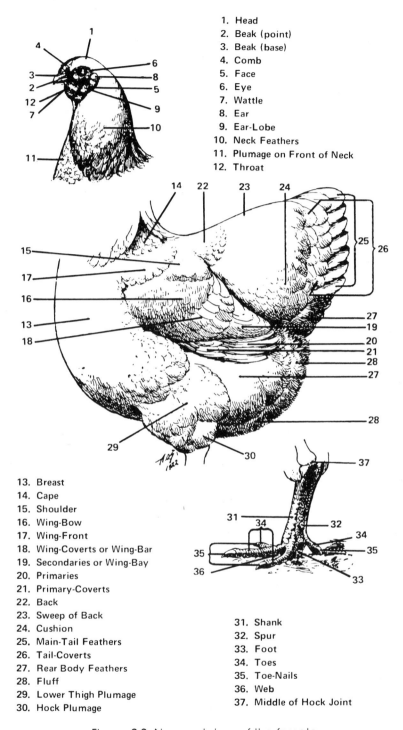

1. Head
2. Beak (point)
3. Beak (base)
4. Comb
5. Face
6. Eye
7. Wattle
8. Ear
9. Ear-Lobe
10. Neck Feathers
11. Plumage on Front of Neck
12. Throat

13. Breast
14. Cape
15. Shoulder
16. Wing-Bow
17. Wing-Front
18. Wing-Coverts or Wing-Bar
19. Secondaries or Wing-Bay
20. Primaries
21. Primary-Coverts
22. Back
23. Sweep of Back
24. Cushion
25. Main-Tail Feathers
26. Tail-Coverts
27. Rear Body Feathers
28. Fluff
29. Lower Thigh Plumage
30. Hock Plumage

31. Shank
32. Spur
33. Foot
34. Toes
35. Toe-Nails
36. Web
37. Middle of Hock Joint

Figure 2-2. Nomenclature of the female.
(Courtesy of American Poultry Assoc.)

chicken most of the reptilian scales changed to feathers. Both scales and feathers are chiefly composed of the same protein, *keratin*.

Feathers serve many purposes such as

1. aiding in flight
2. providing insulation from temperature extremes
3. dispelling rain and snow
4. causing predatory camouflage
5. helping attract others of the same species

Parts of a feather. A feather is composed of a root called the *calamus*; a long *quill* or *shaft*, known as the *rachis* to give rigidity; *barbs* extending from the quill; *barbules* extending from the barbs; and *barbicels* extending from the barbules. All parts except the quill tend to mesh together in the flat portion of the feather. Meshing is not pronounced at the base of the feather and the loose construction gives rise to *fluff*, often different in color than the main web.

How feathers are replenished. When the chick hatches, it has almost no feathers. Except for the wings and tail, it is covered with *down*. Soon the down grows longer, and most of the particles develop a shaft. Within a few days the shaft erupts, and the web of the feather makes its appearance. By the time the chick is 4 to 5 weeks of age it is fully feathered. The first feathers are soon molted, and a new set is grown by the time the bird is 8 weeks old. The third set is completed just prior to the time the bird reaches sexual maturity, and it is the first mature plumage.

Feathers make up between 4 and 8% of the live weight of the bird, the variability being related to age and sex; older birds and males have a lower percentage.

The annual molt. Because adult feathers wear away, are broken, or are pulled out, nature has provided the adult chicken with a method of renewing all its feathers once a year: it drops its remaining feathers and grows a new set. The process is known as *molting*. In the wild, the feathers are dropped intermittently in a consistent pattern so the bird is never void of feathers; it has some old and some new. The normal process of dropping the old feathers and growing a new set requires 3 to 4 months.

Molting and the growth of new feathers are under hormonal control. To molt, a chicken must initiate new growth in the buds at the base of the feathers that in turn forces the old feathers out.

The hormone levels that induce egg production and cause broodiness inhibit feather-bud growth. Consequently, hens that are molting are seldom producing eggs. If egg production is curtailed by artificial means, such as reducing the feed intake, the molt may be precipitated in a more rapid and complete manner (see Chapter 19).

Shape of the feather. Not only do feathers vary greatly in size over the surface of the body but certain shapes are associated with sex. Gonadal hormones play an important part in this sex variation. They increase the length and narrow the width of certain feathers of the male bird, including the hackle, saddle, sickle, and lesser sickle feathers.

Feather tracts. Feathers do not cover the body of the bird uniformly, but grow in rows to produce tracts or areas over the body. The ten major feather tracts are

shoulder	abdomen
thigh	leg
rump	back
breast	wing
neck	head

The order and time of the appearance of feathers are as follows:

Shoulder and thigh	2 to 3 wk
Rump and breast	3 to 4 wk
Neck, abdomen, and leg	4 to 5 wk
Back	5 to 6 wk
Wing coverts and head	6 to 7 wk

Color of feathers. There are many feather colors and color patterns. In many instances differences in color vary according to the location of the feathers on the body. Feather patterns are usually different on the male and female. Feather colors and feather patterns are the result of genetic interaction plus the presence of male or female sex hormones (see Chapters 21 and 22-B).

Waxy coating on feathers. The *uropygial gland* or *preen gland* is located on the dorsal area of the tail, and is the only secretory gland located on the surface of the chicken. It secretes an oily wax that the bird spreads over its feathers with its beak. The material makes the feathers water resistant; they do not absorb water, and water quickly runs from coated feathers.

Head

The head of the chicken is represented by the following parts:

Comb. There are several types of combs, but only the first three of the following list are common. The comb types are

single	strawberry
rose	walnut
pea	''V''
cushion	buttercup

Comb type is the result of gene interaction, but comb size is associated with gonadal development and the intensity of light, either natural or artificial. The lower the light intensity, the larger the comb.

Eyes.   Chickens have color vision and show a preference for violet and orange. They are slightly farsighted.

Eyelids

Eye rings.   Inner margin of eyelids.

Eyelashes.   Bristle feathers composed of a straight shaft.

Ears.   Hearing is equal to that of mammals.

Earlobes.   Either red or white.

Wattles

Beak

Feet and Shanks

The shanks and most of the feet are covered with scales of various colors. Yellow color is due to dietary carotenoid pigments in the epidermis when melanic pigment is absent. Varying shades of black are the result of melanic pigment in the dermis and the epidermis. When there is black in the dermis and yellow in the epidermis, the shanks have a greenish appearance. In the complete absence of both kinds of pigment, the shanks are white. Important parts of the shank and foot are

Hock

Shank

Toes.   Most chickens have four toes on each foot, but there are a few breeds with five.

Skin

Most of the chicken is covered with a thin skin. With the exception of the *uropygial gland* (preen gland) the skin is void of glands. The absence of sweat glands makes it impossible for the bird to sweat to lose moisture.

The skin has a different texture in the area of the comb, wattles, earlobes, beak, scales, spurs, and claws. Except for certain specialized areas, the color of the skin is either white or yellow. The density of the yellow color is correlated with the amount of xanthophylls in the feed consumed and the intensity of egg production (see Table 33-14).

2-B. SKELETON

The skeleton is the framework that supports the body and to which the muscles are attached. The rib cage protects some of the vital organs. Close scrutiny shows that the bones found in the skeleton of mammals are also found in the skeleton of chickens. However, some of the bones in the latter are fused or elongated. Figure 2-3 shows this relationship.

The skeleton of the neck is long and freely movable, but the remaining portion of the vertebral column is rigid, containing many fused bones. Several of the thoracic vertebrae are united to form a firm base for the attachment of the wing and its muscles. There is a heavy keel. The hip bone is solidly fixed to the ileum and the pelvic bones do not join vertically.

The wings correspond to the arms and hands of human beings. The legs contain the same bones as found in the legs of man. The bones of

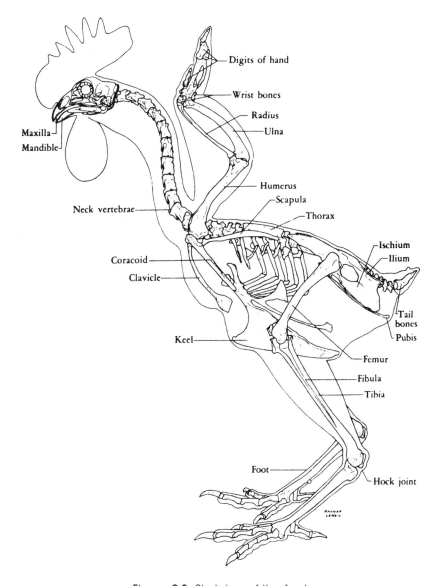

Figure 2-3. Skeleton of the fowl.

the metatarsus, common to the human foot, have been fused and elongated to form the shank.

Bones found in the skull, humerus, keel, clavicle, and some vertebrae are hollow and connected to the respiratory system, with air continually moving in and out of these specialized bones. Most bones are light in weight, yet very strong. There is also a soft, spongy bone material known as *medullary bone* present in varying amounts in the femur, sternum, ribs, ulna, tibia, and certain other bones of the skeleton of females in egg production. This medullary bone is used as a part of the source of calcium for eggshell production. The storage capacity of this specialized bone is highly variable, depending on the length and rate of egg production. Most of the calcium needed for the production of eggshells comes directly from the feed eaten each day.

2-C. MUSCLES

The muscles that motivate the bird are especially important, yet those that control the action of the heart, blood vessels, intestines, and other vital organs cannot be overlooked. Muscles that move the wings are unique to those birds that fly, and are attached to the keel of the *sternum* (breastbone) that also supports the vital organs of the abdominal cavity. These muscles are especially well developed in most birds. Development has increased through genetic selection as evidenced in the modern turkey and meat-type broiler strains of chickens even though these domesticated birds can fly only short distances.

Chickens are endowed with white muscle and red muscle, giving rise to light and dark meat. More fat and myoglobin, an iron- and oxygen-carrying compound, are found in red meat than in white. Usually the activity of the muscle determines its color. In the chicken, those of the leg are darker than those of the breast because there is constant stress on the leg muscles to keep the body upright when the bird stands. In wild flying birds the breast muscle is darker because greater stress is placed on it during flight. Broiler-type chickens have muscle fibers that are larger in diameter and lighter in color than those of layer-type.

2-D. RESPIRATORY SYSTEM

The respiratory system of chickens consists of

nasal cavities	bronchi
larynx	lungs
trachea (windpipe)	air sacs
syrinx (voice box)	air-containing bones

Lungs of the chicken are small compared with those of mammals. They expand or contract only slightly, and there is no true *diaphragm*. The lungs are supplemented by nine air sacs and the air-containing bones. They have four pairs of air sacs divided equally into *thoracic* and *abdominal air sacs*, plus one singular *interclavicular sac*.

Air freely moves in and out of the air sacs, but the lungs are responsible for most of the respiration. Both the lungs and air sacs function as a cooling mechanism for the body when moisture is exhaled in the form of water vapor.

The respiratory rate is governed by the carbon dioxide content of the blood; increased levels increase the rate, which varies between 15 and 25 cycles/min in the resting bird.

2-E. DIGESTIVE SYSTEM

Figure 2-4 shows the digestive system of the chicken. The various parts are discussed below.

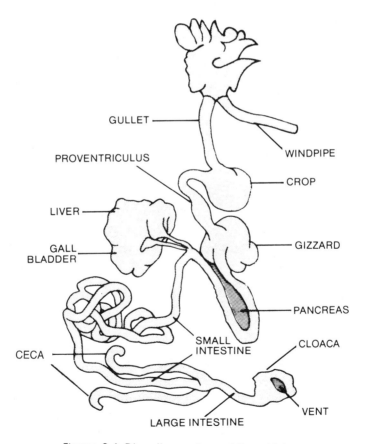

Figure 2-4. Digestive system of the chicken.

Mouth

The chicken has no lips, soft palate, cheeks, or teeth, but there are an upper and lower horny mandible to enclose the mouth, the upper being attached to the skull while the lower is hinged. The hard palate is divided by a long narrow slit in the center that is open to the nasal passages. This opening and the absence of a soft palate make it impossible for the bird to create a vacuum to draw water into the mouth; the bird has to scoop up water when drinking, elevate its head, then let the water run down the gullet by gravity. There is no swallowing.

The two mandibles are referred to as the *beak*. The dagger-like *tongue* has a very rough surface at the back to help force food into the esophagus. Saliva, with its enzyme *amylase,* is secreted in the glands of the mouth, but the primary value of saliva is as a lubricant to make it easier for food particles to pass the mouth where food leaves so rapidly that there is little chance for digestion.

A chicken has fewer taste buds than mammals, the human being having about 9,000. About 24 are located in the several areas of the mouth and beneath the tongue of the chicken, but its ability to taste feeds is relatively high.

Esophagus

The *esophagus* or *gullet* is the tube through which the food passes on its route from the back of the mouth (pharynx) to the proventriculus.

Crop

Just before the gullet enters the body cavity it extends on one side into a pouch known as the *crop,* which acts as a storage place for food. Little or no digestion takes place here except for that involved with the salivary secretion of the mouth, which continues its activity in the crop.

Proventriculus

An enlargement of the gullet just prior to its connection with the gizzard is known as the *proventriculus,* sometimes called the *glandual stomach* or *true stomach.* It is here that gastric juice is produced. *Pepsin,* an enzyme to help protein digestion, and hydrochloric acid are secreted by the glandular cells. Because the food passes quickly through the proventriculus there is little digestion of food material here, but the secretions pass into the gizzard where the enzymic action takes place.

Gizzard

The gizzard, sometimes called the *muscular stomach*, lies between the proventriculus and the upper limit of the small intestine. It has two pairs of very powerful muscles capable of exerting great force, and a very thick mucosa, the surface of which is constantly being eroded and sluffed off. The gizzard remains quiet when empty, but once food enters the muscular contractions of its thick walls begin. The larger the particles of food, the more rapid the contractions. Since the gizzard usually contains some abrasive material, such as grit, rock, gravel, etc., the food particles are soon ground or reduced to small particles capable of being taken into the intestinal tract. When fine material enters the gizzard it leaves in a few minutes, but when the food is coarse it will remain in the gizzard for several hours. There is no difference in the digestibility of feeds in coarsely or finely ground forms.

Small Intestine

The small intestine is about 62 in. (1.5 m) long in the average adult chicken. The first part forms a loop known as the *duodenal loop*. Imbedded in the loop is the *pancreas* that secretes pancreatic juice containing the enzymes *amylase, lipase,* and *trypsin*. Other enzymes are produced by the wall of the small intestine and these further the digestion of protein and sugars.

Ceca

Between the small and large intestines lie two blind pouches known as the ceca. Each cecum is about 6 in. (15 cm) long in the normal healthy adult bird, and soft feed material passes in and out. The exact function of the ceca is not known, but evidently they have little to do with digestion; only minor water absorption and a small amount of carbohydrate and protein digestion plus some bacterial action takes place.

Large Intestine

The large intestine is a relatively short rectum in the chicken, being only 4 in. (10 cm) long in the adult bird, and about twice the diameter of the small intestine. It extends from the end of the small intestine to the cloaca. The large intestine is a place for water resorption to increase the water content of the body cells and maintain the water balance in the bird.

Cloaca

The bulbous area at the end of the alimentary tract is known as the cloaca. Cloaca means ''common sewer,'' and into the cloaca empty the digestive, urinary, and reproductive canals.

Vent

The vent (anus) is the external opening of the cloaca. Its size varies greatly in the female, depending on whether or not she is producing eggs.

Supplementary Digestive Organs

Certain organs are closely associated with digestion because their secretions empty into the intestinal tract and aid the processing of food material.

Pancreas. The pancreas lies within the duodenal loop of the small intestine. It is a gland that secretes pancreatic juice that is then passed into the duodenum by way of the pancreatic ducts where its five powerful enzymes aid in the digestion of starches, fats, and protein. The pancreatic juice neutralizes the acid condition of the proventriculus.

Liver. The liver is composed of two large lobes. Among its functions is the secretion of *bile,* a slightly sticky yellow-green fluid containing the bile acids that, when entering the lower end of the duodenum, aids digestion, particularly of the fats. The bile contains no digestive enzymes. Its chief function is to neutralize the acid condition of the canal and initiate the digestion of fats by forming emulsions.

Gallbladder. The chicken has a gallbladder, but some birds do not. Two bile ducts transfer the bile from the liver to the intestines. The right duct is enlarged to form the gallbladder, through which most of the bile passes and is temporarily stored. The left duct does not enlarge; therefore a small amount of bile passes through it directly to the intestines.

2-F. URINARY SYSTEM

The urinary system is basic to the two kidneys that are located just behind the lungs. A single ureter connects each kidney with the cloaca. The urine of chickens is mainly uric acid, the end product of protein metabolism, which is mixed with the feces in the cloaca and evacuated in the droppings as a white pasty material.

2-G. CIRCULATORY SYSTEM

The purpose of the circulatory system is to transfer blood from the heart to the cells of the body, and return it. The heart of the chicken has four chambers as in mammals: two atria and two ventricles. It beats at a comparatively rapid rate of about 300 pulsations per minute. The smaller the bird, the more rapid the contractions. Chicks show an increased rate as they age. Birds in bright light have a faster beat. The beat of individual chickens is highly variable and may often double as the result of excitement alone.

Composition of blood. Blood is composed of plasma, salts, and other chemicals, plus *erythrocytes* (red cells) and *leucocytes* (white cells). In the chicken the erythrocytes contain a nucleus in contrast to the unnucleated erythrocytes of mammals. The blood of a chicken contains about 3 million erythrocytes per cubic millimeter.

The *spleen* serves as a reservoir for the erythrocytes, and expels its contents into the circulatory system. Blood constitutes about 12% of the weight of a newly hatched chick, and about 6 to 8% of the mature chicken.

Function of blood. Blood has numerous functions, including the following:

1. It moves O_2 to the body cells and removes CO_2 from them.
2. It absorbs nutrients from the alimentary tract and transports them to the tissues.
3. It removes the waste products of cellular metabolism.
4. It transports hormones produced by the endocrine glands to various sections of the bird.
5. It helps regulate the water content of the body tissues.

Blood pressure. Blood pressure of chickens of all ages is measured as mmHg. Even the pressure of the developing embryo may be recorded. As with human beings, there are two measures: (1) *systolic pressure* (arterial), and (2) *diastolic pressure* (as the blood returns to the heart).

Following are the recognized blood pressures of adult chickens:

	Systolic Pressure mmHg	Diastolic Pressure mmHg
Adult female chicken	140–160	130–134
Adult male chicken	180–195	145–150

2-H. NERVOUS SYSTEM

The nervous system regulates all organs and consists of many parts. The brain represents highly concentrated nerve cells, the basis for all nerve stimuli. Hearing and sight are well developed, with the chicken being able to distinguish colors; but the ability to smell is of low magnitude. The sensitive taste buds enable the bird to develop a liking for certain feed flavors that in part determines the type of food the bird eats.

Chickens have an ability to learn; they can be trained to follow certain physical procedures. Furthermore, they learn to recognize large numbers of pen mates at a young age, and their ability increases with age (see Chapter 22-M).

2-I. HORMONE-PRODUCING GLANDS

Within the body are certain *endocrine glands,* or cells of specific organs, that produce chemical products known as *hormones.* Hormones pass directly into the bloodstream and have a controlling effect on cells and organs in other parts of the body. The hormones represent a variety of chemical substances such as proteins, steroids, etc. Some increase the activity of certain organs; others depress the activity; some have an effect on metabolic processes; some do not.

The glands producing hormones include

thyroid	pineal
parathyroids	adrenals
testes	ultimobranchial body
ovary	islets of Langerhans
pituitary	pancreas
hypothalamus	

In addition to the glands, hormones are produced by the gastric and intestinal mucosa.

The function and interaction of the hormones are varied and great in number. Thyroxine, produced by the thyroid, helps regulate the metabolic rate. Parathormone from the parathyroids influences calcium and phosphorus metabolism. The hormones of the pituitary, a small gland at the base of the brain, are many. Some aid in growth; others affect the thyroid and parathyroids, while others have a pronounced effect on ovulation, the oviduct, broodiness, and egg laying.

Hormones of the ovary influence fat deposition, increase the release of calcium from the medullary bone, and precipitate ovulation. Chemicals produced by the adrenals aid retention of glycogen by the liver and affect mineral metabolism. The islets of Langerhans and some cells of the pancreas produce insulin and glycogen, which regulate the utilization of glucose and its level in the bloodstream. Hormones of the gastrointestinal tract increase the production of gastric juice, pancreatic juice, and bile.

2-J. REPRODUCTIVE SYSTEMS

Male

The male reproductive system consists of two testicles in the dorsal area of the body cavity, just in front of the kidneys. The many ducts of the testes lead to the *vas deferentia* and *vas deferens*, which carry the semen from the testicles to the *papillae* in the dorsal area of the cloaca, and then to the copulatory organ located in one of the folds of the cloaca. The process takes 4 days.

Normally, semen is stored in the vas deferens. Here it is diluted with lymph fluid, and both are ejaculated as a mixture during copulation.

The penis of the male chicken is quite small, but waterfowl have a well-developed, long, and twisted organ. Lymph enters the penis to form a mild erection, but it does not penetrate the cloaca. Rather, during mating, the cloaca of the female only opens to expose the end of the oviduct, and semen is deposited over this opening, then finds its way up the oviduct.

Spermatozoa from the male chicken show a long pointed headpiece, followed by a long tail. The pH of semen is between 7.0 and 7.4. The volume of semen ejaculated during one mating may be as high as 1.0 cm^3 at the beginning of the day, but decreases to as little as 0.1 cm^3 after many matings (see Chapter 17-M).

Female

The female reproductive system consists of an ovary and oviduct. These are described in detail in Chapter 3.

2-K. HOW A CHICKEN GROWS

The body of the chicken consists of innumerable cells that are about the same size in all breeds regardless of ultimate mature body weight. Most early embryonic increases in growth occur as the result of cell multiplication: 1 cell divides into 2, 2 into 4, 4 into 8, 8 into 16, and so on. But this rhythmic increase does not continue indefinitely. Soon there is cell specialization to form different body components and the growth rate and rate of division of some cells is retarded after the chick hatches. The older the bird, the lower the daily increments of increased body weight.

After hatching, when the number of muscle fibers (single cells) no longer increases, growth of muscle and nerve cells is the result of cell enlargement rather than cell division. Muscle fibers have a maximum dimension, controlled mainly by the genetic makeup of the bird, but can decrease or increase in size with varying amounts of activity. Both protein synthesis and protein degradation are involved. Better feed is more responsible for protein degradation, while synthesis is due more to genetics. Both synthesis and degradation operate simultaneously, the net result determining whether muscles increase or decrease in size. The muscles of the breast are exceptionally well developed in birds because these muscles are used to motivate the wings during flight.

The degree of fatness of a chicken rests entirely on the number of fat-containing cells. Some breeds and lines of chickens have a greater number of fat cells than others, probably as an indirect consequence of breeding birds for larger size and plumper carcass. Fat cells reach their maximum number in the growing period. The ability of a broiler to gain weight rapidly is principally the result of fat deposits in the fat cells rather than increases in the growth of the skeleton or muscle fibers.

2-L. BODY CHANGES DURING EGG PRODUCTION

During the time female chickens are laying and during the time they are molting certain changes occur in the appearance of the birds.

Molting

Some layers may produce a few eggs after the molt begins, but generally cessation of egg production preempts the molt. The length of the molting period varies. Good egg producers molt late in the season, and rapidly; poor egg producers molt early, and slowly.

Order of the molt. During the molt, feathers are dropped from the various parts of the body in a definite order, that is,

1. head	5. fluff
2. neck	6. abdomen
3. breast	7. wings
4. back	8. tail

Many times a flock will become involved with some type of temporary stress as the result of disease, high or low ambient temperature, reduced feed or water consumption, etc., causing a partial molt of the feathers from the head and neck and a few feathers from the wings. If the cause can be corrected, it should not interfere with the primary annual molt.

Yellow Pigmentation

The yellow color in the skin and shanks of yellow-skinned chickens is due to several *xanthophylls* (hydroxycarotenoid pigments) that are deposited in the fat layer below the skin. The bird's only source of these xanthophylls is from the feed it eats. The more xanthophylls in the feed, the denser the yellow skin color. Xanthophylls in this fatty layer continually undergo chemical breakdown, but are replenished through the feed.

Bleaching. Xanthophylls are also responsible for the yellow color of egg yolks. However, when a pullet starts laying at a fast rate most of the xanthophylls in the feed go to the egg yolks. Not enough is left to replenish those being chemically lost in the skin, and it begins to bleach. The longer a bird lays, the greater the bleaching. When the bird has laid about 180 eggs the skin will have a blue-white color. For a more in-depth discussion of xanthophylls see Chapter 31-I.

Minor Changes Resulting from Egg Production

There are some other changes in the bird during the course of egg production, namely

1. Vent becomes large and moist.
2. The pubic bones become thinner.
3. Space between the pubic bones increases.
4. Distance between the pubic bones and end of keel bone increases.
5. Skin on the skull becomes much thinner.

3

Formation of the Egg

The avian egg consists of a minute reproductive cell quite comparable with that found in mammals. But in the case of the chicken, this cell is surrounded by yolk, albumen, shell membranes, shell, and cuticle. The ovary is responsible for the formation of the yolk; the remaining portions of the egg originate in the oviduct.

3-A. OVARY

At the time of early embryonic development, two ovaries and two oviducts exist, but the right set atrophies, leaving only the left ovary and oviduct at hatching. Prior to egg production the ovary is a quiet mass of small follicles containing ova. Some ova are large enough to be visually seen; others require microscopic magnification. Several thousand are present in each female chicken, many times the number that will eventually mature into full-size yolks necessary for egg production during the life of the bird.

Formation of the Yolk

The yolk is not the true reproductive cell, but a source of food material from which the minute cell (*blastoderm*) and its resultant embryo partially sustain their growth.

When the pullet reaches sexual maturity, the ovary and the oviduct undergo many changes. About 11 days before she is destined to lay her first egg, a sequence of hormonal activity takes place. The *follicle-stimulating hormone* (FSH) produced by the anterior pituitary gland causes the ovarian follicles to increase in size. In turn, the active ovary begins to generate the hormones *estrogen, progesterone,* and *testosterone* (sex steroids). Higher blood plasma levels of estrogen initiate development of the medullary bone, stimulate yolk protein and lipid formation by the liver, and increase the size of the oviduct, enabling it to produce albumen proteins, shell membranes, calcium carbonate for the shell, and cuticle.

The first yolk to begin maturity does so because major amounts of the yolk material produced in the liver and transported by the blood go directly to it. A day or two later, the second yolk begins to develop, and so on, until at the time the first egg is laid from five to ten yolks are in the growth process. About 10 days are required for an individual yolk to mature. Deposits of yolk material are very slow at first and light in color. Eventually the ovum reaches a diameter of 6 mm at which time it grows at a greatly increased rate, the diameter increasing about 4 mm per day. A greater number of yolks are under development at one time in the broiler breeder hen than in the egg-type hen, but the broiler breeder hen does not have the viability to produce as many complete eggs; therefore, she produces fewer.

The coloring matter of the yolk is xanthophyll, a carotenoid pigment derived from the food the chicken eats. The pigment is transferred first to the bloodstream, then quickly to the yolk. Consequently, more is deposited in the yolk during the hours the hen is eating than during dark hours when she is not. This gives rise to deposits of dark and light layers of yolk material depending on the dietary pigment available. From seven to eleven layers are produced per yolk. Yolk formation is rather uniform and the total thickness of both dark and light deposits during 24 hours is about 1.5 to 2.0 mm.

Egg yolk is composed mainly of fats (lipids) and proteins, which combine to form lipoproteins of which two-thirds are the low density fraction (LDF) and known to be synthesized by the liver through the action of estrogens. In the laying hen, LDF is removed from the blood plasma as intact particles for direct deposition in the developing ovarian follicle.

What influences growth rate of the yolk.   Although the addition of fat and protein to the laying ration has been shown to increase the size of the developing yolk, nothing has been found that is economical and practical.

Ova vary greatly in size, not only those produced by an individual chicken but those produced by the various hens in the flock. Size is not associated with rate of lay, but more probably with the length of time required for the ova to reach maturity. Those from an individual hen increase in size the longer she is in production. Furthermore, the first egg laid in a clutch usually will contain a larger yolk than the remaining ones.

Location of the germinal disc.   The yolk material is laid down adjacent to the germinal disc that continues to remain on the surface of the globular yolk mass. Once the egg is laid, the yolk rotates so the germinal disc is uppermost.

Ovulation

At maturity the ova are released from the ovary to enter the oviduct by a process known as *ovulation.* Each ovum hangs on the ovary by a narrow stalk containing the artery that supplies the blood to the developing yolk. This artery undergoes much branching in the surface membranes of the yolk and the follicle appears highly vascular except for the *stigma,* a narrow band surrounding the yolk that is almost void of blood vessels.

When an ovum is mature, the hormone *progesterone,* produced by the ovary, excites the hypothalamus to cause the release of the *leuteinizing hormone* (LH) from the anterior pituitary, which, in turn, causes the mature follicle to rupture at the location of the stigma to release the ovum from the ovary. The yolk is then surrounded by only the *vitelline membrane* (yolk membrane).

Delaying first ovulation. Sexual maturity, as indicated by the first ovulation, may be accelerated or retarded. Restricting feed or restricting the length of the light day during the pullet's growing period are the two main procedures used, but there are many others (see Chapters 14-J and 18-B).

What initiates ovulation. It is not known what sets the hour for the bird's first ovulation, but both the nervous system and hormonal secretions are of primary importance. The second ovulation is regulated by the *laying* of the first egg, and occurs about 15 to 40 minutes after the first egg passes through the vent. Further ovulations occur with the same time period after eggs are laid.

Eggs laid in clutches. Chickens lay eggs on successive days known as clutches, after which they will not lay for one or more days. The length of the clutch may vary from 2 days to more than 200 before a day is missed, but most commercial egg-type chickens produce between three and eight eggs per clutch. The length is quite consistent with individuals: poor producers have shorter clutches; good producers have longer clutches. Once the clutch length is reached, the hen skips a day or more of ovulation and resultant egg production, then produces another clutch. Poor egg producers have a longer rest period between clutches than do good producers.

Time necessary to produce an egg. The time necessary for an egg to transverse the oviduct varies with individuals. Most hens lay successive eggs with time intervals of 23 to 26 hours. If the time is greater than 24 hours, each successive egg will be laid later in the day, and the ovulation of the yolk for the next egg will also occur later in the day. Eggs laid in the afternoon have spent several more hours in the oviduct than those laid in the morning. Eventually eggs are laid so late that the rhythm is broken and an ovulation is skipped.

Time of ovulation. Hens that produce long clutches lay their first egg of a clutch early in the day, an hour or two after the sun rises or the artificial lights go on. Ovulation of the next yolk comes quickly after an egg is laid the next day, with only a slight time lag. Those hens with shorter clutch lengths lay their first egg of the clutch later in the day, ovulation of the next yolk is slower, and the time lag for laying is greater. Most ovulations occur during the morning hours; it is not natural that ovulations occur in the afternoon.

Egg production at start of lay. During the first week of lay ovulation is quite irregular; the hen's hormonal mechanism is not in balance. Often, only two to four eggs will be produced. But by the second or third week, ovulation is progressing at its peak rate, only to drop slowly each week throughout the remainder of the laying cycle.

Light and ovulation. Light, either natural or artificial, has a stimulating effect on the pituitary gland, forcing it to secrete an increased quantity of the FS hormone, which in turn, activates the ovary. Both duration and intensity of light are important. The procedure for correctly lighting a flock of laying hens is complicated and has been detailed in Chapter 18.

Nesting as an indication of ovulation. On most occasions the hen seeks a nest about 24 hours after ovulation, and many scientists further the theory that nesting is a better indication of ovulation than egg laying itself. Evidently, the presence of a fully formed egg in the cloaca has nothing to do with the hen's endeavor to seek a nest. Some hens ovulate and the ovum does not reach the oviduct, yet these hens will seek a nest a day later.

Double ovulation. Normally, only one yolk is ovulated per day, but occasionally two may be released and on rare occasions there may be three. If two are ovulated at the same time only one may enter the oviduct, but if both are picked up simultaneously a double-yolked egg will result. About two-thirds of the double-yolked eggs are the result of ovulations within 3 hours of each other. If there is a great difference in ovulation time, two eggs may be produced on the same day, but usually the second is soft-shelled.

Double-yolked eggs are more common during the first part of the egg production period because of an overactive ovary, and are more often associated with meat-type strains than with egg-type ones. The incidence is due in part to genetic reaction since some birds produce higher percentages of double-yolked eggs than others.

Spring- and summer-housed pullets produce a greater number of double-yolked eggs than fall- and winter-housed pullets.

Defective Eggshells

When the normal interval of about 23 to 26 hours between ovulations is broken, more eggs are produced with defective shells, including those with sandpaper texture, white bands, calcium splashing, and chalky white deposits. The occurrence is greater in meat-type breeds than in egg-type ones. From 5 to 7% of the eggs produced by meat-type hens may have defective shells. Regardless of the type of hen, more eggs are laid with defective shells when the birds are kept in cages than when they are kept on a litter floor.

Yolk Size Affects Egg Size

The size of the completed egg is more closely associated with yolk size than with any other factor, although variations in albumen secretions in the oviduct have some influence. The yolk-albumen relationship changes throughout the laying period. Eggs produced at the beginning of the laying period have yolks that comprise about 22 to 25% of the total weight of the egg; while yolks make up about 30 to 35% with hens well advanced into their laying period. As egg size increases, yolk size increases more than the quantity of albumen. When egg size is small, increasing the protein in the diet may increase the total weight up to 1.5 oz/doz (3.5 g/ea) or vice versa.

Blood Spots and Meat Spots

Often, when the yolk sac ruptures along the stigma, small blood vessels near the area are broken, leaving a clot of blood attached to the yolk to be enclosed as the egg is completed in the oviduct. The hemorrhage is related to many things: genetics, feed, age of the hen, and others. Blood spots are more common in brown-shelled eggs than in white-shelled.

Any tissue sluffed from the follicular sac or the oviduct will probably be included in a part of the developing egg as it passes through the oviduct. These bits of tissue darken with age and are known as meat spots. Many blood spots darken too, and are often incorrectly classified as meat spots.

3-B. PARTS OF THE OVIDUCT

The *oviduct* is a long tube through which the yolk passes and where the remaining portions of the egg are secreted. Normally, it is relatively small in diameter, but with the approach of the first ovulation its size and wall thickness expand greatly. The segments of the oviduct and their purpose are summarized below and are illustrated in Figure 3-1.

Infundibulum

The funnel-shaped upper portion of the oviduct is the *infundibulum*. When functional, its length is approximately 3.5 in. (9 cm). Normally

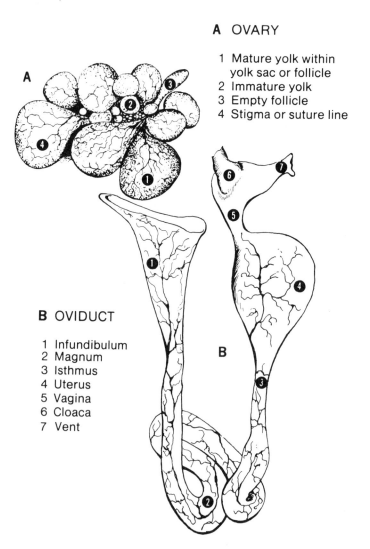

A OVARY

1 Mature yolk within
  yolk sac or follicle
2 Immature yolk
3 Empty follicle
4 Stigma or suture line

B OVIDUCT

1 Infundibulum
2 Magnum
3 Isthmus
4 Uterus
5 Vagina
6 Cloaca
7 Vent

Figure 3-1. Ovary and oviduct.

inactive except immediately after ovulation, its purpose is to search out and engulf the yolk to cause it to enter the oviduct. After ovulation the yolk drops into the ovarian pocket or the body cavity, from which it is picked up by the infundibulum. The yolk remains in this section for only a short period of about 15 minutes, then is forced along the oviduct by multiple contractions.

Malfunction of the infundibulum. To be completely functional, the infundibulum should pick up all the yolks dropped into the body cavity. However, it has been found that an average of 4% are not drawn into the infundibulum, but remain in the body cavity where they are absorbed within about a day. The percentage varies with strains of chickens, some of which retain up to 10% of their yolks in the body cavity. Meat-type birds are more often affected than egg-type ones.

Internal layers. Sometimes the infundibulum loses its power to pick up a high proportion of the yolks, and they accumulate in the body cavity faster than they can be absorbed. Such hens are known as "internal layers," although the term does not well define the condition. The abdomen becomes extended, and the hen stands in an upright position.

Magnum

The magnum is the albumen-secreting portion of the oviduct, and is about 13 in. (33 cm) long in the average laying hen. It takes approximately 3 hours for the developing egg to pass through the magnum.

The albumen in an egg is composed of four layers. The names and percentages are

Chalazae	2.7%	Dense white	57.0%
Liquid inner white	17.3%	Outer thin white	23.0%

All four are produced in the magnum, but the outer thin white is not completed until water is added in the uterus.

Chalazae. Upon breaking an egg, one notices two twisted cords, known as chalazae, extending from opposite poles of the yolk through the albumen. The chalaziferous albumen is produced when the yolk first enters the magnum, but the twisting to form the two chalazae seems to occur much later as the egg rotates in the lower end of the oviduct. Twisted in opposite directions, the chalazae tend to keep the yolk centralized after the egg is laid.

Liquid inner white. As the developing egg passes through the magnum only one type of albumen is produced, but the addition of water plus the rotation of the developing egg gives rise to the various layers, one of which is the liquid inner white.

Dense white.   The dense white makes up the largest portion of the egg albumen. It contains mucin that tends to hold it together. The amount of thick white generated in the magnum is large, but the breakdown of mucin and the addition of water as the egg moves through the oviduct tends to reduce the amount of thick white while increasing the amount of thin white. At the time the egg is laid, it has about one-third of its original content of thick white, but what remains still comprises over half the albumen in the egg.

Egg quality deterioration.   After laying, there is a constant change in the interior contents of the egg. The thick white does not retain its viscous composition and its volume decreases, while the thin white becomes more watery, and the amount increases.

Isthmus

Next, the developing egg is forced into the isthmus, a relatively short section approximately 4 in. (10 cm) in length, where it remains for about 75 minutes. Here the inner and outer shell membranes are formed in such a manner as to represent the final shape of the egg. The contents at this time do not completely fill the shell membranes, and the egg resembles a sack only partially filled with water.

The shell membranes are a papery material composed of protein fibers. The inner one is laid down first, followed by the outer shell membrane, about three times as thick as the inner. The two are held closely together as one prior to the time the egg is laid; then at some section of the membranes the two separate to produce an *air cell.* The area of separation is usually in the large end of the egg, but it may be "misplaced" and occur in the small end or on the side in a small percentage of the eggs.

Air cell is important.   When the egg is first laid there is no air cell. However, it soon appears and increases in diameter to about 0.7 in. (1.8 cm). As the egg ages, the interior contents dehydrate, and the air cell increases in diameter and depth. But increases in size may be altered according to conditions under which the egg is kept: High surrounding temperature and low humidity increase the size of the air cell and vice versa.

Shell membranes act as a barrier.   The shell membranes act as a barrier to penetration of outside organisms such as bacteria. Eggs laid by young hens have thicker shell membranes than eggs laid by older hens.

Uterus

The uterus is primarily the shell gland, and is approximately 4.0 to 4.7 in. (10 to 12 cm) long in the laying hen. The developing egg remains in the uterus about 18 to 20 hours, much longer than in any other section of the oviduct.

Outer thin white deposited after shell membranes. When the egg first enters the uterus, water and salts are added through the shell membranes by the process of osmosis to plump out the loosely adhering shell membranes and to liquify some of the thin albumen to form the fourth layer, the *outer thin white.*

The shell. Eggshell calcification begins just before the egg enters the uterus. Small clusters of calcium appear on the outer shell membrane just before the egg leaves the isthmus. These are the *initiation grains* for calcium deposition in the uterus. Their number is probably inherited and plays a part in the amount of calcium deposition later. They disappear a short time after the egg enters the shell gland.

The first shell is deposited over the initiation sites to form the inner shell, a *mammillary* layer composed of calcite crystals, a spongelike material. This layer is followed by the addition of the outer shell made up of a layer of hard calcite crystals, chalky and about twice as thick as the inner shell surface. The longer the calcite columns, the stronger the shell. The completed eggshell is composed almost entirely of calcite ($CaCO_3$), with small deposits of sodium, potassium, and magnesium.

Source of calcium for eggshell. There are only two sources of calcium for eggshell production: the feed, and certain bones. Normally, most of the calcium for egg formation comes directly from the feed, but some comes from the calcium reservoir, the *medullary bone*, particularly at night when the bird is not eating, yet eggshell is being deposited.

Formation of calcium carbonate. The calcium carbonate of the shell is formed when calcium ions are furnished through the blood supply, while the carbonate ions come from both the blood and the shell gland. Anything that reduces the supply from the blood interferes with maximum $CaCO_3$ deposits of eggshell, and shells of poor quality will result. It is thought that high environmental temperatures may cause such a reduction inasmuch as eggshells are thinner during hot weather.

Poor shell quality. Many factors may cause a deterioration in eggshell quality, and their influence may or may not be due to an inadequate supply of carbonate ions. Shell quality is definitely correlated with shell strength (thickness). Several factors lower eggshell quality, for example,

1. Quality is reduced the longer the bird continues to lay, evidently because the hen cannot produce adequate calcium carbonate to cover the larger eggs produced during the latter part of the laying cycle.
2. Increased environmental heat.

3. Eggs laid in the morning have poorer shell quality than those laid in the afternoon.
4. Stress of the birds in the flock.
5. Practically all misshapen eggs and eggs with body checks are laid between 6:00 and 8:00 a.m.
6. Certain poultry diseases (bronchitis, Newcastle disease, etc.).
7. Certain drugs.

Calcium requirements are high during production.   The demand of the laying hen for calcium is extremely high. A 4-lb (1.8-kg) hen producing 250 2-oz (56.7-g) eggs per year requires about 1.25 lb (0.56 kg) of calcium. Since this is about 25 times the amount of calcium in the bird's skeleton, it is evident that the dietary need for calcium is great. Most laying rations contain from 3 to 4% calcium to meet the demand (see Chapter 31-H).

Pores in the eggshell.   Both the inner and outer shell layers contain small openings called *pores.* There may be 8,000 per egg. Through these pores air finds its way into the egg to supply oxygen to the developing embryo, and remove carbon dioxide and moisture. In the freshly laid egg, the pores are almost completely closed, but as the egg ages the number of open pores is greatly increased.

Color of eggshell.   Eggshells are predominantly white or various shades of brown, but a South American chicken, the Araucana, produces eggs with green or blue shells. Pigments produced in the uterus at the time the shell is produced are responsible for the color. The shade of coloring is quite consistent for each bird, the density being a derivative of the genetic makeup of the individual. Some strains of birds lay eggs with very dark brown shells, while others may vary all the way to pure white. The brown pigment in eggshells is *porphyrin,* uniformly distributed throughout the entire shell.

The cuticle.   Laid down on the outside of the shell in the uterus is the last of the concentric layers of egg formation. It is known as the *cuticle,* and is composed primarily of organic material. Containing a high percentage of water, it acts as a lubricant during the laying process. But once the egg is laid the cuticle material soon dries, blocking off many of the pores of the eggshell to help prevent too rapid an exchange of air and moisture and also to aid in preventing bacteria from entering the egg contents.

Vagina

The next section of the oviduct is the vagina, about 4.7 in. (12 cm) in length in a bird during egg production. Here the cuticle is deposited on the shell to fill many of the shell pores. Normally, the egg is held in the vagina for only a few minutes, but if necessary it may be held for several hours.

Eggs laid large end foremost. Although the egg transverses the oviduct small end first, if the hen is not molested or frightened the egg will rotate horizontally just prior to oviposition and it will be expelled large end first. The rotation requires less than 2 minutes, and makes it possible for the uterine muscles to exert greater pressure on more surface area during oviposition. However, if something disturbs the bird prior to rotation the egg will be laid quickly, and forced through the vent small end foremost.

The parts of the newly laid egg are shown in Figure 3-2.

3-C. SHAPE AND SIZE OF THE EGG

Shape

Although most eggs are ovoid in shape, the exact shape of the egg is generally due to inherited genetic factors. Each hen lays successive eggs of the same shape, that is, pointed, long, thick, etc.

Specifications for a standard egg.

Weight (oz)	2.0
Weight (g)	56.7
Volume (cm^3)	63.0
Specific gravity	1.09
Long circumference (cm)	15.7
Short circumference (cm)	13.7
Shape index	74.0
Surface area (cm^2)	68.0

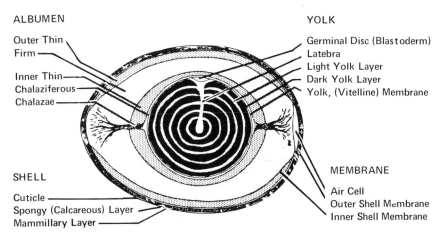

Figure 3-2. The parts of the newly laid egg.

Imperfections in Egg Shape

Some hens continually lay eggs with shape imperfections. These come under several categories: wrinkled, ridged, flat sided, pointed tip, etc. Similar imperfections will be found on each egg the hen lays. Some are of genetic origin; others are probably due to abnormalities of the oviduct.

Egg Size

Eggs from a flock of chickens vary in size (or weight) because of many circumstances. The exact cause of some of the variations is not known, but much has been established regarding others. Some of these variations are as follows:

1. Some hens lay eggs that are larger or smaller than those laid by other hens. Obviously, this difference is due mainly to genetic factors that have an effect on the length of the growth period of the ova. In turn, larger yolks produce larger eggs, while smaller yolks give rise to smaller ones.
2. The first eggs a hen lays are smaller than those laid later. Egg size gradually increases as the hen continues to lay, but the increments of increase are not uniform. Egg size increases rapidly during the first part of the egg production period, but only gradually thereafter.
3. The sequence of eggs within a clutch affects egg size. In most instances, the first egg of the clutch is the heaviest, with each succeeding egg being proportionately smaller. In such cases the yolk size is reduced, but the decrease in size of subsequent eggs in the clutch is also due to a decrease in the amount of albumen produced.
4. Some feed components will affect egg size. For example, egg size may be increased by increasing the protein content of the feed.
5. Hot weather affects the flock, causing a decrease in egg size.

3-D. COMPOSITION OF THE EGG

Water makes up about 65% of the shell egg; the contents contain about 74%. The albumen is high in water content; the solid portion is almost entirely protein with a small amount of carbohydrate. The yolk is composed of about half water, but the solid portion is made up of a high amount of fats, proteins, vitamins, and minerals. The composition of the average egg is given in Table 3-1.

Table 3-1.   Composition of the Newly Laid Egg

Component	Egg with Shell %	Egg Contents without Shell %	Yolk %	Albumen %	Shell and Shell Membranes %
Whole egg	100	—	31.0	58.0	11.0
Water	65	75.0	48.0	87.0	2.0
Protein	12	12.0	17.5	11.0	4.5
Fat	11	11.0	32.5	0.2	—
Carbohy-drate	1	0.5	1.0	1.0	—
Ash	11	1.5	1.0	0.8	93.5

Age of hen affects egg composition.   As the laying flock ages there is an increase in egg weight, dry weight, and the percentage of yolk, while the percentage of shell, albumen, and albumen solids decreases.

Altering egg contents.   Although the composition of the egg has remained rather constant over the years, it is possible to make slight changes in energy content, amounts of certain vitamins, and trace minerals by altering the diet. Likewise, through genetics, certain portions of the egg contents may be altered.

Composition of Araucana eggs.   On an equal weight basis, typical Araucana eggs had 23% more yolk, 9% less albumen, 9% less shell membrane, 10% more whole-egg dry matter, and 0.6% less moisture than White Leghorn eggs according to Robert Simmons, III and Ralph Somes, Jr. (1982. *Poultry Sci. 61,* 1777–1781).

Cholesterol Content of Eggs

Cholesterol is a crystalline alcohol with the formula $C_{27}H_{45}OH$, belonging to a group of lipid (fatlike) compounds known as *sterols.* It is synthesized in the body of all animals, birds, and human beings, and is essential to cellular metabolism. It is not associated with the vegetable kingdom. Cholesterol is found in the blood of humans, and high levels have been associated with many types of arterial diseases because it reduces the size of the blood vessels, thereby increasing the blood pressure. In 1974, the U.S. Department of Agriculture declared that a large egg (2oz. [56.7g]) contained 274 milligrams of cholesterol, a high amount, which implicated eggs as one of the fatty substances associated with heart disease in humans.

In 1987, several large egg producers in the United States began selling eggs with a cholesterol content of less than 210 milligrams. This action aroused the industry and the U.S. Department of Agriculture, and in 1988, a massive test was conducted and analyzed eggs from samples collected across the United States. Another similar sampling was made of eggs collected in the spring of 1989. A standard quantitative test for cholesterol, using direct saponification for extractions, and gas chromatography/capillary column for analysis, showed that eggs contained 22% less cholesterol than previously thought.

USDA results showed that the average large egg produced in the United States contained 213 milligrams of cholesterol. Extra-large eggs contained about 230 milligrams, while medium-sized eggs contained about 180 milligrams; or 399 mg/100 g of whole egg, or 1,280 mg/100 g of egg yolk.

Total fat in a standard-sized egg was found to be 5.1 g instead of 5.7 g. Fatty acid composition was not significantly different, either proportionally or in total numbers. An average large egg contained 1.61 g of saturated fatty acids, 2.17 g of monosaturated fatty acids, and 0.71 g of polyunsaturated fatty acids.

The USDA has now set the new standards as official. Governmental chemists have stated the changes are due to better testing methods, along with different feeds, husbandry, and breeding practices.

4
Development of the Chick Embryo

Processes involved with modern artificial incubation of chicken eggs should be prefaced with a brief description of the development of the chick embryo. Unlike that of the mammal, the avian embryo develops from the food material stored in the egg rather than from nutrients derived from the blood supply of the mother. Furthermore, most of the embryonic growth takes place outside the body of the mother, and development is more rapid than in the case of the mammalian embryo.

4-A. FERTILIZATION

Normally, fertilization is a natural process, but artificial insemination is also practiced.

Natural Fertilization

The male initiates sexual activity in chickens through a process of *courting.* Chickens are polygamous, but certain males and females mate together regularly.

During the course of normal mating between a male and female chicken the male ejaculates from 1.5 billion to 8 billion sperm cells, with more being produced at the beginning of the day than after many matings by the end of the day. First ejaculates average about 1.0 ml, but after several matings they will be reduced to 0.5 ml or less.

A male may mate from 10 to 30 or more times a day, depending on the availability of females and competition from other males. As matings continue, the volume of semen and the number of sperm cells decrease.

But seldom will an ejaculate contain less than 100 million sperm, the minimum requirement seemingly necessary for good fertility.

The copulatory organ.   Male chickens have a small phallus that becomes engorged with lymph to produce a slight erection and to form a copulatory organ, although there is practically no penetration at the time of mating. Ducks, geese, and some other birds have well-defined copulatory organs.

Movement of sperm cells.   The sperm are immediately carried to the uterovaginal and infundibular spermatozoa storage glands in the upper part of the oviduct of the hen, the infundibulum. If no egg is in the oviduct, the movement takes only 30 minutes. Within 15 minutes after an ovulation a few sperm cells find their way to the area of the pronucleus on the surface of the egg yolk. Three or four sperm cells may enter, but only one unites with the female egg cell to form a new individual, the *zygote*. After a single mating fertile shell eggs may be produced within 20 hours, but normal maximum fertility from a flock of hens will not be attained until about the third day.

Ambient temperature and fertility.   A cold room temperature depresses the normal activity of the male testes and thereby reduces fertility. The optimum ambient temperature is in the neighborhood of 66°F (19°C) for both males and females.

Fertility after removal of males from flock.   If the males are removed from the flock of hens, some fertile eggs will be produced for as long as 4 weeks but the percentage of fertile eggs decreases each day after removal of the males, the decrease being more rapid after the fourth or fifth day.

Viability of newly produced sperm.   New sperm cells are more viable than old ones. Because of the increased viability, new sperm cells are more likely to unite with the female egg cell. Thus, if the males are removed from a flock, and new ones substituted on the same day, practically all the fertilization after 3 days will be from the sperm produced by the new males.

Fertilization and time of mating.   With natural mating, the time of day that copulation occurs does not seem to affect the degree of fertilization of the egg cell.

Body type no indication of fertility.   During natural matings, fertility is poorly correlated with male body weight at 8, 12, 20, and 24 weeks of age, breast angle, or hip width.

Parthenogenesis

Parthenogenesis is reproduction by an unfertilized gamete—that is, no male is involved. It is quite common among invertebrates. Although a rare phenomenon in chickens, it does occur. However, most embryonic development lasts only a few days; then the embryo dies. But occasionally a very small percentage hatch, all of which are diploid males with small testes, and capable of living only a short time after hatching.

Artificial Insemination

It is possible to obtain semen artificially from the male chicken to inseminate females. The soft part of the abdomen below the pelvic bones of the male is massaged to protrude the papillae and the semen is gently squeezed (milked) out, beginning closest to the body and collected in a vial. The semen is then transferred to a syringe, diluted with special diluents, and about 0.025 to 0.035 ml are forced into the oviduct of the hen to a depth of about 1 to 2 in. (2.5–5.0 cm), depending on the size of the bird. The semen must be fresh, and inseminations must be repeated every 5 to 7 days to maintain optimum fertility. Fertility through artificial insemination is better in Leghorns than in meat-type strains, but the reason is unknown.

Fewer males needed.   For natural mating, one male is usually required for about every ten females, but with artificial insemination, one male can produce enough semen to fertilize 100 to 150 females on a weekly basis.

Semen not stored easily.   Avian semen is difficult to store, but new diluents show hope of extending the viability of storage. Males should be ejaculated about three times weekly, although fertility will not be impaired if semen is collected as often as once a day, but the semen volume will be lower. Insemination must be made quickly after the fresh semen is collected. It will not withstand freezing, for when frozen and thawed, fertility will drop to about one-half.

Time of semen collection.   Semen collected in the morning will have greater volume, greater sperm motility, and slightly higher sperm concentration than that collected in the afternoon.

Insemination should be completed in late evening.   In a test at Auburn University, meat-type females inseminated at 9:00 p.m. gave the best fertility. Mid-morning and mid-afternoon inseminations produced no difference in fertility, but both were lower than at 9:00 p.m. There was no difference between young pullets and hens.

Increased cost the economic deterrent.   Artificial insemination of chickens under present procedures is costly. It requires one man-hour to collect the semen from 145 males. About 200 to 260 females can be inseminated per man-hour, but besides this cost there is the expense of keeping the females in two- to four-bird cages and the males in single-bird cages. The procedure has more economic merit with meat-type birds than with egg-type birds.

4-B. PREOVIPOSITAL EMBRYONIC DEVELOPMENT

The first embryological development takes place within the body of the hen at her body temperature between 105° and 107°F (40.6° and 41.7°C). About 4.5% of the total time consumed in embryological development takes place in the oviduct. On the average, the total incubation process requires about 22 days: 1 day in the hen and 21 days in the incubator.

Preoviposital development is initiated in the infundibulum about 15 minutes after the yolk is ovulated at which time a sperm cell enters the female egg cell to form the *zygote*, a one-celled individual. About 5 hours later, the new zygote enters the isthmus portion of the oviduct, and the first cleavage (cellular division) takes place and two daughter cells arise. About 20 minutes later, the two new cells divide and four cells are present. Eight cells have developed by the time the forming egg leaves the isthmus to enter the uterus. After 4 hours in the uterus the developing embryo has grown to 256 cells, all by geometric progression.

During the above divisions of the cells the blastodisc is formed, each cell clinging tightly to its neighbors in a single layer in contact with the yolk. As cellular divisions continue, several layers of cells develop and make up the blastoderm. Soon the cells in the center of the blastoderm become detached from the yolk to form a pouch or cavity. It is in the center of this cavity that further embryonic development takes place.

While the developing egg is still within the body of the hen the blastoderm develops into two layers by what is called *gastrulation*. The upper layer of cells is called the *ectoderm*, and the lower, the *entoderm*. Soon a third layer, the *mesoderm*, develops between the ectoderm and entoderm. From these three layers all the organs and parts of the body develop. The ectoderm gives rise to the nervous system, parts of the eyes, the feathers, beak, claws, and skin. From the entoderm come the respiratory and secretory organs along with the digestive tract. The mesoderm is responsible for the development of the skeleton, muscles, blood system, reproductive organs, and the excretory system. By the time the egg is laid the developing embryo will be composed of thousands of cells.

4-C. POSTOVIPOSITAL EGG-HOLDING PERIOD

In a fertile egg just laid the embryo should be developed past the gastrula stage and well suited to a cessation of development prior to being placed in the incubator. But to arrest all development during the holding period the egg should be maintained at a temperature of 65°F (23.9°C) when held for less than 5 days. When held for a longer period, hatching eggs should be kept at a temperature of 55°F (12.8°C).

4-D. DEVELOPMENT OF THE EXTRAEMBRYONIC MEMBRANES

Because the embryo has no anatomical connection with the mother's body, nature has endowed it with certain membranes necessary to utilize the food material contained in the egg. These membranes are as follows:

Yolk sac.   Enveloping the yolk, this membrane secretes an enzyme that changes the yolk contents into a soluble form so that the food material may be absorbed and carried to the developing embryo. The yolk sac and its remaining contents are drawn into the body cavity just prior to hatching to serve as a temporary source of food material.

Amnion.   The amniotic sac helps the young embryo during development as it is filled with a transparent fluid in which the embryo floats.

Allantois.   This membrane serves as a circulatory system, and when fully developed completely surrounds the embryo. The allantois is initiated on the third day and is fully developed by the 12th day. It has the following functions:

> *Respiratory.* It oxygenates the blood of the embryo and removes the carbon dioxide.
>
> *Excretory.* It removes the excretions of the embryonic kidney and deposits them in the allantoic cavity.
>
> *Digestive.* It aids in the digestion of albumen and in the absorption of calcium from the eggshell.

Chorion.   This membrane fuses the inner shell membrane with the allantois, and helps the latter in completing its metabolic functions.

4-E. DAILY CHANGES DURING EMBRYONIC GROWTH

During the incubative process, moisture is lost from the egg through the shell. This drying reduces the size of the egg contents and increases the size of the air cell. After 19 days of incubation the air cell usually occupies one-third of the egg. It is deeper on one side than on the other. The development of the chick embryo is a complicated process. The main changes follow, but remember there has been about 24 hours of embryonic development prior to the time the egg is laid.

First Day

Several embryonic processes are in evidence during the first 24 hours of incubation in the incubator.

4 hr.   Heart and blood vessels start to develop.

12 hr.   Heart starts to beat. Blood circulation begins with the joining of the blood vessels of the embryo and the yolk sac.

16 hr.   First sign of resemblance to a chick embryo through the development of the somites, blocklike structures that develop on both sides of the spinal cord. From these, bones and muscle develop.

18 hr.   Appearance of the alimentary tract.

20 hr.   Appearance of the vertebral column.

21 hr.   Origin of the nervous system.

22 hr.   Head begins to form.

24 hr.   Eyes originate.

Second Day

25 hr.   Beginning of formation of the ear.

Third Day

60 hr.   The nose is initiated.

62 hr.   The legs begin their development.

64 hr.   Beginning of the formation of the wings. The embryo begins to rotate so that it lies on its left side. The circulatory system rapidly increases during the third day.

Fourth Day

The tongue begins to form and now all body organs are present. The vascular system is clearly evident to the naked eye.

Fifth Day

The reproductive organs differentiate, and sex is developed. The heart begins to take a definite shape and the vascular area of the yolk sac covers two-thirds of the yolk. The face and nasal portions of the embryo begin to take a lifelike appearance.

Sixth Day

The beak and the eggtooth begin to take a normal form. Some voluntary movement of the embryo may be noted.

Seventh Day

The body begins rapid development, more so than the head. Body organs are visible.

Eighth Day

Feather germs, the origin of the feather tracts, appear.

Tenth Day

The beak begins to harden. Toes, as well as scales on the legs, start to appear.

Eleventh Day

The walls of the abdomen appear, and the intestines may be seen in the yolk sac.

Thirteenth Day

Chick down is present, the skeleton is beginning to calcify, and most organs are differentiated with only final growth necessary.

Fourteenth Day

The embryo rotates to position itself parallel to the long axis of the egg with the head normally toward the large end.

Seventeenth Day

The head turns so that normally the beak is under the right wing and toward the lower part of the enlarged air cell.

Nineteenth Day

The yolk sac begins to enter the body cavity through the umbilicus. This yolk material is used as food during the first few days of the chick's life. The chick finds a position necessary for pipping the shell.

Twentieth Day

The yolk sac has completed its entrance into the body cavity. The embryo occupies all the area within the shell except for the air cell. The umbilicus is beginning to close. Next the beak of the chick penetrates the inner shell membrane and enters the air cell. Slowly the chick inhales some air, and pulmonary respiration begins. Next it pips the shell to gain entrance to outside air. At this time the lungs become fully functional, and the chick is under critical stress, the second in its life.

Twenty-first Day

After first pipping the shell, the chick rests for several hours, then cuts a circular line around the eggshell in a counterclockwise direction by striking the shell with its eggtooth. Normally, if the chick has positioned itself correctly prior to hatching, this pipping is near the large end of the egg. From the time the shell is first broken until the chick is able to liberate itself, exhausted and wet, 10 to 20 hours have elapsed.

4-F. EMBRYONIC COMMUNICATION

It has been known for some time that eggs from some species of birds tend to hatch at the same time under natural incubation even though the eggs have been laid over a period of several days and those laid first have been incubated longer than those last laid, inasmuch as there is some incubation when the bird sits on the nest in the course of laying the remaining eggs in her clutch.

The phenomenon is the result of the developing embryos "talking" to each other by emitting a clicking noise. The speed of the clicking is the factor responsible for accelerating or retarding embryonic growth. Slow rates tend to accelerate development, while fast rates retard it. Artificial clicking has been shown to increase the rate of growth of the chicken embryo resulting in eggs hatching earlier, yet fast artificial clicking has had little effect, if any, on retardation.

4-G. EMBRYONIC METABOLISM

The developing embryonic chick requires carbohydrates, fats, proteins, minerals, vitamins, water, and oxygen as food materials to complete its normal development.

Energy

Energy for the embryo comes from proteins, carbohydrates, and fats, but the source varies with the age of the developing chick. Carbohydrates furnish the supply during the first 4 days but thereafter both carbohydrates and proteins are utilized as evidenced by the formation of urea, a by-product of protein metabolism. Fat from the egg yolk is the probable source of energy during the latter part of the incubation period, but it cannot be surmised that the processes are so well established that one or another does not carry over into a different time period.

Minerals

Calcium is the most important mineral involved in embryonic metabolism. It is transferred from the shell to the embryo. In fact, the calcium content of the egg contents and embryo rises markedly beginning with the 12th day; and the increase in both is parallel, although the amount in the embryo is always greater than that of the egg contents. The fact that the amount of calcium in the shell membranes increases during incubation offers further proof of the transfer of calcium from the shell to the shell membranes. Additional proof is noted in the fact that infertile eggs, incubated with fertile eggs, show no transfer of calcium from the shell to the interior contents or shell membranes.

Other minerals are also needed by the young growing embryo. The embryonic source is the egg contents. In many instances, deficiencies in the diet of the hen laying the egg will result in inadequate amounts of minerals available to the embryo. Many of these, as well as the importance of other essential embryonic nutrients, are discussed in Chapter 32.

Vitamins

All the vitamins are needed for embryonic growth just as they are for the growth of the chick after it hatches. However, the requirements for the individual vitamins are different for development during the embryonic stage. Rations for breeders producing hatching eggs are, therefore, different from those used for the production of market eggs (see Chapters 8-L and 32).

5

Chick Hatcheries

Chick hatcheries are modern buildings that provide separate rooms for each hatchery operation, but each room has its individual requirements. The following pages describe these requirements.

5-A. SIZE AND LOCATION

Determining the Size of the Hatchery

The size of the hatchery is computed as follows:

1. Egg capacity of the setters and hatchers
2. Number of eggs that can be set each week
3. Number of chicks that can be hatched from each setting
4. Number of chicks that can be hatched each week

Table 5-1 shows the relationship between the above measurements when eggs are set twice weekly.

Because of the importance of hatchery isolation with modern MG (*Mycoplasma gallisepticum*) and MS (*Mycoplasma synoviae*) disease-control programs, proper location of the hatchery is important (see Chapters 37-H and 37-I). The hatchery should be situated at least 1,000 ft (305 m) from houses containing chickens, but even this distance is not enough to ensure that there will not occasionally be horizontal transmission of disease-producing organisms from the chicken houses to the hatchery. The hatchery area should be a separate unit with its own entrance and exit, unassociated with those of the poultry farm.

Table 5-1. Methods Used for Determining Hatchery Capacity (Chicks Hatched Twice Weekly)

Egg Capacity of Incubators (Including both Setters and Hatchers)	Number of Eggs That Can Be Set		No. of Chicks That Can Be Hatched at 85% Hatchability	
	Per Setting	Per Wk	Each Hatch	Each Wk
100,000	16,666	33,333	14,166	28,333
200,000	33,333	66,667	28,333	56,667
400,000	66,667	133,333	56,667	113,333
600,000	100,000	200,000	85,000	170,000
800,000	133,333	266,667	113,333	226,667
1,000,000	166,667	333,333	141,667	283,333

5-B. PERSONNEL FLOW THROUGH HATCHERY

To maintain an MG-free and MS-free hatchery it is essential that all persons entering the premises shower and change into clean clothing in an adjoining room. They may leave only through this same room where they change back into street clothing. Thus, this shower room becomes an integral part of hatchery planning. It is the only entrance and exit, and the hatchery becomes an isolated unit as far as human beings are concerned. All other doors must be kept locked to eliminate the human element in the transfer of infectious diseases.

The shower room must be carefully constructed so that those entering the hatchery may gain entrance only through the water from the shower; they must not be able to walk around it. Also, there must be a room or area for disrobing and another for dressing in clean work clothes. Some method of heating the room during cold weather should be provided. An individual locker for each employee is advantageous.

5-C. EGG-CHICK FLOW THROUGH HATCHERY

Hatcheries should be constructed so that the hatching eggs may be taken in at one end and the chicks removed at the other. In other words, eggs and chicks should flow through the hatchery from one room to the one next needed in the hatching process. There should be no backtracking. Such a flow affords better isolation of the rooms and there is less human traffic throughout the building. Care in designing the floor arrangement of a hatchery will result in fewer steps. A diagrammatic flow system is shown in Figure 5-1.

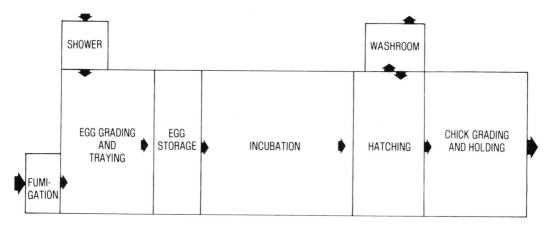

Figure 5-1. Egg-chick flow through hatchery.

Modern Hatchery Design

Most large hatcheries are T-shaped in order to use floor space effectively. An example of the floor layout of such a hatchery with a setting capacity of about 1 million eggs is shown in Figure 5-2.

Getting Hatching Eggs into the Hatchery

Employees delivering hatching eggs to the hatchery must not enter the hatchery in the course of their duties. Eggs should be delivered to the door of the fumigation cabinet. A hatchery employee takes the eggs from the unloading dock into the fumigation room. Electric buttons to ring a bell inside the hatchery should be installed on the outside door to notify people that someone on the outside is seeking admittance.

Removing Chicks from the Hatchery

Similarly, employees responsible for loading chicks into chick delivery vans must not enter the hatchery building. Hatchery employees should deliver chick boxes to the truck delivery door where they are taken by the truck driver and loaded into the truck. Under no condition should the hatchery employee walk through the delivery door, or the truck driver enter the hatchery through the chick delivery exit.

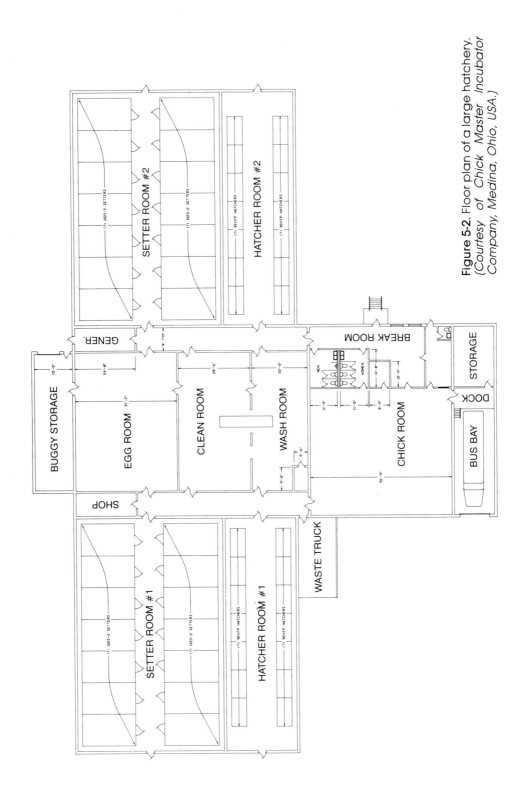

Figure 5-2. Floor plan of a large hatchery. (Courtesy of Chick Master Incubator Company, Medina, Ohio, USA.)

5-D. HATCHERY CONSTRUCTION

Hatchery buildings should be intricately designed, properly constructed, and adequately ventilated. An architect should be employed to draw the details and write the specifications. Generally, the incubator manufacturer can supply details for building construction. Only brief general points will be considered here.

Truss Design

Unless the hatchery building is extremely wide, trusses should be used for supporting the roof. Since the hatchery must be ceiled, almost any type of roof design may be used—flat, gable, monitor, or shed. Posts should be avoided if possible. If it is necessary to use posts, a floor design should be drawn showing the location of all hatchery equipment. Posts may be used in construction when they will not interfere with the equipment or hatchery chores.

Width of Hatchery

The width of the setter and hatcher rooms will be determined by the type of incubator used. Find the depth of the incubators; then allow space for the working aisles, behind the machines (if necessary), and the walls. The total width of the room may then be determined. Other rooms in the hatchery must be built "around" these two rooms to provide proper egg flow, and to present an overall pleasing design and appearance.

Ceiling Height

Most commercial hatcheries are built with forced-draft ventilating systems. Thus, there is no need for high ceilings. The recommended height is 10 ft (3.1 m).

Walls

Fireproof material should be used in constructing as much of the hatchery building as possible. Because the interior of the hatchery building is continually being washed and disinfected, the inside walls should be covered with a glazed hard nonabsorbent finish. This finish also prevents the growth of molds common to walls that are porous and absorbent.

Concrete blocks lend themselves very well to wall construction. They may be painted with a material that seals the pores of the blocks and leaves a hard glazed surface. Inside walls between rooms should not be wood, because the water remaining after each washing eventually rots them. If wood has to be used, it should be pretreated and waterproofed.

Ceiling Material

Most hatchery rooms have a high humidity, and during cold weather condensation of moisture on the ceilings is common. Thus, it is impractical to construct the ceilings of any material that will deteriorate in the presence of water. Plaster should not be used. The best materials are waterproof pressed woods or metals. Insulation of the ceiling will eliminate a good share of moisture condensation. Adequate ventilation in the room will also help, as will negative pressure.

Doors

Most common doors are 7 ft (2.1 m) in height, but these are not adequate for a hatchery. Because chick box racks and other high equipment must be moved through the hatchery doors, openings should be 8 ft (2.4 m) high and at least 4 ft (1.2 m) wide, and doors double-swinging. Those into the washroom and those at the chick box exit should be much wider. Where the opening is wide, a sliding door may be more practical.

Floor Construction

All floors must be concrete, preferably with imbedded steel to prevent cracking. The concrete must be given a glazed finish. There must be no high or low spots, for floors are to be washed almost daily. All water must drain away; none can be left standing.

Slope of the floor. Floor drains should be numerous and the slope of the floor to the drain should never be greater than 0.5 in. (1.3 cm) in 10 ft (3 m). Otherwise, hatchery equipment will not roll over the floor easily.

Floor drains. Tremendous quantities of water are used in the hatchery. No smaller than 6-in. (15 cm) trap-type floor drains should be used. Be sure these have a flat top so hatchery equipment can be rolled over them.

Another method is to construct a gutter about 6 in. (15 cm) wide and 6 in. (15 cm) deep in the concrete floor in such a way that water will run to the end of the gutter and into the sewer system. Cover the gutter with a steel plate in which holes have been drilled or with slotted cast iron covers built for this purpose.

Sewers

Sewers should be larger than those used in most industrial buildings. This requirement is necessary because

1. A very large amount of water is involved.
2. Broken eggshells settle in the sewer lines if the water does not flow fast enough.

Electric Lines

Electric lines may be laid below the concrete floor in waterproof conduits, or over the ceiling. If many changes in the location of electrical equipment are contemplated, the electric lines should be over the ceiling.

Electricity

In most countries commercial electricity is 60 cycles, but is 50 cycles in others. Be sure you secure the correct type of electrical equipment.

Water Lines

Water lines are best placed below the concrete floor. Water from those laid over the ceiling is warmer and, as many incubators use water to cool the interior, cool water is better than warm water. Large amounts of water will be used in the hatchery for washing hatching trays and cleaning, as well as in the incubators. Be sure the incoming water lines are adequate in size, and the water pressure is 50 PSI (lb/in.^{2}) (3.51 kg/cm^2) at the automatic valves in both setters and hatchers.

Docks

Unloading eggs from trucks and loading chicks into trucks will be implemented if a dock constructed at truck height is used. The top of the dock should be level with the floor of the hatchery, constructed of concrete, with a drain in the middle.

Caution: Water that has been used to wash docks should not be allowed to run back into the hatchery or on the ground by the dock. Construct drains to carry the water away.

Daily Water Requirement

On the average, about 800 U.S. gal (3,028 L) of water per day will be used by the entire hatchery for each 100,000 incubator egg capacity, but the figure is highly variable, depending on the hatchery equipment used.

Truck Bay

An inside loading dock or chick truck bay is very advantageous. The truck is warmer in the winter when chicks are being loaded, and protected from the sun's rays in the summer. A dock or bay may also be used as a truck fumigation room provided it can be sealed and a blower installed to remove the fumigant and fumes from the truck motor.

5-E. HATCHERY ROOMS

Hatchery rooms must be adequate in size. It is better to have them too large than too small. Usually, hatcheries of medium size will hatch chicks twice a week, but large hatcheries will hatch as many as six times a week in order to equalize the daily work and equipment load. Consequently, hatching schedules will affect the size of some rooms in the hatchery. Table 5-2 shows the general recommendations for floor space of the various hatchery rooms when there are two hatches per week.

Table 5-2.   Floor Space for Hatchery Rooms (Two Hatches per Week)

Room Type	Per 1,000 Eggs Incubator-Hatcher		Per Case of Eggs Set per Setting (360 Eggs/Case)		Per 1,000 Straight-run Chicks Hatched per Hatch	
	ft²	m²	ft²	m²	ft²	m²
Egg-receiving room	2.00	0.19	4.32	0.40	15.00	1.39
Egg-storage room[1]	0.33	0.03	0.71	0.06	2.47	0.23
Chick-holding room	4.00	0.37	8.64	0.80	30.00	2.79
Wash room	0.80	0.07	1.73	0.16	6.00	0.55
Storage room	0.70	0.07	1.51	0.14	5.25	0.49

[1]When cases are stacked four high, or racks of similar size.

Fumigation Room

The fumigation room should be as small as possible in order to reduce the amount of fumigant used. It should be the size necessary to hold one pickup of eggs. A fan should be used to circulate the air and exhaust the fumigant.

Egg-Holding Room

Proper construction of the egg-holding (egg-cooler) room is important if the quality of the hatching eggs is to be maintained. It should be about 8 ft (2.5 m) high, insulated, slowly ventilated, with *complete* air movement, cooled, and humidified. The walls should have the following *R* values (see Chapter 11-E):

> *Walls:* R value of 12
> *Ceiling:* R value of 16

Refrigeration. The room must be refrigerated to maintain a temperature of 65°F (18.3°C). A forced-air type of refrigeration unit is required in order to keep a uniform temperature throughout the room.

Measuring unit size. Capacity, or size of the refrigeration unit, is measured in Btu/hr, which is the rate of heat removal. Sometimes the unit of measurement is in tons. A ton of refrigeration is equivalent to 12,000 Btu/hr. At other times the units may be rated according to the size of the motor that operates the compressor: 1 hp, 2 hp, etc.

Calculating size of refrigeration unit. Room insulation, outside temperature, and other factors will determine the size of the unit needed to adequately cool the egg-cooler room. The following calculation will provide an estimate of the size required:

Item	Btu Heat Removed per Hr
Floor area in ft² × 3.00	_____
Wall plus ceiling area in ft² × 4.00	_____
Number of dozen eggs cooled per day × 5.5	_____
Miscellaneous: 35 Btu for each 10 ft² of floor space	_____
Total Btu required	_____

Refrigeration units vary in size and cooling capacity. Specifications for some are given in Table 5-3.

Setting and hatching rooms. The size of these two rooms will depend on the make of equipment used. The manufacturer will be able to supply these dimensions and other necessary construction details. The incubating equipment takes relatively little floor space. The

Table 5-3.   Size and Ratings of Refrigeration Units for Egg-holding Rooms

Dozen of Eggs per Da	Room Size[1]		Refrigeration Unit		
	ft	m	hp	ton	Btu
800	11 × 12 × 7	3.3 × 3.7 × 2.1	1/2	1/2	6,000
1,200	12 × 21 × 7	3.7 × 6.4 × 2.1	3/4	3/4	9,000
1,600	16 × 21 × 7	4.9 × 6.4 × 2.1	1	1	12,000
3,100	21 × 28 × 7	6.4 × 8.5 × 2.1	2	2	24,000

[1]Necessary to set twice weekly.

exact room size involves the aisle and working area necessary to move the eggs and chicks in and out of the machines.

5-F. HATCHERY VENTILATION

Forced air should be used to ventilate the hatchery, but each room must be considered a separate entity inasmuch as each has a different requirement for temperature, humidity, and air.

Each room should be ventilated as a separate unit with displaced air exhausted outside the building. However, if the air is filtered, as much as 80% may be recirculated within a room, but this should not alter the fresh air requirement given in Table 5-4 because of the minimum need for oxygen.

Incoming air should be heated in the winter and cooled in the summer. It should be humidified if necessary. More air should be moved through the hatchery rooms during warm seasons than cool ones. Therefore, rheostats should be installed on all ventilating fans to provide increased or decreased airflow to help control room temperature.

Table 5-4.   Airflow per Minute through Hatchery Rooms

Outside Temperature			Per 1,000 Eggs			Per 1,000 Chicks
°F	°C		Egg-holding Room	Setting Room	Hatching Room	Chick-holding Room
10	−12.2	(ft³)	2.00	7.00	15.00	30.0
		(m³)	0.06	0.20	0.43	0.86
40	4.4	(ft³)	2.00	8.00	17.00	40.0
		(m³)	0.06	0.23	0.48	1.14
70	21.1	(ft³)	2.00	10.00	20.00	50.0
		(m³)	0.06	0.28	0.57	1.42
100	37.8	(ft³)	2.00	12.00	25.00	60.0
		(m³)	0.06	0.34	0.71	1.70

Air Movement Through Hatchery Rooms

Table 5-4 shows the amount of air that should flow through the hatchery rooms according to the outside temperature.

Type of Ventilating System

Positive air pressure is created in a room when the volume of air coming in is greater than that going out. With the reverse, a negative room pressure is created. Positive or negative pressure in a room can often alter the operation of the ventilating system within the incubator; therefore, some incubator manufacturers specify positive pressure in the room; others, negative pressure. But regardless, the pressure inside the room and inside the machines will be the same. The exact room pressure will, of course, be affected by the air pressure outside the building, but the increase or decrease of room pressure should never be greater than $\frac{1}{8}$ in. (0.32 cm) static water pressure. Devices may be purchased or made that will regulate and record the static pressure.

Basics of Hatchery Ventilation

Ventilation is required to

1. Supply oxygen.
2. Remove carbon dioxide.
3. Remove heat from the incubator. Most incubators operate best when the heating, cooling, and ventilating devices of the room function only part of the time. Continuous operation is harder on equipment.

    Heat produced by an egg under incubation starts at about 0.4 gram calorie per day, but on the 19th day will increase to approximately 90 gram calories, giving an average of about 45 when the incubator contains eggs with varying lengths of incubation.
4. Provide incubators with air of proper texture. Most setters operate with a relative humidity of 50 to 60%, but the relative humidity of the air of the incubator room will be quickly reduced once it enters the incubator where the temperature is higher, and humidifiers will be needed in the machines to restore the moisture content of the air. Room relative humidity should be kept at about 50%. Outside air will often vary greatly from

this percentage, and must be conditioned prior to entering the incubator.

Air entering the machines should have a temperature of about 75°F (24°C). When the outside air is colder or warmer it will have to be heated or cooled in the incubator room prior to entering the incubators.

5. Remove heat produced in the hatcher and chick rooms. Each newly hatched chick will produce at least 2 Btu of heat per hour, most of which must be removed from the building by the ventilating system. This heat production compares with about 30 Btu per hour for a 3-lb (1.4-kg) bird.

Air Delivered by Ducts

With the negative pressure system of ventilation, the air must be moved by ducts to provide uniform distribution in the room. These also reduce the speed at which the air travels. Although there may be one main duct for taking air from an inlet to the various hatchery rooms and another main one for exhausting it, there must not be a transfer of air from room to room.

Some ventilating systems employ supplementary air movement within rooms containing incubators. A fan picks up air from the room and forces it through a duct running over, or in front of, the incubators. Holes are cut in this duct to allow the air to escape throughout the room. This recirculated air provides a better air mixture and more rapid circulation, but does not alter the ventilation requirement for fresh outside air.

Air Movement Within Hatchery

Regardless of the place where air enters the hatchery, the air within the building should always move toward the washroom, the most contaminated room in the hatchery. Thus, each room into which air moves should have a lower static pressure (air pressure) than the room from which it came. If it does not, the air will flow in the opposite direction.

Exhaust Air Through a Waterbath

Air from the hatcher room and chick-holding room should be exhausted through a waterbath to prevent chick down from these two rooms from being blown into the open air, then redrawn into the intake of the ventilating system and recirculated to all rooms in the hatchery. The watertank should be located in the attic to prevent freezing in the winter. A disinfectant should be added to the bathwater to help prevent the dissemination of disease-producing organisms.

Down collectors. Some hatchers have down collectors that reduce the amount of down ejected, greatly alleviating the need for a waterbath.

Negative Pressure in Hallways

Temperature, humidity, and air pressure in hallways will adjust to an average of the conditions in the rooms adjacent to the hallways as door openings are not tight. Conditions in hallways will usually be satisfactory except in very cold climates when ceiling heaters will be necessary.

Capacity of Electric Fans

The approximate capacities of fans to move air are shown in Table 5-5. Width, angle, cleanliness, and shape of the fan blades, as well as loose fanbelts and high static pressure in the room, will affect the amount of air delivered. Normally, fans move about 90 to 95% of the ft^3/min (m^3/min) shown in the table.

Table 5-5. Capacities of Electric Exhaust Fans

Motor		Fan Blades			Air Capacity @ "0" Static Pressure	
			Diameter			
hp	r/min	No.	in.	cm	ft^3/min	m^3/min
1/8	1,725	4	12	30.4	1,650	46.7
1/4	1,725	4	18	45.7	2,900	82.1
1/4	1,140	4	18	45.7	1,800	50.9
1/3	1,140	5	18	45.7	3,600	101.9
1/2	1,140	5	24	60.1	5,300	150.0
1/3	630	4	24	60.1	6,200	175.5
1/3	473	4	30	76.2	6,300	178.3
1/2	412	4	36	91.4	12,000	339.6

5-G. COOLING THE HATCHERY

During periods of hot weather it becomes necessary to cool the hatchery rooms. Certain rooms require more cooling than others. For example, the chick room is usually the first to show temperature increases because of the buildup of heat from the chicks. Next is the hatching room. The most economical method of reducing the temperature in hatchery buildings is by evaporative cooling.

Principle of Evaporative Cooling

The theory of evaporative cooling is based on the fact that when water evaporates a cooling effect is produced. For example, when one steps out of a shower one feels cold because body heat is evaporating the moisture on the skin, which in turn is lowering the surface temperature. Once the water is evaporated or removed with a towel, one feels warm again because there is no more evaporation.

For the same reason, to evaporate water vapor in air, heat is required, and with the use of an evaporative cooler this heat comes from the heat in the outside air. This hot outside air is drawn through a moisture-laden pad, where the heat evaporates the water in the pad, thus lowering the air temperature. But the amount of evaporation is regulated by the amount of water vapor in the incoming outside air. If the air is dry, more evaporation takes place to cause more cooling. When the incoming air is moist, less evaporation and cooling take place.

The amount of moisture is measured as the percentage of moisture compared to air completely saturated (100% relative humidity). The *absolute* temperature to which air may be cooled can be determined by taking an outside reading on a wet-bulb thermometer. Such a thermometer uses both air temperature and humidity for its reading.

The Evaporator-Cooler

A practical application of the above phenomenon is through the use of commercial evaporator-coolers. These come in varying sizes from 2 to 20 ft (0.6 to 6.0 m) on a side. The larger the size, the more air moved.

In the evaporator-cooler, moisture is provided by a moisture-laden pad. Air is sucked through the pad, usually by a squirrel-cage fan; moisture is absorbed; the air is cooled and forced throughout the building in ducts, then out of the rooms through an exhaust opening. A slight air pressure will build up in the ducts. The usual back pressure is in the neighborhood of 1/20 to 1/10 in. (0.12 to 0.25 cm) of static water pressure. To reduce this as much as possible, the outlet area in the room or building should be about three times the area of the cooler exhaust area. A more common method is to use exhaust fans. The slight back pressure will offer resistance to the fan in the evaporator-cooler causing the fan to move only about 85% of its maximum amount of air.

Fans are used to exhaust air. To maintain an equilibrium in the room between incoming and outgoing air, the capacity of the exhaust fans should be about 10% greater than that of the intake fans, thus creating negative pressure in the rooms. There are automatic devices to control the outlet fans so as to maintain any desired static pressure.

Location of the cooler. Smaller coolers should be installed on the side of the building and the air blown directly into small rooms. For larger installations, however, where the air is ducted inside the building, the most common place for the cooler is at, or near, the peak of the roof.

How much can air be cooled? The amount of cooling by an evaporator-cooler depends on the temperature and relative humidity of the incoming air. Using Table 5-6, it is possible to calculate the cooling effect when the temperature and percentage of relative humidity of the outside air are known. In Table 5-6, to arrive at the temperature reduction, first find the outside temperature in the left-hand column, then go horizontally to the right until the relative humidity of the outside air is found. Follow this column to the bottom of the table where the temperature reduction is given.

Table 5-6. Temperature Reduction That May Be Accomplished with Evaporative Cooling When Dry-bulb Temperature and Relative Humidity Are Known

Outside Dry Bulb		Outside Relative Humidity %											
°F	°C												
70	21.1	86	77	68	59	51	44	36	29	22	15	9	3	0
72	22.2	86	77	69	61	53	45	38	31	24	18	12	6	0
74	23.3	86	78	69	61	54	47	39	33	26	20	14	8	3
76	24.4	87	78	70	62	55	48	41	34	28	22	16	11	5
78	25.6	87	79	71	63	56	49	43	36	30	24	18	13	8
80	26.7	87	79	72	64	57	50	44	38	32	26	20	15	10
82	27.8	88	80	72	65	58	51	45	39	33	28	22	17	12
84	28.9	88	80	73	66	59	52	46	40	35	29	24	19	14
86	30.0	88	81	73	66	60	53	47	42	36	31	26	21	16
88	31.1	88	81	74	67	61	54	48	43	37	32	27	22	18
90	32.2	89	81	74	68	61	55	49	44	39	34	29	24	19
92	33.3	89	82	75	68	62	56	50	45	40	35	30	25	21
94	34.4	89	82	75	69	63	57	51	46	41	36	31	27	22
96	35.6	89	82	76	69	63	58	52	47	42	37	32	28	24
98	36.7	89	83	76	70	64	58	53	48	43	38	34	29	25
100	37.8	89	83	77	70	65	59	54	49	44	39	35	30	26
102	38.9	90	85	78	72	67	62	56	51	46	42	36	32	28
104	40.0	90	85	78	72	67	62	56	52	47	43	38	33	29
106	41.1	90	85	78	73	67	62	57	52	47	43	39	34	30
108	42.2	90	85	78	73	67	62	57	53	48	44	40	35	32
110	43.3	91	85	79	73	68	63	57	53	49	45	41	37	33

Potential Cooling for a Given Temperature and Relative Humidity[1]

						°F							
3	5	7	9	11	13	15	17	19	21	23	25	27	
						°C							
1.7	2.8	3.9	5.0	6.1	7.2	8.3	9.4	10.6	11.7	12.8	13.9	15.0	

[1]This amount is theoretical. Approximately 80% of this change is accomplished in usual practice. For example, when the dry-bulb temperature is 100°F and the relative humidity is 30% the table shows a 25°F potential drop. In practice, 80% of this, or 20°F, can be readily accomplished.

Specifications for evaporator-coolers.   Specifications for representative evaporator-coolers are given in Table 5-7.

Evaporator-coolers in humid areas.   Air expands during the hotter hours of the day, as during the afternoon, the relative humidity of the air is reduced, and the air is able to absorb more moisture than during the cooler hours at night. Therefore, when temperatures are high, evaporator-coolers will reduce the temperature more than when they are low. Thus, coolers may be used in rather humid climates; their use cannot be confined to arid areas. In fact, the relative humidity during afternoon hours may be as much as 20% less than during the nighttime hours.

Table 5-7.   Evaporator–Cooler Specifications

Width in.	Depth in.	Height in.	Outlet Velocity ft/min	Air Delivery ft³/min
27	27	25	1,300	2,300
34	28	30	1,750	3,100
34	42	42	2,500	5,600
38	45	42	2,400	7,700
56	56	60	1,980	9,100

Heating and Cooling in the Same Ventilating System

When ducts are used to transfer air in the hatchery rooms it is common practice to install both heating and cooling units in the same ventilating system. During cold weather, the heating unit operates; during hot weather, the cooling unit functions. A thermostat is installed in the respective rooms and adjusted so that the transfer from one unit to the other will take place automatically.

Ability of Air to Hold Moisture

When air is heated it expands, and its capacity to hold moisture increases. Therefore, when air temperatures increase, the relative humidity decreases. The opposite occurs when air temperatures decrease. Figure 5-3 is a nomograph used for calculating the change in relative humidity when air temperatures increase or decrease.

Question.   If the air temperature is 70°F (21.1°C), and the relative humidity is 75%, and the air temperature is raised to 100°F (37.8°C), what will be the relative humidity of the air?

Calculation.   Using Figure 5-3, lay a ruler across 70°F (21.1°C), line B, and 75% relative humidity, line C, and find the intersecting point

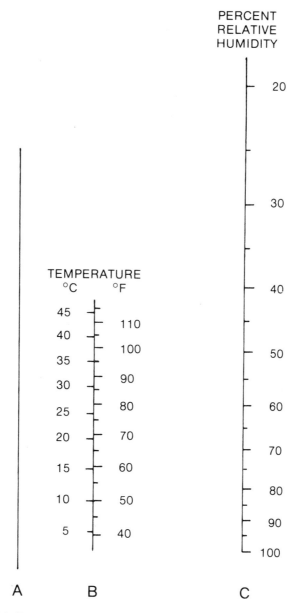

Figure 5-3. Temperature change and its effect on relative humidity.

on line A. Hold the point on line A and move the other end of the ruler up to 100°F (37.8°C), line B. The ruler intersects line C at 30% relative humidity.

5-H. TEMPERATURE, HUMIDITY, AND AIR PRESSURE

The recommendations for temperature, humidity, and air pressure of various hatchery rooms are given in Table 5-8.

Table 5-8. Temperature, Humidity, and Air Pressure by Rooms

Room	Temperature		Rh %	Air Pressure	
	°F	°C		%	Pressure
Egg-holding	65	18.3	75	0	Even
Setter	75	23.9	50	0	Even
Hatcher	75	23.9	50	5	Negative
Chick-holding	75	23.9	65	10	Negative
Wash	60–70	15.6–21.1	60–75	10	Negative
Hallways	60	15.6	65–70	5–10	Negative
Clean	70	21.1	50–60	5	Positive

Hatchery Room Specifications

Egg-holding room. Use a 75% relative humidity to prevent eggs from drying too rapidly during the preincubation period.

Setting and hatching rooms. Setters and hatchers will operate more uniformly and economically if the relative humidity of these two rooms is maintained at 50%.

Washroom. Here the chicks are removed from the hatching trays and boxed. Use an exhaust hood over this area to reduce cross-contamination. After the chicks are boxed they are moved to the chick-holding room, and the trays are washed in a tray washer in the washroom.

Chick-holding room. Maintain a 65% relative humidity to prevent excessive chick dehydration.

Clean room. After the trays are washed they are placed in their buggies and moved to the adjacent clean room to dry. The clean room should have more static pressure than the washroom so air always flows from the former to the latter. Sometimes the clean room can be made airtight and used as a fumigation room if ceiling exhaust fans are used to remove the fumigant.

5-I. HATCHERY DEBRIS

Sanitation plays an important part in the operation of a hatchery. It is a must. Floors, walls, and air must be kept clean, incubators scrubbed and disinfected, and debris collected and either removed or incinerated. As the amount of residue from hatcheries is tremendous, the question of how to handle it is important. Some key points are as follows:

1. Keep the material in a damp condition when floors and incubators are being cleaned in order to keep as much of it out of the air as possible.

2. Do not sweep debris; vacuum it. Keep each room isolated. All material should be placed in a sealed container before moving it through rooms.
3. From the standpoint of disease organisms, the washroom is probably the most contaminated room in the hatchery. Do not track from it to freshly cleaned rooms.
4. Use chick-down collectors of one type or another over the exhausts from the chick hatchers and from the chick-holding room.
5. There are two alternatives for waste removal, namely
   a. Use a commercial eggshell vacuum.
   b. Place the waste in bags of plastic or similar material and remove it from the hatchery.
6. Clean and sanitize the hatchery thoroughly. When in doubt, sanitize again.

5-J. OTHER HATCHERY ROOMS

Depending on the type of hatchery, other rooms are often found, including

small office
lounge and lunch room
small laboratory
rest rooms
tool room

emergency generator room
electrical control room
box storage room
electrical monitoring room

6

Hatchery Equipment

Good equipment plays an important part in increasing hatchery profits. Not only will the hatchability of the eggs and chick quality be improved, but labor costs will be lowered. All are involved with an increased financial return on the investment.

Hatchery equipment cannot be standardized because many factors enter into a proper choice, namely

1. Size of the hatchery
2. Number of hatches per week
3. Type of disease-control programs
4. Whether breeder chicks or commercial pullet chicks are hatched
5. Type of incubator
6. Hatchery services provided, such as beak, comb, and toe trimming, vaccination, inoculation, and so on

6-A. WATER EQUIPMENT

Water Softeners and Filters

An analysis of the water should be made. Excessive minerals will cause lime deposits on humidity controls, spray nozzles, and jets, making them inoperative. Valves will not seat properly, and they will leak. Proper filters may be necessary. A water softener may be needed (see Chapter 39-C).

Water Heaters

The demand for hot water in the hatchery is great. Industrial-type heaters with a large capacity should be installed.

6-B. EGG-HANDLING EQUIPMENT

Reducing the cost of labor calls for many labor-saving devices and is a major concern. Although hatching eggs were originally cased and delivered to the hatchery, this process necessitated that the eggs be handled several times. A more modern method is to place the eggs in the setting trays at the farm, then slide the trays into buggies and transfer them to the hatchery. With some incubators, the buggies may be wheeled directly into the fumigation room and setters.

Hatchery Carts

Egg cases, egg flats, and chick boxes should not be handled any more than necessary. Carts of several types may be used to expedite their transfer.

Wheeled carts.   These have four wheels, no platform, and come in several sizes.

Semilift carts.   These are larger carts with two wheels and a capacity of 20 to 25 30-doz cases of eggs or 1,200 lb (544 kg).

Hand trucks.   These lightweight trucks are capable of handling four or five cases of eggs. The capacity usually is 250 lb (115 kg).

Pallets.   Cases of eggs, egg cases, and chick boxes may be placed on pallets, to be moved later with a forklift.

Conveyors

Moving cases of eggs, boxed chicks, and materials short distances may be expedited by using conveyors rather than carrying them. Conveyors are of great help in reducing the hatchery labor by making it possible to rapidly move a large number of items with little physical effort.

6-C. EGG GRADING AND WASHING EQUIPMENT

In many hatcheries it is necessary to grade eggs by size before they are set; in others, washing may sometimes be a practical procedure.

Vacuum Egg Lifts

To expedite the removal of eggs from egg flats, vacuum egg lifters should be used. Most of these lift the eggs by suction. The lifters are operated by a vacuum pump that creates a suction on rubber cups, the number varying from 12 to 48. Most of those used in hatcheries are 6 × 6. That is, they lift eggs from filler flats that are six rows long and 6 rows wide, lifting 36 eggs at a time. Some are larger (6 × 8), lifting 48 eggs. The size of the lifter is determined by the size of the egg flat used for collecting eggs and delivering them to the hatchery.

Special lifters. Special lifters are employed to transfer eggs from egg flats to some incubator trays. These pick up the eggs, then narrow the distance between the rows of eggs so that the rows fit the incubator trays.

Graders for Hatching Eggs

If eggs are to be graded for size, automatic graders should be used. There are many of these on the market, of various types and capacities. One should be secured that will do the job adequately, in a limited amount of time, and with little egg breakage.

How egg graders are rated for size. Most egg graders are rated according to the number of 30-doz cases of eggs that can be run through them in 1 hour. These ratings are for continuous operation, a feat not generally accomplished. Under practical usage 85% of the rating would be a more accurate capacity. Some indication of the size necessary may be found in Table 6-1.

Table 6-1. Hours Necessary to Grade Eggs

Number of Eggs to Be Graded	Capacity of Grader in 30-Doz Cases per Hr				
	30	50	100	150	200
	Hr Necessary to Grade Eggs				
50,000	4.6	2.8	1.4	0.9	0.7
100,000	9.3	5.6	2.8	1.9	1.4
150,000	13.9	8.3	4.2	2.8	2.1
200,000	18.5	11.1	5.6	3.7	2.8
250,000	23.2	13.9	6.9	4.6	3.5
300,000	27.8	16.7	8.3	5.6	4.6

Hatching Egg Washers

Hatching eggs can only be satisfactorily washed in a commercial egg washer. First, hold the eggs at 65°F (18.3°C) for 8 to 12 hours, then wash, using a balanced cleaning compound. The washwater should have a temperature of 110°F (43°C) to which 200 ppm of chlorine has been added. Completely drain the washwater often and use new cleaning compound each time.

Detergents used for washing. Most washing compounds have a detergent as their base. There are many on the market. Do not use laundry detergents. As detergents have a tendency to foam, a compatible antifoaming solution is often added to the washwater.

Hardness of water determines amount of detergent. The more minerals in the water, the larger the amount of detergent necessary. Water samples should be taken and analyzed to determine the exact quantity needed. Water softeners may help in the presence of high percentages of minerals (see Chapter 39-C).

Disinfectants sometimes added to washwater. To lower the level of disease-producing organisms in washwater reused in the washing process, disinfectants are generally added. A solution that releases chlorine slowly is an example. If such disinfectants are used, care must be taken to keep them at a uniform level in the washwater. Many are easily destroyed in the presence of organic material. Some egg washers rinse the eggs with a chlorine solution after they have been through the washing section.

Internal egg temperature rises during washing. The water used to wash hatching eggs should have a temperature of about 110°F (43°C). Washing increases the internal temperature of the egg as much as 7° to 13°F (4° to 7°C). Several factors influence the actual number of degrees.

1. Temperature of the eggs prior to washing.
2. Size of the eggs.
3. Temperature of the washwater.
4. Length of time in the washwater.

Plastic egg flats. Some egg flats are made of plastic. They are more durable than those made of fiber, and may be washed, disinfected, and reused. Some egg washers are made to accommodate eggs held on plastic flats. Thus, there is no need to transfer the eggs from flats to the loading tray of the washer. The trays and the eggs on them go through the washer together. Furthermore, some incubators are manufactured so that plastic flats with eggs on them may be inserted in the incubator. Again, there is no transfer of eggs at the time they are set. Thus, a combination of washing and incubation of eggs may be accomplished with the eggs on the flats, without transfer. Such a procedure reduces egg breakage because of less handling, and there is a material reduction in the cost of hatchery labor.

6-D. EMERGENCY ELECTRIC STANDBY PLANTS

When there is a failure in the outside electrical supply, the hatchers must have a quick source of other electric power. This is furnished by a standby plant, situated in or adjacent to the hatchery building.

What Type Plant?

When the standby plant is to be used only in case of a failure of regular electric service, purchase a plant that matches the commercial power in voltage, phases, and number of wires. Consult an electrical engineer and your incubator manufacturer before you buy.

Automatic or Manual?

With the automatic plant the transfer from the outside electric supply to the standby unit is automatic. When the outside power fails, the motor used to operate the generator starts automatically, and switches transfer the load to the new supply of power.

When the changeover is manual, the switches must be moved by hand, and the motor on the generator must be hand-started. For plants with manual controls, an *alarm system* should be installed. The alarm bell will ring automatically when the regular electric current ceases.

Calculate the Electric Load Required

Before purchasing a generator, calculate the electric load necessary when power fails. Remember that most incubators have auxiliary heaters that operate after the incubators have been without electricity for a few minutes. This increases the load, often doubling it. There is also the increased load involved with starting motors. Thus, the total starting load is much greater than the normal operating requirement. Calculate the maximum load required; then add 10% as a safety factor. Another suggestion is to start a few machines at a time, thus reducing the electrical load.

Wiring the Standby Plant

If power failures are of short duration, the standby generator may be wired to operate only the hatchers since eggs in the setter section (1 to 19 da) can withstand several hours with no electricity to operate the electrical fans, heaters, and humidifiers. However, an electrical failure of only a few minutes in the hatchers will prove disastrous.

6-E. INCUBATING–HATCHING EQUIPMENT

Through the years the forced-draft incubator has undergone a variety of changes to make it a highly specialized piece of equipment: lighter cabinet materials; floorless machines; easier cleaning; improved thermostats; automatic egg turners; separate hatchers; more accurate humidifiers; and better cooling devices. In the past few years, ingenious new ideas have been incorporated to make the incubator a superb electronic machine. Some of these new developments are included in the following list:

High-speed electric fans.   These create a more even temperature and humidity throughout the cabinets of certain setters and hatchers.

Automatic and manual egg-turning devices.   Trays can be engaged either way, at any angle, or stopped in any position.

Spray-nozzle humidifiers.   A high-pressure vaporlike spray introduces moisture into a mixing chamber when more humidity is needed.

Removable assembly.   Fans, heaters, and coolers are joined and can be removed in one piece for cleaning and for better cleaning of the cabinet.

Plastic hatcher trays.   These not only allow more air movement through them but they may be stacked and placed on carts (dollies), replacing the use of buggies.

Solid-state controls.   Incorporated now in many machines are solid-state controls to activate egg-turning devices, humidifiers, heaters, coolers, and fans.

Digital readouts.   Dry- and wet-bulb incubator temperatures are read by large digital numbers on the front of the machines. Some show the *relative humidity* (rh), rather than the wet-bulb thermometer reading.

More safety devices.   These include indicator lights on a panel at the front of the machines to show when power is on or off, high and low cooling, heaters on and off, fans on and off, egg turners connected, humidity control on and off, and so on.

Improved down collectors.   An improved design has increased the amount of down removed from the hatchers.

Remote monitoring.   Some machines are capable of electronically transmitting a number of setter and hatcher readings to a monitoring room, remote from the setters and hatchers.

Modern Chicken Egg Setter
and Hatcher

A chicken egg setter is shown in Figure 6-1. It has a capacity of 93,312 eggs. Note the method of turning the eggs, outside device for recording temperature and humidity, control panel with digital readout, auxiliary

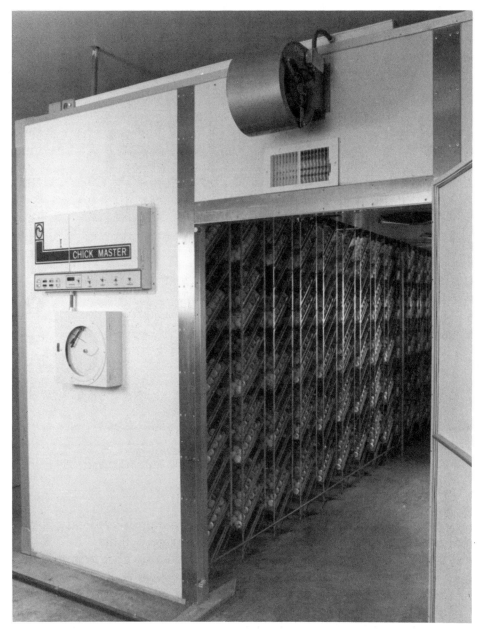

Figure 6-1. Chicken egg setter with 93,312-egg capacity.
(Courtesy of Chick Master Incubator Company, Medina, Ohio, USA.)

front-mounted fan, and hatchery floor designed to serve as the setter floor. This machine is designed to set 15,552 eggs twice weekly.

A chicken egg hatcher is shown in Figure 6-2. It has a capacity of 15,552 eggs, one-sixth the capacity of the matching setter. Note tiered plastic trays on roll-out dollies sitting directly on the hatchery floor, roll-

Figure 6-2. Chicken egg hatcher with 15,552-egg capacity.
(Courtesy of Chick Master Incubator Company, Medina, Ohio, USA.)

out assembly containing fans, heaters, and cooling coils for easy cleaning, digital outside temperature recorder, and safety lights.

Ten matching setters and hatchers will give an incubator capacity of more than 1 million eggs, hatching about 270,000 chicks a week.

6-F. OTHER HATCHERY EQUIPMENT

There are many other items necessary to operate a hatchery efficiently and to furnish services necessitated by customer request.

Egg Candlers

·In some instances eggs are candled during the incubation period. Eggs that are infertile, and those in which the embryos have died, are removed. At times there is a market for the infertile eggs, as food for human consumption (see Chapter 9-D).

Test Thermometers

Several highly accurate test thermometers should be a part of the hatchery equipment. These may be used for checking the accuracy of the thermometers in the setters and hatchers (see Chapter 8-D).

Movable Chick Service Turntable

When day-old chicks are graded, beak trimmed, comb trimmed, vaccinated, and so on, a rotating service table will be of help (Figure 6-3). Chicks are elevated to a round turntable, employees sit around it, and in turn complete a necessary chore. Then the chicks are placed in chick boxes.

Chick Box Racks

Although chick boxes filled with day-old chicks may be placed on dollies, some prefer to place the boxes in racks. These are built of 1-by-2-in. (2.5-by-5 cm) lumber or metal and constructed so that each box has its own compartment with an air space between it and the neighboring boxes. Racks may be constructed to accommodate 6 to 8 tiers of boxes, and are generally four boxes long. Casters are used on the bottom of the legs to facilitate movement.

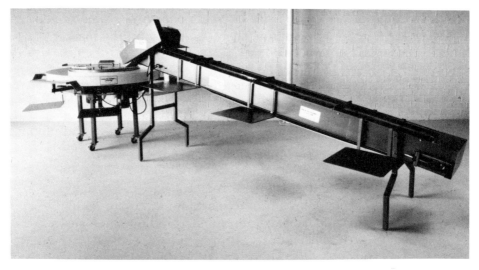

Figure 6-3. Movable chick service turntable.
(Courtesy of Chick-Go-Round, Inc., Farmville, North Carolina, USA.)

Chick Counters

Automatic equipment is available for counting chicks at the rate of 50,000 per hour, tumbling 25 into each of the four compartments of the chick box.

Sexing Equipment

There are four common methods of sexing chicks at day old. The method used will determine the equipment necessary, as follows:

1. *Vent sexing.* Most chick sexers furnish their own equipment that consists of only a bright light.
2. *Machine sexing.* These are specialized sexing machines capable of visually observing the sexual organs of the chicks. These may or may not be supplied by the hatchery.
3. *Color sexing.* No equipment is needed other than suitable tables for holding the chick boxes, or movable belts, or movable turntable.
4. *Feather sexing.* Chicks may be feather sexed as they are picked from chick boxes, or as they circle a round service turntable.

Vacuums

Hatchery dust should be vacuumed, not swept. Large, industrial-type vacuums should be used. The size and number will be determined by the size of the hatchery operation. There are many on the market. Variations in their makeup are

1. *Self-contained, or conversion-covered, machines.* Self-contained units are complete; conversion units are made to be placed on top of a steel barrel.
2. *Dry or wet-and-dry units.* Some vacuums pick up only dry material; others may be used with either wet or dry debris, and are preferable.
3. *With or without dust collectors.* Dust collectors are usually bags that move with the machine, and are easily cleaned. Hatchery vacuums should employ some system of dust collection.
4. *Portable hand units.* These are either smaller vacuums that may be carried, or those with smaller suction

heads. They are helpful in cleaning the inside of incubators and other hatchery equipment.

5. *Eggshell vacuums.* Such vacuums are used to remove the debris left on the hatcher trays after the chicks have been removed. The debris goes into a holding tank, and is later removed to a truck.

Pressure Pumps

Increased water pressure is necessary to do a thorough job of washing the floors, walls, incubators, and hatching trays. A pressure pump should be installed in the hatchery to give the necessary water pressure. These come in various sizes and capacities; some are portable and some must be permanently installed.

Tray Washers

Cleaning hatching trays is laborious when done by hand. To facilitate the procedure, an automatic tray washer may be used. Many are on the market. In some instances washers are matched with the make and size of the incubator tray used.

Comb Trimming Shears

These are small manicure scissors. They are used for trimming the combs of day-old chicks.

Beak Trimming Equipment

An electric automatic or semiautomatic machine is used for trimming the beak. Although the procedure is recommended when chicks are 6 to 9 days of age, customers sometimes request that the trimming be done in the hatchery, particularly with broiler chicks.

Toe Trimming Equipment

Although a beak-trimming machine may be used to trim the ends of the toes of day-old chicks, a special toe trimmer is more practical. The toes should be trimmed only to the outer joint.

Automatic Syringes

Some hatcheries inject day-old chicks with vaccines or solutions involved with disease prevention. To facilitate the procedure the syringes are automatic; that is, they are capable of refilling automatically, and may be adjusted to allow varying dosages.

Marek's Disease Vaccination Equipment

Most chicks are vaccinated against Marek's disease in the hatchery. Several types of vaccine are available and the equipment used for each is different (see Chapter 37-D).

Mycoplasma Gallisepticum Test Equipment

As a part of a PPLO S-6 negative program the rapid serum plate test should be completed in the hatchery, using culled chicks or chicks that have pipped the shell, in order to determine if antibodies are present. This requires

    1. Small test tubes for holding the blood
    2. Testing plate and cabinet

Some testing cabinets use absorbent paper instead of glass. The test paper may be dried and filed for future reference. This is not possible when a glass plate is used (see Chapters 9-J and 37-J).

Refuse Disposal Equipment

Although not a part of the hatchery building, equipment for the disposal of hatchery waste material is an integral part of the hatchery requirement (see Chapter 9-H).

Incineration seems to be the most efficient means of disposing of the hatchery debris, most of which is unhatched eggs and eggshells. Incinerators will reduce the material to ash, having about 15% of its original volume. Hatchery waste may be dehydrated, ground, and legally used as a poultry feed in most areas (see Chapter 25-D).

7

Maintaining Hatching Egg Quality

Maintaining the hatching potential of the newly produced egg is of vital importance. But a lot can happen to a hatching egg between the time it is laid and the time it enters the incubator. As a result of faulty egg handling during this critical period much of the inherent ability of the egg to hatch well and to produce a quality chick may be lost. Every technique of good egg care and handling must be put into practice.

7-A. MAINTAINING EGG QUALITY IN THE CHICKEN HOUSE

Nesting Material

Many eggs are broken in the nest as the result of inadequate cushioning of the nesting material. But more important is the prevention of stained and dirty eggs as a result of unclean and wet nesting material. The ability to absorb moisture is an important quality to consider in selecting a good covering for the bottom of the nest.

Nesting materials should be absorbent; durable; coarse, so they will not be blown easily from nests; dust-free; porous, so eggs will cool more quickly; of good cushioning quality; and inexpensive. Common nesting materials are

extruded volcanic ash	dried sugar cane
shavings	chopped corn cobs
peat moss	straw, hay
rice hulls	excelsior nest pads

peanut hulls

carpet remnants

shredded paper

oystershell mixed

with shavings

Astroturf®

Training Birds to Use the Nests

Birds should be trained to use the nests rather than be allowed to lay on the floor where a high percentage of the eggs will be stained and broken. The following suggestions will help induce hens to lay in the nests rather than on the floor:

1. Have the lowest nest perch no more than 27 in. (69 cm) above the floor of the house.
2. Place the nests in the pen before the birds start to lay.
3. Put the nesting material in the nests when the nests are placed in the pens. Keep the nesting material clean before egg production starts as hens may refuse the nest if the nesting material is dirty, dusty, or soiled.
4. Supply adequate nesting material. If it has been blown from the nests, or worn out, and the bare surface of the nest is exposed, birds are not likely to lay in them.
5. Provide adequate nest ventilation to help keep the nesting material dry and the birds comfortable.
6. Provide one nest for every four hens. If birds cannot get into the nests to lay they will be forced to find a "nest" on the floor.
7. When pullets first start to lay, pick up floor eggs six to eight times a day. Other birds lay more floor eggs when they see eggs on the floor.

Collecting Hatching Eggs

Normally, hatching eggs are picked up from the nest four times a day. However, during periods of extremes in temperature, either high or low, five or six collections will be necessary.

Nest eggs, covered by hens in the course of laying, are undergoing preincubation, which shortens the regular incubator incubation period, reduces hatchability, and lowers chick quality.

When eggs are picked from the nests four times a day, some eggs have been in the nest for 3 to 4 hours, and some only a few minutes. These preincubation differences exert themselves in an equivalent amount of variability in hatching time, hatchability, and chick quality.

Important: Eggs laid late in the day should be collected the same day rather than left in the nests until the next morning. Hatching eggs left in the nests overnight will lose some of their hatching qualities. Shell sanitization of these eggs the day after laying rather than immediately after laying will be of little value in preventing the entrance of bacteria into the egg.

Automatic egg pickup. Eggs may be collected by a sloping nest floor that delivers eggs to a belt that moves them to the end of the house as fast as they are laid. This setup reduces preincubation time to a minimum and improves eggshell disinfection, because eggs may be fumigated or sprayed faster after laying.

Close the Nests at Night

Hens should not be allowed to sit in the nests overnight. Close the openings after the day's egg production has been completed, first removing any birds inclined to want to remain in the nests. Open the nests early in the morning, before laying begins.

Hatching Egg Containers

The type of container used for holding hatching eggs collected from the nests is important. Once laid, hatching eggs should be cooled as quickly as possible. They should be taken to a specialized cooling room designed for this purpose. Even though the temperature of the egg-holding room in most cases should be 65°F (18.3°C), it will be several hours before the interior contents of the egg will reach such a temperature.

Eggs are best collected on flats because they allow the maximum amount of air circulation around the eggs, which is beneficial not only for cooling but also for fumigation. Unless the hatching eggs are to be transported long distances, they should not be put in egg cases, even if holes are cut in the sides. It will take several days for eggs in cases to cool, but only a few hours with flats. With some setters, provision is made to leave the eggs on the same flats during washing, fumigation, and incubation.

Remove nonhatching eggs. While gathering the eggs separate the extra large, cracked, and soiled ones. Many hatcheries do not grade commercial hatching eggs for size, thinking that the benefits do not offset the expense involved. Removing the obvious nonhatching eggs in the hen house will be helpful when eggs are not to be graded for size in the hatchery.

Plastic flats better than fiber. Litter and dirt accumulate on fiber flats and they cannot be washed. Select a plastic flat that will allow any nesting material to fall through openings cut in them.

Egg baskets not advised. Piling one egg on top of another in wire baskets (or buckets) will only increase breakage. Furthermore, eggs eventually must be removed from the basket to another container or tray. The less handling the better.

7-B. REDUCING BACTERIAL CONTAMINATION OF EGGS

Most hatcheries sanitize hatching eggs, but the wrong program can do little good and may give a feeling of false security. Shell contamination starts early. There is no such thing as a sterile eggshell. Even eggs removed from the oviduct will be covered with many bacteria. Then there is major infestation as the eggs pass the cloaca where the urinary and intestinal canals also enter. From 300 to 500 organisms may be found on the shell at the time the egg is laid. Even though only a few may be pathogenic, there are enough to cause trouble. Most involved are the *Salmonella, Pseudomonas, Escherichia coli,* and Arizona organisms. In the presence of adequate heat and moisture, within 15 minutes after the egg is laid these numbers will have increased to between 1,500 and 3,000. In another hour, 20,000 to 30,000 will be present.

To add to these numbers are those bacteria laid on the shell from material the newly laid egg comes in contact with, such as floor litter, dirty nesting material, and intestinal filth. The number of organisms found on the average egg at the time it is picked from the nest or floor can be shown as follows:

Clean eggs	3,000 to 3,400 organisms
Soiled eggs	25,000 to 28,000 organisms
Dirty eggs	390,000 to 430,000 organisms

One can easily account for the high numbers on dirty eggshells because fecal material is the main contributor, and 1 gram of this material contains from 2 billion to 6 billion bacteria.

Unless the weather is unusually hot, the egg contents begin to cool, shrink, and produce internal suction immediately after the egg is laid. Suction and shell penetration of bacteria are greatest immediately after oviposition. Of all eggs laid by a flock, 15% will have bacterial shell penetration in 15 minutes, 21% in 30 minutes, 25% in 60 minutes, but only 33% in 24 hours.

Bacterial Penetration of the Egg Coverings

Nature has endowed the egg with several coverings to prevent the majority of surface bacteria from gaining entrance to the young developing embryo through the yolk sac. Ranking first in penetration prevention is the cuticle, but because it is highly variable in thickness it is highly variable in its ability to prevent bacteria from gaining entrance to the shell pores.

Second best of the four egg coverings to resist bacterial penetration is the shell. Although a 2-oz (56.7-g) egg will contain about 8,000 pore openings, most of these are too small for bacteria to pass. But there is a small percentage of large malformed pores and they are several times the diameter of most bacteria. Through these pores hundreds of organisms pass to get to the two shell membranes. Usually the organisms pass easily, but many remain viable and reside in the pores for up to 3 weeks before passing.

E. A. Sauter and C. F. Peterson showed that shell thickness has a major effect on the ability of bacteria to pass through the eggshell. As specific gravity is directly correlated with shell thickness, it was used as the criterion for classifying eggshells and the resultant bacterial penetration of each group as shown in Table 7-1.

The figures show that shell quality, rather than time, is more important to shell penetration. For example, 21% of the bacteria penetrated the shells of good quality after 24 hours, while 34% penetrated the shells of poor quality after 30 minutes.

Table 7-1. Shell Quality and Bacterial Penetration of Eggshells

Specific Gravity of Eggs	Shell Quality	% Bacterial Penetration of Shell		
		After 30 min	After 60 min	After 24 hr
1.070	Poor	34	41	54
1.080	Average	18	25	27
1.090	Good	11	16	21

Source: Sauter, E. A., and C. F. Peterson, 1974, Poultry Sci. 51, 2,159.

Methods of Sanitizing Eggshells

It is obvious that poultrymen interested in reducing the bacterial penetration of eggs must have a program of producing eggs with thick and clean shells. But even then, huge numbers of organisms will have gained entrance to the egg contents long before any shell sanitizing can be administered. Shell sanitizing is only effective in destroying the remaining bacteria on the shell. With any program, time is of the essence; the more quickly the sanitizing is done, the better the results.

There are several methods of sanitizing eggshells, but each has its shortcomings. The one chosen must fit the mechanics of the operation and must be administered quickly after the eggs are laid.

1. *Quaternary ammonia.* This is a shell bactericide to be sprayed on eggs as they lie on flats. It is mixed with lukewarm water at 200 ppm.

2. *Quaternary ammonia-formalin.* A better mixture for spraying eggs as they lie on flats is a mixture of quaternary ammonia and formalin added to lukewarm water at 156 ppm. Use the following formula:

Ingredient	U.S. Customary	Metric
Formalin (40%)	1 oz	29.6 cm^3
Quaternary ammonia	1 oz	29.6 cm^3
Water (lukewarm)	1 gal	3.8 L

Caution: Wear rubber or plastic gloves when handling sprays containing chlorine or formalin. Do not breathe the fumes. See Chapter 9-I regarding legal concentrations of formaldehyde gas in air breathed by humans.

3. *Ozone (O$_3$).* Ozone may be generated at low cost and administered at 100 ppm to eggs held in a tight cabinet; consequently it is not too effective because some eggs have been laid from 1 to 5 hours. However, it can be administered to eggs in the setter during the first day of incubation, according to Brian Sheldon, North Carolina State University. As an eggshell disinfectant it is nearly equal to formaldehyde.

4. *Formaldehyde gas.* Formaldehyde gas is generated at 3× strength in a tight egg-holding cabinet or room (see Chapter 9-J). The gas provides excellent bacterial destruction on contact and it is very easy to fumigate large numbers of eggs. Hatching eggs should be fumigated on the farm for 30 minutes after each egg pickup. Formaldehyde's only deficiency is that it cannot be administered until several hours after some eggs have been laid.

Note: For best results, hatching eggs should be sprayed or fumigated while they are still warm. When eggs are cooling, the egg contents are shrinking and drawing in any disinfectant that is on the outside of the shell. Once the eggs are cooled, this process stops.

Sanitizing eggs delivered by belt. With automatic belt pickups, eggs are delivered continuously to the end of the house. This system has a sanitizing benefit, because eggs can be sprayed as fast as they are delivered, negating the delay when eggs are sprayed in the house after each egg gathering.

7-C. TRANSPORTING HATCHING EGGS

The poultry breeding farm or farms are often located some distance from the hatchery, making it necessary to transport eggs a considerable distance. In such instances, eggs are most easily kept in conventional egg cases containing egg flats or in incubator trays placed on dollies.

Optimal truck-holding requirements. The temperature in the truck should be 65°F (18°C) and the relative humidity should be 70 to 80%. Although most eggs are transported by truck, occasionally eggs are moved long distances by air. Although such transportation will not normally affect the hatchability of eggs through jarring, shaking, or altitude, the time involved in making the transfer from one point to another and any variation in the ambient temperature outside the limits of normalcy will affect the hatchability.

7-D. EGG SELECTION AND HATCHABILITY

Hatching eggs of poor quality do not hatch as well as eggs of good quality. The term *quality* refers to the condition outside the shell, the condition of the shell itself, and that of the contents. Eggs with inferior characteristics should be eliminated rather than placed under incubation. Some obvious physical differences of eggs and their effect of hatchability are described below.

Egg Size

Within a given group of eggs the very small ones and the extremely large ones do not produce as satisfactory a hatch as do those of normal size. Extremely large eggs, eggs with double yolks, and extremely small eggs should not be set. The incidence of these will vary according to the length of time the flock has been producing eggs. More double-yolked eggs will be produced at the beginning of the laying period than later. Hatching eggs that are used for the production of breeding stock should be more carefully selected than those used to hatch commercial broiler chicks and commercial layer chicks.

Egg Size and Chick Size

Chick size is related to egg size, as well as to the relative humidity of the air surrounding the eggs prior to and during incubation. Chicks also lose weight rapidly after hatching because of dehydration, so day-old chick weight varies greatly. Average relationships are found in Table 7-2.

Mature broiler weight associated with egg size.   One can expect 0.0275 lb (12.5 g) additional weight per straight-run broiler at market time for each oz per doz (2.36 g/ea) of egg weight.

Table 7-2.   Egg Weight as It Affects Day-old Chick Weight

Egg Weight		Chick Weight		
oz/doz	g/ea	oz/ea	lb/100	g/ea
22	52.0	1.19	7.4	33.8
23	54.3	1.25	7.8	35.3
24	56.7	1.30	8.1	36.9
25	59.1	1.35	8.4	38.4
26	61.4	1.41	8.8	39.9
27	63.8	1.46	9.1	41.5
28	66.2	1.52	9.5	43.0

Eggshell Imperfections

Eggs closest to ovoid shape hatch best. Excessively long, thin, or completely round eggs do not hatch well. Many eggs have shell imperfections such as ridges, pointed ends, etc., and do not hatch satisfactorily. Some of these imperfections are inherited; therefore, such eggs should not be placed under incubation in order to reduce their incidence in the next generation. Poor interior quality of the egg also affects hatchability. Table 7-3 shows the effect when eggs with abnormalities are set.

Table 7-3.   Hatchability of Abnormal Broiler Breeder Eggs

Description of Abnormality	Hatchability %
Normal (control)	73.9
Misshape (slight deviations due to ridges)	65.0
Slightly round	63.2
Small	62.4
White (no pigment)	49.3
Obviously round	47.8
Pimpled (rough shell)	18.8
Wrinkled (obvious)	12.7
Dark top (rough area)	7.6

Source: Brake, J. L., 1987, North Carolina State University study.

Shell Color and Thickness

The density of the pigment in brown-shelled eggs is often correlated with hatchability. When a group of brown-shelled eggs from a single flock of birds is being hatched, those with darker shells will hatch better than those with lighter shells. However, because hatchability is a genetic factor, strains of chickens may be developed that produce high or low hatchability irrespective of eggshell color. Thus, strains of chickens laying eggs with light-brown shells would not necessarily involve poor hatchability.

Not only is shell quality an inherited factor, but it is partially dependent on the nutrition of the dam and the ambient temperature at which she is kept. Diets low in calcium and vitamin D, and continuous ambient temperatures above 80° to 90°F (27° to 32°C), are more conducive to the production of eggs with inferior shells. Furthermore, the longer a hen remains in egg production, the greater the deterioration in shell quality. This deterioration occurs because the daily production of calcium carbonate by the uterus for eggshell formation remains about the same, and as eggs get larger the longer a bird lays, the calcium carbonate is spread over the egg in a thinner layer.

Shell thickness. Eggshell thickness is important to hatchability. Few eggs with a shell thickness less than 0.27 mm will hatch. For best results eggshells should be between 0.33 and 0.35 mm in thickness. Special calipers are available for measuring the thickness of eggshells.

Candling hatching eggs. Many people candle eggs prior to setting in order to remove the checks and cracks. If this procedure is used, cool the eggs overnight before examining them. Twice as many cracks will be observed after cooling.

Interior Quality

Some eggs are laid with tremulous air cells; others incur them later through jarring and improper handling. These air cells present one of the greatest depressors of hatchability.

Caution: Hatching eggs should be handled carefully.

Haugh units. The higher the reading of Haugh units (see Chapter 16-U) for albumen quality, the better the hatchability of the eggs. Best hatches are secured when the Haugh units of fresh eggs are 80 or over. The reading decreases as eggs are held prior to incubation because of lower albumen viscosity and hatchability will be reduced.

The loss of viscosity is accompanied by a decrease in ovomucin, a protein essential for embryonic growth. Therefore, eggs with high Haugh unit readings hatch better than those with low readings.

7-E. HANDLING EGGS PRIOR TO INCUBATION

Once eggs are laid they must be held for a day or more to fit the setting schedule of the incubator. On occasion, as with breeding farms, hatching eggs may not be set for 1 or 2 weeks after they are laid. The conditions under which eggs are held have a great bearing on helping to maintain as much of their original quality as possible.

Embryonic Threshold

Although the optimum temperature for embryonic development in the forced-draft incubator is in the neighborhood of 99.5°F (37.5°C), this does not mean that there is no embryonic growth when the temperature is below this figure. There is a *threshold* temperature of 75°F (24°C) above which embryonic growth commences and below which it ceases.

Inasmuch as the embryo of a newly laid egg is slightly cold-blooded (similar to reptiles), one may alter its environmental temperature above or below the threshold area several times before the embryo is completely killed. However, each time the temperature goes above or below the threshold the embryo grows weaker and its chance for hatching decreases.

Egg-Holding Room Temperature

After hatching eggs are laid they should be cooled to a temperature well below the threshold of embryonic development and kept at this temperature until shortly before being placed in the incubators. Temperature in the egg-holding room should be 65°F (18.3°C) to curtail embryonic development completely. When eggs are set at intervals of less than 5 days hatchability will be reduced if the holding temperature is less than this. (See Table 7-4.)

Long-time holding temperature. Although the above is a practical application of correct temperature, there is evidence that when eggs are held for more than 5 days holding at 51°F (10.5°C) will produce better hatches.

Table 7-4. Time Necessary to Cool Eggs to 65°F (18.2°C) from 100°F (37.8°C)

Sealed Egg Cases	Egg Cases with Holes in Side	Wire Baskets	Incubator Egg Trays
4–5 da	1–2 da	1 da	3/4 da

Egg-Holding Room Humidity

Moisture held within the contents of the egg is continuously lost by evaporation through the shell. The rate of this process is governed in part by the relative humidity of the air surrounding the egg; when the relative humidity is low, evaporation of the egg contents is more rapid; when it is high, evaporation is less rapid.

During the holding period, evaporation of the interior portion of the egg should be held at a minimum by increasing the humidity of the air in the egg-holding room. To completely suppress egg evaporation by this method seems impractical because the humidity would have to be raised to near the saturation point. Fiberboard egg cases then would become saturated with moisture and would be impossible to handle, and mold growth could become a problem.

Correct humidity in egg-holding room. Regardless of the type of container or tray used to hold the eggs, the relative humidity in the egg-holding room should be 75%. This level materially reduces egg evaporation and does not cause deterioration of egg cases.

Effect of Holding Hatching Eggs

When hatching eggs are held at a temperature of 65°F (18.3°C), embryonic development is fully arrested. However, hatchability decreases for each day the eggs are held. Eggs held for less than 5 days show little perceptible reduction in their hatchability or in the quality of the chicks hatched from them. When the period of holding is longer than 4 days, hatchability will drop materially with each additional day.

Under commercial conditions eggs are set at least twice a week. When set twice, some of the eggs would be 3 days old, some 2, and some 1; and there would be little discernible loss in hatchability of the group. However, on occasion eggs are held for longer periods before placing them in the incubators. Such holding not only reduces the hatchability as shown in Table 7-5 but increases the length of the incubation period.

Rule of thumb: Hatching time is delayed 30 minutes and hatchability is reduced 4% for each day eggs are held or stored after 4 days.

Age of Breeder Hens, Shell Quality, and Hatchability

Eggs with good shell quality from older breeders hatch almost as well as those from younger breeders. Hens having the least gain in egg weight throughout their laying year have the least decline in shell quality. Increases in egg weight are more detrimental to hatchability than decreases in shell quality.

Table 7-5.   Effect of Egg Storage on Hatchability
and Incubation Period

Days of Storage[1]	Hatchability of Fertile Eggs %	Hatching Times as a Delay from the Normal hr
1	88	0
4	87	0.7
7	79	1.8
10	68	3.2
13	56	4.6
16	44	6.3
19	30	8.0
22	26	9.7
25	0	

[1]Storage at 65°F (18.2°C).

Age of Egg Affects Broiler Weight

As the age of the hatching egg affects day-old chick quality, so does it affect mature broiler weight. Table 7-6 shows the results of a test.

Eggs from a young breeder flock can be held longer than those from an old breeder flock. Eggs from older flocks should be set as soon after laying as practical.

Table 7-6.   Length of Egg-Holding Period and
Broiler Weight

Egg Holding Period[1] da	Mature Broiler Weight	
	lb	kg
1–7	4.11	1.86
8–14	4.06	1.84
15–21	3.95	1.79

Source: McDaniel, G. R., 1972, *Poultry Dig.,* July.
[1]Eggs held at 65°F (18.2°C).

Holding Eggs for Long Periods in Plastic Containers

To prevent rapid loss of moisture from hatching eggs stored for long periods, a liner of plastic (or similar material) may be used inside the egg cases. These liners are usually sealed to make them air tight. For further preservation of egg quality, nitrogen gas may be flushed into the egg area before sealing the container. (Do not use carbon dioxide as it depresses hatchability.) After a few hours in lined cases, moisture escaping from the eggs will raise the humidity. This, in turn, aids in slowing further egg evaporation. Such a method will

1. Prolong the hatching quality of the eggs
2. Increase the hatchability
3. Not be detrimental to the chicks hatched from such eggs
4. Not alter first-year production from birds hatched from eggs that have undergone the treatment

Procedure

1. Disinfect the shell with an eggshell sanitizer.
2. Cool the eggs thoroughly to 65°F (18.3°C). Both (1) and (2) are necessary to help prevent mold from growing on eggs stored in humid airtight plastic containers.
3. Place eggs in egg cases with plastic container, flush with nitrogen gas, and seal.
4. Store eggs at 65°F (18.3°C).

Position of Eggs During Holding Period

During a holding period of less than 10 days, eggs should be placed small end down in the trays or on the flats. When held longer than 10 days, hatchability will be improved if eggs are held small end up.

Turning Hatching Eggs During the Holding Period

When hatching eggs are held for less than 1 week before being set there seems to be no need for turning them during this period. However, in the case of certain poultry breeding and genetic farms, it may be necessary to hold eggs for rather long periods. Rotating eggs from side to side over a 90° angle will then improve hatchability.

How to turn the eggs during the holding period. Place the eggs in an egg case and place a 10-in. (25-cm) block under one end. The next day remove the block and place it under the other end.

Moisture Condensation on Eggshell

When eggs are removed from a cold room to a room with a higher temperature, as moving eggs from the cool egg-holding room to the egg-traying room, moisture will often condense on the shells; this picks up additional bacterial organisms floating in the air and increases shell contamination. Furthermore, such moisture makes eggs hard to handle and they are easily soiled.

Table 7-7 shows the temperature and humidity at which eggs will "sweat" (condense moisture on the shell).

Remedy for moisture condensation. If moisture is condensing on the shells when the eggs are moved from the egg-cooler room to the egg-traying room, there are two remedies:

1. Decrease the relative humidity in the egg-traying room. This may be difficult, and even impractical.
2. Increase the temperature in the egg-traying room. When temperature is increased, the relative humidity decreases, thus lowering the condensation of moisture.

Caution: Never fumigate moisture-laden eggs with formaldehyde gas. All eggs *must be dry* before the fumigation process is initiated.

Table 7-7. Effect of Humidity and Temperature on Moisture Condensation on Eggshells

Egg-traying Room Temperature		Temperature of Eggs (or Temperature of Cooler)		
		55°F (12.8°C)	60°F(15.6°C)	65°F(18.3°C)
		Eggs Will Sweat if Relative Humidity in Egg-traying Room Is Higher Than		
°F	°C	%	%	%
60	15.6	82	—	—
65	18.3	70	85	—
70	21.1	58	71	83
75	23.9	50	60	71
80	26.7	42	51	60
85	29.4	36	44	51
90	32.2	30	37	43
95	35.0	26	32	38
100	37.8	22	28	32

7-F. GRADING AND TRAYING HATCHING EGGS

If it is necessary to sort hatching eggs by size, the process should be completed after the eggs have cooled in the egg-holding room. The usual procedure is to grade and tray eggs as close to setting time as possible. When there is a uniform daily labor supply in the hatchery this may not be practical, and eggs may have to be trayed daily and ahead of setting time. If this is the case the trayed eggs should be placed in cabinets, and the cabinets covered. Do not allow hatching eggs to be located in free-flowing air as in front of fans. Increasing the flow of air around eggs increases the rate of egg evaporation, and thus dries out the contents more rapidly.

Egg grading increases cracked eggs. Each handling increases the number of cracked eggs. Do not grade eggs unless absolutely necessary.

Most broiler hatching eggs not graded. With integration in the broiler industry there is little advantage in making a close selection of hatching eggs for quality and weight. Integrators feel it is more profitable to set eggs without grading than to go to the expense of the necessary handling. But this does not necessarily mean that some *minimum* weight of hatching eggs should not be established.

Minimum Egg Weight

The size of the newly hatched chick is directly related to the size of the egg from which it hatches. Because egg size increases as the hen continues to produce eggs, there is a marked variation in the size of eggs coming into a hatchery. Most hatcheries set up a minimum egg size for hatching purposes, dependent on the use to which the chicks are to be put—broiler, Leghorn commercial, breeder, etc. Variations of the minimum will be between 22 and 24 oz per doz (52.0 and 56.7 g/ea).

As customers do not like small chicks, particularly when they are mixed with those of normal or large size, weight of the eggs placed in the incubator may become especially important, as Table 7-2 shows.

7-G. WARMING EGGS PRIOR TO INCUBATION

It may be advisable to warm eggs prior to placing them in the incubator.

Warming Eggs

Hatching eggs should not be removed from the cool holding room and placed directly in the setters. Rather, they should be warmed to room temperature first, but not to any temperature above 75°F (23.9°C) or embryonic development will be initiated. The warming process may take from 4 to 6 hours depending on the temperature of the egg-holding room. Placing cold eggs in the setters usually reduces the temperature within the machine until the freshly set eggs reach the incubating temperature. This cool environment delays the hatching time of the newly set eggs and lowers the hatchability of the eggs already in the incubator. However, some setters are equipped with an extra set of heaters that engage until the setter temperature is brought up to normal.

Preincubation of Hatching Eggs

In some instances hatching eggs are preincubated to increase the percentage of hatchability. The increase is from 1 to 2%. That is, they are subjected to a temperature of 101°F (38.2°C) for 6 to 8 hours, then cooled to room temperature before being placed in the incubators. Under some conditions this may have merit, but any increase in hatchability is usually offset by the increased cost of preincubating the eggs.

Incubation process cumulative. That the incubation process is cumulative and any preincubation at near normal incubating temperatures shortens the regular incubation period by an equivalent time is shown in Table 7-8.

Table 7-8. Effect of Preincubation Time on Length of Regular Incubation Period

Hours Held After Laying at 101°F (38.2°C) Before Cooling	Hours of Regular Incubation Period	Total Hours of Incubation	Hatch of All Eggs Set %
0	532	532	80
6	523	529	82
12	519	531	81
18	514	532	56
24	498	522	52

8

Factors Affecting Hatchability

The actual process of hatching a chick is complicated and there are a great many factors affecting the normal procedure. Although optimums for temperature, humidity, and air have been determined, and incubators are capable of operating within narrow limits, some hatches are poor, and there is little doubt that hatchability could be improved in most hatcheries. This chapter deals with the many factors that actually cause the variability.

Hatchability may be measured by two formulas:

1. The number of chicks hatched as a percentage of *all* eggs set
2. The number of chicks hatched as a percentage of the *fertile* eggs set

From a commercial standpoint the first definition is popular, but to distinguish variability in the fertility of eggs from hatchability, the second definition is used when scientists examine hatchability data critically.

8-A. FERTILITY

The ability of the female breeder to produce fertile eggs depends on factors in the laying pen. Good viable breeding males and healthy normal breeding females are a requisite. Therefore, fertility is the result of laying house management rather than hatchery management. The difficulty lies in the fact that most commercial poultrymen talk of hatchability as the percentage of chicks hatched from all eggs set, rather than those from the fertile eggs, which leads to incorrect assumptions as Table 8-1 shows.

Table 8-1.   Percentage of Total Hatch When Percentage of Fertile Eggs and Percentage of Hatch of Fertile Eggs Vary

Fertility %	% Hatch of Fertile Eggs					
	95	90	85	80	75	70
	% Hatch of All Eggs Set					
95	90.2	85.5	80.0	76.0	71.3	66.5
90	85.5	81.0	76.5	72.0	67.5	63.0
85	80.8	76.5	72.3	68.0	63.8	59.5
80	76.0	72.0	68.0	64.0	60.0	56.0
75	71.3	67.5	63.8	60.0	56.3	52.5
70	66.5	63.0	59.5	56.0	52.5	49.0

Example.   With a fertility of 95% and with 75% hatch of fertile eggs, the hatchability of all eggs set is 71.3%. But also when the fertility is 75%, and hatchability of fertile eggs is 95%, the hatchability of all eggs set is 71.3%.

From Table 8-1, there are many combinations of fertility and hatchability of fertile eggs that give the same figure for hatchability of all eggs set.

Fertility or hatchability problem?   Always be sure to separate fertility and hatchability and consider each as a separate entity. First see where the problem lies—in fertility or hatchability.

Fertility Cannot Be Predetermined

It would be advantageous to be able to differentiate between fertile and nonfertile eggs prior to incubation, yet no system has been devised to make this classification. Specific gravity, egg shape, air cell, and shell texture are not indicators of fertility. The only acceptable practice is to incubate the eggs for several hours or days, then place them before a bright light (candle) to observe embryos (living or dead), or lack of embryos. Special lighting systems make it possible to complete this technique after only a few hours of incubation. With commercial candlers several days of incubation are required.

Eggs Must Be Broken to Determine True Fertility

Candling is a crude method of determining infertility, giving only approximations as observed through the shell. If an accurate differentiation between fertility and infertility is desired, incubated eggs should be broken out and examined. With this procedure it may be observed that many embryos have died in the very early stages of development—and these eggs appear clear under the candling light. Such eggs must be classified as fertile. The problem lies with very early embryonic mortality, often before the egg is laid.

Fertility Inherited

To some extent, fertility is an inherited factor. For example, some strains of chickens produce better fertility than others. Furthermore, individual males and females vary in their ability to produce viable embryos. Certain mutations are correlated with infertility. Homozygosity of the ''R'' gene for rose comb is associated with poor fertility in males, but not in females. Similarly, it is known that fertility in Cornish birds is less than in other breeds, yet when artificial insemination is used, fertility is normal. It is also common knowledge that by continuous selection within a strain of chickens over a period of years it is possible to increase or decrease fertility.

Improved hatchability indirectly improves fertility. Genetic selection, generation after generation, to improve hatchability as defined as the percentage hatch of *total* eggs set, has obviously involved both fertility and hatchability. But the modern geneticist in developing his poultry lines will breed for each separately.

8-B. SEX OF CHICKS

Impossible to Predetermine Sex

There is no method for determining the sex of the living embryo at the time the egg is laid or at any time thereafter until hatching. Shape of the egg, position of the air cell, specific gravity of eggs, or other factors are not correlated with the sex of the developing chick.

Sex Ratio

In all probability the ratio of males to females is nearly equal at the time the ova are fertilized (primary sex ratio), but unequal mortality during embryonic development usually causes a varying preponderance of males over females at the time of hatching (secondary sex ratio). The following are causes of variability in the secondary sex ratio:

1. *Genetics.* Varieties and strains of chickens differ.
2. *Lethals.* Some lethal genes are associated with sex, reducing the hatchability of one sex more than the other.
3. *Physical factors.* Evidently one sex is better able to adjust itself to the environmental conditions of incubation.

4. *Time of egg laying.* Sex ratios change according to the egg production period—time of day, time of year, higher proportion of males hatched during hot weather, etc.

5. *Correct sex ratio.* There is no such ratio because of the many factors involved. Besides, the sex ratio varies throughout the year as egg size changes and as birds continue through their egg production period.

8-C. METABOLISM OF THE CHICK EMBRYO

In part, the metabolic rate of a chick embryo is the result of the temperature at which it is incubated. Higher temperatures accelerate growth; lower temperatures retard growth. As the rate increases, the cells of the embryo must have more oxygen to metabolize the fats, carbohydrates, and dozens of other cellular components, which in turn produce more by-products: water and carbon dioxide (CO_2). As embryonic metabolic rate is decreased, less oxygen is needed, and less water and CO_2 are produced as by-products.

When these reversible chemical reactions are in balance, the incubating temperature is correct. In the modern setter, this optimal temperature is about 99.5°F (37.5°C), but recording devices are often in error, so to be sure the temperature is correct the recording devices must be checked occasionally.

Importance of Water Loss from the Egg

To correctly control the loss of metabolic water from the egg the surrounding air in the setter must have a relative humidity of 50 to 60%. When the humidity is too high, the metabolic water will not be exhausted through the shell as fast as it is produced. Embryos will be "drowned" in their own by-product, which is water.

Inasmuch as oxygen enters the egg through the shell pores at a rate proportional to that of water being eliminated, when water leaving the egg is reduced entrance of oxygen is lowered and embryos "suffocate." With too much water and too little oxygen, the percentage of *early dead* embryos increases, the chicks hatch later, and they are large and weak.

When the relative humidity of the air surrounding the egg is too low the metabolic water will be removed from the egg contents faster than the developing embryo is producing it, and water will also be removed from the yolk and albumen. Early dead embryos increase, the chicks hatch early, and they are smaller and weaker.

Research has shown that when eggs of average size and quality are incubated in air with 50 to 60% relative humidity, they will lose approximately 12% of their initial weight in 19 days of incubation. This percentage may be converted to 0.632% per day. Although the weekly egg weight loss is slightly greater at the end of the incubation period, the difference is not great enough to warrant a change in the above figures.

Egg Weighing to Determine Weight Loss

The only way to determine if incubating eggs are losing weight according to schedule is to calculate the weight loss. Hygrometers and other such recording devices only register the relative humidity of the air, not the egg weight loss.

The practice of weighing trays of eggs and calculating the percentage of weight loss at various periods of incubation must be made a part of the hatchery program. In most cases, the first calculation of egg weight loss will show that the relative humidity of the setters has been too low; eggs have been evaporating too rapidly and have been losing too much weight.

Tray calculations may also show a high variability. A close study of the tables in this chapter should make it possible to determine the cause and make the necessary adjustments in the relative humidity.

How to calculate egg weight loss. Use the following procedure for several trays of eggs, calculating each tray separately:

1. Weigh an empty incubator tray when eggs are set. Fill the tray with eggs and weigh the tray and the eggs.
2. Subtract the weight of the empty tray from the weight of the eggs and tray to obtain the net weight of the eggs. Calculate the average weight of the eggs in ounces per dozen or grams each.
3. After several days of incubation, replace cracked and broken eggs with comparable eggs from another tray; then weigh the tray and the eggs and subtract the weight of the tray, arriving at the net weight of the eggs.
4. Calculate the egg weight loss as a percentage of the original weight of the eggs. Divide the percentage by the number of days of incubation.
5. Check the calculation against the recommended percentage daily weight loss from Table 8-3 according to egg weight.

8-D. TEMPERATURE DURING INCUBATION

The living embryo has an optimum temperature at which it completes its growth best. This does not mean that growth will not be initiated at temperatures below and above the optimum. It does mean that when temperatures are different than the optimum, growth rate will be impaired and the embryo will be weakened.

Physiological Zero

Physiological zero is that temperature below which embryonic growth is arrested, and above which it is initiated. It has been difficult to determine the figure exactly; there are too many disturbing factors. Furthermore, it varies with strains and varieties of chickens. Most research work of late has established the physiological zero at about 75°F (23.9°C).

Optimum Temperature for Incubation

When incubated in a forced-draft incubator, some chicken eggs will hatch provided the temperature is between 95° and 105°F (35° and 40.5°C). However, there is an optimum temperature, somewhere between the two figures, at which the embryo develops best. Research work has shown that the optimum temperature during the first 19 days of incubation is somewhat higher than that required during the last 2 days.

The exact optimum will vary with different makes of incubators; and as each manufacturer has accurately established the temperature at which hatchability and chick quality are best, the directions should be followed explicitly.

When incubating temperatures deviate from the optimum, hatchability declines and the incidence of malformed chicks increases. The ambient temperature at which the egg is incubated affects the length of the incubation period. As temperature increases above the optimum, the incubation period is shortened; as it decreases, the incubation period is lengthened. This does not mean that it is advisable to raise or lower the temperature from the established optimum. Doing so only weakens the embryos and results in chicks of poor quality.

The optimum incubating temperature is not the same for all eggs. The following factors influence it:

1. Size of the egg
2. Shell quality
3. Genetics (including breed and strain of chicken)

4. Age of the egg when it is set
5. Humidity of the air during incubation

As eggs vary greatly in the above characteristics, optimum incubating temperature has been established for "average" eggs. In most instances eggs are not segregated to take advantage of different optimum temperature for eggs in the five categories, but there may be occasions when it is practical.

Three Optimum Temperatures

Embryonic growth may be divided into three phases, each requiring a different temperature, as follows:

1. *Prior to egg laying.* The body temperature of the broody hen fluctuates between 105° and 107°F (40.6° and 41.7°C). As the new embryo completes many cellular divisions during the 22 hours between the time of the union of the sperm and egg cell and the time the egg is laid, the optimum temperature for embryonic development during this period must be that of the body temperature of the hen.
2. *During the first 19 days of incubation.* Although varying slightly according to the make of forced-draft incubator, the temperature is about 99.5°F (36.7°C).
3. *During the 20th and 21st days of incubation.* In forced-draft incubators, best hatchability occurs when the temperature is lowered from that during the first 19 days to between 98° and 99°F (36.7° and 37.2°C).

These variations indicate that the developing embryo is quite critical of its environment, and because of the narrow confines in the temperature at which it develops best after artificial incubation begins, all incubators must be capable of having their temperature regulated within small fluctuations.

Adjusting Incubating Temperatures

Because optimum incubating temperatures have been developed using "average" eggs, hatchability could obviously be improved if the optimum were known for every type and size of hatching egg under varying weather and other environmental conditions, and adjustments made in incubating temperature accordingly. Regardless of the manufacturer's recommendations, incubators should undergo some experimentation to determine the exact optimum incubating temperatures under local conditions and with the type of hatching eggs involved. Even small corrections will usually show some improvement in hatchability or chick quality.

Embryonic Effects from Overheating

Exposure of 16-hour embryos for 24 hours to a temperature of 104°F (40.0°C) caused no major detrimental effect on hatchability. But exposure for 6 hours to a temperature of 110°F (43.3°C) caused a decrease in hatchability with a severe decline after 9 hours of exposure. Heating to 115°F (46.1°C) for 3 hours or 120° (48.6°C) for 1 hour killed all embryos. Chicks hatched following severe heat stress had a high incidence of clubbed down, wiry down, and exhibited unsteady gait (J. H. Thompson et al., 1976, *Poultry Sci.*, pp. 892–894).

Embryonic Effects from Cooling

Under natural conditions, hens leave the nest many times each day during the incubation period. Cooling the eggs during the hens' short absences from the nest is evidently not detrimental to hatchability under such conditions.

With artificial incubation there are many times when electric failures cause a reduction in the environmental temperature within an incubator that may last for a short period of time or for several hours. During the first 3 days of incubation embryos are quite resistant to low incubating temperatures. After 3 days there is a linear reduction in this resistance up to 19 days. Hatched chicks in the hatcher are more resistant to cold than to heat.

Cooling lengthens the incubation period.   Because incubation is a cumulative process, any reduction in the incubating temperature will increase the length of the incubation time. However, the incubation time will be longer by only about half as much as the cooling period itself because incubator temperatures are seldom reduced below room temperature of 65° to 70°F (18° to 21°C). Even then, the contents of the egg will remain above room temperature for several hours.

Cooling increases the incidence of malpositions.   Cooling incubating eggs during the first 19 days of incubation increases embryonic malpositions (see Chapter 8-I). The lower the temperature, the higher the incidence.

Cooling during the 20th and 21st days of incubation.   Although the embryo can withstand short drops in incubating temperature during the first 19 days, disaster generally results when the temperature is lowered more than 2°F during the last 2 days. Electric interruption during this period of incubation, even if only for a short period, is critical.

What Happens When Electric Power Fails

When electric power fails, fans or blades used to distribute the heat uniformly in the incubator stop. Hot air rises to the top of the compartment, and eggs in this area become overheated, while those in the lower section become chilled. When chicks are hatching, this excess heat at the top of the cabinet is usually lethal; the chicks suffocate.

Standby electric equipment necessary. No hatchery should be operated unless there is assurance that there will be no cessation of electric current. Continuous power is an absolute necessity during the hatching period (20 to 21 days). Consequently, a standby electric generator must be installed to take over when the regular source of electricity falters (see Chapter 6-D).

Thermometers Should Be Checked Often

All thermometers used in the incubators should be checked for accuracy occasionally. First secure a good test thermometer. Heat a pail of water to approximately 100°F (38°C). Place the test thermometer and one or more incubator thermometers in the water and stir to keep the water temperature uniform. While thermometers are still in the water, check the recording thermometer against the test thermometer.

Reuniting separated mercury. When the mercury in the thermometer separates, temperature readings are inaccurate. To reunite the mercury

1. Place the thermometer in the freezer section of a refrigerator for about 30 minutes. Remove the thermometer and shake the mercury down into the bulb, or
2. Place the thermometer in water warm enough to force the mercury into the bulb at the top of the tube, being careful not to have the water too hot. Remove and shake the mercury down as it cools.

8-E. HUMIDITY DURING INCUBATION

For an embryo to develop properly and to transform into a chick of normal size, the egg contents must evaporate at an established rate. When the egg contents dry out too rapidly the chick will be smaller than normal; when they do not evaporate fast enough the chick will be larger than normal. In either case the embryo is weakened, resulting in lowered hatchability and reduced chick quality. To regulate the evaporation of the egg contents the amount of moisture in the air surrounding the egg must be controlled, since this outside moisture determines the rate of the egg weight loss. High humidity reduces egg evaporation; low humidity increases it.

Measuring Relative Humidity of the Air

Relative humidity may be calculated by comparing the temperatures recorded by wet-bulb and dry-bulb thermometers. The dry bulb records the normally known temperatures of the air. The wet-bulb thermometer is an ordinary thermometer in which the bulb has been covered with a water-moist wick. When air is forced around this bulb and wick, there is a cooling effect produced by the evaporation; and the more cooling, the lower the wet-bulb temperature.

As temperature determines the amount of moisture air will hold at maximum capacity, a table must be used in order to determine the percentage of relative humidity of the air being measured (see Table 8-2). As the temperature of the air in the setter and hatcher is maintained within close limits some hygrometers have been converted to percentage of relative humidity and read direct at 99.5°F (37.5°C). There are also instruments that show the relative humidity regardless of the incubator temperature.

Table 8-2. Percentage Relative Humidity as Determined by Wet-bulb and Dry-bulb Thermometer Readings

Wet-bulb Temperature		Dry-bulb Temperature			
		98.0°F 36.7°C	98.5°F 37.0°C	99.0°F 37.2°C	99.5°F 37.5°C
°F	°C	Relative Humidity (%)			
80	26.7	46	45	44	43
82	27.8	51	50	49	48
84	28.9	56	55	54	53
86	30.0	61.5	60.5	59	58
88	31.1	67	66	65	63.5
90	32.2	73	72	71	69.5
92	33.3	79	78	77	75.5

Importance of Correct Humidity

To ensure proper dehydration of the egg contents the relative humidity of the air in the setter during the first 19 days of incubation must be confined to within rather narrow limits. Depending on the make of the incubator, these limits are 50 to 60%, but hatchery operators should experiment to determine a precise percentage.

Increasing the humidity of the setter (1 to 19 days) lengthens the incubation period; reducing it shortens the incubation period. Generally, too much humidity during the first 19 days of incubation will cause chicks to hatch later than normal and they will be larger and soft in the abdomen. Too little humidity will produce the opposite effects, as well as dehydrated shanks.

Egg Size and Its Effect on Egg Weight Loss

Hatching eggs weighing 24 oz/doz (56.7 g/ea) and having good shell quality should lose approximately 12% of their weight during the first 19 days of incubation, but there are many factors influencing the figure, one of which is egg size. Table 8-3 shows the egg weight loss when eggs of different sizes are incubated, but the humidity remains the same.

Table 8-3. Daily Weight Loss of Hatching Eggs of Various Sizes (Relative Humidity of 50–60%)

Avg Beginning Egg Weight		Egg Weight Loss, 1–19 Da of Incubation	Avg Daily Egg Weight Loss
oz/doz	g/ea	%	%
23	54.3	12.25	0.645
24	56.7	12.00	0.632
25	59.1	11.80	0.621
26	61.4	11.60	0.611
27	63.8	11.45	0.603
28	66.2	11.30	0.595

Shell Area and Egg Weight Loss

The *surface area* of the shell is indirectly correlated with the weight of the egg contents. That is, large eggs have less shell area per unit of interior egg weight than do small eggs. As egg evaporation depends mainly on the area of the shell, and the resulting number of shell pores through which moisture is lost, smaller eggs lose a larger *percentage* of their weight during incubation than larger eggs (see Table 8-4).

Furthermore, smaller eggs produce smaller chicks not only because the eggs are smaller but chicks hatched from these eggs are even smaller because smaller eggs evaporate at a faster rate. With larger eggs, the reverse is true: Larger eggs produce larger chicks, and the eggs from which they came evaporate less.

Most flock-run eggs will vary as much as 5 oz/doz (2 g/ea) in weight, and it is obvious that far from all are losing 12% of their weight during

Table 8-4. Relative Humidity and Egg Size as They Affect Incubating Weight Loss

Relative Humidity in Setter %	Original Weight of Eggs				
	oz/doz				
	22	24	26	28	30
	g/ea				
	52.0	56.7	61.4	66.2	70.9
	Loss of Egg Weight, 1 through 19 da of Incubation				
	%	%	%	%	%
70–79	10.6	10.3	10.0	9.8	9.6
60–69	11.5	11.1	10.7	10.4	10.2
50–59	12.5	12.0	11.6	11.3	11.1
40–49	13.7	13.1	12.6	12.2	11.9
30–39	15.0	14.3	13.8	13.4	13.1

19 days of incubation. Table 8-5 shows the relative humidity of the setter necessary for eggs of different weights to lose 12%. The best one can do is calculate the average weight of the eggs and use this table.

Remember.   As egg size of the breeding flock increases the longer the flock is in production, the percentage of relative humidity in the setter will have to be lowered or egg evaporation will not be great enough.

Table 8-5. Egg Size as It Relates to Relative Humidity

Original Weight of Eggs		Relative Humidity in Setter for Eggs to Lose 12% Weight in 19 Da %
oz/doz	g/ea	
22	52.0	58–62
23	54.3	56–60
24	56.7	53–57
25	59.1	51–55
26	61.4	49–53
27	63.8	47–51
28	66.1	45–49

Shell Quality Affects Humidity Requirement

Moisture moves more freely through shells of poor quality than shells of good quality. Thin chalky porous shells are instrumental in increasing evaporation of the egg contents, thus producing chicks smaller than normal for the size of the eggs involved. Chicks from eggs with thick dense shells tend to be larger than normal, because there has not been as much egg evaporation during the incubation process. Table 8-6 shows these variations.

Table 8-6. Influence of Shell Quality on Egg Weight Loss During Incubation (57% Relative Humidity)

Egg Weight		Shell Thickness	Weight Loss 1 through 19 Da of Incubation %
oz/doz	g/ea		
24	56.7	Average	12.0
24	56.7	Thin	14.0
24	56.7	Thick	10.5

Causes of Poor Shell Quality

Although there are many things that lower hatchability, poor shell quality probably has the greatest effect. Shell quality from young breeder flocks is usually good and hatches are high, but as the birds continue through their laying year shell thickness and shell quality deteriorate and hatchability drops. This difficulty can be laid to the bird's inability to increase calcium deposition on the egg to compensate for the increase in egg size during the latter part of the production period. A dietary deficiency of calcium during the time the eggshell is being formed in the oviduct aggravates the situation.

In some instances, hatching eggs laid early in the morning have poorer hatchability than those laid later in the day. These early eggs will have poorer shell quality, usually because the egg has passed through the shell gland at night when the hen is not eating. Therefore, the amount of calcium for eggshell formation is below optimum. To remedy the situation two-thirds of the oystershell in the mash should be from large-sized flaked oystershell, which dissolves and digests more slowly, so more calcium is available at night (see Chapter 31-G).

Since shell thickness and specific gravity of the egg are closely correlated, the break point between good and poor hatchability is at a specific gravity of 1.080, regardless of the age of the breeding flock. Eggs from young and old breeder flocks should produce about the same hatchability when the specific gravity of the eggs is the same (see Chapter 16-U).

Cracked Eggs

From a study of commercial hatcheries, up to 2% of all eggs set were cracked prior to hatching. On the average, 1.1% were cracked when trayed and 0.9% were cracked at transfer, but in many hatcheries the numbers were twice these figures. Hatcheries with a lower number of cracked eggs had about one-third these figures. Cracked eggs are a great economic loss, but the above figures show that the number can be reduced.

Temperature and Humidity in the Setter

There is an interaction between temperature and humidity in the setter. As the temperature has usually been set by the manufacturer, the only adjustment necessary is in the relative humidity. Whether the humidity is correct or incorrect depends only on the moisture loss through the eggshell by evaporation. A calculation should be made at least three times during the incubation period because humidity varies from changes in egg weight, shell thickness, shell quality, age of the breeding flock, season of the year, and nutrition.

Temperature and Humidity in the Hatcher

No matter what its age, the chicken cannot withstand high temperature and high humidity at the same time. Its body is just not built to handle both. This rule applies not only after the chick has hatched, but during the period of embryonic development.

During the last 2 days of incubation (20 and 21), while the eggs are in the hatcher, the humidity must be increased but only within certain limits. Correct moisture prevents the beak of the chick from sticking to the newly pipped shell and allows for freer movement of the chick's head at the time of pipping. Too little moisture at the time of hatching will produce chicks smeared with egg or shell, stuck down, and partially dehydrated. Too much humidity during this period will cause the chicks to be smeared with egg, and they will have a navel that has not closed properly.

The relative humidity of the hatcher should be about 65% when the eggs are transferred to it, an increase of 5 to 10% from that of the setter. A temperature reduction of from 0.5° to 1.5°F (0.3° to 1.0°C) is necessary. But remember that when the air temperature is reduced there is an automatic decrease in relative humidity as cooler air holds less moisture. A temperature reduction of 1.0°F (0.6°C) will decrease the relative humidity by 2.5%.

Some chicks will have pipped the shell when the egg transfer is made, so changes in temperature and humidity must not be delayed. As the hatch progresses, the chicks will usually exhale enough moisture to increase the relative humidity in the hatcher to about 75%, the optimum requirement.

8-F. AIR REQUIREMENTS DURING INCUBATION

The main components of air are oxygen (O_2), nitrogen (N_2), carbon dioxide (CO_2), and water vapor (H_2O). The free movement of these through the pores of the shell and the shell membranes is important; the developing embryo must have a constant supply of oxygen and exhaust carbon dioxide and moisture.

Oxygen in the Air

Approximately 21% of the air at sea level is oxygen, and it is impossible to increase the percentage appreciably in incubators unless pure oxygen is introduced.

Although the oxygen content of the air in a commercial incubator is not often altered, there may be some variation in the hatcher where large amounts of carbon dioxide are being liberated by the newly hatched chicks. In such cases, hatchability drops about 5% for each 1% that the oxygen content of the air drops below 21%.

Air Supply Generally Adequate

As the embryo advances in age its oxygen requirement increases and more carbon dioxide is given off. Each process is speeded up approximately 100 times between the first and 21st day of incubation, as Table 8-7 shows.

On the 18th day of incubation, 1,000 eggs would require 143 ft^3 (4.1 m^3) of fresh air (oxygen in the air at 21%). An incubator holding

Table 8-7. Gaseous Exchange During Incubation per 1,000 Eggs

Day of Incubation	Absorption of Oxygen ft^3	Expulsion of Carbon Dioxide ft^3
1	0.5	0.29
5	1.17	0.58
10	3.79	1.92
15	22.70	11.50
18	30.00	15.40
21	45.40	23.00

Source: Romanoff, A. L., 1930, *J. Morphol. 50*, 517–525.

40,000 eggs would need 5,720 ft³ (162 m³) of fresh air, or approximately 238 ft³ (6.8 m³) per hour. This would mean that the air in the incubator would have to be changed about eight times a day or once every 3 hours. Necessarily these are minimums, and although oxygen is necessary for the embryonic process the amount of fresh air needed in an incubator is relatively small. As air intakes in most machines are generally more than adequate, care should be taken to see that overventilation does not become a problem.

Carbon Dioxide Tolerance

Carbon dioxide (CO_2) is a natural by-product of metabolic processes during embryonic development beginning during gastrulation. In fact, CO_2 is being released through the shell at the time the egg is laid, the result of the pH of the egg contents changing to an alkaline condition.

Carbon dioxide concentrates in the air within the setter and hatcher when there is insufficient air exchange to remove it. Young embryos have a lower tolerance level to CO_2 than old ones when measured as the concentration in the air within the machines. The tolerance level seems to be linear from the first day of incubation through the 21st day. During the first 4 days in the setter, the tolerance level of CO_2 is 0.3%.

Carbon dioxide levels above 0.3% in the setter reduce hatchability, with significant reductions at 1.0%, and completely lethal at 5.0%. Chicks hatching in the hatcher give off more CO_2 than embryos contained in eggs, and the tolerance level in the hatcher has been set at 0.75%. Recording devices are available for measuring the CO_2 content of the air, and some incubators have them as standard equipment.

Speed of Airflow

There is no evidence to show that variations in the velocity of air flowing past the eggs in an incubator have any effect on hatchability. The important factor seems to be the ability to maintain enough air movement to provide a uniform temperature throughout the incubating cabinet. Many innovations in devices to circulate the air in incubators have been perfected; paddles, blades, and fans are used, each creating an airflow, or movement, from very slow to rapid.

Deduction.   As long as the hatchability of eggs incubated in all sections of the setter and hatcher is uniform, one may surmise that airflow is adequate. If nonuniformity occurs, anemometers and temperature sensors should be employed to monitor conditions in all sections of the machine.

8-G. INCUBATION AT HIGH ALTITUDES

In 1944, North reported that eggs incubated at an altitude of 7,200 ft (2,195 m) produced a greatly reduced number of chicks compared with similar eggs incubated at an altitude of 700 ft (213 m). This warrants some discussion, as many hatcheries around the world are located at high elevations.

Air varies in its density according to elevation; the higher the altitude, the less dense it becomes. Because air weighs less at higher altitudes it exerts less barometric pressure. Also, when air expands, as at high altitudes, a cubic volume contains less oxygen. Figures are shown in Table 8-8.

Research has shown that hatchability of chicken eggs is reduced as the altitude at which they are incubated is increased. However, for altitudes under 2,500 ft (760 m), the reduction is so slight as to be seldom noticed. But when the altitude is over 3,500 ft (1,067 m) the loss in hatchability becomes an acute problem. Figure 8-1 shows the reduction in hatchability at increased altitudes based on a hatchability of 80% at sea level.

Lower hemoglobin values the probable cause.   It has been found that lower hemoglobin, particularly at about 13 to 14 days of incubation, is the primary cause for the high embryonic mortality at higher elevations. But many embryos mature and hatch inasmuch as some are able to generate enough hemoglobin to further normal embryonic development. Lower hemoglobin production, or a delay in its production, delays the time of hatch. Along with a diminished supply of oxygen, embryonic growth is also retarded.

An interesting fact associated with high-altitude incubation is that lack of oxygen alone may not be the sole contributing factor. Although some scientists feel it is a lack of oxygen others believe it is

Table 8-8.   Relationship Between Altitude, Oxygen Content of Air, and Barometric Pressure

Altitude Above Sea Level		Barometric Pressure in. Hg	Reduced Weight of Air (or Oxygen) %
ft	m		
0	0	29.92	0
2,000	609	27.82	5.1
4,000	1,217	25.84	11.2
6,000	1,829	23.98	16.4
8,000	2,438	22.22	21.4
10,000	3,048	20.58	26.2
12,000	3,658	19.03	30.7

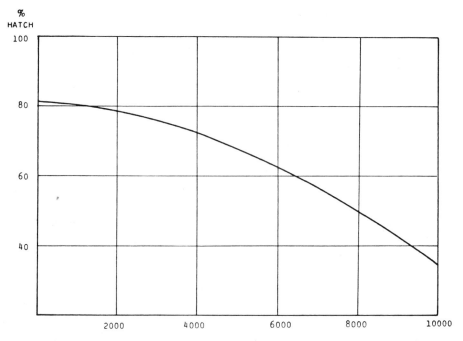

Figure 8-1. Relationship between altitude in feet and hatchability.

a reduction of pressurized oxygen, while others indicate that the amount of carbon dioxide in the incubating air is also important. Embryos developing at high altitudes are quite sensitive to carbon dioxide concentrations outside the limits shown in Chapter 8-G.

Increasing air pressure restores normal hatch. By increasing the air pressure to sea level normalcy during incubation it is possible to produce satisfactory hatchability. This indicates at least one method, though probably not always economical, of overcoming the ill effects of high-altitude incubation. But some hatcheries do provide pressurized atmosphere within the incubation rooms.

Oxygen injection a better method. Another and more practical method of restoring normal hatchability at high altitudes is to inject oxygen directly into the cabinet in which the eggs are being incubated. Concentrations of oxygen at 23 to 23.5% will increase hatchability materially.

How to inject oxygen. Oxygen is introduced into both the setter and hatcher compartments by a tube from oxygen cylinders that have a pressure regulator valve and flowmeter. A gas-analysis apparatus is used to determine the percentage of oxygen in the mixed air within the incubator cabinets. Readings must be taken several times a day to maintain the required ratio of oxygen.

Other Factors Involved
with High-Altitude Incubation

Of importance are factors involved with the practical aspect of high-altitude incubation and maintaining breeding flocks as follows:

1. Eggs from different strains have different hatchabilities.
2. Incubation time is increased as the elevation increases, evidently due to a decrease in the carbon dioxide content of the air rather than the change in the oxygen content.
3. Eggs produced by breeder hens kept at high altitudes seem to produce normal hatches if they are incubated at low altitudes.
4. Altitude has no effect on the fertility of eggs produced by the breeder flock.
5. If one rears the chicks produced from eggs laid by breeder hens at high altitudes and the eggs incubated at high altitudes, the mortality of the growing chicks is higher than normal.
6. Some of the hatchability may be restored over a period of years if genetic selection is practiced in the breeding program, but it cannot be restored to normal.

8-H. POSITION OF THE EGG DURING INCUBATION

It is necessary that eggs be kept in their proper position and turned regularly during days 1 to 19 of incubation. Eggs under artificial incubation should be held with their large ends up. It is the natural procedure for the head of the chick to develop in the large end of the egg near the air cell, and for the developing embryo to orient itself so that the head is uppermost. Most of this rotation occurs during the second week of incubation, and is most easily completed when the large end of the egg is kept higher than the small end. When eggs are incubated with the small end up, about 60% of the embryos will develop with the head near the small end. Thus, when the chick is ready to hatch, its beak cannot break into the air cell to initiate pulmonary respiration.

About 10% of the eggs set with the small end up will fail to hatch, and chick quality of those hatching will be reduced. Most of the reversals when eggs are set small end up are accidental or because it is difficult to tell the large end of the egg from the small, particularly with those eggs from older hens.

Turning Eggs During Incubation

The yolk of the newly laid egg has a specific gravity that causes it to settle in the thin albumen, but once the egg is placed under incubation the specific gravity lessens and the yolk rises in the thin albumen to come in contact with the outer thick albumen if the egg is not turned. If the egg is not rotated the two layers of thick albumen, normally separated by a layer of thin albumen, come in contact, and the embryo usually dies.

Under natural incubation the hen turns the eggs many times a day. With artificial incubation eggs are set large end up and rotated back and forth along their long axes to produce the turning process. Eggs should not be turned in a circle, as this ruptures the allantoic sac with resultant embryonic mortality. Most eggs are turned to a position of 45° from vertical, then reversed in the opposite direction to a similar position. One incubator turns them to a position of 90° from vertical, then reverses them to the opposite position. Less rotation than 45° is not adequate for high hatchability as Table 8-9 shows.

Interval of turning. During the first 19 days that eggs are under incubation they must be turned regularly and often. Table 8-10 shows the percentage of hatchability of eggs turned from two to ten times a day.

Although other experiments have shown that turning eggs as often as every 15 minutes is not detrimental to hatchability, nothing is to be gained by turning them more than six times a day when eggs are rotated back and forth along their long axes. Most commercial incubators provide for turning the egg automatically every 1 to 3 hours.

Period of turning. Table 8-11 shows the effect of various periods of turning hatching eggs during incubation. The results indicate that turning the first week is the most important, and the second week, next. Turning the last week seems to be of questionable value. Under practical conditions, eggs of various ages are intermingled in most setter compartments, so all have to be turned together.

Important: When eggs are turned the process should be completed

Table 8-9. Effect of Angle of Turning Eggs During Incubation

Angle Turned to Each Side of Vertical	Hatch of Fertile Eggs %
20°	69.3
30°	78.9
40°	84.6

Table 8-10. Effect of Turning Eggs on Hatchability

Times Turned Daily	Hatch of Fertile Eggs %
2	78.1
4	85.3
6	92.0
8	92.2
10	92.1

Table 8-11. Effect of Turning Hatching Eggs at Various Periods

Period Turned During Incubation da	Hatch of Fertile Eggs %
No turning	28
1–7	78
1–14	95
1–18	92

quickly, then the eggs allowed to remain stationary and resting until the next turning. Hatchability is lowered when eggs are kept in a constant back-and-forth motion.

Transferring Eggs to the Hatcher

With modern incubators eggs are transferred from the setter to the hatcher at about the end of 19 days of incubation. But the time is not constant: many factors affect it.

Length of incubation period varies. Several things affect the length of the incubation period—breed, sex, age of eggs, size of eggs, shell quality, etc. Eggs with shorter incubation periods should be set later than those needing longer periods of incubation. When setting procedure is correct, all eggs should hatch within a short time period.

Important: Do not transfer eggs too early or too late. Difficulties arise when eggs from egg-type and meat-type breeder birds are set at the same time in the same machine, as eggs from breeds such as Leghorns have a shorter incubation period than those from heavy breeds.

Females hatch before males. There is evidence that when fresh eggs are incubated the females hatch as much as 3 hours before the males. However, the spread decreases the longer the eggs are held prior to incubation and completely disappears when eggs are held for 14 days (Christine Mather and K. F. Laughlin, 1976, *British Poultry Sci. 17*, p. 471).

When to transfer eggs. Eggs should be transferred to the hatcher when approximately 1% of the eggs are slightly pipped. If eggs from Leghorns and meat-type birds are to be set in the same machine or hatchery, the Leghorn eggs should be set so they hatch last. A Leghorn chick from a 24-oz/doz (56.7-g/ea) egg will dehydrate faster after hatching than one from a meat-type egg of similar size. Hatching Leghorn chicks last will shorten the period from the hatchery to the farm.

Position of Eggs During Hatching

Most commercial incubators provide for keeping the eggs in a horizontal position during the last 2 days of incubation in the hatcher. Although they will hatch as well if kept upright with the large end up, this method has not been practical because of the additional space needed by the chicks in the trays once they liberate themselves from the shells.

Turning eggs the last 2 days of incubation has no value, and may be injurious to the embryo. There is no evidence to show that changing the position of the eggs at the time they are transferred to the hatchers is detrimental to hatchability, provided the transfer is not made too early. Starting with the 17th day of incubation the embryo begins to position itself for hatching, and the process may take more than 24 hours. Embryos should not have their routine changed during this period.

The incentive to hatch.   The beak of the chick first breaks the inner shell membrane of the air cell and enters the air cell where pulmonary respiration is initiated. But the air in the air cell contains only 15 to 16% oxygen compared with 21% for normal air, and about 4% more CO_2. In the chick's endeavor to get more oxygen and less CO_2, it pips the shell to get outside normal air, but it will be 10 to 20 hours later that it will be able to liberate itself from the shell.

8-I. OTHER FACTORS AFFECTING HATCHABILITY

Hatchability is a varying thing as anyone who has operated a hatchery well knows. There are a variety of factors affecting the number of chicks produced from a given number of eggs. Although many are of minor significance, they are cumulative, and several minor ones add up to greatly reduced hatchability.

Egg Laying Pattern
and Hatchability

The first eggs a breeder hen lays do not hatch well. Usually they are held in the hen for a period longer than normal, and the preincubation is detrimental to hatchability. Neither do the chicks from eggs laid during the first 2 weeks of egg production live satisfactorily. Under normal conditions, hatching eggs produced during the first 2 weeks of egg production are not set, not only because of poor hatchability and chick growth but because such eggs are small and produce small chicks.

Eggs produced at the end of the laying year do not hatch as well as those laid earlier. Normally, there is a pattern of increased hatchability from the first eggs set until about the 12th to 13th week of egg production, after which hatchability gradually decreases the longer the hen is producing eggs. The yolks from such eggs contain a higher percentage of fat, as do the day-old chicks hatched from them.

Eggs from birds with a high rate of lay hatch better than those from birds laying at a medium or low rate. There is evidence not only that eggs laid in longer clutches have higher hatchability, but those laid near the end of the clutch hatch better than those laid at the first of the clutch.

Weather Affects Hatchability

Lower hatchability from eggs laid during periods of extremes in environmental temperatures is common. Continuous days of hot or cold weather are likely to cause a drop in hatchability because heat and cold affect the breeders producing the eggs, but periods of short duration (1 or 2 days) will not. Hot weather during the summer months is a deterrent to good hatchability. In a study of large commercial hatcheries in the United States, the hatchability of eggs during the months of July, August, and September was about 5% lower than during the remainder of the year. The decreases encountered were the result of

1. Decreased feed consumption by the breeding flocks, which caused embryonic nutritional deficiencies
2. Deteriorated hatching egg quality during the preincubation (holding) period

Factors Affecting the Length of the Incubation Period

Although the normal average incubation period for chicken eggs is 21 days, this figure is highly variable. In fact, the variations may become so great at times as to affect the normal routine of hatchery labor and to lower chick quality. The following are some of the causes of this variation:

1. Certain diseases and stresses in the breeder flock will lengthen the period of incubation.
2. The longer a hen is in egg production, the longer the incubation period of her eggs.
3. The longer an egg is held in the body of the hen prior to oviposition, the more the early embryonic growth, which reduces the incubation time in the incubator. Embryos which are just past the gastrula stage when the egg is laid hatch better.
4. Leghorn eggs have a shorter incubation period than those produced by meat-type birds.

5. Eggs produced in the warmer season have a shorter incubation period than those laid in the colder season. This is the result of more preincubation during the warmer periods.
6. The smaller the breed, the shorter the incubation period.
7. Eggs with thick shells require a longer incubation period than those with thin shells.
8. The longer an egg is held at a temperature above 75°F (23.9°C) prior to setting, the shorter the incubation period.
9. The longer eggs are held in the cooler room prior to setting, the longer the incubation period.
10. Small eggs hatch sooner than large eggs.
11. Some procedures used in egg dipping delay the hatching time.
12. Embryos with certain malpositions require a longer time to emerge from the shell.
13. Eggs warmed prior to setting will require a shorter incubation period.

Noise and Hatchability

Does excessive noise during incubation affect hatchability? To answer this question, eggs were incubated near an airport and subjected to 600 sonic booms of overpressures of 0.5 to 4.8 lb per in.^{2} in the test area, but surprisingly did not exceed background noise from fans within the incubators. Such sonic booms did not affect the hatch (J. M. Heinemann, 1969, *Symposium on Extraauditory Effects of Audible Sound, Boston*).

Age of Breeder
Affects Hatchability

As breeders age, hatchability drops. Their eggs become much larger and are held in the oviduct longer, thereby increasing the length of the preoviposital incubation period. This places the embryo in a more advanced state of development at the time the egg is laid, a period not conducive to holding prior to incubation. To add to this difficulty, the eggshell of older hens is always thinner, especially in hot weather.

These large eggs laid by the older hens show a higher incidence of embryonic deaths at the time they are placed in the incubator when embryonic growth is reinitiated. These deaths come so early that they are often not noticed and are usually classified as infertiles. Hatching eggs from older hens should be *gathered more often* than those from younger hens.

Position of the Embryo in the Egg

Normally, the chick embryo develops with the head in the large end of the egg (near the air cell) and with its head under its right wing. But there are many embryos that do not develop in this position. These are called *malpositions,* and have been classified and described. The more common ones are

Classification	Description of Malposition
I	Head between thighs
II	Head in small end of egg
III	Head under left wing
IV	Head not directed toward air cell
V	Feet over head
VI	Beak above right wing instead of under

Some chicks in many of the above malpositions will hatch; some will not. Of all embryos at 18 days of age, approximately 1 to 4% will be malpositioned. An examination of the dead-in-shell will be necessary to determine the percentage and the type involved.

Abnormal Embryos

There are many embryos that develop abnormally and they should be culled during chick grading. A partial list of these variations is as follows:

small head	short beak	curled toes
popeye	crooked neck	wingless
one eye	twisted spine	clubbed down
no eyes	thickened hocks	short down
parrot beak	extra leg	dwarf
crossed beak	unabsorbed yolk	spraddle legged

8-J. EMBRYONIC MORTALITY PATTERNS

There are four periods during the development of the embryo when mortality may be excessive and thereby offer some indication of the cause of poor hatches.

Period I (Preoviposital Mortality)

Gastrulation represents the first critical point of embryonic growth. When eggs are held in the hen too long, embryonic development has advanced too far past the gastrula stage, and embryonic mortality during egg holding after the egg is laid increases markedly. There is also an increase in embryonic mortality during egg holding if the period of gastrulation has not been completed when the egg is laid.

Several factors have a bearing on the length of time it takes the egg to move through the oviduct. Large eggs take longer than small eggs. Eggs with thick shells take longer than those with thin shells. Hens whose eggs gain the least amount of weight through the laying period lay eggs that hatch better.

Eggs may be prematurely laid, which shortens the preoviposital incubation period. These eggs are generally characterized by thin shells, or in the case of brown-shelled eggs, by a lighter shell color. Certain respiratory diseases in the breeder flock cause eggs to be laid early.

The eggs of poorer producers require a longer period in the oviduct, sometimes as long as 27 hours and embryo growth has advanced too far when the egg is laid. This is one reason why better egg producers usually have higher hatchability.

Period II (Early-Dead Embryos)

These are embryos that die during the first 3 days of incubation. Some do not reinitiate development once the eggs are placed in the setter. Many may be the result of poor egg-holding conditions between the time the egg is laid and the time it is placed in the incubator, which lowers embryo vitality. But as most early-dead embryos are classified by the presence of coagulated blood, they cannot be identified until after the blood system begins to develop. If the vascular system is advanced far enough when the young embryo dies, the blood will circulate to the outer edges of the blood vessels and coagulate there, leaving a blood ring. Too much formaldehyde fumigation during the egg-holding period can cause increased mortality during Period II.

Period III (8- to 18-day Mortality)

Daily embryonic mortality during Period III *should* be very low, but sometimes it is high. During this period most nutritional deficiencies in the breeder diet have their greatest effect on the embryo, although too little vitamin A may cause excessive embryonic mortality in Period II since vitamin A aids in the development of the blood system.

Not only is mortality increased during Period III when there are nutritional limitations in the breeder diet but more embryonic abnormalities appear. Clubbed down, curled toes, dwarfing, micromelia (shortening of the long bones), parrot beak, crooked keel and beak, malpositions, blood clots, and edema occur.

Period IV (Nineteenth, Twentieth, Twenty-first Days of Mortality)

These 3 days represent another critical stage. Many changes occur in the developing chick during Period IV (see Chapter 4-E). Most of the embryonic mortality during Period IV is caused by factors of long duration. Of those chicks that fail to hatch, 50% will be found in a position other than normal and the most important cause of embryonic malpositions is that eggs are not set with the air cell up.

Normal Mortality Pattern

With good commercial hatcheries the following pattern is indicative of those eggs that fail to hatch. Variations from these figures will help define the mortality problem.

Infertile	5.0%
Period I mortality (preoviposital)	0.6%
Period II mortality (1–7 days)	2.0%
Period III mortality (8–18 days)	0.6%
Period IV mortality (19, 20, 21 days)	3.0%
Pips	0.8%
Total	12.0%

Normal embryonic mortality during Period IV should be about 50% greater than during Period II. Any variation in these two critical stages is indicative of a serious mortality problem.

Field Test of Embryonic Mortality

A team of scientists examined the hatching trays of nine North Carolina (USA) hatcheries, involving 70 large breeder flocks, with observations as shown in Table 8-12.

Table 8-12. Classification of Eggs and Chicks at Hatching Time

Item	Results of All Eggs Set		
	Avg %	Best Hatch %	Poorest Hatch %
Cracked eggs	1.28	0.81	2.93
Infertile eggs	5.10	3.54	4.63
Early dead embryos[1] (Period II)	2.12	0.78	5.96
Late dead embryos (Periods III and IV)	3.89	2.54	6.03
Pipped	1.11	0.97	1.12
Cull chicks	0.50	0.48	0.73
Hatchability	86.06	90.89	78.58

Source: North Carolina Poultry Dept.
[1]Includes preoviposital mortality

8-K. HATCHING AND HOLDING TIME AFFECT CHICK WEIGHT

The effects of hatching time and holding time in the incubator after hatching on subsequent broiler chick body weights were studied by J. E. Hager and W. L. Beane of Virginia Polytechnic Institute. Hatch weights of chicks at 20½ days of incubation were significantly greater than at 21½ days. Starting weights were dramatically reduced when the chicks were held in the incubator for 18 hours or longer after hatching.

The chicks held in the incubator for 18 hours or longer after hatching had significantly lower body weights through 3 weeks of age. Those held for 36 hours had lower weights through 4 weeks of age.

8-L. NUTRITIONAL EFFECTS ON HATCHABILITY

Nutritional deficiencies or toxic materials are likely to affect both egg production and hatchability, the difficulty increasing gradually as the involvement becomes more acute. Sudden drops in either egg production or hatchability are more apt to be the result of disease in the flock or incubator failure.

With nutritional deficiencies embryonic mortality arrives at an earlier age. For example, mortality that normally occurs at 18 to 21 days will be found at 15 to 19 days. A summary of some nutritional deficiencies is shown in Table 8-13.

Table 8-13. Nutritional Deficiencies of the Breeders That Affect Hatchability

Deficient Items	Embryonic Description
Vitamin A	Failure to develop normal blood system. Embryonic malpositions.
Vitamin D$_3$	Rickets. Lack of phosphorus. Stunted chicks and soft bones resulting from improper calcification of eggshells.
Vitamin E	Reduced fertility. Bulging eyes. Inadequate embryonic vascular system. Exudative diathesis (edema). Embryonic mortality 1 to 3 days.
Vitamin K	Prolonged embryonic blood clotting time. Hemorrhages and blood clots in embryo and extraembryonic blood vessels. Hemorrhagic syndrome of embryos and newly hatched chicks.
Riboflavin	High mortality 9 to 14 days. Edema, atrophied leg muscles, clubbed down, curled toes, enlargement of the sciatic nerve sheaths, and dwarfing are indicative of too little riboflavin. Reduced hatchability 2 weeks afte breeder rations become deficient.
Pantothenic Acid	Abnormal feathering. Subcutaneous hemorrhages in embryo. Chicks hatch in a weak condition and most fail to survive.
Biotin	Perosis. Short long bones (micromelia). Shortened and twisted bones of the feet, wings, and skull. Webbing between third and fourth toes. Parrot beak. Excessive mortality between 1 to 7 days.
Vitamin B$_{12}$	Embryonic malposition with head between legs. Edema. Short beak. Poor muscle development. High embryonic mortality 8 to 14 days.
Folacin	Similar to biotin deficiency. Chicks die shortly after pipping the shell.
Vitamin B$_6$	Reduced hatchability.
Calcium	Rickets. Reduced hatchability. Short and thick legs, wings, and lower mandible. Pliable beak, legs, and neck. Edema.
Phosphorus	Rickets. Soft legs and beak. High embryonic mortality between 14 and 16 days.
Manganese	Skeletal abnormalities. Chondrodystrophy (short wings and legs, abnormal head and parrot beak). Imperfect development of inner ear. Retarded growth. Edema. Abnormal down.
Zinc	Micromelia. Skeletal deformities (absence of rump, wings, legs, and toes). Underdeveloped eyes. Tufted down. Newly hatched chicks are weak and cannot stand, eat, or drink. Chick mortality increased soon after hatching.
Selenium	Subcutaneous fluid. Exudative diathesis (edema). Degeneration of the pancreas. Reduced hatchability. Selenium deficiencies are enhanced when the breeder diet is low in Vitamin E.

Excessive Items	
Selenium	Selenium is very toxic at high levels. Edema. Crooked toes. High embryonic mortality.
DDT	Affects chick growth, egg production, and livability, but not hatchability.
PCBs (polychlorinated biphenols)	Toxic to developing embryos. Hatchability drastically reduced, but egg production and fertility not affected.
Nicarbazin	Brown eggshells lose their pigment. Hatchability decreases up to 32%.

8-M. DISEASE AND HATCHABILITY

Several poultry diseases that affect the parent breeder flock have an effect on the developing embryo, hatchability, and chick quality; other disease organisms establish themselves in the hatchery and incubators to infect future hatches. Many of the pathogenic organisms produce similar conditions: high embryonic mortality, weak chicks, and whitish diarrhea. Therefore, it is almost impossible to differentiate the source of infection by observations of the dead embryos or the newly hatched chicks. Only a laboratory examination will determine the organism involved.

Many of these diseases are discussed in Chapter 37. The important ones involving incubation and chick quality are as follows:

pullorum disease
Arizona disease
fowl typhoid
paratyphoid
aspergillosis
omphalitis
Escherichia coli infection

infectious bronchitis
Newcastle disease
avian encephalomyelitis
Mycoplasma gallisepticum infection
Mycoplasma synoviae infection
aflatoxosis (toxin poisoning)
laryngotracheitis

8-N. ANALYZING POOR HATCHABILITY

Observation	Possible Causes
Eggs exploding	Bacterial contamination of eggs Dirty eggs Improperly washed eggs Incubator infection
Clear eggs	Infertile Eggs held improperly Too much egg fumigation Very early embryonic mortality
Bloodring (embryonic death, 2–4 days)	Hereditary Diseased breeding flock Old eggs Rough handling of hatching eggs Incubating temperature too high Incubating temperature too low
Dead embryos second week of incubation	Inadequate breeder ration Disease in breeder flock Eggs not cooled prior to incubation Temperature too high in incubator Temperature too low in incubator Electric power failure Eggs not turned Too much CO_2 in air (inadequate ventilation)

Observation	Possible Causes
Air cell too small	Inadequate breeder ration Large eggs Humidity too high, 1 to 19 days
Air cell too large	Small eggs Humidity too low, 1 to 19 days
Chicks hatch early	Small eggs Leghorn eggs versus meat-type eggs Incorrect thermometer Temperature too high, 1 to 19 days Humidity too low, 1 to 19 days
Chicks hatch late	Variable room temperature Large eggs Old eggs Incorrect thermometer Temperature too low, 1 to 19 days Humidity too high, 1 to 19 days Temperature too low in hatcher
Fully developed embryo with beak not in air cell	Inadequate breeder ration Temperature too high, 1 to 10 days Humidity too high, nineteenth day
Fully developed embryo with beak in air cell	Inadequate breeder ration Incubator air circulation poor Temperature too high, 20 to 21 days Humidity too high, 20 to 21 days
Chicks pipping early	Temperature too high, 1 to 19 days Humidity too low, 1 to 19 days
Chicks dead after pipping shell	Inadequate breeder ration Lethal genes Disease in breeder flock Eggs incubated small end up Thin-shelled eggs Eggs not turned first 2 weeks Eggs transferred too late Inadequate air circulation, 20 to 21 days CO_2 content of air too high, 20 to 21 days Incorrect temperature, 1 to 19 days Temperature too high, 20 to 21 days Humidity too low, 20 to 21 days
Malpositions	Inadequate breeder ration Eggs set small end up Odd-shaped eggs set Inadequate turning
Sticky chicks (albumen sticking to chicks)	Eggs transferred too late Temperature too high, 20 to 21 days Humidity too low, 20 to 21 days Down collectors not adequate
Sticky chicks (albumen sticking to down)	Old eggs Air speed too slow, 20 to 21 days Inadequate air in incubator Temperature too high, 20 to 21 days

Observation	Possible Causes
	Humidity too high, 20 to 21 days Down collectors not adequate
Chicks too small	Eggs produced in hot weather Small eggs Thin, porous eggshells Humidity too low, 1 to 19 days
Chicks too large	Large eggs Humidity too high, 1 to 19 days
Trays not uniform in hatch or chick quality	Eggs from different breeds Eggs of different sizes Eggs of different ages when set Disease or stress in some breeder flocks Inadequate incubator air circulation
Soft chicks	Unsanitary incubator conditions Temperature too low, 1 to 19 days · Humidity too high, 20 to 21 days
Chicks dehydrated	Eggs set too early Humidity too low, 20 to 21 days Chicks left in hatcher too long after hatching completed
Mushy chicks	Unsanitary incubator conditions
Unhealed navel, dry	Inadequate breeder ration Temperature too low, 20 to 21 days Wide temperature variations in incubator Humidity too high, 20 to 21 days Humidity not lowered after hatching completed
Unhealed navel, wet, and odorous	Omphalitis Unsanitary hatchery and incubators
Chicks cannot stand	Breeder ration inadequate Improper temperature, 1 to 21 days Humidity too high, 1 to 19 days Inadequate ventilation, 1 to 21 days
Crippled chicks	Inadequate breeder ration Variation in temperature, 1 to 21 days Malpositions
Crooked toes	Inadequate breeder ration Improper temperature, 1 to 19 days
Spraddle legs	Hatchery trays too smooth
Short down	Inadequate breeder ration High temperature, 1 to 10 days
Closed eyes	Temperature too high, 20 to 21 days Humidity too low, 20 to 21 days Loose down in hatcher Down collectors not adequate

9

Operating the Hatchery

The operation of a chick hatchery involves the production of the largest number of quality chicks possible from the hatching eggs received at the hatchery. In addition, chicks must be produced economically. Ever-increasing competition and integration within the industry have made hatchery operation a business of small unit margins, and managers must be ever-cognizant of the importance of the little things that produce top profits.

9-A. SECURING HATCHING EGGS

Most hatcherymen in the United States and many other countries have their own flocks of breeder chickens, or at least their own guaranteed source of hatching eggs. But elsewhere there are still many variations in the source of supply.

Source of Eggs

Hatchery owns the breeder hens. The breeders may be on hatchery-owned farms, or on farms owned by the flockowner who has entered into an agreement with the hatchery to produce hatching eggs on contract. In the latter case the hatchery actually owns the breeders; the flockowner is paid a contracted amount to produce the eggs (see Chapter 10-A). This system is particularly applicable when the complete poultry enterprise is integrated.

Hatchery secures eggs from flockowners. In many sections of the world flockowners exist who own both the breeders and the poultry farm.

Although most are involved with some type of contract to produce eggs for a hatchery in predetermined quantities, the hatchery does not own the birds. In many cases, however, the hatchery is involved in financing flockowners; that is, the hatchery may lend money to help finance the poultry breeder operation, and the flockowner agrees to pay off his debt by having the hatchery make weekly deductions from his egg payments.

Other hatcheries supply eggs. On occasion one hatchery may have an oversupply of hatching eggs while another does not have enough. Under these circumstances the hatchery with the surplus will sell eggs to the other. Such a procedure is quite common with large integrators where one person or company may own and operate several hatcheries, all hatching the same strain of chickens.

Hatching egg-producing companies are sources of eggs. In certain countries there is still another method of securing hatching eggs. Eggs are produced by an egg-operating company that has a large number of poultrymen under contract to produce hatching eggs for it. These egg-operating companies, in turn, have an agreement to sell a given number of hatching eggs each week to unattached hatcheries that are directly involved with chick hatching. In many instances the eggs may be shipped to a hatchery in another country.

How Eggs Are Delivered to the Hatchery

Hatching eggs usually arrive at the hatchery by truck, although they may arrive at some point in the general locality by rail or air. As many hatcheries are on a program of producing MG-negative or MS-negative chicks, care must be taken to prevent the entrance of these disease-producing organisms into the hatchery. In these cases, the hatching eggs received would be negative, and they must not be contaminated. The following rules must be rigidly observed:

1. Truck drivers and their helpers must shower and change into clean clothes, headgear, and footwear before entering any trucks involved with the egg transport.
2. All trucks must be disinfected and fumigated with formaldehyde gas before eggs are placed in them (see Chapter 9-I).
3. Only eggs that are MG-negative or MS-negative may be placed in the truck.

Containers to Hold Eggs

Eggs are collected on paper or plastic flats, placed in cases or racks, and delivered to the hatchery. The type of egg washer, incubator, and other equipment will often determine the type of container.

Egg Delivery Record

A system of identifying hatching egg receipts must be established. This system not only offers a record for the flockowner or trucker but is of value to the hatchery manager and the hatchery bookkeeping department.

The delivery record should include

1. Source of eggs (flockowner, trucker, etc.)
2. Date eggs are received
3. House, pen, or flock numbers
4. Breed or line of chickens involved
5. Number of cases or number of dozens of eggs received

9-B. HANDLING HATCHING EGGS

The handling of hatching eggs between the time they are laid and the time they are placed in the incubator requires certain procedures to give some assurance that hatchability will be high.

Cooling Hatching Eggs

Prior to setting, eggs should be cooled by placing them in a room with a temperature of 65°F (18.3°C), and at 75% relative humidity (see Chapter 7-E).

Grading Hatching Eggs

Some eggs are graded for quality and size in the hatchery; others are not. The decision depends on the type of eggs, breed, and whether they are the product of an integrated operation.

Broiler-type hatching eggs. Eggs from which commercial broiler chicks are to be hatched may or may not be graded. The following situations are involved:

1. *When broiler-chick production is part of an integrated operation*, there is little to be accomplished by egg grading.

The process is costly, and most integrators would rather the expense be eliminated, even though the resultant chicks may be of slightly inferior quality and some smaller in size.

2. *When broiler chicks are sold to poultry customers*, chick quality is of utmost importance. Small chicks will be objectionable to the customer; therefore, many of the hatching eggs must be graded and the small ones removed. This process is particularly applicable during the first few weeks the breeder hens are producing hatching eggs, for the egg size is naturally smaller during this period.

Egg-type hatching eggs.   Hatching eggs that are to be used for the production of commercial laying-type pullet chicks generally are graded for size and quality. Most such chicks are not a part of an integrated program and must be sold to poultrymen who are in the business of producing commercial eggs. Like begets like, and any imperfections in the exterior quality of the hatching egg—shape, shell imperfections, or shell quality—are likely to be reproduced by the next generation.

Breeder-type hatching eggs.   Certain poultrymen are involved with developing and merchandising breeder-type chicks. Many times pedigree hatching is involved. Under these circumstances egg weight and quality are of the greatest importance and eggs must be graded carefully.

How Hatching Eggs Are Graded

Hatching eggs are graded automatically for weight. There are many machines on the market that sort eggs according to designated sizes. Some are capable of grading only a few 30-doz cases per hour; others will handle up to 200 per hour. Some egg graders are equipped with egg candlers that may be used for detection of eggs with thin shells, cracked shells, and other imperfections.

Traying Hatching Eggs

Eggs may be placed in the incubator setting trays either at the time they are gathered or just before they are cooled. Usually they should be trayed as close to setting time as possible. However, this procedure places a heavy load on the labor requirement during a relatively short period of time. A better method is to tray eggs each day, placing the trays in special holding racks designed for this purpose. After traying, the racks are returned to the cooler room for further storage of the eggs at 65°F (18.3°C). The eggs should be covered with a plastic "shirt" to reduce egg evaporation.

Recording the source of eggs. At the time the eggs are trayed, a record should be made of the source and type involved. Small cards that fit into a card holder on the tray should be used. All the necessary data regarding the eggs should be written on the cards. If egg settings are made twice weekly, these cards usually are numbered 1 through 6 to correspond to the 6 settings of eggs in the machines at one time.

Use the number 1 card for the first setting, number 2 for the next, etc. These numbers make it easier to select the correct egg trays when the eggs are transferred to the hatcher after 19 days of incubation.

Warm Eggs Prior to Setting

Approximately 6 hours prior to placing eggs in the setter they should be moved from the egg-cooler room to a room with a temperature of about 72°F (22°C). It is advantageous to warm eggs before placing them in the incubator. Placing cool eggs in the machine lowers the incubating temperature for several hours, thus delaying the hatching time of the eggs already in the machine.

Setting Hatching Eggs

The time the chick trucks are to leave the hatchery will determine the time that the eggs should be set. Chicks should be at the customer's farm about 12 hours after the hatch is "pulled." Chicks will have to be removed from the hatcher at 9:00 to 10:00 p.m. inasmuch as chicks should be delivered early the following morning. Eggs should be set at a time that will allow chicks to hatch and dry prior to 9:00 to 10:00 p.m. This will mean the average egg setting time will be about 5:00 to 6:00 p.m.

Remember: Some eggs, as those from Leghorns, require less than 21 days to hatch; others, depending on age, shell quality, size, and other factors will require more or less than 21 days of incubation.

Important: When eggs from different breeds or with different incubation periods are being set, they should be placed in the incubator so that all chicks hatch at the same time; e.g., eggs from Leghorns should be set later than those from meat-type breeders.

Allow time for hatchery chores. Certain hatchery jobs are time-consuming. Toe, comb, and beak trimming, sexing, and other tasks require many hours of labor and cannot be done simultaneously. When planning the hour to set eggs, ample time should be allowed for those chores after the chicks hatch.

Variability in Hatching Time

Unfortunately, all chicks do not hatch at the same time. Even under the best of incubation procedures, about 32 to 35 hours elapse between the time the first and last chicks hatch. Therefore, some chicks are almost one and one-half days of age before the last ones hatch.

Most scientists agree that a chick should drink no later than 36 hours after hatching, but because some chicks are 32 to 35 hours old when removed from the hatcher, they will be more than 58 hours old when they reach the farm. From a practical standpoint, chicks should be delivered to the farm about 12 hours after the entire group is removed from the hatcher. Even then, some will be nearly 2 days old.

9-C. FILTERING RECYCLED AIR FROM THE HATCHER

Microorganisms from chick down and hatchery debris are usually disseminated throughout the hatchery as the air from the hatchers is recirculated. Some of these organisms may be pathogenic, causing morbidity and mortality in the chicks after they are placed in the brooder house.

Filtering the recirculated air in the hatcher will remove almost all of the bacteria in that area. In one test 99% were removed. This test and others have led to the use of filters in commercial incubators and in hatchery rooms themselves.

In another test, the hatch of fertile eggs was significantly greater in the filtered hatcher. Broilers from the filtered hatcher were heavier at 4 weeks of age than chicks from a nonfiltered hatcher and the mortality was lower.

9-D EGGS CONTAINING EMBRYOS FOR HUMAN CONSUMPTION

In many countries, particularly those in the Far East, embryonated eggs are considered a delicacy among foods for humans. Duck and chicken eggs are the most common.

The eggs of the chicken are incubated for 12 to 14 days and sold as "Balut" or "Embryo Eggs." Such eggs must be kept refrigerated if any storage is involved.

In the United States some states require that hatcheries producing such eggs declare that they are intended for sale as poultry embryos at the time the eggs are placed in the incubators. When marketed, the containers in which they are sold must be marked with the word ''EMBRYOS'' in one-half-inch high letters together with the species from which the eggs came (chicken, duck, etc.).

9-E. AFTER THE CHICKS HATCH

The process of removing the chicks from the hatcher is often called *pulling the hatch.* It involves the following:

Drying the Chicks

Dehydration is a stress and a problem with newly hatched chicks. Excessive drying in the hatcher should be avoided. Chicks should be removed from the hatcher as soon as all are hatched and about 95% are dry. Further drying and hardening should be confined to chick boxes.

Chick-holding Room Temperature and Humidity

When the chicks are removed from the hatcher and boxed they should be moved to the chick-holding room with a temperature of 75°F (23.9°C) to reduce the danger of chilling, and with a relative humidity of 75% to reduce dehydration.

Chick Box Sizes

Chick boxes vary in size and construction. The number of chicks placed in a box, the outside temperature, and the shipping distance will determine the dimensions. Various sizes holding 100 chicks are as follows:

Box Type	Inches	Centimeters
Standard winter	22 × 18 × 6	56 × 46 × 15
Standard summer	22 × 18 × 6 or 7	56 × 46 × 15 or 18
Oversize	24 × 20 × 6	61 × 51 × 15

Many boxes of patented design are on the market. Some call for the use of staples to hold them together; others have intricate corner folds that can be snapped together without stapling.

In most cases, a box must be separated from the box above it. Originally, this was accomplished by gluing a piece of wood to the cover, but now the boxes have extensions of the dividers that protrude through and above the lid. These act as separators, keeping the boxes apart.

Although most chick boxes are constructed of corrugated fiber, some are made of plastic. These, unlike those of fiber, may be washed, fumigated, and reused. When cold weather is a factor use flexible plastic boxes because they will not crack. For short hauls, a lidless box is used.

Chick Box Pads

Although each box holding 100 chicks is usually divided into four compartments to prevent the chicks from piling in one corner, some material to which they may clamp their toes should cover the bottom of the box. Many are used, but the excelsior pad or rough, absorbent mat predominate. These come in sizes necessary to fit the box being used. As each is one-fourth the size of the bottom of the chick box, four pads are necessary for each box. However, some boxes do not have dividers; there is but one compartment, and the boxes are satisfactory for short hauls.

Calculating the Number
of Chicks Hatched

It is recommended that chicks be ''scooped'' from the trays to expedite the procedure, but it is still necessary to get an accurate count of the number hatched. The procedure is as follows:

Scoop all chicks except the obvious culls. Reference is made to the tray card on which will be recorded the number of eggs set in the tray. Next, count the number of eggs left on the tray (infertile, dead-in-shell, and pipped), plus the number of dead or cull chicks on the tray. Subtract this figure from the number of eggs set. This will give the number of chicks hatched, which can be recorded on the tray tag. After all the chicks have been placed in boxes, the total number of chicks hatched may be calculated by adding the number of chicks hatched as recorded on each tray tag. By subtracting the ''extras'' the number of salable chicks may be calculated. Divide by two if only pullet chicks are sold.

Hardening the Chicks

When chicks are first placed in the chick boxes they are soft in the abdomen, are not completely fluffed out, and do not stand well. They must be ''hardened'' by leaving them in the boxes for 4 or 5 hours. Such hardening makes it easier to grade the chicks for quality, and the chicks are more easily vent-sexed.

Grading the Chicks

A definite standard of chick quality must be outlined. No chick below the minimum standard must be allowed to go to a customer. The standard should be the same for all breeds, and functional in all seasons of the year. Do not cut quality when the hatch is poor.

Some standards for quality are

1. No chick deformities
2. No unhealed navels
3. Above a minimum weight
4. Not dehydrated
5. Down color representative of the breed
6. Stand up well; are lively

Sexing the Chicks

The method of sexing depends on the breed. Leghorns are vent or feather sexed; most brown egg varieties are color sexed, and broilers are color sexed or feather sexed (see Chapter 22-K and 22-L).

Extra Chicks

For many years it has been customary to guarantee live delivery of chicks to the customer. So as not to complicate any adjustments after they have been delivered, the hatcheryman has added "extra" chicks to the order to replace any death losses prior to the arrival of the chicks at the farm. The percentage of extras may vary from 1 to 4, with 2 being about average. The procedure is to place 102 chicks in a box, and invoice for 100. Thus, two chicks are a direct charge against the cost of production.

Record the Data

The following data should be recorded for each group of hatching eggs set:

1. Breed
2. Number of eggs set
3. Number of quality chicks hatched
4. Percentage of total hatchability
5. Number of "grade-outs" (culls)
6. Percentage of "grade-outs"

7. Percentage of ''extra chicks'' given to customer
8. Percentage of chicks invoiced of those hatched

9-F. CHICK DELIVERY

Safe and sanitary delivery of the day-old chick is the last of the many hatchery operations. Worldwide, most deliveries are by truck, although other means of transportation are sometimes used. These would include rail and air. In some instances the customer may pick up his chicks at the hatchery, using his own means of transportation. But most hatchery-men feel they must deliver their chicks in a manner acceptable to their standards of disease prevention, quality control, and in their own vehicles.

Chick Trucks

Specialized trucks are used for the delivery of day-old chicks. These may be custom-built or secured from a manufacturer. They must have adequate ventilation and a means of enabling the boxes to be stacked and separated.

Size and type.   The size of the compartment in which the chick boxes are placed will vary. Some trucks hold only 10,000 chicks; others as many as 50,000 or more. The size of the truck will be determined by the size of the hatchery, and of course some large hatcheries have several trucks. Usually larger operations hatch chicks on several days of the week, making it necessary to have fewer trucks than if hatching is confined to two days a week.

There are three main types of truck bodies:

1. *Cab-over.* In these the cab and the body are one unit, enabling the driver to sit in the same area in which the chicks are placed. Usually these are smaller units holding a maximum of 20,000 chicks.
2. *Box-type.* Larger units have a box built separately from the cab in which the driver rides. These may hold 50,000 or more chicks. In order to safeguard the birds, special controls that can be read in the cab are installed.
3. *Converted buses.* School buses and those that are similar sometimes are used as the framework for chick trucks. They may be either new or reconditioned. Since these are equipped with special ventilators, and sometimes air-conditioners, they lend themselves very well to the delivery of chicks.

Ventilation system. Several thousand chicks generate a great deal of heat, and when chick boxes are confined to a small area, as in a truck body, there must be adequate ventilation to remove the heat. Special fans, air intakes, and exhausts must be provided.

Heated in winter. In exceptionally cold climates the chick compartment must be heated. Water, heated by the truck engine, is the usual source of heat, although electric heaters are sometimes used.

Cooled in summer. In the hot months of the year the flow of air through the van will not produce adequate cooling. A refrigeration system must be installed to lower the temperature.

Caution: Most of the ventilating fans used for moving air through the chick area are powered by electricity generated by the truck motor. When the motor fails, the supply is cut off. Supplementary units, powered by gasoline engines, must be installed. A load of chicks is too perishable not to take adequate provisions. Even refrigeration units are best powered by a separate engine.

Length of haul. The length of the chick delivery may determine the amount of truck cooling needed. If the hauls are short, and can be completed during the early-morning hours, adequate ventilation, rather than refrigeration, may be all that is needed.

Delivering Chicks to the Farm

Baby chicks should reach the customer's farm early in the morning. Not only is the weather cooler during this part of the day but the early arrival allows a full day for close observation of the chicks by the caretaker.

Log the truck driver. Truck drivers should keep a running log during chick deliveries, especially those involving long hauls. They should record the time they leave the hatchery, all stops, and time of arrival at the farm. Not only does this provide a permanent record but it keeps the drivers more cognizant of their duties to provide fast and safe arrival of the chicks.

Receipt for chick condition at arrival. Every customer should complete an ''Arrival Condition'' form supplied by the hatchery. This should include

1. Number, sex, and breed of chicks delivered
2. Time of arrival at farm
3. Number of chicks dead on arrival
4. Condition of chicks
5. Condition of the brooding facilities on the farm
6. Other comments

It should be in duplicate, and signed by both the customer and the driver. The original is returned to the hatchery; the copy is left with the customer.

Shipping Chicks by Air

Many chicks are shipped long distances by air. Special recommendations and instructions are necessary.

Chick box for air shipments. No smaller than 18 × 24 × 7 in. (46 × 61 × 18 cm) boxes should be used. These have a volume of 2,800 in.[3] (45,884 cm³), which is important for air shipments. Punch out all the holes in the box. When the air temperature is above 70°F (21°C), pack 85 chicks to a standard box; when below 70°F (21°C), pack 100 to the box.

Timing the air trip

1. Get to the airport 3 hours before the plane is scheduled to leave.
2. If airline transfers are to be made enroute do not schedule less than 6 hours for the transfer. Book direct flights if possible.

Type of air equipment used in flight.

1. *Passenger flights.* Passenger carriers vary in their ability to transport chicks. Both flights and type of equipment need to be checked prior to making arrangements to ship chicks by air on passenger flights.
2. *Freighters.* Freighter lines differ in their ability to load and carry chicks. Check the carrier previous to making any shipments. Some factors involved are:
   100 boxes represent maximum loads.
   Pallets are used for loading.
   Igloos are sometimes not acceptable for shipping chicks by air.
   Check the carrier; ask where the chicks are to be carried in the plane. Belly compartments may not be ventilated. Never load chicks in the same compartment with dry ice (frozen CO_2).

Instructions at the airport. Probably more chicks die or are damaged at the airport than in the plane. Extreme precautions must be taken. Comply with the following:

1. The captain of the airliner must be informed that chicks will be on board.
2. Keep chicks in the shade.
3. Keep chicks away from drafts.

4. Do not allow chick boxes to stand outside in cold weather, or in the sun at any time.
5. Do not cover chick boxes with a tarpaulin. Never place boxes in a corner of a room. Keep in a well-ventilated room inside the cargo building.
6. Don't stack boxes over eight high. The heat buildup is too great.
7. Never allow the boxes to become wet. They will collapse.
8. Don't stack other cargo on top of chick boxes. Leave air space around the boxes. This is a must.
9. Keep boxes level at all times.

9-G. HATCHERY SANITATION

Hatching eggs must be produced by healthy breeding stock, the eggs incubated under sanitary conditions, and the chicks delivered in clean vehicles. A sanitary hatchery is necessary to produce high hatchability and good chick quality. The latter implies that chicks not only appear healthy at the time they are hatched but are free of many disease-producing organisms that may attack the very young bird.

Disinfectants for the Hatchery and Poultry Farm

Surface disinfectants are most effective in the absence of organic material. Disinfecting is not a substitute for cleanliness; it is a means of destroying microorganisms, but is effective only when things start out relatively clean.

Specifications for disinfectants. All disinfectants used in the hatchery should be

1. Highly germicidal
2. Nontoxic to humans and animals
3. Effective in the presence of moderate amounts of organic material
4. Noncorroding and nonstaining
5. Soluble in water
6. Capable of penetrating materials and crevices
7. Unassociated with pungent odors
8. Readily available and inexpensive

How Surface Disinfectants Work

Surface disinfectants differ in their action according to the type of organism involved.

Bacterial organisms. Most surface disinfectants are bactericidal (kill), but some chemicals only disrupt the life cycle of the bacterium (bacteriostatic). A large number will suppress the growth of gram-positive bacteria, while others will inactivate the gram-negative. A disinfectant can be bactericidal at one concentration and bacteriostatic at lower levels.

Viral organisms. Viruses pose a special problem for a surface disinfectant inasmuch as viruses are inanimate until housed in the host cell. Most disinfectants only partially inactivate them. To inactivate a virus, the agent must disrupt the nonnucleic acid protein coat around the nucleic acid (gene material) contents. When only the coat, and not the contents, of the virus is destroyed, the viral nucleic acid can enter the host cell and become reactivated.

Fungal organisms. Surface disinfectants having fungicidal or fungistatic properties for use in hatcheries or on poultry farms must be capable of destroying or inactivating the fungal spores. But the fungus organism is difficult to destroy, and most fungicides only poison the fungus rather than destroy the fungal cells.

Protozoal organisms. Some, but not all, protozoal parasites spend part of their life cycle outside the host. Sanitizers effective against such organisms therefore must make their attack during the quiescent life stages, which in themselves vary and generally represent the most resistant phase of the organisms' lives.

There are few surface disinfectants specific for protozoal organisms affecting poultry. Even though several chemical compounds are applicable, their concentration and cost make them unacceptable.

9-H. CHEMICALS USED FOR DISINFECTING

The chemicals used for surface disinfecting are many and their values are highly variable. Sanitizers may be grouped according to their base ingredient, but there are many things that affect their potency, all of which must be understood to cause the disinfectant to produce normal effectiveness.

Cresols and Cresylic Acid

Cresols and cresylic acid are liquid yellow or brown coal tar derivatives. They have a strong odor, irritate the skin, and turn milky when water is added, but they have excellent germicidal action. Many types are available; some are used in conjunction with a detergent. Their odor may injure day-old chicks.

They are effective against gram-positive and gram-negative bacteria, most fungi, and some viruses. They are compatible with anionics (charged ionic compounds) but not with nonionics, and work best in an acid pH. They have greater use in the poultry house than in the hatchery. Cresol ($C_6H_4CH_3OH$) (methyl phenol) may be used for disinfecting floors or room surface areas, equipment, and footbaths.

Phenols

These also are coal tar derivatives with a base of carbolic acid (C_6H_5OH). Synthetic related compounds also fall in this category. Phenols have a characteristic odor, turn milky in the presence of water, and are effective germicides. The most common types used as disinfectants are the synthetic arylphenols, the simple akylphenols, the halogenated phenols, and the nitrophenols.

Most phenols are not compatible with the nonionic (neither anionic or cationic, but produce neutral colloidal properties) surfactants and should not be used with them. They are compatible with anionics (charged ionic compounds) and are more active in an alkaline pH because of their greater solubility. Their action is rapid.

Such disinfectants are effective against fungi, gram-positive and gram-negative bacteria, but not effective against bacterial spores. They will control some viruses. In high concentration, phenols act as a protoplasmic poison, penetrating and disrupting the cell wall and precipitating the cell proteins, but in low concentrations only the essential enzyme systems of the cell are disrupted. Synthetic phenols may be used for egg dipping, hatchery and equipment sanitation, and footbaths.

Phenol coefficient. This may be defined as the ratio of killing power of a disinfectant to the killing power of pure phenol. It is used to compare the efficiency of disinfectants, using phenol as the base. But the intricacies of the laboratory comparison are many; the test must be regulated and conducted carefully.

Iodine

Iodine compounds are available as iodophors, which are combinations of elemental iodine and an organic solubilizing agent, usually a nonionic surface action agent that is soluble in water. The compounds react only with nucleic acid of the cell contents. They are good disinfectants in an acid situation (2 to 4 pH), but activity diminishes in an alkaline pH and in the presence of organic material. They are effective against gram-positive and gram-negative bacteria, attacking the nucleic acid of the organisms. Iodophors are also effective against fungi and some viruses.

Chlorine

Chlorine is an effective constituent of certain disinfectants. Included are the powder forms of sodium or calcium hypochlorite combined with hydrated trisodium phosphate, liquids containing sodium hypochlorites and organic chloramines.

Chlorine compounds are good disinfectants when free chlorine is available in abundance (200 to 300 ppm). When elemental chlorine or hypochlorites are added to water they produce hypochlorous acid (HOCl), which produces the bactericidal action. But in the presence of organic material the chlorine combines with the organic material to form stable compounds, and therefore reduces the free chlorine in solution.

Chlorine is effective against bacteria and fungi, and when coming from hypochlorites it attacks both the protein coat and the nucleic acids of viruses. Bound chlorines, as in chloramines, have poor activity. Chlorine solutions are much more active in acid solutions than in alkaline, and in a warm rather than in a cold mixture. Sodium hypochlorite is very active, but its disinfecting life is short. In contrast, calcium hypochlorite is less active, but its disinfecting quality persists for a long period. Chlorine compounds are somewhat irritating to the skin and corrosive to metal.

Quaternary Ammonium

These compounds are cationic (have positively charged ions), odorless, clear, generally nonirritating, have a deodorizing and detergent action, and are quite effective as surface disinfectants. The most common is alkyl dimethyl benzyl ammonium chloride.

Quaternary ammonium compounds are extremely water soluble, but cannot be used in soapy solutions or where there is a residue of soap or anionic detergent. Their germicidal properties are reduced in the presence of organic material.

These chemicals are effective against gram-positive organisms, moderately effective against gram-negative, and will control some fungi and some viruses. Their effectiveness is increased by the addition of sodium carbonate inasmuch as this increases the alkalinity of the solution, and are quite effective in mild acid carriers and with hard waters.

A solution of 500 ppm quaternary ammonium, 200 ppm, EDTA (ethylene diamine tetra-acetic acid) and sodium carbonate added to adjust the pH to about 8.0 (about 200 ppm) is an excellent hatchery disinfectant. It may be used on floors, walls, and incubator trays.

Caution: The use of quaternary ammonium compounds on poultry farms has met with resistance in some countries because of the injurious effects on human beings when foods are involved. It should be withdrawn five days before the birds are marketed.

Formaldehyde (HCHO)

Formaldehyde will react with both the nonnucleic acid protein coat and the nucleic components within the particle, but time, potency, pH, and nature of the water affect the rate of diffusion. Although a powerful disinfecting agent in 10% formalin-water solution, formaldehyde's primary value lies in the use of the gas as a fumigant. Probably no other hatchery disinfectant is used as much, and this warrants special consideration and discussion.

9-I. FORMALDEHYDE FUMIGATION

Formaldehyde is commercially available as a 40% solution in water (37% by weight) known as *formalin*, and as a powder, *paraformaldehyde*, containing 91% formaldehyde. When either is heated, formaldehyde gas is liberated. Human skin is very sensitive to formalin and there should be no direct contact.

Legal Standards for Use of Formaldehyde Gas

Humans are very susceptible when they inhale formaldehyde gas, so much so that in the United States the Occupational Safety and Health Administration has set maximum concentrations in air content as a protection of employees against injury. (Some foreign countries have similar regulations.)

Peak allowable for 15 minutes. The allowable legal concentration of formaldehyde gas for this short period is 2 ppm, which is very low.

Peak allowable for 8-hour day. The allowable legal limit for those who are exposed to formaldehyde gas for more than 15 minutes, or an entire day, is 0.5 ppm.

Formaldehyde products covered. These maximums apply to all forms of occupational exposure to formaldehyde, including gas, solutions, and materials such as trioxen. Employees should wear face masks.

Directions for Use of Formalin

The heat necessary to release formaldehyde from formalin is best produced by mixing potassium permanganate ($KMnO_4$) with it. Use an enamelware or earthenware vessel of large capacity because of the boiling foaming splattering action when the two are mixed. Do not use vessels that will crack. Place the vessel and the permanganate in the area to be fumigated; then add the formalin.

Caution: Never add the permanganate to the formalin. Great heat is generated when the two chemicals are combined, and care should be taken. Formaldehyde gas is generated quickly; do not allow the fumes to get into the eyes.

Recommendations for Mixing Permanganate and Formalin

Two parts (by volume) of formalin are mixed with approximately one part (by weight) of potassium permanganate. This will produce complete expulsion of the gas. When the reaction is complete, a dry brown powder will be left. It the residue is wet, not enough permanganate was used; if the residue is purple, too much permanganate was added.

Varying concentrations of formaldehyde gas are necessary to fumigate under different conditions. The normal, or single strength concentration, is obtained by mixing 1.35 fluid oz (40 cm^3) of formalin with 0.71 oz (20 g) of potassium permanganate for each 100 ft^3 (2.83 m^3) of space. This is sometimes known as 1× concentration. Other concentrations are 2× (twice the above amounts), 3×, and 5×. Table 9-1 shows the amount of the two chemicals needed to produce the desired concentration, along with the amount of paraformaldehyde needed to produce a comparable amount of formaldehyde gas.

Table 9-1. Chemical Levels to Fumigate 100 ft^3 (2.83 m^3)

| Strength | To Produce Formaldehyde Gas (Mix Together) | | | | To Produce Equivalent Amount of Formaldehyde Gas | |
| | Formalin | | Potassium Permanganate | | Paraformaldehyde Powder | |
	fluid oz	cm^3	oz	g	oz	g
Single (1×)	1.35	40	0.71	20	0.36	10
Double (2×)	2.71	80	1.41	40	0.71	20
Triple (3×)	4.06	120	2.12	60	1.06	30
(5×)	6.76	200	3.53	100	1.77	50

Directions for the Use of Paraformaldehyde

Paraformaldehyde, a powder, should be placed in an electric skillet or pan with the thermostatic control set at 450°F (232°C). For single-strength concentration, use 0.36 oz (10 g) of paraformaldehyde for each 100 ft³ (2.83 m³) of space. See Table 9-1 for other concentrations. Heated in a skillet, 1 lb (454 g) of paraformaldehyde will release its gas in about 20 minutes.

   Important: An electric switch to turn off the generator should be on the outside of the compartment or room, as it is usually impossible to get into the area during fumigation.

Temperature and Humidity for Maximum Formaldehyde Disinfection

The efficacy of formaldehyde gas is increased in the presence of heat and moisture. It is impossible to get maximum results unless the temperature of the area to be fumigated is 75°F (24°C), or higher, and the relative humidity is 75%, or more.

Neutralizing Formaldehyde Gas with Ammonium Hydroxide

In certain instances it is necessary to stop the action of formaldehyde gas after the required period of fumigation. Usually this may be accomplished by opening the air intakes and exhausts in the incubators or rooms, but often this method is too slow. Ammonium hydroxide (NH_4OH) may be used to expedite the process. A solution of 26 to 29% ammonia should be used. Check to see how much formalin was used; then sprinkle the same amount of ammonium hydroxide on the floor of the area fumigated.

Use 0.14 oz (4 cm³) of ammonium hydroxide for each gram of paraformaldehyde used. In each case the ammonium hydroxide will neutralize the remaining formaldehyde gas.

Strength of Formaldehyde Fumigation

Varying concentrations of formaldehyde gas are needed under different conditions, and the length of the fumigation period is variable. The gas is detrimental to the living embryo and to the newly hatched chick, and care must be exercised that the concentration of the gas and the length of the fumigation period meet the requirements to kill all the pathogenic organisms possible, yet not be lethal to the embryo or chick. Table 9-2 gives these recommendations.

Table 9-2. Formaldehyde Fumigation Concentration Recommendations

Fumigation of	Concentration of Fumigant	Time to Fumigate min	Neutralizer (Ammonium Hydroxide)
Hatching eggs immediately after they are laid	3×	20	No
Eggs in setter (1st day only)	2×	20	No
Chicks in hatchers	1×	3	Yes
Incubator room	1×, 2×	30	No
Hatcher, between hatches	3×	30	No
Hatcher room, chick room, between hatches	3×	30	No
Washroom	3×	30	No
Chick boxes, pads	3×	30	No
Trucks	5×	20	Yes

Special Directions for Fumigation

Fumigating eggs soon after laying. Fumigating with 3× concentration of formaldehyde for 20 minutes will kill about 97.5 to 99.5% of the organisms on the shells of brown eggs, and about 95 to 98.5% of those on white eggs, the difference probably being due to the fact that brown eggs have a thicker cuticle which absorbs more gas.

Because the population of organisms on the shells of eggs is greater in the summer than in the winter, the destruction by gas will vary. A higher percentage will be killed in the summer. However, the types of organisms are different during the two seasons.

Fumigating the setters. Normally eggs need not be fumigated in the setters (1 to 19 days). However, if "blowups" are a problem, fumigation will be necessary. Some hatcherymen, however, make it a regular practice to fumigate the setter section of the machines about once a week.

Formaldehyde gas is toxic to developing embryos, particularly between 24 and 96 hours of age, and during the time the chicks are pipping the shell. The first is a very critical period in the life of the embryo. To prevent the embryo from being weakened, eggs should be fumigated *only* as soon as they are placed in the incubators. With multiple settings, this is the only time some eggs in the machines will not be between 24 and 96 hours of age. Eggs should be fumigated at 2× strength for 20 minutes with vents closed.

Fumigating chicks. Fumigation of chicks generally is not recommended. However, with omphalitis outbreaks in the hatchery it may be necessary in order to get the disease under control.

Formaldehyde fumigation colors the down of the chicks a deep orange, often noticeable by the customer. Chick fumigation should be considered an emergency measure only.

Fumigating between hatches. After the chicks have been removed from the hatchers, and the trays washed, the hatchers, trays, hatcher room, and wash room should be fumigated with formaldehyde gas at the concentrations given in Table 9-2.

Fumigating chick trucks. Because it is difficult to raise the temperature and humidity in the chick truck at the time of fumigation, the concentration of formaldehyde gas should be increased to 5×. Thoroughly clean the chick truck first. It may be advisable to construct a ''tent'' that will quickly cover the truck and chassis. Fumigation can then be completed inside the tent, disinfecting the complete van. Some hatcheries are constructed with an inside truck bay that can be sealed and used as a fumigation room for trucks (see Chapter 5-D).

9-J. EFFICACY AND USE OF DISINFECTANTS

Properties of Disinfectants

Table 9-3 shows the efficacy of five groups of disinfectants on various pathogenic organisms to kill or subdue them.

Table 9-3. Properties of Disinfectants

Property	Chlorine	Iodine	Phenol	Quaternary Ammonium Compounds	Formaldehyde
Bactericidal	+	+	+	+	+
Bacteriostatic	−	−	+	+	+
Fungicidal	−	+	+	±	+
Virucidal	±	+	+	±	+
Toxicity	+	−	+	+	+
Activity with organic matter	+ + + +	+ +	+	+ + +	+

Source: Canadian Dept. Agr., Hatchery Sanitation, 1970.
Notes: + indicates a positive property (degree of affinity); − indicates a negative property; ± indicates limited activity.

Use of Disinfectants in the Hatchery

Table 9-4 gives the uses of the various disinfectants in the hatchery.

Table 9-4.   Hatchery Use of Disinfectants

Use area	Chlorine	Iodine	Phenol	Quaternary Ammonium Compounds	Formaldehyde
Hatchery equip- ment	+	+	+	+	+
Water disinfec- tion	+	+	−	+	−
Personnel	+	+	−	+	−
Egg washing	+	−	−	+	+
Floor	−	−	+	+	+
Footbaths	−	−	+	+	−
Rooms	±	+	±	+	+

Source: Canadian Dept. Agr., Hatchery Sanitation, 1970.
Notes: + indicates a positive property (degree of affinity); − indicates a negative property; ± indicates limited activity.

9-K. THE HATCHERY'S PART IN DISEASE-CONTROL PROGRAMS

Certain pathogenic organisms that produce several chicken diseases may be passed from an infected breeder hen through her eggs to her resultant chicks. To prevent transmission by this route it is necessary that the blood of breeder hens (and males) be tested to determine if they have the disease or have had the disease. When reactors to the test are found, it is necessary that no more eggs be incubated from the entire flock.

Even though no reactors may be found in the breeding flocks, and the hatching eggs are not carriers of the disease, there is still the possibility that unsanitary conditions in the hatchery will lead to contamination of developing embryos in the setter and of newly hatched chicks in the hatcher. Therefore, the hatchery becomes an integral part of any disease-prevention program. The following diseases are involved.

Pullorum Disease

The causative bacterium is *Salmonella pullorum* (see Chapter 37-A). There is a blood test for identifying any carriers in the breeding flock, but incubators and all other hatchery equipment must be fumigated with formaldehyde (gas) at regular intervals to prevent hatchery contamination. Likewise, all hatching eggs coming into the hatchery should be fumigated or sanitized.

However, since all prime breeders sell only pullorum-free eggs and chicks, and if the hatchery is kept free of the causative pullorum organism, there is little need to test the birds in the breeder flocks, especially if they are part of an integrated operation. But hatchery sanitation is still necessary.

Fowl Typhoid

Caused by an organism similar to *Salmonella pullorum,* the fowl typhoid bacterium is *Salmonella gallinarum* (see Chapter 37-B). When breeder birds are tested for pullorum, they are also tested for typhoid at the same time, so all rules for blood testing apply to both diseases. A sanitary hatchery also applies to both.

Mycoplasma Gallisepticum (MG)

This disease was formerly known as PPLO (pleuro-pneumonia-like-organism) or CRD (chronic respiratory disease), and the causative organism is *Mycoplasma gallisepticum* (see Chapter 37-H). Over 20 serotypes have been discovered, and the one involved with this disease is known as S-6. The hatchery is directly involved with the program of hatching MG S-6-free chicks.

There are many breeders offering MG-free chicks, and the best way to get started is to secure such chicks, then maintain a rigid quarantine program, gradually increasing the number of chicks hatched and the size of the flocks.

When all the requirements of the (U.S.) National Poultry Improvement Plan are met, the poultry products may be sold as "U.S. M. Gallisepticum Clean" (see Chapter 1-F).

Mycoplasma Synoviae (MS)

This disease is often known as synovitis, and is due to the organism, *Mycoplasma synoviae,* similar to *Mycoplasma gallisepticum* (see Chapter 37-I). There is but one serotype, and the chicks are distressed because of an inflammation of the joints and tendon sheaths. The program for producing MS-clean chicks is the same as that for MG-clean chicks, except that an antigen specific for MS must be used for all blood tests.

When all requirements of the (U.S.) National Poultry Improvement Plan are met, the poultry products may be sold as "U.S. M. Synoviae Clean."

Control Programs

U.S. Pullorum Clean Hatchery Program
U.S. Typhoid Clean Hatchery Program

U.S. M. Gallisepticum Clean Hatchery Program
U.S. M. Synoviae Clean Hatchery Program

The accepted method for the control of pullorum, typhoid, MG, and MS diseases is by the elimination of breeder flocks containing carrier birds. One of several authorized blood tests is employed to identify the carrier birds. To accompany the blood test is a sanitation program in the hatchery to assure that there is no reinfection of the hatching eggs and the chicks hatched from them.

The hatchery program is quite similar for all four diseases and should encompass the following:

1. Secure disease-negative hatching eggs.
2. Hatch only disease-negative chicks.
3. Use only new or disinfected chick boxes.
4. Prior to entering the hatchery, all personnel must shower and change into clean clothing, including footwear and caps.
5. Fumigate or otherwise disinfect almost everything in the hatchery between hatches except the setter room.
6. Have only one entrance and exit to the hatchery. Keep all other doors locked.
7. Place a large deep disinfecting pan and scrub brush inside the entry door to clean footwear. Renew disinfectant often.
8. Moisten all debris to settle dust before removing it from the hatchery to the incinerator.
9. Construct a fence around the hatchery area, at least 100 feet from the hatchery building. Post security signs.

Inactivating Mycoplasma Organisms Within Hatching Eggs

There are methods for inactivating the mycoplasma organisms located inside the shell. A procedure is used where the breeder farm has some contamination, but it is not practical to initiate a disease-free program. Neither of the programs below will give assurance that all mycoplasma organisms within the eggshell will be inactivated.

Egg Dipping to Inactivate Mycoplasma

There are two methods as follows:

1. *Temperature differential (TD) system of egg dipping.* Hatching eggs are warmed for 3 to 6 hours at 100°F (37.8°C).

The heat expands the egg contents and the air is forced out. Next, the eggs are dipped in an antibiotic solution for 15 minutes at a temperature of 40°F (4.4°C). Not only does the antibiotic destroy the mycoplasma organisms on the surface of the eggshell but some of the antibiotic is drawn through the shell pores as the interior contents shrink.

2. *Pressurized differential (PD) system of egg sanitizing.* Hatching eggs are placed in a tank with a tight lid and filled with antibiotic solution at 55° to 60°F (12.8° to 15.6°C). A vacuum pump is connected to the tank and the inside pressure of the tank is reduced to 10.3 to 15.2 in. (25.7 to 38.1 cm) pressure of mercury. At this pressure air is sucked out of the egg, but when the pressure is restored to normal some antibiotic solution enters the egg contents.

Immunization to MG Infection

Several states recognize the vaccination of potential breeder chickens against MG infections during growing period, using a vaccine by way of the drinking water. Although fairly successful for individual flocks, there have been failures. Therefore, vaccination is not recognized as a method of developing an MG-free flock and hatchery.

Marek's Disease Vaccination in the Hatchery

There are four methods of vaccinating day-old chicks for Marek's disease in the hatchery (see Chapter 37-Q). Determine the necessary procedure required; then secure the proper vaccine and train personnel to do the vaccinating.

9-L. CLEANING THE HATCHERY BETWEEN HATCHES

Cleaning the hatchery between hatches is of primary importance. The process must be complete. Except for the setters and setter room, every piece of equipment must be thoroughly vacuumed, scrubbed, disinfected, and fumigated. Clean the setter room, *but do not fumigate.* Developing embryos in the setter are sensitive to formaldehydes (see Table 9-2).

Cleaning the Hatchers

1. Remove all racks, trays, carts, and roll-out assemblies and clean.

2. Vacuum the inside and outside of hatchers.
3. Wash the inside and outside of hatchers.
4. Scrub the inside walls with a suitable disinfectant.
5. Return all racks and clean trays from the clean room to the hatchers and fumigate with formaldehyde at 3× strength.

Cleaning the Hatcher Room and Chick Room

1. Vacuum all debris from the floor and walls.
2. Wash the floors and walls; then disinfect.
3. Wash and disinfect remaining equipment.
4. Fumigate rooms and equipment with formaldehyde at 3× strength.

Cleaning the Hatcher Trays

All hatcher trays, carts, and racks should be moved to the washroom, washed thoroughly, dipped in a disinfecting solution, moved to the clean room to dry, then moved to the hatchers where they are to be fumigated at 3× strength.

Cleaning the Washroom

1. After all hatching trays and portable equipment have been washed and disinfected and taken out for fumigation, remove all debris from the washroom. Empty the drain trap. Either incinerate the material removed, or place it in plastic (or similar) bags and remove it from the hatchery.
2. Next, wash and disinfect the ceiling, walls, and floors. Do not forget that the washroom probably is the most infectious area in the hatchery.
3. Fumigate the room with formaldehyde at 3× strength.

9-M. HATCHER WASTE DISPOSAL

With the small hatchery, most hatchery waste (infertile and non-hatched eggs, plus dead and cull chicks) may be disposed of easily enough, but with the large hatchery the volume creates a problem (see Chapter 40-C).

There are four main methods of disposing of the material.

1. *Incineration.* Incinerators come in many sizes, depending on the amount of wet material processed per hour. Regulations generally require that incinerators be at temperatures above 1,600°F (870°C) with a resulting ash of less than 0.2 lb (91 g) per 100 lb (45.4 kilos) burned. In some states and countries it is necessary to get a permit to operate an incinerator.

2. *Landfall disposal pit.* Usually the debris coming directly from the hatchery must be covered with no less than 6 in. (15 cm) of soil. But in some states and countries ecological laws have made it impossible to dump such refuse in open areas, even when covered. Other methods of disposal must be considered.

3. *Lagoon system.* Basically the process uses a grinder to pulverize the material, an aerated lagoon, and a "polishing pond." The system requires about 0.3 gal (1.25 liters) of water for each "no-hatch" egg run through it.

   Floating aerators are moved over the pond to increase aeration. The final liquid material is pumped from the polishing pond and sprinkled over nearby land, using a regular overhead irrigation system (see Chapter 40-D).

4. *Dehydration.* Equipment has not been designed specifically for handling dehydration of hatchery waste. The triple-pass rotary dehydrator or the tube-type flash dryer are the primary types of equipment employed. Fat may be left in the final ground product, or it may be removed.

   Because of variations in hatchability of eggs, dried poultry waste is highly variable. When hatchability is high, there is a decrease in the protein level and an increase in the calcium percentage. Even the breed or

Table 9-5. Nutrient Profile for Three Types of Poultry Egg Waste When Heated and Dehydrated

Item	Broiler Hatchery Byproduct %	Egg-type Hatchery Byproduct %	Eggshell Meal %
Protein	22.2	32.3	7.61
Calcium	24.6	17.2	36.4
Phosphorus	0.3	0.6	0.12
Fat	9.9	18.0	0.24

Source: Vandepopuliere, J. M., 1975, Arkansas Nutrition Conference, *Ark. Poultry Times,* Oct. 17.

strain of chicks hatched will influence the composition.
Table 9-5 gives an indication of the makeup of three
products originating from hatchery waste.

Nutritive value. Research work indicates that hatchery waste byproduct meal can replace at least 6% of the meat scraps or soybean oil
meal in the growing ration, while the shell meal can be substituted
totally for other calcium in the feed. In the layer ration 16% of the
total feed can be composed of dried hatchery waste (J. M. Vandepopuliere et al., *Poultry Science, 53,* 1985).

10

Hatchery Management

There is no doubt that good hatchery management means good *cost management*. The manager who can reduce his costs to the minimum will have accomplished one of the greatest of feats. The large size of present-day hatcheries means that more people will be necessary to oversee the many segments of the operation and the manager's duties are accentuated: breeder flocks, disease control, hatching egg transport, hatchery operation, chick delivery, and customer service.

But there is more than just knowing what is involved with each segment: There must be a system to bring reports to the attention of the manager on a regular basis. Appraising these reports and having the ability to make corrections and adjustments in the program when necessary are the prime functions of a capable manager.

10-A. SECURING HATCHING EGGS

A constant supply of quality hatching eggs is a requisite for a profitable hatchery operation. Because hatching eggs usually can no longer be purchased on the open market, a careful program must be projected to assure that flockowners will be able to furnish a definite number of hatching eggs at specified times. The supply of eggs must be adequate to produce all the chicks that are needed in an integrated operation, or sold to those producing their own hatching eggs, yet there must not be an excess of eggs that would have to be liquidated as market eggs at less-than-cost price.

Determining Hatching Egg Need

A sales or use projection must be made. This projection involves charts and graphs showing the number of chicks, by breeds and sexes, that will be sold or needed during the next 18 months. It must be kept current and extended regularly to encompass an 18-month period. Many operations use computers to compile these data. Such a projection is based on

1. *Past history of weekly sales or need.* Such a record shows the sales or need during the past year.
2. *Cyclical demand for chicks.* Are there seasonal demands for chicks? Although the weekly demand for egg-type pullet chicks is much more uniform than in years past, more chicks may be needed during certain periods. With meat-type straight-run broiler chicks the weekly demand is now very uniform because sales of dressed broiler chicks have few weekly variations, but there are still some seasonal variations.
3. *Whether or not your hatchery is a part of an integrated operation.* Chick sales projections will be on a different basis when the hatchery is a part of integration. In these cases the number of chicks needed each week will be determined by the number of broilers to be processed, or the number of commercial, egg-type pullets necessary to produce a given quantity of market eggs.

Securing Hatching Eggs from Hatchery-Owned Farms

When the supply of hatching eggs comes from hatchery-owned farms, there is only the need to house as many breeder pullets as will be necessary to produce the required number of eggs. However, determining the number is an intricate project—one that will require many hours of work if there is to be assurance that the supply of hatching eggs will meet the demand and that there will be no overproduction. When many farms are involved, the process becomes more complicated because only birds of one age may be practical at each unit; birds of other ages will have to be kept on other farms. Many commercial operations employ the use of a computer to make the calculations.

Securing Hatching Eggs Through Contract

When the hatchery does not own a poultry farm or farms it must make some arrangement with other poultrymen for its egg supply. Such poultrymen usually will provide the buildings, equipment, and labor necessary to raise the pullets and cockerels and keep the pullets in egg production. The hatchery usually will furnish the day-old breeding chicks, along with the feed and medications necessary to grow the birds. Under these conditions the hatchery and each flockowner will enter into a contract. Under such an agreement the hatchery will own the birds at all times; the flockowner only contracts to furnish the houses, equipment, and labor necessary to care for the birds.

10-B. CALCULATING NUMBER OF SALABLE CHICKS

The number of salable chicks available for sale from a hatch of a given number of eggs set is determined from the figures for percent hatchability, percent grade-outs (culls), and percent extra chicks.

Example: If 100,000 eggs were set, and the total hatchability was 80%, 80,000 chicks would be hatched. However, with 2% ''grade-outs'' (culls), and 2% ''extra chicks'' the number of ''salable chicks'' would be 76,863.

How to make the computation:

Step 1. Calculate the total number of chicks hatched and subtract the number of grade-outs; i.e., if 2%, then subtract 2% from the number of chicks hatched, as

100,000 eggs set
  80,000 chicks hatched (80% total hatch)
  <u> 1,600</u> subtract 2% grade-outs (2% of 80,000)
  78,400 chicks available after grading

Step 2. For every 100 chicks sold (invoiced), 102 chicks are to be shipped. Thus, the chicks available after grading must equal 102% of those sold (invoiced), as

78,400 chicks available after grading
76,863 equals number salable chicks.
    Divide 78,400 by 102, and multiply by 100.

10-C. CHICK PRODUCTION COSTS

Differences in hatching egg cost, percent hatchability, hatchery operating cost, and delivery cost make the total chick cost highly variable. Table 10-1 represents chick production costs at this writing, but they will certainly change. However, the table may be used as a guide for calculating local costs.

Table 10-1.   Hatchery Costs to Produce a Chick

Item	Cost per Salable Chick Hatched				
	Commercial Egg-type Pullet Chick		Commercial Straight-run Broiler Chick		
	Actual U.S. Cents	% of Total	Actual U.S. Cents	% of Total
Egg procurement	0.74		1.5	0.38		1.5
Hatching eggs	29.40[1]		58.3	18.23[3]		72.8
Hatchery operation[2]	9.96		19.8	4.11		16.4
Delivery expense	2.41		4.8	0.71		2.8
Subtotal		42.51			23.43	
G&A expense	3.68		7.3	1.63		6.5
Total		46.19			25.06	
Sales expense[4]	3.38		6.7			
Sales G&A expense	0.81					
Total sales expense		4.19	1.6			
Total procurement, hatchery, and sales expense		50.38	100.0		25.06	100.0

[1]Eggs at U.S. $1.50/doz; 85% hatch.
[2]Includes chick sexing, vaccination.
[3]Eggs at U.S. $1.75/doz; 80% hatch.
[4]No sales expense for broiler chicks.

Cost to Produce, Sell, and Deliver a Chick

As a starting point in discussing the economics of chick production, an example of the various costs involved is given in Table 10-1. Hatchery costs are broken down into the major components of the operation. It is impossible to manage a hatchery unless these data are known on a regular basis. Certain figures, such as overhead, General and Administrative expense, and sales expense, may be estimated for making the weekly calculations, but must be accurate when computing the month-end breakdown.

Hatching Egg Procurement Cost

Procurement cost should involve a bookkeeping cost center, with the following charged to this account:

1. Labor, including all benefits
2. Truck depreciation
3. Truck operating expense
4. Other expense (parking, tolls, uniforms, etc.)

Calculate costs on a per-dozen basis. Procurement costs must have a unit for cost analysis. Inasmuch as total egg costs are to be on a one-dozen basis, any procurement cost should be the same.

Hatching Egg Cost

The cost of hatching eggs fluctuates, and the percentage of the eggs that will produce quality chicks also is variable. Because the two factors work together to produce a final effect, the "hatching egg cost to produce a quality chick" is a necessary calculation. Some of these relationships are shown in Table 10-2. A study of this table will show the wide variability in the egg cost to produce a chick. Thus, in an analysis of these costs, egg cost and hatchability must be considered separately.

How hatching egg costs are treated. Hatching eggs are purchased, or they come into the hatchery from flocks connected with an integrated poultry enterprise. When they are purchased, the cost is the price paid for the eggs. When received from an integrated operation, hatching eggs are usually priced into the hatchery operation (cost center) at their cost of production; no farm profit is added. In an integrated project there is only one profit, when the final product is sold.

How hatching costs are treated by the accountant. Table 10-3 shows how variations in hatchability affect the cost to produce a chick. How-

Table 10-2. Egg Cost and Hatchability as They Affect Egg Cost per Salable Broiler Chick Hatched

Salable Hatch %	U.S. $ per Dozen Hatching Eggs												
	0.80	0.90	1.00	1.10	1.20	1.30	1.40	1.50	1.60	1.70	1.80	1.90	2.00
	Egg Cost per Salable Chick Hatched in U.S. Cents												
70	9.5	10.7	11.9	13.1	14.3	15.5	16.7	17.9	19.0	20.2	21.4	22.6	23.8
72	9.3	10.4	11.6	12.7	13.9	15.0	16.2	17.4	18.5	19.7	20.8	22.0	23.2
74	9.0	10.1	11.3	12.4	13.5	14.6	15.8	16.9	18.0	19.2	20.2	21.4	22.5
76	8.8	9.9	11.0	11.9	13.2	14.3	15.4	16.5	17.5	18.7	19.7	20.8	21.9
78	8.6	9.6	10.7	11.8	12.8	13.9	15.0	16.0	17.1	18.2	19.2	20.3	21.4
80	8.3	9.4	10.4	11.5	12.5	13.5	14.6	15.6	16.7	17.7	18.8	19.8	20.8
82	8.1	9.2	10.2	11.2	12.2	13.2	14.2	15.2	16.3	17.3	18.3	19.3	20.3
84	7.9	8.9	9.9	10.9	11.9	12.9	13.9	14.9	15.9	16.9	17.9	18.9	19.9
86	7.8	8.7	9.7	10.7	11.6	12.6	13.6	14.5	15.5	16.5	17.4	18.4	19.4
88	7.6	8.5	9.5	10.4	11.4	12.3	13.3	14.2	15.2	16.1	17.0	18.0	18.9
90	7.4	8.3	9.3	10.2	11.1	12.0	13.0	13.9	14.8	15.7	16.7	17.6	18.5
92	7.3	8.2	9.1	10.0	10.9	11.8	12.7	13.6	14.5	15.4	16.3	17.2	18.1
94	7.1	8.0	8.9	9.8	10.6	11.5	12.4	13.3	14.2	15.1	16.0	16.8	17.7
96	6.9	7.8	8.7	9.5	10.4	11.3	12.2	13.0	13.9	14.8	15.6	16.5	17.4

Table 10-3.   How Hatchability Affects Broiler Chick Cost (in U.S.$)

Number eggs set	125,000
Number dozen eggs set	10,417
Price per doz eggs set	$1.75
Expenses	
Hatching egg cost	$18,230
Egg procurement	380
Hatchery operating	4,110
Delivery	710
G&A	1,630
Total	$25,060
75% Hatchability	
Number salable chicks hatched	93,750
Cost per salable chick hatched	$0.267
80% Hatchability	
Number salable chicks hatched	100,000
Cost per salable chick hatched	$0.251
85% Hatchability	
Number salable chicks hatched	106,250
Cost per salable chick hatched	$0.236

ever, the procedure outlined in this table is never used in keeping the financial books of the company. Rather, cost is recorded. The total cost for a given period is then divided by the number of salable chicks hatched from these eggs. This gives the cost per chick. The greater the percentage of chicks hatched, the lower the cost per chick.

Example.   Table 10-3 shows the variations in cost to produce a chick when three groups of eggs have the same cost, but different hatchabilities.

Hatchery Operating Cost

Many costs other than egg costs are involved in the actual operation of a chick hatchery. Usually, these costs are segregated into various categories by the company accountant. Although differing slightly according to the type of hatchery operated, the following represent the main divisions of these accounts:

1. Labor, including all benefits
2. Heat, light, and power
3. Depreciation
4. Containers (chick boxes)

5. Repairs and maintenance
6. Consumable supplies
7. Other hatchery expenses (miscellaneous)
8. Services (beak trimming, sexing, comb trimming, etc.)
9. General and Administrative (office expense, telephone, management expense, taxes, insurance, interest, etc.)

Important: G&A expenses are not usually considered a direct cost of hatchery operation, but are an indirect cost, and most items add to the expense of producing a chick.

Delivery Cost

One of the obligations of most hatcheries is to deliver the chicks to the customer's farm in a satisfactory manner, which is sometimes costly, depending on the distance. In some cases, chick deliveries are nearby and the expense is almost trivial. But delivery costs must be separated from other hatchery expenses and calculated independently on a "per chick" basis. Delivery is not always by truck; in some countries rail and air are used to a large extent.

The account ledger should include the following breakdown of delivery expenses:

1. Labor, including all benefits
2. Vehicle operating expense
3. Vehicle depreciation expense
4. Air or rail delivery costs (if the hatchery is reimbursed for a part of these expenses, the account should be credited for these reimbursements)
5. Other delivery costs (miscellaneous) (tolls, parking, driver expense, clothing, etc.)
6. General and Administrative expense

Sales Cost

Last of the items of expense in hatchery operation is sales expense. On a chick basis this figure is highly variable. Hatcheries selling egg-type chicks usually have a high sales expense, while an integrated broiler-chick hatchery would have practically no sales expense, its endeavors encompassing chick placements rather than chick sales.

Where a definite sales program is involved, the sales expense ledger might be broken down into categories as follows:

1. Labor and commissions
2. Vehicle operating expense
3. Vehicle depreciation expense
4. Salesmen's travel expense
5. Advertising expense
6. Other expenses (miscellaneous)
7. General and Administrative expense

Calculating Costs in Egg-Type Hatcheries

When egg-type (Leghorn, etc.) chicks are sold, only the pullet chick is involved; the cockerel is a byproduct and is usually destroyed. Thus, the cost of producing a pullet egg-type chick is twice that of a straight-run (nonsexed) chick, since only half the chicks produced are pullets.

Some hatcheries calculate their hatchery costs on the basis of a straight-run chick, then multiply any final figures by two to arrive at the pullet cost. Hatchability is always measured on the basis of straight-run chicks.

10-D. FACTORS AFFECTING COST OF HATCHING A CHICK

A managerial analysis of the hatchery operation is important. There are many items that affect the cost of producing a chick, and a constant watch must be kept on these to keep production costs at a low level. Some that affect the operational costs are

1. *Labor efficiency.* Automation and labor efficiency are instrumental in reducing labor costs. One index used to measure labor efficiency is "chicks hatched per hatchery employee." Although the figure can hardly be used to compare one hatchery with another, it does offer a method of determining the weekly or monthly variations with a single operation over a period of time.
2. *Wage rate.* The hourly cost of hatchery labor is an important criterion of hatchery costs. In many instances it will be necessary to improve efficiency to offset increases in the wage rate.
3. *Managerial efficiency.* Just how good is the manager? Can he direct people and conduct good business procedure? Is he cost conscious?
4. *Utilization of incubator capacity.* If there are cyclical de-

mands for chicks, what happens to costs during the
''off-season''? Do they increase to such an extent that
profits during other parts of the year are consumed?

5. *Hatchability of the eggs.* Probably no single factor is as
responsible in determining chick costs as hatchability.
It is difficult to be competitive when hatchability
drops.

6. *Size of operation.* Generally, the cost of hatching a chick
is lower in larger hatcheries. This is particularly true
in hatcheries producing egg-type, commercial pullet
chicks, where seasonal variations in the demand for
chicks is greater than in those hatcheries selling meat-
type broiler chicks.

7. *Age and condition of hatchery.* Both of these affect results, either efficiency of operation, or hatchability.
Old-fashioned operations are apt to be less sanitary;
chick quality may be impaired.

8. *Depreciation costs.* Hatcheries should be well equipped,
clean, and efficient. Luxuries in construction are not
necessary; they only add to the depreciation cost.

9. *Discounts on purchases of supplies.* Every little bit helps.
Discounts should be taken. Generally, it is economical
to borrow money when necessary to take advantage
of cash discounts on purchases.

10. *Utility rates.* Large amounts of electricity are used in a
hatchery. The power rate is important. Wire the
hatchery to take care of power (commercial) and volume rates; incubators and motors should have a separate circuit and separate meter.

10-E. HATCHERY MANAGEMENT RECORDS

To manage a hatchery efficiently, and to keep costs to a minimum, the
manager must have certain records after each hatch, each week, and at
the end of each month. Good management is the result of pinpointing
inefficiencies and correcting them.

Manager's Hatch Report

Figure 10-1 is a form containing the data that should be supplied to
the manager after each hatch. The manager should not be confused with
too many reports. He needs only those necessary to make major decisions; minor decisions will be the result of discussions with his employees. The Manager's Hatch Report is initiated immediately after every

Flock		No. Cases Set	No. Eggs Set	No. Total Chicks Hatched	% Total Hatch		No. Grade-outs	% Grade-outs	% Extras	Salable Chicks	
Number	Wks. in Prod.				Act.	Std.				No.	%
TOTAL											

Hatch date _____

Egg cost per dozen eggs   _____
Egg cost per salable chick   _____
Other costs per salable chick (est.)   _____
TOTAL cost per salable chick   _____

Total chicks sold   _____
Prime product chicks destroyed   _____
TOTAL salable chicks hatched   _____
Estimated number salable chicks
    next hatch   _____

Figure 10-1. Manager's hatch report.

hatch; it must be current to be effective. Naturally, all cost figures, other than the egg cost to hatch a chick, are best given as estimates based on actual costs incurred the previous month.

Manager's Monthly Hatchery Report

At the end of the month the manager should be supplied with an accurate report of his operating costs, along with a month-end report of factors involving hatchability. There also may be a cumulative report for the year.

What the month-end report should include.   The report at the end of the month should be in three segments: (1) hatchability data, (2) cost analysis, and (3) other data. The following figures under each category should be on the report:

1. *Hatchability data*
   hatchery egg setting capacity for the month
   number of eggs set during the month
   percentage of hatchery egg capacity utilized
   total chicks hatched for month
   percentage of total chicks hatched for month
   standard hatchability for month
   percentage of chicks culled (grade-outs)
   percentage of extra chicks
   total salable chicks hatched
   percentage of salable chicks hatched
   number of salable chicks destroyed
2. *Cost analysis*
   total procurement cost per case of eggs
   egg cost per salable chick hatched
   total cost to hatch one salable chick

total cost to deliver one salable chick
total cost to sell one chick
TOTAL cost to hatch, deliver, and sell one chick
NOTE: These data should be completed for each breed involved. It is impossible to segregate problems if chicks hatched from various breeds are grouped together.

3. *Other data*
total procurement employees
total hatchery employees
total sales employees
chicks hatched per hatchery employee
chicks delivered per delivery employee
number of hatches for the month

11
Poultry Housing

Chickens, being warm blooded (*homeothermic*), have the ability to maintain a rather uniform temperature of their internal organs (*homeostasis*). However, the mechanism is efficient only when the ambient temperature is within certain limits; birds cannot adjust well to extremes. Therefore, it is very important that chicks be housed and cared for so as to provide an environment that will enable them to maintain their thermal balance.

11-A. BODY TEMPERATURE CONTROL

The confines of the internal body temperature of birds show more variability than mammals, so much so that there is no absolute body temperature. In the adult chicken this variability is between 105° and 107°F (40.6° and 41.7°C). Some variations may be observed as follows:

1. The temperature of the newly hatched chick is about 103.5°F (39.7°C), then rises daily until it attains a constant "adult" figure at about 3 weeks of age.
2. Smaller breeds have a higher body temperature than larger breeds.
3. Male chickens have a slightly higher body temperature than female, probably the result of a higher metabolic rate and more muscular use.
4. Activity increases the body temperature. For example, the temperature of birds on the floor is higher than that of birds kept in cages.

5. Molting birds have a higher temperature than those fully feathered.
6. Broody birds have a lower temperature than non-broody. Evidently the metabolic rate is lower in broody hens, and there is less muscular activity.
7. After food enters the digestive tract, body temperature increases.
8. The body temperature of the chicken is higher during periods of greater light intensity than during periods of darkness.
9. There is a tendency for the body temperature to rise as the ambient temperature rises.

How Heat Is Lost from the Body

Even though there are many contributing factors to a slight increase in the deep body temperature, the rise would be excessive if it were not possible for the bird to dissipate excess heat from the body. The chicken is continually producing heat through metabolic processes and muscular activity, and the heat lost from the body must equal the heat produced or the body temperature will rise. There are several methods of heat liberation:

Radiation. When the temperature at the bird's surface is greater than the adjacent air, heat is lost from the body by radiation, and ceases when the temperature of the surrounding air is reduced to, or below, the temperature of the bird's surface area.

Conduction. Loss of heat by conduction is caused when the surface of the bird comes in contact with any surrounding object, either air or some solid material, as when the bird sits on a cool floor. However, air is a poor thermal conductor, either when its moisture content is high or low, and generally, heat lost from the bird's body by conduction is very low.

Convection. When cool air comes in contact with the surface of the bird the air is warmed. The heated air expands, rises, and heat is carried away as the warmer air moves on. When the speed of air moving over the body is increased, as by fans, the amount of heat lost from the bird by convection increases. In most mammals, this body heat is water-laden because the sweat glands continually exude moisture, which evaporates, producing an even greater cooling. But chickens have no sweat glands and skin moisture is not a factor.

As the ambient temperature rises, heat loss by convection decreases, and when it reaches body temperature there is little loss by this method. In still air, there is none. Only 10 to 25% of all heat lost from the surface of the body is by convection; increased air speed over the bird increases the elimination. When the air temperature is higher than the surface temperature, both convection and conduction operate to cause heat removal.

Vaporization of water. As a replacement for moisture lost through sweat glands in most mammals, the chicken uses a process of evaporative cooling by the vaporization of moisture from the damp lining of the respiratory tract. Heat lost in this manner is a major method of heat elimination from the body of the bird when the ambient temperature is high.

Fecal excretion. A small amount of heat leaves the body with the fecal excretions.

Production of eggs. Heat lost with the laying of an egg is evident, but of minor importance.

Lethal Body Temperature

When the heat produced by the bird is greater than that dissipated through the various processes of elimination, the deep body temperature rises. When it gets to a certain point the bird dies from heat prostration. This is said to be the upper lethal temperature, and is about 116.8°F (47°C), but it is not absolute.

Mechanisms to Maintain Body Temperature

At about 70°F (21°C) 75% of all heat generated by the bird is lost through radiation, conduction, and convection. But the rate of loss is influenced by the ambient temperature. When the weather is cool these systems do their job well, but when the environmental temperatures are at, or near, the body temperature of the bird, they operate only a little or not at all.

The hen's ability to dissipate heat is influenced by the skin temperature rather than by the deep body temperature. As the temperature of the air surrounding the bird decreases, the blood vessels in the skin contract, thus reducing the flow of blood, which in turn acts to reduce the amount of heat lost from the body. When the temperature of the surrounding air increases, the blood vessels dilate, increasing the flow of blood, thereby increasing the amount of heat lost.

Panting necessary at high environmental temperatures. When radiation, conduction, and convection are unable to transfer all the heat produced, the next mechanism is called upon. This is *panting*, which is a means of bringing more outside air in contact with the membranes of the respiratory tract. Heat is removed from the body by the incoming air itself, and because the outside air has a lower humidity, more moisture is absorbed from the bird, along with its content of heat. This is known as *insensible heat loss.*

At an average humidity, birds will begin panting when the ambient temperature reaches 85°F (29.4°C). As the outside temperature increases above this figure, so will the respiratory rate of the bird (panting), and more heat will be eliminated from its body.

Panting and dehydration.   The increase in the breathing rate is accompanied by an increase in the loss of moisture from the body. To compensate for this loss, the bird drinks more water to avoid dehydration. Eventually, the bird drinks more water than it can exhale and the surplus is excreted through the droppings. The amount of moisture in the ambient air (humidity) also affects the panting rate; the higher the humidity the more rapid the respiration.

High temperatures and high humidity.   Chickens cannot withstand concurrent high temperature and high humidity, regardless of their age. When the surrounding air is moist, it cannot absorb as much moisture from the lungs; consequently the bird must pant faster. Similarly, when the outside temperature and high humidity are present, the bird may not be able to pant fast enough to remove the heat from its body. Prostration and death occur when the body temperature rises above the physiological maximum.

Heat production and feed consumption.   As the body temperature begins to rise during increases in outside temperature, the bird also makes other adjustments to keep its body temperature normal. Feed consumption is reduced as the ambient temperature rises; it is increased as the temperature lowers. In turn, growth and egg production are decreased.

Activity affected.   As the outside temperature changes, so does the activity of the bird. Movement is lessened during hot weather in the bird's endeavor to generate less heat. The bird rests more, as evidenced by less eating, less mating, and more sitting rather than standing or walking, and the wings are extended to expose more body surface. More molting of surface feathers occurs. Conversely, when the air temperatures are low the bird induces greater production of body heat through increased activity, and greater feed consumption. The bird conserves heat by fluffing its feathers.

11-B. HEAT PRODUCTION AND LOSS

Good poultry housing tends to alleviate the extremes in the bird's environment to make it easier for the bird to compensate for its surroundings. This means a more optimum ambient temperature in the house, along with proper and adequate removal of moisture. A discussion of adequate housing must include a study of the contributing factors of heat and moisture production.

Btu defined. The output of heat produced by a bird is measured in British thermal units (Btu). One Btu is the amount of heat required to raise the temperature of 1 lb of water 1 degree Fahrenheit at or near 39.2°F, the maximum density of water.

Heat production by birds. The data on bird heat production are highly variable because of the contributors involved—type of bird (broilers, growing pullets, layers, etc.), caloric value of the feed, ambient temperature, relative humidity, etc. It would require many tables to produce accurate figures showing the influence of each. But inasmuch as heat production is discussed as a means toward proper house ventilation, and ventilation requirements are not too well established, a composite table (Table 11-1) has been constructed to show averages for all types of birds kept under all types of conditions.

From the table the data show that a 4-lb (1.8-kg) bird produces about 40 Btu of heat per hour. On a unit of body weight basis this is only one-half as much as produced by a 1-lb (0.45-kg) bird. Thus, it must be kept in mind that heat produced per unit of weight decreases as a bird becomes larger and heavier.

In ventilating a poultry house during hot weather the additional heat produced by the birds must be removed from the building to reduce the ambient temperature, but during cold weather a high percentage of the generated heat must be kept in the house to raise the ambient temperature.

The following rule-of-thumb figures may be used for Btu production of laying pullets:

Standard Leghorn layer	40 Btu per hour per bird
Brown-egg layer	45 Btu per hour per bird
Meat-type layer	55 Btu per hour per bird

Egg-type bird/broiler relationship. Broilers of the same weight as egg-type birds will produce a greater quantity of heat because broilers grow faster and consume more feed per unit of weight, both of which increase the body heat production.

Sensible and Insensible Heat Production

The above discussion regarding heat loss from the body has included all heat lost. However, this total heat may be divided into two categories:

Sensible heat loss. Sensible heat is heat that causes the temperature of the surroundings (air or physical medium) to increase. Such heat is one endpoint of muscular activity and metabolism of the feed con-

Table 11-1. Heat, Moisture, and Fecal Production by Chickens at 70°F (21°C) by Weight Variations

Avg Body Weight		Heat Output per Hour			Moisture Output per 100 Birds per Day						Fecal Output per Day per 100 Birds	
		Per lb Body Weight	Per kg Body Weight	Per 100 Birds	Fecal Output		Respiratory Output		Total Output			
lb	kg	Btu	Btu	Btu	lb	kg	lb	kg	lb	kg	lb	kg
1	0.5	20.0	44.0	2,000	15.9	7.2	5.2	2.4	21.1	9.6	9.6	4.4
2	0.9	14.5	31.9	2,900	20.2	9.2	10.7	4.9	30.9	14.1	18.1	8.2
3	1.4	11.5	25.3	3,450	23.1	10.5	15.5	7.1	38.6	17.6	25.0	11.4
4	1.8	10.0	22.0	4,000	25.3	11.4	19.3	8.8	44.6	20.3	30.8	14.0
5	2.3	9.0	19.8	4,500	27.2	12.4	20.7	9.4	47.9	21.8	35.7	16.2
6	2.7	8.2	18.0	4,920	29.2	13.3	22.3	10.2	51.6	23.5	39.2	17.8

sumed and this heat raises the temperature of the air surrounding the chickens in the house.

Insensible (latent) heat loss.   This is the heat lost from the body through the elimination of respiratory moisture. To vaporize moisture in the lungs and air sacs, heat is required and eliminated. This heat does not increase house temperature.

Relationships between sensible and insensible heat loss.   Two factors, ambient temperature and ambient humidity, affect the rate of sensible and insensible heat lost from the body. At temperatures below 70°F (21°C) the body processes are capable of equalizing about 75% of the body temperature through the processes of radiation, conduction, and convection (sensible heat loss), but as temperatures rise above the thermoneutral range the processes of vaporization of moisture increase. Therefore, as the ambient temperature increases, a greater percentage is insensible heat loss, while a lesser percentage is sensible heat loss (see Table 11-2). However, humidity of the air is more of a factor than is air temperature, for it governs the rate of evaporation of moisture from the linings of the respiratory tract.

Temperature influences sensible and insensible heat.   Not only does body weight affect the amount of heat produced but the ambient temperature alters the proportion of sensible and insensible heat produced. Lowering the air temperature increases the proportion of sensible heat. It is also greater during the day than during the night. All these factors have a bearing on the ventilation of a poultry house (see Table 11-2).

Table 11-2. Sensible and Insensible (Latent) Heat Production as Influenced by Ambient Temperature (White Leghorn Hens)

Ambient Temperature		Sensible Heat %	Insensible Heat %	Output of Sensible Heat per Hour	
°F	°C			Per lb Body Weight Btu	Per kg Body Weight Btu
40	4.4	90	10	9.0	19.8
60	15.6	80	20	7.9	17.4
80	26.7	60	40	6.1	13.4
100	37.8	40	60	4.3	9.5

11-C. WATER PRODUCTION AND LOSS

The quantity of water produced by a bird depends on the body weight, type of feed, density of the feed, salt content of the feed, and air temperature and humidity. Water is eliminated mainly through the fecal material of which about one-fourth comes from the urine and three-fourths from the intestinal tract, by respiration, and in the production of the egg. The amount lost must be determined before an adequate ventilating system can be designed for the poultry house.

At normal ambient temperature and humidity, the moisture lost by respiration by a 4-lb (1.8-kg) bird about equals the amount lost through the feces, but at lower body weights the proportion of water excreted in the feces decreases (see Table 11-1).

On a weight basis, the total amount of water eliminated through respiration and fecal discharge decreases as the size of the bird increases. It is about two times as much per unit of weight for a 6-lb (2.7-kg) bird as for a 1-lb (0.45-kg) bird (see Table 11-1). This factor is to be reckoned with in poultry house ventilation because most ventilating formulas are based on the number of pounds (kg) of chickens in the house.

Amount of water in the feces. This amount is very highly variable, as one would expect. It is associated with the ambient temperature, which in turn affects the amount of water being consumed. Younger (smaller) birds have less moisture in their droppings than older (larger) birds. Most adult chickens in a normal environment and consuming a normal amount of feed will produce feces containing 75 to 80% water (see Table 11-1).

Feed and water consumption affect water production. At 70°F (21°C) a chicken will normally drink twice the weight of water compared to the feed consumed, but as ambient temperatures rise, feed consumption decreases while water intake increases. But the feed decreases and water increases are not uniform with temperature variations. They are less at cool temperatures than at hot ones. The relationships are shown in Table 11-3. Notice that a bird will drink twice as much water at 100°F (37.8°C) as at 70°F (21°C).

Respiratory and fecal elimination of water. With a 4-lb (1.8-kg) bird at a temperature of 70°F (21°C) about 43% of the moisture is lost through the fecal material, while only 25% is lost in this manner by a 1-lb (0.45 kg) bird.

More moisture is lost from the body than is consumed in the form of water as a result of the moisture in the feed plus the additional metabolic water produced and eliminated. As an example, 100 4-lb (1.8 kg) birds held at a temperature of 70°F (21°C) will drink about 44 lb (20 kg) of water per day, but will produce about 45 lb (21 kg) of respired and fecal moisture (Table 11-3). Approximately 2.5 lb (1.1 kg) will be consumed in the feed and 6.0 lb (2.8 kg) voided in the eggs. From the same table it may be computed that 100 birds will produce about 5.3 U.S. gal (20 L) of water per day, all of which must be removed from the poultry house.

Ambient temperature is a very important consideration in determining total water output per bird. Although temperatures below 70°F (21°C)

Table 11-3. Ambient Temperature as It Affects Feed and Water Consumption and Water Elimination (100 4-lb Leghorn Laying Pullets per 24 Hours)

Item	Mean Daytime House Temperature						
	°F						
	40	50	60	70	80	90	100
Feed consumed (lb)	26.0	25.5	24.1	22.0	19.1	15.3	10.5
Pounds of water con- sumed per pound of feed consumed	1.3	1.4	1.6	2.0	2.9	4.8	8.4
Water consumed (lb)	33.9	35.7	38.6	44.0	55.4	73.5	89.2
Water consumed (U.S. gal)	4.1	4.3	4.7	5.3	6.7	8.9	10.8
Droppings produced (lb)	36.4	35.7	33.7	30.8	26.7	21.4	14.7
Fecal water (lb)	28.8	28.7	27.3	25.3	22.2	18.0	12.5
Respired water (lb)	4.7	6.4	11.2	19.3	33.6	56.0	76.8
Fecal and respired water (lb)	33.5	35.3	38.5	44.6	55.8	74.0	89.3
	°C						
Item	4.4	10.0	15.6	21.1	26.7	32.2	37.8
Feed consumed (kg)	11.8	11.6	11.0	10.0	8.7	7.0	4.8
Kg of water consumed per kg of feed con- sumed	1.3	1.4	1.6	2.0	2.9	4.8	8.4
Water consumed (liters)	15.5	16.3	17.8	20.1	25.4	33.7	40.9
Droppings produced (kg)	16.6	16.2	15.3	14.0	12.1	9.7	6.7
Fecal water (kg)	13.1	13.0	12.4	11.5	10.1	8.2	5.7
Respired water (kg)	2.1	2.9	5.1	8.8	15.3	25.5	34.5
Fecal and respired water (kg)	15.2	16.0	17.6	20.3	25.4	33.7	40.2

do not greatly affect the amount of water consumed, temperatures above rapidly increase water consumption (see Table 11-3).

11-D. THE ENVIRONMENTAL PROBLEM

Adequate housing, from an environmental standpoint, is what is necessary to meet the optimum requirements for best bird growth, freedom from stress, good egg production, high fertility, and the most efficient utilization of feed. Briefly, adequate housing must provide the flock with optimum air quality and temperature conditions so that performance may be optimized.

11-E. INSULATING THE POULTRY HOUSE

The question of whether or not to insulate a poultry house must be analyzed for each individual situation. This is a requisite for open-sided houses as well as for those that are environmentally controlled. Most insulation is confined to the roof, since this area has the greatest heat loss during cold weather and it is the area where the sun's rays strike.

Insulation is beneficial during the winter months as a means of conserving heat and thereby improving feed utilization. It is also beneficial during the summer months when it can effectively exclude the high radiant heat load from metal roofing. Excluding heat not only improves flock comfort but also results in improved performance.

R-values for Insulating Efficiency

The efficiency of any insulating material, or combination of materials, or type of construction, is rated by its ability to resist the transfer of heat through it. There are many materials or combinations of materials used to insulate poultry houses, and the resistance of these materials to the transfer of heat has been given a practical term known as *R-value*, or thermal resistance. These are given in Table 11-4.

Vapor Barrier

To be effective, insulating material must be dry, since moisture can conduct both heat and cold. To prevent moisture from penetrating the outside wall or roof and wetting the insulation, a dead-air space between layers of material may be provided. This space is known as a *vapor barrier*. However, during the last few years new materials that take the place of the dead air space have come on the market. These materials are porous and do not conduct moisture. Therefore, they may be placed directly against other material, which makes them easy to install. Sometimes these insulating materials are placed on the underside of the rafters, thus leaving an air space between them and the roof sheathing or covering.

How Much Insulation?

Obviously, there should be more insulation in cold climates than in warm or hot. But the average should show the following R-values:

Type of Climate	R-value for Roof and Ceiling	Walls
Hot climates	4	2
Medium climates	8	2.5
Cold climates	12–14	8–10

Table 11-4.   R-values of Various Building Materials

Item	Thickness in.	Thickness cm	Resistance Rating
Insulation per 1 in. (2.5 cm) of thickness			
Blanket bat	1	2.5	3.70
Balsam wool (wood fiber blanket)	1	2.5	4.00
Cellulose fiber	1	2.5	4.16
Expanded polystyrene, molded (bead board)	1	2.5	3.50
Expanded polystyrene, extruded (Styrofoam®)	1	2.5	5.00
Urethane foam	1	2.5	6.60
Fiberglass (glass wool)	1	2.5	3.70
Palco wool (redwood fiber)	1	2.5	3.84
Rock wool (machine blown)	1	2.5	3.33
Rock wool (blanket)	1	2.5	3.33
Foam glass	1	2.5	2.50
Glass fiber blanket	1	2.5	3.33
Mineral wool	1	2.5	3.33
Insulation board	1	2.5	2.37
Vermiculite (expanded)	1	2.5	2.05
Wood fiber	1	2.5	3.33
Sawdust or shavings (dry)	1	2.5	2.22
Straw	1	2.5	1.75
Materials (thickness as indicated)			
Air space, horizontal	0.75+	1.8+	2.33
Air space, vertical	0.75+	1.8+	0.91
Asbestos cement	0.12	0.3	0.03
Building paper			0.15
Concrete	8.00	20.3	0.61
Concrete block	8.00	20.3	1.11
Hardboard	0.25	0.6	0.18
Plywood	0.25	0.6	0.32
Plywood	0.50	1.2	0.63
Surface, inside			0.61
Surface, outside			0.17
Siding, drop	0.75	1.9	0.94
Sheathing	0.75	1.9	0.92
Metal siding			0.09
Glass, single			0.61
Shingles, asbestos			0.18
Shingles, wood			0.78
Roofing (roll, 55-lb)			0.15
Vapor barrier			0.15

Determining the R-value of Walls and Roofs

Because each type of wall or roof covering has an R-value, the sum total of the R-values of the materials used will give the R-value for the wall or roof. Using Table 11-4, an example of the resistance value of a wall has been calculated on the following page.

Wall Insulation Item	R-value
Outside surface	0.17
Metal siding	0.09
Vertical air space	0.91
2-in. fiberglass	7.40
1/4-in. plywood	0.32
Inside surface	0.61
Total resistance rating of wall	9.50

Types of Insulation

Each type of insulation has its advantages and disadvantages. The blanket or bat is usually made from some type of mineral, such as glass fiber. If it has a foil back be sure to install it so it is tight and there is little vapor loss. When such materials are moisture laden they lose their insulating value.

Some insulating materials are extremely flammable while others have flame retardants added to them. But even with retardants they emit a lot of smoke when burned. Most materials are tested and rated according to their flame spread and smoke development. In some locations the building code or insurance companies require a spread rating of 20 to 25, or less.

Attic Insulation

By running stringers across the house on top of the studs, then lining the bottom of the stringers with some sort of insulating material, it is possible to construct an attic in the house. This has the advantage of creating a large vapor barrier or air space in the gable. But the size of the air space is not correlated with any insulating value. A 4-in (10-cm) dead-air space probably would provide as much insulation as an air space 4 to 5 ft (1.2 to 2 m) in depth. Furthermore, these attics become extremely warm, and although this has some advantage in the cold months of the year, it is a disadvantage during the hot summer months. To remove the hot air from the attic some sort of ventilation must be provided, either with suction cupolas on the roof or exhaust fans in the gables of each end of the house.

In environmentally controlled poultry houses, the air entering the pens during the cold winter months is first brought into the attic, warmed, then drawn into the area where the chickens are. During the summer months, air is admitted directly from the outside, bypassing the attic.

11-F. MOISTURE IN THE POULTRY HOUSE

Moisture in the house constitutes one of the greatest problems of adequate poultry housing. This moisture is due to fecal elimination of water, spillage, and to that contained in the respired air. About the only way it may be removed from the house is to increase the movement of air through the building. But the relative humidity of the incoming air governs the amount of moisture it will absorb. When the relative humidity is low, the air will take up more house moisture than when it is high (see Chapter 5-F). But all moisture in the poultry house is not in the air; most of it is in fecal material and/or the litter. Fresh fecal material will contain from 75 to 80% water, but the amount of water in the litter is more variable. In dry climates it may be as low as 5 to 10%; in wet, it may rise to 70 to 80%. The optimum for growing birds is between 20 and 40%. For older birds it is between 10 and 30%. When the relative humidity of the air in the house is 50%, the moisture content of the litter will approach 25%.

Water in Fecal Material

At 70°F (21°C) a chicken will drink about 2 lb (or 2 kg) of water for each pound (or kilo) of feed it eats, and 60 to 65% of the water consumed will appear in the fecal material. At higher ambient temperatures, the amount of water consumed, and the amount excreted, will increase greatly.

Elimination of moisture from the poultry house becomes much more of a problem during cold weather than in warm weather because it is necessary to retain the heat in the poultry house by reducing the flow of air through the building. Soon the air contains more moisture, evaporation is reduced, and the litter becomes wet (see Table 11-3).

Amount of Fecal Material

At 70°F (21°C) the amount of fecal material will be about 40% greater than the weight of the feed eaten. Ambient temperature will alter this percentage. A breakdown of feed and water consumed by 100 4-lb (1.8-kg) Leghorn hens and the composition of their droppings is shown in Table 11-3.

Although many data relative to the amount and composition of the fecal material have been reported, they have been highly variable because there are so many factors involved in the production of the fecal matter. Humidity, temperature variations between night and day, composition of the feed, and many other factors influence the figures. Table 11-3 presents data that are the result of average conditions. Of material significance in this table is the great variation in feed and water consumption when the environmental house temperature changes.

Moisture Buildup in the Poultry House

Unless means are provided to remove it, moisture will increase in the poultry house when:

1. *Water consumption by the birds increases.*
2. *House temperature decreases.* Decreasing the temperature reduces the ability of the air to hold moisture; therefore, less fecal moisture will be removed from the litter. The air holds less respired moisture; the excess is absorbed by the litter. When air in the house is warmer it will absorb more moisture, so the heat generated by the birds is of great help in removing moisture from the building.
3. *Humidity of the air in the house increases.* As the humidity increases, the air takes up less water vapor, and more fecal and respired moisture are absorbed by the litter.
4. *The salt content of the diet increases.* Birds drink more water as the salt content is increased.
5. *The energy value of the feed is reduced.*
6. *A feed is pelleted.* Water consumption is increased when feed is compressed, either as crumbles or pellets.
7. *Drinking water is contaminated by microorganisms.*
8. *Birds are kept in cages.* Caged birds consume more water than those kept on the floor.

11-G. TOXIC GASES

Several gases are found regularly in poultry houses. Many are lethal at high concentrations, but there is a maximum practical level of each for chickens as shown in Table 11-5.

Table 11-5. Lethal and Practical Limits of Gases in a Poultry House

Gas	Lethal Limit %	Practical Limit	
		%	ppm
Carbon dioxide	Above 30	Below 1	Below 10,000
Methane	Above 5	Below 5	Below 50,000
Hydrogen sulfide	Above 0.05	Below 0.004	Below 40
Ammonia	Above 0.05	Below 0.0025	Below 25
Oxygen	Below 6		

11-H. AMMONIA CONCENTRATION

Ammonia in a poultry house can become troublesome when the concentration is high. It is nauseating to the caretaker, irritates the eyes, and affects chickens.

Ammonia is measured in parts per million (ppm). Normally, 15 ppm will prove uncomfortable for human beings; 50 ppm for 8 hours is considered the maximum allowable concentration.

Tolerance level for chickens. Continuous high concentrations of ammonia lessen the activity of the cilia of the respiratory tract of chickens. With laying birds, 30 ppm are probably slightly injurious, affecting egg production and the general health of the birds, while 50 ppm produce serious trouble, particularly growth. But much higher concentrations of ammonia can be tolerated, perhaps as high as 100 ppm for short periods. However, this amount produces a higher incidence of breast blisters, and water consumption is higher. Thus, for practical purposes, the ammonia concentration in the poultry house should not be over 25 ppm.

How to measure ammonia concentration. A kit is available for measuring ammonia concentration in ppm. A special impregnated paper is moistened and held in the poultry house. It turns various colors from orange to blue, depending on the concentration of ammonia in the atmosphere, and the color can be correlated with the concentration of ammonia present.

Reducing ammonia fumes. Ammonia in the poultry house may be reduced by

1. Increasing the amount of air flowing through the building.
2. Replacing the litter.
3. Reducing the pH of the litter below 7.0. This may be accomplished by adding phosphoric acid or superphosphate. Ammonia release is rapid when the pH is 8 or above (see Chapter 20-F).

11-I. OPEN-SIDED POULTRY HOUSE

Most of the poultry houses in the world are conventional or open-sided (see Figure 11-1); that is, they rely on the free flow of air through the house for ventilation. Certain requirements must be met if such a ventilated house is to provide an adequate environment. Care in following these rules during the course of construction will avoid pitfalls later.

Width of house. The width of the open-sided poultry house should be about 30 ft (9.8 m) and no more than 40 ft (12.2 m) wide. Houses that are wider will not provide ample ventilation during hot

Figure 11-1. An open-sided high-rise cage house with side curtains.

weather. Wide houses also require additional interior supports that
may interfere with equipment or manure removal. This width recommendation is basic for growing birds, broilers, and laying hens.

Height of house.   Most open-sided houses have a stud that is 8 ft
(2.4 m) long. The stud represents the distance from the foundation
to the roof line. In areas where the temperature is exceptionally high
throughout the year, the stud length should be increased to 10 ft
(3 m). High-rise houses, with manure storage areas below the cages
or slats, should be as high as 14 ft (4.3 m) or more at the eaves.

Length of house.   Poultry houses may be almost any convenient length.
The terrain on which they are to be built often determines the
length; rolling land means more grading before construction can
start. Because automatic feeding equipment will limit the length of
the poultry house, the equipment manufacturer should be consulted about the optimum length of the feeding system. Many times
the feed hopper is placed in the center of long houses to provide
better use of automatic feeders.

Shape of roof.   Practically all poultry houses built today have a gable
roof, the pitch varying from one-quarter to one-third. A good overhang should be provided to protect the inside from driving rains
and to afford interior shade.

Roof exhausts.   Houses should be equipped with a covered exhaust
area at the peak of the roof to allow excess heat to escape. Various

systems are available to close the exhaust during the colder months in order to conserve heat.

Insulation. Even with the conventional poultry house, it is well to provide some type of insulation. The roof may be insulated, using special products for this purpose, or an attic, or partial attic, may be installed. Attics should be ventilated with suction cupolas, or by vents.

Building materials and construction. Open-sided and environmentally controlled houses use a variety of building materials. The choice is dependent on the structural strength required, the insulative characteristics of the material, material availability, and material cost. Galvanized steel or aluminum are most commonly used for roofing and siding. Framing is usually done with wood or steel and some houses constructed recently have used the tilt-up concrete wall method of construction. Cages and other equipment can be supported either from the roof trusses or from the ground. However, most of the newer multiple-deck cage units are supported from the ground because of their weight.

Foundation. A solid and adequate foundation should support the building. Concrete, concrete blocks, bricks, or other permanent and termite-proof material should be used. Evenness of the foundation is important, for it will determine the evenness of the completed structure.

Floor. With certain disease-control programs, a concrete or similar floor is mandatory. It is also necessary when the soil is very dense and can absorb and transfer moisture from lower subsoil, but in certain areas, where the soil is sandy, and where commercial broilers or commercial layers or breeders are kept, a concrete slab is not used when birds are placed on the floor. Cage houses usually have concrete walks to facilitate the movement of hand egg collection carts and mobile feed carts. The area beneath the cages may or may not be paved depending on the manure removal program and method.

Doors. Doors at the end of the house should be large enough for a truck, tractor, or manure-handling equipment to pass through. Such equipment will be used when the house is cleaned.

Orientation. Houses must be oriented in a direction to take advantage of prevailing airflow patterns. Orientation must also be considered relative to solar heat transfer into the building from exposed roofs or side walls. Pullet-rearing areas should always be located upwind from adult birds.

Sides

With this type of house most of the side areas are open. The height of the opening will be determined by climatic conditions, and by the type of bird being housed, as follows:

Broilers and young chicks. From one-half to two-thirds of each side is left open, the exact amount being determined by summer and winter temperatures. When both heat and cold are to be dealt with, the size of the opening should be medium. Where heat is continuous, the opening should be larger; sometimes almost all of the side is left open.

Growing birds and layers. The opening size is greater for older birds. They should be provided with more air because bird density is greater and more ventilation is necessary.

Cage houses. Houses equipped with cages necessitate the greatest amount of air movement. The bird density is the greatest of any type of flock. Sides should be almost completely open.

Curtains during cold weather. Young chicks and older birds should be given some protection during periods of cold weather and extreme winds. This protection is usually provided by curtains made of some durable and plasticlike material. They are installed down the length of the building and hung so that the entire curtain may be rolled up or down by cables and a winch located at one end of the building or by thermostatically controlled automatic winches. This construction makes it easy to regulate the size of the opening according to weather conditions—an almost indispensable provision.

Cooling the Open-Sided House

As long as there is some air moving, the open-sided house works well in warm weather, but in hot weather the warm breezes may be detrimental to the birds. It is when the wind stops that trouble begins. The buildup of heat within the building is quick, and at inside temperatures of 95°F (35°C) and above, the birds are distressed, and suffocation begins as their body temperature rises above the point of toleration. One or more methods of comforting the birds must be employed.

1. *Sprinkle the house roof.* Circulating sprinklers may be installed on the roof. In order for these to have maximum effectiveness, it is important that the entire roof be wet.
2. *Sprinkle the ground area outside the house.* This tends to cool the air around the house, but it also increases the humidity, a detriment to bird cooling.
3. *Use foggers in the poultry house.* A variety of in-house fogging systems are available, ranging from those producing large droplets with considerable wetting of birds and equipment to high pressure systems that reduce temperatures through evaporative cooling. The latter systems can reduce temperatures by more than 20°F (11°C) in regions of relatively low humidity.

4. *Use fans outside of or in the poultry house.* Undoubtedly, the natural movement of air over the birds helps to lower their body temperature. It dissipates body heat more quickly, and removes the exhaled moisture. Increased air movement usually is necessary during hot weather, particularly when natural air movement ceases. Fans may be placed on the windward side of the poultry house to increase the velocity of air as it blows through the building. But when the outside temperature is unusually high, it may be better to place the fans inside the house, to blow the air lengthwise of the building. High-speed fans are better than low-speed, regardless of where placed.

The Wind-Chill Factor

Cold is relative to a human being. How cold one feels is determined by the loss of heat from the surface of the body. Obviously, the lower the ambient temperature the more heat lost from the body, and the colder the human body feels. But if the surrounding air is blown over the surface of the body one not only feels colder because heat is dissipated more rapidly but because moisture is evaporated more quickly from the skin. For instance, if one stands in still air at 35°F (1.7°C), he or she feels chilly. But if the air movement over the person is increased to 10 miles per hour, it feels as though the temperature were 21°F (−6.1°C). If the air is speeded to 35 miles per hour, it feels the equivalent of still air at 3°F (−16.1°C).

Although these facts are known for human beings, the chicken has no sweat glands; thus there is little, if any, evaporation from the surface of the body. But we do know that the movement of air over the birds, regardless of how it is produced, makes them more comfortable in hot weather, as heat loss from the body surface is increased. Similarly, the birds must feel colder during cold weather when air speeds are increased. Thus, increasing air movement during extremes of heat is an important factor in ventilating the open-sided house (see Table 11.6).

11-J. CONTROLLED-ENVIRONMENT HOUSE

A controlled-environment house is one in which inside conditions are maintained as near as possible to the bird's optimum requirements. Doing so usually necessitates a completely enclosed insulated house with no windows. Air is removed from the house by exhaust fans and fresh air is brought in through intake openings. Artificial light, rather than natural daylight, is used to illuminate the interior. Where high outside

Table 11-6. Chicken-Cooling Effect of Air at Various Speeds.

Air Speed		Cooling Effect	
ft/min	m/min	°F	°C
20	6	0.0	0.0
50	15	1.0	0.5
100	30	3.0	1.7
250	76	6.0	3.4
500	152	10.0	5.6

Source: Tilley, M. F., Col. Agric. Engineering, England.

temperatures are involved, some method of controlling the temperature inside of the house is provided. The houses are not heated except for brooders. The heat from the birds is used to keep inside temperature within the range required for maximum feed efficiencies. Figure 11-2 illustrates typical controlled-environment housing.

Much of the structural makeup of the environmentally controlled poultry house is similar to that of the house with open sides. It should have a good foundation and a gable roof. Insulation is a must; both the sides and the top should be given protection. The overhang of the roof need not be as great because the sides are completely covered. But venti-

Figure 11-2. An environmentally controlled high-rise cage house with feed bins.

lating a completely enclosed house is difficult. Details must be worked out so that air movement is adequate and evenly distributed during both hot and cold weather, a complicated procedure.

Width of House

Because air is mechanically exhausted from the environmentally controlled house, rather than removed by natural air movement as in the open-sided house, the width of the enclosed house may be greater. These houses are often 40 to 50 ft (12.2 to 16.5 m) wide. Most ventilating systems will adequately remove the air from a house of this width, but there may be difficulty with those that are wider.

11-K. MECHANICAL MOVEMENT OF AIR

Air must be moved through the environmentally controlled poultry house to replenish the oxygen, to remove moisture and ammonia, and to keep an optimum temperature. Furthermore, the movement of air in the house must be uniform from top to bottom, and from side to side. This calls for special ventilating methods.

Air is brought into the house through inlets usually located in the eaves, through light traps located in the walls or ends of the building, or through cooling pad panels. The amount of air to be moved will determine the size and number of fans necessary. But be sure to keep the fan blades and shutters clean. When they are allowed to get dirty, airflow is reduced materially.

Typical ventilating fans used in the poultry house are 36 in. (0.9 m) in diameter, operated with a one-half horsepower motor. Such fans deliver about 10,000 ft³/min (283 m³/min) of air. Larger fans of up to 54 in. (1.4 m) in diameter are available and these can deliver up to 30,000 ft³ (850 m³) of air per minute.

11-L. CALCULATING VENTILATION NEEDS FOR NEGATIVE PRESSURE

In the environmentally controlled poultry house, the amount of air exhausted should be slightly more than the amount of air coming into the building. A negative pressure is created within the building and is regulated by the *exhaust system*. The amount of air to be drawn out of the poultry house by the exhaust fans is determined by the type, age, and size of the birds, the insulation characteristics of the house, the desired inside temperature, and the outside temperature.

When the houses are exceptionally wide, air circulation within the house will not be adequate with the above type of ventilation. To remedy this inadequacy, sometimes the intakes are placed at the front and back of the house and the air is exhausted through a cupola by fans placed in the center of the ceiling.

Heat Production and Heat Removal

Each 50 ft³ (1.4 m³) of air will remove 1 Btu of heat for each 1°F (0.56°C) rise in temperature. Practically, this means that if the house temperature were increased by 5°F (2.8°C), 50 ft³ (1.4 m³) of air leaving the building would remove 5 Btu of heat. Although this approximation is rough, it is used for determining the number of fans necessary to remove the heat from the poultry house during hot weather.

Static pressure reduces fan efficiency. The following shows the effect of static pressure on the delivery of air by fans, based on a fan delivering 10,000 ft³/min with no static pressure at 38°N. latitude.

Static Pressure in. H₂O	Air Delivery by Fan ft³/min
0.00	10,000
0.05	9,400
0.10	8,600

How to Compute Necessary Airflow

There is a simple method used to compute the necessary airflow through a chicken house. Computation includes environmental temperature and size (weight) of the birds.

Rule of thumb. Provide 0.012 ft³ of airflow per minute per pound of body weight of the chickens in the house for each 1°F of temperature. Typical examples for 30 to 60% relative humidity are as follows.

Air Temperature	
°F	ft³ of Air per min/lb Body Wt
40	0.48
60	0.72
80	0.96
100	1.20
110	1.32

Fan Operation

Fan capacity must be that necessary for maximum temperature and maximum stocking density. The total fan capacity of the poultry house should be what is necessary to remove the heat from the building when the outside temperature and flock heat production is the highest. There are two methods of lowering the airflow through the house when less air is needed at lower temperatures:

1. *Rheostats on the fans.* Some are hand-operated, and some operate automatically as the house temperature changes. Rheostats are capable of altering the speed of the motors driving the fans.
2. *Operate only a part of the fans.* Since fans must be located at intervals along the length of the building, some may be stopped when the air requirement is less. In other instances the fans are installed in banks of three fans placed side by side. Usually one of these—a smaller one—runs continuously; the others cut off one at a time as less air is needed. These latter fans should be run intermittently; e.g., 10 minutes on, 5 minutes off.

Fan Ventilating Efficiency Rating (VER)

VER is the ratio of the volume rate of air movement (ft^3 or m^3 per minute) by a fan to the rate of electrical consumption (watts). When comparing fans, the variation has been found to be as much as 50% in the cost to move air when calculated by the VER. Therefore, the cost to move air is a very important consideration when selecting ventilating fans.

11-M. THE AIR INTAKE

When the air in a building is not moving, the cold air, being heavier, will drop to the floor and the hot air, being lighter, will move to the ceiling. Air in the house must be kept moving so that all areas have the same temperature.

The air intake is as important as the air exhaust in the environmentally controlled house. In the first place, slightly less air must be admitted than exhausted. The difference should be about 0.04 in. (0.1 cm) of static pressure. This will allow the fans to function at about 85% of capacity and the air to circulate well in the house because of air velocity.

Air Intake Computation

Although many types of air intakes are used with modern poultry buildings, the recommendation for the size of the opening is as follows:
Rule of thumb:

> Allow 1 in.^{2} of intake opening for each 4 ft^3/min of air exhaust. When light traps are used over the fan openings, increase the intake to 1.25 in.^{2}
>
> or
>
> Allow 6.5 cm^2 of intake opening for each 0.113 m^3/min of air exhaust. When light traps are used over the fan openings, increase the intake to 8 cm^2.

Shape of air intake opening affects air distribution in house. For the air to be distributed well throughout the building, it must enter at a gushing rate; therefore, it should enter the building through some type of narrow entrance. This is called the *intake slot.* The width of this slot should be about 2 in. (5 cm) when the outside temperature is 70°F (21°C). In most instances this will make it necessary for the slot to run almost the length of the house and in high-density houses, along both sides, to get adequate air intake and velocity.

Adjustable slot intake. Since it is necessary to exhaust more air from the poultry house when temperatures are high and less when they are low, the air intake must be adjusted to maintain 0.04 in. of static pressure within the building. For this reason the slot intake too must be adjustable. This may be accomplished by hand; or a winch and cable system may be used so that the slot can be opened or closed the entire length of the house from a central location. Automatic devices also are available, some operating by static pressure.

Location of the slot intake. In practically all cases, the slot intake should be as high as possible on the wall near the eaves. Air may be brought directly into the building from the outside, or may enter from the attic, where it will have been warmed during cold weather.

Baffle to direct incoming air. An adjustable baffle should be used to direct the amount and direction of the air coming in through the slot. At normal temperatures the baffle should be almost horizontal. This will mix the incoming air adequately with that in the house. When more incoming air is needed, the inside of the baffle edge is dropped so as to deflect the air toward the floor.

Velocity of incoming air through the slot. When all details of the ventilating system are operating at average, the velocity of the air coming through the slot will be between 700 and 750 ft (213 and 228 m) per minute, or about 8.2 miles (13.2 kilometers) per hour. However, the speed of the incoming air soon diminishes as it flows into the house when there is no static pressure (see Table 11-7). For this reason,

Table 11-7. Speed of Air at Inlet and Varying Distances from It

Air Speed at Inlet ft/min	Distance Away from Inlet		
	10 ft	15 ft Air Speed	20 ft
	ft/min	ft/min	ft/min
500	56	41	31
750	84	61	47
1,000	112	83	62
	Distance Away from Inlet		
	3.1 m	4.6 m Air Speed	6.1 m
m/min	m/min	m/min	m/min
152	17.1	12.5	9.5
228	25.6	18.6	14.3
305	34.1	25.3	18.9

Source: Tilley, M. F., Col. of Agric. Engineering, England.

the air flow should be as short as practical to ensure evenness of temperature and air quality. Air flow within cage houses must provide for evenness of distribution in each cage row.

Caution during hot weather. Many people make the mistake of only opening the air intakes wide during hot weather. This disrupts the circulation of air in the house, and is not the solution to cooling the birds. The correct procedure is to increase the amount of exhausted air first, then open the air inlet slots enough to maintain the designed static pressure in the house.

Caution during cold weather. It is just as important to keep the air gushing through the slot intake during winter as during summer. A high speed of air at the intake does not necessarily cool the house; it only provides better dissemination of air within the room.

In both situations the air velocity must be held constant, which is accomplished by adjusting the number of fans operating while at the same time adjusting the size of the air inlet openings. More fans operating will require a larger air inlet area while fewer fans will require a reduction in inlet size.

11-N. PRESSURIZED SYSTEM OF VENTILATION

A second, but less popular, method of ventilating the environmentally controlled poultry house is the *pressurized system* or *positive pressure sys-*

tem. Fans force air into the poultry house, and the outlets are so regulated as to give a slight positive pressure in the building by exhausting slightly less air than is brought into the room. Usually, the air is distributed through ducts running the length of the house.

Large, Low-Speed Fans Used

To make this system work it is necessary to move a large volume of incoming air at a slow rate of flow. Thus the ducts that distribute the air throughout the house must be large. Large, low-speed fans should be used to bring in the air. Such a system reduces drafts in the house, and improves the distribution of air.

Location of Ducts

The ducts should be located about 12 in. (30.5 cm) below the ceiling of the house to induce the best air circulation. In narrow houses they may be placed near the leeward wall; in wide houses they should be placed at the middle of the house. Both solid-walled and collapsible plastic ducts are used.

Formula for Fan Velocity

When ductwork is used in a ventilating system with positive pressure, the static-pressure losses through the ducts reduce the amount of air moved by the fans. Most air velocities in ducts are between 600 and 1200 ft/min (183 and 366 m/min) corresponding to head pressures of 0.022 and 0.88 in. of water. The expected fan velocity can be estimated from the following formula:

$$V = \sqrt{h}$$

where V = feet per minute and h = inches of water.

11-O. SELECTING THE RIGHT TYPE OF HOUSE

Several innovations in house design have evolved over the years. Their importance is based on many factors including a reduction in floor space per bird, less labor, higher fertility with breeding birds, better disposition of droppings, and improved sanitation. Each of these houses can use either the open-sided or environmentally controlled systems. The primary categories of each of these house designs are

1. Cage house
2. Slat-and-litter house
3. All-slat house
4. High-rise house

11-P. CAGE HOUSE

Several types of houses and roofs lend themselves to cage housing. Most variations are the result of climatic conditions. In cool to cold climates, the environmentally controlled house is almost a must. In mild climates, only a roof over the cages seems to be necessary. Special cage housing is discussed in detail in Chapter 16-B and 16-H, but several of the following house types may be used to hold cages.

11-Q. SLAT-AND-LITTER HOUSE

The slat-and-litter house is constructed so that a part of the floor area is covered with slats. Although built primarily for those birds producing hatching eggs, particularly meat-type breeders, the house may also be used for growing birds, but they must be trained to use the slats when they are young.

In some instances the slats are replaced with welded wire fabric, but wire is a poor substitute. This type of house is best suited to broiler-breeder parents (see Figure 11-3). The advantages and disadvantages are as follows:

Advantages

1. More eggs can be produced per unit of floor space when slats are used inasmuch as the floor space requirement per bird is less.
2. Fewer floor eggs are produced with the slat-and-litter house than with the all-slat house.
3. There is less labor per bird.
4. There is no difference in mortality, cracked eggs, or hatchability.
5. Fertility is better with the slat-and-litter house than with the all-slat house.

Disadvantages

1. Housing investment is higher with the slat-and-litter house than with the all-litter house.
2. Birds lay fewer eggs with the slat-and-litter house than

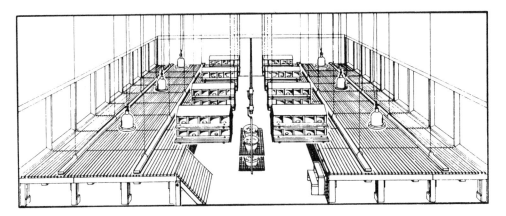

Figure 11-3. View of slat-and-litter house.
(Courtesy of Arbor Acres Farm, Inc., Glastonbury, Connecticut, U.S.A.)

with the all-litter house, and even fewer with the all-slat house.

3. Crowding birds on slats increases the incidence of dirty eggs.
4. The separation of the birds from the manure beneath the slats commonly results in fly problems.

Amount of slats (or wire). Slats should cover about 60% of the floor area; 40% should be covered with litter. Although an all-slat floor may be used for commercial laying birds and egg-type breeders, meat-type breeders mate better if some litter area is available. Fertility may be reduced when the breeders are kept on an all-slat or all-wire floor.

Slat size and spacing. Slats should be about 1 to 2 in. (2.5 to 5 cm) wide and spaced about 1 in. (2.5 cm) apart, and run lengthwise of the building. *Do not* run them crosswise of the house because birds cannot brace themselves to eat from the feeders.

Wire size. When wire is to be used in place of slats, extreme care should be used in construction. The wire should be heavy enough not to sag. This will call for supports about every 12 in. (30.5 cm). Welded wire fabric should be used to provide greater support. It should have a mesh size of 1 by 2 in. (2.5 by 5 cm) with the long part of the mesh running crosswise of the building.

Constructing the slats (or wire). Slat or wire floors should be constructed in sections so that they may be removed when it is necessary to clean the droppings from under them, or when the house is cleaned.

Location of the slats (or wire). The slats may be located in the house in two ways:

1. *Down the edges of the house.* One-half of the slats are placed against one wall, and one-half are against the opposite wall. This placement has an advantage in open-sided houses as any driving rain falls on the slats rather than on the litter. Furthermore, all chores may be handled from the central litter area (see Figure 11-3).
2. *Down the center of the house.* This system leaves one-half the litter area on one side of the house and one-half on the opposite side, with the slats in the center. This construction has some advantages in the environmentally controlled house, because the feeders and waterers placed on the slats are closer together. With the open-sided house, there are disadvantages because rains blow in and wet the litter, and it is difficult for a person to cross over the slats to get to the opposite side of the house.

Height of the slats. The top of the slats should be about 27 in. (68 cm) above the floor. This will allow space below the slats for a year's droppings to accumulate. Wire mesh should be used on the sides of the slat supports to provide more air movement across the droppings in an endeavor to keep them dry. When slats are placed against the outside wall of an open-sided house it will be necessary to construct a moveable curtain or cover for the area below the slats to keep air from blowing through during cold weather and to allow air to enter during hot weather.

11-R. ALL-SLAT HOUSE

Commercial laying birds may be kept on an all-slat floor. The advantage is that it requires less floor space per bird than when the birds are kept on a litter floor. On litter, commercial laying pullets will require about 2 ft^2 (0.18 m^2) of floor space per bird. When they are kept on an all-slat floor, 1 ft^2 (0.09 m^2) is ample. The requirement for constructing the slats, their size, and spacing are the same as for the slats used for the slat-and-litter combination.

11-S. HIGH-RISE HOUSE

Poultrymen have long looked for new and more economical methods of collecting and disposing of poultry manure. The Leghorn hen will produce 90 to 100 lb (41 to 45 kg) of fresh droppings containing 75 to 80% moisture during 52 weeks of lay. More droppings will be produced during cold weather than during hot weather. With litter on the floor the droppings will be mixed with the litter, but with cages they remain intact, and are difficult to dispose of until partially dried.

To overcome many of the evils of conventional handling of wet manure, the high-rise house (sometimes called the deep-pit house) has become increasingly popular. It provides for in-house drying and sheltered accumulation of the droppings until the flock is sold or until disposal of the manure can be arranged.

A high-rise house is essentially a two-story house. The top floor is for the birds in cages or on a slat floor. The bottom floor, with no ceiling, is directly underneath and is used for the accumulated manure. Each is about 7 ft (2.1 m) high.

Ventilating the high-rise house. Most such houses are completely enclosed and environmentally controlled. Air is brought in through a slot-type opening at the eaves of the upper unit holding the layers. The air is warmed by the birds, then is pulled down to the manure pit and over the droppings by large exhaust fans installed in the walls of the lower pit area. Warm air entering the pit has the capacity to absorb moisture from the droppings before it is exhausted from the building.

Bacterial action. Bacterial action will decompose the droppings after most of the water is removed. However, there is little or no action with wet material and the manure will release ammonia, thus reducing its fertilizing value. Therefore, it is imperative that no excess water get into the pit, either from outside seepage or from overflowing waterers. The floor of the pit section should be about 18 in. (45 cm) above the ground level to help prevent water seepage, and should be covered with 6 to 12 in. (15 to 30 cm) of cinders, sand, or gravel.

Determining ventilation needs. The basic requirements for air through this type of house are the same as for the one-story environmentally controlled house. However, during cool or cold weather it is very important that the amount of air coming into the building be reduced as the rule provides. This is necessary in order to retain heat in the upper section where the birds are located. But this reduction does have a detrimental effect during cool weather, for not as much air enters the droppings pit, and consequently the manure does not dry as rapidly.

Supplementary fans in the pit. When the droppings dry slowly, as during cold weather, the process may be expedited by using supplementary fans in the pit. These blow air down the front of the pit and back on the opposite side, creating a circular movement over the droppings. The fans should be hung from the top of the pit, and be reversible. According to D. C. Sprague, of the New York State College of Agriculture, the air should move over the droppings at a velocity of 400 to 600 ft/min. One 10,000-ft^3/min (283-m^3/min) fan should be supplied for each 2,000 ft^2 (186 m^2) of floor space.

Manure accumulation in the pit. Most poultrymen prefer to clean their high-rise houses after each flock but some elect to store the manure in the house for periods up to 7 years. A complete clean-out is commonly practiced but it is advisable to leave a 6- to 12-in. (15- to 30-cm) dry pad of manure to aid in the drying of new manure and to preserve beneficial insect populations. Manure removal with the flock in the house is not considered to be a good practice.

The rate at which manure accumulates in the pit depends on the density of birds, the configuration of cages, and the rate of decomposition and drying.

> *Note:* Although the above design is for a house equipped with laying cages and the area below the birds is open at the ceiling except for service walks, slat floors may be used as the floor of the upper section. The birds are placed on the slats rather than in cages. The high-rise house may also be used for growing pullets kept in cages or on slat floors.

11-T. COOLING THE HOUSE

Moving more air through the poultry house when the outside temperature gets above 85°F (29.4°C) is not the solution to providing a comfortable environment for the birds. There are four accepted methods of cooling the birds:

1. *Low-pressure fogging system.* Fogging nozzles that operate at regular water pressures are installed throughout the house or over the birds in cages.
2. *Pad-and-fan system.* Exhaust fans in the house draw incoming air through a wet pad where the evaporation of moisture from the pad reduces the temperature of the incoming air.
3. *Fog-and-fan system.* This is similar to the pad-and-fan system except that incoming air is drawn through a hood in which high-pressure foggers have been installed. As air is drawn through the fog, its temperature is reduced.
4. *High-pressure fogging system.* This special nozzle converts water from liquid to vapor form. This change has a great cooling effect on the air in which it comes in contact.

Pad-and-Fan System

The principles and recommendations for this method are given in Chapter 5-G since the system is also used to cool chick hatcheries. There are two types of this system used for cooling poultry houses:

1. *Pressurized system.* Evaporative coolers are placed outside the house and air is sucked through the evaporative pads of the cooler, then forced into the poultry house. Openings are provided in the room through which the air is exhausted from the building. A slight pressure is built up in the room.
2. *Vacuum system.* This is a variation of evaporative cooling used in poultry houses. The evaporative pad is placed in the wall at one end of the house and exhaust fans at the other. The exhaust fans cause air to be sucked through the evaporative pad, thus reducing the temperature of the incoming air. The principles and amount of cooling are the same as those involved with the pressurized system.

Pad and Fan Specifications

The five types of evaporator pads commonly used are

1. 2-in. (5.0-cm) aspen
2. 1-in. (2.5-cm) concrete-coated bagasse
3. 4-in. (10-cm) corrugated-fluted cardboard
4. 1-in. (2.5-cm) rubberized hogshair
5. Cellulose and glass fiber pad (Humid-Kool®)

Selection of pad.   Each of the five types of pads has its own unique set of characteristics. Selection of a pad material should be based on its relative cooling efficiency, life expectancy, ease of cleaning, and freedom from problems. Cooling efficiency is a measure of the amount of the cooling potential that is actually achieved. When the air from an evaporative cooling pad reaches the wet bulb temperature, 100% efficiency is obtained.

Location of the pads.   Unless the house is more than 200 ft (61 m) long, the pad may be placed in one end of the building, the fans in the other. In the case of longer houses, pads may be placed in both ends of the house and the exhaust fans in the center, or the pad in the center and the fans at each end.

Most pads are installed in a vertical position. Water is usually distributed at the top of the pad and allowed to trickle down to the bottom. The thinner, more flexible pads (aspen and hogshair) can also be installed horizontally and water is applied as a spray. This technique usually results in cleaner pads without sagging.

Exhaust Fan Requirements

Exhaust fans should be located in the opposite wall or end from the evaporator pads. Fans that are large in diameter and that revolve slowly, yet move large volumes of air, should be used.

The amount of air to be moved through the house having pad-and-fan cooling will depend on:

| number of birds | type of building insulation |
| age and weight of birds | maximum outside temperature |

Houses with this method of cooling require the movement of more air through them than houses with only air circulation. The required amount of air to be moved through the house will vary from 1.4 to 2.0 ft^3/min per pound of bird (0.04 to 0.06 m^3/min per kg of bird) in the house.

Rule of thumb:

> Provide 1.75 ft^3 of airflow per minute per pound of live birds in the house.
>
> or
>
> Provide 0.11 m^3 of airflow per minute per kilo of live birds in the house.

Example:

> A house holding 10,000 Leghorn pullets, weighing 4 lb each, represents 40,000 lb of birds. At the rate of 1.75 ft^3 per minute per pound, 70,000 ft^3 per minute of air would need to be exhausted. This figure is equivalent to 7 ft^3 per minute per bird.
>
> or
>
> A house holding 10,000 Leghorn pullets, weighing 1.82 kg each, represents 18,200 kg of birds. At the rate of 0.11 m^3 per minute per kilogram, 2,000 m^3 per minute of air would need to be exhausted. This figure is equivalent to 0.2 m^3 per minute per bird.

Heat buildup within the building. When the cool air originates at one end of the building, flows over the birds, then out the other end of the house, the heat from the birds warms the air as it flows through the house. This buildup should not be greater than 8°F (4.5°C). If it is greater, redesigning of the system should be considered. In most cases, this is an indication that an insufficient pad area or amount of air is available. Care must be taken to assure that the pads are clean and completely wet. Dirty pads will greatly reduce airflow and dry pads will not cool the incoming air.

Pad Requirement

The air should come through the pad at a speed of about 150 ft (46 m) per minute. To calculate the area of the pad, take the amount of air moved per minute by the exhaust fans at 0.1 in. (0.3 cm) static pressure and divide by 150.

Remember: The air to be exhausted must first be accurately calculated, then the pad area computed.

Wet Pad Not Used at All Times

Pad-and-fan cooling is not used at all hours, day and night, but only during hot hours. When the water going to the pad is shut off, the ventilating fans continue to run to keep air moving through the house. When the pad is dry, the ventilating system becomes a conventional one except for the fact that there is a very long distance from the entrance of air to the exit; therefore, a faster air movement is necessary. During the winter months, pads should be partially closed with plastic curtains or sheets.

Thermostat on water inlet. Water flowing to the pad should be controlled automatically rather than manually by using a thermostat to control the activity of the pump. The thermostat should be set at 80°F (27°C). When the temperature in the house reaches this point the pump starts; when it drops below this temperature, the pump stops.

Thermostats on Exhaust Fans

With most installations, two or more exhaust fans will be required. Several small fans are better than one large one. There should be a thermostat on each fan, but the temperature at which the fans start and stop should be varied. All fans should be operating when the house temperature reaches 80°F (27°C).

Pad-and-fan assembly kit available. Several manufacturers produce the materials necessary to install pad-and-fan cooling equipment. The items included are the pads, troughs, water pump, float, and thermostats. Exhaust fans and thermostats are also available from many manufacturers and distributors.

High-Pressure Fogging System for Cooling

There is a second method of cooling poultry houses. A hood is constructed over air inlets, and high pressure fogger nozzles are placed in a pipe just below the lower limit of the hood so as to emit a water mist into the hood that is then sucked into the house by incoming air created by the exhaust fans. The nozzles in the water line operate under about 500 lb of pressure to produce the necessary fog. No wet pad is used.

A slot at the top of the wall and at the top of the hood is used to give a better mix of the fogged air after it enters the house. Exhaust fans are located on the opposite wall, and covered with a hood.

The system calls for filtered water to prevent sediment from disrupting the flow of water through the nozzles and a high-pressure centrifugal pump to produce 500 lb of pressure. A thermostat stops and starts the system as the house temperature changes.

Painted Roofs for Additional Cooling

Whitewashing the roof of the uninsulated house will be of great help in lowering the interior temperature because the sun's rays will be reflected rather than absorbed.

Suitable whitewash formulas are

1. 20 lb (9.1 kg) hydrated lime
   5 U.S. gal (18.9 liter) water

2. 20 lb (9.1 kg) hydrated lime
   20 lb (9.1 kg) white cement
   6 U.S. gal (22.7 liter) water
   ½ cup (118 cc) bluing

3. 20 lb (9.1 kg) hydrated lime
   5 U.S. gal (18.9 liter) water
   1 U.S. qt (946 cc) polyvinyl acetate

11-U. LIGHT CONTROL IN THE ENVIRONMENTALLY CONTROLLED HOUSE

Although the use of light in the poultry house is discussed in Chapter 18, the environmentally controlled house must be lightproof; that is, no outside light should be allowed to enter the building. Where fans are installed in the side, light seeps through the fan openings, and the house is not lightproof. For this reason a *light trap* must be installed. This is accomplished by building a hood on the outside of the building over the fan opening. It should extend down far enough to prevent light from entering, yet impair the movement of air as little as possible. The air requirement is higher when light traps are used inasmuch as they do decrease the speed of the air.

11-V. VENTILATING AND COOLING THE BROILER HOUSE

There is now a new system for ventilating and cooling the broiler house (see Chapter 20-C).

12

Poultry House Equipment

Good equipment is a requisite of good poultry management. In all probability, most equipment that was modern a few years ago is now out of date, impractical, and uneconomical. Today, there is more demand for automation in the poultry house to lower the hours of labor required to care for the birds and to reduce labor costs. Figure 12-1 illustrates a modern egg-production complex.

This chapter deals with equipment for houses in which the birds are kept on the floor. Chapter 16 discusses cages and cage house equipment; Chapter 20 has to do with broiler house equipment.

12-A. BROODING EQUIPMENT

The so-called brooding age of chickens refers to the first 5 to 6 weeks. Chicks are small during the early part of this stage, and require equipment that is small in size, most of which cannot be fully automated. Supplementary heat is also needed.

The Brooder House

Brood-grow house. Until recently the accepted practice among poultry-men was to brood the chicks in a separate *brooding house.* Now, however, separate brooding houses are seldom used; chicks are kept in the same house from day 1 until sold or point-of-lay.

Brood-grow-lay house. In many instances the birds remain in the same house throughout their lives—from the first day until the end of their laying year. Thus, the house becomes a brooding house with

Figure 12-1. A large egg-production complex.

supplemental brooding heat; a growing house must accommodate birds of various ages.

Small-room brooding. When the brood-grow or the brood-grow-lay system of housing is used the house is much larger than that needed to brood the chicks. To reduce the size of the brooding area in such houses a plastic, or similar, curtain is placed crosswise of the house to segregate one section for brooding. This reduces the heat necessary to warm the brooding section, an economic saving. As the chickens grow older, the space is gradually increased until the entire house is occupied.

Brooder Fuel Supply

Brooding heat may be supplied by a variety of fuels. Almost any available product may be used to furnish the fuel as heating devices have been manufactured to use the one selected. These fuels are as follows:

Gas. Gas may be used for heating individual stoves or central heating units. It is available in three forms.

1. *Natural gas.*
2. *Liquefied petroleum:* Two types are available: (a) propane or (b) butane.
3. *Methane gas:* Methane gas is now being produced by digesting dried poultry waste.

Kerosene. This fuel is used for heating individual stoves or central heating installations.

Coal. Anthracite rather than bituminous coal is used for brooder stove fuel where it is available. It burns with a minimum of volatile matter and smoke.

Oil. Fuel oil is commonly used in some sections of the world to heat brooder stoves or central heating systems.

Wood. Wood is used to heat brooding rooms, thereby reducing the fuel requirement of brooder stoves. A blower moves oxygen to the firebox of the furnace and another blower moves air into the house.

Electricity. Electricity is used as the fuel supply for a variety of brooder heating devices, such as heating rings, electric light bulbs, infrared light bulbs, etc.

Solar energy. There are now many innovations used to convert the heat of the sun's rays to other forms of energy. In some instances the solar energy is converted to heat poultry buildings; in others, it is used to operate special space heating devices that supply the necessary supplementary brooding heat.

Btu Values of Various Fuels

Although somewhat variable, particularly with natural gas and coal, the following show the heat values of some common fuels:

Gasoline	124,000 Btu per U.S. gal
Kerosene	134,100 Btu per U.S. gal
Diesel fuel (#2)	140,000 Btu per U.S. gal
Propane	92,000 Btu per U.S. gal
Butane	117,750 Btu per U.S. gal
Natural gas	1,025 Btu per cu ft
Fuel oil (#2)	138,500 Btu per U.S. gal
Coal (anthracite)	12,950 Btu per lb
Coal (bituminous)	11,800 Btu per lb
Wood	7,700 Btu per lb
Electricity	3,412 Btu per KWH

Methods of Supplying Brooder Heat

Chick brooders are units that furnish the heat necessary to keep the young bird warm. Usually some provision is incorporated that will deflect the heat downward toward the chicks. There are many types of brooders or brooding arrangements:

1. *Hover-type:* These represent the most common type of brooder. The heat unit is covered with a round or angular piece of metal to deflect the heat toward the floor. The brooder unit is usually suspended from the ceil-

ing by a cord or cable in such a manner that the brooder may be raised or lowered. When not in use it is drawn as high as possible, out of the way, and left there until it is needed for the next group of chicks; or it may be removed from the house. There are several methods of heating hover-type brooders and these give rise to their classification.

Conventional gas.   These are 6 or 8 ft (1.8 or 2.4 m) in diameter with a gas burner located in the top of the dome. By changing the gas jets they operate on either natural or LP (liquefied petroleum) gas. They accommodate from 500 to 750 day-old chicks, and have an input rating of 12,000 to 20,000 Btu per hour, depending on their size.

"Pancake" or flat-top.   This hover utilizes an almost flat top and a specialized burner that produces radiant heat. The canopy (hover) is about 4 ft (1.2 m) in diameter, and is hung about 24 in. (61 cm) above the floor; 500 chicks are usually placed under each brooder. The Btu input capacity varies from 25,000 to 50,000 per hour. The amount of heat required for this type of brooder stove, according to the room temperature, is shown in Table 12-1. Radiant heat cannot be measured with a thermometer. Only the comfort of the chicks can be used as an indicator of the correct supplementary heat.

Catalytic.   There is a gas brooder in which a catalyst is employed to produce a chemical reaction and heat production. Catalytic combustion produces a clean, flameless heat, and dust and moisture do not affect its operation. The heating device has a low surface temperature and will not ignite dust or litter. No hover is used, the heat being deflected from the "burner." It has a Btu rating of about

Table 12-1.   Heat Requirements for Pancake Brooders at Varying Room Temperatures

	Ambient Temperature		
	°F		
	40	50	60
	°C		
	4.4	10.9	15.6
Age in Days	Btu per sq in. per hour		
1	1.33	1.08	0.85
7	0.93	0.74	0.55
14	0.80	0.61	0.42
21	0.61	0.45	0.28
28	0.34	0.30	0.12
35	0.24	0.15	0.05
42	0.20	0.12	0.04

Source: Griffin, J. G., and H. A. Stephens, 1971, *Poultry Meat,* Apr.

22,000 per hour, and such burners are said to use about 20% less gas than the conventional gas brooders.

Infrared. These are heated with a special burner under a tile refractory that produces infrared rays when heated. There are several types of these brooders. Some employ a canopy; others do not. The Btu rating is relatively low.

Conventional kerosene. A special burner, using kerosene as the fuel, and covered with a large canopy, is used in many areas where other fuels are not available.

Conventional electric. Where electric power is comparatively inexpensive, electricity may be used to produce brooder heat. Electric heaters are placed under a canopy, with a thermostat to turn the current on and off. In some models a small electric fan is placed in the canopy cone. It draws fresh air from the outside at the top, and circulates it down and over the chicks on the floor. An electric brooder may require up to 2,500 watts of power.

2. *Hot-water brooders:* Hot-water pipes are placed about 12 in. (30 cm) above the floor to supply the heat. A boiler at one end of the house heats the water in the pipes. A thermostat is placed on the boiler to turn the burner on and off so as to maintain a uniform water temperature in the boiler. An additional thermostat is placed under the pipes in the middle of the house. This controls a pump at the location of the incoming water line at the boiler so that when the brooding temperature drops, hot water is drawn through the pipes by the pump until the brooding temperature reaches a desired figure. From four to eight hot-water pipes are placed down the middle of the house and an insulated cover is placed over them to hold the heat near the floor.

Important: Never place the water pipes near the back or front wall of the house; they should be in the middle to provide ample ventilation under the cover.

3. *Slab heating:* Rather than supply heat from above the chicks it is possible to furnish the heat from below. A concrete slab (floor) is heated by running pipes through the concrete and forcing warm water through the pipes. A thermostatically controlled boiler is located at one end of the house. In some instances the slab is heated with electric wires imbedded in the concrete. Generally no litter is used on the heated slab.

Caution: Do not heat the entire floor of the brooder house; heat only a section from 6 to 8 ft (1.8 to 2.4 m) wide in the middle and extending the entire length of the brooding area. Chicks feather poorly if all the floor is heated.

4. *Room heating:* When environmentally controlled houses are used for brooding chicks, it is possible to heat the entire house to furnish the necessary brooding temperature. Normally, this temperature is lower than that necessary under a canopy-type brooder. Room heating requires

a floor temperature of about 85°F (29°C) for starting chicks. Heat necessary to warm the entire house may be supplied by a central heater, then forced throughout the house through a duct. Another method is to suspend regular brooder stoves about 6 ft (1.8 m) above the floor and use them to heat the entire room, but usually this is not economical. Room heating is particularly adaptable to the partial house brooding concept.

Attraction Lights

Training day-old chicks to go to the heat supply when they are cool is difficult. To teach them the location of the brooder heat, a small light may be placed under the canopy or at the location of the heat. This light may be supplied by one 7½-watt white light bulb for each brooder stove or equivalent area. After 2 or 3 days the chicks will learn the heat source and the light may be turned off.

Brooder Guards

Some material must be placed around the brooder stoves to prevent the chicks from straying too far away from the heat supply until they learn the source of heat. These are called *brooder guards* and circle the brooder heat area at a distance of about 30 in. (76 cm). As the chicks grow older the area inside the *ring* is increased to give the birds more room. The material used for the brooder guards may be solid or wire mesh. Solid is preferred during cold weather and wire mesh is preferred during hot weather.

Solid guards may be constructed of some flexible material such as Masonite. Strips from 16 to 24 in. (40 to 61 cm) wide are placed around the brooder stove and the sections held together with large "clothespins." Corrugated cardboard also may be used, but as it cannot be washed and cleaned it must be discarded after the guards are no longer needed.

Waterers for Young Chicks

Waterers for young chicks are confined almost entirely to the pan-and-jar type. Since water must be easily accessible, several *small founts* must be placed around each brooder stove, inside the brooder guard. The pan-and-jar waterer is the most practical. There are several innovations: (1) jar and pan, (2) plastic, (3) all glass, (4) metal, and (5) plastic and metal. Each fount should hold approximately 1 gal (3.8 L) of water. Several small founts are better than a few large ones.

Waterers After One Week of Age

As soon as the chicks learn to drink, and the brooder guards are expanded or removed, larger waterers should be substituted. Most of these are of a type that can be used until the birds reach sexual maturity; some may be used for laying birds as well. These waterers and watering systems usually are automatic. There are many types:

Automatic troughs. The trough is usually V-shaped, and adjustable for height from 2 in. (5 cm) to about 16 in. (41 cm) above the floor, and 8 ft (2.4 m) in length. There are three types, the differentiation being in the type of valve used to turn the water on and off.

1. *Suspension valve.* One end of the trough is suspended from the frame; the other hangs on a valve. The weight of the water in the trough turns the water on and off.
2. *Float valve.* The trough is mounted solid to the frame. At one end of the trough is an enclosed pan in which there is a float valve that operates according to the height of the water in the pan and the trough.
3. *Electric valve.* An electric valve governs the flow of water to the trough.

Hanging waterers. These are round waterers constructed of plastic or metal that hang from the ceiling. A bell-shaped dome encloses a valve to maintain a designated level of water in the circular pan. They may be raised or lowered by changing the length of the suspension cord, cable, or chain.

Cup waterers. Cup waterers are small drinking cups, from 2 to 6 in. (5 to 15 cm) in diameter, and from 1 to 3 in. (2.5 to 7.6 cm) deep. They may be classified according to the manner in which water is admitted to them.

Suspension-type. A relatively large cup is attached to the end of a vertical pipe or hose. A valve opens and closes according to the weight of the water in the cup.

Trigger-type. Usually these cups are clamped onto the top of a horizontal pipe running the length of the chicken house. A valve is situated in the bottom of the cup with a "trigger" that opens and closes the intake valve. The "trigger" is operated by the bird. Cups on this type of waterer are quite small. The horizontal pipe may be raised or lowered by cables and a winch; thus the height of all of the cups may be altered at one time according to the age of the birds.

Note: To operate cup waterers a pressure equalizer must be installed in the incoming water line. Cup systems operate on extremely low pressures.

Drip nipples. These are small valves operated from below by the chicken, and allow the water to run down the throat as the bird extends its neck. They cannot be used for chicks before they are 1 week old.

First Feeders

So that the chicks learn to eat at an early age, a large area of feed must be supplied the first few days. Usually the first feed is spread over a large, flat container with a shallow edge 1 to 2 in. (2.5 to 5 cm) in height. Such a container may be supplied through one of the following:

1. Chick box lids.
2. Feeder lids—the same size as chick box lids but manufactured especially for this purpose.
3. Plastic feeders—similar in size to chick box lids but made of plastic or some similar material. An advantage is that they may be washed and reused.

Second Feeders

When the chicks are 5 days of age the feeder lids are no longer practical; larger feeders must be substituted. In most instances these new feed containers should be large enough to handle the birds until they are fully grown. Some are automatic; some must be filled by hand. There are several types.

Hand feeders

Trough feeders. Troughs that are 4 to 6 ft (1.2 to 1.8 m) in length are filled by hand. A grill or reel over the trough will help prevent feed wastage and keep the birds out of the troughs.

Tube feeders. These are large tubes about 8 to 16 in. (20 to 40 cm) in diameter and about 2 ft (0.6 m) long. At the bottom a large pan is suspended into which the feed flows from the tube, and from which the birds eat. Tube feeders are usually suspended from the ceiling.

Automatic feeders. Automatic feeders consist of a pan or trough from which the birds may eat, and a mechanism for automatically transferring feed from a central hopper to the pans or troughs. There are many types, but the main classifications are as follows:

Trough and chain. A continuous trough goes around the poultry house and a special chain with cross-cleats drags the feed the length of the trough. A hopper at one end of the house or pen acts as a reservoir for the feed.

Conveyor-and-pan system. An auger or chain pushes feed through a tube or trough. Openings are made at intervals in the tube so that the feed drops out into circular pans. Some automatic method is employed to stop the flow of feed once the pans are full.

Tube and trough system. Feed is moved along a tube that has a feed trough attached to it. Holes in the tube allow the feed to be pushed into the trough at the bottom.

Tube and tube feeder. An auger or cable with discs forces feed around the house through a tube. Other tubes are attached to the main tube at regular intervals. Through these smaller tubes the feed drops into hanging feeders.

Type of feeder and age of birds. The "second feeder" must be of a size and type that will be satisfactory for chicks as young as one week of age. If the trough is too large it will be necessary to use another feeder between the time the feeder lids are removed and the growing feeders are used. This will increase the investment in equipment. Chicks will get in the trough to eat when they are young rather than eat from the sides, which is not a disadvantage and should be encouraged. As the flock grows, the feeding system should be raised to maintain the feed level at the same height as the back of the chickens. An electric wire can be attached above the trough to prevent roosting on the feeder.

Feeder Time Clocks

Rather than operate the automatic feeding equipment continuously, it may be connected with a time clock that runs the feed-moving mechanism intermittently. Time clocks are used to start and stop the equipment. This procedure lengthens the life of the feeding equipment, particularly the tubes and troughs.

Artificial Light

Regardless of the type of poultry house, some method of providing artificial illumination will be necessary. The amount and type of lighting is a detailed subject, and is covered fully in Chapter 18.

Carrier

In many instances a *carrier* in the poultry house will facilitate the movement of equipment and other items throughout the length of the poultry house. Such a carrier will consist of an overhead track with a suspended platform. Correct placement of the carrier track in the house will be determined by the jobs for which it is to be used. In the brood-grow-lay house it should be located near the nests so that it may be used to transfer eggs.

Bulk Feed Tank

Bulk feed has replaced bagged feed in most commercial poultry areas of the world. As a reservoir for feed, tanks are set next to the poultry building. Their size should not be governed by the feed consumption of young chicks, but rather by the amount eaten by the oldest birds the house will contain. Each tank should hold one week's supply of feed, plus about two days' reserve. It is advisable to have two small tanks rather than a single large one to allow time for cleaning and aeration.

Feed Scales

Most modern methods of feeding chickens involve weighing the feed consumed each day. Many programs call for weighing the daily feed allocation. These requirements necessitate scales that are accurate and practical. Some scales are automatic; some are semiautomatic. Scales may be associated with the feeder system or with the bulk feed tank.

Water Meter

Water meters are used on many poultry farms to measure the amount of water being drunk by the chickens each day. There are a variety of styles, some with recording devices.

12-B. GROWING EQUIPMENT

Most of the equipment used during the bird's growing period will be the same as that provided during the laying period. Careful selection of equipment should be made to make certain that as little as possible of the equipment will be duplicated. There are, however, several items used for growing birds that should be discussed.

Waterers

Most watering systems manufactured today are suitable for birds that are growing and those that are producing eggs. However, any system that uses a deep pan or deep trough will not be satisfactory for young or small growing birds because chickens cannot reach the water. This is particularly true of the arrangement using a pan and float valve. Long, shallow watering troughs through which water runs continuously also are not satisfactory for young chicks, but are acceptable for growing birds and layers.

Feeders

In most instances the same feeder will be used for both growing and laying birds, especially when automatic feeders are used. But some systems pose a problem. Many times the trough or pan used for the growing birds is too small for adult birds, and a larger one must be substituted. Select a type that may be used for both ages.

Specialized feeders for controlled feeding. It is becoming commonplace to restrict the feed intake of growing birds. This is often difficult with automatic feeding equipment as the feeders run and stop alternately and automatically. When the feeder starts, and fresh feed begins to flow through the system, the birds near the feed reservoir get the first feed and eat longer than those at the end of the house. The difficulty is further complicated when there are several pens in the house, operated with one feeder.

To alleviate this problem, some manufacturers of automatic feeding equipment have an arrangement whereby feed flows through a tube until it is completely filled; then the feed drops into pans in all sections of the house at the same time; therefore, all birds have an equal period of feed consumption.

Other special feeders. Tube-type feeders are used for feeding cockerels when males are kept with the females during either the growing or laying period. These feeders are filled by hand and are kept at a height so that males may eat from them, but not the females. They are known as *cockerel feeders.*

Other similar tube feeders are used for oystershell or grit, and are known as *shell and grit feeders.* These should be hung at a low level to enable both males and females to eat from them.

Roosts Not Needed

As a general rule, roosts are not needed in a poultry house, either for growing or laying birds. In fact, it is advisable to provide some ''antiroosting'' device over certain equipment to prevent the birds from roosting on it.

Water Proportioner

A water proportioner is an automatic mechanical device, usually a pump, that draws a medicament, chemical, or vaccine from a container and injects it into the incoming water line. Most of these inject 1 oz (30 cm³) into every 1 gal (3.79 L) of incoming water. By adjusting the dilution of the medicament in the stock solution, more or less of the medicament can be made to enter the water supply.

12-C. LAYING HOUSE EQUIPMENT

In general, the feeding and watering equipment used for the growing birds should be ample when they are producing eggs. This is particularly true in the brood-grow-lay house. But there will be some equipment pertinent to laying birds only, which is described below.

Nests

Nests should be adequate in size, ample, easily cleaned, well ventilated, and dark.

Individual nests.   The floor should be 12 in. (30 cm) wide for egg-type pullets and 14 in. (35 cm) wide for meat-type. Both nests should be about 14 in. (35 cm) deep. Provide one nest for every four hens. Most are of metal, constructed in batteries five nests wide and two or three tiers high with a sloping roof to prevent birds from roosting on them. The bottoms should be removable for easy cleaning. If hot weather is to be encountered, wire mesh across the back is better than solid material. The jump roosts in front should be hinged to make it possible to raise them to close off the nests at night.

Community nests.   These are unpartitioned boxes about 2 ft (0.6 m) wide and 8 ft (2.4 m) long, with an opening at each end through which the birds enter and leave. Each has a sloping cover that is hinged so that it may be opened for gathering the eggs. If the weather is hot, the back side should be covered with wire mesh.

Roll-away nests.   The mechanical collection of eggs involves the roll-out (roll-away) nest. It has a wire or plastic floor, sloped so the eggs will roll to the back and onto a moving belt that delivers the eggs to the end of the house. Usually two rows of nests are placed back to back with a 4-in. (10 cm) belt between them. Astroturf® mats should be placed in the bottoms of the nests; otherwise hens may refuse them. The nests are a single compartment.

Trapnests.   Individual conventional nests are used, but each opening has a trap door that closes when a hen enters. She must be liberated by hand. Trapnests are used for recording the actual number of eggs a hen lays. Each hen is given an identification band or badge with her number on it. Her number is written on the egg when pedigree hatching is involved. Provide one trapnest for every three hens.

Artificial Illumination

The use of artificial light in the laying house is mandatory. This is true not only in the environmentally controlled poultry house but in the open-sided house because it is necessary to increase the length of the light day during the months when hours of natural daylight are less than adequate to sustain normal egg production. See Chapter 18 for complete details.

Broody Coops

It is the natural inclination for laying hens to go ''broody'' after they lay their first clutch of eggs. Broodiness is an inherited characteristic, more pronounced in meat-type breeds than egg-type. A broody hen may be returned to her normal condition if she is placed in a coop with a wire or slat floor. Such broody coops were a necessity years ago. However, the broody factor has been almost eliminated in most strains of chickens through genetic selection and modern breeding methods, and because of this broody coops are no longer used.

Manure-Collecting System

When birds of any age are raised on slats or wire, or a portion of the poultry house floor is covered with slats or wire, large quantities of droppings build up below them. This manure may be allowed to accumulate until the birds are removed from the house, when it is removed; or the manure may be collected regularly by specialized collecting equipment.

Collecting equipment. One method of removing the manure from under slats or wire while the birds are in the house is by the use of a *drag*. At the time the house is constructed, a concrete pit is built under the slats or wire. A steel, motorized drag is used to remove the droppings from the pit. The drag is made to go back and forth pulling the droppings to the end of the house where they are removed.

Litter Removal Equipment

Although it is difficult to remove litter from the house when there are chickens in it, except by manual methods, special equipment may be employed when the house is vacant. Such items as a skip loader, vacuum, and belt-and-loader are common.

12-D. ELECTRICITY ON THE POULTRY FARM

The present demand for electricity on the poultry farm is great. Automation has increased the use of electrical equipment, and environmentally controlled housing has increased the need for artificial illumination and forced air movement by electrically operated fans. In many instances special control panels and heavier wiring will be necessary. Some electrical terms and requirements are given below to provide a better understanding of the factors involved with electricity on the poultry farm.

Definitions

Alternating current.   Alternating current (AC) is current that reverses direction rapidly and regularly. In AC, any one wire is first positive, then negative, then positive, etc. The cycle is completed 50 or 60 times every second, with usual electric power, thus giving rise to 50-cycle or 60-cycle terms for current. Electric motors must match the cycle involved.

Ampere.   The unit of measurement of electric current is the ampere. It is the rate at which current is caused to flow through a resistance of one ohm by a pressure of one volt.

Current.   Current is the flow of electricity in a circuit. Current is measured in amperes.

Direct current.   Direct current (DC) is current that flows in one direction only. One wire is always positive; the other, negative.

Horsepower.   One horsepower (hp) is equivalent to 746 watts.

KVA.   This is an abbreviation for kilovolt-ampere, the product of volts times amperes, divided by 1000.

KW.   Kilowatts (KW) are a measure of electric power. A kilowatt equals 1000 watts.

Kilowatt hour.   The amount of electric power represented by 1000 watts for one hour.

Ohm.   The ohm is the amount of resistance that will permit current to flow at the rate of one ampere under a pressure of one volt.

Single phase.   A single-phase, alternating-current system is one having a single voltage in which the reversals of that voltage occur at the same time and are of the same alternating polarity throughout the system.

Three phase.   A three-phase alternating current system has three single phase circuits (or groups of circuits), each so timed that the alternations of the first are 1/3 cycle (120°) ahead of those of the second and 2/3 cycle (240°) ahead of those of the third.

Voltage. Voltage is the force, pressure, or electromotive force (EMF) that causes electric current to flow in an electric circuit. Its unit of measurement is the *volt*, which represents the amount of electric pressure that will cause current to flow at the rate of one ampere through a resistance of one ohm.

Watt. The watt is the unit of measurement of electric power; 746 watts are equivalent to one horsepower. The watt represents the rate at which power is expended when a pressure of one volt causes current to flow at the rate of one ampere.

Current Requirement of Electric Motors

The running current requirements of electric motors are given in Table 12-2. Motors are rated by horsepower (hp). Ampere and wire sizes are also shown.

Table 12-2. Running Current Requirements of Electric Motors

| Motor Horse-power | 115 volt 1 phase, AC | | 230 volt 1 phase, AC | | 230 volt 3 phase, AC | |
	Current Load[1] Amp	Wire Size[2] Gauge	Current Load[1] Amp	Wire Size[2] Gauge	Current Load[1] Amp	Wire Size[2] Gauge
1/6	4.4	14	2.2	14		
1/4	5.8	14	2.9	14		
1/3	7.2	14	3.6	14		
1/2	9.8	14	4.9	14	2.0	14
3/4	13.8	12	6.9	14	2.8	14
1	16.0	12	8.0	14	3.5	14
1½	20.0	10	10.0	14	5.0	14
2	24.0	10	12.0	14	6.5	14
3	34.0	6	17.0	10	9.0	12
5	56.0	4	28.0	8	15.0	10

Source: National Electric Code.
[1] Full rated load current in amperes (starting load much higher).
[2] Minimum wire size. Larger sizes required when there is a voltage drop in the line.

Wire Size and Electric Load

The relationship between wire size, electric load, and fuse size is shown in Table 12-3.

Table 12-3. Relationship Between Wire Size, Electrical Load, and Fuse Size

Wire Gauge Size	Maximum Watts		Fuse or Breaker Amps
	at 120 V	at 240 V	
14	1,725	3,450	15
12	2,300	4,600	20
10	3,450	6,900	30
8	4,600	9,200	40
6	6,325	12,750	55
4	8,050	16,100	70
2	10,925	21,850	95
0	14,375	28,750	125

12-E. STANDBY ELECTRIC PLANTS

Standby electric plants may be needed on many poultry farms to operate feeders, pumps on water systems, lights, etc. (see Chapter 6-D).

12-F. EQUIPMENT CONTROLS

Historically, equipment was either controlled manually or by thermostats and timers. Today, computer programs are available to monitor house temperatures, feed intake, water consumption, body weight, ammonia levels, etc. These programs can regulate the number of fans operating, the timing of feeding and manure removal, the lighting program, the synchronization of mechanical egg collection from different houses, and the summarization of flock performance records on a daily basis.

12-G. ALARM SYSTEMS

With more mechanization and less frequent flock visitation, management must depend on various alarm systems to notify them of equipment failures. Such systems are used to warn of temperature extremes, power failure, inoperative fans or feeders, and a variety of other problems. The alarm consists of a siren or bell and may be tied into a telephone system with automatic dialing to two or more individuals.

13
Brooding Management

This chapter deals with management factors having to do with the brooding period, usually defined as the first 5 to 6 weeks of a chick's life. Floor management only is discussed here; cage management is treated in Chapter 16; broiler management is discussed in Chapter 20. However, there are many problems discussed in this chapter that have a bearing on cage and broiler operations; it is impossible to separate them completely.

13-A. THE BROODER HOUSE

Production of a good pullet is one of the most important requisites of good management, for how well the pullet is grown will greatly determine how well she will produce in the laying house. Similarly, well-managed growth in a male will exert an influence on his behavior in the breeding pen.

Isolation of the Brooder House

Chicks should be brooded in a house that is not located near other poultry. There is just too much danger of disease transmission to do otherwise. At least 300 ft (91 m) should be allowed between such houses, and a greater distance is preferable. The brooder house must be isolated both by distance and by management. Prevailing air movement must be from the brooder area to other poultry areas and never in the opposite direction. It should be enclosed with a fence at least 100 ft (30.5 m) from the house, and the gate to the enclosure should be kept locked except when in use.

With modern disease-control programs it will be necessary for everyone entering the enclosure to shower and change into clean clothing (see Chapter 37-H, 37-I). Visitors must be kept out of the enclosure unless they, too, shower and change clothing. Contact with other poultry is not permitted.

All-In, All-Out System

Although more than one brooder house may be within the fenced enclosure, the chicks should be of a similar age, the oldest being no more than 7 days older than the youngest. In addition, flocks should originate from a single source whenever possible, meaning from a common parent flock located on one farm. Vaccination and other programs are more complex when chicks are not of a similar age and source. All the chicks should be started the same week; later all should be removed from the house at approximately the same time. This program gives rise to the term, "all-in, all-out," meaning that all chicks are placed in the house (or houses) at one time, and some time later, all are moved out at one time. Another group of chicks is not to be placed in the house (or houses) until all the older birds are moved out and the premises cleaned. This provides a period when there are no chicks within the enclosure, thus breaking any cycle of disease infection.

Field research has shown that up to 20 more eggs per hen were realized with the all-in, all-out system compared with multiple age flocks during laying but a lot will depend on the variables involved (Fred C. Price et al., *Univ. of California Extension Bulletin*).

13-B. PREPARING FOR THE CHICKS

Everything within the fenced enclosure must be cleaned thoroughly before placing a new group of chicks in the brooder house or houses.

Cleaning the House and Equipment

Immediately after the fenced brooding enclosure is depopulated of birds, the house or houses should be cleaned, disinfected, and readied for another group of chicks. Immediate preparation is necessary so that the buildings may lie empty for one or two weeks prior to placing new chicks in them. Disinfection and fumigation will kill most of any disease-producing organisms; an empty house will break the life cycle of most of those remaining. The cleaning process must involve the following:

1. *Remove all old litter.* Used litter should be removed from the poultry house, then hauled away from the premises.
2. *Clean and scrub the house.* All loose debris must be taken from the building, and the floors scraped clean. Any slats or wire should be similarly scraped. Next, a pressure sprayer should be used to thoroughly wash the interior of the house. Following this, it will be necessary to disinfect the entire inside of the building. Use the most powerful concentration of disinfectant, as recommended on the label. Allow the house to dry.
3. *Clean the equipment.* All equipment must be scraped, washed, and disinfected. Dip the smaller items in a disinfecting solution. If possible, the equipment should be moved outside the house to an area inside the fenced enclosure to complete the cleaning process. After cleaning, the equipment should be moved back into the house.
4. *Fumigate.* If the house is tight, or curtains can be closed to keep the fumigant in, the house and equipment should be fumigated, using $3 \times$ concentration of formaldehyde gas (see Chapter 9-J).
5. *Clean and fumigate bulk feed bins.* These should be washed and disinfected, then fumigated at $3 \times$ strength formaldehyde.
6. *Treat dirt floors.* Spray dirt floors with an oil-and-disinfectant mixture, or some commercial product suitable for this purpose.
7. *Clean the grounds.* Remove all weeds and debris from the area outside the house, burn feathers, mow the grass, and make any necessary road repairs. Driveways should be properly surfaced to enable trucks to move over them easily during periods of inclement weather. Spray the area adjacent to the house with an oil-and-disinfectant mixture or a commercial product. If a truck dip vat is involved, it should be emptied, thoroughly cleaned, and fresh disinfectant added.

13-C. PROPER NUMBER OF CHICKS

There are several computations necessary to determine the number of chicks to order for any individual house. Care should be exercised to work out every detail according to the type of management program, breed and sex of bird, etc. These variations are as follows:

Management Program

Will the house be used only for brooding, or will the chicks be brooded and grown in the same house, or will the brood-grow-lay program be followed? The answer will certainly influence the number of chicks that may be started on a floor of a given area. The minimum requirement for floor space for the oldest birds kept in the house will determine the number of chicks started, for older birds require more floor space than younger birds.

Commercial Pullets or Breeder Birds

As a general rule, all ages of birds to be used for breeding purposes should be given more floor space than those used for the production of commercial eggs. Furthermore, males are involved with flocks to be used for breeding purposes, and these need even more floor space. In most instances the cockerels are raised in pens separate from the females until 6 to 10 weeks of age; at times both sexes are kept in the same pen. Thus, the amount of floor space necessary will vary according to the use to which the birds are to be put.

Will the Layers Be Kept on the Floor or in Cages?

Although practically all chicks are brooded under similar conditions during the first 5 or 6 weeks of their lives, the handling of the birds after this period, particularly during the laying period, will determine the number of chicks to be started and housed. Although the more common practice today is to brood and grow commercial laying birds in the same house, the age at which they are moved to their permanent laying quarters will vary. If they are to be kept on the floor during their laying period, they probably will stay in the growing house until about 20 weeks of age. If they are to be transferred to cages they may be moved at any time between 14 and 20 weeks of age. However, the tendency today is to transfer them at an early age, usually about 18 weeks of age. As less floor space is required for 18-week-old birds than for older ones, this alters the bird capacity of the growing house.

In the case of birds that are being raised for future breeding purposes and the production of hatching eggs, there are two ages at which the birds may be housed in their laying quarters.

> 1. *Transfer at 10 weeks of age.* More and more breeder birds are being kept under a MG- or MS-clean status. Breeders are carefully blood-tested at regular intervals to de-

termine their freedom from the disease, but the most likely time for flocks to become infected or *break* is when the birds are coming into egg production. For this reason it is much better to ''house'' the birds far in advance of this age, to avoid as much stress as possible at laying age. Usually the birds are moved at 10 weeks of age.

2. *Brood-grow-lay system.* This management method has become popular with those poultrymen involved with a Mycoplasma program, because the birds are never moved from house to house, thus avoiding any moving stress, and because the poultry house may be completely isolated during the entire life of the birds.

Either of these two alternatives will influence the number of breeder-type chicks that may be placed in the house at 1 day of age, and also the requirements for feeder and waterer space.

Floor Space Requirements During Brooding

The amount of floor space necessary for each chick during the first 5 or 6 weeks of its life is given in Table 13-1. The figure varies according to the type of bird involved, and the purpose to which it is to be put in the laying or breeding pen.

Table 13-1. Floor Space Requirement During the Brooding Period

Type of Bird	Floor Space per Bird		
	ft^2	m^2	Birds/m^2
Mini-Leghorn, egg-type pullets	0.60	0.056	17.8
Leghorn, egg-type pullets	0.75	0.070	14.3
Medium-size, egg-type pullets	0.85	0.079	12.7
Leghorn breeder pullets	0.85	0.079	12.7
Leghorn breeder cockerels	1.00	0.093	10.8
Medium-size breeder pullets	1.00	0.093	10.8
Medium-size breeder cockerels	1.00	0.093	10.8
Meat-type breeder pullets	1.00	0.093	10.8
Meat-type breeder cockerels	1.25	0.116	8.6

Pens and Floor Space

Except for birds to be used as breeders, pens probably are not advisable. Although pens were widely used several years ago, the procedure today is to house birds in larger groups; several thousand may be kept together. Breeder-type birds, however, seem to do better when the unit size is small, with 400 to 500 adult birds being placed in one pen. However, if necessary, up to 1,000 may be brooded together even though the stress factors increase. It is best to keep the breeder males and females in separate pens during the early brooding period because the procedures for culling and handling are different. The use of pens within a house does not alter the floor space requirement shown in Table 13-1.

Effects of Crowding

Chicks should be given adequate floor space during the brooding period; they should not be crowded. The amount of floor space given in Table 13-1 does not involve crowding. Although the amounts vary according to the type of bird being brooded, some birds or sexes require more room in which to move about. Certain chicks, as meat-type cockerels, are costly and small compared with other birds. They necessarily should have more floor space.

A decrease from the recommended floor space usually means an increase in mortality and reduced rate of growth. Although the latter is not of primary importance with birds to be kept for egg production purposes, any increase over "normal" death loss becomes costly. Not only is there a monetary loss involving the cost of the chick but the value of the feed, labor, and other items necessary to grow a chick until the time of death is a direct loss too. There is also the loss of profit that would have been derived from the birds had they lived. A reduction in body weight below that recommended by the breeder is generally associated with poor lay-house performance and must be avoided.

How much mortality during brooding? Normally, chick mortality during the first week in the brooder house is greater than any week thereafter during the growing period, but it should not be over 1%. Losses the second week should be slightly less. Beginning with the third week, deaths should be at a relatively low weekly level and run rather uniformly until the end of the growing period. Most replacement pullets are successfully reared with less than 5% loss to point of lay.

Order Chicks Well in Advance

Chicks should be ordered or planned for weeks, even months, in advance of the time they are to be placed in the brooder house. This gives ample time for the hatchery to plan for the order, and for the customer to have a high assurance that he will get the chicks when he wants them.

First, calculate the number of chicks needed. Usually, the quantity of day-old, female chicks needed is determined by the number of pullets necessary at sexual maturity. To this figure is to be added the number of cull and death losses during the growing period. Also to be considered is the number of *extra* chicks given by the hatchery.

    Example: Number of pullets required at sexual maturity (10,000)
        Number of chicks ordered if net losses are expected to be 1%
        (4% extras delivered minus 5% loss)

$$\frac{10,000}{0.99} = 10,100$$

Growing mortality a factor. In the above example, 95% of the birds started are to be *housed* at sexual maturity. Certainly, the goal of every poultryman should be at least this, but many will have a much higher figure, particularly with egg-type lines. However, others will not do as well.

Extra chicks. Hatcheries have a custom of giving free *extra chicks* to make up for those lost in delivering the order plus a normal amount of early death loss. Between 2 and 4% extra chicks are usually included. However, mortality rates must be calculated on the total number delivered.

When cockerel chicks are needed. If the flock is to be used for breeding purposes once they reach maturity, it will be necessary to calculate the number of cockerels necessary according to the number of pullets, then order the required number of male chicks. Cockerels missed during sexing (usually less than 0.5%) are generally removed as soon as comb development is observed.

13-D. BROODING REQUIREMENTS

A few days before the chicks are due to arrive at the farm the caretaker must begin to ready the poultry brooder house. Everything must be in order.

Litter

There are many types of litter, and an evaluation of these is found in Chapter 20 and Table 20-6. Cover the floor with about 2 in. (5 cm) of the material. The particles should be large enough not to pass a ¼-in. (0.6-cm) sieve.

Treated litter should not be used. Do not use a litter that has been treated with an insecticide, herbicide, preservative, or other chemical.

Delivery warning. Remember that the fenced enclosure and the brooder house have been thoroughly disinfected. Any truck used to transport litter to the house should also be clean before it enters the enclosure. Do not undo a good job of cleaning the house by admitting dirty trucks to the premises. Install a "dip vat" at the entrance. Trucks must drive through a disinfecting solution held in the dip vat.

Litter management.   During the first 3 weeks of the chick's life the litter should be only slightly moist; after that it should contain about 25% moisture. Do not start the brooder stoves until the day before the chicks are to arrive as this tends to dry the litter too much. When chicks are placed on exceptionally dry litter there is a tendency to increase their dehydration. After the chicks arrive, the droppings add moisture to the litter. If the litter becomes too wet, increase the amount of air moving through the house. If increasing the air movement does not dry the litter, more litter should be added, mixing the new with the old. But be careful. Often molds grow under wet litter; stirring or turning only exposes more moldy material to the chicks. Remove caked or wet litter before adding more.

Feathering and litter condition.   There must be some humidity in the poultry house. Chicks do not grow or feather well in a dry atmosphere, and the moisture in the litter tends to influence the amount of dampness in the air. However, the area near the heat supply has a tendency to get wet because the young birds frequent the area more, and because the warmed area is relatively small; but other parts of the house may be too dry. Both may become problem areas.

Brooder Guards

Brooder guards should enclose the heated area, either the brooder stoves, the hot-water pipes, or the heated slab. The type of canopy brooder will determine the distance the guards should be from the edge of the hover-type stove. The guards will need to be farther from the blast-type stove than the conventional, but normally the distance should be about 30 in. (76 cm) in winter and 36 in. (91 cm) in summer.

Increase the area within the guards.   As soon as the chicks learn the source of the supplementary heat, the guards must be expanded to allow a greater area inside them. Begin increasing the area on the third day. Guards should be used for 6 to 9 days, after which they may be removed. Often the guards may encircle two or more brooder stoves after the fourth or fifth day and larger waterers and feeders placed inside the rings. Many pullet chicks are beak trimmed and some males are toe-clipped at 6 to 10 days of age; if this is practiced the chicks should remain in the guards until these jobs are completed.

Guards with room-type heating.   Although chicks cannot move away from warmth when the entire brooder house is heated, they may not be able to find the feed and water if they are allowed to stray over the entire floor area. It is just impossible to provide enough feeders and waterers. Therefore, guards will be necessary to restrict the chicks to a small area where the feed and water are more easily available.

Attraction Lights

A 7½-watt white or red light bulb should be used under each hover. The presence of brooder guards does not remove the need for such a light. It should be used for 2 or 3 days only; chicks will learn the source of heat during this time.

House Ventilation for Hover-type Brooding

There need be but little air movement through the brooder house during the first few days; three to five air changes per hour will be ample. The house temperature should be kept at about 75°F (24°C) the first 4 or 5 days, after which it should be lowered to 65° to 70°F (18–21°C). Chicks do better in a relatively cool environment if there is a place they can go to get warm.

Ventilating the open-sided house. Curtains should have been installed on this type of house. When chicks are placed in the house, the curtains should be closed. As the chicks grow older the curtains may be opened more during the warm hours of the day and closed at night. Weather conditions and the age of the chicks will determine the daytime and nighttime opening size. There must be no drafts in the house.

Avoid disaster: Ventilation in the open-sided house is difficult to control. But regulating the amount of air going through the poultry house is most important. Changing the curtains according to the outside and inside temperature is a prime factor in raising good birds. Its importance cannot be overestimated. Alterations may be necessary several times a day. Just opening the curtains in the morning and closing them at night is not enough. Improper ventilation and house temperature will only lead to difficulties in the flock, sometimes to disaster. Mechanical curtain controls operated by a thermostat are much more effective than manually operated ones.

Use plastic curtains inside the house: When the outside temperature gets very low it may be impossible to keep the temperature of the brooder house high enough for chick comfort. Even the surplus heat from the brooder stoves will not warm the house to an optimum temperature. A remedy in some instances is to hang two clear plastic curtains from near the ceiling to near the floor, extending the length of the house. One is hung on each side of the area in which the stoves are located, and just outside the guards. In this manner only about half of the floor area is used until the chicks get older and can withstand cooler conditions. Be careful, however, that the ventilation inside the curtains is adequate. Leave some air space at the top and bottom of the curtains. Remove the curtains when the chicks are 2 to 3 weeks of age.

Half-room brooding. If the brooders are placed in the middle of the poultry house, leaving the ends empty, only about half of the house will be used for brooding. Reducing the brooding space decreases the fuel necessary to brood the chicks. Reductions of up to 35% have been demonstrated.

Ventilating the environmentally controlled house. After the chicks are a few days old, air movement through the brooder house will need to be increased to furnish a proper environment. The size and number of birds in the building are the factors used in determining proper airflow (see Chapter 11-K).

Light During the First 48 Hours

In the brooder house, artificial light must be used to supplement natural daylight during the first 48 hours after the chicks arrive on the farm. Chicks should receive a total of 23 hours of light at 3.5 fc (35 lx) of illumination at the floor during this period. This amount will be supplied by approximately 3.5 watts of light bulb for each 4 ft^2 (0.37 m^2) of floor space when the bulb is approximately 8 ft (2.4 m) above the floor, and under a good clean reflector. This is very bright illumination. It should reflect off the drinking water to attract the chicks.

A 23-hour program is preferable to 24 hours because it acquaints the flock with periodic periods of darkness. Power failures can frighten a flock not used to total darkness. The first 2 days is a critical period in the flock's life and it is absolutely essential that all chicks learn to drink and eat as soon as possible.

Light After the First 48 Hours

After the first 2 days, the intensity of the light in the brooder house should be reduced. At floor level it should be about 1 fc (10 lx). Supply approximately 1 watt of bulb for each 4 ft^2 (0.37 m^2) of floor space when the bulb is 8 ft (2.4 m) above the floor, and under a good, clean reflector.

Brooder Capacity

The brooder capacity is related to the area of the floor that is warmed by the heat unit. The size of the canopy has little to do with it, as some heating devices deflect the heat farther than others. The type of chicks being brooded will also alter the number that can be placed under one unit, as follows:

Commercial and meat-type pullets. Most canopy-type brooders have a capacity of 500 chicks, and this is the number usually placed under each unit, although certain brooders will handle up to 750. Do not place more pullet chicks than the brooder will handle at 4 or 5 weeks of age.

Egg-type breeding cockerels: Place about 400 cockerel chicks per stove.

Meat-type breeding cockerels: Place about 350 cockerel chicks per stove.

Central heating. Allow approximately 6 in.^{2} (38 cm^2) per pullet chick and 8 in.^{2} (52 cm^2) per cockerel chick under hot-water hovers. The recommendation will vary according to the height of the pipes above the litter. The shorter the distance, the more space required per chick because the nearness to heat tends to drive the chicks out from under the hover and they will not use the entire area below the warm pipes.

Slab heating. When a warmed slab of concrete is used as the source of supplementary heat, provide about 20 to 25 in.^{2} (129 to 163 cm^2) of heated slab per chick to be brooded.

Brooding Temperature

It is difficult to recommend any brooding temperature applicable to all types of brooders and all conditions. Usually, however, a temperature of 85° to 90°F (30° to 32°C) at a point of 6 in. (15 cm) outside the canopy and 2 in. (5 cm) above the top of the litter is satisfactory for chicks at 1 day of age. Use the higher temperature for meat-type, breeder, cockerel chicks. As chicks grow older the temperature may be reduced at the rate of about 5° to 7°F (2.8° to 3.9°C) per week. But a thermometer is a poor tool for measuring chick comfort. The chicks themselves should be the indicator. At night they should bed down just outside the edge of the hover and completely circle the brooder. If they are too far out, the temperature is too high; if too far in, the temperature is too low. Thermometers should be used before the chicks are placed under the hovers, but after the first 2 days, thermometers may be removed and stored. Chicks should be fully feathered before supplementary heat is removed.

Slab brooding temperature. The usual starting recommendation for slab brooding and warm-room brooding is about 85°F (27°C) at a point 2 in. (5 cm) above the top of the litter.

Caution on warm-room brooding. Extreme care should be taken to reduce the brooding temperature in warm-room brooding. With heat in the entire house, the chicks have no way of getting away from it if they are too warm. Furthermore, they have no place to go if they are cool and require additional heat. If the temperature is correct, the birds will spread out evenly over the floor or in small congregations at night; they will not huddle or extend their wings.

Guards relate to brooding temperature. Although brooder guards are necessary to confine day-old chicks to the heated area, and to concentrate the feeders and waterers within a small area, often the guards are responsible for chick injury. Although solid guards prevent a great deal of draft over the chicks during cool weather, the birds may develop a custom of congregating against them the first day to keep warm, rather than going to the heated area of the brooder stove. During hot weather there is a tendency for the stove to become too hot, forcing the chicks away from the heat and against the guards, where they pile in their endeavor to get away from the heat. For this latter reason mesh wire, rather than solid material, is recommended during hot weather. Be careful to adjust the brooder heat so the chicks stay away from the guard.

Watch the chicks. During the first 2 weeks of life, chicks are easily stressed. Sudden changes in temperature, light, ventilation, water and feed supply, and noise will produce observable deterioration in the quality of the birds and in some cases may lead to piling and mortality due to suffocation (see Chapter 39-A). Frequent visits to the brooder house are necessary during this period. Watch every detail; do not make a casual observation. Check the house at night and see that everything is normal.

Increase brooder heat if needed. Be ready to increase the brooder heat if chicks appear chilled during a stress created by vaccination, beak trimming, etc. Keep the birds comfortable at all times.

13-E. WATER

Chicks must learn quickly to drink and eat. Although they can get along without water and feed for up to 3 days after hatching, such a delay will be detrimental. Any postponement only dehydrates and weakens them, and weak chicks do not learn to drink and eat as rapidly. For best results, chicks should be given water about 24 hours after they are hatched.

Function of Water

Water is important for it serves many functions in the chicken, some of which are

1. It helps to cool the bird by evaporation through the lungs and air sacs.
2. It makes up a high percentage of the body.
3. A high percentage of the egg contents is water.
4. It aids in softening the feed in the crop and forms a carrier during its passage through the alimentary tract.
5. It aids in certain digestive processes.
6. Water is an important part of the blood and lymph.

Availability and Temperature of Water

Number of waterers. Provide two fount-type chick waterers for every 100 pullet chicks the first 2 weeks. With 500 chicks to a stove, this will mean ten waterers per brooder. The waterers should be placed just *outside* the edge of the hover and on the litter so that the water level will be convenient to the chicks. After 2 days the founts should be placed on stands about 1 in. (2.5 cm) high to keep litter from getting in them. Availability of drinking surface is the criterion with day-old chicks rather than the amount of water in the founts; several small waterers are better than a few large ones. There are also a small cup and trough that may be placed at floor level.

Water temperature. Fill the waterers about 4 hours before the chicks arrive. This allows time for the brooder heat to warm the water. The water temperature should be 65°F (18°C), and over. Do not use stale water. Water must be fresh.

Provide Clean Drinking Water

Clean and disinfect the waterers. Chick founts and other waterers should be cleaned with a scrub brush each day. Then the old water should be removed from the building. A disinfectant must be added to the scrub water to facilitate destruction of microorganisms and to aid in preventing mold growth.

Sanitizing the water. Many commercial poultrymen use a sanitizer in the drinking water for varying periods of time. The procedure may or may not have merit in reducing losses from disease. But such a practice is detrimental if water-type vaccines are to be used as water sanitizers reduce the efficacy of the vaccine, sometimes causing it to become completely ineffective (see Chapter 39-C).

Sugar-water. The addition of sucrose (sugar) to the first drinking water has been shown to reduce growing mortality but not to alter chick weight. An 8% sugar solution is usually provided for the first 15 hours after the chicks are placed in the brooder house. To get an 8% solution, dissolve 3 lb (1362 g) of sugar in 3 U.S. gal (11.3 L) of water.

Water with soluble vitamins and electrolytes. If chicks are stressed when arriving at the farm, water-soluble vitamins and electrolytes may also be added to the drinking water for the first 3 or 4 days. Contact your supplier for such a mixture.

Give Water Before Feed

Water should be provided for about 3 hours before feed is given. It is best that chicks drink before they eat inasmuch as this procedure reduces dehydration.

Be sure the chicks are drinking. Just to provide an adequate number of waterers is not enough; chicks must drink soon after being placed under the brooder. Be sure *all* chicks are drinking. Too often it appears that they are getting water, but close observation will reveal that some are not getting to the founts. If this is a general problem on the farm, increase the number of waterers and check to see if the light intensity during the first 2 days is great enough.

Minimum Waterer Space

Once the jug-type waterers or "mini-troughs" are removed, Table 13-2 shows the waterer space required during the remainder of the brooding

Table 13-2.   Waterer Space for Pullet and Cockerel Chicks First 6 Weeks

| | Automatic 8-ft (2.4-m) Trough per Bird | | Number Founts per 1,000 Birds | | |
| | | | 8-ft (2.4-m) Troughs | 13-in. (32.5-cm) Dome Type | 1- to 2-in. (2.5- to 5-cm) Automatic Cups or Nipples |
Type and Sex	in.	cm			
Egg-type pullets or cockerels	0.60	1.5	3	12	75
Meat-type breeder pullets	0.75	1.9	4	15	100
Meat-type breeder cockerels	1.00	2.5	5	20	125

period (5 to 6 weeks). The space requirements for trough waterers assumes both sides are accessible.

Water Consumption of Young Leghorn Pullet Chicks

The daily water consumption of 1,000 Leghorn pullet chicks at 70°F (21.1°C) for 14 days is given in Table 13-3. This table is also of help in preparing and administering water vaccines and water medicaments. These figures vary greatly as the house temperature changes as shown in Chapter 14 and Table 14-3.

Table 13-3. Water Consumption per 1,000 Leghorn Pullet Chicks per Day

Age in Days	Water Consumption		Age in Days	Water Consumption	
	U.S. gal	L		U.S. gal	L
1	2.2	8.3	8	5.2	19.6
2	2.5	9.5	9	5.9	22.3
3	2.8	10.6	10	6.7	25.4
4	3.2	12.1	11	7.6	28.8
5	3.6	13.6	12	8.7	32.9
6	4.1	15.5	13	10.0	37.9
7	4.6	17.4	14	11.5	43.5

Metering Water Consumption

Small meters are available for determining the water consumption of a flock of chickens and are to be installed on the incoming water line of each house. Water consumption is important and these meters offer an excellent means of determining the daily water intake of the birds. A sudden drop in water consumption is often the first indication of trouble in the flock. A large increase in water usage is usually associated with leakage problems.

13-F. FEED

Although the subjects of nutrition and feeding are covered in detail in Chapters 24 through 33, there are certain items that are best discussed in this section. Most poultrymen have little to do with the formulation of the feed they give to their chickens, but they do have a direct responsibility to see that the feed is supplied clean and in amounts adequate to furnish the nutritional needs of the bird. This in itself is a very scientific subject and involves many factors.

Feed delivery. Fresh feed is essential. It should be delivered to the poultry house at least once a week. Bulk feed tanks should be of a size that will handle 7 days of feeding with about 2 days of reserve supply. Tanks should be located outside of the house so that truck drivers do not have to enter the house.

Clean feed trucks. Many programs of poultry production are built around complete cleanliness and isolation of the birds. Anyone entering the premises must first shower and change to clean clothing. But how to properly disinfect a feed truck at the time feed is delivered is one of the most baffling of management problems, as well as one of the most important. To disinfect the tires, the truck may be driven through a dip vat containing water to which a good disinfectant has been added that still retains its disinfecting properties. Although it is probably impossible to completely disinfect the trucks that deliver the feed, extreme precautions should be taken to do the best job possible, for feed trucks represent an important avenue for the entrance of disease-producing organisms into the poultry house.

The first feed. For the 6-week brooding period, chicks need to have feed available at all times. During the first few days of their lives it is important that the feed be *easily* available. For this reason, large, flat containers make the most suitable feeders. Inverted chick box lids or similar trays should be used. Supply one such feeder for every 100 chicks. Some use clean egg flats. Provide two of these for each 100 chicks. Give the first feed approximately 3 hours after the chicks arrive, sprinkling it over the entire area of the feeder lid or container. Feed should be fresh. Feed little and often the first few days.

All chicks must eat. Be sure that each chick is eating at the start. Provide plenty of light on the feed to make it easier for the young birds to see. Keep the brooder heat high enough so the chicks do not have to stay under the hovers—away from the feed—to keep warm. Place the feeder lids outside of the area covered by the brooder canopy so as to provide better light. This procedure also prevents feed from drying as much as when the feed is placed under the hovers and closer to the heat supply.

Watch for "starve-outs." Some chicks will have difficulty in learning to eat, resulting in what is known as "starve-outs." Examine some of the seemingly timid chicks the first day to see if their crops are full. After a few hours on feed the first day a chick should have a well-filled, distended crop. If not all the chicks are eating, see where the trouble lies, and correct it.

How to help prevent vent pasting. On certain farms and during some seasons of the year chicks will develop a laxative condition leading to *vent pasting* during the first few days after they begin eating. Often this may be avoided if the chicks are given some cracked corn (or cracked wheat or cracked milo) as part of their first feed. Preferably, the corn should be cracked with a roller, then screened to keep only the smaller particles while eliminating the powdery material. It should be sprinkled on top of the regular first feed. Feed the cracked corn for only 2 days at the rate of about 10 lb (4.5 kg) for each 1,000 chicks.

13-G. WHEN THE CHICKS ARRIVE

Preparedness is of great importance in getting ready for the chicks, but once they arrive at the farm many management factors are involved. This list is long but necessary; a complete understanding of each item is the manager's responsibility.

Lock the House

During the cleanup period and thereafter, the fenced enclosure and the house must be kept locked. No visitors should be allowed unless they shower and use clean clothing. Poultry house doors should be self-locking once the door is closed. These have the advantage of keeping other people out once the attendant is inside the building.

Chick Delivery

Chick trucks go from farm to farm. Normally, they are cleaned and fumigated at the hatchery after every delivery. But make sure; ask the driver. Furthermore, the cleanliness of the driver should be an issue. Is he dressed in clean clothing? If there is any doubt, he should not be allowed in the fenced enclosure; chicks should be taken to the house by farm personnel who have showered and put on clean clothing.

Destroy the chick boxes. Unless the chick boxes are made of plastic, or some similar material, and are reusable, they should be destroyed on the farm. Nowadays, with so much dependent on disease-control programs, it is poor practice to return fiberboard boxes to the hatchery.

Chick Arrival Time

Chicks should arrive at the farm early in the morning so they will have the entire day to learn to drink and eat and be under close observation. Request this of the hatcheryman. During hot weather it is important that the chicks be transported during the cooler hours of the night, arriving at the farm soon after sunrise.

Examine the chicks. Make an examination of several boxes of birds, checking on mortality, quality, and thriftiness. Ask the driver for a delivery form and make any notations as to number and condition of the chicks at the designated space on the form. Both the driver and the customer should sign.

Dump the chicks quickly. Remove the lid and tip the boxes over quickly, dumping the chicks near the brooder heat.

Day-old vaccination. Some day-old vaccination may be needed. Usually this is done in the hatchery; but it may be done in the brooding house. This will necessitate a slow process of removing chicks from the boxes, vaccinating them, and transferring them to the brooder area—another reason for an early arrival of the chicks. If vaccinating crews come on the farm to do the work, be certain that they shower and dress in clean clothing before entering the premises.

> *Caution:* Dispose of all empty vaccine vials after vaccinating is completed. Put them in a solution containing a disinfectant or burn them.

13-H. LARGER EQUIPMENT

Chick founts and flat-type feeders are essential for starting chicks, but their usefulness is short-lived; soon larger equipment must be substituted.

Make changes gradually. Chickens are creatures of habit; they dislike having their daily routine changed. Any variation in a management procedure must be made over a period of time. When changing equipment within the poultry house, leave the old equipment at its usual location for several days after the new is added, or slowly move the old equipment to the location of the new.

Larger Waterers

Leave the small jug-type chick founts or "mini-troughs" inside the guards until the birds are 7 to 10 days of age; then begin taking a few out each day and substitute larger ones so that all small founts will be removed by the time the chicks are 10 to 14 days old.

Waterer space. Water space requirement during the brooding period is shown in Table 13-2.

Waterer height. The second waterers should be placed first on the floor, then gradually raised as the chicks grow older. Finally, adjust them so that the water level is even with the tops of the backs of the birds. Birds have a tendency to regurgitate their feed when the water level is lower.

Larger Feeders

When the chicks are 3 days of age, add larger feeders. Then gradually remove the feeder lids, taking a week to complete the job. In most cases automatic feeders may be used after the third day by placing the brooder guards just outside them, but if this is not possible, it will be necessary to use small trough-type feeders until the guards have been removed and the chicks have access to the automatics.

If hanging feeders are to be used during the growing period, they should first be set on the floor inside the brooder guards. Regardless of the type of larger equipment used, chicks will get into the feeders at the start, but this is not to be discouraged until the birds are older. At the start the feeder trough or pan should be kept full of feed; later, the feed depth should be reduced so as to keep only the chain covered in automatic troughs with chain drags. In pan-type automatic feeders, keep a minimum of feed in the pans. About 1 in. (2.5 cm) is ample. When the feed is too deep, more is wasted as the birds beak it over and out.

Feeder space requirement. Provide the following feeder space, remembering that "space" is for one side of the trough only. Allow 2 in. (5 cm) of space per chick for the trough. About 20% more chickens can eat from a circular pan than from a straight-line trough of the same distance. For round pans, provide 1.6 in. (4 cm) of feeder space per chick.

> 12-in. (30.5-cm) diameter pans: Supply four for each 100 birds.
>
> 16-in. (40.6-cm) diameter pans: Supply three for each 100 birds.

Location of the feeders. Feeders, as well as waterers, should be distributed uniformly throughout the house. No bird should have to go over 10 ft (3 m) to get either feed or water.

When the slat-floor house is used. In many instances, houses will have a part of floor covered with slats. This raises a problem as to where to place the feeders and waterers for the young birds. Older birds are usually fed and watered from automatic equipment placed on top of the slats. But inasmuch as young chicks will not use the slats until they are several weeks of age, it will be necessary to provide second feeders and waterers that can be placed on the litter floor.

Check for feed wastage. Regularly examine the litter around the feeder to see if feed is being beaked from the feeder, and wasted. A small percentage lost increases the feed cost by a similar percentage. If feed is being wasted, lower the level of feed in the troughs or pans or raise the feeder system.

Grills and reels on feeders.   Many feed troughs have either a wire grill or a reel over them to prevent birds from getting into the feeder. Although most of these serve their purpose, care should be taken that they do not restrict the feed intake.

13-I. BEAK TRIMMING

Cannibalism is prevalent among chickens of all ages and some method of preventing this vice should be used. The common procedure is to trim their beaks, but the age at which this is best done is controversial even though the operation may be completed at any time prior to the onset of egg production, but not always with comparable results. The criteria for a good beak trimming job are twofold:

1. Create as little stress as possible.
2. The beak should not grow out again.

Beak trimming must be considered a precision operation and experience is a great asset in doing it properly. Too often it is done carelessly, creating more stress than necessary when the beak is cut too short, or by not removing enough, which allows the beak to grow and eventually regain a near-normal length.

Advantages and Disadvantages of Beak Trimming

There are advantages and disadvantages to beak trimming, but certainly the advantages far outweigh the disadvantages. Cannibalism in a flock, either through feather pulling, picking, vent nipping, or fighting sometimes leads to disaster. The vice gets to be habit-forming; a few cannibalistic birds can instill the character in others and soon the entire flock is affected. The advantages of beak trimming are as follows:

1. Toe picking is reduced.
2. There is less stress in the flock.
3. It helps to prevent feather picking and cannibalism.
4. Feed efficiency is improved as a result of less wastage.
5. Livability is better, with fewer culls.
6. There is more uniformity of the birds in the flock.

The disadvantages are

1. Birds lose weight for 1 to 2 weeks after trimming.
2. Growth rate is reduced for a long period; it will take from 10 to 20 weeks for a bird to attain the weight of a similar nontrimmed bird.
3. Beak trimming may slightly delay sexual maturity, reduce body weight at sexual maturity, reduce egg production rates, and may reduce egg size.

Beak Trimming with a Hot Blade

To properly trim the beak of a bird, a part of the upper and lower mandible is removed with an electrically controlled cauterizing blade having a temperature of 1500°F (815°C). With some procedures the machines are automatic to the extent that when the beak of the bird is held against a trigger, the blade drops down to make the cut. The hot blade not only severs the beak but acts as a cauterizer, destroying the tissue responsible for generating beak regrowth. With any method of hot beak trimming, it is extremely important that the chick's tongue not be burned. Follow instructions carefully.

Beak-Trimming Methods

There are several methods and several ages when beaks can be effectively trimmed. Often the age will determine the procedure selected.

Block trimming at 6 to 8 days. This is one of the better procedures for pullets to be used for egg production, but is not equally effective among various Leghorn strains. The chicks are easy to handle at this age, and the procedure is rapid. But the method calls for extreme refinement, and is often termed *precision beak trimming.* Considerable experience is needed to do a good job and many have had failures. However, when the beak is cut and cauterized correctly it will not grow back; there is no need for a later touch-up.

The chicks are trimmed with an electric beak-trimming machine having a guide plate with an $^{11}/_{64}$-in. (0.44-cm) hole. A guide with a larger hole should be used if the chicks are larger than normal. The chick is held with the thumb on the back of the head and the forefinger under the throat. The closed beak is inserted in the hole and light pressure exerted on the throat to pull back the tongue. The beak hits the trigger which allows the hot blade to drop down and automatically make the cut. There should be the thickness of a nickel (U.S. five-cent coin) (2 mm) between the cut of the upper beak and the nostrils. The lower beak should be slightly longer than the upper, accomplished by tilting the chick's head downward at the time the beak is inserted in the hole. The severed beak must be kept in contact with the blade for exactly 2 seconds. This is accomplished by a timed cam built into the trimming machine. If held longer, the beak may be burned and can be permanently damaged; if held shorter, the beak may grow back.

Trim accurately, and not more than 15 chicks per minute. Change blades every 3,000 chicks. Keep the guide holes clean so that each chick receives the same treatment. Be sure not to sear the eyes when trimming.

When the birds mature, the upper and lower beaks will be well rounded with the upper beak slightly shorter than the lower. This procedure can only be used with chicks between 6 and 10 days of age.

Side-type beak trimming at 6 to 8 days. With this type of trimming a special cauterizing blade is used. The bird is held on its side and both beaks are cut and cauterized simultaneously and with an inward slant so that the inner portion is deeper than the upper and lower. Keep the blade in contact with the beak for 1 second. The theory behind this method is that the inability of the bird to close its beak tightly at the end makes it difficult to pull feathers or become cannibalistic. Acceptance of the procedure has been slow, mainly because it offers little, if any, advantage over block trimming. The method takes more time and body weight recovery is usually longer.

High-speed trimming at 1 day of age. This method uses a high-voltage arc across two electrodes to burn a hole in the upper beak. Up to 2,000 chicks can be handled per hour. In a few days, the front of the upper beak begins to drop off, and is entirely gone by the tenth day, and healed by the fourteenth. Only that portion between the burned hole and the head remains.

Notch-type Trimming at 1 day of age. A special method of hot-notch trimming has been developed that is used mainly for broilers at 1 day of age at the hatchery. Rather than sever the beak, a special blunt blade burns an area near the tip of the upper beak. The "blade" has been designed to leave a thin part to the inside of the upper beak. This makes it possible for the chick to eat inasmuch as the beak does not have a sensitive raw tip. The end of the upper beak will gradually drop off without apparent injury to the chick, leaving a trimmed upper beak and a normal lower one.

Beak trimming between 6 and 12 weeks of age. Many producers of egg-type pullets are unable to obtain lifetime control of cannibalism by only one early trimming and have added a second or final trimming at 6 to 12 weeks of age. Trimming at this age results in a more permanent cutback of the beak, but it is also generally more stressful in terms of its effects on body weight.

Each method of trimming has a different effect on body weight, but the merits of one program over another must be evaluated on its successful reduction of picking and cannibalism and how well the flock performs in the laying house. California research (Table 13-4) shows significant improvements in performance when birds are trimmed at 6 weeks of age compared with 12 weeks.

How to do it: Use an electric beak-trimming machine. For birds between 6 and 12 weeks of age, cut the upper and lower beaks about 3/16 in. (.45 cm) in front of the nostrils, cutting the lower beak slightly longer than the upper. Cut one beak at a time. Cauterize the area and round the corners thoroughly.

18-week trimming. As a last resort, birds may be trimmed 2 weeks prior to egg production, but do not remove as much of the beak. Trim each mandible separately and the lower beak of pullets should be no more than 1/8 in. (.3 cm) longer than the upper beak.

Touch-up trimming. If the early trimming has not been done properly, many beaks will partially grow back by the time the pullets are 8 weeks of age, or older, and will need to be touched up. This procedure is common and is usually done on a bird-by-bird basis.

Beak trimming cockerels. Cockerels to be used for breeding purposes, as adults, should be trimmed also, but the procedure is somewhat different, according to age.

Table 13-4. Beak Trimming Age and Layer Performance

	Age of Beak Trimming							
	6 weeks				12 weeks			
	Strain of White Leghorn							
Measurement	A	B	C	Avg	A	B	C	Avg
Hen-day production (%)	74.5	73.0	74.3	73.9	72.6	72.4	72.0	72.3
Eggs/hen-housed	221	221	222	221	212	219	209	213
Avg egg weight (g)	62.2	61.8	62.5	62.2	62.8	61.5	62.8	62.4
Daily feed intake (g)	109	107	110	109	108	104	106	106
Total egg mass (kg)	13.8	13.6	13.9	13.8	13.3	13.5	13.1	13.3
Adult mortality (%)	7.5	5.2	6.3	6.3	8.3	4.0	9.5	7.3

Source: University of California, 1986.

6 to 8 days: Block trim, the same as for pullets, except that both the upper and lower beaks should be cut the same length.

2 to 10 weeks: Remove no more than ⅓ of the beak, cutting the upper and lower the same length.

18 weeks: Cut the two beaks back to the quick only. Do not trim mature male birds in the breeding pen.

Avoid Trimming Troubles

Don't trim beaks when birds are under stress. Don't trim when other stresses are present. Never combine beak trimming with moving birds or with vaccinations.

Note that sulfa causes profuse bleeding. Certain sulfa drugs, particularly the sulfonamides, are known to prolong bleeding after beak trimming. If there has been such trouble in the past, give water-soluble vitamin K in the drinking water for 2 days prior to trimming.

Trimming machines are source of infection. Beak-trimming machines must be cleaned and disinfected regularly to prevent cross-contamination among chicks. Although the hot cauterizing blade obviously kills all organisms with which it comes in contact, there are other areas of the machine that soon get contaminated. Mortality in the chicks is increased about 10 to 12 days later. This is particularly true when trimming is done at 1 day of age. Organisms isolated from trimming machines not sanitized have been Salmonella, Staphylococcus, Aspergillus, and Coliform. It is well to check the residue on trimming machines for bacteria and other organisms. Instigate a monitoring procedure.

Watering devices may not be satisfactory. Newly trimmed chicks will have difficulty tripping the valves of trigger-type waterers and drip-type waterers. If these are to be used as second waterers, be sure the chick founts remain in the pens for several days after trimming. Cups may have to be hand-triggered for a few days.

Increase depth of feed in troughs. This prevents the newly trimmed bird from striking the bottom of the feed trough with its cauterized, tender beak when feeding. Striking the tender beak would certainly be a deterrent to normal feed consumption.

Caution: When birds are older than 10 days, proceed as follows: Do not trim during a vaccination period; supply feed immediately; increase the level of both feed and water; trim during the cool part of the day; add vitamin K to the ration during hot weather, or if profuse bleeding occurs, start 2 days prior to trimming; catch a small number of birds at a time; do not move the birds to new facilities.

> *Note:* Crews must be carefully monitored. Beak trimming is a very tedious task and procedures must be followed properly. Blades must be sharp and maintained at proper temperatures. The amount of beak removal must be consistent from bird to bird. Cauterization procedures must be uniformly applied. Pullets should be evaluated constantly during the operation and subsequently in the laying house.

13-J. OTHER JOBS DURING THE BROODING PERIOD

Toe Clipping Males

The inside and back toes of all breeding males should be clipped to prevent tearing the backs of the females during mating. A special piece of equipment is best suited for this purpose (see Chapter 6-F). Clip the toes at the outer joint, just above the toenail. Although the clipping is best done in the hatchery, toes may be cut when the 6- to 8-day beak trimming is done.

Toe Clipping Females

Few female chicks are toe clipped, but research work has shown that the procedure of removing the front, middle, and back toes at the outer joint at day old may have merit under certain conditions. When the laying pullets were housed in cages, toe clipping at 1 day of age increased egg production by 15 eggs per bird per year. However, a great deal depends on the results obtained with a particular strain of birds under certain housing conditions. If trouble has been common, one might expect some remedial effect; if there has been no trouble, it is obvious the improvement in egg production will not be as great.

Dewinging

Cutting off the end of the wing at the carpohumeral joint when chicks are one day of age is not recommended. Although a common practice with turkeys to prevent flying, it produces severe stress in chickens, reducing the rate of lay and income over feed and chick cost, while increasing growing and laying mortality, and feed per dozen eggs.

Dubbing

Removing the comb, or *dubbing*, is a practice followed by many poultrymen. It prevents the comb from being injured during fighting and picking, provides better vision, and lowers the damage done by frostbite. It also reduces injuries incurred when the comb comes in contact with feeder and waterer grills and reels, and with wires in cages. Normally, all cockerel chicks should be dubbed. On a farm where comb damage in pullets has been excessive, they too should have their combs removed. Chickens should not be dubbed in warm climate areas since the comb functions to eliminate excess body heat.

How to do it: Dubbing is best done when the chicks are 1 day of age, although it may be completed during the first few weeks of their life, but hemorrhaging is often severe after the first day.

With a pair of manicuring scissors, cut the comb off close to the head of the day-old chick, running the shears from the front to the back of the comb. The concave side of the scissors should be up.

Vaccinating During the Brooding Period

There are several vaccinations that may be given before the chicks are 6 weeks of age. Programs for vaccinating are discussed in Chapters 37 and 39, according to specific diseases.

13-K. PROBLEMS DURING BROODING

Stress

Stress may be brought on by a variety of conditions at any time of the bird's life. It must be dealt with promptly (see Chapter 39-A).

Vaccination stress.   Most vaccinations cause some stress; sometimes the vitality of the flock is lessened materially. This is particularly so when birds are handled. Feed intake will decrease, and the birds will huddle near the brooder heat. It may be necessary to increase the brooding temperature to keep the birds comfortable. Often it is practical to add medicaments to the drinking water to treat a stress condition. Sometimes the addition of soluble vitamins to the water will help alleviate any condition brought on by inadequate feed intake. Be careful about vaccinating when a stress from another cause is evident, or when the flock is being medicated. Adding one stress to another only increases flock problems.

Coccidiosis Control

Most chicks are on some type of coccidiosis-control program during the brooding period. As coccidiosis continues to be one of the most important diseases during the first weeks of the chick's life, care should be taken to provide a good coccidiostat in a quantity adequate to completely suppress the multiplication of the oocysts (see Chapter 37-R). If the CocciVac program of coccidiosis control is being used, follow the directions carefully. Regardless of the program, be on the lookout for any evidence of coccidiosis: ruffled feathers, droopiness, bloody droppings, and so forth. Treat quickly.

Unabsorbed Yolk

The presence of unabsorbed yolk during autopsy the first 2 weeks has often been associated with bacterial disease or severe stress. Diseases that raise the body temperature prevent utilization of the yolk material in young chicks. However, the popular procedure of feeding chicks soon after hatching also causes a slower absorption of yolk material, and its presence during the first 14 days should not be considered of importance. Overheating of chicks during the first 2 days under the brooder also lowers yolk absorption. There is no correlation between unabsorbed yolk color and the presence of disease. Yellow, green, and orange unabsorbed yolks are found in healthy chicks.

Hot Weather

Although young chicks can tolerate higher temperatures than older birds, environmental heat may create a severe stress. Birds eat less and drink more water. Be sure there is adequate feed and water space available. It may be necessary to increase the amount of watering space if the chicks are crowding the waterers.

During hot weather, litter problems may be accentuated. Most of the additional water consumed will be deposited in the litter through fecal elimination. Increased air movement through the house will help alleviate the condition, but if this does not improve the litter condition, add 5 lb (2.27 kg) of superphosphate to each 100 ft² (9.3 m²) of floor space. Stir and mix the phosphate with the litter. *Do not use lime.*

Cold Weather

Cold weather problems should be minimized. Better housing has done a great deal to provide a higher temperature within the poultry building. During the brooding period, supplementary heat is added to increase the warmth under the hover, or where the chicks bed down. But the area in the rest of the house needs attention too. Provide heat to warm the area to about 65°F (18°C).

Decreasing the Brooder Heat

The brooder heat should be decreased regularly after the first week, usually at the rate of about 5°F (2.8°C) per week. Birds should not need supplementary brooding heat longer than 4 weeks in warm weather, and 5 weeks in cold weather.

Beginning with the third week, the brooder hovers should be raised regularly until they are about 3 ft (0.9 m) above the birds, depending on the type of brooder used. But it is this period when the brooder heat is being withdrawn, and after, that leads to trouble in cold houses. Respiratory diseases increase, birds eat more feed in their endeavor to warm their bodies, and houses are harder to ventilate. Always maintain a condition that assures adequate chick comfort.

14

Growing Management on Floors

The growing period follows the brooding period and concludes with sexual maturity. Probably no other age of chickens commands the respect of management more than this period. How well a bird is grown will greatly determine how well it does in the laying or breeding house. One's ability to raise a good pullet or cockerel is a requisite of a profitable poultry flock. Those who can't raise a high-quality pullet will be plagued with difficulties during the laying period. Poor-quality pullets at maturity will always perform below breeder standards. Egg production will be poor, egg quality and size inferior, feed conversion suboptimal, and many other factors will be below expectations.

14-A. THE GROWING HOUSE

Formerly, young chickens were moved from the floor brooding house to the floor growing house at about 6 to 8 weeks of age, but with many modern disease-control programs this is no longer the predominant practice. Many commercial egg-type pullets are kept in the same house from their first day until they are to be moved to their laying quarters, either cages or litter floor houses. More and more growing birds, to be used later for breeding purposes, are involved with a program for the eradication of *Mycoplasma gallisepticum* organisms and *Mycoplasma synoviae* organisms. For this reason the birds are left in the brooding house until they are 10 weeks of age, when they are moved to the growing-laying house, or they remain all their lives in one house, where brooding, growing, and egg production are completed.

Brooding House as Growing House

Although the actual brooding period is usually confined to 6 weeks, birds today often remain in the brooding house until they are 8 to 10 weeks of age. Thus the brooding house becomes one for brooding and the early stages of growing. Floor space and equipment requirements have been given in Chapter 13-C for chicks up to 6 weeks of age, but when chicks remain in the same house for the first 10 weeks, care must be taken to provide more floor space, more feeder space, and more waterer space, the latter being especially important during hot weather when there will be a great increase in water consumption.

Although most brooding houses have adequate air movement to suffice for 6 weeks, many may be poorly ventilated during the next 4 weeks. After the first 6 weeks, birds are heavier, grow faster, and occupy more house space; all require an increase in airflow to remove moisture and ammonia from the building. Be sure that the necessary increase is provided.

Even though an additional move is required, many replacement pullet growers prefer the two-house rearing program with the pullets being moved to grow cages at 6 to 10 weeks of age. Both houses are better suited to the age and size of birds, equipment needs can be better tailored to the flock, pullets can adjust at an earlier age to layer-type equipment, and dust, disease, and parasite problems can be reduced.

Breeders involved most. Keeping the chicks in the brooding house for 10 weeks is more important to poultrymen keeping breeding flocks than to those keeping commercial pullets. The reasoning behind this is that it is much better to move the birds to their permanent laying quarters at a young age in order to prevent stress and possible "breaks" in the MG-clean or MS-clean status of the flock. Such breaks occur more often when older birds are moved, often just when egg production is starting. Thus, the birds are kept in the brooding house longer, but moved to the laying house earlier.

The Brood-Grow House

Many poultry producers raising commercial Leghorns or medium-size egg-type pullets have found it more advantageous to use the same house for brooding and growing. This eliminates one moving of the birds, creating less stress. But the brood-grow house has found acceptance because a great many commercial laying pullets are moved to cages rather than to floor operations. Pullets may be moved to cages at a younger age than when they are moved to a floored laying house. Some operators will transfer the growing birds to laying cages when the birds are as young as 14 weeks, although more will keep them until they are 18 weeks of age.

Floor space, feeder space, and waterer space should be adequate for birds just prior to the time they are moved from the house. Do not allow crowding at the feeders or waterers as the birds grow older.

Grow-Lay House

When birds to be used later as breeders are moved to the permanent laying quarters at 10 weeks of age, the laying house becomes a growing house for the first 12 to 14 weeks. In order not to duplicate any feeding and watering equipment it should be of a size adequate for laying birds. So-called small-size growing equipment is not to be used in these houses.

Number of birds to place in house. First determine the number of mature pullets and cockerels the house will accommodate; then add the number of culls and dead birds expected between 10 weeks of age and maturity. This total will represent the number of birds to be placed in the house at 10 weeks of age.

Remember: Do not move birds from house to house or from pen to pen. Birds placed in the building at 10 weeks of age should remain in the same pen until the end of their laying period.

Brood-Grow-Lay House

Another accepted housing practice, particularly among those poultry-men keeping breeding birds, is to confine the birds to the same house from 1 day of age until the end of the laying year. Such a house is known as the *brood-grow-lay* house, to be used for brooding, growing, and laying. Necessarily, equipment must vary according to the age and requirements of the birds as they grow older and space will be underutilized during the brood-grow stages.

Ventilating this type of building must have special consideration because of the different needs associated with bird age. Usually this presents few problems, as the house must be designed for proper ventilation when filled with mature birds. As a result, the air movement can be adjusted for younger and smaller fowl.

14-B. FLOOR

The type of floor in the growing house will vary the management recommendations. Litter, part slats, part wire, all slats, all wire, and other floor innovations are used for growing pullets and cockerels, and alter certain procedures.

Litter Floor

If the floor is completely covered with litter, the equipment should be evenly distributed throughout the house. Consumption of both feed and water is greater when the distance a bird has to go to eat and drink is short and stress is less. Furthermore, birds congregate in groups within the house or pen and each group does its eating and drinking within a relatively small area; birds dislike going to an area with which they are not familiar (see Chapter 22-M).

In the narrow house, the automatic feeders should be placed lengthwise of the house, with the waterers in between. In the wide house, the equipment should have greater distribution. Any short, conventional troughs should be placed crosswise of the building.

Slat-and-Litter Floor

When a portion of the floor is covered with slats, young birds must be trained to frequent the slat area. Even after they can jump on the slats, they may not use them, particularly if the feeders and waterers remain on the litter part of the floor. In most houses of this type, however, the automatic feeders and automatic waterers are placed on the slats in order to concentrate the droppings in this portion of the house. Prior to about 8 weeks of age all feeding and watering must be done on the litter portion of the floor. When the birds reach 8 weeks of age the feeders and waterers on the slats should be placed in operation. As the birds begin to drink and feed from the automatic equipment, gradually remove the feeders and waterers that are on the litter. But do it slowly. There will be a few backward birds that will not easily learn the new location of the new equipment. Similar recommendations are made when wire replaces the slats. At first the birds will bed down on the litter at night, but as they get older all should roost on the slats (or wire).

All-Slat Floor

In many instances, particularly with egg-type commercial pullets, the growing birds are moved from the brooding, or brood-grow unit, to a house having slats over the entire floor. Usually these houses do not present as many management problems as those with a slat-and-litter floor.

Cages

Cage management is discussed fully in Chapter 16, which must be read thoroughly if birds are grown conventionally on the floor and then moved to some type of cage.

14-C. FLOOR SPACE FOR GROWING BIRDS

The amount of floor space needed by different strains and ages of birds is highly variable. Breeders selling the respective lines should be consulted regarding their recommendations. However, averages for birds on litter floors are given in Table 14-1.

The floor space suggested in this table is based on the requirement at growing maturity, or for the week designated. Although birds require less floor space when younger, it is unwise to plan for anything but maximum needs. However, in instances where birds are to be moved from the house at an early age, e.g., at 10 weeks, more birds may be raised in a given area.

Slat-and-litter-house floor space. Floor space requirements for birds grown in a house with a slat-and-litter floor are about 70% of those given in Table 14-1.

All-slat-house floor space. When an all-slat house is used for growing, the floor space requirements are approximately 60% of those given in Table 14-1.

 Remember: All-slat houses are not recommended for some breeds and lines of chickens.

Table 14-1. Floor Space Requirement for Growing Birds (Litter Floor)

Line and Sex	Floor Space per Bird		Birds per m²
	ft²	m²	
Mini-type Leghorn pullets	0.8	0.07	14.3
Leghorn egg-type pullets			
to 18 weeks	1.0	0.09	11.1
to 22 weeks	1.5	0.14	7.1
Medium size, egg-type pullets			
to 18 weeks	1.2	0.11	9.1
to 22 weeks	1.7	0.16	6.3
Leghorn breeder pullets	1.7	0.16	6.3
Leghorn breeder cockerels	1.7	0.16	6.3
Medium-size breeder pullets	1.9	0.18	5.6
Medium-size breeder cockerels	2.1	0.20	5.0
Mini-meat-type breeder pullets	1.7	0.16	6.3
Meat-type breeder pullets	2.5	0.23	4.3
Meat-type breeder cockerels	3.0	0.28	3.5

14-D. GROWING EQUIPMENT

Larger birds require more and larger equipment. In the commercial operation the requirements for automation have an important bearing on watering and feeding.

Waterer Space

As chicks grow older the waterer space must be increased. The requirements during the growing period are governed by the need when the birds reach sexual maturity. These requirements are given in Table 14-2. It must be remembered, however, that when older growing birds are being raised during periods of extremely hot weather, these recommendations may not be ample; Leghorn pullets drink twice as much water per day at 100°F (38°C) as they do at 70°F (21°C), for example. Be certain that there is no crowding around the water containers.

Table 14-2. Waterer Space for Growing Pullets and Cockerels

| Type and Sex | Automatic Troughs or Regular Troughs per Bird | | Per 100 Birds | | | |
	in.	cm	8-ft (2.4-m) Troughs	Large Pan[1]	Cups	Drip-Type Nipples
6–20 Weeks of Age, Moderate Temperature						
Mini-Leghorn pullets and cockerels	0.6	1.5	0.31	0.6	6	9
Leghorn pullets and cockerels	0.75	1.9	0.39	0.7	7	10
Medium-size pullets and cockerels	0.85	2.2	0.44	1.1	8	11
6–22 Weeks of Age, Moderate Temperature						
Mini-meat-type pullets and cockerels	0.85	2.2	0.44	1.1	8	11
Meat-type pullets	1.0	2.5	0.52	1.3	9	12
Meat-type cockerels	1.25	3.2	0.65	1.6	10	13

[1]A large pan has a circumference of 50 in. (127 cm).

Waterers

Although most poultry waterers are of the automatic type, operating from a valve to keep a constant water level, some poultrymen use running water. In such cases a V-shaped trough is installed the length of the house, and the trough should have a slight drop so that a stream of water runs through it constantly. They must be constructed so that the birds will not bump the trough and disrupt the flow of water. In some instances, to give added protection they are placed outside the building, and the birds are allowed to place their heads through slats or "feeder fence" to drink.

Water depth. With all V-shaped troughs, the depth of the water should be very low, perhaps not over 0.5 in. (1.3 cm). When there is little water in the trough the birds will keep it clean as they move their beaks in the trough to get a drink. Similarly, they will also keep the feed out of the troughs. When the water is too deep, feed and debris accumulate at the bottom of the trough, and molds may grow on the sides.

Water Consumption Varies with House Temperature

Table 14-3 shows weekly water consumption of growing Leghorn pullets at four house temperatures, along with body weight, feed consumption, and pounds of water consumed per pound of feed consumed. Notice that growing Leghorn pullets will drink about twice as much water per day at 100°F (37.4°C) as they do at 70°F (21°C), but only slightly less at 50°F (10°C). Relative figures are the same for breeds other than Leghorns.

Feeder Space

As with waterer space, feeder space requirement during the growing period must be adequate for birds approaching sexual maturity. However, if the chickens are moved from the house prior to maturity, the necessary feeder space is proportionately less.

Feeder space and feed restriction. More and more, poultrymen are using a controlled feeding program that involves restricting the feed intake during the growing period. The program calls for more feeding space, since it is necessary that all the birds have room enough at the feeders to eat once the day's feed allotment is provided. Thus, the recommendations found in Table 14-4 are greater than those made a few years ago. When feeder space is inadequate, the birds will lack uniformity at maturity because the aggressive birds will have eaten more feed than the timid ones.

Feeder Height

During the growing period, adjust the height of the feeder so that the bottom of the trough is even with the backs of the birds. When the level is lower, birds will beak out more feed—a direct waste and expense. Adjustable legs, suspension-type chains, or cables will be necessary. Many suspended feeders are quickly regulated with a cable and winch, thereby moving all feeders in the house at one time.

Table 14-3. Feed and Water Consumption of Leghorn Growing Pullets

| Week of Age | 70°F Average Daytime House Temperature | | | | U.S. Gallons of Water Consumed per 1,000 Pullets per Day | | | |
| | Avg Live Weight per Pullet lb | Water Consumed per Day as % of Live Weight lb | Avg Feed per 100 Pullets per Day lb | Water Consumed per lb of Feed Consumed lb | Average Daytime House Temperature | | | |
					50°F	70°F	90°F	100°F
1	0.15	18.3	2.9	0.94	3.3	3.3	3.3	5.0
2	0.26	25.5	3.6	1.84	7.7	8.0	9.3	12.6
3	0.39	31.9	5.4	2.31	13.5	15.0	20.5	29.2
4	0.53	32.2	7.0	2.42	18.6	20.4	31.2	40.3
5	0.68	29.3	8.0	2.47	19.2	23.8	39.7	48.2
6	0.83	27.0	8.4	2.65	21.8	26.8	44.7	54.4
7	1.04	23.4	8.6	2.82	23.7	29.2	48.7	59.2
8	1.17	22.1	9.0	2.86	25.1	31.1	51.7	62.9
9	1.36	19.8	9.5	2.87	26.3	32.5	54.2	66.0
10	1.54	18.4	10.0	2.84	27.8	33.1	57.1	69.4
11	1.68	17.1	10.5	2.74	28.1	34.7	57.8	70.3
12	1.87	15.8	11.0	2.68	28.9	35.5	59.2	72.0
13	2.03	14.8	11.5	2.61	29.3	36.2	60.3	73.3
14	2.18	14.0	12.0	2.54	29.8	36.8	61.2	74.5
15	2.32	13.4	12.5	2.48	30.3	37.4	62.2	75.7
16	2.44	12.9	13.0	2.43	30.9	38.0	63.4	77.2
17	2.55	12.5	13.5	2.37	31.3	38.5	64.3	78.1
18	2.65	12.2	14.0	2.31	31.6	39.0	65.0	79.0
19	2.74	11.9	14.5	2.26	32.0	39.4	65.9	80.0
20	2.82	11.7	15.0	2.20	32.2	39.8	66.3	80.6
21	2.89	11.6	15.5	2.15	32.6	40.2	67.0	81.4
22	2.94	11.5	16.0	2.10	32.8	40.6	67.7	82.1

Roosts

Under most circumstances growing birds do not require roosts, regardless of the breed or type. But the lack of conventional roosts may pressure the birds into using other objects to sit on. Avoid high objects if possible. Place antiroosting guards where the birds may have occasion to sit.

14-E. CLEANING THE GROWING HOUSE

As with the brooding house, all growing houses containing birds of the same age must be isolated. For better security, a fence should surround the house or houses. Separate caretakers should be employed in each unit; no one should be allowed to go from one unit to another without showering and changing into clean clothing.

Regardless of the age at which the birds are to be moved from their first home, the new quarters should be completely cleaned and disin-

Table 14-4. Feeder Space for Growing Pullets and Cockerels

Type of Bird	When Cockerels and Pullets Grown Separately			When Cockerels Grown with Pullets		
	in. per Bird	cm per Bird	Large Tube Feeders per 100 Birds[1]	in. per Bird	cm per Bird	Large Tube Feeders per 100 Birds[1]
Mini-Leghorn, egg-type pullets	2.0	5.0	2.5	—	—	—
Leghorn, egg-type pullets	2.5	6.4	3	—	—	—
Medium-size, egg type pullets	3.0	7.6	4	—	—	—
Leghorn breeder pullets	2.5	6.4	3	2.5	6.4	3
Leghorn breeder cockerels	3.0	7.6	4	2.5	6.4	3
Medium-size, breeder pullets	3.0	7.6	4	3.0	7.6	4
Medium-size, breeder cockerels	3.5	8.9	5	3.0	7.6	4
Mini-meat-type breeder pullets[2]	4.0	10.0	6	4.0	10.0	6
Meat-type breeder pullets[2]	6.0	15.0	8	6.0	15.0	8
Meat-type breeder cockerels[2]	8.0	20.0	10	6.0	15.0	8

[1]A large, tube feeder has a pan with a 50-in. (127-cm) circumference.
[2]These will be on restricted feeding and all birds should be able to eat at one time.

fected prior to the time the birds are admitted. Follow the cleaning directions for poultry houses as given in Chapter 13-B.

Add clean litter to the house to which the birds are to be moved.

Be careful: New litter contains no coccidia oocysts; therefore, a watchful eye for symptoms of coccidiosis will be necessary after the birds are placed in the new house. Although coccidiosis should not be evident in the new house unless the birds are in the midst of an attack when moved, oocysts soon build up in the new quarters, and the disease may hit suddenly. Be ready to medicate immediately.

14-F. VENTILATING THE GROWING HOUSE

Growing birds must have an ample supply of fresh air, without drafts. House ventilation has been discussed in detail in Chapter 11-I and 11-J, but it should be brought out here that growing chickens do not do well in environmental extremes. They feather poorly, growth is reduced, birds are not uniform, and feed conversion deteriorates—each a costly factor.

Hot Weather Problems

When the temperature gets above 80°F (27°C), chickens begin to suffer. The higher the temperature, the greater the stress. Usually the first hot period of the season is the most disastrous; birds are not accustomed to the heat and cannot adjust to differences in water and feed consumption rapidly enough to offset the high temperatures. But be ready for the first heat wave; do not wait until it is in progress. Equipment for cooling the house and the birds should be operational when the heat strikes. Fans, foggers, and roof sprinklers, used to help cool the open-sided house, should be ready. Provide plenty of cool, fresh water. If the number of waterers was adequate for only cool weather, provide more as the weather begins to warm, before any extremely hot day.

If environmentally controlled houses have been constructed properly, the advent of hot weather should not cause serious repercussions. Thermostats should force the fans to operate more, causing a greater amount of air to flow through the building. In many instances, however, the intake opening, or intake slot, will have an adjustment for hot weather. Be sure the change is made prior to any heat wave. Any cooling device, such as evaporator-pads, should be started with the advent of warm weather, and be in full operation when the weather gets hot. Fogger and sprinkler nozzles should be cleaned and operable.

Cold Weather Problems

During cold weather, proper ventilation of the growing house may become difficult. It will be necessary to conserve heat within the building without creating too high a level of moisture and ammonia. Insulation and dry litter aid in keeping the environment as near optimum as possible.

The open-sided house will need curtains that can be closed when necessary. Body heat must be conserved, but not at the expense of poor air quality.

The environmentally controlled house offers a better means of creating a good environment for growing chickens during cold weather. It will be tight and adequately ventilated to conserve heat, yet still maintain a low level of humidity and ammonia. But because growing birds increase in weight each week, the ventilating requirements will vary accordingly. Be sure to make the necessary adjustments.

14-G. LITTER MANAGEMENT FOR COCCIDIOSIS CONTROL

The poultryman who can manage his litter will prevent a great many problems. During the growing period litter should contain between 20 and 30% moisture. There are several reasons, as follows:

1. Feathering is better.
2. Growth is closer to normal.
3. Feed conversion is improved.
4. The coccidiosis problem is more easily controlled.
5. Ammonia in the house is reduced.
6. Flies and parasites will breed in moist litter.

Determining litter condition. To determine if the litter contains the correct amount of moisture, pick up a handful and squeeze it tightly; then open the hand. If the condition of the litter is correct, crevices will form in the compressed material; it should not form a cohesive ball or fall away in a pile. Gauges are also available for determining the moisture percentage in the litter.

Causes of Changes in Litter Condition

Many things affect the condition of the litter:

1. relative humidity and temperature of the outside air
2. relative humidity and temperature inside the poultry house
3. number, age, and weight of the birds in the building
4. amount of air moving through the house
5. water consumption of the birds
6. makeup and form of the feed
6. stress in the birds
8. disease and mold conditions

Litter Condition for Coccidiosis Control

Probably the biggest reason for keeping a required percentage of moisture in the litter has to do with the control of coccidiosis (see Chapter 37-F).

Wet Spots Around Waterers

Birds will spill some water when drinking and the litter around the waterers may become extremely wet. Oocyst sporulation is increased in this area of the house, and may be great enough to cause an increase in the incidence of coccidiosis. Remove the wet material and check to see that the water level in the waterers is low. Be sure all automatic water valves are working properly and that water is not overflowing.

Coccidiosis Control When Layers Are to Be Caged

When birds are grown on a litter floor, but are to go into cages or on a wire floor during the last part of their growing period or at sexual maturity, some change in the coccidiosis program will be necessary. Immunity should be developed prior to the time the birds are placed on the wire. If it is not possible to do this, or the degree of immunity is not known when the birds are caged or placed on a wire floor, continue to feed some coccidiostat for 3 or 4 weeks. Too often poultrymen feel that caged birds will not get coccidiosis, but this is true only when the chickens are kept in cages from 1 day old. Birds transferred from a litter floor to a wire floor will be loaded with sporulated oocysts, and if the feed used after the birds are placed on the wire contains no coccidiostat, or immunity is not complete, coccidiosis will develop, sometimes seriously.

Coccidiosis When Birds Are on Slats

Birds may develop severe attacks of coccidiosis when they are kept on slats. An outbreak can occur because the birds have access to enough sporulated oocysts from the droppings that stick to the boards. Manipulate the cocciodiosis-control program accordingly, using a coccidiostat in the feed; then withdraw it gradually to develop immunity.

14-H. PREVENTING CANNIBALISM

Although 6- to 10-day beak trimming is gaining more acceptance for laying strains, other ages and methods are used by many poultry producers to help prevent cannibalism. See Chapter 13-I for details of beak-trimming ages other than 6 to 10 days.

14-I. SANITATION, DISEASE, AND RODENT CONTROL

Sanitation during the growing period is just as necessary as during any other age of the bird. Keeping things clean requires a daily vigil. Isolation of the growing house, showers, clean clothing, separate caretakers, daily cleaning of waterers, proper litter, clean feed, and so forth, are all parts of the sanitation program. Daily routine to keep the premises sanitary should be detailed and written in advance, and the program should be carried out to the letter. Half a job is little better than no job at all. Stress and disease do not operate in this manner; usually they are either all or nothing. See Chapter 39 for recommendations for preventing disease.

MG and MS Testing Procedure

A high proportion of those farms engaged in the raising of breeding stock are now on a program to prevent *Mycoplasma gallisepticum* organisms or *Mycoplasma synoviae* organisms from gaining entrance to the birds. Flocks are blood-tested every 4 weeks during the growing period, beginning when the chicks are 8 weeks of age and continuing until they enter the breeding-laying pens. About 5% of the birds are to be tested in each house or pen every 4 weeks. Local regulations will determine whether the tube or plate test is to be used. Complete isolation of the growing house, along with strict sanitation, should allow the flock to remain free of the disease (see Chapter 37-J, 37-K).

What to do when reactors appear. When reactors to the MG or MS blood test appear, the flock has broken and the birds have had the disease. Many will remain as ''carriers,'' transmitting the organism through the hatching egg to the day-old chick. If the hatchery to which the hatching eggs from this flock were to be shipped has a clean status, eggs from this flock cannot be used for incubation. If all other flocks on the farm are clean, the diseased flock should be moved off the farm immediately to prevent it from contaminating clean birds in other houses. But before any infected birds are moved, take some to the laboratory for a confirmation of the blood test.

Infected flocks may be used for the production of eggs for human consumption, but the birds should be moved to a new location if there are other birds with a clean status on the original farm.

Disease-Control Program

Some vaccination programs begin in the hatchery; others begin during the brooding period and still others are initiated during the growing period. The program should have been determined before the arrival of the day-old chicks at the farm. Consult specialists and ask them to outline the correct vaccination program for your operation. These will vary from country to country, from farm to farm, from flock to flock, and with the season of the year. Then, do not neglect the necessity of vaccinating correctly, and on schedule.

Most vaccination failures are managerial mistakes; the vaccine was old, improperly kept, administered incorrectly, at the wrong time, or was not necessary. See Chapter 39 for suggestions.

Check for immunity status.   Just to vaccinate is not enough. An appraisal of the immunity developed should be made. This usually involves determining titers of the blood to see if an adequate number of antibodies developed as the result of the vaccination (see Chapter 35). Contact a laboratory that can make these determinations and give you information you need for your disease-control program.

Disease Diagnosis

In spite of all the precautions taken on poultry farms, disease and other problems will still occur. In many instances, a medication program is available to treat the difficulty. But be sure you know what the difficulty is. Take five or six typically involved birds to a good disease diagnostic laboratory. Once the trouble is identified, discuss remedial measures and act promptly; problems are more difficult to cope with the longer treatment is delayed.

Double-check medicament dosages.   Read the instructions carefully, particularly as they apply to dosages. Many mistakes have been made because the birds were overtreated or undertreated. Be sure the drug has been legally approved for the specific disease involved and for the dosage and method of administration. Know the withdrawal period for the medicament, i.e., how long its use must be stopped before the birds are to be slaughtered for consumption.

Weekly Culling During Growing Period

Inferior, crippled, injured, and deformed birds should be removed from the flock during the growing period. In all probability these birds will do poorly when egg production begins, and to keep them in the growing pen only adds to the expense of producing good pullets. But do not catch the entire pen or house of birds to remove these few. Use a catching hook and remove them quietly one day each week. Be sure to make an accounting of the number removed on the house record sheet.

Worm Control

If worms have become a problem, some procedure for keeping them under control should be initiated (see Chapter 38-A). But do not spend money on unnecessary worm-control programs. Be sure you have an accurate diagnosis of the type or worm involved. Consult a recognized veterinarian and discuss treatment with him.

Rodent Control

Rats and mice cause a heavy economic loss. They consume and con- taminate a great deal of feed, destroy building materials, harbor disease and parasites, and create a general nuisance. They are extremely prolific; the premises soon can be overrun with them. When the presence of ro- dents is obvious, begin a program of elimination immediately. See Chap- ter 38-E for details for the control of rats and mice on a poultry farm.

14-J. FEEDING DURING THE GROWING PERIOD

Although the accepted practice is to full-feed (self-feed) egg-type birds during the growing period, restricted feeding may be required to regu- late body weights. These feeding programs may call for more feeding equipment inasmuch as there must be feeder space for all birds at one time. Full details are given in Chapters 30 and 32.

Feed Equipment Changes During Growing

Chickens dislike changes in their day-to-day routine, and care should be taken to see that the birds adjust to any new equipment or surround- ings. Moving birds from house to house creates a stress, but new types of waterers or feeders add to the difficulties. In the new house be sure that feed and water are easily available, and not in some location to which the birds will not readily go. In some cases it may be advisable to provide open pans for a few days. Place water in some; feed in others.

14-K. LIGHTING DURING THE GROWING PERIOD

Growing birds are susceptible to cannibalism, but the vice may be par- tially eliminated by using light intensities of less than 0.5 fc (5 lx) at bird level. This is easily accomplished in the lightproof house, but impossible in the open-sided house.

Light, either natural or artificial, not only makes it easy for the birds to see to eat and drink but it also affects the pituitary gland at the base of the brain, and the stimulation causes mature pullets to begin the pro- duction of eggs. The fact that the hours of natural daylight differ during the seasons of the year causes more pituitary activity during some months than others. When natural daylight is short, artificial light must be added. The procedure is complicated and is discussed in more detail in Chapter 18-A, 18-B, 18-C, and 18-D.

In-Season and Out-of-Season Birds

Pullets tend to come into egg production at a younger-than-normal age if they are grown under natural daylight during the time when daylight hours are lengthening. When the days get shorter, the onset of production is delayed. Those grown during the period when most of the days have decreasing light are known as *in-season birds*. Those grown when most of the days have increasing hours of light are known as *out-of-season birds*. These latter ones present the following growing difficulties.

1. Brooding costs are higher in most locations.
2. More feed is usually required because the birds are grown during periods of colder weather.
3. There are more production problems between 6 and 20 weeks of age.
4. It is more difficult to use a controlled feeding program during the growing period.
5. Pullets mature earlier and lay smaller eggs at the start of egg production.
6. Pullets lay at a lower rate and over a shorter period after they begin their production cycle.
7. Mortality rates during early stages of egg production are generally higher due to an increased amount of prolapse.

All the above factors necessitate special artificial lighting procedures to try to compensate for the deficiencies of natural daylight. Feed limitation also affects the age of sexual maturity. Thus, the feeding and lighting programs must be correlated so that a quality pullet will be produced—one that not only is good physically but one that will begin to lay at a normal age.

Both the lighting and feeding programs are discussed in later chapters.

14-L. EGG-TYPE STARTED PULLETS

There are many compexities involved in growing a good replacement pullet for the production of commercial eggs. Some farms do not have facilities for brooding and growing acceptable pullets. At others the growing houses are too close to the laying units to provide ample isolation. These and other factors have led certain poultry producers to specialize in the brooding and growing of commercial, egg-type pullets, known as *started pullets*. Pullets are sold to those requiring mature birds for their egg production program. There are advantages and disadvantages for both the producer and the recipient (flockowner) of such pullets.

Advantages to the Flockowner

1. The producer can specialize in layer flock management.
2. Pullets are raised by a specialist.
3. Birds are more likely to have been grown on a single-age farm.
4. Outside crews are not required for vaccination, beak trimming, and moving.
5. Maximum rearing-house utilization is not an issue.

Disadvantages to the Flockowner

1. Started pullets are usually more expensive.
2. Lighting, nutrition, beak trimming, and disease-control programs may not suit the buyer.
3. Moving stress may be a complicating factor.
4. Disaster during growing could prevent delivery of pullets on schedule.
5. A large cash outlay is necessary at the time pullets are delivered. The cash outlay is much less if the flockowner raises his own pullets; depreciation and some other factors are not cash expenses.

Advantages to the Started Pullet Grower

1. The grower is allowed to specialize.
2. The grower can make more efficient use of his or her buildings.
3. A better distribution of labor is possible.

Disadvantages to the Started Pullet Grower

1. Guaranteed delivery of quality birds at a previously specified price is required.
2. Excessive mortality may offset any profit from the venture.
3. Increased feed and labor costs will lower profit.
4. The customer may refuse to take the mature birds for one reason or another; the grower is left with them.

Have a Written Contract

To protect both the grower and the flockowner, a written agreement should be prepared and signed in advance by each party. It should contain

1. Strain and number of pullets to be delivered
2. Date and age of pullets at delivery
3. Type of vaccinations, along with name of vaccine manufacturer
4. Special services such as beak trimming, dubbing, and so forth
5. Feeding program involved during growing
6. Price at time of specified delivery, and alternative prices for making delivery early or later
7. Inspection of pullets by flockowner before delivery
8. Type of crates used for delivery
9. Who is to furnish delivery and placement of the pullets

Weekly written progress reports should be provided to the buyer relative to mortality, body weights, and problems. Upon delivery, a complete summary should include vaccination and lighting schedules, feed intake, and body-weight graphs.

Cost of Producing a Started Pullet

The cost of producing a salable pullet is highly variable and differences as high as 40% are not uncommon. Feed price, labor cost, age at delivery, and mortality are the factors largely responsible for these variations.

Delivery age—an important cost factor. Most started pullets are delivered when they are between 18 and 20 weeks of age. Flockowners like to get pullets in their permanent laying quarters several weeks in advance of egg production and for this reason 18 weeks is now the most common age for housing. To avoid fluctuations in prices, it is now common to contract for the growing of pullets exclusive of the chick, feed, and vaccine costs.

Livability—an important cost factor. The ability to raise a high percentage of the chicks started is an important cost item, and differences create great variations in the total cost of raising a pullet. Pullets of poor quality at maturity are not easily sold, and even if they are, a distress price may have to be arranged. One should certainly be able to deliver 90% of those started; many deliver 95%.

Evaluation of pullet quality. Body weight and frame measurements are used to estimate the progress of pullet development (see Chapter 16-G).

Transporting Started Pullets

Started pullets are transported over distances up to several hundred miles with little, if any, difficulty. They are placed in wood or plastic coops or in specially designed wire racks, then hauled by trucks. A tarpaulin should be placed over the front of the coops during warm weather to act as a windbreak. During cold or wet weather it should be extended over the top and down across the sides to give added protection.

Careful handling.   Pullets should be handled carefully at the time they are moved. Broken bones will permanently damage a replacement pullet. Particular care must be provided when pullets are removed from or placed in cages.

Transportation diarrhea.   Most started pullets develop a laxative condition when they are moved. The wet droppings in the crates soil the plumage during moving, giving the birds a very unsightly appearance. Tranquilizers will stop most of this; careful handling will help control it further. Eliminate all possible stress.

Special moving precautions

1. Do not mix birds of different lines and ages.
2. Do not create a stress by beak trimming birds during the period 10 days before to 10 days after moving.
3. Provide extra waterers and feeders when pullets are placed in permanent laying quarters. Beware of changes in types of waterers.
4. Get an accurate count of the birds delivered.
5. During hot weather remove tarpaulins if truck should stop enroute. Give the birds plenty of air. Allow a space in the center of the load, between the tiers of coops.
6. Move pullets at night when the temperature is lower. Trucks should arrive at the flockowner's farm early in the morning.
7. Some states have laws governing the production, sale, and moving of started pullets. They may require a health certificate.
8. Do not overcrowd moving coops or racks. Place no more than 3 pullets per square foot of space.

Should You Grow Pullets or Purchase Them?

There probably is no direct answer to this question. Many factors are involved. Although the grower is entitled to a profit for his endeavors and a return on the money he has invested, a started pullet necessarily will usually cost more than the flockowner's expense of raising his own. But the flockowner may decide that he could make more on his investment if the money were spent on his laying program rather than on his growing operation. Furthermore, it may be worth more to the flockowner to be able to get good, uniform pullets when he needs them. The pullet grower has additional costs that are usually not considered by a flockowner, such as advertising, management, risk, transportation, and profit.

14-M. RECORD KEEPING

Keeping adequate records during the brooding and growing periods is an essential part of flock management. Careful records inform one of what has happened in the past, and help one to plan for the future. The house record should include the following:

1. Line and source of chicks
2. Vaccinations and medication (see Chapter 39-H for examples of forms for recording these data)
3. Feeding program used
4. Feed consumption, by days and weeks
5. Lighting schedule
6. Body weight and uniformity by week after 4 weeks
7. Frame measurements at 12 weeks
8. Mortality by days and weeks
9. Beak trimming description and age
10. Record of problems and observations

14-N. GROWING COST MANAGEMENT

The only reason for growing a pullet is that she may produce eggs. Thus, any pullet-growing costs become a direct charge against egg production costs. The greater the pullet expense, the greater the cost of producing eggs. Reducing pullet production cost lowers egg cost.

Example (1): If the Leghorn pullet growing cost is reduced by U.S. $0.22, the expense of producing a dozen eggs will be lowered by U.S. $0.01 per dozen, figuring on the basis of 22 dozen eggs laid per hen housed.

Example (2): If meat-type breeder pullet growing cost is reduced by U.S. $0.30, the expense of producing a dozen *hatching* eggs will be lowered by U.S. $0.02 per dozen, figuring on the basis of 15 dozen hatching eggs laid per hen housed.

Estimated Costs of Growing a Pullet

When the poultryman owns his own house and equipment and takes care of the birds, an estimate of the cost breakdown for Leghorn, medium-size commercial pullets, and meat-type, breeder replacement pullets is found in Table 14-5.

If contract growing is involved, the cost breakdown of a growing pullet is shown in Table 14-6, where it is assumed that the grower is paid a predetermined amount for the use of buildings and labor. Although the table shows the costs through the 20th week, most pullets are moved to their permanent laying quarters earlier than this, sometimes several weeks. But regardless of the moving date, growing costs accumulate until the birds are at least 20 weeks of age, and sometimes longer.

Table 14-5. Estimated Cost to Produce a Pullet to 20 Weeks (140 Days) of Age, Pullet-Housed Basis (in U.S. Dollars)

Cost Item	Standard Leghorn Pullet	Commercial Medium-size Pullet	Breeder, Meat-type, Replacement Pullet[7]
Day-old chick	$0.50	$0.52	$0.67
Feed	1.13[1]	1.24[2]	1.50[3]
Labor	0.32[4]	0.40[5]	0.56[6]
Medicines and vaccines[8]	0.15	0.15	0.20
Consumable supplies	0.04	0.04	0.05
Vehicles	0.04	0.04	0.04
Maintenance and repair	0.03	0.03	0.04
Depreciation	0.10	0.12	0.14
Other expense	0.04	0.04	0.05
Mortality	0.11	0.13	0.21
Total	2.46	2.71	3.46

[1]15 lb @ $0.075/lb (6.8 kg @ $0.165/kg).
[2]16.5 lb @ $0.075/lb (7.5 kg @ $0.165/kg).
[3]20 lb @ $0.075/lb (9.1 kg @ $0.165/kg).
[4]4 min @ $0.08.
[5]5 min @ $0.08.
[6]7 min @ $0.08.
[7]No males included.
[8]Expense for breeder birds is larger.

14-O. ANALYZING GROWING RECORDS

Variations in Growing Costs

Although only three lines of birds are found in Tables 14-5 and 14-6, there are other types of chickens that will involve different estimated computations. Furthermore, itemized costs will vary according to the re-

Table 14-6. Estimated Contract Cost to Produce a Pullet to 20 Weeks (140 Days) of Age, Pullet-Housed Basis (in U.S. Dollars)

Cost Item	Standard Leghorn Pullet	Commercial Medium-size Pullet	Breeder, Meat-type, Replacement Pullet
Day-old chick	$0.50	$0.52	$0.67
Feed	1.13[1]	1.24[2]	1.50[3]
Growout contract	0.40	0.51	0.53
Medicines and vaccines	0.15	0.15	0.20
Vehicles (supervisor)	0.06	0.06	0.08
Supervision, records	0.06	0.08	0.08
Other expenses	0.04	0.04	0.05
Mortality	0.11	0.13	0.21
Total	2.45	2.73	3.32

[1] 15 lb @ $0.075/lb (6.8 kg @ $0.165/kg).
[2] 16.5 lb @ $0.075/lb (7.5 kg @ $0.165/kg).
[3] 20 lb @ $0.075/lb (9.1 kg @ $0.165/kg).
[4] No males included.

gion where the birds are grown: feed price, feed consumption, labor cost, contract agreement, breeding or commercial flock, investment, chick cost, mortality rates, vaccination programs, and many other factors will affect the total cost. Each operation will need a different set of figures, but when similar tables are constructed under different conditions, estimated costs must be based on a good operation. They need not be goals, but they should represent what could be accomplished under good management.

Male Production Costs

When replacement breeding pullets are involved, either egg-type or meat-type, the cost to raise the accompanying males must be considered. Although the procedure for handling these costs is probably not the best, male costs are usually added to the cost of raising the breeder pullets. Thus, the pullet-growing cost under this system is not an accurate interpretation of actual pullet cost. However, Tables 14-5 and 14-6 show the cost for meat-type pullets only.

Cash vs. Accrual System of Bookkeeping

Most smaller poultrymen keep their books on the cash system; larger operators usually go to the accrual system, but not all. In many instances the cash system of accounting has an advantage from a tax standpoint, but the procedure will never give a true picture of the cost of producing a pullet. Costs must be broken down into their various components regardless of the type of bookkeeping system employed.

Cash system of accounting. Cash expenditures and cash income, along with a few other items are used to determine cash costs, cash income, and cash profit. Many important costs do not appear in the records, and there is no correlation between the time of the cash expenditure and the time of actual use. For instance, feed may be purchased at the end of one month, but not fed until the next month.

Accrual system of accounting. This system gives the producer an exact cost of his expenses, income, and profit. But it may have a disadvantage from a tax standpoint. However, some type of accrual bookkeeping should be employed even if the cash system is used. This is necessary to get a picture of the actual monthly cost and profit. At times, only an accrual *record* will be necessary on the smaller poultry farm. Such a record should contain

1. Cost of actual feed consumed by the birds
2. Actual labor cost (family labor included)
3. Bird inventory value increase or decrease
4. Consumable supplies cost (actually used)
5. Depreciation cost (buildings and equipment)
6. Other actual costs

Interest Expense

From a cost accounting standpoint, interest is not a *production expense*. In bookkeeping procedures it is broken out of production costs and placed in the G&A account. This is necessary, because interest is highly variable. One flockowner could have borrowed a great amount of money resulting in a high interest charge; another might have no borrowed money, and no interest obligation.

Interest on investment. This is a rather fictitious expense figure, and means little. Such an investment figure, determined as the value of what the invested money would draw if placed on loan, is added as a direct cost of production. No other business uses this procedure. Profit should not be determined in this manner. Rather, it is better to determine the actual profit, then calculate the profit as a percentage of the value of the fixed assets (investment in land, buildings, and equipment), or as a percentage of the total money invested.

Capitalizing Pullet Costs on the Accrual System

When the *accrual system* of bookkeeping is used, it is necessary to capitalize the cost involved with growing the pullets to be used for commercial egg production (or for pullet and cockerel costs when a breeder replacement flock is involved). As capitalized, pullet-growing costs are offset by inventory valuations; there is no profit or loss involved with the growing process. Once capitalized, however, the amount must be reduced during the period of egg production, usually by a procedure of charging an equal amount to each dozen of commercial eggs or hatching eggs produced.

Charging an equal amount to each dozen is the preferred method if projected and actual production are similar. Any problems with mortality or production, though, will result in significant error. Some producers prefer to capitalize their pullet expenditures on a time basis. This procedure allocates a share of the pullet cost to the eggs produced during a given week. This system results in a fluctuating cost per dozen as production varies.

Age to capitalize. If flocks are to be capitalized, there can be a great difference in the age of the birds when the accounting capitalization takes place. When it is early, such as at 18 weeks of age, the flock will not be fully grown, and remaining growing costs will be charged to the laying period, adding to the expense prior to sexual maturity when no egg income is forthcoming. If capitalization is delayed until after the birds are producing eggs, some "laying costs" get into growing costs. When growing costs are capitalized at sexual maturity when egg production starts, book financial losses will occur for several weeks after capitalization because egg income will not be sufficient to offset the costs of maintaining the laying flock. For this reason most people like to capitalize the commercial flock when it reaches about 50% hen-day egg production. This will be about 1 month after egg production begins. Thus, if a flock of commercial pullets were to lay its first eggs at 21 weeks of age, the flock would be considered a "growing flock" until it was 25 or 26 weeks of age. At this time the income from eggs produced would offset the cost of maintaining the laying flock. Of course the delay increases the "growing costs" and adds to the amortization cost per dozen eggs produced during the laying period. The *amortization cost per dozen eggs* is the "pullet cost" at capitalization less the salvage value at the end of the laying year, divided by the number of dozen eggs the bird will lay during her production cycle after her amortization date.

Capitalizing started pullets. Regardless of the age at which the started pullets are received on the laying farm, in most instances the pullets should not be capitalized until they are 24 to 25 weeks of age. Costs involved at the laying farm prior to this age should be charged to "growing costs."

Reducing Growing Costs

Cost management involves the reduction of growing costs to their minimum without impairing the quality and performance of the flock in the laying house. There are many avenues open to such a reduction.

Reduced feed cost. Feed is the largest of the growing cost items; thus it becomes one of the most vulnerable. Excessive feed cost should be examined critically.

1. Is the feed the right feed?
2. Are the protein and energy values correct?
3. Is the ration being fed correctly?
4. Is the feed priced right?
5. Is there feed wastage?
6. Should feed restriction be practiced?
7. Is feed consumption too high?

Labor cost. In most cases this is the third largest of the pullet-growing costs. Labor efficiency has become an important item in this age of high labor rates. If pullet-growing costs are to be reduced, the farm must be automated as much as possible and there must be enough birds to keep the labor force busy. Mass vaccination techniques should be used wherever advisable. In no case, though, should labor be eliminated, which would lead to the improper supervision or care of the flock.

Other cost items. There are many other cost items. Some are open to efficiency reductions; little can be done about others. For example, cost of medicines, vaccines, consumable supplies, and so forth must be kept under constant scrutiny, while little, if any, reduction can be made in depreciation, light and power rates, and so on.

Mortality as a cost item. Although not a direct cost, mortality plays an important part in the cost of raising a pullet. One should "house" at least 95% of the pullets started, except where selection pressure is necessary for breeding flocks. The cost of keeping a bird until it dies must be charged to those living, thus raising the growing cost of the live birds. Probably nothing is more costly than "excessive" mortality. Not only is there a bird loss, but the profit the birds would have produced is lost.

Quality of grown pullet important. The productivity in the laying house or cage is due in great part to the quality of the pullet at the time she reaches sexual maturity. Cost management involves quality management. No program of reducing growing costs should impair the quality of the mature bird.

15

Layer Management
on Floors

Although the trend has been toward housing commercial laying strains in cages, there is still a good percentage of commercial layers kept on a floor of litter, slats, or wire. This is particularly true in the lesser-developed countries.

This chapter deals with laying strains such as Leghorns, Mini-Leghorns, and egg-type medium-size pullets kept on floors for the production of commercial eggs. Cage management is to be found in Chapter 16; breeder management is discussed in Chapter 17. Both of these chapters should be read as they contain material pertinent to floor operations.

15-A. DEFINITIONS

Through the years certain poultry terminology has been standardized, but there are still some misconceptions. The following is offered to help clarify the situation.

Age of Moving to Permanent
Laying Quarters

Because of the variations in growing management procedure, pullets are now moved to permanent laying houses at ages between 14 and 21 weeks. Therefore, the laying quarters are used for "growing" for some period before the pullets reach egg production. However, another management change confuses the picture even more when laying rations replace growing rations just before the birds drop their first eggs. Thus, housing age, age to change to a laying ration, and start of the laying period do not coincide. Each must be treated independently.

Housing Time

The term *housing time* or *housed* is also indefinite in poultry nomenclature. Formerly, it meant the time when birds were placed in the laying house just before the onset of egg production. Through the years, the concept of hen-housed production required that a specific age be accepted and 20 weeks is the commonly chosen definition. Thus, even though commercial flocks now commonly commence egg production at less than 20 weeks of age, hen-housed egg production is still based on the 20-week count. Eggs laid prior to 20 weeks are credited to the 20-week flock.

Laying Period

The *laying period*, sometimes called *laying cycle*, or *biological year*, is difficult to define. There are many ways to describe the length of time involved. But in this text it starts when the birds reach 5% egg production on a hen-day basis and continues to the end of the bird's laying period, and the birds are sold as "spent hens." In some cases, where the birds are molted, this may involve one or two molts (see Chapter 19). However, in most discussions, it involves a normal laying "year" (without a molt) of about 12 to 14 months.

Often laying birds are not capitalized on the books of the company until they reach 24 to 26 weeks of age, when egg production has reached 50% or higher. Thus, there is no similarity between the production and the bookkeeping period. Each should be considered a separate entity, and this distinction must not be confused (see Chapter 14-O).

15-B. THE LAYING HOUSE

Type and style of laying houses with a litter floor are detailed in Chapter 11-I and 11-J. As will be noted, these houses may be either

1. Open or curtain-sided conventional
2. Environmentally controlled

Equipment used in these laying houses is described in Chapter 12-C. House management as it applies to egg production is discussed in this chapter, and for breeders in Chapter 17.

15-C. PREPARING FOR THE PULLETS

When birds are to be moved from a growing house to a laying house just prior to sexual maturity, the usual routine of cleaning the house and equipment as outlined in Chapter 13-B must be made a part of the management program. Young pullets must be given a clean start. However, if they have been reared on litter in the laying house, no cleaning will be necessary. The use of the old litter should be continued and the daily sanitation methods should be followed.

Laying House with Litter Floor

If the house has been void of birds for some time and has been cleaned, it will need new litter. Provide a litter that is common to the area, free of mold, dry, clean, and economical. Add about 3 in. (8 cm) during summer months and 4 in. (10 cm) during winter months.

Laying House with Slat or Wire Floor

If birds have been grown on litter, then transferred to a house with a slat or wire floor, be on guard for an outbreak of coccidiosis. Clean the slats or wire thoroughly before placing pullets on them. If a coccidiostat has been in the growing feed, continue to feed a coccidiostat to the pullets on the slats or wire for 2 or 3 weeks, then gradually withdraw it.

Nest Preparation

Nests should be in the laying house, and open, about a week before the first eggs are laid. This allows time for the pullets to get accustomed to them prior to egg production. In this manner fewer floor eggs will be laid, and the first eggs will be cleaner. Use nesting material that is clean and of a type that will prevent as much egg breakage as possible. Close the nests at night, even before egg production starts. It is a bad management habit to allow pullets to remain in the nests overnight.

Automatic Equipment

Be sure that all automatic equipment is in proper working condition. When an environmentally controlled house is involved, a standby generator is essential and should be run and tested before receiving the birds, and each week thereafter during the laying period.

Automatic waterers should be inspected and carefully repaired. Be sure the valves are working properly, and are not leaking. If waterers are in a different location from what the birds were accustomed to during growing add pans of water for a few days. Take the first reading on the water consumption gauge. If the birds are not accustomed to the feeder location, place pans of feed on the floor at first.

15-D. SELECTING PULLETS FOR EGG PRODUCTION

Regardless of whether the pullets are moved from growing to laying houses just prior to sexual maturity, or whether they remain in the same house, some selection should be made at this time to remove the inferior birds. If the pullets are to be moved at this time, each bird should be individually handled. If they are not moved, cull the inferior birds by catching them with a hook. Cull pullets that are

runty	blind
crippled	injured
emaciated	diseased

15-E. SPACE REQUIREMENTS FOR LAYERS ON FLOORS

Type of Floor

When it is to be covered with litter, the floor may be either dirt or concrete. Concrete certainly has its advantages and is to be recommended where the soil is not porous or sandy, or if it does not have good drainage. Concrete floors are easier to keep clean and maintain. There is less likelihood that disease organisms will be carried from one group of birds to the next when concrete is used and cleaned before a new group of birds is placed in the house. Floors may also be slat-and-litter, wire-and-litter, all-slat, or all-wire.

Floor Space Requirements

The larger the bird, the more floor space needed. The type of floor also affects the space necessary. Layers on slats or wire require less space than those on a litter floor because of the separation of the birds from their droppings. Requirements are given in Table 15-1.

Conditions affect floor space requirement. The required floor spaces given in Table 15-1 are average, and are generally recommended. However, many things affect the space needed by each bird. Crowding, per se, up to certain limits, does not seem to affect the general health of the birds. However, it is difficult to reduce the floor space and

Table 15-1. Floor Space Requirements for Layers[1]

Type of Floor	Mini-Leghorn			Standard Leghorn			Medium-Size		
	ft²	m²	Birds per m²	ft²	m²	Birds per m²	ft²	m²	Birds per m²
All-litter	1.25	0.11	8.6	1.50	0.14	7.2	1.75	0.16	6.2
Slat-and-litter[2]	1.00	0.09	10.8	1.25	0.12	9.0	1.50	0.14	7.2
Wire-and-litter[3]	1.00	0.09	10.8	1.25	0.12	9.0	1.50	0.14	7.2
All-slat	0.75	0.07	14.4	1.00	0.09	10.8	1.25	0.12	9.0
All-wire	0.75	0.07	14.4	1.00	0.09	10.8	1.25	0.12	9.0

[1]Floor space requirements must be a happy medium between what is required for maximum egg production and what produces wet litter and poor house ventilation.
[2]Approximately 40% litter; 60% slats.
[3]Approximately 40% litter; 60% wire.

maintain optimum conditions in the poultry house. The temperature rises and the litter becomes wetter. Therefore, the bird density usually may be increased as long as environmental conditions in the house do not drop below optimum. It must be remembered, however, that as bird density increases, the productivity of the flock will decrease and mortality will increase. The three factors must be considered when studying the most economical amount of floor space for each bird. Sometimes one can sacrifice some productivity in order to reduce the housing cost per bird.

Room size. The present tendency is to house layers in larger groups than formerly. However, extremely large populations have disadvantages, and for best results the pens should be constructed to hold a maximum of 1,000 layers.

15-F. EQUIPMENT REQUIREMENTS FOR LAYERS ON FLOORS

Special equipment must be provided for laying pullets. Some of it is required to provide only for handling the eggs.

Feeders

The feeders for layers should be at least the same type as those employed for growing birds, but the amount of feeder space necessary for each bird should be greater. Keep the bottom of the feeder the same height as the backs of the birds. Remember that about 20% more birds can eat from the same feeder space provided by a round pan as compared with that provided by a straight trough. Feeder space requirements are given in Table 15-2.

Table 15-2.　Feeder Space Requirements for Layers

Item	Mini-Leghorn		Standard Leghorn		Medium-Size	
	in.	cm	in.	cm	in.	cm
Trough space[1]	3.0	7.5	3.5	8.75	4.0	10.5
	Number of Pullets per Pan or Tube Feeder					
Pans[2]	16		14		12	
Tube feeders[3]	21		18		15	

[1]Space on one side of trough only.
[2]Approximately 12 in. (0.3 m) in diameter. Usually found on certain automatic feeders.
[3]A pan with a circumference of 50 in. (1.27 m) or a diameter of 16 in. (40.6 cm).

Waterers

Many types of waterers are used in floor-type laying houses. Some are automatic, trough-type; some are circular; others have running water. Pans, cups, and nipples are used too, but are less common. Table 15-3 gives the waterer space requirements for layers of different sizes during hot weather. Don't forget that pullets will drink much more water when temperatures are high than when they are low. When you plan for waterer space, plan for maximum need.

Table 15-3.　Waterer Space Requirements for Layers

Item	Mini-Leghorn		Standard Leghorn		Medium-Size	
	in.	cm	in.	cm	in.	cm
Trough[1]	1.0	2.5	1.0	2.5	1.25	3.1
	Number of Pullets per Pan, Cup, or Drip Valve					
Pans[2]	25		25		20	
Automatic cups	8		8		6	
Drip valves	8		8		6	

[1]Space on one side of trough only.
[2]A pan approximately 10 in. (25.4 cm) in diameter.

Nests

Single-compartment nests.　The single-compartment nest is preferred by most poultrymen keeping commercial layers on the floor. Provide one nest for each four pullets in order to have an ample number during the height of egg production. Sufficient nests will aid in the prevention of floor eggs. Hens will use nests of this type better if the nests are placed crosswise of the house. Since nests are usually manufactured in tiers, place the floor of the lowest tier 24 in. (0.6 m) above the floor.

Community nests. On occasion community nests are used. One community nest 2 ft (0.6 m) × 8 ft (2.4 m) should be provided for every 60 pullets. The bottom of the nest should be 24 in. (0.6 m) above the floor (see Chapter 12-C).

Roll-away nests. When roll-away nests are used in a house having a litter floor the bottoms should be 24 in. (0.6 m) above the floor. When an all-slat or all-wire floor is used, there will be fewer "floor eggs" (laid on slats or wire) if the nests are set very close to the floor. Provide one such nest for every four pullets.

Other House Equipment

Light-timing devices. Artificial light is a necessity in the laying house, and electricity to operate the bulbs must be turned on and off according to a predetermined schedule. Automatic time clocks of a capacity ample to operate all the light bulbs are required.

Shell hoppers. Although most laying feeds contain ample calcium, on occasion supplementary feeding of oystershell or other form of calcium carbonate will be necessary. These may be fed in the automatic feeder, or mixed with the mash when the birds are hand-fed. If neither of these methods is suitable, shell hoppers may be used. These are usually hanging tube feeders. Supply one for every 250 pullets in the pen.

Grit hoppers. Provide one hanging tube feeder for each 250 birds in the pen. Feed 1 lb (454 g) of grit per 100 birds per week, and feed the week's allowance on one day. Note that this is a larger amount than that for caged layers. However, if large groups of pullets are kept on an all-slat or all-wire floor, the grit recommendation is the same as for those kept in cages (see Chapter 31-N).

Automatic equipment. Automatic feeders are usually run intermittently; a time clock is used to start and stop the feeder according to a required schedule. Practically everyone uses some form of automation to replenish the water supply in the waterers. Although automatic egg gathering is increasing in favor, this type of automation is still limited to a small percentage of laying houses.

Many systems have been devised to keep the poultryman informed as to what is going on in the chicken house even though he is not there. Alarm systems on motorized equipment may be installed to notify the caretaker at some remote point that something has failed to operate. Closed-circuit television units are being used to show how the birds are behaving in the laying house.

Scales for weighing feed. It is essential that a record of daily feed consumption be kept. This is difficult when bulk feed is fed unless scales are used to weigh accurately each day's supply. In some instances, a controlled feeding program will be used. In this, an allotted amount of feed is provided each day, and it becomes essential to weigh the feed. Several types of scales are available. Some scales are attached to the bulk bin system; others are a component of the feeder system.

Bulk feed bins. Bulk feed bins should hold a week's supply of feed, plus about 2 days' reserve. Feed consumption over a 7-day period will vary according to the density of the feed used, the size of the birds involved, the environmental temperature, the feeding program, and other factors. Many growers prefer to have two bins per house to allow for cleaning and inventory control.

Catching equipment. The good poultry manager will keep a record of the average body weight of his laying hens. A sample weighing should be taken every 4 weeks during the laying cycle. To do the job quickly and efficiently, catching fences, hooks, and suitable scales will be needed.

Dead bird disposal container. Dead birds should be put in a closed container as soon as they are picked up from the house. Off-farm disposal of dead birds should be within 7 days of collection. These containers must be cleaned and disinfected regularly.

Music in the chicken house. Flightiness is common with certain lines and strains of laying birds. Often it leads to hysteria. Sudden, unusual sounds are a contributing factor. In many instances musical sounds tend to calm the birds and reduce any piling or flightiness when the "unusual" occurs. It is certainly far from uncommon to find radios playing loudly in the laying house. Some poultrymen have recorded the noises made by "happy" hens in the course of laying or feeding, and play these recordings continuously throughout the day.

Table 15-4. Flock Size and Egg Cooler Requirements (Eggs Cooled and Cased Twice Daily)

Number of Layers	Number of 15-dozen Baskets or Stacks of Flats to Be Cooled at One Time	30-dozen Cases to Be Stored, Eggs Picked up Twice Weekly	Minimum Size of Cooler		Btu of Cooling Required
			ft	m	
5,000	10	44	6 × 8	1.8 × 2.4	4,500
10,000	20	88	9 × 9	2.7 × 2.7	7,500
20,000	40	176	10 × 14	3.1 × 4.3	12,000
30,000	60	264	14 × 16	4.3 × 4.9	20,000

Egg storage. Market eggs must be cooled as soon after laying as is practical. In many instances a cooling room is constructed at one end of the poultry house, off the feed and service room. The room must be kept at 45° to 55°F (7.2° to 12.8°C), with 80% relative humidity. If eggs are to be shipped to a central grading station, eggs will be transported from the farm daily, or at least twice a week.

Flock size and cooler requirements. The size of the cooler room and the cooling equipment may be correlated with the size of the laying flock. Recommendations are given in Table 15-4.

15-G. HOUSE TEMPERATURE AND LAYING PERFORMANCE

The details of adequate and proper housing are discussed in Chapter 11. Providing an optimum laying environment starts with a good house. It must be capable of protecting the birds from climatic variations encountered with normal day-to-day changes in temperature, for egg production falters when the pullets are subject to temperatures above or below the optimum.

How Temperature Increases Affect the Bird

As the ambient temperature rises, the laying pullet undergoes many changes.

Rising temperatures increase
water consumption
respiration rate
body temperature
stress

Rising temperatures decrease

oxygen consumption	bird weight
blood pressure	egg production
pulse rate	egg weight
thyroid size and activity	eggshell quality
blood calcium level	shell thickness
feed intake	interior egg quality

In-house temperature and laying performance. Just how much the house temperature affects egg production, egg size, and feed consumed per dozen eggs is shown in Table 15-5. Normally egg production does not start to decline until the average house temperature reaches 80°F (27°C), while egg size decreases at temperatures over 75°F

Table 15-5. Influence of In-House Temperature
on Layer Performance

Average House Temperature		Relative Egg Production	Relative Egg Size	Relative Feed/Dozen Eggs
°F	°C	%	%	%
60	15.6	100	100	100
65	18.3	100	100	96
70	21.1	100	100	93
75	23.9	100	99	90.5
80	26.7	99–100	96	88.5
85	29.4	97–100	93	87
90	32.2	94–100	86	86

Source: Shaver Focus, Jan. 1975.

(24°C). Feed efficiency improves as temperatures increase above 60°F (16°C). The effects of high temperature are more pronounced when feeds are not adjusted to compensate for lower consumption. Absolute nutrient intake levels must be maintained.

Cold Weather Problems

Although cold weather must be compensated for, warming the poultry house is much easier than cooling it. All heat in the building is supplied by the birds, and the amount of air moving through the house must be reduced to conserve this heat. Insulation, draft-proof walls, curtains, and reduced fan speed have their place in conserving heat.

Moisture buildup. As the amount of air flowing through the poultry house is reduced to conserve heat during cold weather, less moisture is moved out of the building. Litter becomes damp or wet and creates a difficulty. Although dry litter may be added and mixed with the wet to improve its consistency, such a program is too expensive to be carried out over a long period of time. The manager's ability to regulate the movement of air to remove most of the moisture from the house while conserving as much heat as possible provides the normal solution to the problem.

House temperature below freezing. When the outside temperature gets very low, it may become impossible for the birds to generate enough heat to keep the temperature inside the house above freezing. Water in the waterers may freeze, pipes burst, and the combs and wattles of the birds may freeze. Egg production suffers, and if the cold weather continues for some time the birds may undergo a partial molt. When laying birds are to be kept in such a climate, better house construction is the only positive answer to the difficulty. Although inside freezing results in bird disaster, the physical well-being of the layers diminishes when the house temperature drops below 55°F (12.8°C). The optimum house temperature is between 65° and 75°F (18.3°C and 23.9°C).

Inadequate feed consumption. As temperatures drop, the birds eat more feed in an endeavor to maintain their body temperature. But in some instances layers are fed a restricted or controlled amount of feed each day. If the feed requirement to maintain body temperature and egg production is greater than the feed given, the layers will reduce their production of eggs in order to use the feed for body heat rather than for eggs. The good manager will make adjustments in the feed allocation when cold weather strikes.

The Challenge of Hot Weather

At temperatures above 80°F (27°C) laying pullets begin to suffer and performance begins to diminish. At 100°F (38°C) things become serious. Egg production drops drastically and many birds may die from heat exhaustion. The skilled manager is one who can cope with high outside temperatures by reducing the temperature inside the house. Too often nothing is done; the manager just takes the heat as an "act of God," and offers no remedial measures.

Handling the conventional house. Even with the open-sided house many programs can be used to reduce the house temperature and to make the laying pullets more comfortable.

1. Insulate the roof or ceiling.
2. Increase ventilation.
3. Move the air faster by providing fans.
4. Lower the humidity.
5. Use foggers.
6. Sprinkle the roof—run sprinklers intermittently.
7. Wet the area outside and around the house.
8. Provide cool nests—open the backs.
9. Give cool, fresh water.
10. Increase the waterer space.
11. Give fresh feed during the morning and evening cool hours.
12. Avoid direct sunlight on the birds.
13. Keep activity in the house to a minimum.

Handling the environmentally controlled house. The totally lighttight house that uses forced air for ventilation may present hot-weather problems greater than those incurred with the conventional house. Certainly the house should be well insulated to reduce the penetration of heat from the sun's rays. Fans should operate at their maximum capacity. Foggers may be used in the building. Be cognizant of the fact that birds on slats or wire are hotter than those on a floor, because the floor is quite cool compared with the air, and birds can "bury" themselves in the litter to get the coolness from the floor. When they are on slats or wire, they are completely surrounded with hot air.

There is no doubt that the solution to cooling the environmentally controlled house lies with evaporative cooling. It is the best system known today, and is being installed in more and more laying houses (see Chapter 11-T).

15-H. MANAGING THE LAYING FLOCK

The day-to-day management of the laying flock on a litter, slat, or part-slat (or part-wire) floor taxes the ability of most poultrymen. It is during the period of egg production that profits are made or lost. Maximum egg production is essential, and the value of daily flock care cannot be overemphasized.

Changing the Lighting Program at Sexual Maturity

Most growing pullets will have been on a light-control program during the growing stage. The hours of natural plus artificial light per day must be increased when the birds reach sexual maturity, but the procedure is at the discretion of the managing poultryman. It must also be tied in with an increase in feed allocation at this time. Both are discussed in Chapter 18.

Management. Follow these procedures:

1. Clean the light bulbs regularly. A good practice is to clean one-third of the bulbs and reflectors each week.
2. Replace burned-out bulbs daily.
3. Watch the length of the light day. As daylight hours increase or decrease, the period of artificial lighting must be adjusted accordingly.
4. Do not decrease the total light day during the laying period. The light day must remain constant or increase as the production year progresses.
5. Be sure the light intensity at floor level is correct. Either too little or too much light is to be avoided. Use a light meter if in doubt (see Chapter 18-E).

Nest Management

Proper nest management will be a great aid in producing quality eggs, and with less breakage.

Time of day eggs are laid. Egg production is not uniform throughout the hours that light is available. Although time of egg laying is not associated with this type of house, type of nest, density of the birds, or ambient temperature, the time of first light in the morning does produce an effect. Pullets will start laying eggs about 1 or 2 hours after the light is bright. The proportion of eggs laid during each hour thereafter is given below:

Hour after Bright Light Begins	% Daily Total of Eggs Laid
1	Few
2,3	40
4,5	30
6,7	20
8,9	10
10,11	Few

Close the nests at night. To prevent pullets from remaining on the nests overnight, any birds found on them at the end of the day should be removed, and the nests closed. Be sure to open them in the morning before the birds are ready to lay. Having nests empty when not needed at night will keep the nest litter cleaner, help prevent broodiness, and keep the eggs cleaner.

The market-egg pickup. Market eggs should be gathered at least two times a day during cool weather and three times during hot weather. Egg breakage will be reduced if there are few eggs in the nest at any period of the day. Do not leave eggs in the nest overnight. Pick up any remaining eggs when the nests are closed at the end of the day.

Nesting material. Any good, dust-free, dry material will suffice as a nesting material. See Chapter 7-A for a list of these. Replenish the material as needed to keep the floor of the nest covered and the nesting material clean.

How to Prevent Floor Eggs

Eggs laid on the floor, rather than in the nests, are a costly expense. Usually they are dirty, many are broken, and it is laborious to gather them. Pullets must be trained to use the nests when they begin to produce eggs. Once they lay on the floor it becomes difficult to change their habit. There are several helps to get the birds in the nests early.

1. Fence off the corners of the pen. Corners are the most likely place for a floor "nest."
2. Place the nests in a darker part of the house.
3. In houses with part-slat floors, place the nest sections at right angles to the house. One end of the nest section may rest on the slats.
4. Use darkened nests. Cover the upper part of the nest opening, and upper part of any back opening, to darken the nests. But be careful during hot weather. This procedure may cause excessive heat in the nests.
5. Open the nests and have nesting material in them a week before any eggs are produced. Let the pullets get acquainted with them.
6. Have an adequate number of nests. No pullet should be required to lay on the floor because she cannot find an empty nest.
7. Keep the nesting material clean and ample.
8. Community nests increase the incidence of floor eggs.
9. Sometimes roll-away nests prove difficult, as the nest floor is usually wire, and the birds may refuse them at first. Covering the wire with coarse straw, a nest pad, or Astroturf® for the first week or two will induce the pullets to lay in the nests rather than on the floor, but some eggs will need to be hand gathered from the nests at first.

Culling

Individual unproductive and inferior pullets should be removed from the pens during the laying period. Usually it is unwise to believe that these will be able to produce a profit. Remove the birds with a hook about once a week. Do not catch the entire group to remove the culls. Such handling will generally reduce egg production in the entire flock. Pullets that have not started to lay by 30 weeks of age should be removed. Culling should be intensified during the latter stages of the laying year.

Broodiness

Broodiness, an inherited factor, is due to the pituitary hormone, prolactin. Today, however, the genetic factor has been eliminated from practically all Leghorn lines, but some broodiness may be evident in those birds producing brown-shelled eggs. Small cages, with a wire or slat floor, may be constructed in the poultry house. When broody hens are removed from the nests and placed in these, the broody instinct will disappear in 2 or 3 days, and the pullets may be returned to the floor. Supply feed and water to those birds in the broody coop.

15-I. FEEDING THE LAYERS

The feeding of commercial egg-type layers can be complicated. Such factors as phase-feeding, controlled-feeding, and self-feeding may enter into the program. Both the feeding method and feed formula are important and must be altered according to the weather. Breed and age of the birds dictate changes in feeding methods. All are detailed in Chapter 31.

Coccidiostat in the Laying Feed

When mature pullets reach the age of 21 to 22 weeks they should have developed immunity to coccidiosis through a program for cocciodiostat removal or by other means. But be careful. Laying feed does not usually contain a coccidiostat; if the growing feed contained this ingredient, and the laying feed doesn't, and the birds do not have full immunity, an outbreak of coccidiosis may occur in the laying quarters. At this age, the disease may prove disastrous; egg production will be delayed, and a less than normal number of eggs will be produced.

15-J. HANDLING MARKET EGGS

Oiling Market Eggs at Pickup

When eggs are to be washed several hours after gathering, they may be oiled at the time they are removed from the nest to preserve initial interior egg quality (see Chapter 16-P).

Cooling Market Eggs

As soon as eggs are laid, their quality should be preserved by cooling. Some means of refrigeration is necessary on most poultry farms. Any laying house of size should have a refrigerated room where the temperature may be kept at optimum (see Chapter 16-P).

Preventing Egg Breakage

Refer to Chapter 16-P for details.

Market Egg Quality

Refer to Chapter 16-T and 16-U for details.

16
Cage Management

This chapter deals with those procedures necessary to grow and keep layers in cages. The system of management is not new, although there have been numerous changes through the years. As early as the 1930s, chickens were kept in cages. The first held one laying bird; sometime later multiple-bird cages were introduced.

It is now estimated that about 75% of all the commercial layers in the world are kept in cages. In the United States, over 95% of the layers and 50% of replacement pullets are in cages. Many of the recommendations given in Chapter 15 have a bearing on cage management, and this chapter should be read first.

16-A. BROODING-GROWING CAGES

Not all cage operations keep the pullets on a wire floor from one day of age until they reach sexual maturity. There are four combinations for using a litter floor and a wire floor during this period:

1. Wire brooding (to 6 weeks), litter growing
2. Litter brooding (to 6 weeks), wire growing
3. Two-house, wire brooding and growing
4. Brooding-growing continuous cage

Brooding Cage Specifications

These specifications vary widely according to the manufacturer of the equipment and the procedures used for brooding and growing. Usually, however, the chicks are started in one battery cage containing a heating unit. Later, some of the growing birds are moved to one or more cage batteries, thus allowing more floor space as the birds become older.

Brooding cage size.   Although brooding cages usually are for growing birds as well as for young chicks, and are about 14 to 16 in. (31 to 41 cm) high, the size of the floor varies. Some common sizes are

> 22 in. wide × 24 in. deep (55.9 × 61.0 cm)
> 24 in. wide × 24 in. deep (61.0 × 61.0 cm)
> 24 in. wide × 27 in. deep (61.0 × 68.6 cm)
> 24 in. wide × 36 in. deep (61.0 × 91.4 cm)

Floor material.   The floors of the brooding cage are made of

> *Welded wire fabric:* The mesh size usually is ½ × 2 in. (1.3 × 5.1 cm) or 1 × 1 in. (2.5 × 2.5 cm). The wire size is about 14 gauge. When the mesh size is greater than ½ in. (1.25 cm), the floor will need to be covered with paper for the first 2 weeks.
> *Plastic:* This material is either entirely plastic or plastic-covered wire.

Slope of floor.   The floors of most cage brooding units do not slope, but some slope slightly upward at the front near the feeder.

Front of cage.   In most instances, feeding is done at the front of the brooding cage, nearest the aisle. The front of the cage is adjustable; this provides access to the feeders while preventing the chicks from getting out of the cages. By adjusting the size of the openings through the wire, birds of various sizes may use the feeders.

Gates.   Gates are built into the brooding cage to make it possible to put the chicks into the unit and to take them out. The gates may be on the front or on the top of the cage.

Heating units.   Unless warm-room brooding is used, it will be necessary to provide some supplementary brooding heat. Although there are several arrangements, a hot-water pipe is a common installation. The pipe runs lengthwise of the house, above each cage unit.

Waterers.   Cups, drip-nipples, troughs, and other arrangements are used to supply water. Regardless of the type used, the height should be adjustable to allow for the growth of the birds.

Feeders.   The trough type of feeder is in predominate use during the brooding period. It may be filled manually or automatically. Some method of adjusting the height is preferable (see Figure 16-1).

Growing Cage Specifications

Because brooding cages are usually used for growing as well as brooding, the specifications are similar. In some instances, however, there are

Figure 16-1. Replacement pullets in growing cages.

separate growing units that are installed in buildings different from the brooding houses.

Cage size. In most instances, this is the same size as the brooding cage. Because of the age and size of the birds, however, fewer birds are placed in each unit. Space per bird is generally doubled.

Floor. The mesh size of the wire on the floor of the cage should be large enough to allow the droppings to fall through easily. The floor should be flat, or nearly so.

Waterers. The types of waterers are those employed with brooding units, but their height is greater.

Feeders. For growing birds, trough-type feeders are the most common. They may be filled by hand or automatically. In the latter case, the troughs need not necessarily be outside the cage; some are in the center, allowing the birds to eat from both sides. Troughs are about 5 to 6 in. (13 to 15 cm) wide, with a good lip that helps prevent feed wastage.

Two-House, Wire Brooding and Growing

With this system, one house or unit is used for brooding; the second and larger house is devoted to growing. Some producers use one brooder to serve two grow houses on an alternate basis. There are advantages and disadvantages to this system of management.

Advantages

1. The system utilizes the smallest amount of floor space, particularly when multideck units are used.
2. Housing, cages, and equipment can be better tailored to the age of the chickens.
3. Brooding heat costs are low because of the high bird density in the brooding house.

Disadvantages

1. Length of time the houses are empty is greater between broods, often increasing the growing cost.
2. Close scheduling of facilities to reduce downtime (vacant time) may lead to difficulties in providing time for cleaning between groups of birds.
3. There is the added stress and cost of moving birds from one house to another.

Brooding–Growing Combination Cages

With this program, the birds are kept in the same house or unit from one day of age to the time they are moved to the laying location. The cages are large enough for growing birds, but some of them are used for brooding. Again, there are advantages and disadvantages:

Advantages

1. The birds are not moved to another house; therefore, the stress associated with the transfer is reduced.
2. The downtime usually is shorter.
3. Labor costs may be lowered, because some of the vaccinating and beak trimming may be done when the chicks are transferred to larger cages, and one move is eliminated.

Disadvantages

1. In most cases the unit may be more expensive than with the two-house system, but a lot depends on the location of the brooding units in the combination house.
2. The house is cooler during the brooding period because of its larger size.
3. Heating equipment is poorly utilized during the growing period and feeding and watering systems are usually a compromise between the needs of day-old chicks and 18-week-old pullets.

Number of decks. Some brooding-growing systems employ the use of one deck, some two decks, and some three or more.

1. *Single-deck:* Most of the new installations do not use the single-deck, brooding-growing units. The housing cost is greater because of the reduced bird density in the building, but handling of the birds is usually much easier, air quality is generally better, and the handling of droppings is simplified.
2. *Two-deck:* Chicks are started on one deck. Supplementary heat is supplied to this deck or to the entire room. At 4 to 6 weeks of age, half the chicks are moved to a cage above or below the starting cages.
3. *Three-deck:* Chicks are placed in a heated cage at starting time. At about 4 weeks, one-third of the birds are moved to a cage above or below, and when they are about 6 weeks of age half the remaining chicks in the brooding unit are moved to the empty cage.

Manure disposal. Droppings fall through the mesh bottom of the cage to the floor below. Although this is a simple procedure when there is but one deck of cages, and with two decks when the upper deck is not directly over the lower, there is a problem with the three-deck cage, as the upper decks will overlap the deck below. To prevent the droppings from falling on the birds in such cases, a sloping droppings board is inserted between the decks. In most cases it slopes toward the center of lines of cages, and the droppings fall to a common place in the middle of the floor.

Removing manure from the floor. Although a concrete floor is not necessary under cages, it does aid when it becomes necessary to scrape up the manure from the floors. Droppings may be removed by a power cart or small tractor or with mechanical scraping devices installed under the cages in a shallow pit. If mobile equipment is used, there should be no cage-supporting posts to interfere with the procedure. For this reason, many cages are hung rather than supported by posts.

Deep-pit system of collecting droppings. The pit system for cage operations is becoming popular. Walks are constructed between the cages for service work, but the area under the cage is open, allowing the droppings to fall into a deep pit that may be easily cleaned (see Chapter 11-S).

16-B. BROODING–GROWING CAGE HOUSES

The width of the brooding-growing house must be determined by the dimensions of the cages that go into it. In some mild climates, little housing may be required; "housing" is confined to a suitable roof with curtains on the sides. These curtains are to be used only during the first few weeks and during inclement weather.

High Chick Density

It is obvious that double-decking and triple-decking increases the density of the birds in the house. There must be ample ventilation (see Chapter 11-I and 11-J). In moderate or cold climates, the environmentally controlled house is the best means of providing a constant flow of fresh air, and the elimination of moisture and ammonia.

Because pullets grown on wire floors in cages require much less house floor space than do those raised on a litter floor, there may be from 3 to 5 times as many birds in the house, depending on the age of the birds and the type of cage construction. With so many birds in the cage house there usually will be difficulty in keeping the house temperature at optimum. Even with the open-sided house, there may be problems. In the environmentally controlled building the use of evaporative cooling is the best means of producing lower house temperatures (see Chapter 11-T).

16-C. CAGE BROODING MANAGEMENT

Preparing the cage house for new pullets follows the procedures outlined for the house with a litter floor. Cleanliness and a good start are important. Isolation of the brooding unit should be made a part of the management program. All chicks should be the same age. Practice the all-in, all-out system.

Floor, Feeder, and Waterer Space Requirements

Mini-Leghorns, standard Leghorns, and medium-size pullets, raised for the production of commercial eggs, are the lines most common for cage operations. The size of the cage and the age of the bird will determine the number of pullets that can be kept in it, and the floor-space requirements per bird are given in Table 16-1 during three growing ages. Feeder and waterer space are also given.

Table 16-1. Space Requirement per Pullet During Cage Brooding, Growing, and Laying

Item	Period					
	0–5 Wk		6–18 Wk		19 Wk and Over	
	in.^{2}	cm^2	in.^{2}	cm^2	in.^{2}	cm^2
Floor space						
Mini-Leghorn	20	129	36	232	48	310
Standard Leghorn	24	155	45	290	60	387
Medium-size[1]	28	181	54	348	70	452
	in.	cm	in.	cm	in.	cm
Feeder space						
Mini-Leghorn	1.6	4.1	2.0	4.1	2.4	6.1
Standard Leghorn	2.0	5.1	2.0	4.1	3.0	7.6
Medium-size[1]	2.2	5.6	2.7	6.9	3.3	8.4
Waterer space (trough)						
Mini-Leghorn	0.60	1.5	0.80	2.0	1.25	3.2
Standard Leghorn	0.75	1.9	1.00	2.5	1.50	3.8
Medium-size[1]	0.80	2.0	1.20	3.1	1.70	4.3
	Number of Pullets per Nipple or Cup					
Waterer space (nipple)						
Mini-Leghorn	20		13		10	
Standard Leghorn	15		10		8	
Medium-size[1]	12		8		6	
Waterer space (cup)						
Mini-Leghorn	33		24		19	
Standard Leghorn	25		15		12	
Medium-size[1]	19		13		10	

[1]Producing brown-shelled eggs for market.

Supplemental Heat

When the heat from a hot-water pipe is used for supplying supplementary heat to the brooding unit, the water in the pipe should be kept at a temperature of about 180°F (82°C). Often one pipe will supply heat for two cages when they are placed back to back. Heavy, reflector-type paper may be placed over the top of the pipe to deflect heat into the brooding cage. Some units have a deflector built into them. Maintain a temperature of between 82° and 85°F (27.8° and 29.4°C) about 2 in. (5 cm) above the floor and in the locality of the heated area. Higher temperatures will increase chick dehydration. Chicks should not huddle if the temperature is correct.

Paper on Cage Floor

Paper is often laid on the wire floor and kept there for about 2 weeks. It should be heavy, rough, and highly moisture-proof. This paper serves several purposes.

1. It provides a solid floor when the chicks are young.
2. It allows for the use of wire with larger openings.
3. It keeps the brooding area warmer.
4. It serves as an area for the first feed.
5. Rough paper allows the chicks to move about more freely.

Watering

Chicks should drink quickly once they are placed in the cage brooder. Have water available in the founts when the chicks arrive at the farm. Many times the watering devices are not easily used by day-old chicks; the birds have difficulty in learning to drink from nipples, certain cup founts, and others. It may be necessary to trigger these waterers by hand the first several days or to supplement with small jug waterers. Watering devices should be adjustable so they can be lowered when the chicks are young, and raised as they grow older.

Water sanitizers. Sanitizers may be used in the drinking water, but take the usual precautions to provide clean water and clean founts when administering any vaccine through the drinking water (see Chapter 39-C).

Waterer space. See Table 16-1.

Room Temperature

Cage brooding requires a relative high room temperature. It should not be allowed to drop below 60°F (16°C). This promotes chick comfort and reduces the cost of providing the supplementary brooder heat. But do not allow the room temperature to get too high. Remember that chicks do not normally feather as well on wire as they do on litter, and high temperatures induce cannibalism. Provide adequate ventilation and house cooling when necessary to minimize the effects of high outside temperatures.

Light

It is much more difficult to provide uniformity of light intensity in cage operations than in floor systems. This problem is increased with multidecked cages, because the light intensity from a single bulb hung overhead is greater at the top deck than at the lower. Furthermore, shadows produced by the cages cut off a great deal of the light supply. In some specialized operations in mild climates, natural daylight may suffice for the birds to see to eat, but it cannot be used during a light-control program (see Chapter 18-E).

Several factors affect the proper use of artificial light:

1. Provide bright light continuously for the first 4 days the chicks are on feed so that they may get off to a good start by learning quickly to eat and drink. The intensity of light should be 3.5 fc (35 lx) at bird level.
2. After the chicks are 4 days of age, gradually change the length of the light day (natural plus artificial light) according to a predetermined schedule (see Chapter 18). The light-control schedule is necessary to regulate the age at which the birds reach sexual maturity. There are several lighting programs, but each poultry producer must follow the one that will produce the best results under his system of housing, management, and geographic location.
3. After 4 days, the intensity of the light for growing pullets should be about 0.3 fc (3 lx) at the location of the lowest cage deck or 0.5 fc (5 lx) if there is but one deck. This will be adequate for normal feed consumption and will prevent a great deal of cannibalism. Many poultry producers have installed rheostats or solid-state dimmers in their artificial lighting systems to regulate intensity accurately and easily.

Vaccination and Medication

There must be a satisfactory vaccination program with caged birds, as with those reared on the floor. Because of the house density, some diseases are more prevalent in cages. Respiratory diseases are especially common. See Chapter 39-E for suggestions regarding a vaccination program.

Medication too is important not only to treat disease outbreaks but to handle certain nonspecific types of cage difficulties. Keep a record of all medications. Coccidiosis is almost completely eliminated when pullets are kept on wire from 1 day of age until they complete their laying cycle. Remember, however, that such birds do not build any immunity to the disease, as they have never come in contact with an ample number of sporulated oocysts. If the pullets are moved from wire to litter, there is a great chance that coccidiosis will develop. Be sure some means of handling the coccidiosis problem is provided in such cases. Be ready to medicate quickly.

Dubbing

Some strains of chickens are endowed with large combs that are difficult for the birds to handle in cages. In some other cases there may have been severe injury to the combs of some of the birds while in the laying cages. Pullets so involved should be dubbed at hatching time (see Chapter 13-J).

Beak Trimming

Cannibalism is more common when birds are confined to cages, either during growing or laying. There are several methods of reducing the incidence; of these, beak trimming is the best and most commonly used (see Chapter 13-I). Although laying-type pullets may be beak trimmed at any age prior to reaching sexual maturity, there are three ages most often involved: (1) 7 to 10 days, (2) 6 to 8 weeks, and (3) 12 weeks.

Toe Clipping

Flightiness and nervousness of caged layers often results in bird injury. One contributor to the difficulty is long toenails. Toe clipping the center toe or the front three toes at 1 day of age has been shown to alleviate the difficulty and may actually improve annual egg production. However, some strains are more nervous than others, so improvement is relative. Recent research has demonstrated that shorter nails will result when an abrasive strip is available inside the cage.

16-D. FEEDING DURING THE BROODING PERIOD

Correct nutrition is detailed in chapters later in this book. However, there are several feeding practices necessary for young chicks that should be discussed here.

First Water

Chicks should have fresh, warm water but no feed when first placed under the brooder. Adding sucrose (sugar) to the first water will improve early growth and livability. If chicks are stressed when delivered to the farm, water soluble vitamins plus electrolytes may be added to the water (see Chapter 13-E).

Form of Feed

Before receiving any chicks, the poultryman must make a decision regarding the form of feed to be used. One has the choice of either mash or crumbles, but most chicks are started on mash. Keep in mind that compressing a feed, as in the manufacture of crumbles, induces more cannibalism in cages.

First Feed

Three hours after the chicks have been given water, supply them with feed. Many flockowners prefer to place paper or shallow cardboard containers on the wire floor adjacent to the cage feed trough for the first feed where the chicks can easily find it. The cage feed trough is usually filled at this time as well.

Feed Consumption

Keep a weekly record of feed consumption. This will necessitate weighing the feed. Many factors, such as the strain of the birds, composition of the feed, and the ambient temperature, influence the amount of feed consumed during the first 6 weeks. As a guide, however, figures for daily consumption are given in Table 30-5.

16-E. CAGE GROWING MANAGEMENT

The growing period often depends on the age at which the mature pullets are placed in their permanent laying quarters. It may vary between 16 and 22 weeks under most conditions, but occasionally pullets are placed in laying cages as young as 14 weeks of age.

Floor, Feeder, and Waterer Space Requirements

The floor space, feeder space, and waterer space requirements for cages are in Table 16-1. Floor space requirements also vary with the ambient temperature as shown in Table 16-2.

Table 16-2. Effect of Pullet Cage Space and Temperature on 20-Week Body Weights

Temperature		Floor Space per Pullet		
°F	°C	43 in.^{2}	38 in.^{2}	35 in.^{2}
		(Weight in Pounds)		
70		3.06	2.90	2.86
90		2.71	2.63	2.46
		276 cm^2	248 cm^2	225 cm^2
		(Weight in Kilograms)		
	20	1.39	1.32	1.30
	32	1.23	1.15	1.12

Source: Ralston Purina Co., 1980.

Growing Waterers

As with the waterers used during the brooding period, there are many types of small waterers capable of furnishing water in growing cages. Troughs with running water, cups, and nipples are the usual methods of providing water.

Growing Feeders

Feeding is done by hand or by automatic equipment. Feed troughs are a part of most installations, and automatic devices force or drag the feed through the building, filling the feeders. Feed carts are used to semiautomate "hand" feeding. These are motorized, and feed is scooped up from a feed tank and added to the troughs by hand; or an auger and elevator lift the feed from the tank, after which it flows to the troughs by gravity.

Light

The length of the light day is very important during the growing period. It must be regulated to cause the pullets to begin egg production at a desired age. There are several programs; all are explained in Chapter 18.

Vaccination and Medication

Vaccination and medication programs must be carried out carefully during the growing period. Most of these start during the brooding phase, but continue as the pullets grow older. Consult a trained poultry pathologist or your vaccine and medicine supplier for programs that are necessary in your area. Some suggestions are given in Chapter 39.

Cannibalism

When birds are given limited space, as in cages, there is a tendency for many to become cannibalistic. Some means of preventing this vice must be initiated before trouble begins. Beak trimming is the most accepted practice. If the birds were not trimmed when they were 7 to 10 days of age, their beaks should be cut back some time during the growing period and prior to an increase in picking.

Music to quiet the pullets. As with pullets grown on a litter floor, radio music offers a possibility for quieting the growing birds. They are less apt to become frightened from unusual noises and the movements of the caretaker.

Crowding in Growing Cages

In *Poultry Tribune,* January 1969, results of experiments having to do with crowding birds in growing cages and the effect on body weight were reported. Various numbers of Leghorn pullets were placed in 24 × 24 in. (61 × 61 cm) cages at 1 day of age and weighed at different ages. Floor space per pullet varied between 96 in.^{2} (619 cm^2) and 29 in.^{2} (187 cm^2). Body weight at 16 weeks was not materially affected until the pullets were kept in less than 58 in.^{2} (374 cm^2) of floor space, after which weight decreased as floor space was reduced. From an economical standpoint, it seemed evident that a 24 × 24 in. (61 × 61 cm) cage would accommodate 14 Leghorn pullets to 16 weeks of age (Table 16-3).

Sample Weighing of Birds

During the course of keeping chickens it becomes necessary to weigh the birds to establish mean flock weights. But it is not necessary to weigh all the birds; only a small percentage need be weighed, but the percentage must be great enough to be representative of the entire flock. Addi-

Table 16-3.   Effect of Cage Floor Space on Standard Leghorn
16-Week Body Weight

	Data for Each 24 × 24-in. (61 × 61-cm) Cage							
Pullets per Cage	6	8	10	12	14	16	18	20
In.² floor space per pullet	96	72	58	48	41	36	32	29
Cm² floor space per pullet	619	464	374	310	265	232	206	187
Body weight, 16 wk (lb)	2.79	2.64	2.70	2.57	2.50	2.45	2.38	2.26
Body weight, 16 wk (kg)	1.27	1.20	1.23	1.17	1.13	1.11	1.08	1.02

Source: Poultry Tribune, Jan. 1969.

tional samples should be taken if the flock is housed in two or more houses or locations. The more birds weighed, the greater the accuracy. Table 16-4 shows the minimum numbers according to flock size.

Instructions for sample weighing

1. The sample should be representative of the flock.
2. Use an accurate scale capable of weighing in ounces (grams), or less.
3. Weighing should start at 3 weeks of age.
4. Weigh each week during the growing period and monthly during the laying period.
5. When pullets are on a litter floor and in a large pen weigh the birds from several locations. Segregate a small group with a catching fence and weigh each bird separately.
6. Cage pullets should be weighed from various locations in the room or line. Mark the cages and weigh the same birds either weekly or monthly.
7. Weigh at the same time each weighing, preferably in the afternoon.

Table 16-4.   Number of Birds to Be Sample Weighed
According to Flock Size

Flock Size	Number of Birds to Be Weighed	Flock Size	Number of Birds to Be Weighed
Under 500	60	4,000–6,000	150
500–1,500	80	6,000–8,000	175
1,500–3,000	100	8,000–10,000	200
3,000–4,000	125		

8. When growing pullets are on a skip-a-day feeding program (see Chapter 30-M), weigh them on a no-feed day in the afternoon.

Classifying body weights. Two classifications are necessary as follows:

1. Average (mean) weight of the sample. Compare the figure with the standard for the same age.
2. Sort the individual pullet weights by intervals of 10% from the *standard* flock body weight. Total each group and calculate the percentage of each group. Compare these percentages with those in Table 16-7 to show the uniformity of the flock.

These two measurements must be considered together since two flocks may have the same average weight but marked differences in uniformity, and two flocks can be equally uniform but one can be light in weight and the other heavy. Thus, comparison of weights must be with the standard and not the average of the sample.

16-F. GROWTH STANDARDS FOR LAYING-TYPE PULLETS

Weekly growth standards for egg-type pullets as given by the primary breeder are highly variable because each breeder has determined a mature body weight (20 weeks) for his strain in order to produce the best results in the laying house. These mature weights may vary 10% from the figures shown in Table 16-6. Some breeders recommend restricted feeding during the growing period in order to meet these weights; others recommend full feeding. The figures in Table 16-5 are an average of those recommended by the primary breeders.

16-G. BODY WEIGHT AT SEXUAL MATURITY

Practically all primary breeders publish an optimum body weight guide for each strain of commercial pullets they produce. Evidence points to the fact that egg production is greater when these optimum weights are attained at sexual maturity. Those flocks that are too light and those that are too heavy do not produce as many eggs.

Factors Affecting Body Weight

Chick weight. The heavier the day-old chick, the heavier the pullet at 12 to 18 weeks, but the correlation is less at the older age. At 20 and 21 weeks of age, factors other than initial chick weight show their effect on body weight and variations in body weight cannot be as

Table 16-5. Body Weight Standards for Egg-Type Growing Pullets

Week of Age	Standard Leghorn		Medium-size[1]	
	lb	kg	lb	kg
1	0.14	0.065	0.30	0.13
2	0.27	0.121	0.40	0.18
3	0.41	0.186	0.60	0.27
4	0.58	0.262	0.80	0.36
5	0.74	0.335	1.00	0.46
6	0.94	0.427	1.30	0.59
7	1.13	0.513	1.50	0.68
8	1.31	0.593	1.70	0.77
9	1.48	0.671	1.90	0.86
10	1.66	0.754	2.10	0.95
11	1.83	0.828	2.30	1.04
12	1.99	0.904	2.50	1.14
13	2.13	0.968	2.70	1.23
14	2.27	1.030	2.90	1.32
15	2.41	1.092	3.00	1.36
16	2.55	1.157	3.20	1.45
17	2.67	1.211	3.30	1.50
18	2.78	1.259	3.40	1.54
19	2.89	1.311	3.60	1.64
20	3.00	1.362	3.70	1.68

[1]Producing brown-shelled table eggs.

Table 16-6. Effect of Hatch Date on 18-Week Body Weights

Month	%	Month	%	Month	%
January	105	May	96	September	104
February	98	June	97	October	101
March	97	July	96	November	105
April	96	August	99	December	105

Source: University of California, 1981.
Note: % is percent of annual average weight.

easily associated with day-old weight. However, pullet weight uniformity at sexual maturity is closely associated with uniformity of weight at day old.

Season affects body weight. Most weight recommendations furnished by the breeders are for pullets raised during the cooler months, but some are an average of birds grown during all seasons. Fall-hatched Leghorns, which are grown during the cooler months of the year, will be about 10% heavier than those raised during the warmer months. Although some breeders recommend some degree of feed restriction during the cooler months in order to reduce the mature pullet weight, the difficulty during hot periods is getting the birds to eat enough to bring them up to the recommended mature weight (see Table 16-6).

Growing houses that are environmentally controlled tend to offset many of the difficulties encountered with summer-grown pullets. In the first place, the length of the light day may be controlled. Also, if the house has evaporative cooling, the ambient temperature may be lowered, which will materially affect the 20- to 21-week body weight. However, if the birds are grown in open-type houses, little can be done to increase body weight except by following those practices that will increase feed consumption.

Weekly body weights must be maintained. Not only do prime breeders recommend an optimum body weight at 20 to 21 weeks for each strain of their birds but they also give weekly weights during the entire growing period. These bear special significance, for if a certain mature weight is to be attained, it is necessary that weekly growth proceed according to a prescribed formula. Too many times the flockowner will wait until the last part of the growing period to make the weight adjustments. If the birds are light in weight and on full feed, it may be impossible to get them to eat more feed at this time; if they are overweight, excessive reduction in feed intake to reduce weight quickly may be detrimental and lead to stresses and increased mortality.

Problems of light body weights are usually corrected by increasing protein and energy levels in the diet, reduction in ambient temperatures, elimination of crowding, and changing to an earlier beak trimming age.

How much change in feed consumption? In order not to make body weight adjustments too rapidly, there must be some regulatory guide.

Rule of thumb: For each 1% over or under the optimum weekly body weight, decrease or increase the feed allocation by 1% respectively.

Example: If the recommended weight at 11 weeks of age is 2 lb (907 g), and the flock average weight is 2.1 lb (953 g), the birds are overweight by 5% and the feed allocation should be reduced by 5%.

Uniformity of Growing Pullets

Genetic uniformity of birds within a flock is an indication of an increase in homozygosity (see Chapter 22-C). Close breeder selection and a higher degree of inbreeding accomplish this. Uniformity varies by strains of birds, but uniformity also depends on the type of management.

Crowding, stress, age, nutrition, mortality, and so forth are factors involved.

Uniformity is most often measured as a percentage of the birds in the flock that are within 10% of the mean weight of all the birds. Table 16-7 shows a scorecard for degree of uniformity for pullets based on flock deviation from the mean weight of the flock. When the mean weight is equal to the breeder standard, the uniformity measurement is much more meaningful.

Variations in uniformity.   Flock uniformity is one of the best indicators of pullet quality. The better the uniformity of growing birds, the better the future egg production. The flock score (Table 16-7) is variable, however, and the following factors affect it. Therefore, the percentages in the table should be adjusted accordingly.

Scale weight interval affects the percentage of pullets falling within 10% of the mean flock weight. The larger the interval, the higher the percentage, as shown in the following example:

Scale Weight Interval	% of Birds within 10% of Mean for "Average" Flock
5g	68
1 oz (28 g)	73
0.1 lb (1.6 oz) (45 g)	78

For this reason, never compare uniformity results when different scales are used.

Flocks of normal health are more uniform at 18 weeks of age than younger or older flocks. However, stressed flocks lose their uniformity the longer the stress is in effect. Therefore, in many instances, younger flocks show the best uniformity, but it decreases the older the flock becomes.

Sexual maturity affects uniformity. Pullets beginning to lay increase

Table 16-7.   Percentage of Pullets Within 10% of the Average Weight of the Flock

Score	Percentage of Pullets within 10% of Average Flock Weight[1]
Superior	91 and above
Excellent	84–90
Good	77–83
Average	70–76
Fair	63–69
Poor	56–62
Very poor	55 and below

[1]With scale weight interval of 1 oz (28 g), or less.

greatly in body weight. Because all pullets do not reach sexual maturity at the same age, flocks express their poorest uniformity, and figures at this time must be considered of lesser value.

Table 16-8 shows the percentage of birds expected by body weight categories for the seven grades, superior, excellent, good, average, fair, poor, and very poor based on a normal curve. That is, the percentages in each weight category below the mean flock body weight are the same as those above. However, any stress in the group tends to distort the curves because there are more individuals in the lighter weights and this decreases the percentages of those in the heavier weights.

Body Frame Measurements

Body weight is by far the most common method used to evaluate pullet quality. In recent years, though, frame measurements have been adopted by at least one major breeder to help evaluate pullet flocks.

Breeders recommend measuring the shank of a representative sample of pullets every other week beginning at 4 weeks of age until the flock is housed at maturity. Because the skeleton is practically fully developed by 10 weeks, this procedure provides a very useful method to evaluate pullet quality at an early age.

16-H. LAYING CAGES

The use of laying cages for commercial egg production has become exceptionally popular. But cages are not a panacea for all the problems arising; there are advantages and disadvantages.

Advantages

1. It is easier to care for the pullets; no birds are underfoot.
2. Floor eggs are eliminated.
3. Eggs are cleaner.
4. Culling is expedited.
5. In most instances, less feed is required to produce a dozen eggs.
6. Broodiness is eliminated.
7. More pullets may be housed in a given house floor space.
8. Internal parasite problems are eliminated.
9. Labor requirements are generally much reduced.

Disadvantages

1. The handling of manure may be a problem.
2. Generally, flies become a greater nuisance.

Table 16-8. Percentage of Standard Leghorn Pullets in Each Weight Category at Sexual Maturity[1] (Mean Flock Weight of 3.0 lb (1.36 kg))

Body Weight Groupings		Percentage Within 10% of Mean Flock Weight						
lb	kg	91% +\nSuperior	84–90%\nExcellent	77–83%\nGood	70–76%\nAverage	63–69%\nFair	56–62%\nPoor	55%–\nVery Poor
4.0 +	1.78 +	—	—	0.25	0.5	1.25	2.0	2.75
3.7–3.9	1.65–1.77	0.25	0.75	1.25	2.5	3.75	5.0	6.25
3.4–3.6	1.51–1.64	2.75	6.75	8.5	10.5	12.0	13.5	15.0
3.1–3.3	1.37–1.50	47.0	43.5	40.0	36.5	33.0	29.5	26.0
2.8–3.0	1.24–1.36	47.0	43.5	40.0	36.5	33.0	29.5	26.0
2.5–2.7	1.10–1.23	2.75	6.75	8.5	10.5	12.0	13.5	15.0
2.2–2.4	0.97–1.09	0.25	0.75	1.25	2.5	3.75	5.0	6.25
2.1–	0.96–	—	—	0.25	0.5	1.25	2.0	2.75

[1]With scale weight interval of 1 oz (28 g), or less.

3. The investment per pullet may be higher than in the case of floor operations.
4. There is a slightly higher percentage of blood spots in the eggs.
5. The bones are more fragile and processors often discount the fowl price.

Laying Cage Size

Although the height of most laying cages is quite similar at 16 in. (40.6 cm) at the rear of the cage, the size of the floor area is highly variable. Some common floor dimensions (width × depth) are as follows:

in.	cm
10 × 16	25 × 41
12 × 16	31 × 41
12 × 18	31 × 46
12 × 20	31 × 51
14 × 16	36 × 41
14 × 18	36 × 46
16 × 18	41 × 46
16 × 20	41 × 51
24 × 18	61 × 46

The Reverse Cage (Shallow Cage)

There is an indication that when 2-, 3-, or 4-bird laying cages are reversed, that is, the longer section of the cage is across the front rather than on the side, improvement is made in the egg productivity of the laying pullets. Results from experimental work have not been consistent, however, and some tests have shown little or no improvement in some production categories. The results may be summarized as follows:

1. Annual hen-housed egg production is increased from 2 to 3%, and the birds gain more weight.
2. Housing and cage investment is higher.
3. Mortality is not affected.
4. The percentage of cracked eggs is reduced.
5. Egg size is the same.
6. Feed consumption per bird is slightly higher.
7. Egg production cost is similar.

The benefits of the reverse cage are attributable to the following conditions:

1. With up to 50% more feeder space per bird, the competition for feeding space is reduced.
2. The birds move around more.
3. The egg roll-out distance is shorter; therefore, eggs roll at a slower velocity and fewer cracked eggs are produced.

There are indications from the North Carolina Random Sample Test that performance in reverse cages may differ between housing types. Results for five consecutive tests between 1981 and 1986 are given in Table 16-9.

Table 16-9. Effects of Cage Shape and Housing Type

Performance Trait	Controlled Environment		Open (curtain)	
	Shallow Cage Type	Deep Cage Type	Shallow Cage Type	Deep Cage Type
White-egg layers				
Eggs/hen-housed	241.8	239.0	243.6	236.5
Daily feed intake (lb)	0.229	0.227	0.237	0.232
(g)	103.9	103.0	107.5	105.2
Feed/doz (lb)	3.55	3.55	3.66	3.70
(kg)	1.61	1.61	1.66	1.68
Egg weight (oz/doz)	23.7	23.7	24.7	24.6
(g/egg)	56.0	56.0	58.4	58.1
Large eggs (%)	63.3	63.5	75.7	74.9
Egg mass/hen-housed (lb)	30.3	30.0	31.7	31.2
(kg)	13.8	13.6	14.4	14.2
Cracked eggs (%)	2.9	2.7	2.6	3.7
Mortality (%)	9.5	10.5	8.4	9.5
Feed:Egg ratio	2.36	2.36	2.33	2.36
Brown-egg layers				
Eggs/hen-housed	229.8	224.4	227.1	220.4
Daily feed intake (lb)	0.253	0.249	0.258	0.251
(g)	114.8	112.9	117.0	113.9
Feed/doz (lb)	4.13	4.18	4.28	4.33
(kg)	1.88	1.90	1.95	1.97
Egg weight (oz/doz)	25.7	25.7	26.8	26.7
(g/egg)	11.7	11.7	12.2	12.1
Large eggs (%)	82.7	82.9	88.4	88.6
Egg mass/hen-housed (lb)	31.3	30.5	32.1	31.2
(kg)	14.2	13.9	14.6	14.2
Cracked eggs (%)	3.0	2.7	3.2	4.2
Mortality (%)	10.8	10.0	9.3	7.6
Feed:Egg ratio	2.53	2.55	2.51	2.55

Source: North Carolina Random Sample Tests 1981–1986.

Types of Laying Cages

Since the advent of laying cages, not only have the floor sizes fluctuated but the method of arranging the cages in the house and the number of birds per cage have become highly variable.

Cage systems. These may be categorized according to the number of birds in a cage, as follows:

1. *Single-bird cages:* The first laying cages were suited to one bird, but because the cage investment is high, single-bird cages are seldom used today.
2. *Multiple-bird cages:* These laying cages hold two or more pullets, but usually no more than eight or ten, with three or four being the most common.
3. *Colony cages:* These are large laying cages suitable for holding between 20 and 30 pullets.

Cage arrangement. To conserve space, thereby reducing the investment in the house in which the laying cages are placed, many methods have been originated to get more cages in a given area. This has led to the following general classification of cage arrangement (Figure 16-2).

1. *Single-deck:* Placing but one tier of cages in a house produces a high housing investment. The arrangement is practical only in areas with a warm climate where the "house" consists of nothing but a roof.
2. *Double-deck:* These are popular because the upper deck is offset, allowing the droppings to fall through the wire mesh to the house floor without touching the lower deck. The offset arrangement is often called the *stair-step system.* Another version of this system incorporates droppings boards or belts between the decks and the cages are not offset.
3. *Triple-deck:* To conserve house space still further, three decks are common. However, the upper two decks are only partially staggered or not staggered at all. To prevent the droppings from falling on the birds below, tilted droppings boards installed below the top cages cause the manure to fall into one area.
4. *Four-deck:* Same as triple-deck, except that four decks are used.
5. *Five-deck:* Same as triple-deck except that five decks are used.
6. *Flat-deck:* These are also called wall-to-wall houses. Although these are single-deck cage installations, the

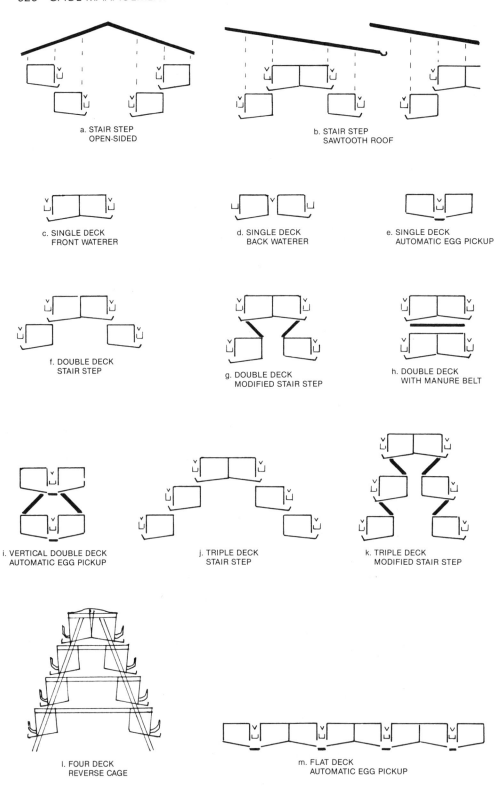

Figure 16-2. Some cage arrangements.

cages are placed close together without a walkway for servicing the pullets. All such work is done from a motorized catwalk that moves back and forth over the cages.

Laying Cage Floors

Most material for laying cage floors is welded wire fabric, although there are some plastic floors in use. All laying cage floors slope so that the eggs roll to a collection area or onto a movable belt.

Cage wire. Most cage floors are constructed of 14-gauge wire to give the necessary strength with a mesh size of 1 × 2 in. (2.5 × 5 cm). The uppermost wire must run at right angles to the length of the house, or from the back to the front of the cage, so that the eggs will roll off the cage floor easily.

Egg collection. Laying cages are constructed with a sloping wire floor that causes the eggs to roll to the front or back of the cage, the front being closer to the service aisle. The wire should slope about 2.4 in. (6.1 cm) for each 12 in. (30.5 cm) of cage depth.

When eggs are to be collected by hand, the wire floor should be extended past the front of the cage and rounded so as to act as a collection basket for the eggs. In some instances the wire at the point of collection is machine-formed, to slow down the speed of the eggs as they roll to the front, thus avoiding some breakage. A rubber or plastic bumper is sometimes added at the front of the collection area to further reduce breakage.

Perches not recommended. Some have installed perches in cages, but these have not proved practical. More cracked egs are produced as some birds lay while perched.

Floor construction and egg breakage. There is evidence to show that the steeper the slope of the wire, the higher the percentage of cracked eggs. Furthermore, the farther the eggs have to roll, the greater the breakage. Heavier gauge wire has been shown to result in more egg breakage.

Automatic egg collection. To reduce the labor requirement of egg gathering, automatic devices have been developed (see Figure 16-3). Although there are some variations, the procedure is standard; a movable belt, about 2 to 4 in. (5 to 10 cm) wide, is used to transfer the eggs from the cage to a collection area at the end of the house or to a crossbelt or conveyor, which in turn transports the eggs to a processing plant or packing unit. Belts must be kept clean to prevent the eggs from becoming dirty during transport.

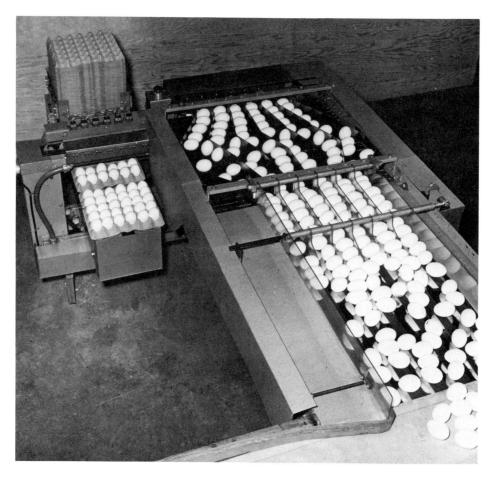

Figure 16-3. Mechanized egg collector with packer head.

Waterers for Caged Layers

Water is supplied by troughs running the length of the cage unit, cups, or drip nipples. When the long trough is used, water runs continuously in a small V-shaped (or similar) trough. Cups or drip nipples may be placed in each cage; or sometimes they may service two adjoining cages, depending on the number of birds in a cage.

Location of the trough waterer. The long, continuous water trough is commonly placed above the feed trough outside of the cage. This is to prevent any spilled water from dripping on the eggs in the egg tray, the feed trough acting as a catch basin.

Feeders for Caged Layers

Long, continuous troughs are used almost universally for feeding caged layers. Since the troughs are usually located on the outside of the cage, the birds must reach through a "feeder fence" to feed. The troughs may be filled by hand or automatically, but the method changes the arrangement of the cages within the house.

Hand feeding. When the birds are fed by hand there must be a service aisle between the rows of cages. Feed troughs are hung on the front of the cages, next to the aisle. The distance between the aisle sides of the troughs should be at least 28 in. (71 cm), inasmuch as motorized equipment is commonly driven through the aisles, and this equipment is built for this width. The long troughs extending the length of the cage units should have no partitions, as feed in the troughs is often supplied by a traveling hopper or is stirred by running a paddle inside the trough, and any partitions would interfere with the ease of operation. Most troughs are made of metal, but some are plastic. Feed troughs should be installed so that the inner lip is 8 to 9 in. (20 to 23 cm) above the floor.

It is important that the entire feed trough be accessible to the birds. Cage fronts are usually designed with 2-in. spacing between vertical wires to accommodate eating. Newer styles are designed with horizontal bars, thus providing no restrictions to bird movement up and down the feed trough.

Mechanical feeding. Most new cage houses are equipped with mechanical feeders. These are most commonly installed at the front of the cage and one feeder serves two troughs in a back-to-back configuration. When houses have both mechanical feeders and egg collectors, aisle space may be reduced.

16-I. CAGE LAYING HOUSES

Caged layers require protection from the weather. In mild climates, with no freezing weather during the winter months, this may be no more than a roof over the cages. But where the weather is colder, complete housing must be provided. Because of the high bird density in cage operations, adequate ventilation must be provided to bring fresh air into the building and to exhaust the heat and ammonia fumes. Environmental control is becoming an accepted practice in house construction. Totally dark houses with cooling devices are used extensively (see Chapter 11-J).

Cage House Width

The single most important factor affecting the width of a house relates to ventilation. For this reason open houses are usually constructed in widths of 30 ft (9.1 m) or less. Environmental houses, on the other hand, are seldom more than 50 ft (15.2 m) wide. Distribution of air is extremely difficult in houses of greater width.

Cage House Length

The major limiting factors relative to the length of a poultry house are associated with the proper functioning of equipment. Mechanical feeders, egg collectors, and manure removal devices are designed to operate efficiently within specific length constraints. Most manufacturers limit this length to about 500 ft (152 m).

Other length constraints include ease of moving birds in and out, ventilation pathways, number of eggs per row when eggs are gathered by hand, the amount of manure per row when manure is removed frequently, and employee safety in case of emergencies. Environmentally controlled houses over 200 ft (61 m) in length should be provided with emergency exits on each side for employees and to provide some natural ventilation in case of power failure.

Shape of the Roof

In mild climates, where only a roof is needed to supply protection from the sun and the rain, the gable roof is common. The width of the roof will be determined by the number of cage rows to be placed beneath it. In cold climates where environmental protection is needed, the gable roof is used almost exclusively.

In milder climate regions, the sawtooth style roof may be used. This roof has a series of sloping sections with an opening every 10 to 15 ft. It is important that the openings be on the side with the prevailing winds since proper ventilation is dependent on drawing the air out of the house. Extremely large sawtooth houses are not recommended as air intake areas relative to the large roof are limited to the house perimeter.

Environmentally Controlled Houses

The basic rules as outlined in Chapter 11 are used in the construction of environmentally controlled cage houses. However, the amount of air to be moved through the house is greater, and this calls for more ventilating fans.

The High-Rise Cage House

Because of problems with manure removal and resultant pollution, many new cage operations in colder climates incorporate a deep pit in their house construction; only walkways serve as the floor, the droppings falling directly into the pit (see Chapter 11-S).

16-J. CAGE REQUIREMENTS FOR LAYERS

Although the handling of layers in cages seems an easy procedure, many management factors have a bearing on the economy of egg production.

Cage versus Floor Operations

There is quite a variation in the behavior of layers in cages compared with those kept on a litter floor. Although the differences are highly variable depending on the density of the birds within a single cage, a comparison of cage versus litter floor operations shows the following:

1. The cage investment per bird is 50 to 100% higher.
2. Caged layers generally produce fewer eggs but production is highly dependent on space allowances.
3. Eggs from birds kept in cages are slightly heavier.
4. Mortality is usually lower in cages.
5. Some strains adapt themselves to cages better than others.
6. Eggshell quality deteriorates more rapidly when layers are kept in cages.
7. Interior egg quality drops off more quickly in the case of caged layers.
8. Labor requirements for cage operations are about 50 to 80% of those for floor management. The difference depends on the degree of automation.
9. Caged pullets weigh more at end of laying year, but their market value is less.

Types of floors compared. Most of the research comparing floor houses and cages for laying hens was conducted prior to 1970. Much of this research compared small floor pens with 50 or less chickens per pen with cages. The California Random Sample Test results for 1957 to 1963 are listed in Table 16-10.

Cage Floor Space Requirements for Layers

Undoubtedly, this subject has been the basis of more research than any other cage management factor. But the results still are inconclusive. Differences of opinion occur because there must be an optimum floor

Table 16-10. Floor Versus Cage Performance (California, 1957–1963)

Performance Trait	Floor	Cages
Hen–day egg production (eggs)	276	249
Mortality (%)	8.9	10.2
Egg weight (oz/doz)	24.5	24.7
(g/egg)	57.9	58.4
Blood spots (%)	5.8	8.5
Shell thickness (in.)	0.0145	0.0147
(microns)	368	374
Total egg mass (lb)	35.16	31.99
(kg)	15.98	14.54
Feed consumption	Not measured	

Source: California Random Sample Test.

space to produce cage eggs at the lowest cost. Many tests have been conducted only on the basis of egg production, livability, and egg quality. All these decrease as the birds are given less cage floor space. Investment in housing and equipment is also involved, as crowding reduces the investment per pullet. The size of the cage is an economic factor. Regardless of the differences of opinion, some general conclusions can be reached regarding cage floor space requirements. Those shown in Table 16-11 represent a happy medium.

Complexities of cage floor space requirement. Although test after test to show the requirement for floor space in laying cages could be given here, the results would be difficult to summarize. There are so many variables: cage size, feeder space per pullet, colony size, strain of chicken, housing type, number of cage decks, lighting programs, and dozens of other factors enter into any analysis of the problem. Obviously, crowding reduces performance, but most poultrymen feel they can sacrifice some performance in order to realize a greater profit on their investment. Just where the break point is has not been determined, largely because of the variables involved. Some have said that the amount of feeder space available to each bird in the cage is a better criterion of maximum cage density than the actual floor space available.

Table 16-11. Cage Floor Space Requirements per Laying Pullet

	in.^{2}	cm^2
Mini-Leghorn	48	310
Standard Leghorn	60	387
Medium-size[1]	70	452

[1]Producing brown-shelled table eggs.

In 1977, University of California research published the results of an experiment comparing 1, 2, 3, and 4 birds in a 12-in. (30 cm) (wide) × 18-in. (45 cm) (deep) cage. Data were collected from 20 to 76 weeks of age (56 weeks of lay), and some of the results are found in Table 16-12.

The comparison shows that for each additional pullet kept in a 12- × 18-in. cage, hen-day egg production was decreased by 2.5%, hen-housed egg production was decreased by 13.5 eggs, and average egg weight was increased by 6 g/doz (0.5 g/ea).

The economic evaluation of cage crowding represents the ultimate criterion of how much crowding in laying cages is economical. The results of this experiment (Table 16-12) favor three birds per cage, with a return of U.S. $1.56 each. Total profit per cage was lowered when four instead of three birds were kept in a cage, but the margin would be less if the price of market eggs increased, and greater if it decreased.

The Random Sample Test at North Carolina State University compared three and four hens per 12-in. (30 cm) wide × 18-in. (45 cm) deep cage over five consecutive tests. White and brown strains were compared in environmentally controlled and curtain-sided houses. Table 16-13 lists the results of these tests.

Hen-housed egg production consistently favored the lower density. Housing type and strain or breed appeared to have very little influence on the differences noted. In general, the three-bird cages outperformed

Table 16-12. Leghorn Bird Density in a 12 × 18-in. (30.5 × 45.7-cm) Cage (Standard Leghorn, 56 Weeks of Lay)

Item	Hens per Cage			
	1	2	3	4
In.² floor space per bird	216	108	72	54
Cm² floor space per bird	1417	697	464	348
Hen-day egg production (%)	73.2	69.3	68.6	65.0
Eggs per hen housed	284	267	264	241
Mortality (%)	4.4	5.7	6.7	15.7
Avg egg weight (oz/doz)	24.95	25.04	25.24	25.64
Avg egg weight (g/ea)	58.9	59.1	59.7	60.5
Avg feed per day per bird (lb)	0.231	0.223	0.222	0.221
Avg feed per day per bird (g)	105	101	101	100
Feed per dozen eggs (lb)	3.80	3.87	3.88	4.09
Feed per dozen eggs (kg)	1.73	1.76	1.76	1.86
Egg income[1] minus feed and pullet cost per bird	$1.99[2]	$1.65	$1.61	$1.08
Total income minus total cost, per cage	$1.12	$2.73	$4.68	$4.37
Total income minus total cost, per bird	$1.12	$1.36	$1.56	$1.09

Source: University of California, 1977.
[1]Large egg value at $0.45/doz.
[2]All in U.S. dollars.

Table 16-13. Effects of Cage Density and Housing Type

Performance Trait	Controlled Environment		Open (curtain)	
	3 Hens per Cage[1]	4 Hens per Cage[1]	3 Hens per Cage[1]	4 Hens per Cage[1]
White-egg layers				
Eggs/hen-housed	245.0	235.6	246.4	233.9
Daily feed intake (lb)	0.233	0.224	0.239	0.229
(g)	105.7	101.6	108.4	103.9
Feed/doz (lb)	3.58	3.52	3.69	3.66
(kg)	1.63	1.60	1.68	1.66
Egg weight (oz/doz)	23.7	23.7	24.7	24.6
(g/egg)	56.0	56.0	58.4	58.1
Large eggs (%)	63.7	62.8	76.1	74.8
Egg mass/hen-housed (lb)	30.7	29.5	32.2	30.4
(kg)	13.95	13.41	14.64	13.82
Cracked eggs (%)	2.3	3.2	3.0	3.4
Mortality (%)	8.1	11.0	7.7	10.3
Feed:Egg ratio	2.38	2.34	2.36	2.34
Brown-egg Layers				
Eggs/hen-housed	234.4	220.4	230.6	217.3
Daily feed intake (lb)	0.256	0.245	0.260	0.249
(g)	116.1	111.1	117.9	112.9
Feed/doz (lb)	4.15	4.16	4.30	4.33
(kg)	1.89	1.89	1.95	1.97
Egg weight (oz/doz)	25.8	25.6	26.7	26.7
(g/egg)	61.0	60.5	63.1	63.1
Large eggs (%)	83.8	81.7	88.6	88.4
Egg mass/hen-housed (lb)	32.0	29.8	32.6	30.7
(kg)	14.55	13.55	14.82	13.95
Cracked eggs (%)	2.4	3.2	3.3	4.0
Mortality (%)	7.9	11.8	7.2	9.7
Feed:Egg ratio	2.53	2.56	2.52	2.54

Source: North Carolina Random Sample Tests, 1981–1986.
[1]Cages are 12 × 18 in. (30 × 45 cm). Half are shallow and half are deep.

the four-bird cages by about 12 eggs per bird, consumed 4 to 5% more feed, and produced eggs of the same weight, but experienced 3% less mortality during the laying period.

Egg producers use higher densities to increase total production and to reduce overhead costs per dozen eggs. When economic margins are low, the effects of high densities often result in lower total profits for a farm even though the number of birds may be 33% more.

Crowding Layers in Cages

When determining the floor space per bird in laying cages, the following must be considered:

1. Crowding increases mortality.
2. Crowding decreases hen-day egg production.
3. Crowding decreases hen-housed egg production.
4. Crowding reduces eggshell quality.
5. Crowding reduces net profit per bird.
6. Crowding may reduce total farm income.

Feeder Space

When cage dimensions are reversed (see Chapter 16-H) the feed trough is then located along the long dimension of the cage. This may increase feeder space per bird by up to 50%. Researchers at Kansas State University have demonstrated that practically all of the benefits observed in reverse cages are associated with increased feeder space.

California researchers in 1980 compared four different two-bird cage configurations with feeder space allowances of 5 or 6 in. (12.7 or 15.3 cm) per bird. Results are given in Table 16-14.

Both wide cages gave significant improvements in egg production, lower mortality, and fewer cracked eggs over the narrower cages. Even though the space per hen was identical in the 10- × 12-in. and the 12- × 10-in. (25 × 30 cm and 30 × 25 cm) cages, egg production per hen improved by 13 eggs when the feeder space allowance increased from 5 in. per hen to 6 in. (12.5 to 15 cm).

Table 16-14. Effect of Feeder Space on Performance in Two-Bird Cages[1]

Performance Trait	Cage Size (width × depth)			
	10 × 10 in. 25 × 25 cm	10 × 12 in. 25 × 30 cm	12 × 10 in. 30 × 25 cm	12 × 12 in. 30 × 30 cm
Eggs/hen-housed	256	258	271	269
Daily-feed intake (lb)	0.236	0.239	0.240	0.240
(g)	107	108	109	109
Feed/doz (lb)	3.60	3.62	3.51	3.55
(kg)	1.64	1.65	1.60	1.61
Egg weight (g/egg)	59.3	59.2	58.3	58.4
Large eggs (%)	78.2	74.1	72.2	74.8
Egg mass/hen-housed (lb)	33.3	33.6	34.7	34.6
(kg)	15.2	15.3	15.8	15.7
Cracked eggs (%)	3.5	3.7	1.7	3.3
Mortality (%)	8.3	5.0	3.0	2.2
Feed:Egg ratio	2.29	2.31	2.28	2.29

Source: University of California, 1980.
[1]20 to 68 weeks of age, White Leghorns.

Egg Production in Various
Cage Decks

If light intensity is good at the feed container, there should be no difference in the egg production of hens in the various decks. But in many cage houses, this is impossible because the light source is above the top deck and the birds in the lower decks do suffer from inadequate light. Both egg production and uniformity of production are poorer the farther from the light source.

16-K. MOVING PULLETS TO LAYING CAGES

Pullets to be placed in laying cages may have been raised on litter, slats, wire floor, or in a growing cage with a wire floor. Care should be taken at moving time to handle the birds carefully to lower the incidence of stress.

Age When Moved

Pullets may be transferred to the laying cages at any age between 14 and 20 weeks. Seventeen to 18 weeks seems to be the optimum so far as the birds are concerned. The birds are in the laying cages sufficiently early to recover from moving stresses prior to the time egg production begins. But growing facilities may dictate moving at other ages. Sometimes the facilities may not have enough space to allow the birds to remain in the growing quarters until 17 or 18 weeks of age; in other cases, the laying quarters may not be available until the growing pullets are older.

Sorting Pullets by Weight

Obviously, all pullets are not of equal body weight or sexual maturity when they are placed in the laying cages; some will begin egg production earlier than others. Since body weight is a good criterion of sexual maturity within a given flock, some poultry producers have placed smaller birds together within a cage, birds of average size in others, and the largest pullets in other cages. Researchers at the University of California conducted an experiment in which Leghorn pullets were segregated into five weight classifications and maintained for a year of production. Results showed that the smallest and the largest birds, when caged separately, produced the lowest number of eggs. The smallest pullets came into egg production last. The best egg production came from the birds closest to the average weight of the entire group.

Table 16-15. Effects of Restricted Feeding Light and Heavy Halves of a Layer Flock Separated at 18 Weeks of Age[1]

| Performance Trait | Body Weight Group | | | | | |
| | Light Half | | Heavy Half | | Average | |
	Full Feeding	Restrict Feeding	Full Feeding	Restrict Feeding	Full Feeding	Restrict Feeding
18-wk Body weight (lb)	2.25	2.25	2.67	2.67	2.46	2.46
(kg)	1.02	1.02	1.21	1.21	1.17	1.17
64-wk Body weight (lb)	3.95	3.59	4.39	4.04	4.17	3.82
(kg)	1.79	1.63	1.99	1.83	1.89	1.73
Eggs/hen-housed	200	190	203	201	201	196
Daily feed intake (lb)	0.218	0.203	0.232	0.217	0.225	0.210
(g)	99	92	105	99	102	95
Feed/doz (lb)	3.75	3.69	3.99	3.80	3.87	3.74
(kg)	1.70	1.68	1.81	1.73	1.76	1.70
Egg weight (oz/doz)	24.4	24.3	25.2	25.2	24.8	24.8
(g)	57.6	57.5	59.5	59.6	58.6	58.6
Large eggs (%)	68.0	67.6	77.1	80.4	72.6	74.0
Egg mass/hen-housed (lb)	25.3	24.0	26.6	26.4	26.0	25.3
(kg)	11.5	10.9	12.1	12.0	11.8	11.5
Mortality (%)	15.0	11.3	10.0	8.3	12.5	9.8
Feed:Egg ratio	2.46	2.43	2.53	2.41	2.49	2.42
Egg income minus feed Cost/hen-housed (U.S. $)	3.25	3.16	3.19	3.42	3.21	3.29

Source: University of California, 1977.
[1]Fed from 6 to 8 a.m. and 6 to 8 p.m. versus full access to feed.

A second experiment studied the effects of adult feed restriction with the heavy and light halves of a commercial flock. Restriction consisted of two 2-hour feedings per day. Significant benefits of restricting the heavy half and normal feeding of the light half were demonstrated (Table 16-15).

Sorting pullets by body weight with separate feeding programs can significantly improve profits. The savings of feed more than compensates for the additional labor involved in sorting the flock.

16-L. CONSUMPTION OF WATER BY LAYERS

Invariably laying pullets drink more water when they are on wire than when they are kept on a litter floor. This fact, along with more water being drunk during periods of hot weather, may cause difficulties. More moisture is eliminated through the droppings, and they become wet and soggy. In many cage operations the manure becomes difficult to scrape up and remove from the cage building. In pit houses the excess moisture is not easily removed through normal ventilating procedures; extra fans must be used in the pit.

Egg Production Affects Water Consumption

Approximate water requirement at varying percentages of egg production is shown in Table 16-16.

Table 16-16. Egg Production and Water Consumption (Standard Leghorns)

Hen-Day Egg Production %	Water Consumption per 1,000 Pullets per Day at 70°F (21.1°C) in Cages	
	U.S. Gal	L
10	40	151
30	42	159
50	46	174
70	53	201
90	63	239

Temperature Affects Feed and Water Consumption

Table 16-17 shows average laying cycle figures of standard Leghorns for feed and water consumption at seven different house temperatures.

Note: Of interest is the pounds of water consumed per day as a per-

Table 16-17. House Temperature as It Affects Feed and Water Consumption of Standard Leghorn Laying Pullets in Cages

Item	Average Daytime House Temperature						
	40°F 4.4°C	50°F 10.0°C	60°F 15.6°C	70°F 21.1°C	80°F 26.7°C	90°F 32.2°C	100°F 37.8°C
Lb feed consumed per 100 pullets per day	26.0	25.5	24.1	22.0	19.1	15.3	10.5
Kg feed consumed per 100 pullets per day	11.8	11.6	11.0	10.0	8.7	7.0	4.8
Lb of water consumed per lb of feed consumed (or kg per kg)	1.3	1.4	1.6	2.0	2.9	4.8	8.5
Lb of water consumed per 100 pullets per day	33.9	35.7	38.6	44.0	55.4	73.5	89.2
Kg of water consumed per 100 pullets per day	15.4	16.2	17.6	20.0	25.2	33.4	40.6
Water consumed per day as % of live body weight	9.7	10.2	11.0	12.6	15.8	21.0	25.5
U.S. gal. of water consumed per 1,000 pullets per day	40.8	43.0	46.6	53.0	66.8	88.3	107.5
Liters of water consumed per 1,000 pullets per day	154.4	162.8	176.4	200.6	252.8	334.2	406.9

centage of live body weight of the Leghorn layers at different house temperatures.

Important: Daily temperatures used throughout this text are average daytime house temperatures. These are considerably higher than the average daily temperature for 24 hours. They are also lower than the maximum daily temperature (usually midafternoon). However, the average daytime house temperature is the most consistent figure for recording daily house temperature variations. One approximation of these three temperatures on the same day is as follows:

Average daytime house temperature	90°F
	(32.2°C)
Average daily temperature (24 hr)	80°F
	(26.7°C)
Maximum daily temperature	105°F
	(40.6°C)

Regional fluctuations in day and night temperatures and the amount of environmental control will result in different temperature relationships.

Water Consumption During the Laying Cycle

Table 16-18 shows the variations in water consumption throughout the laying period for standard Leghorns. The figures associate themselves with the varying body weight and rate of egg production during various weeks of the laying cycle.

Water Restriction

Water consumption highest late in the day. Caged layers drink about 25% of their daily water consumption during the 2 hours immediately preceding the time the lights go off or the sun sets regardless of the house temperature. Consumption the remainder of the day shows a fairly uniform hourly intake.

Reducing water consumption. Evidently pullets in cages consume more water than they actually need causing the droppings to become wet. Intermittent watering will eliminate some of the problem, but do not restrict water prior to the peak of egg production or during hot weather. Water restriction is not practiced in cup or nipple systems because of leakage problems.

The practical amount of water restriction during moderate or cool weather is shown in Table 16-19. In this test, comparisons were

Table 16-18. Feed and Water Consumption of Standard Laying Leghorn Pullets in Cages

Week of Egg Production	Feed Consumed per 100 Pullets per Day at 70°F (21.1°C) lb	Average Daytime House Temperature						
		40°F 4.4°C	50°F 10.0°C	60°F 15.6°C	70°F 21.1°C	80°F 26.7°C	90°F 32.2°C	100°F 37.8°C
		U.S. Gallons of Water Consumed per 1,000 Pullets per Day						
1	18.0	31	33	36	41	51	68	83
2	18.4	38	41	45	51	67	84	103
3	18.7	42	45	49	55	74	93	112
4	19.1	44	47	51	58	77	96	117
5	19.4	46	49	53	60	81	101	122
6, 7	20.0	48	51	55	63	85	105	127
8–12	21.0	46	49	53	60	82	101	122
13–18	23.0	44	47	51	58	78	96	117
19–38	23.0	42	45	49	55	75	93	113
39–49	23.0	40	43	47	53	70	88	108
50–60	23.0	38	41	45	51	67	84	103
	kg	Liters of Water Consumed per 1,000 Pullets per Day						
1	8.2	117	125	136	155	193	257	314
2	8.4	144	155	170	193	254	318	390
3	8.5	160	170	186	208	280	352	424
4	8.7	167	178	193	220	292	363	443
5	8.8	174	186	201	227	307	382	462
6, 7	9.1	182	193	208	239	322	397	481
8–12	9.5	174	186	201	227	310	382	462
13–18	10.5	167	178	193	220	295	363	443
19–38	10.5	160	170	186	208	284	352	428
39–49	10.5	151	163	178	201	265	333	409
50–60	10.5	144	155	170	193	254	318	390

Table 16-19. Average Egg Production, Feed Efficiency, Feed Consumption, Livability, and Manure Moisture as Influenced by Restricting Watering Time (Standard Leghorns)

Waterings per Day (15 min ea)	Hen-Day Production %	Feed per Dozen Eggs		Feed per 100 Birds per Day		Livability %	Manure Moisture %
		lb	kg	lb	kg		
15	61.5	4.99	2.27	25.6	11.6	92.1	78.2
8	61.6	5.04	2.29	25.8	11.7	92.1	78.5
4	62.2	4.90	1.22	25.4	11.5	91.3	75.6

Source: Maine's Timely Topics, March 1976.

made between 4, 8, and 15 15-minute watering periods per day. Four such periods produced the best results.

Watering Devices

Waterer space.   Waterer trough space is usually more than adequate in cages because any trough runs across the entire front of the cage, but the minimum is 1.5 in. (3.8 cm) of waterer space per pullet, which will be ample in hot weather.

One cup every other cage is adequate except when cages are exceptionally large and contain a great many birds. The amount is similar with drip nipples, but remember that one cup will accommodate more birds than one nipple (see Table 16-1).

Water temperature.   It is important that drinking water is cool during the warmer months. Chickens will back away from hot water; feed consumption will decrease and performance will suffer. Closed-pipe systems using cups or nipples can have water temperatures as high as the house temperature. Water pipes must be protected from the sun and should not be placed in hot attics or in areas where heat buildup is possible. The benefits of cold water were demonstrated by Canadian researchers (see Table 16-20).

Trough slope for continuous water troughs.   Such troughs, having running water, should have a slope of 3 in. per 100 ft (7.6 cm per 30.5 m) of trough.

Table 16-20.   Effect of Drinking Water Temperature on Layer Performance at High Environmental Temperatures (95°F, 35°C)

	Water Temperature	
Performance Trait	95°F (35°C)	36°F (3°C)
Hen-day egg production (%)	81	93
Egg weight (oz/doz)	20.7	20.6
(g/egg)	49.0	48.5
Daily feed intake (lb)	.141	.168

Source: University of Guelph, Canada.

16-M. ENVIRONMENTAL TEMPERATURE IN CAGES

What to Do in Hot Weather

Hot weather can cause depressed feed consumption and can cause extremely high mortality in caged birds. They cannot withstand hot climatic conditions as well as birds on a litter floor; caged birds are completely surrounded by hot air, and have no way to get away from the heat. Some hot weather tips are

1. Shade the outside row of cages in open-sided houses.
2. Install foggers over the birds. These foggers may be automatic; that is, they will turn on automatically when the temperature reaches a certain point and turn off at a lower temperature. In many instances a pressure system will be needed to make the foggers work efficiently. *But be careful:* If there is an electric failure at the time of high temperatures, the automatic device that operates the water intake valve to the system or the compressor pump will not work.
3. Sprinkle the roofs of the buildings.
4. Install circulation fans.
5. Provide cool water. Cool water on hot days increases feed consumption, which helps to maintain normal egg production.
6. Do not practice water restriction.
7. Increase the nutrient levels in the feed. Chickens eat less at high temperatures, thereby reducing daily nutrient intake.
8. Open the air exhausts and intakes in environmentally controlled houses.
9. Clean fans and pads in environmentally controlled houses.
10. Be sure pads are fully saturated with water and that water outlets are free-running.

What to Do in Cold Weather

When laying cages are constructed within an insulated and environmentally controlled house, there should be little difficulty during periods of cold weather. But in many mild areas the cages are not enclosed; only a roof protects the birds from the elements. Cold wind can be particularly harmful to chickens.

Wind protection. With open housing, the birds in the outside row of cages are subjected to blowing winds. Some protection should be furnished. Windbreaks should be constructed about 5 to 7 ft (1.5 to 2 m) away from the outside of the house. They may be composed of lath fence, lattice, fly screen, or louvers. Adjustable curtains or temporary plastic sheeting will also protect the flock.

Do not let pipes freeze. Before the temperature falls below freezing drain all exposed pipes and water equipment.

16-N. ARTIFICIAL ILLUMINATION

Some method of maintaining a constant length of light day must be provided. Supplemental artificial light will be needed during the season of short hours of natural daylight. Light-proof houses will need artificial illumination entirely. See Chapter 18 for complete details of lighting programs.

Light and cannibalism. It is a well-known fact that the greater the intensity of light, the higher the incidence of cannibalism in laying cages. The amount of light necessary to stimulate egg production is much less than many people think. Furthermore, when cages are constructed with three or more decks the upper deck gets the most light, the lower, the least. There is no simple solution to this problem. The only alternative is to provide the birds in the lower deck with the minimum amount of light, so as not to overilluminate those in the upper deck, where the light is the brightest. Cannibalism will be most prevalent in the outside rows in an open-sided house, as these rows are exposed to more bright natural daylight.

16-O. FEEDING CAGED LAYERS

The nutrition of caged layers is detailed in later chapters, particularly Chapter 31.

Prevent Feed Wastage

Feed beaked out of a cage feed trough is a complete waste; there is no way the birds can get to it. A study of the amount of feed wastage when layers were kept in cages was made at Oregon State College. The findings showed that the more pullets there were in a cage, the less the feed loss. Seemingly, more birds at the trough allows less space for picking and billing the feed. Lower light intensity will also reduce feed wastage. Do not put much feed in the troughs at one time; provide fresh feed oftener. The results of the Oregon test were as follows:

Number of Birds per Cage	Feed Wasted per Bird per Year	
	lb	kg
1	4.52	2.05
2	2.01	0.91
3	1.08	0.49

Excessive feed depth is a major contributor to feed wastage. Care must be taken, though, to provide enough feed for all birds. Poorly beak-trimmed birds may require greater feed depths.

16-P. EGG HANDLING

Uncollectable Eggs

Uncollectable eggs are those broken prior to gathering that pass through the floors of the cages. Many are soft-shelled. Following are some causes and possible reasons for their being uncollectable:

Body checks. These vary between 0.5 and 3.0% of all eggs laid. Most are laid prior to 10:00 a.m. Many are uncollectable, but not all (Table 16-21).

Misshapen eggs. Up to 15% of all eggs produced are misshapen. Only a few are laid after 10:00 a.m.

Length of egg production period. Young pullets produce some uncollectable eggs (shell-less, ultra-thin, and thin-shelled), but the incidence

Table 16-21. Body-Checked Eggs, Oviposition Time and Cage Density

Time of Oviposition	Body Checks %
0600 to 0800 hr	16.7
0800 to 1000	4.4
1000 to 1200	1.7
1200 to 1400	0.5
1400 to 1600	0.6
1600 to 1800	0.5
Number of Birds per Cage	(eggs laid between 0600 and 0730)
2	8.3
3	14.6
4	33.8
Lighting Program (lights on)	
0400 to 0600 hr	1
0400 to 0800	7.7
0400 to 1100	18.3

Source: Auburn University, 1982.

increases as the birds get older. On the average, one-half of the un-collectable eggs are shell-less.

Number of eggs uncollectable.   Up to 7% of all eggs laid are uncollectable. Such eggs are never credited to the total egg production of the bird; they are a complete loss.

Molt.   Birds lay about 75% fewer uncollectable eggs after molting than they do before.

Gathering Eggs

Egg collection is either manual or automatic, but which is the better system is highly debatable among poultrymen. It is not difficult for an efficient operator to pick up the eggs by hand from 30,000 layers and still have time for other chores. To facilitate the procedure, carts are moved through the aisles and the eggs placed in flats on the carts. Some carts are motorized, either with electric batteries or by small gasoline-driven engines. But automatic egg-gathering devices are being installed in many new cage operations. Eggs roll from the sloping floor of the cage onto a movable belt, which delivers them to a service room at the end of the building or directly into a processing plant for cleaning, grading, and packing.

Sorting during collection.   Although it is inefficient to sort eggs accord-ing to size at the time of collection, any excessively large, broken, or unusually dirty eggs or those with poor shells should be removed as the eggs are gathered. The procedure eliminates difficulties later.

Egg breakage.   Most egg producers do not realize the seriousness of the number of cracked eggs they get from their flocks. Although most say the percentage is low, recent tabulations at the grading station showed that typically 5% of the eggs received showed cracks. During washing and grading more eggs were cracked. Indi-vidual problem flocks often have 20% or more of their eggs cracked. When 5% of the eggs are cracked through faulty handling by the producer, there would be an annual loss of U.S. $5,000 for every 20,000 hens in production.

Causes of Cracked Eggs

It has been shown that the quality of the eggshell is closely related to the incidence of cracked eggs. Some of the influencing factors are as follows:

Genetics.   Some strains of birds have the ability to produce eggs with better shells. However, as strains are bred for better egg production, there is a tendency toward a reduction in eggshell quality.

Position of egg within a clutch.   The first eggs of a clutch possess better shell quality than those laid later in the clutch. As high-producing hens have longer clutches, the correlation between clutch length and poorer shell quality is obvious.

Length of lay.   The longer the period of egg production, the poorer the shell quality becomes.

Temperature.   The higher the environmental temperature, the poorer the quality of the eggshells. When high temperatures occur at the end of the laying period, both factors work to decrease shell quality.

Disease.   Certain respiratory diseases, such as infectious bronchitis and Newcastle disease, have a marked effect on shell quality.

Humidity.   Recent research work has shown that the moisture level of the air—not the temperature of the air—is the critical factor affecting the breaking strength of eggshells. Market eggs should be held and handled in a room with as low a humidity as possible.

Classification of Cracked Eggs

Body checks.   These are eggs in which cracks develop during the time the egg is in the uterus and are subsequently partially repaired. They may be caused by bird fright, nutritional failure, stress, disease, and are more of a problem with increasing age.

Cracked during oviposition.   The position of the bird when she lays an egg is partially responsible for such egg cracking. The more upright the bird stands, the longer the drop, and the more eggs cracked. The material on which the egg is dropped also affects the incidence.

Cracked during rolling on cage floors.   When eggs roll down the cage floor to the collection area, they bump each other and some are cracked as are many when they hit the bent-up portion of the collection tray. This is the principle reason for recommending twice-a-day or more gathering. Bumper guards at the tray's edge will help to prevent much of this breakage.

Cracked during gathering.   Handling of eggs increases the number cracked. Some mechanical collection devices contribute to greater egg breakage because of more bumping. Excessive speed, too many eggs per hand, and poor egg roll-out design are major contributors to this problem.

Cracked during processing.   Remove cracked eggs prior to processing. The procedures of washing, grading, and cartoning increase egg breakage by about half.

Season of year and age of flock.   More eggs are cracked during the summer months because the shells are thinner during hot weather. Twice as many eggs are cracked from hens over 40 weeks of age compared to younger flocks because of thinner and weaker shells (see Table 16-22).

Table 16-22. Relationship of Flock Age to Egg Breakage[1]

Age of Flock wk	Cracked Eggs %
20 to 29	2.01
30 to 39	2.62
40 to 49	3.56
50 to 59	4.40
60 to 69	6.18
Average	3.75

Source: University of California, 1981–1982.
[1]On the collection racks prior to processing (hand-candled).

Egg gathering rates.   These are highly variable from farm to farm depending on the experience of the gatherer, the frequency of gathering, cage configurations, egg quality, and the type of gathering equipment. Experienced gatherers can gather 5,000 or more eggs per hour. At this rate, a full-time person could gather the eggs from a 50,000-hen flock with a rate of lay of 75% in an 8-hour day with time to spare. Mechanized egg collection systems with automatic flat packers can triple this rate of collection and fully automated in-line systems completely eliminate the need for the in-house gatherer.

Reducing Egg Breakage

Eggshell breaking strength.   When shell quality is good, it will require from 6 to 8 lb (2.7 to 3.6 kg) of pressure to break the eggs. But when shell quality is poor, 5 lb (2.3 kg) of pressure will break them, and there is a problem. Normally, eggs laid at the end of the laying cycle will break with 5 lb (2.3 kg) or less of pressure.

How to reduce egg breakage.   Reducing egg breakage should be a real project on most poultry farms. First determine what the percentage of breakage is; then put on a campaign to reduce it. Keep a record and plot the progress on graph paper. Some methods of reducing the incidence of cracked eggs are as follows:

1. Some strains of egg-type layers produce a higher percentage of cracked eggs. Make a careful choice.
2. Handle eggs more carefully at the end of the laying period.
3. Start an employee-education program to reduce breakage. Candle eggs gathered by different personnel and compare results.

4. Provide a cushion bumper at the front of the egg collection area of the cages.

5. Try some cages with fewer birds. Crowding increases egg breakage. It may be that an excessive number of cracked eggs is canceling the additional income from more birds per cage (see Table 16-12).

6. Collect eggs more often. Eggs rolling down the cage floor have a greater chance of being broken if there are eggs in the receiving area. It is estimated that at least one-half of the breakage is due to this (see Table 16-23).

7. Collect eggs more often during the summer months and from older flocks.

8. Collect eggs on flats to avoid extra handling and to minimize breakage. Do not stack flats more than six high. If baskets must be used, do not fill them over half full.

9. Consider changing the ration if poor shell quality is a continuous problem.

10. Reduce every stress possible; stresses lower shell quality in most cases and therefore increase egg breakage.

11. Prevent cannibalism. It often causes birds to pick at freshly laid eggs. Check on the light intensity.

12. Reduce bird fright. The jumping of birds in the cages can only create more cracked eggs.

13. Be sure any automatic egg-gathering equipment does not produce an increase in egg breakage. Check the belt material, the speed of the belts, and any angling or corner devices.

14. Reduce house activities during the egg laying period.

Table 16-23. Egg Collection Frequency and Egg Breakage

Age of Flock wk	Twice/day			Once/day
	% Breakage			
	10 A.M.	2 P.M.	Avg	12 noon
30	5.6	9.8	8.1	7.6
30	3.1	5.8	4.6	7.3
58	0.8	2.0	1.5	3.1
58	1.4	2.6	2.0	3.8
106	4.5	7.3	5.9	8.0
Average	2.7	5.2	4.0	5.6

Source: University of California, 1979.

15. Constantly check egg-processing equipment for its contribution to egg breakage. Poorly maintained equipment breaks an excess of 5% of the eggs. The data in Table 16-24 illustrate the effects of egg washer maintenance on egg breakage.

Table 16-24. Eggshell Damage During In-line Washing

Breakage in the Washer	Number of Washers
Less than 0.5	25
0.5 to 1.0	39
1.0 to 1.5	23
1.5 to 2.0	7
2.0 to 2.5	9
2.5 to 3.0	3
3.0 to 3.5	6
3.5 to 4.0	2
4.0 and more	3
Average (1.24%)	

Source: University of California, 1976.

Dirty Eggs

Egg cleanliness is one reason for the acceptance of laying cages. However, there will be times when eggs seem to get dirty in cages. Moisture and dust will collect on cage floors and dirty rings will be left on the shells as the eggs roll down the wire floors. Try to eliminate as much dust as possible. Periodically, brush the egg roll-outs. This will help to reduce stained and dirty eggs. Do not allow the cage wires to rust. Some wire floors are constructed of material that is rust-resistant. Keep automatic egg-gathering belts clean. One broken egg can dirty several clean ones.

Oiling Eggs at Pickup Time

Most market eggs are washed in specialized equipment before they are placed in cartons or cases. Oiling the newly laid eggs will facilitate the washing process and help preserve initial egg quality and egg weight prior to processing. A thin layer of mineral oil should be sprayed over the eggs as they are placed on flats at gathering. Two or three grams of oil per flat of 30 eggs is sufficient. However, if eggs are handled quickly, it is questionable if egg oiling of cage eggs at this time is economically justified. Note: Most market eggs are not washed.

Egg-Cooling Room

Each cage laying farm must be provided with an egg-cooling room. If the poultry houses are large it is best to construct such a room in each house (see Chapter 15-I).

16-Q. FACTORS AFFECTING EGG PRODUCTION

Age of Pullet at First Egg

Annual average egg weight is correlated with the age of the pullet when she lays her first eggs. Since age at first eggs may be controlled by lighting programs and increasing or decreasing the feed intake, the management program must be one that gets the bird to sexual maturity at the most opportune age. Although the egg-size relationship is indirect, that is, eggs are larger because the bird is older at the time she goes through her laying cycle, the relationship is important. But forcing pullets to delay the onset of egg production increases the cost of raising a pullet, which of course raises the cost of producing a dozen eggs. Thus, the poultry manager must weigh the added pullet costs against the greater cash return from larger eggs. The most economical age for first eggs varies between strains of chickens and between countries depending on egg pricing methods relative to egg weight. But initiation of egg production between 19 and 20 weeks is now considered optimum. At about 5 weeks after the first egg is produced the flock should attain 50% egg production on a hen-day basis depending on the flock's uniformity.

Mortality During Egg Production

Excessive mortality during the laying period is an expensive management failure. Not only is there a loss of the bird herself, but the profit she would have generated does not materialize. There probably is no such thing as "normal" mortality; thus there is no method of determining "excessive" mortality. Studies of large cage laying farms in southern California have shown that monthly average death losses have been from 0.5 to over 2%. However, individual flocks have had losses as low as 0.3% per month. Some strains of layers have a lower incidence of mortality than others, so all excessive mortality cannot be the result of poor management. Yet one should certainly feel that anything over 1% per month is attributable to management.

Culling the Cage Laying Flock

Culling should be a continuous process in laying cages, day-by-day, removing all injured and weak birds. Besides this, a detailed culling should be made at the following times:

> 15–20 weeks of egg production
> 30 weeks of egg production
> 45 weeks of egg production

Remove all birds that are nonproducing, sick, molting, cannibalistic, and those with prolapse at these time periods.

Pullet Replacement Schedule in Cages

Egg producers today attempt to optimize the length of the laying cycle to maximize average returns. This age is commonly between 75 and 80 weeks except where producers elect to molt their flocks (see Chapter 19). Rearing programs are coordinated to replace flocks at regular intervals, thereby making the most efficient use of facilities and providing the farm with a balance of egg sizes and a steady flow of production. One-age houses or farms are totally depopulated and cleaned before new pullets are placed in the house.

Typical losses of less than 12% during the laying year result in an average housing occupancy of 94% or more. Consolidation of flocks is rarely done except in countries with bird quota schemes or when flocks are molted. Most producers feel that mixing different flocks is not worth the effort and that it risks disease exposure and disrupts established social orders.

Variations in Pullet Egg Production

At sexual maturity most flocks of egg-type pullets are pretty much lookalikes. But it soon becomes obvious that one cannot predict the future of a group of pullets by simply looking at them at this age. To a large extent it is how they were raised that determines how well they will produce eggs. The hidden factors that cause variations will include, but not be limited to, the following:

Strain differences. There is great variability among strains of egg-type pullets in regards to egg production, egg size, mortality, and other factors.

Hatching egg size. Hatching egg size is directly correlated with chick size. This weight difference in day-old chicks remains throughout the pullet growing period. It is estimated that 30% of the variation in mature body weight of the pullets in the flock is due to variation in hatching egg size.

Dubbing. Dubbing may increase or decrease egg production. In warm or hot climates the comb is of special value in dissipating heat from the body, and dubbing may be a detriment. In cool climates, dubbing will help maintain a better social order, and is an advantage.

Toe clipping. This will improve future egg production up to ten eggs per pullet, depending on how docile the strain is. Toe-clipped pullets are significantly heavier at 20 weeks of age.

Beak trimming. Type and age of trimming influences flock behavior. Different programs may affect egg production by 20 or more eggs per year.

Season of the year. Where there are hot summers and cold winters, the date of hatch will affect performance as shown in Table 16-25.

Body weight at maturity. Most prime breeders publish the optimum flock average body weight of their birds at sexual maturity. It has been shown that production will be decreased by two eggs per pullet per year for each 0.1 lb (45 g) above or below this optimum mature weight figure.

Body weight uniformity. Probably the best indicator of pullet quality at sexual maturity is *uniformity* of body weight (see Chapter 16-G). The number of eggs produced by flocks with varying percentages of

Table 16-25. Month of Hatch, Egg Production, and Mortality (Standard White Leghorn Flocks, U.S. Data)[1]

Month of Hatch	Number of Flocks	Egg Production			Mortality %
		4-Week Peak %	Average Hen-Day %	Hen-Housed Eggs	
Jan.	91	85.8	75.9	206.1	8.8
Feb.	68	86.5	75.3	204.1	8.0
March	80	86.9	75.0	203.5	8.1
April	71	87.0	75.2	203.0	7.9
May	81	86.9	74.1	200.9	8.7
June	93	88.1	75.3	202.2	9.1
July	89	87.1	75.6	203.7	10.1
Aug.	79	88.5	76.8	206.4	9.5
Sept.	112	88.9	77.5	207.1	9.1
Oct.	89	88.1	76.7	206.4	8.9
Nov.	80	87.3	76.2	205.7	8.6
Dec.	85	86.9	75.6	206.1	8.3
Av/Total	1018	87.3	75.8	204.6	8.8

Source: University of California.
[1]1985 and 1986 hatch dates, 21 to 60 weeks of age.

birds within 10% of the average flock body weight are shown in Table 16-26.

At sexual maturity, most flocks average from 90 to 105% of the breeder's recommendation for body weight, but at the end of the laying year most flocks fall between 80 and 100%. Extra light and extra heavy pullets at sexual maturity do not do as well as layers (see Table 16-27).

Table 16-26. Body Weight Uniformity at Sexual Maturity and Egg Production Variation

% of Pullets Within 10% of Average Flock Body Weight at Sexual Maturity[1]	Difference in Average Number of Eggs Produced per Pullet When All Hens Sold at Same Time
91 and above	+ 10
84–90	+ 7
77–83	+ 4
70–76	0 (base)
63–69	− 4
56–62	− 8
55 and below	− 12

[1]Scale weight interval of 1 oz (28 g), or less.

Table 16-27. Body Weight Groups and Annual Performance

Body Weight		Avg Hen-Day Egg Production	Mortality	Feed Consumed per Doz Eggs		Egg Size	
lb	kg	%	%	lb	kg	oz/dz	g each
3.3+	1.50+	62.5	9.9	5.79	2.63	27.0	63.8
3.0–3.1	1.36–1.41	64.0	5.7	5.07	2.31	26.7	63.1
2.7–2.8	1.23–1.27	64.6	7.3	4.76	2.16	25.6	60.5
2.4–2.5	1.09–1.14	64.6	9.6	4.67	2.12	25.5	60.2
2.2–	1.00–	55.1	18.5	4.54	2.06	24.8	58.6

Source: University of California, 1968.

16-R. PRODUCTION STANDARDS

Every line of chickens has a bred-in potential for each genetic *quantitative* character. For instance, genes that affect egg production are present. When management is correct, there will be full expression of these genes. But management is not always good enough to get the full gene potential; there will be all sorts of variation in the number of eggs a bird will produce. All the geneticist can do is make a statement that under good management, his line of birds will produce a calculated number of eggs. Those poultrymen who get more eggs from their birds than the calculated number will have done a fine job; those who get fewer will have done a poor job. These calculated figures may be termed "standards." They are not averages of what the birds will accomplish under field conditions, but are somewhat higher. Neither are they goals that can be attained only on rare occasions.

Importance of Standards

Most commercial poultrymen have the problem of not being able to determine whether their birds are doing well or poorly. The underlying cause of this difficulty is that there are day-to-day changes in the expression of most quantitative characters in the chicken. Birds get larger; egg production first increases and then decreases during the laying year; egg size generally increases the longer a bird lays; hatchability varies; and so forth. Standards set forth certain figures that indicate optimum behavior for many of these characters. If the standards are given on a weekly basis, they offer a means for the flock owner to compare the weekly productivity of this flock with figures that are standard for the particular line of birds involved. He then knows whether the flock is producing at standard, above standard, or below; he knows whether he is doing a good or a poor job of management.

Figures Are a Guide Only

These standard figures follow. They can be offered only as a guide, as each strain of chickens differs in its inherited ability to produce at a given rate. However, the figures in the tables are quite representative of averages of all lines. For the poultryman who wishes data for the particular strain of birds he is using, the primary breeder should be contacted.

Age at Early Egg Production

There are four common dates used for computing early egg production as shown below. Approximate average ages when the flock reaches these four categories are given, but they are highly variable. Most management recommendations are tied to age at 5% hen-day egg production. These ages would be lower with out-of-season flocks and higher with in-season flocks. They would be lower with flocks highly uniform for body weight and higher for flocks with lower uniformity.

Age of flock when first egg is laid	140 days
Age of flock at 5% hen-day egg production	150 days

| Age of flock at 50% hen-day egg production | 168 days |
| Age of flock at peak hen-day egg production | 210 days |

Production Indices

Common to the industry are two methods of measuring daily or weekly egg production. Each has its fallacies as an index, but each is a good rule of thumb.

Hen-day egg production for one day.   The following formula is a measure of the egg productivity of the live hens on any given day.

$$\frac{\text{Number eggs produced}}{\text{Number live hens}} \times 100 = \text{\% hen-day production for 1 day}$$

> *Example:*   There are 1,000 hens alive on a certain day and they produce 750 eggs that day. Their *hen-day egg production* is 75%.

Hen-housed egg production for one day.   The following formula is a measure of the egg productivity in relation to the number of hens (housed) at the beginning of the laying period.

$$\frac{\text{Number eggs produced}}{\text{Number hens housed}} \times 100 = \text{\% hen-housed production for 1 day}$$

> *Example:*   1,200 hens were "housed" at the beginning of the laying "year." Today they laid 750 eggs. Their *hen-housed egg production* is 62.5%.

Hen-day egg production for a long period.   This may be calculated by first computing the number of hen days in the period by totalling the number of hens alive on each day of the period. Then calculate the number of eggs laid during the same period. Proceed by using the following formula:

$$\frac{\text{Number of eggs produced during period}}{\text{Number of hen-days in the period}} \times 100 = \text{\% hen-day egg production for the period}$$

Hen-housed egg production for a long period.   First compute the average number of eggs laid per day during the period. Then use the following formula:

$$\frac{\text{Avg daily number of eggs produced}}{\text{Number of hens housed}} \times 100 = \text{\% hen-housed egg production for the period}$$

Difficulties with Hen-Day and Hen-Housed Interpretations

Although *hen-day egg production* is an excellent indicator of how well the live birds are laying, it does not consider egg size or egg quality. Since these factors help determine egg income, hen-day egg production is often misleading from a profit standpoint. Neither does the index account for mortality. Theoretically, one could lose all but 4 from a flock of 1,200 hens housed, and if he got three eggs on one day the hen-day production would be 75%. However, it is the best production index available to the industry. Almost everyone uses it.

The other index, *hen-housed production*, is not reliable either, as it includes both egg production and cumulative flock mortality. Therefore, there are many combinations of egg production and death loss that could give an identical hen-housed production figure. But from a cost-of-egg production standpoint, this percentage is good because it includes both production and mortality. It tells what the flock has done and is doing. However, as in the case of hen-day production, there is no indication of egg size or egg quality. Hen-housed production has almost no meaning for a short period. It is used primarily to summarize the flock's performance over an entire laying cycle. When so used, it is a much better measure of economic performance than hen-day production because it measures the yield based on the original flock size and not on the current count.

Relationship Between Production Indices

Some indication of the variations occurring with different methods of measuring flock egg productivity are given below (from Table 16-28). The only assumption involved in constructing the data is that flock mortality is uniform throughout the year (365 days) at 0.2% per week. Hen-day production is always larger than hen-housed production for the same flock.

Eggs produced per hen to 76 wk of age, hen-housed	288.9
Percent egg production per hen to 76 wk of age, hen-housed	73.7
Eggs produced per hen to 76 wk of age, hen-day	305.8
Percent egg production per hen to 76 wk of age, hen-day	78.0

In this example, hen-housed production is a better measure of profitability because all eggs are credited to the original flock. Eggs produced on a hen-day basis give the production of the average live hens regardless of mortality.

How to Use the Standards

Because the standards are indicative of what a good strain of birds should do under good management practices, they offer a means of comparing the egg production of an actual flock of birds with the standard figures. If actual egg production is below the standard, there is a problem; when the actual figures are at or above the standards, management is fulfilling its requirements. As the egg production standards are given on a weekly basis, the comparison may be made at any time during the production cycle.

The first week of egg production must start when the flock average hen-day egg production for the week is 5%. In an attempt to standardize record-keeping systems, 20 weeks is usually considered the starting point.

Production Standards for Commercial Layers

Table 16-28 gives the production standards for Standard Leghorn and medium-size (laying brown-shelled eggs) layers, respectively. The production figures given in Table 16-28 are highly variable. In-season flocks would be expected to produce more eggs, out-of-season flocks would produce fewer. Strain, temperature, stress, mortality, number of birds per cage, cage floor space, and many other factors would affect the production percentages. Figures are a guide only.

Table 16-28. Production Standards for White Leghorn and Medium-Size Commercial Layers in Cages

Week	Hen-Day Egg Production %	Hen-Housed Egg Production %	Cumulative Egg Production per Hen Housed
21	10.0	10.0	0.7
22	23.0	22.9	2.3
23	40.0	39.8	5.1
24	60.0	59.5	9.3
25	76.5	75.7	14.6
26	84.5	83.5	20.4
27	87.0	85.8	26.4
28	89.5	88.1	32.6
29	91.2	89.6	38.8
30	92.5	90.7	45.2
31	92.0	90.0	51.5
32	91.5	89.3	57.7

Table 16-28. (*continued*)

Week	Hen-Day Egg Production %	Hen-Housed Egg Production %	Cumulative Egg Production per Hen Housed
33	91.0	88.6	63.9
34	90.5	88.0	70.1
35	90.0	87.3	76.2
36	89.5	86.7	82.3
37	89.0	86.0	88.3
38	88.5	85.4	94.3
39	88.0	84.7	100.2
40	87.5	84.1	106.1
41	87.0	83.4	111.9
42	86.5	82.8	117.7
43	86.0	82.1	123.5
44	85.5	81.5	129.2
45	85.0	80.9	134.8
46	84.5	80.2	140.4
47	84.0	79.6	146.0
48	83.5	78.9	151.5
49	83.0	78.3	157.0
50	82.5	77.7	162.5
51	82.0	77.1	167.9
52	81.5	76.4	173.2
53	81.0	75.8	178.5
54	80.5	75.2	183.8
55	80.0	74.6	189.0
56	79.5	74.0	194.2
57	79.0	73.3	199.3
58	78.5	72.7	204.4
59	78.0	72.1	209.4
60	77.5	71.5	214.5
61	77.0	70.9	219.4
62	76.5	70.3	224.3
63	76.0	69.7	229.2
64	75.5	69.1	234.1
65	75.0	68.5	238.9
66	74.5	67.9	243.6
67	74.0	67.4	248.3
68	73.5	66.8	253.0
69	73.0	66.2	257.6
70	72.5	65.6	262.2
71	72.0	65.0	266.8
72	71.5	64.4	271.3
73	71.0	63.9	275.8
74	70.5	63.3	280.2
75	70.0	62.7	284.6
76	69.5	62.1	288.9
Eggs per hen	305.8	288.9	288.9
Average %	78.0	73.7	

16-S. EGG SIZE STANDARDS

Period of Laying Year and Egg Size by Percentages

The first eggs laid during the production period are smaller than those laid later, egg size gradually increasing as the pullet continues to lay. Since eggs are marketed by weight classifications, the producer is extremely interested in standards for the percentage of eggs that fall into each size as the bird continues production. Egg size definitions in the United States vary from state to state. In some instances, every egg graded large must weigh at least 2 oz (56.8 g). In other states and in the USDA standards, an individual egg weighing $1^{11}\!/_{12}$ oz (54.2 g) is acceptable in a carton of large eggs.

This definition problem causes considerable confusion within the egg industry. The percentage of eggs at different ages for each size category and size definition is shown in Table 16-29. *Note:* Care must be taken with both definitions to avoid egg weight loss. Eggs graded to precise definitions may fall below the legal definitions, resulting in rejections.

Each strain of birds will vary, and many management, nutritional, and environmental factors will affect the percentages. The table exemplifies a strain of birds whose average egg weight throughout the first laying cycle is 25.4 oz to the dozen (60 g ea). This egg weight is about average; it is neither large nor small. Strains of birds laying larger eggs would produce different percentages of the grade sizes listed, as would those laying smaller eggs.

Period of Lay and Number of Eggs Produced by Egg Size

In Table 16-29, the percentages of eggs laid in each weight classification are given. Table 16-30 shows the standard figures as a guide to determine actual number of eggs that would be produced. The standards involve birds that have the capacity to produce 289 eggs to 76 weeks of age with an average egg weight of 25.4 oz per dozen (60 g/ea). The percentage of large-and-over eggs is 79.4% (based on the $1^{11}\!/_{12}$-oz large definition), which is about average for all strains of Leghorns, although some are known to produce larger eggs than this; but there are many that produce smaller eggs (see Table 16-30).

Egg size variations. One factor of great importance in variation in egg size is the temperature of the air surrounding the birds. High temperatures decrease egg size as a result of a decrease in the daily nutrient intake of the flock (principally energy and protein). This distorts the percentage of eggs in each grade. The standards shown in Table 16-29 make no compensation for such changes, as evi-

Table 16-29. Standard Relationship Between Flock Age and Egg Size Percentages (White Leghorns)

24 oz Minimum Weight for Large

Age in Weeks	Peewee, Small Under 21 oz[1] Under 49.6 g[2]	Medium 21–23.9 oz 49.6–56.6 g	Large 24–26.9 oz 56.7–63.7 g	Ex. large, Jumbo 27 oz and over 63.8 g and over
	Percentage of Total Eggs Produced			
25	57	38	2	3
30	12	65	22	1
35	3	41	50	6
40	1	24	60	15
45	0	16	59	25
50	0	12	56	32
55	0	10	52	38
60	0	9	50	41
65	0	8	49	43
70	0	7	48	45
75	0	7	47	46

23 oz Minimum Weight for Large

Age in Weeks	Peewee, Small Under 20 oz 20–22.9 oz Under 47.3 g	Medium 20–22.9 oz 47.2–54.2 g	Large 23–25.9 oz 54.4–61.4 g	Ex. large, Jumbo 26 oz and over 61.5 g and over
	Percentage of Total Eggs Produced			
25	33	58	6	3
30	4	50	43	3
35	1	22	62	15
40	0	11	57	32
45	0	6	49	45
50	0	4	42	53
55	0	3	38	59
60	0	3	35	62
65	0	3	33	63
70	0	2	32	65
75	0	2	31	66

Note: Weighted average egg weight: 25.4 oz/doz (60.0 g/ea).
[1]Per dozen.
[2]Each.

denced by the fact that the percentage of the larger egg sizes increases until the end of the laying year.

Ambient Temperature and Egg Size

High house temperatures reduce egg size in two ways:

1. *High temperature during growing.* When pullets are grown during hot weather the bird weight at 20 weeks

Table 16-30. Standard Relationship Between Age and Egg Size by Numbers (White Leghorns)

24 oz Minimum Weight for Large

Period (age)	Peewee, Small Under 21 oz[1] Under 49.6 g[1]	Medium 21–23.9 oz 49.6–56.6 g	Large 24–26.9 oz 56.7–63.7 g	Ex. large, Jumbo 27 oz and over 63.8 g and over	Total
	Number of Eggs Produced per Hen Housed				
21–25	10.00	3.71	0.11	0.65	14.47
26–30	7.99	18.36	3.60	0.61	30.56
31–35	1.58	15.84	12.53	1.04	30.99
36–40	0.40	9.00	17.14	3.24	29.78
41–45	0.18	5.44	17.06	6.05	28.73
46–50	0.04	3.67	15.59	8.17	27.47
51–55	0.04	2.77	14.08	9.47	26.36
56–60		2.30	12.85	10.15	25.40
61–65		1.98	11.88	10.37	24.23
66–70		1.76	11.05	10.30	23.11
71–76		1.94	12.38	11.99	26.31
Total	20.23	66.77	128.27	72.04	287.41
%	7.0	23.3	44.6	25.1	100.00

23 oz Minimum Weight for Large

Period (age)	Peewee, Small Under 20 oz 47.3 g	Medium 20–22.9 oz 47.3–54.2 g	Large 23–25.9 oz 54.2–61.2 g	Ex. large, Jumbo 26 oz and over 61.4 g and over	Total
	Number of Eggs Produced per Hen Housed				
21–25	7.02	6.37	0.47	0.65	14.51
26–30	3.35	18.11	8.35	0.83	30.64
31–35	0.40	9.90	17.71	2.92	30.93
36–40	0.07	4.28	17.96	7.52	29.83
41–45		2.20	14.94	11.48	28.62
46–50		1.40	12.28	13.90	27.58
51–55		1.01	10.44	14.94	26.39
56–60		0.79	9.14	15.37	25.30
61–65		0.72	8.21	15.34	24.27
66–70		0.58	7.56	15.01	23.15
71–76		0.65	8.22	17.22	26.09
Total	10.84	46.01	115.28	115.18	287.31
%	3.8	16.0	40.1	40.1	100.00

Note: Weighted average egg weight: 25.4 oz/doz (60.0 g/ea).
[1]Per dozen.
[2]Each.

will be up to 20% lower than during cool weather, thereby causing the pullets to produce eggs lighter in weight at the start of egg production. Generally, such birds do not attain their ideal body weight during production and continue to lay eggs smaller than normal.

2. *Laying house temperatures above 80°F (26.7°C).* These depress egg production, egg size, and shell quality. The longer the house temperatures are excessive, the lower the values in these categories. High environmental temperatures at the end of the laying period produce more depression than those earlier.

The combined effect of these two factors expresses itself to the extreme when April and October hatch dates are compared (see Table 16-31).

Table 16-31. Effect of Season of Housing on Egg Weight

| Age in Weeks | Month Housed | | | | | |
| | April | | October | | Annual Average | |
	oz/doz	g/egg	oz/doz	g/egg	oz/doz	g/egg
25	21.2	50.0	20.6	48.6	20.9	49.4
30	22.9	54.2	22.9	54.2	22.9	54.2
35	24.2	57.2	24.2	57.2	24.2	57.2
40	25.1	59.4	25.1	59.4	25.1	59.4
45	25.9	61.1	25.5	60.3	25.7	60.8
50	26.5	62.5	25.7	60.8	26.1	61.7
55	26.9	63.6	25.9	61.1	26.5	62.5
60	27.2	64.2	25.9	61.1	26.6	62.8
65	27.2	64.2	26.0	61.4	26.7	63.1
70	27.2	64.2	26.2	61.9	26.7	63.1
Average	25.4	60.0	24.8	58.6	25.3	59.7

Source: University of California, 1988.

Month of Maturity, Egg Production, and Egg Size

Pullets reaching sexual maturity in the spring lay more large eggs than those maturing in the fall as a result of their larger body weights at point-of-lay. This effect produces larger eggs throughout the life of the flock.

Other Factors Affecting Egg Weight

Strain of birds. Egg size is a genetic factor; thus it is possible to develop strains of chickens laying large, medium, or small eggs. Egg size is highly correlated with body weight.

Age at first egg. The older a pullet is when she lays her first egg, the larger her eggs during her laying period. Inasmuch as it is possible to delay the onset of egg production, the age association becomes an important economic consideration (see Chapter 18-E).

Environmental temperature. As ambient temperatures rise, egg size decreases. At times hot weather will create a major egg-size problem.

Egg size when birds kept in cages.   Normally, and under similar circumstances, pullets in cages produce eggs that are slightly larger than those laid by pullets kept on a litter floor.

Laying ration.   Certain components of the laying ration will affect egg size. Increases in protein percentage usually are associated with increases in egg size (see Chapter 31-G).

Size of pullets in the flock.   The larger the pullet *within a given flock,* the larger the eggs. As the larger birds also produce more eggs, body weight becomes important. However, the birds in the flock are never uniform in weight. They go from small to medium to large. But the more uniform the body size, the more uniform the egg size.

Table 16-32.   Effects of Age and Strain on Egg Weights (White Leghorns)

Age in Weeks	Strain	No. of Flocks	Average Egg Weight		Range in Wts. Among Flocks		% Eggs over	
			oz/doz	g/egg	oz/doz	g/egg	23 oz/doz 54.3 g/egg	24 oz/doz 56.7 g/egg
24	A	25	21.6	51.0	2.9	6.9	18	10
	C	20	20.1	47.5	2.6	6.1	6	3
	E	33	20.6	48.7	3.5	8.3	9	6
	ALL	110	20.8	49.1	4.0	9.5	10	6
30	A	23	24.3	57.4	1.8	4.3	76	54
	C	21	23.1	54.6	1.7	4.0	49	27
	E	35	23.4	55.3	3.1	7.3	56	36
	ALL	111	23.5	55.5	3.1	7.3	59	38
40	A	25	25.7	60.7	1.7	4.0	93	83
	C	21	24.9	58.8	1.5	3.5	88	70
	E	35	25.0	59.1	2.2	5.2	86	71
	ALL	113	25.2	59.5	2.8	6.6	89	75
50	A	25	26.6	62.8	2.7	6.4	96	90
	C	21	26.0	61.4	1.9	4.5	96	88
	E	35	25.9	61.2	2.6	6.1	94	84
	ALL	113	26.1	61.7	2.6	6.1	95	86
60	A	22	27.3	64.5	3.4	8.0	97	93
	C	19	26.8	63.3	2.4	5.7	98	93
	E	33	26.4	62.4	3.2	7.6	95	87
	ALL	103	26.7	63.1	3.8	9.0	96	90
80	A[1]	21	27.2	64.3	2.8	6.6	97	93
	C	16	27.0	63.8	1.6	3.8	98	94
	E	29	26.6	62.8	2.7	6.4	96	89
	ALL	100	26.9	63.6	3.1	7.3	97	91
100	A	14	27.6	65.2	2.1	5.0	98	95
	C	12	27.6	65.2	2.1	5.0	99	96
	E	26	26.9	63.6	2.9	6.9	96	92
	ALL	74	27.2	64.3	3.4	8.0	97	93

Source: University of California, 1982.
[1]80- and 100-week-old flocks are molted.

Egg weight variation throughout the laying period. The various factors affecting egg weight have been discussed. Field studies in California during 1980 to 1982 compared egg weight in 113 commercial flocks under a wide variety of management conditions. Table 16-32 lists the average weights and range in weights observed at different ages.

16-T. EXTERIOR MARKET EGG QUALITY

Pimply eggshells. Such imperfections vary with the strain of chickens and the age of the flock. Individual hens lay eggs with the same degree of pimpling, day after day. The incidence is not associated with any disease-producing organism but rather to particles of calcium carbonate adhering to the shell. Such eggs are downgraded because of the pimple or the hole produced in the shell. High levels of vitamin D_3 in the feed have been shown to increase the amount of eggshell pimpling, but different calcium levels appear to have no effect.

Windowed eggs. Translucent streaks or "windows" in eggs are prevalent on occasion. These eggshells have a lower breaking strength and specific gravity. Some are associated with blind checks (very small cracks). Water moves out through these areas more rapidly, thereby decreasing the strength of the shell in the region. There is no known method of reducing the incidence of windowed eggs.

Misshapen eggs. Eggs that are ridged and misshapen are generally an economic loss. The number is directly related to the number of birds in a cage; the more, the higher the incidence. Most of these eggs are laid early in the morning. Misshapen eggs are a common result of infectious bronchitis and Newcastle disease infections. Soft-shelled eggs are more numerous when the birds are kept in cages than when maintained on a litter floor, and most are laid at night.

Ridged eggs. Birds that are crowded in cages while the shell hardens in the oviduct are more likely to lay eggs with ridges. The hardening process takes about 12 hr, and crowding produces more pressure on the sides of the birds and the egg itself. The largest diameter of the egg is affected most and may crack or buckle, later to be sealed or covered with more shell material. Avoid unnecessary bird activity during the period when the shell is being formed and hardened.

Declining shell quality during the laying cycle. Declining shell quality with increasing age of the hen occurs in all breeds and strains of chickens due to the following factors:

1. The amount of calcium in the medullary bone decreases with the age of the hen.
2. The amount of shell (weight) material produced re-

mains rather constant week after week during the egg production cycle. As eggs continually get larger during this period, the shells must necessarily become thinner to cover the larger egg contents. There is some difference in the thickness of shells from white- and brown-shell eggs; white shells are usually somewhat thicker.

Specific Gravity of Eggs and Eggshell Quality

Even fresh eggs vary in many ways. Shell strength, condition of the albumen, yolk color, and incidence of blood and meat spots differ. These, along with other quality factors, deteriorate as the eggs age.

Specific gravity of eggs. Specific gravity of an egg and shell thickness are related: Higher specific gravities are indicative of greater shell thickness. An egg is placed in solutions of different specific gravities until the solution is found where the egg floats. The specific gravity of each solution is given a score and these figures are those commonly used. They are as follows:

Specific Gravity of Solution	Specific Gravity Score
1.068	0 (thinnest shell)
1.072	1
1.076	2
1.080	3
1.084	4
1.088	5
1.092	6
1.096	7
1.100	8 (thickest shell)

Any specific gravity score above 5 indicates good eggshell quality, but annual averages of all eggs laid by the flock may vary between 3 and 5. The specific gravity score decreases as birds continue to lay. A typical example would show a score of 6 the first month, decreasing uniformly to 2.5 by the end of 12 months. Birds with the highest scores at the first part of the laying cycle have the highest at the end.

Procedure for calculating specific gravity. Secure a hydrometer that will measure the specific gravity of solutions between 1.060 and 1.100. Then get nine 10-gal (37.8 L) plastic pails and add 7 U.S. gal (26.5 L) of water and 7 lb (3.2 kg) of salt (NaCl) to each pail. Dilutions of one lb (454 g) of salt to one gal (3.8 L) of water will give a specific gravity of about 1.079. Using the hydrometer, add salt or water to each container to bring the solutions to the nine desired specific gravities as shown above.

Place 50 eggs in a plastic-coated egg basket and lower the container and eggs into the solution having the lowest specific gravity. Remove the floating eggs, identify the number, and repeat the procedure for the pail having the next highest specific gravity, and so on, continuing through the remaining pails. The temperature of the eggs and the solutions should be standardized for the most accurate results and to prevent contamination of the eggs.

Calculate the percentage of eggs floating in each solution and calculate the average specific gravity of the entire lot. By repeating the procedure, using another 50 eggs, 100 eggs will be used and the figures will convert easily to percentages. When making comparisons always use eggs laid during the same time of the day.

Specific gravity and cracked eggs. Research has shown that specific gravity of eggs is correlated with the probability that eggs will crack during processing. Table 16-33 shows the relationship.

Archimedes' method for measuring specific gravity of eggs. An alternate method of measuring specific gravity (SG) of eggs is to measure the weight of each egg in air, immerse the egg in water and weigh it, and then estimate the specific gravity by Archimedes' principle as

SG = (weight in air) ÷ (difference between weight in air and water)

Although the Archimedes' method is more accurate under some conditions, the flotation method to measure egg SG should be satisfactory for most applications (B. K. Thompson and R. M. Hamilton, 1982, *Poultry Sci. 16,* 1599).

Table 16-33. Relationship of Processing Cracks to Specific Gravity of Eggs (Leghorns)

Specific Gravity of Eggs	Cracks %
1.065	27.3
1.070	21.0
1.075	11.1
1.080	7.5
1.085	2.4
1.090	0.7

Source: Holder, D. F., et al. 1974. *Poultry Sci. 58,* 250.

Cracked egg percentages.   Dr. David Roland of Auburn University stated that for every 100 hard-shelled eggs collected from standard laying Leghorns, there are an additional 7.8 eggs that are uncollected as the result of being lost through the bottom of the cage. These eggs are classified as shell-less, ultra-thin-shelled, and thin shelled. Even though these eggs are never credited to the flock as eggs produced, they still represent an investment in feed and must be included in calculating theoretical nutrient requirements.

Much of this loss can be prevented by reducing disturbances in the laying houses during the period of maximum oviposition. The percentages vary with the season of the year as shown in Table 16-34.

Table 16-34. Number of Uncollectable Eggs per 100 Collected

Age of Layers Months	Winter	Summer
8	2.4	2.9
11	5.1	3.6
14	10.6	6.8
17	14.7	16.1
Seasonal avg	8.2	7.4
Yearly avg	7.8	

16-U. INTERIOR MARKET EGG QUALITY

Air Cell

Air cell and egg quality.   Normally the two shell membranes are separated at the large end of the egg to form the air cell. In the newly gathered egg this cell is approximately ¾ in. (2 cm) in diameter and ⅛ in. (0.32 cm) in depth. As the egg ages the diameter and depth of the air cell increase, the speed depending on temperature and composition of the gases surrounding the egg. In the United States, the size of the air cell is used to estimate the age of the egg. USDA standards for Grade AA eggs require that the air cell be no more than ⅛ in. (0.32 cm) deep.

Yolk

Yolk shadow. When an egg is candled the yolk creates a definite shadow. It is more pronounced in eggs with white shells than in those with brown shells. In fresh eggs the shadow is light, as the thick albumen tends to keep the yolk centralized within the shell.

But as the egg ages, the albumen becomes thinner, allowing the yolk to approach the shell, and thereby creating a darker shadow. The intensity of the yolk shadow is an indication of the quality (age) of an egg.

Yolk color. The density of the yellow pigment in egg yolk is closely related to the xanthophylls in the ration. Consumers vary in their choice of yolk-color density; some prefer light-colored yolks, some prefer dark. In most cases the diet is altered so as to produce egg yolks of the correct color density. Many eggs find their way to egg-breaking plants where the yolks and whites are separated and either dried or frozen, later to be used in food preparation. As many yolks are used for the production of noodles and similar products, there is a necessity for an intense yellow color. In many instances yolk-color intensifiers are added to the feed (see Chapter 31-H).

Mottled yolks. Mottling of yolks appears as small, fatlike globules on the surface of the yolk, produced by movement of water from the albumen through the vitelline membrane into the yolk. Mottling increases during the summer months. The cause of mottling is indefinite. Nicarbazin, ammonia fumes, gossypol, and piperazine may be contributors, along with many others.

Double-yolked eggs. The frequency of two yolks in an otherwise normal egg may be due to a variety of things (see Chapter 3-A). Some scientists have said that double-yolked eggs at the start of egg production may be due in part to too much light stimulation, while those occurring late in the laying period are due to the lack of enough light, either intensity or length of the daily light period.

Albumen

Haugh units to measure albumen quality. There is great variability in the albumen quality of freshly laid eggs when the albumen is measured by its ability to remain viscous. The age of the egg also influences the quality of egg albumen.

Today's most widely used measure of albumen quality is the Haugh unit, a measure of the height of the albumen after correcting the reading for difference in egg size. Prof. R. R. Haugh, who developed the correction factor, noted that ''observed'' albumen quality in the broken-out egg as a logarithmic rather than a linear function of albumen height.

> *Example:* A change in height from 10 mm to 9 mm (a difference of 1 mm) is relatively less important from the standpoint of albumen appearance than a change from 3 mm to 2 mm (also a difference of 1 mm).

Haugh proposed that albumen height as a measure of quality be improved upon by taking its log and multiplying by 100 to correct to whole numbers. He added a correction factor for egg weight for he found that if two eggs of unequal size have the same albumen height, the smaller egg had better apparent quality. Eventually, Haugh's formula became very complicated, but conversion charts are available to simplify the computation. There is also a micrometer that reads the height of the albumen and converts the figure to Haugh units (HU). The egg weight must be set on the dial before the albumen height is measured. The range can extend from a high of 100 or better to a low of 20 or less. The higher the value, the better the albumen quality.

Factors affecting Haugh units. The Haugh unit has been used to determine the interior quality of table eggs. It not only measures the relative freshness of the egg but also the age of the hen that produced it. At one time, it was used by the USDA in its voluntary grading program. There are many factors that can affect this measurement such as age of egg, age of flock, ambient temperature, strain of chicken, and egg-handling procedures.

Pink albumen. Pink albumen may be caused by gossypol from cottonseed meal.

Watery white. This is often the result of high ambient temperature or ammonia in the house, eggs from older hens, sulfa drugs, or disease.

Blood and Meat Spots

Blood spots in eggs. Blood spots are classified according to their size. Small spots are less than $\frac{1}{8}$ in. (0.32 cm) in diameter; large ones are greater than this. Leghorn strains vary in the number of eggs they lay with blood spots; usually the quantity is about equally divided between large and small spots with a total averaging about 2.0%. Eggs from brown-egg layers will usually show a higher incidence with a typical average of around 5.0%.

Meat spots in eggs. Meat spots are grouped according to their size in the same manner as blood spots. Meat spots are usually seen very infrequently in eggs from white-egg layers. The average number is usually less than 1.0% and less than half are larger than $\frac{1}{8}$ in. (0.32 cm) in diameter. On the other hand, brown-egg layers are very prone to produce large numbers of meat spots. The average is about 20% but some strains can go as high as 30%. Table 16-35 lists the amount of blood and meat spots observed in the North Carolina Random Sample Test over four separate tests.

Table 16-35.   Incidence of Blood and Meat Spots in Chicken Eggs

Strain Code	Blood Spots		Meat Spots	
	Large %	Small %	Large %	Small %
White-egg strains				
A	0.7	0.8	0.2	0.3
B	0.9	1.0	0.2	0.4
C	1.1	1.2	0.2	0.2
D	1.1	0.8	0.3	0.6
E	0.5	0.7	0.3	0.5
Avg	0.9	0.9	0.2	0.4
Brown-egg strains				
F	2.0	2.9	9.7	10.2
G	2.4	3.1	9.9	10.6
H	2.2	3.5	13.5	14.2
Avg	2.2	3.2	11.0	11.7

Source: North Carolina Random Sample Tests, 1980–1984.

Oiling Eggs to Preserve Quality

Oiling eggs by spraying them with a thin coat of mineral oil after they have been washed is an accepted practice to reduce the transfer of air and carbon dioxide through the shell, thereby prolonging egg quality. Oiling has the added benefit of reducing the loss of egg weight due to evaporation of moisture.

16-V. PROBLEMS IN CAGED LAYERS

Cage layer fatigue.   Birds kept in cages develop brittle bones, the height of the incidence being determined by several factors. Such layers have difficulty in standing, but egg production does not seem to be greatly affected. However, the market value of the spent hens will be seriously reduced, often to the point that the birds cannot be sold. There is no known control measure other than to assure that the flock's daily intake of nutrients, with emphasis on minerals, is adequate for the flock's level of performance (see Chapter 37-Z).

Fatty liver syndrome.   Caged layers are prone to increased fat deposition, sometimes to the extent that the liver is affected. The condition is known as fatty liver syndrome. There are some remedial measures, but at times nothing seems of value (see Chapter 37-U).

Prolapse.   The failure of the female to retract the outer end of the oviduct after oviposition is known as prolapse. The appearance of the oviduct leads to cannibalism, and the problem is difficult to cope with. In many cases, the prolapse becomes so great in individual birds that they are picked to death by their cage mates (see Chapter 13-I).

The difficulty is definitely inherited, and is more prevalent in cages than with regular nests where the birds are not exposed during the time the oviduct is being retracted. Pullets too fat at the onset of egg production is a definite cause. Feeding a high-fiber diet, or doing most anything that will cause reduced egg production will help return the flock to normal.

16-W. CAGE INSECT CONTROL

Flies are a common problem in all cage operations. The separation of the chicken from its droppings allows fly breeding to continue unabated unless the farm operator intervenes. Flies increase when birds are crowded.

The more successful fly control methods emphasize prevention of the problem by keeping the droppings dry, encouragement of beneficial insects that consume or parasitize immature fly stages, and the use of supplementary chemical treatments (see Chapter 38-B).

16-X. CAGE MANURE DISPOSAL

The handling of manure on a cage poultry farm requires management techniques of enormous proportions. Many methods have been tried. House design is an important factor. Not only is the moisture content of poultry manure a problem, but the fly aggravation associated with poultry droppings may take on major proportions.

Wet droppings.   High levels of protein and salt in the ration are instrumental in increasing the amount of moisture in the droppings. As environmental temperatures rise, the pullets drink more water, thereby increasing fecal moisture. Leaky watering devices are a major contributor to excessive water in the droppings collection area.

Droppings odor.   Fumes from the ammonia in the droppings plus those created by bacterial action are obnoxious. The odors are accentuated when droppings are wet. Some odors may be materially reduced by using some commercial products on the market. Regular removal of the manure below cage floors should be made a part of the management program. In pits, exposure to the air and the use of more fans will help dry the droppings.

Manure disposal. There are many mechanical, chemical, and bacterial methods of manure handling that are practical on poultry farms. These are discussed in Chapter 40.

16-Y. COST MANAGEMENT IN CAGES

Producing market eggs at a profit is the end point of a complex program not only through good flock performance but in the egg producer's ability to manage his production costs. Even though prices are not always under his control, the elements of top performance are. These costs are best itemized as those necessary to produce one dozen eggs.

Cost of Producing One Dozen Eggs

Although not always an accurate method of evaluating cost efficiency, because it does not consider egg size and egg quality, the index of "cost of producing a dozen eggs" is one common to the poultry industry. The estimated costs are given in Table 16-36. Obviously, the costs in the table are representative of a good, efficient operation. However, egg production costs are often lower; more often, they are higher. Feed cost may be a determining factor. But in this day of narrow margins of profit from egg production it will be necessary to keep costs somewhere near those given in the table.

Table 16-36.   Estimated Costs to Produce One Dozen Table Eggs (in U.S. Dollars)

Item	Standard Leghorn $	Medium-Size[1] $
Feed	0.263[2]	0.300[3]
Replacement pullet	0.104[4]	0.120[5]
Labor	0.032	0.032
Housing	0.025	0.027
Equipment	0.020	0.022
Other costs	0.050	0.050
(Less hen salvage value)	(0.010)	(0.012)
Total	0.484	0.539

[1]Producing brown-shelled eggs.
[2]3.5 lb/doz @ $0.075/lb (1.59 kg/doz @ $0.165/kg).
[3]4.0 lb/doz @ $0.075/lb (1.82 kg/doz @ $0.165/kg).
[4]$2.50 per pullet amortized over 21 doz eggs.
[5]$3.00 per pullet amortized over 21 doz eggs.

Pullet Growing Cost Per Dozen Eggs

The cost to grow the pullet is a necessary factor in determining the cost to produce a dozen eggs. As it is the second largest of the expense items, it is of great importance. Table 16-37 shows the relationship between pullet costs and egg production costs.

Age at first egg. Pullet growing costs are closely related to the age at which the first eggs are produced. The later the sexual maturity, the longer the growing period and the greater the costs. However, this factor is not responsible for all variations in this item. Efficiency, ability to grow a high percentage of the chicks that are started, freedom from flock disease, and other costs enter into the final figure.

As the age of sexual maturity is delayed, egg size increases, particularly for those eggs produced at the beginning of the laying cycle. However, the extra value of the larger eggs must be balanced against the added cost of keeping the pullet longer in the growing period. In most instances egg production should reach 10% on a hen-day basis when the pullets are 21 weeks of age. When egg production starts later the cost per egg unit increases.

Length of laying cycle. Probably the two most important contributors

Table 16-37. Pullet Cost as It Affects Pullet Cost per Dozen Eggs (in U.S. Cents)

Pullet Growing Cost (in U.S. Dollars)	Number of Eggs Produced per Pullet					
	240	252	264	276	288	300
	Number of Dozen Eggs Produced per Pullet per Year					
	20	21	22	23	24	25
2.00	10.0	9.5	9.1	8.7	8.3	8.0
2.10	10.5	10.0	9.5	9.1	8.8	8.4
2.20	11.0	10.5	10.0	9.6	9.2	8.8
2.30	11.5	11.0	10.5	10.0	9.6	9.2
2.40	12.0	11.4	10.9	10.4	10.0	9.6
2.50	12.5	11.9	11.4	10.9	10.4	10.0
2.60	13.0	12.4	11.8	11.3	10.8	10.4
2.70	13.5	12.9	12.3	11.7	11.3	10.8
2.80	14.0	13.3	12.7	12.2	11.7	11.2
2.90	14.5	13.8	13.2	12.6	12.1	11.6
3.00	15.0	14.3	13.6	13.0	12.5	12.0
3.10	15.5	14.8	14.1	13.5	12.9	12.4
3.20	16.0	15.2	14.6	13.9	13.3	12.8
3.30	16.5	15.7	15.0	14.3	13.8	13.2
3.40	17.0	16.2	15.5	14.8	14.2	13.6
3.50	17.5	16.7	15.9	15.2	14.6	14.0

to pullet-growing costs per dozen eggs are the length of the laying cycle and whether or not the flock is recycled (see Table 16-37). Lifetime production can exceed 35 or more dozen depending on the number and length of cycles. This will greatly reduce the pullet cost per dozen eggs produced but other costs will increase.

Feed Cost Per Dozen Eggs

About 60% of the cost of producing one dozen market eggs is feed cost. It is the largest of the cost items, and consequently represents an important facet for an egg cost analysis. Two items go into a calculation of the cost of feed per dozen eggs:

1. Pounds (or kilograms) of feed necessary to produce one dozen eggs
2. Cost of feed per lb (or kilogram)

Reducing feed cost per dozen eggs. The answer to methods of reducing feed cost does not lie entirely with the price of feed. There are many ways of reducing feed cost per egg unit other than obtaining cheaper feed. Some of these methods are as follows:

1. *Increase average house temperature.* Feed consumption will decrease approximately 0.5% for each 1°F (0.6°C). Nutrient density will have to be adjusted upward to compensate for the reduction in feed intake.
2. *Reduce the pullet body weight.* Pullets reaching sexual maturity at a reduced size will usually remain smaller during the laying period, thus requiring less feed.
3. *Increase egg production.* When egg production increases, the amount of feed necessary to produce a dozen eggs decreases. There are many management factors that increase egg production.
4. *Use controlled feeding during laying period.* Maintaining correct body weight during the laying period through controlled feeding will keep the pullets from consuming more feed than normal and getting overly heavy.
5. *Use phase feeding.* The main purpose behind phase feeding is to lower the cost of the feed consumed by a pullet during her laying period.
6. *Follow a rigid culling program.* Culls consume feed which must be charged to egg production.
7. *Prevent feed wastage.* When as little as 1.0 lb (454 g) of feed is wasted per pullet per year, the cost of producing eggs is increased by about 0.3 U.S. cents per dozen.
8. *Reduce the individual size of eggs produced.*

Other Cost Items

There are many items other than feed and pullet cost that go into the cost of producing eggs. Investment and some overhead items are fixed costs and usually cannot be reduced. Labor is probably the most variable of all costs other than feed and pullets. The most intensive daily labor chore involves egg collection, and costs may vary from as much as 2 U.S. cents per dozen for hand egg collection to practically zero for mechanized collection systems. New fully mechanized egg farms may employ as few as one person per 50,000 or more layers.

Contract Egg Production

The production of market eggs on a contract basis has gained an important foothold in the industry during the past few years as the result of an increased amount of integration. Under such contracts, the PRODUCER furnishes the buildings, equipment, labor and a few incidentals, while the INTEGRATOR furnishes the pullets, feed, and some other items, along with the necessary supervision.

Contract PRODUCER return. There are many methods for paying the PRODUCER for the use of his equipment and for his services. These, too, are variable. One common and simple method is for the INTEGRATOR to pay the PRODUCER a contracted price based on each dozen eggs produced and delivered to the INTEGRATOR. A contract of this type offers the incentive for the PRODUCER to do a good job; the more eggs he produces, the greater his cash return. See Chapter 10-A for an example of an egg production contract.

Records

An adequate record of the performance of the caged layers must be kept. It should contain items such as egg production, egg size, feed consumption, mortality, body weight, egg quality, and medications. In addition to these production records, facts associated with cost management must be recorded, for the ability of the poultryman to produce eggs at the lowest possible cost is the ultimate of attaining the highest profit. Examples of these records and other pertinent information are given in Chapter 23.

17
Breeder Management

Keeping birds for the production of hatching eggs involves practices necessary for the production of commercial market eggs plus practices needed that will produce eggs that are fertile and hatch well. Many of the differences in management begin with the young chick and continue through the growing period. Chapters 13, 14, and 15 should be read first as many recommendations will not be repeated here. The information that follows is broken down to conform with the three types of breeder birds as follows:

1. Those producing pullets that lay white eggs for human consumption.
   a. Standard Leghorns
   b. Mini-Leghorns (see Chapters 1-D and 22-F)
2. Those producing pullets that lay brown eggs for human consumption.
   a. Sex-link (gold male and barred female parents)
   b. Brown egg (gold male and silver female parents)
3. Meat-type broiler breeder parents that produce straight-run broiler chicks.
   a. Non-sexed-linked (progeny cannot be sexed at day old)
   b. Sex-linked (progeny can be sexed at day old)
   c. Mini-meat-type (see Chapters 1-D and 22-F)

17-A. GROWING PROGRAMS FOR BREEDERS

There is a separate growing program for each of the seven types listed above. Male and female programs are a part of each. In many instances each sex is handled differently.

Although breeding flocks will consist of only about 10% as many males as females, the males are responsible for 50% of the *gene* makeup of the progeny and must be given more room, more feed, and special handling.

Meat-line grandparents produce cockerel chicks that are smaller than the grandparent pullet line chicks because the cockerel lines produce smaller and fewer eggs, and smaller eggs produce smaller chicks. These cockerel chicks must be started separately from the pullet chicks.

Biological Isolation

Most breeding birds are now subjected to some type of disease-free program against such diseases as pullorum, fowl typhoid, M. Gallisepticum, and M. Synoviae. As chicks they must have complete isolation; as adults they will be blood-tested to determine if they are free of the disease organism involved. Isolation of the flock or flocks must be complete to prevent entrance of the disease from the outside. Separate personnel must be used; they must shower and put on clean clothing before entering the premises.

Brood–Grow System of Growing

Today, practically all poultrymen with breeding stock use the same poultry house for brooding and growing. Not only does the system remove the need for separate brooding and growing houses but the system makes it easier to carry out the many necessary vaccination programs.

With the brood-grow system about one-third of the house is used during the first 5 weeks when brooder stoves are needed. After this time the chicks should be given access to the entire building until egg-production breeders are about 18 weeks of age, and broiler breeder parents are about 20 weeks of age. They should then be moved to permanent laying houses.

Brood–Grow–Lay System

With this program the same house is used for brooding, growing, and laying. Once the chicks are placed in it, they never leave. This means that the house is really designed for laying birds and cockerels, then temporarily converted to a brooding and growing facility. The number of birds to be placed in it at the start of the laying period will determine the number of chicks placed in it at day old.

17-B. LEGHORN MANAGEMENT DURING BROODING AND GROWING

The poultryman has the option of two programs, the choice being the one best suited to his method of management.

1. *Sexes intermingled.* Start the sexes separately in the same house. They may be in separate rooms or confined by brooder guards to separate stoves within the same room. Intersperse the brooders with cockerels throughout the building, for when the guards are removed the chicks have a tendency to return to the same brooders rather than mingle with all the pullets. When 2 weeks of age the brooder guards should be removed to allow the cockerels and pullets to run together.

2. *Sexes raised separately.* Confine the cockerels and pullets to separate sections of the house by using a high wire fence. When the birds are about 12 weeks of age, mix about 5% of the cockerels with the pullets. At about 18 weeks of age, when culling of the inferior cockerels is completed, intersperse the remaining cockerels with the pullets by moving them during a period of darkness to reduce fighting in the males.

17-C. MEDIUM-SIZE BIRD MANAGEMENT DURING BROODING AND GROWING

These birds are parents of pullets raised for the production of brown-shell eggs for human consumption. Most of these breeding birds are comprised of genetically gold males and genetically silver females. The offspring chicks may be color sexed at day old. Female offspring chicks are gold; male offspring chicks are silver.

17-D. MEAT-TYPE MANAGEMENT DURING BROODING AND GROWING

These are broiler breeder parents. During the growing period meat-type male and female birds will get exceptionally large if fed all they will eat. The ideal male body weight at 24 weeks of age is about 7.4 lb (3.36 kg), and the ideal female weight is 5.5 lb (2.50 kg). However, if given free access to feed, their weights will be about 50% more than these weights. Therefore, feed restriction is the main ingredient of a management program. There are two recommendations:

1. *Sexes raised separately.* With meat-type birds it is recommended that the cockerels and pullets be raised separately until about 21 weeks of age. Then mix the cockerels with the pullets at the ratio of 12 males per 100 pullets. At sexual maturity (about 21 to 22 weeks of age) reduce the number of males to 9 to 11 per 100 females (see Table 17-8).

2. *Sexes intermingled.* Keep the sexes separate the first 2 weeks. During this period the beak trimming will be completed and the smaller cockerel chicks will get off to a good start for they will not be crowded by the larger pullet chicks. When 2 weeks old, place an equal number of cockerel chicks under each brooder containing pullet chicks.

17-E. HATCHERY PROCEDURES FOR BREEDER CHICKS

The hatchery has certain responsibilities that are different when hatching and delivering breeder chicks.

Number of Each Sex to Deliver

If the primary breeder develops both male and female parent lines the hatchery will be responsible for hatching and delivering the required number of male-line cockerels and female-line pullets. These numbers will be as follows:

Leghorns: Deliver 10 to 12 cockerel chicks for each 100 pullet chicks.

Medium-size (the offspring producing brown eggs for human consumption): Deliver 10 to 12 cockerel chicks for each 100 pullet chicks.

Broiler breeder parents: Deliver 12 to 15 cockerel chicks for each 100 pullet chicks.

Toe Trimming and Comb Trimming

To prevent injury to the backs of the females during mating, the toes of day-old meat-type cockerel chicks should be trimmed at the hatchery. Trim at the outer first joint of the back toe and inside toe of each foot. Use an electric beak trimmer or toe clipper.

Adult males in breeding pens do a lot of comb pecking to set up their social order. To reduce comb injury, the combs of male chicks are usually trimmed at the hatchery (see Chapter 13-J).

Sexing Errors

Although the main reason for toe and comb trimming is to prevent body injury, such trimming offers a means of identifying the sexes of male and female parent lines. All parent line breeder chicks are sexed at the hatchery, and the combs of all male-line cockerels to be used as breeders are trimmed. The toes of male meat-line cockerels are also trimmed. The female-line chicks are sexed, but only the pullet chicks are to be used as breeders, and these are not trimmed.

As no sexing procedure is perfect, there are errors, and birds that are sexing errors must be removed from the growing flock as soon as identifiable. If they are left in the flock many factors of genetic excellence in production will be forfeited.

The following shows how the sexing errors are identified:

	Cockerel Chicks	Pullet Chicks
Combs trimmed	Keep	Sexing error
Toes trimmed	Keep	Sexing error
Combs untrimmed	Sexing error	Keep
Toes untrimmed	Sexing error	Keep

Hatchery Vaccination and Medication

It may be part of the disease-control program to have some chick vaccination or medication done at day old at the hatchery. Consult with a trained veterinarian or your project supervisor before embarking on any such program. Read the description of the particular disease in Chapter 37.

17-F. BREEDER GROWING HOUSE

Practically all breeder growing birds are raised in one of the following two types of houses:

    1. All-litter floor
    2. Slat-and-litter floor (or plastic-and-litter)

Roosts. Roosts are not required for growing birds, but when used the number of first floor eggs will be reduced.
Floor space. Floor space requirements are given in Table 14-1.
Feeder space. See Table 14-4.
Waterer space. See Table 14-2.

Water Consumption by Growing Leghorns

Water consumption is highly variable for all types and ages of chickens depending on ambient temperature, humidity, density of the feed, and the amount of feed restriction, but ambient temperature has the greatest effect. Table 14-3 shows how the consumption of water by Leghorns varies with changes in the ambient temperature.

Water Consumption by Meat-type Birds During Growing

The amount of feed restriction weighs heavily on the quantity of water consumed by growing meat-type birds. Table 17-1 shows the amount of water that such growing pullets *should* consume at varying ambient temperatures when feed is restricted daily or on a skip-day feeding program.

Note that pullet water consumption is optimal when the average daytime house temperature is 70°F (21.1°C). At 50°F (10.0°C) a bird should drink about 81% of this amount, 167% at 90°F (32.2°C), and 202% at 100°F (37.8°C).

Water Restriction When Feed Is Restricted

Growing chickens will consume more water when on a restricted feeding program. This is particularly true with meat-type growing pullets because many times they are on a skip-day feeding program (see Chapter 32-G), which means they get no feed on alternate days, when they gorge

Table 17-1. Water Consumption by Standard Meat-Type Growing Pullets, for Females Grown Separately (on Daily or Skip-Day Feed Restriction)

	Water Consumed per 1,000 Pullets per Day							
	Average Daytime House Temperature							
	50°F (10.0°C)		70°F (21.1°C)		90°F (32.2°C)		100°F (37.8°C)	
Age wk	U.S. gal	L	U.S. gal	L	U.S. gal	L	U.S. gal	L
2	15	55	18	68	30	114	36	138
4	20	77	25	95	42	158	51	191
6	23	87	28	107	47	179	57	217
8	26	97	32	120	53	200	64	243
10	29	109	35	134	59	224	72	271
12	32	120	39	148	65	247	79	299
14	35	132	43	163	72	273	87	330
16	38	145	47	179	79	298	95	361
18	42	158	52	195	86	336	104	394
20	45	172	56	212	94	354	113	428
22	49	186	61	230	102	384	123	456
24	53	201	66	248	109	414	133	502

themselves with water in an endeavor to give a satiated feeling to their intestinal tract. The crop is distended and the droppings are very wet.

Because there is no alternative to restricted feeding of growing meat-type pullets an accompanying program of water restriction must be followed for birds between 2 and 20 weeks of age. The directions are different for *feed* days and *no-feed* days as follows:

1. *During feed days* (when the daily feed allotment is restricted): Supply water from 1 hour before the first feed is given until 1 hour after the daily allotment of feed is consumed. Then supply water again for 1 hour at 4:00 p.m.

   Note: Use this water program when feed is restricted, but fed every day.

2. *During no-feed days*

   a. The average daily temperature is below 70°F (21.1°C): Supply water for 1 hour at 7:00 a.m., noon, and 4:00 p.m.

   b. The average daily temperature is between 70°F (21.1°C) and 80°F (26.7°C): Supply water for 1½ hours at 7:00 a.m., noon, and 4:00 p.m.

For all restricted feeding programs. Do not restrict water when the average daytime house temperature is above 80°F (26.7°C) or when the birds are stressed.

17-G. BODY WEIGHT RECOMMENDATIONS

Although each type of breeding bird will have different weekly weights during the growing and breeding periods, and each primary breeder will publish weights for his particular strain, Tables 17-2 through 17-5 represent weight averages.

Table 17-2. Average Body Weights of Standard Leghorn Breeders[1]

Flock Age wk	Female		Male	
	lb	kg	lb	kg
1	0.2	0.09	0.3	0.14
2	0.3	0.14	0.4	0.18
3	0.5	0.22	0.6	0.27
4	0.6	0.27	0.8	0.36
5	0.8	0.36	1.0	0.46
6	0.9	0.41	1.2	0.55
7	1.1	0.50	1.5	0.68
8	1.3	0.59	1.7	0.77
9	1.5	0.68	2.0	0.91
10	1.6	0.73	2.2	1.00
11	1.8	0.82	2.3	1.04
12	2.0	0.91	2.5	1.14
13	2.1	0.96	2.7	1.23
14	2.3	1.04	2.9	1.32
15	2.4	1.09	3.0	1.36
16	2.5	1.14	3.2	1.46
17	2.6	1.19	3.3	1.50
18	2.7	1.23	3.4	1.55
19	2.8	1.27	3.6	1.64
20	2.9	1.32	3.7	1.68
21	3.0	1.36	3.8	1.73
22	3.1	1.41	3.9	1.77
23	3.2	1.45	4.1	1.86
24	3.3	1.50	4.2	1.90
25	3.4	1.55	4.3	1.96
30	3.5	1.59	4.4	2.00
40	3.6	1.64	4.6	2.09
50	3.7	1.68	4.7	2.13
60	3.8	1.73	4.8	2.18
70	3.9	1.77	5.0	2.27
80	4.0	1.82	5.1	2.32

[1]For the production of commercial pullets.

Table 17-3. Average Body Weights of Medium-Size Egg-Type Breeders[1]

Flock Age wk	Female		Male	
	lb	kg	lb	kg
1	0.3	0.13	0.4	0.18
2	0.4	0.18	0.5	0.22
3	0.6	0.27	0.7	0.32
4	0.8	0.36	1.0	0.45
5	1.0	0.46	1.3	0.59
6	1.3	0.59	1.6	0.73
7	1.5	0.68	1.9	0.86
8	1.7	0.77	2.2	1.00
9	1.9	0.86	2.4	1.09
10	2.1	0.95	2.7	1.22
11	2.3	1.04	2.9	1.32
12	2.5	1.14	3.2	1.45
13	2.7	1.23	3.4	1.54
14	2.9	1.32	3.6	1.63
15	3.0	1.36	3.8	1.73
16	3.2	1.45	4.0	1.82
17	3.3	1.50	4.2	1.91
18	3.4	1.54	4.3	1.96
19	3.6	1.64	4.6	2.09
20	3.7	1.68	4.7	2.13
21	3.8	1.73	4.8	2.18
22	3.9	1.77	5.0	2.27
23	4.0	1.82	5.1	2.32
24	4.1	1.86	5.2	2.36
25	4.3	1.96	5.4	2.45
30	4.4	2.00	5.6	2.54
40	4.5	2.05	5.7	2.59
50	4.6	2.09	5.8	2.64
60	4.8	2.18	6.0	2.72
70	4.9	2.23	6.2	2.82
80	5.0	2.27	6.4	2.94

[1]For the production of pullets laying brown-shelled eggs for human consumption.

Table 17-4. Average Body Weights of Standard-Size Meat-Type Breeders

Flock Age wk	Female		Male		Male Weight as % of Female Weight
	lb	kg	lb	kg	
1	0.3	0.14	0.32	0.15	
2	0.5	0.22	0.55	0.25	
3	0.9	0.41	1.01	0.46	
4	1.1	0.50	1.27	0.58	115
5	1.3	0.59	1.5	0.68	
6	1.5	0.64	1.8	0.82	
7	1.7	0.77	2.0	0.91	
8	1.9	0.86	2.3	1.04	119
9	2.1	0.96	2.5	1.13	
10	2.3	1.05	2.8	1.29	
11	2.5	1.14	3.1	1.41	
12	2.7	1.23	3.3	1.50	123
13	2.9	1.32	3.6	1.63	
14	3.1	1.41	3.9	1.77	
15	3.3	1.50	4.2	1.91	
16	3.5	1.59	4.5	2.04	127
17	3.7	1.68	4.7	2.13	
18	3.9	1.77	5.0	2.27	
19	4.1	1.86	5.3	2.40	
20	4.3	1.96	5.6	2.54	131
21	4.5	2.05	5.9	2.68	
22	4.8	2.18	6.4	2.90	
23	5.1	2.32	6.8	3.09	
24	5.5	2.50	7.4	3.36	135
25	5.8	2.63	7.9	3.58	
30	6.0	2.73	8.5	3.86	142
40	6.5	2.96	9.1	4.14	140
50	6.8	3.09	9.5	4.32	140
60	7.0	3.18	9.7	4.41	139
70	7.2	3.27	9.8	4.46	136

In the Northern Hemisphere chicks hatched between July and December should weigh about 5% less than the tables show, and those hatched between January and June should weigh about 5% more. Reverse for the Southern Hemisphere.

Egg-Type Breeders and Body Weight

Body weight can only be controlled by reducing or increasing feed consumption. But Leghorn breeders do not usually put on excess

Table 17-5. Average Body Weights of Mini-Meat-Type Breeder Females (to Be Mated with Standard-Size Meat-Type Males)

Flock Age wk	Female lb	Female kg	Flock Age wk	Female lb	Female kg
1	0.28	0.13	18	3.3	1.50
2	0.46	0.21	19	3.4	1.54
3	0.8	0.36	20	3.6	1.63
4	0.9	0.41			
5	1.1	0.50	21	3.8	1.73
			22	4.0	1.82
6	1.3	0.59	23	4.3	1.95
7	1.5	0.68	24	4.5	2.04
8	1.6	0.73	25	4.7	2.13
9	1.8	0.82			
10	1.9	0.86	30	5.0	2.27
11	2.1	0.95			
			40	5.2	2.36
12	2.3	1.04			
13	2.4	1.09	50	5.4	2.45
14	2.6	1.18			
15	2.8	1.27	60	5.5	2.50
16	2.9	1.32	70	5.7	2.59
17	3.1	1.41			

weight. More often their problem is with achieving standard weight. The best remedy is to increase the caloric content of the feed. However, on occasion they do get too heavy, and it becomes necessary to reduce the daily caloric intake either by changing the feed formula or by reducing the feed intake.

Medium-size breeders producing progeny for the production of brown-shell eggs for human consumption usually put on excess weight if full fed during the growing period. Therefore, daily or skip-day programs to reduce feed consumption must be instigated (see Chapter 13).

Although particular attention is given to pullet body weight during growing, the cockerel must not be forgotten. He too should be on a program of feed control similar to that of the pullet (see Chapter 32-P).

Meat-Type Breeders (Broiler Breeder Parents) and Body Weight

The body weight of the *meat-type breeder* is especially important as meat-type lines are bred for fast body growth in the broiler house and if the pullets are full-fed during the growing period they will be much too heavy at sexual maturity to produce their maximum number of eggs during their laying cycle. There are no set rules for maintaining a recommended weekly body weight for the particular strain involved, but feed consumption must be restricted.

Begin spot weighings as early as 3 weeks of age. Average body weight of the flock should increase each week. Do not allow the growth of the birds to reach a plateau.

How to weigh breeders. Weigh a representative sample (see Chapter 16-E). During the first 6 weeks weigh five birds at a time. This procedure is more accurate and reduces the time used for weighing. After 6 weeks, weigh birds individually to calculate the weight uniformity (see Chapter 16-G).

Response to correct mature body weight of meat-type breeder females. Controlling growth weight of breeder females (and males) so that they reach sexual maturity with good body fleshing but without excess fat produces the following effects:

1. Body weight uniformity is improved.
2. The onset of egg production is delayed.
3. The first eggs are larger.
4. Egg production during the laying cycle is increased.
5. More hatching eggs are produced during the laying year mainly because of the larger egg size.
6. Laying house mortality is reduced.
7. The feed cost of growing a pullet to sexual maturity is usually lowered.
8. The feed cost of producing a dozen hatching eggs is reduced.
9. The fertility of the hatching eggs is increased.
10. The hatchability of the hatching eggs is improved.

Body Weights of Broiler Breeder Parents

Average weights of males and females of meat-type broiler parents are given in Table 17-4, along with the male weight as a percent of female weight.

17-H. SPECIAL HANDLING OF BROILER BREEDER PARENTS

There are some special methods of feeding, watering, and lighting of broiler breeder parents that should be detailed as one or all three may fit into a program of handling the breeder flock.

Daily Controlled Feeding During Growing

Special automatic feeding equipment has been designed so that the required amount of feed is given each day rather than by skipping days. The daily allotment of feed for the house is automatically weighed, then dumped into a hopper from which it is delivered in a tube simultaneously to the many feeding pans, or in a trough with a chain speed of 100 feet per minute, circling a 400-foot house in 8 minutes. All pans or troughs are uniformly kept full until the day's allotment of feed is consumed.

The Blackout Growing House for Meat-Type Birds

Growing pullets and cockerels under a blackout lighting program (see Figure 17-1) has many advantages. However, the sexes must be raised in separate houses inasmuch as the lighting schedules are much different during growing. The houses must be environmentally controlled with forced-air ventilation and capable of being *completely* blacked out.

The program reduces the growing body weight below the figures shown in Table 17-4, and it should be continued into the laying period when the birds should be changed to breeder feeding.

Advantages

1. There is greater control of the age at sexual maturity.
2. Consumption of growing and laying feeds are reduced with a financial saving.
3. Flock uniformity is better.
4. The age at sexual maturity can be delayed so the flock produces large first eggs.
5. Out-of-season flocks begin egg production up to 2 weeks earlier compared with those raised in open-sided houses. Egg production is increased up to 8%.

A regular restricted feeding program may be used, but use daily restriction rather than a skip-day feeding program.

17-I. THE BREEDER HOUSE

Floor space. Breeding females require more floor space than females kept for only the production of market eggs. Because of the many types of floors and the variation in the size of the breeder, plus the fact that males make up a part of the flock, floor space requirements vary greatly as shown in Table 17-6. Figures are on a per bird basis, so the males should be counted along with the females.

Age	Females			Males		
da/wk	Light hr/da		Details	Light hr/da		Details
1 da	24		3.5 fc of light Full feed	24		3.5 fc of light Full feed
6 da	8		1 fc of light Controlled feeding	8		1 fc of light Controlled feeding
19 wk	14			10		Move males to breeder house
20 wk						
21 wk			Move females to breeder house	14		
22 wk			First eggs produced Start breeder feed			
23 wk	15			15		
24 wk			5% HD egg produc- tion			
25 wk			Begin saving hatch- ing eggs			
26 wk						
27 wk to 60–70 wk	16		50% HD egg produc- tion	16		Lights off no later than 8:00 P.M.

Figure 17-1. Lighting program for broiler breeder parents in blackout house.

Floor type. Although floor space requirements have been given for egg-type breeders on wire, wire has not been very satisfactory for meat-type breeders. The difficulty lies in the fact that birds do not mate well on wire; some fertility is sacrificed. Fertility will drop about 2 to 3% with egg-type birds and 5 to 7% with meat-type birds.

When wire is used as the only floor material, plenty of supports should be used to keep the wire rigid and even. Use a heavy-gauge wire. A plastic material is now available to replace the wire, but it may require more supports. It is said that better fertility results with plastic.

Table 17-6. Floor Space Requirements per Breeder Bird During Laying Period

	Type of Floor[4]											
	All-litter			Slat (Wire) and Litter[1]			All-Slat			All-wire[2]		
Type of Breeder	ft²	m²	Birds per m²	ft²	m²	Birds per m²	ft²	m²	Birds per m²	ft²	m²	Birds per m²
Mini-Leghorn	1.5	0.14	7.2	1.25	0.12	8.3	1.0	0.09	10.8	1.0	0.09	10.8
Standard Leghorn	2.0	0.19	5.4	1.75	0.16	6.2	1.25	0.12	8.3	1.25	0.12	8.3
Medium-size, egg-type	2.25	0.21	4.8	2.0	0.19	5.3	1.5	0.14	7.2	1.5	0.14	7.2
Mini-meat-type[3]	2.25	0.21	4.8	2.0	0.19	5.3	1.5	0.14	7.2			
Standard meat-type	3.0	0.28	3.6	2.5	0.23	4.4	2.0	0.19	5.4			

[1]Approximately 60% slats; 40% litter.
[2]Or plastic-type floors.
[3]Mini-meat-type females mated with standard-size males.
[4]For each male and female.

Nests.   Provide one nest for each four pullets in the breeding house. Nests for meat-type breeders should be slightly larger than those for egg-type breeders. A perch should be in front of all nests, and it should be constructed so that it may be used to close the nests at night.

Lights.   An adequate program of lighting is a requisite for maximum hatching egg production as well as for commercial egg production. A complete discussion about lighting programs is given in Chapter 18.

17-J. FEEDER SPACE FOR BREEDERS DURING EGG PRODUCTION

Feeder space for breeders is usually greater than for laying pullets. The requirements are given in Table 17-7, but with broiler breeders on a skip-day feeding program, the feeder space may be reduced by a third if the feed depth is increased by a third.

Table 17-7.   Feeder Space Requirements for Breeder Birds[1] (Litter Floor Operation)

Breed	Trough Space[2]		Number of Birds	
	in.	cm	Pan[4]	Tube Feeder[3]
Mini-Leghorn	3.25	8.2	15	19
Standard Leghorn	3.75	9.4	13	16
Medium-size, egg-type	4.25	10.6	11	13
Mini-meat-type	5.00	12.5	10	12
Standard meat-type	6.00	15.0	8	11

[1]Space for each male and female.
[2]Space for one side of trough only.
[3]Approximately 12 in. (0.3 m) in diameter. Usually found on certain automatic feeders.
[4]A pan with a circumference of 50 in. (1.72 m) or a diameter of 16 in. (40.6 cm).

17-K. WATER FOR THE BREEDER FLOCK

Adequate water is important to the breeder flock and the demand increases greatly as the ambient temperature rises.

Waterer space.   The requirements for waterer space for the breeder flock are given in Table 15-3 and are on a *bird basis;* that is, the figures given are for each male and each female in the pen during periods of maximum water consumption.

Water consumption by the mature breeder flock.   Although the breeder flock involves about 10% males, each male drinks approximately the same amount of water per day as a female. Therefore, the males in the flock should be included with the females when water consumption is calculated. The figures for standard Leghorns are given in Table 16-17. Standard meat-type birds will drink 65 to 75% more, depending on body weight.

17-L. TYPE OF FLOOR AND FLOCK MANAGEMENT

Managing on the Litter Floor

Clean and dry litter will be necessary if the breeders are to be kept on the floor. Not only will this improve the general health of the birds but it will prevent the feet of the pullets from becoming dirty and carrying the dirt into the nests. Hatching egg cleanliness is essential in breeder operations.

Prevent floor eggs. Floor eggs are usually dirty, difficult to clean and sanitize, and a probable cause of "blowups" in the incubator. A high percentage of them are broken, resulting in a direct loss. There must be a low incidence of floor eggs if the breeding operation is to be practical and profitable. The number of cracked eggs should never be over 2% for young layers; 3% for old.

Have an adequate number of nests. No hen should be forced to lay on the floor because there is no nesting space available. Pullets seem to want to use the nests more if the nest sections are placed crosswise of the house.

Managing on Slats and Litter

The poultryman keeping breeding birds, particularly those of meat-type lines, has enjoyed the many benefits of the slat-and-litter house (see Chapter 11-Q). Most new meat-type breeding houses are constructed to use this system. Even though it has some drawbacks, the benefits far outweigh the disadvantages. An evaluation of the slat-and-litter house versus the all-litter house shows the following:

1. The litter area of the slat-and-litter house is difficult to ventilate.
2. Birds lay about five more eggs on the all-litter floor than on the slat-and-litter floor.
3. Nearly three times as many floor eggs will be laid on all-litter as slats-and-litter.
4. Some birds continually lay eggs on slats in preference to nests.
5. Laying house mortality is slightly higher on the slat-and-litter floor than on the all-litter floor.

6. The cost of maintaining a breeder hen is identical.
7. The cost of producing a dozen hatching eggs is the same.

Figure 17-2 shows a house of meat-type breeders on slats and litter. Notice the placement of the nests at right angles over the slats and the suspended waterers and pan feeders, the latter being necessary when birds are on some restricted feeding programs.

Managing on the All-Slat Floor

Most poultrymen keeping meat-type breeders have been discouraged with the all-slat house. Usually fertility is lower, and most of the "floor eggs" laid on the slats are broken. Leghorns and medium-size breeders seem to be better suited to slats, and many such birds are housed in this type of accommodation. Many of these houses are now constructed using the deep-pit principle, thus eliminating routine removal of the droppings.

Flightiness on slats. Pullets—particularly Leghorns—tend to be more flighty on an all-slat floor. They may fly back-and-forth through the

Figure 17-2. Meat-type breeders in a slat-and-litter house.
(Courtesy of Broiler Industry®, Mount Morris, Illinois, U.S.A.)

house if it is not separated by partitions. Light, flexible wire netting may be suspended at intervals crosswise of the house. The bottom edge should be loose and about two feet above the slats to allow room for the birds to walk under it.

Location of nests. The bottom of the nests should be closer to the floor than is the case with litter floors. This will prevent some of the "floor eggs," because the pullets can almost walk into the nest. However, placing the nests this low makes it more difficult to gather the eggs; there must be a happy medium. When eggs are picked up automatically by a belt, this low nest position is no problem.

Body weight on slats. There is some inclination for birds to get heavier on slats than on litter. They seem to exercise less, and the increased density of the birds makes it more difficult for them to move about. Both lead to increases in body weight. Thus, any feed control program to maintain body weight becomes more important when the birds are on slats.

Managing on Plastic Floors

Plastic flooring has become quite popular as a substitute for wooden slats in the slat-and-litter or the all-slat house, or the wire in the wire-and-litter or the all-wire floor. There are several types of plastics on the market, enough to make it possible to secure the proper type. It is said that plastic floor material does not seem to cause as much difficulty with foot and leg problems, or with breast blisters.

17-M. MANAGING THE MALE FOR HIGH FERTILITY

Half the germ plasma of the newly hatched chick comes from the sire, half from the dam. Therefore the few males in the breeding pen represent half the mating. Their importance cannot be overestimated. Not only do they produce their half of the genes that go to make up the new individual but they are responsible for the sperm necessary for hatching egg fertility.

Male Reproduction

The potency of male chicken semen is related to the number of mature spermatozoa. In the early part of the day semen is white and opaque, but as matings are consummated, it becomes clear and watery. Each ejaculation varies from as low as 0.1 cc to as high as 1.0 cc.

Male chickens attempt to mate between 10 and 30 times a day, depending on competition, number of females available, social order, temperature, light, and many other factors. Males may mate several times a day with the same hen. They mate more frequently with pullets in the middle of the social order rather than with the more precocious or timid individuals. Of the attempted matings only about two-thirds will be completed.

It seems to make little difference when matings are made to produce the greatest fertility, but more matings usually occur early in the day. Fertile eggs will be produced for days after the males are removed from the flock, but if males are removed and new ones added the same day, fertile eggs produced after 3 days will be the result of matings by the new males (see Chapter 4-A).

Ratio of Males to Females

Too many males in the breeding pen reduce fertility, as do too few. The correct ratio of males to females depends on the type and size of the birds involved, and is defined on the basis of the number of cockerels per 100 pullets. Although the figures given in Table 17-8 are those usually recommended, a few extra males should be placed in the pens at the time the birds are mated to allow for some early culling and mortality from fighting. Provide more males on slats, and slats-and-litter than on all-litter floors. The male-to-female ratio does not affect the frequency of male mating.

Table 17-8.   Ratio of Males to Females in Breeding Pens

Male of Mating	Female of Mating	Mating Produces	Males per 100 Females	
			On Litter	On Slats-and-Litter
Mini-Leghorn	Standard Leghorn	Commercial Mini-Leghorn pullet	8	9
Standard Leghorn	Standard Leghorn	Commercial Standard Leghorn pullet	8	9
Medium-size	Medium-size	Commercial medium-size pullet (brown eggs)	9	10
Standard meat-type	Mini-meat-type	Commercial broiler	9	10
Standard meat-type	Standard meat-type	Commercial broiler	10	11

Importance of Male Body Weight

It is just as important to grow a male of high quality as it is to grow a female of similar quality. Too often the male is neglected. Excessive body weight at maturity must be avoided; growing weight guides for the line involved must be maintained. See Tables 17-2, 17-3, and 17-4 for suggestions for body weight of growing and breeding males.

Cull the Males Often

During the breeding period, the males must be carefully watched. Any inferior birds must be removed with a hook. There is proof that males mate with certain females, and if a particular male becomes unable to mate, his matching females will not take another male until he is removed.

Handle males carefully. If it becomes necessary to handle males, catch them carefully by both legs or both legs and a wing. Many males will be permanently injured if only one leg is caught.

Exercise the Males

Cockerels should be induced to exercise to prevent their legs from deteriorating. Cockerel feeders create a great deal of exercise; the males must jump to get feed from them. Feeding some grain in the litter in the afternoon induces scratching, a worthwhile exercise.

The Timid Male

Males set up a social order, as do females (see Chapter 22-M). The more timid males must be adequately provided for. Be sure they are getting enough feed to maintain their recommended body weight. If underweight, it will be advisable to add cockerel feeders to the pen to increase the male feed consumption. Normally, when males have access to both conventional feeders and cockerel feeders, they will consume about half of their feed from each.

Length of Time Males Are on Litter

When slat-and-floor housing is used the males must have access to the litter at least 8 hours per day for high fertility. Do not fence off the litter area part of the day.

Enlarged Foot Pads

Some males will develop enlarged foot pads. These often become inflamed, and the male will not mate. Wire and slat floors are instrumental in causing an increase in the difficulty. If a male is seriously affected, he should be removed from the pen; such a bird seldom recovers. In a test at Pennsylvania State University, fertility in meat-type breeders on a sloping wire floor was 9.2% less than in similar birds on a litter floor. It was 4% lower during the second 4-week period and nearly 17% lower during the seventh 4-week period. This reduction was due partly to the sore foot problem with the cockerels, which got progressively worse.

Replacing Males in a Pen (Spiking the Flock)

As birds progress throughout the laying cycle there is a natural reduction in fertility. A serious economic situation may be produced; hatches may become so poor as to be unprofitable. The problem is much greater with meat-type breeds than with egg-type. Some poultrymen replace the old males in a flock with a set of new and younger ones after about two-thirds of the egg production period is over. Although this certainly increases fertility, the procedure is costly and not recommended. Generally the cost of raising a new set of males does not offset the cost of the reduced fertility over such a short period of time as the last one-third of the egg production period.

Males Not Mating

When breeder birds are kept in a house with a slat-and-litter or a wire-and-litter floor, males will have a tendency to remain on the slats or wire, as a sort of "roosting" place. Since most of the pullets will prefer to mate on the litter rather than on the slats or wire, such pullets will not be mated because the males will not leave the slats. Feeding a small amount of whole grain in the litter in the middle of the afternoon will prevent a great deal of this difficulty, as it will cause the males to get off the slats in search of feed. Mating on the litter will follow.

17-N. HATCHING EGG PRODUCTION

The only reason for keeping breeding birds is to produce an abundant number of hatching eggs that will produce a high percentage of quality chicks. Furthermore, the hatching eggs must be produced as economically as possible without impairing chick quality. A good egg usually means a good chick.

Hatching Egg Size

Egg size determined by chronological age of pullets. Birds of a similar genetic makeup that are 26 weeks of age, for instance, lay eggs of a certain size regardless of when they started egg production. The same would be true at any other laying house age. Thus, holding the pullets out of production does not cause them to lay larger eggs; rather, the eggs are larger because the birds are older.

Minimum egg weight for hatching purposes. Hatcheries usually set up minimum weights for the hatching eggs that they will incubate. The weight may be different for various lines and breeds. In some cases a lower weight is allowable during the first few weeks of egg production than is allowed later. Regardless of this, large first eggs are essential so that the poultryman can sell or deliver as many hatching eggs as possible over the life of the breeding flock. But it costs more to produce a pullet when she is kept out of egg production during a longer "growing" period. Thus, there must be a most economical age for the production of first eggs. An analysis is given in Chapter 17-R.

When to start saving hatching eggs. There is evidence to show that chicks hatched from the eggs laid by a pullet during her first 2 weeks of egg production do not live well. But after that, eggs may be used for incubation as soon as they are large enough. The minimum size is determined by the needs of the hatchery using the eggs and by the size of the bird laying the eggs. Although the minimum egg weight is variable, the following data will give some indication.

| | Minimum Egg Weight | | | |
| | First 12 Weeks of Egg Production | | After 12 Weeks of Egg Production | |
Line of Parent Breeders	(oz/doz)	(g/ea)	(oz/doz)	(g/ea)
Standard Leghorn	22	52.0	23	54.3
Medium-size, egg-type	23	54.3	23	54.3
Standard meat-type	21	49.6	22	52.0

In the above data it may be noticed that minimum egg weight suggestions are sometimes smaller during the first 12 weeks of egg production, which is the result of two lines of reasoning as follows:

1. Those hatcheries producing broiler chicks from meat-type breeders can get more hatching eggs from first egg production. Broiler weight shows little correlation with egg size at this time, but there is a greater correlation later in the production period.

2. Eggs produced at the start of lay are smaller than those laid later. Because the same birds are laying them, the first hatching eggs have a genetic potential identical with those produced later. Thus from a genetic standpoint, smaller hatching eggs may be used during the first 12 weeks of *total* egg production than may be used later. Following the minimum size requirements re-

gardless of the length of time the bird is in egg production will, of course, eliminate the smaller eggs through the entire production period.

Shell Quality

Shell quality is associated with hatchability. The longer a bird produces eggs, the poorer the shell quality. Season of the year, strain, temperature, diet, and various other factors also affect the texture. Although the quality of the eggshell remains quite acceptable for a 12-month laying period with egg-type breeds, shell quality of eggs produced by meat lines begins to deteriorate rapidly after 8 or 9 months of lay (see Table 19-7).

Interior Egg Quality

Although several of the factors causing fluctuations in the interior quality of eggs also produce variations in hatchability, the important factor is a condition known as *tremulous air cells*. Some eggs are laid with loose air cells; many more loose air cells are produced during the course of handling the hatching eggs before they are placed in the incubator. Handle eggs carefully, and prevent as much jarring as possible, to prevent the loosening of the air cells. Such eggs hatch poorly; many do not hatch at all.

Preventing Cracked Eggs

It is surprising how many hatching eggs are cracked during the process of laying and getting them to the incubator. A large proportion of the cracking is manmade; practically no eggs are laid with cracked shells. The desire of the caretaker to handle more birds, automation, improper holding containers, too many transfers, and inexperienced poultrymen contribute to the high incidence of breakage. As cracked eggs seldom hatch, and hatching eggs are expensive to produce, the economic loss is high on many farms. Sometimes 5% or more of the eggs laid are cracked between the nest and the incubator. A tolerable maximum is 2%; 1% is an economic goal.

Clean Egg Production

Hatching eggs picked from the nest or delivered by a movable belt should not need cleaning. If a few are dirty, remove the debris with a *dry* towel. Never use a wet towel or sandpaper. A wet towel spreads disease organisms; sandpaper creates a pathogenic dust that moves through the air, and also removes the protective cuticle, opening many shell pores (see Chapter 7-B).

17-O. VACCINATION PROGRAM FOR BREEDERS

Vaccination programs are more involved than those for layers because young progeny chicks partially rely on parental immunity to protect them from certain diseases, necessitating more vaccination of the parents not only during the growing period but during the time they are producing hatching eggs.

The vaccination program for the breeder males is the same as for the breeder females. An example of a meat-type breeder replacement vaccination program is given in Table 39-4. The diseases listed in the table are as follows:

Marek's disease	Bronchitis
Infectious bursal disease	Avian encephalomyelitis
Newcastle disease	Fowl pox

For comparison, an example of a layer vaccination program may be found in Table 39-3. A discussion of each disease is in Chapter 37.

17-P. ARTIFICIAL INSEMINATION (AI)

All turkey breeder hens are artificially inseminated because their large size makes it almost impossible for them to mate naturally. However, with chickens AI is economically unfeasible; natural matings produce almost complete fertility of the eggs. But for other reasons, primary breeders often have to resort to AI to carry out their genetic breeding program, which calls for caging the males and females separately to expedite the procedure.

Meat-type breeders should be grown on a litter floor. If meat-type breeders are to be artificially inseminated during egg production; they should be raised to sexual maturity on a litter floor to prevent them from becoming obese.

Female weight during egg production. There is no problem with excess weight in egg strains, but meat-type males and females get excessively heavy in wire-floored cages. They overeat and exercise less. Feed restriction will be necessary to maintain an optimal body weight (see Chapter 4-A).

17-Q. EGG PRODUCTION STANDARDS FOR BREEDERS

Production standards for breeder females have been segregated according to

1. Standard egg-type breeder hens (Table 17-9)
2. Standard meat-type breeder hens (Table 17-10)

Table 17-9. Production Standards for Standard Egg-type Breeder Hens

Age wk	Week of Egg Production	Hen-day Egg Production %	Cumulative Hen-housed Egg Production	Hatching Eggs %	Cumulative Hatching Eggs per Hen Housed	Total Hatchability %	Cumulative Pullet Chicks Hatched per Hen Housed[1]
22	1	5	0.4	—	—	—	—
23	2	11	1.1	—	—	—	—
24	3	21	2.6	—	—	—	—
25	4	34.	4.9	15	0.4	71	0.1
26	5	50	8.4	35	1.6	74	0.6
27	6	68	13	51	3.8	82	1.5
28	7	82	19	63	7.4	83	3.0
29	8	90	25	72	12	85	4.9
30	9	92	31	78	17	87	7.0
31	10	91	38	82	22	88	9.3
32	11	90	44	84	27	88	11.6
33	12	90	50	86	32	89	13.9
34	13	89	56	87	38	89	16.3
35	14	89	62	88	43	89	18.6
36	15	88	68	89	48	89	21.0
37	16	87	74	90	54	89	23.4
38	17	87	80	90	59	89	25.7
39	18	86	86	51	64	89	28.1
40	19	85	91	91	69	89	30.4
41	20	85	97	91	75	88	32.7
42	21	84	103	91	80	88	34.9
43	22	84	108	91	85	88	37.2
44	23	83	114	91	90	88	39.4
45	24	82	119	90	95	87	41.5
46	25	82	125	90	100	87	43.7

47	26	81	130	90	105	87	45.8
48	27	80	135	90	109	87	47.9
49	28	80	141	90	114	86	49.9
50	29	79	146	90	119	86	51.9
51	30	78	151	90	124	86	53.9
52	31	78	156	89	129	85	55.9
53	32	77	161	89	133	85	57.8
54	33	76	166	89	137	85	59.7
55	34	75	171	89	141	84	61.5
56	35	75	176	89	146	84	63.8
57	36	74	181	89	150	83	65.1
58	37	73	186	89	154	84	66.9
59	38	73	190	88	158	83	68.6
60	39	72	195	88	163	82	70.3
61	40	72	200	88	167	82	72.0
62	41	71	204	88	171	82	73.6
63	42	70	209	88	175	81	75.2
64	43	70	213	87	179	81	76.8
65	44	69	218	87	182	81	78.3
66	45	69	222	87	186	80	79.8
67	46	68	226	87	190	80	81.3
68	47	68	231	86	194	79	82.8
69	48	67	235	86	197	79	84.3
70	49	66	239	86	201	78	85.7
71	50	66	243	86	204	78	87.1
72	51	65	247	85	208	77	88.4
73	52	64	251	85	211	77	89.7
Total		267	251		211		
Avg %		73.1		84.1		84.9	89.7

[1]Chicks hatch 3 weeks later.

397

Table 17-10. Production Standards for Standard Meat-Type Breeder Hens

Age wk	Week of Egg Production	Hen-day Egg Production %	Cumulative Hen-housed Egg Production	Hatching Eggs[1] %	Cumulative Hatching Eggs per Hen Housed	Total Hatchability %	Cumulative St-run Chicks Hatched[2]
25	1	5	0.4	—	—	—	—
26	2	20	1.8	—	—	—	—
27	3	38	4.3	35	1	70	1
28	4	56	8	68	4	75	3
29	5	73	13	80	8	80	6
30	6	84	19	85	13	84	10
31	7	86	25	88	18	86	15
32	8	85	31	90	24	88	20
33	9	84	37	91	29	89	24
34	10	84	42	91	34	90	29
35	11	83	48	92	40	91	34
36	12	82	54	92	45	91	39
37	13	81	59	93	50	91	44
38	14	81	65	93	56	91	49
39	15	80	70	93	61	91	54
40	16	79	76	93	66	91	58
41	17	78	81	94	71	90	63
42	18	77	86	94	76	90	67
43	19	77	92	94	81	90	72
44	20	76	97	94	86	90	76

45	21	75	102	94	91	90	81
46	22	74	106	93	96	89	85
47	23	74	112	93	101	89	89
48	24	73	116	93	105	89	93
49	25	72	121	93	110	89	97
50	26	71	126	93	114	89	101
51	27	70	131	93	119	88	105
52	28	70	135	92	123	88	109
53	29	69	140	92	127	88	113
54	30	68	144	92	131	88	116
55	31	67	149	92	136	87	120
56	32	66	153	92	140	87	124
57	33	66	157	92	144	87	127
58	34	65	161	91	148	86	130
59	35	64	166	91	152	86	134
60	36	63	170	91	155	85	137
61	37	62	174	91	159	85	140
62	38	61	178	91	163	84	143
63	39	60	181	91	166	84	146
64	40	60	185	90	170	83	149
65	41	59	189	90	173	82	152
66	42	58	192	90	176	81	155
67	43	57	196	89	180	79	157
68	44	56	199	89	183	78	159
	Total	209	199		183		159
	Avg %	68		92		87	

[1]22 oz/doz (52 g/ea).
[2]Chicks hatch 3 wk later.

399

Not only are figures given for hen-day and hen-housed egg production but there are standards for percentage of hatching eggs, cumulative hatching eggs, and percentage of total hatchability. As with other standards, these figures are for good flocks given good care. It must be kept in mind, however, that egg production is generally lower with breeder lines than with commercial egg-producing lines. However, breeders are usually kept on litter floors and produce more eggs than layers kept in cages, so for all practical purposes, egg production is about the same. The summary of the figures are as follows:

	From Table 17-9 Egg-type (52 wk)	From Table 17-10 Meat-type (44 wk)
Total hen-day egg production	267	209
Total hen-housed egg production	251	199
Percentage hen-day egg production at peak	92	86
Hatching eggs per hen housed	211	183
Straight-run chicks hatched per hen housed	179	159

17-R. BREEDER FLOCK COST VARIATIONS

Probably no phase of poultry production is as dependent on cost management as the period when the breeder flock is producing hatching eggs. The age at which the pullets lay their first eggs, egg size, length of time the birds are kept in egg production, in-season and out-of-season flock variations, and usage of hatching eggs—all have their effect on production cost and profit. Undoubtedly these factors are of more importance to the flockowner keeping meat-type breeders than to one keeping egg-type breeders. The inability of the meat-type breeder to produce a large number of eggs, the large amount of feed consumed because of its heavy weight, the relatively short egg production period, and the high salvage value of the bird at the end of its laying year materially affect the economies of hatching egg production. The egg-type breeder is kept in egg production for a longer period than the meat-type; it is not as greatly affected by climatic changes; it produces more eggs; feed consumption is lower as the result of a smaller size; and the salvage value is very small.

Whether hatching eggs are being produced by an independent poultryman or by a contract producer does not influence the need for cost management of the breeder flock. There must always be a consistent endeavor to keep costs as low as possible. The effects of several of these factors are shown in the following sections.

Cost to Grow an Egg-Type Breeder Pullet (Male Included)

One of the important costs incurred in the production of hatching eggs is that of raising the sexually mature pullet plus the matching cockerel. These costs must be amortized over the egg production period on the basis of a constant value per dozen hatching eggs. These costs are shown in Table 17-11.

Table 17-11. Cost to Raise a Breeder Pullet, Male Included (in U.S. Dollars)

	Cost Through 23 Weeks		
	Standard Leghorn	Medium Size	Standard Meat-type
Base pullet cost to 20 wk of age	$7.22	$7.02	$3.46
Proportionate male cost to 20 wk of age	0.73	0.45	0.62
Bloodtesting, culling, and other costs at 20 wk of age (male included)	0.42	0.42	0.47
Cost to keep pullet from 20 through 23 wk of age (male included)[1]	0.30	0.35	0.85
Total cost per pullet (male included)	$8.67	$8.24	$5.40

[1]Value of eggs laid through 23 weeks credited to "growing" costs.

Cost to Produce One Dozen Hatching Eggs

Estimated costs of producing a dozen hatching eggs from Leghorn, medium-size, and meat-type replacement females, with the matching male cost included, are given in Table 17-12. Naturally, these costs will vary according to the price and quality of the feed, number of hatching eggs produced, pullet growing cost, mortality, and many other factors.

Best Age for First Eggs from Egg-Type Breeders

Although egg-type, commercial pullets producing market eggs should attain 5% HD egg production when they are about 22 weeks of age, the female parent breeder of most lines produces smaller eggs. Consequently, it is economically advisable to delay their onset of egg produc-

Table 17-12.   Cost to Produce One Dozen Hatching Eggs, Male Included (in U.S. Dollars)

Cost Item	Standard Leghorn[1]		Medium-size[2]		Standard Meat-type[3]	
	Per Pullet	Per Doz Hatching Eggs[4]	Per Pullet	Per Doz Hatching Eggs[5]	Per Pullet	Per Doz Hatching Eggs[6]
Pullet growing cost (including male)	$ 8.67		$ 8.24		$ 5.40	
Breeder feed cost[10]	6.16[7]		7.20[8]		7.36[9]	
Labor	0.91		1.00		1.01	
Other costs	1.57		1.84		2.17	
Total cost	$17.31	$0.98	$18.28	$1.01	$15.94	$1.06

[1]52 wk of lay.
[2]52 wk of lay.
[3]44 wk of lay.
[4]211 hatching eggs.
[5]217 hatching eggs.

[6]183 hatching eggs.
[7]77 lb @ $0.08/lb (35 kg @ $0.18/kg.
[8]90 lb @ $0.08/lb (45.8 kg @ $0.18/kg.
[9]92 lb @ $0.08/lb (42 kg @ $0.18/kg.
[10]Male feed included.

tion until they reach 5% when 23 or 24 weeks of age. Doing so will necessitate controlled feeding and light management during the growing period.

17-S. VARIATIONS IN MEAT-TYPE HATCHING EGG COST

Compared with egg-type strains, the actual cost of producing one dozen meat-type hatching eggs is dependent on many more factors. These variations represent one of the major discrepancies in any analysis of meat-type hatching egg production costs. The major ones are as follows:

1. It is more difficult to raise a good pullet and cockerel.
2. Pullets may be brought into egg production at any reasonable age.
3. Season hatching date has a great effect on egg production.
4. Pullet growing costs vary tremendously.
5. The length of the laying period is highly variable.

6. The rate of egg production among flocks is far from consistent.
7. Bird size is not uniform and affects feed consumption.
8. Being larger, the spent hens are more valuable.

Age at Sexual Maturity and Hatching Egg Cost

By careful manipulation of the length of the light day and feed intake, meat-type pullets may be brought into egg production at any reasonable age between 20 and 30 weeks. One question is: At what age of sexual maturity will a pullet produce hatching eggs at the lowest possible cost? Table 17-13 shows a projection of figures to answer this question.

Regardless of the cost and production figures used to construct Table 17-13, the relative differences in the cost to produce a dozen meat-type hatching eggs will be about the same. On close observation of the table, one will realize the following:

1. The longer the delay in reaching sexual maturity (5% HD egg production), the more it costs to raise a pullet (male included).
2. Pullets have a longer egg production period when brought into egg production at 25 to 27 weeks than earlier or later.

Table 17-13. Relationship Between Age at Sexual Maturity and Hatching Egg Cost (Meat-Type Broiler Breeders)

Age at 5% HD Production wk	Grow Cost per Pullet (Male Incl) U.S. $	Egg Prod Period[1] wk	Total Eggs Produced	Hatching Eggs Prod[2]	Grow/Lay Cost per Pullet (Male Incl) U.S.$	Total Cost of Hatching Doz Eggs U.S.$
20	4.62	40	172	130	13.82	1.31
21	4.74	41	179	143	14.58	1.22
22	4.88	42	182	155	14.96	1.16
23	5.03	43	188	165	15.35	1.12
24	5.20	43	194	175	15.52	1.06
25	5.38	44	199	183	15.94	1.06
26	5.59	44	199	183	16.15	1.06
27	5.81	44	197	179	16.37	1.10
28	6.04	43.5	194	175	16.48	1.13
29	6.28	43	190	169	16.60	1.18
30	6.53	42	185	163	16.63	1.22

[1]First wk at 5% HD production
[2]22 oz/doz (52 g/ea) minimum

3. Pullets brought into egg production at 25 to 27 weeks lay more eggs because the percentage of egg production is higher.

4. Pullets beginning egg production at 25 to 27 weeks of age produce a greater number of eggs meeting the size and quality requirements for incubation: 92% for 25 to 27 weeks; 76% for 20 weeks.

5. Pullets kept for an extremely long period of production lay eggs with poorer shell quality at the end of their production, which reduces the number of eggs acceptable for incubation: 92% for 25 to 27 weeks; 88% for 30 weeks.

6. Production costs per pullet during the laying period are mainly determined by the length of the egg production period: Some pullets produce hatching eggs economically for 44 weeks; some, only for 40.

7. Pullets brought into egg production at 25 to 27 weeks produce a dozen hatching eggs at a much lower cost than those brought in at other ages, earlier or later.

8. Not only is the cost to produce a dozen hatching eggs important in Table 17-13 but the number of hatching eggs produced bears just as much weight. Although the cost per hatching egg is lower at 25 to 27 weeks than at 20 weeks, the pullets produce 183 hatching eggs versus 130, and both require about the same investment in facilities and about the same feed, labor, and other costs. The differences are similar for 30 weeks.

17-T. WHEN TO SELL THE MEAT-TYPE BREEDER FLOCK

At the end of the laying period the meat-type breeder flock must be liquidated (sold). By that time weekly egg production has dropped materially, hatchability of the eggs is lower, and mortality has reduced the number of birds in the flock. Obviously, each item increases the cost of producing hatching eggs of quality. But when should the breeder flock be sold?

Peculiarly, with meat-type pullets, the length of the laying period is short, egg production is low, and production costs are high. Few flocks may be kept in egg production for over 44 weeks, and many "good" flocks have laying periods as low as 38 to 40 weeks.

The cost per hatching egg during the early part of the production year is lower than toward the end. But flocks must be kept in production a few weeks at the end even though the production cost of hatching eggs during this period makes it a losing proposition.

The entire program of producing hatching eggs with meat-type breeders is one of reasonably getting all the eggs possible. The theory is that profits made early in the production period can offset the losses later. There is no formula; there are just too many factors involved. But the period of economic loss cannot be extended too long.

There are many things that happen long before the end of the production year approaches to give an indication of how many hatching eggs will be produced. The following factors will shorten the period of hatching egg production, necessitating that the breeder flock will have to be liquidated at an earlier age:

1. Out-of-season flocks.
2. When pullets are overweight at sexual maturity.
3. When egg production has been low since sexual maturity.
4. Those flocks that do not have a high peak egg production.
5. Flocks that are going to end their production period during the hot summer months.
6. Females that are too heavy during egg production.
7. Bird nonuniformity that has been excessive.
8. Those flocks that have encountered some disease or stress.
9. After the flock has undergone a partial molt.
10. When shell quality has deteriorated sharply.
11. When there are too few good males in the flock.
12. When water consumption has been so great that the litter is wet.

17-U. DUAL FEEDING FOR BROILER BREEDER PARENTS

See Chapter 32-P for a full discussion.

18

Lighting Management

It is a well-known fact that light intensity, the length of the daily light period, and the pattern of daily change produce biological responses associated with egg production. The responses are the result of increased activity of the anterior lobe of the pituitary gland located at the base of the brain. Light stimulation causes the release of the follicle-stimulating hormone (FSH) from the pituitary, which in turn causes an increase in the growth of the ovarian follicles. Upon reaching maturity, the ovum is released by the action of another hormonal secretion from the pituitary, the leuteinizing hormone (LH).

18-A. SOME FACTS ABOUT LIGHT

Natural daylight is supplied by the sun. The intensity of the sun's light rays has a day-by-day variation as the result of the position of the sun, cloudiness, dust and moisture in the air, and other factors. However, the length of the light day varies too. The relative position of the earth to the sun causes differences in the length of the daily daylight period. In the Northern Hemisphere, June 21 is the longest day of the year and December 21 is the shortest. In the Southern Hemisphere, the dates are reversed. Because the earth's surface is curved, daylight occurs from 15 to 30 minutes before sunrise and darkness occurs 15 to 30 minutes after sunset. Thus, the length of the light day is somewhat longer than the hours between sunrise and sunset, but the time between sunrise and sunset is usually considered the light day.

Light Vision

As seen by the human eye, light is only that part of the radiant energy spectrum that is represented by wavelengths between about 400 and 700 millimicrons (nanometers). The limits of the chicken eye are quite similar to those of the human eye. All chickens have color vision. There is some indication that they are able to see better when illumination is by those rays at the long end of the spectrum, such as red, orange, yellow, and perhaps blue. The intensity of light necessary for chickens to see to eat is unusually low. After some training, birds will find their way to the feeders and eat when the light intensity is less than one-fourth footcandle. However, it requires from two to four times this amount of light to stimulate the pituitary and increase egg production.

Terms Used in Connection with Light

Light has various measurements, most of which are very specific. Those used in the course of poultry management are as follows:

Candela: The word is derived from "candle." A candela is the unit of luminous intensity of a light source in a specified direction.

Lumen: The lumen is defined as the rate at which light falls on a square foot area surface which is equally distant one foot from a source whose intensity is one candela. A lumen is a standard of measurement of the light intensity from electric bulbs of various types and sizes.

Lux: A lux (lx) of light intensity is equal to one lumen per square meter, and the term is used regularly in some countries. One lux is equal to 0.0929 footcandle. One footcandle equals 10.76 lux.

Lumen efficiency: The amount of electricity necessary to light a bulb is measured in *watts.* The number of lumens of light per watt of bulb is an indicator of the efficiency of the light source. With incandescent bulbs, the rule of thumb is that one watt produces 12.56 lumens of light. However, the figure varies with the size of the bulb; small bulbs produce more lumens, larger produce less. With fluorescent lamps, the efficiency is much greater.

Footcandle: Illumination on a surface is measured in footcandles. A footcandle (fc) is defined as the intensity of light striking each and every point on a segment of the inside surface of an imaginary one-foot radius sphere with a one candlepower source at the center. Thus, one footcandle equals one lumen per square foot.

Available lumens (light): All light generated by a light bulb is not available to the chicken. Many of the rays of an open bulb without a reflector reach the bird only after reflecting off some surface or object. About 30% of the light is absorbed by the walls, ceiling, equipment, etc. Deteriorations and cloudiness of the bulb reduce the light given off by about 30%. Therefore, only about 49% of the rated lumens are available to the chicken.

Example: One watt equals 12.56 lumens; 49% of 12.56 lumens equals 6.15 lumens per watt, the amount available.

Calculating footcandles: One 60-watt incandescent bulb produces 753.6 lumens of light (60 × 12.56), but only 369.3 lumens of usable light (753.6 × 49%). If spread over 240 sq ft of surface there would be 1.54 footcandles per square foot. Although this is a mathematical conception of light intensity, in the chicken house the distance from the light source to the area at the location of the birds also affects the intensity. As the distance from the light source is doubled, intensity is reduced by one-fourth.

Types of Light

There are three common light types:

1. *Incandescent*
   Cheapest to install
   Low light efficiency
   Necessitates reflectors
   Short bulb life—750 to 1000 hours
2. *Fluorescent*
   Three to four times the efficiency of incandescent light
   Much higher light level for power used
   Fluorescent equipment higher priced than incandescent
   Variable performance in cold weather
   Ten times longer life than incandescent bulbs
   Available in both screw-in (see Figure 18-1) and hard-wired models
3. *Mercury vapor*
   Cannot be used in houses with low ceilings
   Almost as efficient as fluorescent
   Better than fluorescent when temperature changes
   Long life of 24,000 hours
   Requires several minutes to warm up

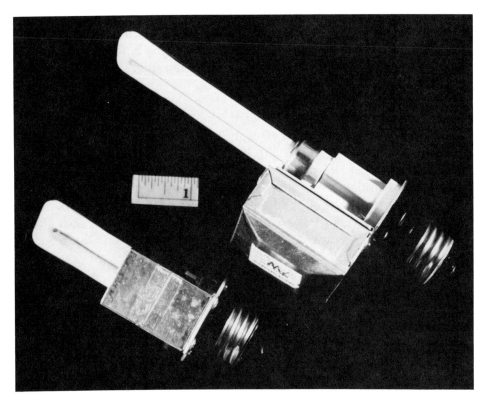

Figure 18-1. Low-wattage fluorescent screw-in tubes for poultry houses.

Color of Light

The color of the light rays has an effect on the productivity of chickens. A part of this difference is thought to be due to the fact that the oil droplets in the retina of the eye filter out some of the shorter rays such as green, blue, and violet. In spite of the improvement in some color categories, white light is used almost entirely, as it represents the best distribution of wavelengths. The relationship between color of light rays and certain production factors are given in Table 18-1. In many cases the relationship is very slight.

Measuring Light Intensity

It is important to know the exact amount of light that is falling on the birds. There are so many factors that influence the intensity of the light rays that any recommendation can only be a guide. Secure a *light meter* and take an exact reading. These may be purchased or secured from a local light and power company. Take readings at several locations in the

Table 18-1. Light Color Relationships

Item	Color of Light Rays				
	Red	Orange	Yellow	Green	Blue
Improves growth				X	X
Depresses feed efficiency			X	X	
Lowers age at sexual maturity				X	X
Increases age at sexual maturity	X	X	X		
Enlarges the eye					X
Reduces nervousness	X				
Lowers cannibalism	X				X
Increases egg production	X	X			
Lowers egg production			X		
Increases egg size			X		
Improves male fertility				X	X
Lowers male fertility	X				

poultry house. Since birds are confined to one location in cage housing, measure intensity halfway between light sources on each tier of cages just above the feed trough.

Comparison of Incandescent and Fluorescent Bulbs

The amount of light produced per unit of electrical power is an economic point to consider when selecting a lighting system. The illuminating power of white incandescent and white fluorescent bulbs when compared on a per-watt basis varies greatly, as can be seen in Table 18-2. However, peak efficiency of fluorescent bulbs can be maintained only

Table 18-2. Lumens of Light from White Incandescent and Fluorescent Bulbs

Bulb Watts	Incandescent Bulbs Lumens	Fluorescent Bulbs Lumens
7		400
15	125	500–700
25	225	800–1,000
40	430	2,000–2,500
50	655	
60	810	
75		4,000–5,000
100	1,600	
150	2,500	
200	3,500	10,000–12,000

when the temperature of the surrounding air is between 70° and 80°F (21° and 27°C) unless the fixtures are enclosed. Efficiency is reduced with temperatures above or below this optimum. At 30° to 40°F (−1.1° to 4.4°C), for example, only about 60% of the maximum output is produced.

> *Recommendation.* In most instances, fluorescent lights are satisfactory for poultry house lighting. When they are used, the intensity of light should equal the recommendations for incandescent lighting. Purchase fluorescent fixtures that are "thermal protected" to help prevent fires. Warm white tubes are preferable to cool white.

Mercury Vapor Lamps

These lamps have been improved and may be used in some poultry houses. The initial investment and operating cost are low. Their disadvantage is that they are slow to reach full brightness and must be mounted at a height of at least 10 ft (3.1 m), but 14 ft (4.3 m) is better, which may preclude their use in some poultry houses. Even at this height light distribution is poor and casts too many shadows, especially in cage houses. With mercury vapor installations it is necessary to supplement with incandescent bulbs to provide uniform illumination over the area.

18-B. LIGHT MANAGEMENT

The manner in which lights are installed in the poultry house has a bearing on their efficiency. There must be a required amount of intensity of light at bird level. Furthermore, the intensity must be uniform throughout the area or areas frequented by the birds.

> *Distribution of bulbs.* In floor operations, a good rule of thumb to follow is the "1 to 1½" ratio; that is, the distance between bulbs should be 1½ times the distance from the bulb to the bird level. If there are more than two rows of lights down the house, the bulbs in each row should be staggered to give better light distribution at floor level. The distance from the bulbs to the outer edges of the house should be only one-half the distance between bulbs.
>
> Distribution of the light bulbs may present more of a problem when cages are used. The bulbs should be placed so that their rays fall on the feed supply and on the birds. When the feeders are on the aisle side, a string of lights placed down the aisle will suffice. But when the feeders are at the back, between two cages, there may be difficulty in getting uniform light distribution, particularly with multideck cages.

Reflectors. In most instances a clean reflector will increase the light intensity at bird level by about 50% compared with no reflector. Avoid inverted cone-shaped reflectors. They confine the light rays to too small an area. Use a flat-type or saucer-type reflector with a rounded edge.

In some instances, bright, reflecting insulation or white paint are used on the inside of poultry buildings. Although not as good as a regular reflector, these materials do reflect some light rays.

Reflector pitch: When reflectors are used, the pitch of the reflector will determine the area illuminated. Although increasing the size of the bulbs will increase the lighted area when no reflectors are used, an increase in the bulb size will not illuminate a greater area when most reflectors are used; only the intensity is increased.

Reflector size: Reflectors should be about 10 to 12 in. (25 to 31 cm) in diameter.

Height of bulbs. Light bulbs should be placed as close to the bird area as is practical. So that the attendant may walk under them, this usually is about 7 to 8 ft (2.1 to 2.4 m) in houses with a litter floor or in cage houses with a service aisle. Avoid hanging bulbs by a cord in open houses. The wind whips them around causing shadows across the birds, often frightening them. The height of corded lights can be regulated in multitiered cage houses with the use of a crank-operated cable system, which allows the movement of the light vertically to provide more light to the lower cages during vaccination and handling operations. Table 18-3 shows the height and size of light bulbs to produce 0.5 and 1 fc at bird level.

Table 18-3. Incandescent Lamp Size and Height to Produce One-Half and One Footcandle at Bird Level

Lamp Size Watts	Height of Lamps above Bird Level							
	To Produce ½ Footcandle at Bird Level				To Produce 1 Footcandle at Bird Level			
	With Reflectors		Without Reflectors		With Reflectors		Without Reflectors	
	ft	m	ft	m	ft	m	ft	m
15	5	1.5	3.5	1.1	3.5	1.1	2.5	0.7
25	6.5	2.0	4.5	1.4	4.5	1.4	3	0.9
40	9	2.7	6.5	2.0	6.5	2.0	4.5	1.4
60	14	4.3	10	3.1	10	3.1	7	2.1
75	15.5	4.7	10.5	3.2	10.5	3.2	7.5	2.3
100	19	5.8	13.5	4.1	13.5	4.1	9.5	2.9

Size of bulbs. The usual recommendation is to supply 1 bulb watt for each 4 ft² (0.37 m²) of floor space to provide 1 fc (10 lx) of light. This is approximate when the bulbs are under a good reflector and 7 to 8 ft (2.1 to 2.4 m) above the floor. Arrange the height and distribution of the bulbs so that it is not necessary to use larger than 60-watt bulbs. When larger bulbs are to be used, the light distribution is less uniform and it requires more electricity to light the house.

Dirty Bulbs

Very dusty bulbs will emit about one-third less light than clean bulbs. Note the following equivalent light intensities:

Type of Bulb and Reflector	Equivalent Light Intensity
Clean bulb, clean reflector	60-watt bulb
Clean bulb, no reflector	40-watt bulb
Dirty bulb, dirty reflector	40-watt bulb
Dirty bulb, no reflector	25-watt bulb

Light bulbs should be cleaned every 2 weeks under normal circumstances, and more often if necessary. Also replace burned-out bulbs each day. This recommendation is particularly true for birds in cages, because the birds cannot move about to find other areas that are lighted.

Time Clocks and Light Controls

Clocks that turn the lights on and off are used in poultry houses. Older models have adjustable on-and-off switches with settings possible in 15-minute intervals. Clocks for open houses require an on-and-off switch for both the a.m. and p.m. Enclosed housing requires only one set of switches: on in the a.m. and off in the p.m. Intermittent programs require a pair of switches for each light period.

Computerized controls are available that can be programmed to the minute, are adjustable for any lighting program pattern, and can control more than one house. Solid-state dimmer controls are especially useful during the rearing stages to control picking. Lights can be turned up when it is necessary to handle the birds for beak trimming, vaccination, or other procedures where good visibility is required.

Photoelectric cells are also used in open houses to turn lights off at sunrise and on at sunset. They will also turn the lights on whenever the light intensity is below a predetermined level.

Warning: Be sure to reset the clocks immediately after a power failure.

18-C. LIGHT EFFECTS DURING GROWING

One of the primary effects of light is the manner in which it alters the age at which pullets reach sexual maturity. It is not the intensity of light that produces all the difference, but rather, the changing length of the light day that alters the time the first eggs are laid. Decreasing the length of the light day as the pullets go through their growing period increases the age at sexual maturity; increasing the length of the light day reduces the age.

Under natural sunlight the length of the light day is continually changing. It is longest on June 21 and shortest on December 21 in the Northern Hemisphere. Therefore, during half the year the light days become longer; during the other half, they become shorter.

> *Light day must not be increased.* With growing pullets the length of the light day should not be allowed to increase. This calls for special lighting procedures with the open-sided house (see Chapter 18-F).

Effects of Natural Daylight

To keep pullets from reaching sexual maturity too early, which will result in an excessive number of small eggs and an increased incidence of prolapse, the poultry producer manipulates his light management program to delay the onset of egg production. With the environmentally controlled dark house, this procedure is relatively easy, as all light is supplied artificially, but under natural daylight the program becomes complicated.

In the Northern Hemisphere, winter-hatched pullets will complete their growing period during the time when the light days are getting longer, causing the pullets to lay their first eggs at a younger age. Spring-hatched flocks are raised during a period in which the length of the natural light day increases until they are about half-grown. During the last half of their growing period, the light day decreases in length, thus delaying the onset of egg production. (Reverse the seasons for the Southern Hemisphere.)

If the requirement is for late sexual maturity, only those birds grown during the period when the natural light day is increasing necessitate a controlled program of lighting, to reverse the increasing light days to decreasing light days.

Artificial Light in the Lightproof House

Many pullets are raised in a lightproof house where no natural daylight is allowed to enter. Fan and other openings must be light trapped to prevent any sunlight from getting into the building. With such a house the length of the light day may be governed at will by the use of artificial lights. There is no problem in following any program of light control.

Results of Light Control

Delaying the onset of egg production by controlled lighting procedures not only increases the age at sexual maturity but also affects other production factors. Although much research has been conducted to determine the various results of light control, the following summarizes most of these findings. All studies, however, have not produced conclusive evidence as listed below. Some of the items represent small changes and thus sometimes are not definite.

1. Regular reductions in the amount of light during the growing period lengthens the time from 1 day of age until sexual maturity.
2. Short day lengths during the growing period increase the number of eggs laid during the first half of the egg production period, but do not greatly increase the total number of eggs laid during the entire period. Egg size is usually smaller than with decreasing light programs.
3. Decreasing patterns of light during the growing period materially increase the size of the first eggs laid because the birds are older with a general increase in the size of all eggs produced during the first 4 or 5 months.
4. The maximum delay in the time for the pullets to reach sexual maturity through a program of reducing the length of the light day is about 3 weeks.
5. Feed restriction during the growing period also delays sexual maturity. The maximum time of delay is about 3 weeks. Meat-type breeders are raised with feed restriction in order to maintain a reduced body weight, and a similar but less drastic program is sometimes used for egg-type lines. Concurrent with the controlled feeding program is light restriction. However, the delay in reaching sexual maturity is not cumulative. Even if both programs are used simultaneously the maximum delay is usually not more than 4 weeks.

All-night lights first 2 days.   So that chicks may quickly learn to eat and drink, continuous 24-hour light should be used during the first 3 days. With brooder cages, continuous light is often used for 4 to 7 days. The intensity at bird level should be 3.5 fc (35 lx).

Continuous dim lights. Do not use continuous light or dim night light for growing pullets.

Light Threshold

The intensity of light during the growing period should be maintained at 0.5 fc (5 lx). Intensities more than this amount may result in picking. If the length of the light day is less than 11 to 12 hours per photoperiod, sexual maturity and egg production will be retarded. If the light day is longer, sexual maturity will be reached at a younger age and egg production will be initiated. This threshold period of 11 to 12 hours is important in developing any satisfactory growing program. It is the criterion used to produce one method of light restriction during this time period.

Age for Maximum Growing Results

Preferably, reduced lighting programs during the growing period should start soon after the brooding period. Some programs start when the chicks are 3 days of age. Maximum results cannot be secured unless a light restriction program is started by the time the growing pullets are 12 weeks of age. To wait longer reduces the length of time during light restriction, and full benefits cannot be realized. However, there is more response to light restriction after 12 weeks than before.

Light and Vices

Light supplied during the growing period should be reduced to a level of 0.5 fc (5 lx) at bird level. Increasing the illumination above this figure will induce more cannibalistic characteristics. Picking and feather pulling will increase; the birds may become highly nervous.

18-D. INTERMITTENT LIGHTING FOR EGG STRAINS DURING GROWING

There are many combinations of intermittent lighting used mainly to conserve electricity and feed. Growing Leghorn pullets with alternating 15 minutes of light and 45 minutes of darkness for a total of 6 hours of light per 24 hours seems the most practical. It will not reduce body weight at sexual maturity, but will reduce growing feed consumption 5% when compared with full feeding. Blackout housing is required.

18-E. LIGHT EFFECTS DURING EGG PRODUCTION

Not only does light affect the growing bird but it stimulates the pituitary gland of layers to secrete the hormones necessary for egg production. Under natural sunlight, maximum egg production is stimulated when light is provided 11 to 13 hours after dawn. During the winter the length of the light day is not normally long enough to foster maximum egg production. To increase the length of the light day in winter, artificial light must be used.

How Light Stimulates the Layer

The light stimulus is initiated when light falls on the eye of the chicken. Light falling on parts of the body other than the eye has no effect on the process.

Light causes the release of LH and FSH hormones from the pituitary, which in turn cause increased growth of the ova in the ovary (see Chapter 3-A).

Red and orange light in the range of the spectrum (6,640 to 7,400 A) are the most effective for increasing egg production. However, artificial light generated by various white bulbs produce enough light within these limits to be effective, and is usually used. Brightness of light has its effective limit, and excessive light provides no additional stimulus.

Light Intensity Threshold

The practical recommendation is that there should be 1 fc (10 lx) of illumination at bird level. In houses with windows or openings, this intensity will be much higher during the hours of natural daylight. Only when artificial lights are used to supplement daylight will the minimum amount of light be necessary. In the light-tight house, 1 fc will be adequate during the entire light period of each day.

Although controlled experiments have shown that 0.5 fc (5 lx) of light is ample for egg production, it must be remembered that a poultry house is full of equipment that creates many shadows that reduce light intensity. This is particularly true of a cage operation. Thus the practical level of light is 1 fc. Table 18-4 shows the egg response from various intensities of light where no shadows are cast.

Light Intensity in Multideck Cages

Maintaining adequate light at all decks in multideck cages is very difficult. There seems to be no practical solution. When a light is hung over-

Table 18-4. Laying Response to Different Levels of Light Intensity in Multideck Cages (Windowless Houses)

Deck	Light Intensity at Each Deck Level		Eggs Produced per Bird (45 wk lay)
	Footcandles	Lux	
Top	3.44	37.0	240
Middle	2.32	25.0	242
Bottom	1.58	17.0	242
Top	0.70	7.5	239
Middle	0.46	5.0	240
Bottom	0.31	3.3	233
Top	0.14	1.5	231
Middle	0.09	1.0	233
Bottom	0.07	0.7	222
Top	0.03	0.3	223
Middle	0.02	0.2	221
Bottom	0.01	0.1	208

Source: Morris, T., Dept. of Agr., Univ. of Reading, Reading, Berks., England.

head, the upper deck receives a greater intensity of light than the lower. In triple-deck cages of normal construction the illumination at the upper deck is from 3 to 3.5 times as great as the lower. If a minimum amount of light is maintained for the lower deck, there is no egg production problem from the increased light at the upper. However, increases in light intensity are correlated with certain vices; picking, cannibalism, prolapse and nervousness are more prevalent.

Table 18-4 shows the effect of various intensities of light in triple-deck cages on egg production. From a practical standpoint, the only recommendation that can be made is that a minimum of 0.5 fc (5 lx) of light must be provided. Remember that dirty bulbs soon reduce the light intensity and that inoperative bulbs can affect egg production in just a few days. Use a light meter.

Length of Light Period

The length of the natural light day varies not only by seasons, but according to one's location on the globe. At the equator the sun is ''overhead'' most of the time and the daily hours of light and dark are more constant; but as one approaches the poles of the Earth the position of the sun changes. This influences the length of the natural daylight hours and the number of daily hours that artificial light must be provided to attain maximum egg production. In the United States the longest day of the year has about 15 hours of daylight; the shortest has about 9. However, the daily variation between regions is as great as 1 hour.

The poultry manager should secure a local schedule of daily sunrise and sunset, along with the length of the daylight hours. Such schedules can be procured from a local weather bureau or similar agency. These figures are very important in planning any lighting program for houses with open sides or windows.

Total hours of light.   Although a light day of greater than 11 to 12 hours will stimulate egg production, the light day must be 14 hours for maximum egg production. Most programs call for 1 to 2 hours more than this to provide a safety factor.

Light day must not be reduced.   During the laying cycle, the length of the light day may be increased, but it should never be reduced. In a house with windows or openings this is very important because the hours of natural daylight vary. Thus, the hours of light supplied, either natural or artificial, must always be as great as the length of the longest day of the year during which the pullets will be kept in egg production.

How to calculate.   If the longest day of the year has 15 hours of daylight, the length of the light day the rest of the year must never be less than 15 hours. If it is, the light day will decrease during a part of the year.

Time of Light During Photoperiod

How to lengthen the light day.   When artificial light is used to supplement the hours of natural daylight in the laying house, there are three practical procedures.

   1. Morning light
   2. Evening light
   3. Morning and evening light

The latter (3) is the best from a convenience standpoint. The hours of light closely coincide with normal working schedules. When morning (1) or evening (2) is used as the time for artificial light, there is also the inconvenience of having to adjust the time clock often to compensate for the daily difference in the time of sunrise or sunset.

Be careful about changing light.   Do not lower the intensity of artificial light or change from white to colored light during the laying period. Be careful about using a temporary program of additional morning and evening light to increase feed consumption during hot weather.

18-F. COMBINATION GROWING–LAYING LIGHT PROGRAMS

To be most effective, the lighting programs for growing birds must be correlated with the lighting programs for layers. There are dozens of such programs, but the common points of all are built around two important features.

1. The length of the light day should never increase for growing pullets.
2. The length of the light day should never decrease for laying pullets.

Light program for entire life of pullet. The type of lighting program used during the growing period will dictate the method of lighting during the laying period; therefore, all programs are growing-laying combinations.

Light Programs During Growing

Lighting programs for growing birds may be classified very briefly according to the time of year the birds are grown and whether the birds are kept in a lightproof house or have access to natural daylight.

1. Reared in lightproof houses.
2. Reared in houses with windows or openings
   A. During a period of *decreasing* length of natural light day, or
   B. During a period of *increasing* length of natural light day.

Using the above broad classifications, one finds some variations under each.

1. *Reared in lightproof houses.* From a management standpoint these programs are easy, as artificial light is the only source of illumination and the length of the light day can be controlled at will. The most practical program is to keep the length of the light day below the threshold in order to delay the onset of egg production. Although 11 to 12 hours of light represents the threshold, the actual length of the light day under these programs is about 8 hours to provide safety. A constant light day of 8 hours is maintained from the time the birds are about 3 days of age until sexual maturity.

2A. *Reared in houses with windows or openings and during a period of decreasing day length.* Birds in these houses have sunlight; the time of the year the day-old chicks are placed in the house will alter the lighting program, because it is almost impossible to start the chicks when all the days in the growing period have decreasing hours of natural light. When birds are given this classification, one can only assume that the length of the light day will be decreasing at least during the last half of the growing period.

   Examples: Referring to Figure 18-2, the following relationships be-tween hatch days in the Northern Hemisphere and the lighting period show:

   June hatch: Each day during the growing period gets shorter.

   March hatch: Each day during the first half of the growing period gets longer, but shorter during the last half.

   Note: Reverse the above and the following for the Southern Hemisphere.

   As a general rule, chicks hatched between March 1 and August 31 in the Northern Hemisphere may be raised in houses with win-dows or openings without any supplementary artificial light. These *"in-season" flocks* are grown during a period when the length of the natural light day is decreasing, at least during the last part of their growing cycle.

2B. *Reared in houses with windows or openings and during a period of in-creasing day length.* When chicks are hatched between September 1 and February 28, and are grown in houses with windows or open-ings, they spend at least the last half of their growing period when the length of the natural light day is increasing.

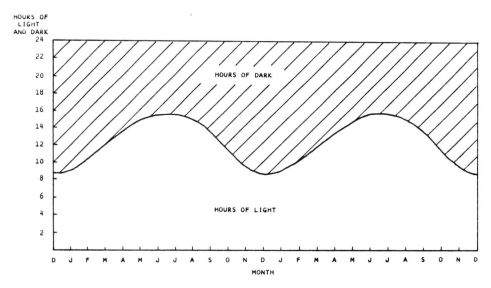

Figure 18-2. Approximate hours of natural daylight in Northern Hemisphere, by months (45°N latitude).

Examples: Referring to Figure 18-2, the following examples are indicative of the picture:

December hatch: Each light day during the growing period gets longer.

October hatch: Each light day during the first half of the growing period gets shorter, but longer during the last half.

These are the problem flocks and are known as *out-of-season flocks.* There is no way that they can be raised under only natural light; artificial light must be added to mask the natural increasing light pattern.

a. *Constant light-day program.* From a local table, determine the number of hours of daylight during the longest day until the flock reaches an average age of 20 weeks for egg lines or 22 weeks for meat lines. When the chicks are 3 days of age provide this amount of total light (natural plus artificial) until the pullets begin to lay. With this system there is never a decrease in the length of light day during the growing period.

b. *Decreasing light-day program.* From a local table, determine the number of hours of daylight when the flock reaches an age of 20 weeks for egg breeds or 22 weeks for meat breeds. Add 7 hours to this figure. The total will be the number of hours of light (natural plus artificial) the pullets are to receive the first week beginning with the third day. Each week thereafter, reduce the length of the total daily light period by 20 minutes until the pullets reach sexual maturity.

Note: The actual weekly reduction in natural light hours will vary according to the distance from the equator.

These programs not very effective. Actually, the 2B programs are a poor substitute for any of the other above methods. At most times during the growing period the total light day is greater than the threshold of light response, which is 11 to 12 hours. Any benefit lies only in the fact that the length of the light day remains constant (a) or decreases (b), rather than increases.

Growing–Laying Light Combinations

Inasmuch as light programs for growing pullets vary in respect to the total length of the light day at the time the birds reach sexual maturity, the method of lighting during egg production must be altered to be consistent with the growing program.

Basic rule for making change to layer program. The basic rule at the time pullets are ready to produce eggs is that the length of the light day must increase. When the daily total hours of light supplied to the growing birds just prior to egg production is less than the threshold of 11 to 12 hours, the number of hours must be increased to above the threshold, usually to about 13 hours as a start. If the pullets reach laying age after a light program during growing that requires more hours of light than the threshold of 11 to 12 hours, the number of hours of light should be increased by 1 hour.

Age for changing to laying lighting program.   The growing light program must be changed to a laying light program at about the time the first eggs are laid. However, flocks differ in the age at which they start egg production, depending on temperature, season of the year, the growing light program, and other factors. The average age to change to the laying light program is 20 weeks for egg-type pullets and 23 weeks for meat-type pullets. The times should be somewhat later for in-season flocks and earlier for out-of-season flocks.

The age to increase light can be varied between 18 and 22 weeks for White Leghorn pullets depending on the season. The earlier age (18 weeks) is preferred during the winter months when body weights are heavier and price differences between egg size categories are small. The later age (22 weeks) is preferred during the summer months because body weights are usually depressed and smaller eggs are severely penalized in price.

Feed intake must increase.   Increases in the length of the light day at sexual maturity must coincide with increases in feed consumption. For the first few weeks after the flock starts to lay, two things occur:

1. Egg production increases rapidly.
2. Body weight increases.

Each of the above changes requires additional feed to produce the desired effect. Many pullets, particularly those in the meat-type classification, are raised on a controlled feeding program in order to lower their body weight at sexual maturity. If additional light is given at the onset of egg production, but feed allocations are not increased, egg production will suffer. If the feed allotment is increased in the absence of additional light, the pullet will gain weight too rapidly because few eggs are being produced, and again, egg production will suffer. Start to increase the feed intake at about 1 week after you change from a growing light program to a laying light program.

Length of Light Day for Layers

In order to get maximum laying response from the layer lighting program, the total daily hours of light should be as follows:

Egg lines: 16-hour light day
Meat-lines: 15- to 15.5-hour light day

Increasing light at sexual maturity. There is some evidence that sudden, large increases in the length of the light day at the time of sexual maturity are conducive to prolapse. Therefore, if a flock reaches sexual maturity with less than 11 to 12 hours of light, the length of the light day at the time egg production starts should never be increased to more than 1 hour greater than the threshold of 11 to 12 hours, or a maximum of about 13 hours. When the growing pullets receive more than 11 to 12 hours of light per day at the time they reach sexual maturity, the increase should not be more than 1 hour. After the first week of the increase, add 15 minutes of light per week until the pullets are receiving their required number of hours according to the program to be followed.

18-G. LIGHT TREATMENT AND PRODUCTION FACTORS

How Light Treatments Affect Production

Two of the ultimate aims of controlling the length of the light day during growing and egg production are to prevent excessively early onset of egg laying and to increase production efficiency. Table 18-5 shows the influence of several different growing-laying light programs on several production factors. Also of interest is the relation between the light threshold and the onset of lay. Gradually reducing the length of the light day during the growing period was not effective unless the light day was reduced below the threshold of 11 to 12 hours. Gradually reducing from 22 to 16 hours was ineffective in delaying sexual maturity, but gradually reducing from 22 to 9 hours was effective.

Early Egg Size Affected by Light During Growing

It is common knowledge that when pullets are delayed in the onset of egg production, the first eggs are larger. The delay may be produced by restricting the growing feed, or by using a proper lighting program. Early egg size is larger because the birds are older (egg size increases the older a bird gets). When she reaches a certain age she lays an egg of a certain size, regardless of the age at which she laid her first eggs. This factor is of great economic importance because first eggs are small in comparison with those laid later in the laying cycle. Thus, delaying sexual maturity increases the percentage of the early eggs that will command a better price because of their larger market weight, or because they are more suitable for hatching purposes in the case of breeding birds. Flocks that

Table 18-5. Influence of Lighting Treatment on Sexual Maturity, Laying House Mortality, and Egg Production[1] (S. C. White Leghorns in Cages)

Light Treatment		Days to Reach 10% Egg Production	Days to Reach 50% Egg Production	Laying House Mortality %	Eggs Produced During 47 Weeks of Lay
Growing Period	Laying Period				
Gradually decreased from 22 hr to 16 hr	Gradually increased from 16 hr to 22 hr	156	172	3.3	225
Gradually decreased from 22 hr to 9 hr	Gradually increased from 9 hr to 22 hr	172	186	3.3	220
Gradually decreased from 16 hr to 9 hr	Gradually increased from 9 hr to 16 hr	171	191	3.8	220
Gradually decreased from 16 hr to 9 hr	Suddenly increased from 9 hr to 16 hr	163	176	5.0	230
Started on constant 16 hr then suddenly decreased to constant 9 hr	Suddenly increased from 9 hr to 16 hr	165	176	4.6	227
Constant 16 hr	Constant 16 hr	156	171	5.0	224

Source: Shutze, J. V., W. E. Matson, and J. McGinnis, 1963. *Poultry Sci. 42,* 150–156.
[1]Average of two tests.

are light in weight at sexual maturity will lay eggs smaller than normal throughout their entire lives. The effects of light stimulation at different ages in White Leghorn laying flocks is shown in Table 18-6.

Table 18-6. Age at Lighting and Egg Size[1]

| | Age at Lighting (wk) | | |
Trait	18	20	22
Average egg weight (g/egg)	57.7	58.8	59.4
Average egg weight (oz/doz)	24.4	24.9	25.1
Percent large and above	65.8	74.2	79.5
Average egg value (U.S.$)	0.467	0.477	0.482

Source: University of California, *Progress in Poultry,* 1982
[1]Lighting increased from 10 hr to 17 hr.

How Light Affects Males

Seemingly, the future fertilizing ability of the males is not affected by light intensities or the length of the light day during the growing period. During the breeding season, however, there is a relationship between light intensity and fertilization: The lower the intensity, the lower the volume of semen produced. However, the point has no practical application, as the intensity necessary for the pullets in the breeding pen to produce an adequate number of eggs also is satisfactory for the production of high semen volumes and good fertilization.

Light for Broilers

Refer to Chapter 20-G for a discussion of the effects of light on broiler growth, feed conversion, and management.

18-H. INSTRUCTIONS FOR GROWING AND LAYING LIGHT PROGRAMS

Explicit directions are given below for the use of the lighting programs outlined above. They are for the Northern Hemisphere and should be reversed for the Southern Hemisphere.

Important: Consult your chick supplier or genetic breeder for detailed instructions for the lighting program specific for his strain of birds.

1. Reared in Lightproof Houses

When pullets are raised in lightproof houses, provide 8 hours of artificial light each day after the third day until egg-type strains reach 20 weeks of age or meat-type strains reach 22 weeks of age. Do not alter the length of the light day during this period.

At 20 or 22 weeks of age, respectively, production begins and the birds may remain in a lightproof house or be moved to a house with open sides or windows. The program for each is as follows:

1A. *Left in the lightproof laying house.* When the laying pullets are to be left in a lightproof laying house, abruptly increase the length of the light day to 13 hours, then add 1 hour per week until meat-type strains receive 15 hours of light and egg-type strains receive 15 min of light per week.

1B. *Moved to laying houses with open sides or windows.* If the pullets are to be moved to houses with open sides or windows at 20 to 22 weeks of age the length of the initial light day will be governed by the length of the natural daylight hours. If they are less than 13 hours, use artificial light to supplement natural daylight so as to provide a 13-hour light day; then add 1 hour more light per week until meat-type strains receive 15.5 hours of light, and egg-type strains receive 15 min per week.

If the length of the natural light day is greater than 13 hours at 20 or 22 weeks of age, respectively, then add 1 more hour of light at this age, and increase 1 hour per week until meat-type strains receive 15.5 hours of light or egg-type strains receive 15 min per week.

2. Reared in Houses with Windows or Openings

2A. *Reared in houses with windows or openings and during a period of decreasing day length.* For chicks hatched between March 1 and August 31, no supplemental artificial light should be given in these growing houses. Only natural daylight should furnish the illumination.

When egg-type strains are 20 weeks of age or meat-type birds are 22 weeks of age, abruptly increase the length of the daily light period to 13 hours; then add 1 hour more per week until meat-type strains receive 15 hours of light in lightproof houses or 15.5 hours in houses with windows or openings. Increase to 16 hours with all egg-type strains at 15 min per week.

2B. *Reared in houses with windows or openings and during a period of increasing day length.* When chicks are hatched between September 1 and February 28, and are raised in these houses, use one of the following light programs:

a. *Constant light day program.* Determine the length of the longest natural light day before the pullets reach 20 or 22 weeks of age. Maintain this period of daily light hours from the third day until the birds are 20 or 22 weeks of age. Supplement natural daylight with artificial light to keep the light day constant.

At 20 weeks of age for egg-type strains and 22 weeks of age for meat-type birds, increase the length of the light day by 1 hour. Make increases each week thereafter until meat-type strains receive 15 hours of light in lightproof houses or 15.5 hours in houses with windows or openings. Increase to 16 hours with all egg-type strains.

   b. *Decreasing light day program.* Determine the number of natural daylight hours when the pullets will reach 20 or 22 weeks of age, respectively; then add 7 hours. This total represents the length of the light day after the third day. Thereafter, reduce the length of the light day by 20 minutes each week. This approximates 7 hours in 20 or 22 weeks, respectively.

At 20 weeks of age for egg-type strains and 22 weeks of age for meat-type birds, increase the length of the light day by 1 hour. Make hourly increases each week thereafter until meat-type strains receive 15 hours of light in lightproof houses or 15.5 hours in houses with windows or openings. Increase to 16 hours with all egg-type strains at 15 min per week.

Altering the Preceding Directions

Although the preceding directions are quite specific as to the hours of total light that should be given on a specific week of age at a time approximating sexual maturity, it should be remembered that under most conditions the laying programs for lighting should be initiated at the time the first egg is laid by the flock. Actually such light increases should precede laying feed increases by 1 to 2 weeks. It will take from 7 to 10 days for additional light to produce its full effect, and birds should be producing well when the feed increase is given. Otherwise bird weight may increase too rapidly.

18-I. INTERMITTENT LIGHT FOR PULLETS PRODUCING EGGS

That the effect of light for laying chickens is not the product of intensity times duration has been shown in several experiments using intermittent light. An experiment by H. V. Biellier, Missouri Agricultural Experiment Station, involved a regular light program for the first 4 months of production, and intermittent light thereafter. The results are given in Table 18-7.

Table 18-7. Intermittent Light Compared with Continuous Light

	Hen-day Egg Production (%)								
Light Treatment[1]	Dec.	Jan.	Feb.	Mar.	Apr.	May	June	July	Aug.
A. 15 hr continuous	89	90	91	92	87	86	79	78	72
B. 1 hr light/3 hr dark, repeated 4 times daily	89	91	93	91	89	86	80	77	72
C. 1 hr light/6 hr dark, repeated 3 times daily	89	90	93	90	87	83	78	69	64

Source: Missouri Agric. Exp. Station, March 1976.
[1]After 4 months of regular lighting.

The table shows exceptional results with only 4 hours (B) or 3 hours (C) of light per day when given intermittently. Four hours of such light proved as productive as 15 hours of continuous light.

Caution: Be sure you have had experience with this lighting system before administering it to flocks large in size. Intermittent daytime lighting programs for layers should not be started until the pullets have been laying for at least 8 weeks. Starting earlier causes a drop in egg production and egg size that may continue throughout the laying year. Some programs call for continuous (24 hr) lighting from the onset of production past the peak before the intermittent lighting program is initiated. Intermittent programs have been shown to reduce the amount of electricity used for the lighting program and lower the quantity of feed consumed, both economic advantages.

Intermittent day-and-night lighting. Scientists at Cornell University have found that a 24-hour lighting program for layers consisting of 8 light, 10 dark, 2 light, and 4 dark periods gave results equal to a continuous lighting program. However, there were still 10 hours of light per day but electricity costs were reduced by almost 40%.

18-J. AHEMERAL LIGHTING PROGRAMS FOR LAYERS

An ahemeral lighting cycle includes any that do not total 24 hours of light and dark. There are two types: (1) a longer day, and (2) a shorter day.

1. *Longer day for egg strains.* Several day lengths have been tested, but the 28-hour day (14 hours of light and 14 hours of dark) seems to offer the most promise.

Advantages: Improved shell quality, particularly in older layers. Shell thickness increased about 9%. Improved egg size as the production year is continued, with a measurable increase in egg albumen.

Disadvantages: Egg production is usually reduced. Requires a fully lightproof house. Light hours are not compatible with a normal working schedule.

2. *Shorter day for egg strains.* The 22-hour day seems the most practical.

Advantages: Yearly egg production is increased by about 2%.

Disadvantages: No increase in shell quality. No increase in egg size. Light hours are not compatible with a normal working schedule.

19

Flock Recycling

Why Poultry Producers Recycle

Although most poultry producers integrate recycling programs into their regular replacement schedules, economic circumstances may require a sudden decision to carry a flock over for a second cycle of egg production. These circumstances include

1. Anticipation of high egg prices
2. Lack of available cash because of depressed egg prices
3. Empty laying houses

In these cases the cyclical effects of the ups and downs of the price the producer receives for his commercial eggs may mean that he has been producing eggs at a cost above their market value. If this has been of long duration, many poultrymen find it financially difficult to accumulate enough ready cash to purchase a new group of chicks and to provide for them until they reach sexual maturity. Many such poultry producers resort to recycling because the cash outlay for the flock is less.

19-A. INDUCED MOLTING TO RECYCLE LAYING FLOCKS

Molting is a natural process of all birds in an endeavor to renew their feathers prior to migration, shorter days, or cooler weather. Normally, wild chickens molt once a year; as they produce but a few eggs, the molt is not associated with the laying cycle. However, domestic chickens have been bred for high egg production, and under ordinary circumstances they do not go through a complete molt until the end of a long and intensive laying period. If nothing is done to alter the normal molting cycle, it requires about 4 months for a hen to drop her feathers and grow a new set. It is possible, however, to speed up the process through a program of inducing pullets to molt rapidly, growing a new set of feathers, then stimulating them to begin producing eggs. The entire artificial program should take no longer than 6 to 8 weeks.

Induced molting is practiced to give the flock a rest at the end of a period of egg production. The ability of a flock to produce eggs at a high rate after the molt can be attributed only to the rest period it receives. Actually then, induced molting is but a procedure for resting the flock so that it may perform efficiently during another cycle.

19-B. INDUCED MOLTING REQUISITES

There are several requisites to a good program of molting. Many programs will do the job well, but as stress must be created to cause the birds to drop their feathers, successful molting programs are those that create the least amount of stress, produce a rapid molt, and get the birds back into egg production quickly. The three main factors involved are described below.

Initiating the molt. All molting programs require that egg production be reduced to zero, which is usually accomplished by fasting (no feed) the flock until or beyond the time production ceases or by limiting critical nutrients such as protein, calcium, or sodium. Some producers include 1 or 2 days of water removal to help get the flock out of production, but this action is not necessary and has certain risks associated with it especially during the summer months. Artificial lights should be turned off in open-sided houses and reduced to no more than 8 hours in environmentally controlled houses.

A fast of 4 days will usually cause a flock to cease egg production. Longer fasts of up to 14 days will usually give superior results, but extreme care must be taken to monitor body weight losses and mortality.

Resting the flock. Once the flock is out of production, it may be held out of production for as little as 1 week to as much as 4 or 5 weeks depending on the intent of the manager and the feeding program implemented during this period.

Short resting periods can get a flock back to its peak of production in 4 or 5 weeks from the initiation of the molt, which is important if egg prices are high at the time of the molt or when the intended second cycle period is to be less than 6 months in length. Short rests, though, are usually associated with lower rates of egg production and poorer eggshell quality.

Longer rests will result in production peaks 10 to 12 weeks after molt inducement. These are associated with higher rates of lay and better eggshell quality. Longer rests should be used if egg prices are low and if the flock is to be kept for more than 6 months.

The length of the resting period can be regulated by the level of nutrition. Low-protein or low-calcium diets will generally keep a flock out of production. Many egg producers have found that cracked grain supplemented with pullet-growing levels of minerals and vitamins provides a suitable diet to keep the flock out of production, for feather regrowth, and to regain body weight.

Returning the flock to production. When the flock is to be returned to production, a layer diet should be fed and lights should be returned to the normal lighting program for layers. A 50% rate of lay should be reached in 2 to 3 weeks, and the peak should follow an additional 2 to 4 weeks later.

Types of Recycling Programs

Egg-producing hens may be molted one or more times, giving rise to two general types of programs.

Two-cycle molting program. This program involves one molt and two cycles of egg production. The hens are molted after about 10 months of egg production, brought back into egg production, and then sold at about 24 months of age.

Multiple-cycle molting program. This program involves two or more molts and three or more cycles of egg production. The hens are first molted after about 9 months of production, then held through successively shorter cycles and sold at 30 or more months of age. These programs are seldom as profitable as two-cycle programs, but are noted for superior egg quality (see Chapter 19-G).

19-C. COMPARISON OF FIRST AND SECOND LAYING CYCLES

Whether or not it pays to routinely recycle laying flocks depends on many factors. Most important is the relative performance of pullet and molted flocks and the cost and prices for eggs, feed, and replacement pullets.

After an induced molt, egg production during the second cycle does not equal egg production during the first cycle. This fact, along with the reduced flock size, would appear to rule out the use of recycling, but reductions in replacement costs usually offset the disadvantages. Therefore, the practice of recycling becomes one of cost analysis. Some of the factors involved are as follows:

1. *Cost to bring to egg production.*   It costs less to molt a hen and bring her back into egg production than it does to grow a pullet from one day of age to egg production. This is a prime cost factor in making the decision of whether to molt or not to molt.

2. *Amortization of bird cost.*   The cost of growing a pullet, or the cost of molting her at the end of her first cycle of lay and returning her to egg production minus salvage value, must be amortized over the number of eggs she produces. Although the molted hen has a lower cost than the pullet, she will not produce as many eggs during her second cycle as during her first.

3. *Mortality comparison.*   The weekly rate of mortality is usually slightly less during successive egg-production cycles. If the weekly mortality rate during the first cycle were 0.20%, the second cycle would be approximately 0.18% and the third cycle would be approximately 0.16%.

4. *Feed consumption.*   Normally feed consumption is practically the same in each cycle of production when measured from peak production to the end of the cycle.

5. *Length of egg production period.*   The profitable period of egg production is longer during the first cycle than during the second. Usually the second cycle lasts only 7 to 9 months. The most profitable two-cycle program is to molt the birds at about 65 weeks of age and sell them 40 weeks later at 105 weeks.

6. *Rate of egg production.*   The rate of egg production is lower during the second period when compared with the corresponding month of the first period. At the peak of egg production the rate will be about 7 to 10% below the first-year peak and will decline at a slightly faster rate in successive cycles.

7. *Egg size.*   Egg size is larger during the second cycle. This becomes an advantage only if there is a market for the larger eggs.

8. *Shell quality.*   The average quality of eggshells is much better during the first cycle than during the second. Although shell quality gradually deteriorates while the bird is in egg production, the molting rest usually restores shell quality to the equivalent of a 10-month-old pullet flock when molting is initiated at 65 weeks of age. Shell quality during the second cycle drops at about the same rate as in the first cycle, but because it starts at a lower level, it reaches the same level earlier.

9. *Interior egg quality.* During the second cycle the contents of the eggs have a slightly lower quality than during the first. About 10% fewer Grade A eggs may be expected during the second cycle compared with a similar production period during the first.

19-D. PROGRAMS FOR TWO-CYCLE MOLTING

There are dozens of molting programs used throughout the egg industry. Each has unique procedures for initiating a flock into a molt, holding it at zero production, and returning it to peak performance. The best programs must be able to get a flock out of production rapidly and uniformly, experience relatively low mortality, be simple to follow, be inexpensive, and postmolt egg production and egg quality must be only slightly poorer than first-cycle results. Of the many satisfactory molting programs, three that feature different basic procedures are given here:

Conventional Molting Program

This program is outlined in Table 19-1. Some additional information is as follows:

1. Self-feed oystershell from the start of molting until 2 weeks after egg production is reestablished; then return to controlled shell feeding.
2. Provide adequate feeding space; all birds must be able to eat at one time.
3. Take 1 to 2 weeks longer to molt and to return breeders to egg production.

California Molting Program

Another method of molting, called the California program, is shown in Table 19-2. It incorporates a longer period of complete feed withdrawal followed by nothing but cracked grain. As water is not restricted, it is a good procedure during periods of hot weather; dehydration is eliminated. Oystershell feeding during the feed withdrawal period is optional. On or about the 7th day without feed the flock will begin to look poorly. However, it should not be fed unless mortality starts to rise. One or 2 days on feed after the 10-day withdrawal period will overcome the poor appearance. Production usually drops to zero on the 5th or 6th day.

Table 19-1. Conventional Molting Program (On-again, Off-again Program)

Day	Feed		Water	Light
1 2	None		None	8 hours
3	Egg-type layers 10 lb (4.5 kg)/100 hens	Meat-type layers 15 lb (6.8 kg)/100 birds	Water	
4	None		None	
5	Same as 3rd day		Water	
6	None		None	
7	Same as 3rd day		Water	
8	None		None	
9	Same as 3rd day		Water	
10 through 55–60	Return to controlled feed restric- tion—about 75% of full-feed intake			
61	Full-feed layer ration	Full-feed breeder ration		14–16 hours

Body weight drops about 25% after 10 days with no feed and original weight is regained in 7 weeks after feeding is restored. Longer feed withdrawal periods of up to 14 days have been successfully used with equal or superior results. Periods of this length, though, must be watched very carefully to assure that mortality does not exceed about 1.25%.

Table 19-2. California Molting Program (No Water Restriction)

Day	Feed	Water	Light
1 through 10	None	Water	Discontinue or provide 8 hours
11 through 28	Full-feed cracked grain		
29+	Full-feed laying mash		16 hours

Source: The University of California, 1988.
Note: The grain feeding period may be eliminated for short molts or lengthened for longer molts.

North Carolina Molting Program

A third program, which is also very popular, was developed at North Carolina State University. This program includes a premolt period of 7 days when artificial lights are turned on during the night to give the flock a 24-hour day before feed is removed. This procedure allows the manager to subject the flock to the simultaneous reductions of light and nutrient intake.

The North Carolina method places a major emphasis on body weight reduction and suggests that weight losses should be 30% for 3.6-lb (1.64-kg) hens, 33% for 3.8-lb (1.73-kg) hens, and 35% for 4.0-lb (1.82-kg) hens. The length of time to achieve this reduction will be 14 or more days.

When the weight reduction is reached, the flock is placed on 0.1 lb (45 g) per day feed intake for 2 days, followed with a high-protein (15–16%) molt ration that contains 2% calcium until the 28th day when the flock is returned to a conventional layer diet. Twelve hours of total light are provided for the first 3 weeks and a minimum of 13 hours on the 21st day. Normal lighting is resumed on the 35th day.

19-E. CONSIDERATIONS INVOLVED WITH TWO-CYCLE MOLTING

The results from recycling are subject to a great deal of variation. Some flocks that are exceptional during their first cycle produce poorly during their second. In other cases, mediocre first-year flocks may have an even better record after being molted. Many poultrymen consistently have poor results, while others find molting very profitable.

Length of First Egg Production Cycle

Normally, flocks are molted after about 10 months of production. This age is about when shell quality deterioration reaches a level that will be a problem in the egg-processing plant. In general, the earlier flocks are molted, the higher the subsequent rate of lay (see Table 19-3).

Table 19-3.   The Effect of Age at Molt on Subsequent Rate of Lay

	Age at Molt in Weeks				
Period	<55	55–59	60–64	65–69	70–74
	(% hen-day rate of lay)				
At peak	82	82	80	80	78
+30 wk	72	72	70	69	67
+40 wk	64	64	63	61	58

Source: University of California, 1988.

Flock Should Be of Good Quality

In spite of the fact that on occasion some flocks that are ordinary during the first cycle do exceptionally well during the second, it is wise to recycle only good first-year flocks. This rule may be difficult to implement because of scheduling and housing requirements. The odds are greater, though, that they will do better than poor first-year flocks.

Check for disease. Before recycling any flock, take a few birds to the laboratory to determine if any diseases are present. If a serious disease is present, it may alter the decision to molt.

Vaccinate. About a week before the hens are to be molted, vaccinate for infectious bronchitis and Newcastle disease. The laboratory technician may also recommend other vaccinations.

Housing Facilities

Housing the molting flock sometimes becomes an expensive item. When recycled hens are to be left in the same house they used during their pullet year of egg production, the house will usually be from 90 to 92% full, thus raising the housing cost per bird and per dozen eggs produced.

As a matter of economics, some producers prefer to regroup molted birds by consolidating houses or by filling in vacant cage space. It should be noted, though, that such practices can introduce disease from one flock to another and can disrupt established social patterns within the flock, resulting in cannibalistic behavior.

Fast Molt Versus Slow Molt

Most conventional molting programs call for returning a flock to 50% egg production in about 8 weeks after molt initiation in order to give the flock a good rest. However one can reduce this time period to 6 weeks to take advantage of current higher prices or for relatively short second cycles. A fast molt program is accomplished by fasting the flock from 4 to 10 days, followed immediately with a layer diet. No rest period is provided.

Body Weight Loss

The present-day theory for a successful molt is one that is brought about by body weight reduction. It is important to get the fat off the reproductive system; reduce it to the minimum. Weight reductions of between 25 and 30% are essential during the molt.

A University of California experiment using the basic California molting program (Table 19-2) held layers off feed for varying lengths of time with results as shown in Table 19-4. Their results showed that the period of feed withdrawal must be at least 10 days to get the necessary weight reduction. Egg production increased with the longer fasting periods.

Weight loss and mortality association. Usually the more severe the molting program, the greater the weight loss, and the greater the mortality. Weight loss is not a problem except that mortality increases when the birds lose too much weight. On most programs feed restriction may be continued as long as the mortality remains under 2%.

Weight loss regained. Once the hens drop to zero egg production and the hens lose about 30% of their weight during molting, the feed allowance should be increased, but only to the extent that it is ample for the birds to replenish their covering of feathers and to gain weight slowly. Hens should regain about half of the weight loss after they have been on the molting program for 25 days. By the time the molting program is completed, the body weight of the hens should approximate their weight at the time molting was first initiated. Once the body weight has returned to normal, the light day should be increased to 14 to 16 hours to stimulate egg production. It is not wise to increase the length of the light day unless weight loss has been regained.

Weigh the birds during molting program. Molting programs should not

Table 19-4. Effect of Length of Feed Removal During Molting on Subsequent Performance

	Days Without Feed[1]			
Performance Trait	6	8	10	12
Hen-day egg production (%)	53.9	51.7	53.7	56.3
Eggs per hen housed	142	141	149	153
Total egg yield (lb/hen housed)	19.4	19.9	20.9	21.5
(kg/hen housed)	8.8	9.1	9.5	9.8
Average egg weight (oz/doz)	26.3	27.2	27.0	27.1
(g/egg)	62.2	64.3	63.9	64.0
Daily feed consumption (lb)	0.209	0.207	0.208	0.212
(g)	94.8	93.9	94.3	96.2
Feed conversion (lb/doz)	4.67	4.85	4.7	4.53
(kg/doz)	2.12	2.20	2.14	2.06
Feed:Egg ratio	2.84	2.85	2.78	2.68
Mortality (%)	10.7	12.5	1.8	7.1
Egg income minus feed cost ($/hen)	1.71	1.63	1.82	1.98

Source: University of California, 1977
[1]Followed with cracked milo (sorghum) to day 28.

be considered to be exact recommendations under all conditions. Variations in the length of the natural light day, environmental temperature, condition of the hens at the time the program is initiated, and many other factors will alter the effects of the program. One good criterion of how well molting is progressing is to weigh regularly a representative sample of the hens. Average flock weights should be taken just prior to molting, and every 2 weeks during the molting program.

19-F. THE ECONOMICS OF RECYCLING LAYING FLOCKS

Analytical Procedures

Historically, egg producers were concerned with the cost to bring a pullet or hen to the point of 5% production. Molting cost was compared to the replacement cost at the same rate of lay. New procedures have been applied in recent years that emphasize maximum returns per unit of time per pullet housed. This approach, when analyzed by computer, can compare an endless number of programs and those with superior monetary returns can be selected. When done properly, the net effect of various performance differences at different prices can be assessed relative to selecting the optimum replacement program.

Number of Eggs Produced

As noted before, egg production in successive cycles tends to be lower. Table 19-5 illustrates the relationship in a flock molted at 65 and 105 weeks of age. In addition, the number of eggs is also a result of the flock size and these percentages will yield fewer eggs as the flock is depleted.

Egg-Size Comparison

Since pullet flocks usually require about 20 weeks of production before they reach 90% or more large eggs compared to a molted flock, which starts at this level, the egg-size advantage belongs to the molted flock. A typical pullet flock may produce 75% Grade A large eggs during its life compared to a three-cycle flock, which produces 80% or more.

Table 19-6 compares the egg-size breakdown during the first, second, and third cycles of production. The major advantage in egg weight for molted flocks is even greater when premium prices are paid for the larger classes or when eggs are sold on a weight basis.

Table 19-5. Comparison of First, Second, and Third Cycles of Leghorn Egg Production

	Hen Day Egg Production During Molting Program				
Age wk	1st Cycle %	Age wk	2nd Cycle %	Age wk	3rd Cycle %
21	10.0	66	37.5	106	35.2
22	23.0	67	0.0	107	0.0
23	39.8	68	0.0	108	0.0
24	60.0	69	0.0	109	0.0
25	76.5	70	9.9	110	9.1
26	84.5	71	24.1	111	22.1
27	87.0	72	39.0	112	35.7
28	89.5	73	53.8	113	49.4
29	91.2	74	66.6	114	61.1
30	92.5	75	75.1	115	68.9
31	92.0	76	80.0	116	73.4
32	91.5	77	85.0	117	78.0
33	91.0	78	84.5	118	77.5
34	90.5	79	84.0	119	76.9
35	90.0	80	83.4	120	76.3
36	89.5	81	82.9	121	75.8
37	89.0	82	82.4	122	75.2
38	88.5	83	81.9	123	74.7
39	88.0	84	81.3	124	74.1
40	87.5	85	80.8	125	73.6
41	87.0	86	80.3	126	73.0
42	86.5	87	79.8	127	72.5
43	86.0	88	79.2	128	71.9
44	85.5	89	78.7	129	71.4
45	85.0	90	78.2	130	70.8
46	84.5	91	77.7	131	70.3
47	84.0	92	77.1	132	69.7
48	83.5	93	76.6	133	69.2
49	83.0	94	76.1	134	68.6
50	82.5	95	75.6	135	68.1
51	82.0	96	75.0	136	67.5
52	81.5	97	74.5	137	67.0
53	81.0	98	74.0	138	66.4
54	80.5	99	73.5	139	65.9
55	80.0	100	72.9	140	65.3
56	79.5	101	72.4		
57	79.0	102	71.9		
58	78.5	103	71.4		
59	78.0	104	70.8		
60	77.5	105	70.3		
61	77.0				
62	76.5				
63	76.0				
64	75.5				
65	75.0				
Avg %	79.5		66.0		59.3

Table 19-6. First, Second, and Third Cycle Egg Size

Egg Weight Classification		Cycle of Production		
Oz per Dozen	G/Each (Approximate)	1st Cycle (21–65 wk)	2nd Cycle (66–105 wk) (% of eggs laid)	3rd Cycle (106–140 wk)
Over 29	Over 68.5	4.1	11.8	12.7
26 to 28.9	61.4 to 68.4	29.1	53.2	53.0
23 to 25.9	54.3 to 61.3	42.4	32.4	31.7
20 to 22.9	47.2 to 54.2	19.7	2.6	2.6
17 to 19.9	40.2 to 47.2	4.5	0.0	0.0
Under 17	Under 40.2	0.2	0.0	0.0

Egg Quality

Egg quality deteriorates with the length of the laying period. By 65 to 70 weeks of age, most Leghorn strains will experience 6 to 8% off-grade eggs. Most of these are due to shell defects. Recycling will improve practically all egg-quality characteristics to a comparable level of a 10-month-old pullet. The earlier the age at molting, the better the recovery.

Egg-quality deterioration after a molt follows a similar pattern as in the first cycle, but it will reach the same objectionable level in a shorter period of time. In most cases, the length of production cycle is determined solely by the reaction of the egg-processing plant manager to the quality of eggs being produced. Table 19-7 illustrates the egg quality associated with three cycles of production and flocks molted at 19 and 31 months of age.

Feed Consumption

Pullet flocks require less feed during the first 10 to 15 weeks of lay because of their lower body weight and egg mass. Most studies indicate that following this period daily feed consumption remains fairly constant unless there are environmental influences.

Feed consumption during the molting period is considerably less than during egg production because of the feed removal process. Upon reaching full production, daily feed consumption is comparable to the corresponding period during the first cycle. Additional savings are achieved during the resting period when a less expensive feed is used.

Costs and Prices

The decision to incorporate molting into a routine replacement program must evaluate all of the performance characteristics previously dis-

Table 19-7.   Egg Quality Changes with Age of Flock Following Molt

Months of Age	% USDA AA[1]	Albumen Height mm	Haugh Units	Shell Thickness microns	Egg Weight oz/doz	Shell Smoothness Score[2]
6	96	9.6	98.7	414	22.7	0.02
7	97	9.2	95.9	411	23.5	0.02
8	96	8.9	94.5	406	24.1	0.03
9	95	8.7	92.5	401	24.8	0.05
10	94	8.4	91.0	399	25.4	0.08
11	92	8.2	88.4	394	25.8	0.13
12	86	7.9	87.2	391	26.4	0.19
13	84	7.7	86.5	389	26.8	0.27
14	79	7.6	85.4	384	27.1	0.36
15	73	7.5	84.4	381	27.3	0.46
16	67	7.3	83.7	381	27.5	0.56
17	60	7.2	82.7	378	27.6	0.68
18	50	7.1	82.0	378	27.7	0.86
19	43	7.1	81.6	378	27.6	0.97
20			Molt 1			
21	84	8.3	89.8	396	27.8	0.14
22	82	8.2	88.6	396	27.9	0.17
23	78	8.0	87.5	394	28.0	0.22
24	75	7.8	86.4	391	28.1	0.29
25	70	7.7	85.5	389	28.1	0.40
26	65	7.6	84.6	389	28.2	0.51
27	58	7.4	83.6	386	28.2	0.64
28	52	7.3	83.0	386	28.2	0.79
29	46	7.2	82.1	386	28.3	0.96
30	38	7.1	81.8	386	28.3	1.14
31	32	7.0	81.3	386	28.2	1.30
32			Molt 2			
33	69	8.0	87.0	406	28.7	0.43
34	68	8.0	87.4	409	28.4	0.40
35	66	8.0	87.2	409	28.3	0.45
36	65	8.0	87.0	411	28.1	0.52
37	61	7.7	85.7	409	27.9	0.68
38	59	7.4	84.3	409	27.7	0.86

Source: University of California, 1963–1965.
[1]Interior quality only.
[2]On a scale of 0 to 3, 0 = smooth, 3 = very rough.

cussed and many important economic factors. The principle justification for recycling is to reduce the cost of a replacement program. Therefore, induced molting will be more profitable when replacement costs are high and the salvage value of the old hen is low.

High egg prices, in general, favor all-pullet programs because of their higher rate of egg production. On the other hand, egg marketing based on weight or premium prices for the larger egg-size categories favor recycling.

Because of the complex interrelationships involved in this question, one cannot conclusively state that recycling is better or worse than the all-pullet program. With accurate performance and pricing information, and the use of the computer, the final answer can be determined for each set of conditions.

Accurate Records Are Essential

Molting and recycling a flock calls for accurate decisions: When should the flock be molted? Is the molt period progressing on schedule? How long should the hens remain in production during the second cycle? What are the costs and income? An adequate record system must be followed. It will also serve as a guide to any future molting program.

19-G. MULTIPLE-CYCLE PROGRAMS

If hens are molted twice instead of once, the laying life of the flock is increased by several months because there are three cycles of egg production rather than two. However, the total production period varies and the weeks in each cycle are at the discretion of the manager. Once again, egg quality and quantity dictate the length of each cycle. Economics dictates whether or not recycling should be done at all. Typical programs are as follows:

2-Cycle	3-Cycle
20 weeks rearing	20 weeks rearing
40 to 50 weeks egg production	32 to 36 weeks egg production
8 weeks molt	8 weeks molt
24 to 32 weeks egg production	28 to 32 weeks egg production
	8 weeks molt
92 to 110 weeks total	24 to 28 weeks egg production
	120 to 132 weeks total

Egg production and egg quality.   Estimated egg production and egg size during 3-cycle molting programs are given in Table 19-6. The earlier birds are molted, the higher the subsequent rate of lay and egg quality.

Economic Evaluation

Extended periods of lay beyond a normal two-cycle program can be analyzed in the same manner as previously discussed. Without regrouping, though, and with even poorer rates of lay, these extended programs are rarely cost efficient.

At the normal age for a second molt, the flock is usually reduced to approximately 85% of its original size, which along with a rate of lay that rarely exceeds 80% (Table 19-5), usually makes these programs uneconomical even though average egg quality can be superior to either the one- or two-cycle programs. Three-cycle production may prove to be economical under circumstances where larger eggs are priced at a substantial premium.

19-H. OTHER METHODS OF PRODUCING THE MOLT

Many feed ingredients and chemical compounds have been used to precipitate the molt. Although some have merit, most have not given as good results as those previously discussed.

Molting by Feeding Zinc

The use of high levels of zinc in the feed has been used to produce a molting pause in layers, and they remain out of egg production only a short time. About 20,000 ppm of zinc is added to the feed. Zinc oxide is superior to zinc sulphate or zinc carbonate. Mix 55 lb (25 kg) of zinc oxide containing 73% zinc to a ton (2,000 lb) of feed that contains 3.5% calcium. Feed the zinc feed mixture for 5 days; then return the birds to a regular laying ration containing 50 ppm of zinc. Reduce the period of lighting during the period the high-zinc diet is fed; then return the birds to a regular lighting schedule.

Hens on the high-zinc program will eat less than 20% of their normal amount of feed. They will lose from 0.75 lb (340 g) to 1.0 lb (454 g) of weight. Egg production should stop by the fifth day after zinc feeding is started. Birds should come back into production about 7 days after the high-zinc diet is removed, and peak at 76 to 80%.

Caution: Some zinc sources may be contaminated with lead, which will be injurious to the flock.

Low Sodium Diet for Molting

This program has had some acceptance. Reduce the sodium content of the ration to 0.04% and turn off the lights in open-sided houses, or provide an 8-hour light day in environmentally controlled houses. After birds are molted (about 6 weeks) return them to the regular laying ration with added sodium and increase the length of the light day.

Low-sodium diets are essentially all-grain diets and may be supplemented with high-fiber feedstuffs. It is best to reduce the calcium and phosphorus levels to those in a typical pullet growing diet to inhibit the flock from returning to production too early. The low-nutrient diet is used extensively in Europe where animal welfare codes prohibit lengthy fasting programs.

Drugs to Produce a Molt

Drugs, such as methalibure, enheptin, progesterone, chlormadinone, iodine, and others have been shown to be effective in producing a molt and have been used experimentally, but most are illegal as feed ingredients. Check with authorities before using.

19-I. RECYCLING THE BREEDER FLOCK

Breeder flocks used for the production of hatching eggs are often molted and recycled. However, the profitability of this program is materially different compared with the molting of table egg flocks because of

1. The added cost of providing for the males
2. The increased value of the hatching egg compared with the commercial egg
3. The higher value of the breeder pullet

Reasons for Recycling the Breeder Flock

The reason for recycling a commercial flock is to make an additional profit over a short period of egg production. However, the reasons for a second period of egg production in the case of breeders are quite different. They are

1. *To supplement normal hatching egg production.* In many instances, particularly in the case of egg-type breeders, there are seasonal demands for day-old chicks that call for an increase in the number of hatching eggs. Often it is more practical to supplement normal, first-year egg production with eggs produced by flocks in their second cycle.
2. *Substitute for high growing mortality.* Most hatcheries have a flock replacement schedule based on no excess of hatching egg production. When a first-year flock has high mortality, there may be a hatching egg deficiency. In these cases, molting an older flock may help to make

up the difference in hatching egg numbers due to the high death loss in another flock.

3. *Unforeseen demand for chicks.* The demand for chicks is difficult to project. Many times it is greater than can be supplied by the available hatching eggs. Recycling certain flocks offers a means of supplementing the egg supply.

Recycling Best with Young Males

Young males produce much better fertility than old males molted, particularly with meat-type lines. Only when the decision to molt the breeding flock comes at a time late in the egg production period should old males be used with the recycled hens.

When young males should be brooded. It will be necessary to determine in advance when the molted hens will return to egg production. Six months prior to this date the new group of matching males should be started in the brooder house. This will allow time for them to grow and reach sexual maturity when the recycled hens begin producing hatching eggs.

Molting breeder males. If old males are to be used with the molted hens, the males should be molted at the same time. This will prevent many males from molting during the second breeding cycle, thus reducing their breeding potential. Use the same molting program for the males as for the females.

Molting Programs for Leghorn or Medium-Size Breeder Hens

Any two-cycle program of molting that produces good results with a commercial flock of hens will be satisfactory for egg-type breeder hens. However, it has been found that it takes from 1 to 2 weeks longer to return a breeder hen to egg production than with a commercial hen. Fertility, hatchability, and chick quality will be better if a slow-molt program is used.

Molting Programs for Broiler Breeder Hens

It will require about 10 to 14 days of fasting (no feed) to get a 25% weight reduction when molting broiler breeder females. Mortality should not be a problem during this period of feed removal but should not be allowed to exceed 3%. Unless there is a 25% weight reduction

Table 19-8.   Production Estimates for Egg-Type Breeders in Second Cycle[1]

| Week of Egg Production | Second Period of Egg Production | | | |
| | Hen-housed Egg Production % | Cumulative Hatching Eggs per Hen Housed | Total Hatch[2] | |
			Young Males %	Old Males %
1	5	0.3	78	73
2	11	1.1	81	76
3	19	2.4	84	79
4	32	4.5	85	80
5	48	7.8	85	80
6	65	12	85	79
7	78	18	84	79
8	86	24	84	78
9	88	30	84	78
10	87	36	83	77
11	86	42	83	77
12	85	48	83	76
13	84	53	82	76
14	83	59	82	75
15	81	64	82	75
16	80	69	81	74
17	79	75	81	74
18	78	80	81	73
19	76	85	80	73
20	75	90	80	72
21	74	95	79	71
22	73	100	79	71
23	72	105	78	70
24	71	109	78	69
25	69	114	77	69
26	68	118	77	68
27	67	123	76	67
28	65	127	76	67
29	64	131	75	66
30	63	135	75	65
31	61	139	74	65
32	60	143	74	64
33	58	146	73	64
34	57	150	73	63
35	55	154	72	63

[1]Leghorn and medium-size breeders.
[2]Chicks hatch 3 weeks later.

Table 19-9. Production Estimates for Meat-Type Breeders in Second Cycle

| | Second Period of Egg Production | | | |
| | | | Total Hatch[1] | |
Week of Egg Production	Hen-housed Egg Production %	Cumulative Hatching Eggs per Hen Housed	Young Males %	Old Males %
1	5	0.3	78	73
2	18	1.5	79	74
3	35	4	80	75
4	51	7	81	76
5	66	12	81	76
6	74	19	81	75
7	75	22	80	75
8	74	26	80	74
9	73	31	80	74
10	72	36	79	73
11	71	41	79	73
12	70	45	79	72
13	69	50	78	72
14	68	54	78	71
15	67	59	78	71
16	66	63	77	70
17	65	67	77	70
18	64	71	77	69
19	63	75	76	69
20	62	79	76	68
21	61	83	75	67
22	60	87	75	67
23	59	91	74	66
24	58	94	74	65
25	57	98	73	65
26	56	101	73	64
27	54	105	72	63
28	53	108	72	63
29	52	111	71	62
30	51	114	71	61
31	49	117	70	61
32	48	120	70	60
33	47	123	69	60
34	46	126	69	59
35	44	128	68	59

[1]Chicks hatch 3 weeks later.

there will not be adequate fat reduction in the reproductive system, which is especially important with meat-type females.

Feather renewal following the molt is especially important with meat-type hens. They will not produce eggs until feather growth is completed. A ration containing 1,250 kcal ME per pound, 16% protein, and 0.9% sulfur amino acids should be fed after the fast to foster better feather growth. Restricted feeding should be practiced to prevent the post-molted birds from getting too heavy. Requirements for new or old meat-type males are the same as for Leghorns.

Estimates of Egg and Chick Production

Estimates of egg production, hatching egg production, and percentage of chicks hatched during the second cycle of egg production are given in Tables 19-8 (egg-type) and 19-9 (broiler breeder parents). It must be remembered that egg size is an important factor with breeding flocks, and post-molted birds have a decided advantage in this respect.

20

Broiler, Roaster, and Capon Management

Broiler production is a business in which volume is necessary to offset the small unit of profit. With margins so small, the producer, whether an individual operator or an integrator, must be aware of the many factors that affect the cost of production. Although each factor exerts only a minor influence, the combined effect of all factors becomes phenomenal.

Through the years, the meaning of these three terms has changed. Originally, *fryer* meant a young frying chicken, but the term has been replaced by *broiler*. Broilers are usually marketed at a live weight between 4 and 4.5 pounds, usually between 6 and 8 weeks of age. In some countries, particularly those where birds are deboned for the consuming trade, live straight-run weights between 5 and 7 pounds are preferred. Roasters average about 6.5 pounds live weight regardless of sex. Because of the similarity between many items in the broiler and roaster growing programs, several recommendations are incorporated together in the following pages.

20-A. BROILER GROWING PROGRAMS

All-In, All-Out System

The most practical program for broiler rearing has been the use of the *all-in, all-out* system, in which only one age of broilers is on the farm at the same time. All the chicks are started on the same day, and later sold on the same day, after which there is a period when no birds are on the premises. This lack of birds breaks any cycle of an infectious disease; the next group of birds has a "clean start," with no possibility of contracting a disease from older flocks on the farm.

Multiple Brooding

Although in the past it has been more profitable to keep one age of broilers on the farm, recent advances in isolation and disease control have made it possible to keep chicks of several ages on the same farm, and there are now many good operations following this procedure. But the system calls for expert management; the program is not for the novice.

Isolation

The broiler growing farm should be isolated. The buildings are best enclosed with a tight fence, and with locks on all entrance gates. Beware of feed and other supply trucks entering the enclosure.

Size of Operation

Broiler growing units have become larger as automation has made its way into the industry. Certainly one person can easily care for 40 or 50 thousand birds with little difficulty. As a contractor, he usually works every day until the birds are sold, then has his days off when the premises are depopulated. In some instances, one caretaker has handled over 100,000 birds with some extra help during the first week when the labor requirement is high.

Broods per Year

The length of the growing period and that of the downtime (time between broods) vary, and in turn these variations affect the number of broods that can be started each year. Normal downtimes range from 7 to 14 days. A short downtime plus a short growing period increases the number of broilers that can be produced in a house during the course of a year. The effects of these two factors on the number of broods per year are shown in Table 20-1.

20-B. BROILER HOUSING

The practices followed in the growing of broilers are very similar to those used in the growing of commercial egg strains. Since housing, equipment, and management have been discussed in earlier chapters, the material included here will be confined to brief outlines of certain procedures, except where some item is especially pertinent to broiler raising.

Table 20-1. Length of Growing Period, Downtime, and Number of Broods per Year

Length of Growing Period da	Length of Downtime in Days							
	7	8	9	10	11	12	13	14
	Number of Broods per Year							
41	7.6	7.5	7.3	7.2	7.0	6.9	6.8	6.6
42	7.5	7.3	7.2	7.0	6.9	6.8	6.6	6.5
43	7.3	7.2	7.0	6.9	6.8	6.6	6.5	6.4
44	7.2	7.0	6.9	6.8	6.6	6.5	6.4	6.3
45	7.0	6.9	6.8	6.6	6.5	6.4	6.3	6.2
46	6.9	6.8	6.7	6.5	6.4	6.2	6.2	6.1
47	6.8	6.6	6.5	6.4	6.3	6.2	6.1	6.0
48	6.6	6.5	6.4	6.3	6.2	6.1	6.0	5.9
49	6.5	6.4	6.3	6.2	6.1	6.0	5.9	5.8
50	6.4	6.3	6.2	6.1	6.0	5.9	5.8	5.7
51	6.3	6.2	6.1	6.0	5.9	5.8	5.7	5.6
52	6.2	6.1	6.0	5.9	5.8	5.7	5.6	5.5
53	6.1	6.0	5.9	5.8	5.7	5.6	5.5	5.5
54	6.0	5.9	5.8	5.7	5.6	5.5	5.5	5.4
55	5.9	5.8	5.7	5.6	5.5	5.5	5.4	5.3
56	5.8	5.7	5.6	5.5	5.5	5.4	5.3	5.2
57	5.7	5.6	5.5	5.5	5.4	5.3	5.2	5.1
58	5.6	5.5	5.5	5.4	5.3	5.2	5.1	5.1
59	5.5	5.5	5.4	5.3	5.2	5.1	5.1	5.0
60	5.5	5.4	5.3	5.2	5.1	5.1	5.0	4.9

Types of Houses for Broilers and Roasters

Broiler houses may be considered as brooder houses. The only variation is that the birds are kept in them for 2 to 4 weeks longer than in the case of chick-type brooder houses. Such houses are described fully in Chapter 11, and it should be reviewed. In the main, there are two general types of housing suitable for broiler production.

1. *Open-sided house.* These houses are open to the extent that at least one-half of the front and back of the house is open. When there are high summer temperatures 75 to 80% should be open. Usually there are curtains that can be rolled up and down over the openings; consequently, these houses are not lightproof.
2. *Environmentally controlled house.* Such houses are lightproof. Artificial light is used in the building and ventilation is by exhaust fans, and both are closely controlled according to the requirement of the birds. Some method of cooling the house is often incorporated.

Floor Space per Broiler or Roaster

Growth and feed conversion are inversely proportional to floor space per bird. The more you crowd broilers and roasters, the poorer the results. However, as floor space is reduced per bird, the greater the weight of broilers produced in the house, and this will, up to a certain point, increase the return on investment. There must be a happy medium. Most broilers are raised under crowded conditions in order to increase the investment return. Producers have become accustomed to sacrificing some growth and feed conversion from crowding in order to market more pounds or kilos of birds. The relationship is shown in Table 20-2. Crowding affects feathering and feed conversion, as well as growth. These relationships are detailed in Table 20-3.

Economics important. An analysis of any floorspace experiment cannot be made solely on the basis of flock behavior. Indeed, limiting the floor space gives poorer results on a bird basis, yet the question has always been and continues to be: What is the least amount of floor space necessary per bird to produce the greatest return on investment? Another fact is that larger birds require more floor space than smaller ones. Thus, the size of the broiler or roaster to be marketed has a bearing on the computation.

Table 20-2. Effect of Floor Space on Weight and Mortality of Straight-run Broilers (40 Days of Age)

Floor Space per Bird		Average 40-day Live Weight			Mortality %	Weight of Birds Raised per Unit of Floor Space		
ft²	m²	lb	kg	% of Base		lb	kg	% of Base
1.0	0.093	4.35	1.88	102	2.0	4.15	1.88	81
0.9	0.084	4.12	1.87	101	2.1	4.58	2.08	90
0.8	0.074	4.09	1.86	100	2.3	5.11	2.32	100
0.7	0.065	4.05	1.83	99	2.6	5.58	2.53	109
0.6	0.056	4.00	1.81	98	3.0	6.67	3.03	131
0.5	0.046	3.94	1.79	96	3.6	7.88	3.57	154
0.4	0.037	3.86	1.75	94	4.5	9.65	4.37	189

Table 20-3. Effect of Floor Space on Feathering and Feed Conversion of Broilers

Floor Space per Broiler		Poorly Feathered %	Feed Conversion	
ft²	m²		Conversion	% of Base
1.0	0.09	0.2	1.73	98
0.9	0.08	0.4	1.74	99
0.8	0.07	1.0	1.75	100
0.7	0.06	2.2	1.79	102
0.6	0.05	4.8	1.84	105
0.5	0.04	8.0	1.91	109
0.4	0.03	14.1	1.98	113

Reducing the floor space will

1. Decrease feed consumption.
2. Decrease the growth rate.
3. Decrease feed efficiency.
4. Increase mortality.
5. Increase cannibalism.
6. Increase the incidence of breast blisters.
7. Increase the percentage of birds with poor feathering.
8. Increase the condemnations at the processing plant.
9. Increase the house ventilation requirement.
10. Increase the pounds of broilers raised in a given house during a 12-month period.

Floor space and broiler or roaster size. In the face of lower mature bird weight, decreased feed efficiency, and increased mortality, approximately the amount of floor space given in Table 20-4 should be allocated to each bird for greatest cash income per house. Decrease the floor space per bird by about 10% during the winter months. In other words, broilers and roasters need more floor space in hot weather than in cold.

Table 20-4. Floor Space Requirement for Broilers and Roasters

Weight of Mature Bird		Floor Space Requirement			Meat Produced	
lb	kg	ft² per Bird	m² per Bird	Birds per m²	lb per ft²	kg per m²
3	1.36	0.5	0.05	20.0	6.0	28.0
4	1.82	0.7	0.06	16.7	6.0	30.3
5	2.27	0.8	0.08	12.5	6.0	28.4
6	2.72	1.0	0.09	11.1	6.0	30.2
7	3.18	1.2	0.11	9.1	6.0	29.0

Size of the Broiler House

There is no formula for determining the size of a broiler house, the dimensions being a combination of many factors. Where commercial production is practiced, few houses hold less than 10,000 broilers and some hold up to 40,000.

Depth and length of the house. The open-sided conventional house should be 30 ft deep (front to back) inasmuch as one loop of an automatic trough feeder will best fit this depth. An open house will not ventilate well when any deeper. It may be any length; a linear foot of house will hold about 40 broilers to maturity.

Environmentally controlled house. In many areas, where outdoor temperatures are highly variable, a large percentage of the broiler houses are environmentally controlled. These are totally tight and dark. All light and air are artificially supplied. Most of these houses are 40 ft deep and 400 ft long, with 16,000 ft² of floor space, holding about 20,000 birds at 0.8 ft² per bird. But there are houses being built that are about 580 ft (176.8 m) long and 56 ft (17.1 m) deep, holding 38,000 to 40,000 birds to maturity, depending on their mature weight.

Floors Versus No Floors

In many locations concrete floors are a requisite to good house construction. However, where the soil is dry, porous, and sandy, many broiler and roaster houses are built so that the soil serves as the floor.

Limited-Area Brooding

During cold weather brood in one area of the broiler house and enclose the area with plastic curtains. The middle is best, leaving the ends of the house unused for 3 weeks. It has been shown that the fuel to operate the brooders may be reduced by one-third when limited-area brooding is practiced. Allow 0.35 ft² (325 cm²) of floor space per chick up to 3 weeks of age. Provide more floor space per chick at 3 weeks of age or growth and feed conversion will suffer.

20-C. ENVIRONMENTALLY CONTROLLED BROILER HOUSE

Chapter 11-A, 11-B, and 11-C should be read first as the information will not be repeated here. Broilers are homeotherms. They have the ability to maintain a rather constant deep-body temperature when the ambient temperature is between 55°F (12.8°C) and 75°F (23.9°C), which is their thermoneutral zone. As ambient temperatures rise or fall outside the thermoneutral zone, birds get to the point where their defenses can no longer maintain an optimal deep-body temperature of 105°F (40.6°C). Thus, changes must be made in housing to keep the birds more comfortable.

Measuring Heat Production

Heat produced by broilers is measured in British thermal units (Btu). Table 20-5 shows the figures for broilers and roasters of different weights.

Notice that the heavier a bird gets, the greater the Btu production, but on a unit of weight bases, the lower it becomes.

Table 20-5. Heat Output by Broilers and Roasters at 70°F (21.1°C)

Live Weight		Bird Heat Output per Hour		
		Per lb Weight	Per kg Weight	Per Bird
lb	kg	Btu	Btu	Btu
1	0.45	26.8	59.0	27
2	0.91	18.5	40.7	37
3	1.36	15.3	33.7	46
4	1.81	13.4	29.5	54
5	2.27	12.1	26.6	61
6	2.72	11.0	24.2	66

Total Heat Produced in the Broiler House

Heat in the broiler house has the following two sources:

1. *Solar rays.* The hot roof and sides of the house transmit heat to the interior. The amount is about 3.25 Btu per hour for each broiler in the house.
2. *Broilers.* Table 20-5 shows a 4-lb broiler produces about 54 Btu per hour at 70°F (21.1°C).

Total heat produced in the house. The total Btu (3.25 + 54) would be 57.25 Btu per 4-lb bird per hour. In a 400- × 40-ft house holding 21,000 4-lb broilers the heat produced would be 1,202,250 Btu per hour, all of which must be removed from the building or the deep-body temperature of the birds will rise.

Negative Pressure Necessary

Air is moved through the broiler house by the use of exhaust fans. Slightly more air is exhausted than enters the building, thereby creating a negative pressure that should be about 0.04 in. static pressure.

With negative pressure, incoming air is brought in through an adjustable slot at the front of the building just below the eaves and runs the entire length of the house. If the static pressure in the building is constant, air drawn into the building remains the same speed regardless of the size of the intake, but increasing the size of the slot opening will allow more air to enter the building.

Cooling the Birds

Evaporative cooling.   There is no way the interior of the building may be cooled by bringing in hot outside air. The temperature of the incoming air must be lowered by evaporative cooling (see Chapter 11-T).

Air is drawn through a moist pad that reduces the temperature about 12°F (6.7°C) when the outside air has a temperature of 90°F (32.2°C) and a relative humidity of 56% (see Table 5-6 and Figure 20-1).

Wind-chill cooling.   Chickens, like humans, feel cooler when air moves over their bodies regardless of the temperature. In the environmentally controlled broiler house the air should enter at one end at a speed of 400 ft per minute (4.6 mph), flow across the birds to the other end, and reach the end at the same air speed. As air flows over the birds they lose heat and the air temperature increases.

In a building 400 ft long, this increase in air temperature should be about 8°F (4.4°C). If it is less, the system is not removing all the excess heat produced by the birds. If it is more, air speed is too slow.

Figure 20-1. Evaporator pad cooling system.
(Courtesy of Cyclone International, Inc., Holland, Michigan, USA.)

Two Ventilating Systems Necessary

The ventilating systems during summer and winter are so different that it is impossible for one system to operate efficiently during both seasons. There must be two systems: one for winter and one for summer.

Winter System

During winter the outside air is cool or cold and the humidity is high. The cool air is brought into the house and warmed by the birds. As the temperature of the air in the house rises, the relative humidity drops, enabling the air to pick up moisture from the litter, thereby keeping it drier.

The speed of the air in the house must be kept at about 40 ft per minute to prevent any wind chill, which would cause serious problems in the birds during the winter months. The only way this speed can be accomplished is to move the air through the intake slot at the front of the house to the back of the house, a distance of 40 ft, by installing some small variable-speed exhaust fans in the back side of the house. If the air is being replenished every minute, it is moving at the rate of 40 ft per minute.

The ventilation requirements are as follows:

Age of Broilers wk	Ft³ per Minute per Bird			
	Avg House Temperature			
	50°F	60°F	70°F	80°F
0–2	0.3	0.4	0.5	0.6
3–4	1.2	1.3	1.5	1.7
5 to market	2.3	2.8	3.2	3.5

Summer System

For this system to operate, both wind chill and evaporative cooling must be used. All air is brought through the pad of the evaporative cooler located at one end and on the front of the house, then drawn 400 ft by the exhaust fans at the other end of the house. The building is ten times as long as it is deep, so to replace all the air in the building in 1 minute the air speed will have to be increased ten times or to about 400 ft per minute.

Air leaves the evaporative cooler at about 12°F (6.7°C) below the temperature of the outside air; then when the speeding air moves across the birds it gives them a wind-chill feeling of an additional 10°F reduction in temperature. Furthermore, the speed of air at this temperature is instrumental in rapidly removing the heat produced by the broilers as fast as they produce it.

Computerizing Ventilation Components

Systems are now available that can be set to control all the changes necessary for proper house ventilation: temperature, air speed, cooling, heating, moisture, etc. Connected to a central computer, the system is fully automatic and takes the guesswork out of making economic changes in ventilation to suit the conditions necessary for greatest broiler profit.

Figure 20-2 shows four 400 × 40 ft totally enclosed broiler houses that are environmentally controlled. Each building houses about 20,000 broilers to maturity near Lavonia, Georgia (USA).

Figure 20-3 shows one end of four 400 × 40 ft totally enclosed broiler houses using special discharge cones on the exhaust fans that operate the evaporative cooling system. These cones increase the air movement by 17%. The four 20,000-bird broiler houses are near Rison, Arkansas (USA).

Figure 20-2. Four totally enclosed broiler houses.
(Courtesy of Aerotech, Inc., Lansing, Michigan, USA.)

Figure 20-3. Air discharge cones on broiler house exhaust fans. *(Courtesy of Aerotech, Inc., Lansing, Michigan, USA.)*

20-D. BROILER GROWING EQUIPMENT

Practically all the equipment used in a commercial broiler growing house is similar to that necessary for keeping meat-type breeder chicks to a similar age (see Chapter 12-A). However, some of the equipment capacities are different.

Brooder capacity. The following shows the capacity of various brooders during moderate weather. Reduce the numbers by 20% in cold weather.

Type of Brooder	Chick Brooder Capacity
30,000 Btu gas brooder	500–650
40,000 Btu gas brooder	600–700
1,000 capacity oil brooder	600–700
1,000 capacity hover-type, coal brooder	600–750
1,000 capacity radiant brooder	600–750
Slab heating	See Chapter 13-D
Central heating system	See Chapter 13-D

Feeder space. Allow 2 in. (5 cm) of trough space through 5 weeks, and 3 in. (7.6 cm) until market time at 7 weeks. When circular pans are used, allow about 20% less feeder space per bird. With 15-in. (38-cm) diameter pans provide one pan for every 33 broilers.

Feeders. An innovation in automatic floor feeders for broilers has been the tube-and-pan feeder. It provides a more uniform distribution of similar feed throughout the house. There is no preferential feeding as there is with a trough-and-chain type feeder. In order to remove some of this difficulty with the trough-and-chain feeder some models have special high-speed motors that move the feed at the rate of 60 ft (18.3 m) per minute.

Waterer space. Provide two chick founts for each 100 chicks at the start of the brooding period. Later, each broiler should have 0.75 in. (2 cm) of waterer space when troughs are used. When circular pans are employed, supply about 20% less space per bird (see Table 20-8).

Water meters. Most modern broiler installations now have meters on the water line going to the pens. Daily water consumption is a good indication of the health of the birds.

Lights. Use continuous bright light at first, then provide 0.35 fc (3.8 lx) of illumination at bird level to supplement natural daylight or in environmentally controlled houses (see Chapter 20-G).

Bulk feed bins. These should be large enough to hold 2,000 lb (900 kg) of feed per 1,000 broilers.

20-E. SELECTING THE CHICKS

Many of the factors associated with economical broiler and roaster growth are inherited. Others are associated with the hatchery in which the chickens are hatched. Decisions that must be made by the grower in selecting the best chick are listed below. In many instances, the integrator is responsible for the choice of chicks; the producer must take the type and quality offered to him as a part of his contract for growing the birds.

1. Which strain of broilers will be most profitable?
2. What vaccination program have the breeders undergone?
3. What is the breeder or hatchery disease-control program?
4. What quality of chicks will be delivered?
5. What is the chick size?
6. Is there need for vaccinating at the hatchery?
7. Are the chicks to be sexed?
8. Will the chicks need to have their beaks trimmed?

Larger Hatching Eggs Produce Larger Broilers

When the average weight of all hatching eggs set is 24 oz/doz (56.7 g each), mature broilers will be heavier by 0.04 lb (18 g) per bird for each 1 oz/doz (2.4 g each) increase in egg size.

Example: If the average weight of mature broilers hatched from eggs averaging 24 oz/doz is 4 lb, 20 oz/doz-eggs would produce birds weighing 3.84 lb; 28 oz/doz-eggs would produce 4.16-lb birds.

Example: If the average weight of mature broilers hatched from eggs averaging 56.7 g each is 1.87 kg, 47.2-g eggs would produce birds weighing 1.74 kg; 66.1-g eggs would produce 1.89-kg birds, and so forth.

Sexing Method Affects Feather Color and Feathering

If it is desired that day-old chicks be segregated by sex, the method of sexing has an effect on the behavior of the growing birds. Two methods of sexing are commonly used to visually segregate male and female broiler chicks in the hatchery. Each calls for an examination based on down color or rate of feathering as a result of special parental mating (see Chapter 22-K and 22-L).

1. *Feather sexing.* Day-old chicks that are rapid and slow feathering may be sex separated by examining the relative length of the primary and covert feathers of the wing. Sexing accuracy is good, with the females carrying genes for fast feathering and the males carrying genes for slow feathering. However, the slow-feathering males often do not feather well in the brooder house, particularly during hot weather, often impairing growth and leading to more cannibalism.

2. *Color sexing.* Usually gold and silver genes are used for color differentiation at day-old. Gold, buff, or red chicks are females; white or yellow chicks are males. They can be sexed rapidly and accurately. However, the color differences are also evident in the mature broiler. Females are gold or brown; males are white. But the pinfeathers are usually white so there should be no processing problem.

20-F. MANAGING BROILERS ON LITTER

Broiler management might be called *mini-management;* it involves a lot of little things, many of which in themselves seem insignificant yet add up to economical production. The manager or caretaker who can care for all the details will achieve success, but the person who is careless, who forgets some ''meaningless'' chore, who does not know the intricacies of the business, or who fails to recognize problems in their infancy, will almost inevitably fail. So many times in the broiler production business the manager makes the difference. Today many flock managers are not as closely associated with their business as they were before integration gained a foothold in the industry, and when they had a greater financial interest in their business.

Preparing for the Chicks

Many of the items regarding broiler chick brooding are the same as, or similar to, egg-type chick care. Read Chapter 13 first.

A certain amount of downtime between broods of broilers is necessary to clean the house and premises and to break a disease cycle (see Table 20-1). Preparation for the new group of broiler chicks during the downtime will encompass the following:

Clean the house and equipment. Do as thorough a job of cleaning as possible. Use a disinfectant when needed.

New litter. In most instances it is best to remove the old litter and supply new. There are many types of litter material, and the most likely to be used will be the most economical. But litters do differ, as Table 20-6 shows.

Qualities of a good litter. A good litter should

1. Be light in weight
2. Have a medium particle size
3. Be highly absorbent
4. Dry rapidly
5. Be soft and compressible
6. Show low thermal conductivity

Table 20-6. Factors Associated with Litter Materials

Litter Material	Grams of Moisture 100 g of Litter Will Hold	Mature Broiler Weight on New Litter	
		lb	kg
Pine straw	207	3.35	1.51
Peanut Hulls	203	—	—
Pine shavings	190	3.38	1.53
Chopped pine straw	186	—	—
Rice hulls	171	3.47	1.57
Pine stump chips	165	3.58	1.62
Pine bark and chips	160	3.40	1.54
Pine bark	149	3.39	1.54
Corncobs	123	3.57	1.62
Pine sawdust	102	3.57	1.62
Clay	69	3.24	1.47

Source: Univ. of Georgia Res. Bull. 75, Poultry Dept., Athens, Ga.

7. Absorb a minimum of atmospheric moisture
8. Be inexpensive
9. Be compatible when sold as fertilizer

Used litter.   In some cases it may be practical to reuse litter, but the procedure is often plagued with problems.

1. Do not reuse litter if the last brood of broilers was diseased.
2. Always depopulate the building when reusing litter; allow for an extra long period of depopulation.
3. Disinfect the building and equipment thoroughly.

Prevent ammonia fumes.   Continuous exposure to 50 ppm of ammonia will reduce 7-week broiler weight by 8%, but feed conversion and livability will not be affected. Phosphoric acid is effective in controlling ammonia fumes the first 10 days of brooding on reused litter, and is superior to superphosphate.

Maintain a low litter pH.   Very little ammonia will be released when the litter is kept below a pH of 7, but is rapid at a pH of 8 or above.

Recommendations for litter treatment

Phosphoric acid: 2 U.S. qt (1.9 L) per 10.5 ft^2 (1 m^2).
Superphosphate: 2.2 lb (1.0 kg) per 10.5 ft^2 (1 m^2).

Clean the bulk feed bins.   Remove unused feed from the bins; then wash and disinfect them.

Have the correct house and brooder temperature.   Both are very important to a proper start of the brooding operation.

Attraction lights.   Where hover-type brooders are used, install attraction lights to attract the chicks to the area of brooder heat during the first few days.

Brooder guards.   With most brooders, guards are necessary to confine the chicks to the heated area.

Warm-room brooding.   Where the entire house is heated instead of using conventional brooding systems, be sure that the building is heated to the correct temperature.

Feeders.   Chick box lids or feeder trays provide the most satisfactory feeder for the first few days.

Chicks should arrive early in the day.   So that the caretaker may have the entire day to closely observe the chicks, they should be delivered early in the morning.

Room Temperature and Brooder Heat Requirement

Inasmuch as brooder heat or radiation is used to supplement the room temperature when brooding chicks, increasing the room temperature will obviously reduce the need for brooder heat. The relationship is shown in Table 20-7, where a base of 100 is given for the heat of energy needed by the brooder stove in a room heated to 70°F (21.1°C) at chick level, by weeks, to give a comfortable environment, and variations are shown for other house temperatures between 40°F (4.4°C) and 80°F (26.7°C).

Example: If room temperature is 40°F it will require only 2.2 times as much heat or energy (fuel) the first week than if the room temperature is 70°F (21.1°C), but seven times as much the fifth week even though the requirement is less.

Table 20-7.   Brooder Heat Requirement Relationship According to House Temperature

Week of Age	House Temperature at Bird Level[1]				
	40°F 4.4°C	50°F 10.0°C	60°F 15.6°C	70°F 21.1°C	80°F 26.7°C
1	220	180	140	100 (base)	60
2	200	160	120	80	40
3	180	140	100	60	20
4	160	120	80	40	0
5	140	100	60	20	0

[1]Avg daytime house temperature.

Brooding Temperature

The brooding temperature at 2 in. (5.1 cm) above the surface of the litter and under the hover should be as follows for broiler chicks:

Week	°F	°C
1	86	30.0
2	86	30.0
3	81	27.2
4	75	23.9

When the Chicks Arrive

Early care is essential; there is nothing like a good start.

Water.   A few hours before delivery of the chicks, fill the founts with water so that the chill will be removed by the time the chicks arrive.

Sugar water.   The addition of 8% sucrose to the drinking water consumed the first 15 hours has been shown to improve broiler growth (see Chapter 13-E).

Use water jugs.   Gallon jugs seem to provide first water best. Provide two for each 100 chicks. Automatic chick waterers may be used.

Water temperature. The water temperature should be about 75°F (24°C) the first few days, then lowered inasmuch as cool water increases feed consumption and helps to reduce mortality.

Dumping the chicks. Invert the chick boxes and dump the chicks near the brooder heat and water founts.

Three hours later give feed. After the chicks have had water for about 3 hours, place chick mash or crumbles on the feeder lids. Keep the surface of the lid covered with feed (see Chapter 13-F).

Should Broilers Have Their Beaks Trimmed?

Beak trimming is used to prevent cannibalism. Trimming at any age creates a stress, reducing feed consumption for several days or longer. It may take 4 weeks for the birds to assume their normal growth pattern. In lighttight houses reduced light intensity should prevent all cannibalism, but in open-sided houses beak trimming may be necessary during the warmer seasons.

Although trimming when the chicks are 6 days old will produce excellent results, there is the added labor involved with catching the birds at this age. A more common procedure when broilers must have their beaks trimmed is to trim them at the hatchery. Use either notch trimming or an electric spark (see Chapter 13-I).

Prevent Feed Wastage

From 0.25 to 0.50 lb (114 to 227 g) of feed per bird can be wasted during 7 weeks if precautions are not taken. This represents 3 to 6% of the total feed. Keep the feed level low in all feeders. If the feeders are two-thirds full, 10% of the feed will be wasted; if one-half full, 3%; when less than one-third full, the waste will be about 1%.

20-G. WATER FOR BROILERS AND ROASTERS

Chicks tend to drink from one waterer only. If it is removed or runs dry, it will be some time before they choose another water source or location.

Waterer Space

Waterers for broilers, roasters, and capons may be trough, dome-type, cup, or nipple. Space requirements are shown in Table 20-8.

Table 20-8. Waterer Space for Broilers and Roasters

Type and Sex	Automatic Trough or Regular Trough per Bird		Per 1,000 Birds			
	in.	cm.	8-ft (2.4-m) Trough	Dome Type	Cups	Drip-Type Nipples
Broilers, Roasters						
0–8 wk	0.8	2.0	4	16	94	94
Roasters						
8 wk to maturity	1.1	2.8	6	22	138	138

Temperature and Age Effects on Feed and Water Consumption of Broilers

The body of a broiler contains up to 70% water, and a bird must have ample water to keep this percentage constant, but an even greater requirement for water is to maintain a constant deep-body temperature during hot weather. Inasmuch as water absorbs heat rapidly its movement through the body is essential to body heat removal. The amount of water necessary to remove this heat increases rapidly as house temperatures rise above the zone of thermal comfort of 65° to 70°F (18.3 to 21.1°C). But below this figure ambient temperatures have but minor effects on water consumption. Accompanying the increases in water intake at high temperatures is a decrease in feed consumption.

Feed and water consumption of broilers at varying house temperatures. Table 20-9 shows feed and water consumption of broilers at varying house temperatures, by weeks. Also shown are feed conversions and live body weights, by weeks.

20-H. LIGHT MANAGEMENT FOR BROILERS

The amount of light for a growing broiler is only that amount necessary to enable the bird to move about and to see to eat and drink. Activity is to be reduced to a minimum. The intensity of illumination at bird level should be about 0.35 to 0.50 fc. In the environmentally controlled house, this requirement is easily provided, but with houses having open sides or windows more than this intensity of light is supplied by sunshine. As illumination above the optimum induces cannibalism, activity, and piling, the open house is a decided disadvantage unless some method can be used to restrict some of the natural daylight. But often this is not possible, which causes a problem.

Table 20-9. House Temperature as It Affects Growth, Feed, and Water Consumption of Straight-Run Broilers by Weeks

Age Wk	\multicolumn{8}{c}{Average Daytime House Temperature}
	50°F	70°F	90°F	100°F	10.0°C	21.1°C	32.2°C	37.8°C
	\multicolumn{4}{c}{Pounds Feed/100 Broilers/Day}	\multicolumn{4}{c}{Kilos Feed/100 Broilers/Day}
1	3.7	3.7	3.6	3.5	1.68	1.68	1.64	1.59
2	9.2	9.1	8.8	8.7	4.54	4.14	4.00	3.96
3	14.7	14.3	13.4	19.0	6.68	6.50	6.09	7.64
4	20.7	19.9	18.4	16.7	9.41	9.05	8.36	8.64
5	26.6	25.3	22.4	20.9	12.09	11.50	10.18	9.50
6	33.0	31.4	27.4	24.7	15.00	14.37	12.46	11.23
7	40.1	37.6	32.1	28.4	18.20	17.09	14.59	12.91
8	44.4	41.4	35.4	30.7	20.20	18.82	16.09	13.96
	\multicolumn{4}{c}{Pounds Body Wt at End of Wk}	\multicolumn{4}{c}{Kilos Body Wt at End of Wk}
1	0.32	0.33	0.32	0.31	0.15	0.15	0.15	0.14
2	0.80	0.86	0.81	0.79	0.41	0.39	0.37	0.35
3	1.43	1.53	1.43	1.38	0.65	0.70	0.65	0.63
4	2.18	2.33	2.13	2.03	0.99	1.06	0.97	0.92
5	2.99	3.20	2.84	2.66	1.36	1.46	1.29	1.21
6	3.88	4.15	3.57	3.28	1.76	1.89	1.62	1.49
7	4.88	5.15	4.26	3.81	2.19	2.34	1.94	1.73
8	5.72	6.12	4.85	4.22	2.60	2.78	2.21	1.92
	\multicolumn{4}{c}{Cumulative Feed Conversion End Wk}	\multicolumn{4}{c}{Cumulative Feed Conversion End Wk}
1	0.81	0.80	0.79	0.78	0.81	0.80	0.79	0.78
2	1.13	1.05	1.07	1.08	1.13	1.05	1.07	1.08
3	1.35	1.24	1.26	1.28	1.35	1.24	1.26	1.28
4	1.55	1.41	1.45	1.48	1.55	1.41	1.45	1.48
5	1.75	1.58	1.64	1.68	1.75	1.58	1.64	1.68
6	1.95	1.75	1.84	1.89	1.95	1.75	1.84	1.89
7	2.15	1.92	2.07	2.15	2.15	1.92	2.07	2.15
8	2.35	2.09	2.33	2.45	2.35	2.09	2.33	2.45
	\multicolumn{4}{c}{US Gal Water/1,000 Broilers/Day}	\multicolumn{4}{c}{Liters Water/1,000 Broilers/Day}
1	8	8	9	10	30	30	34	38
2	12	16	26	48	45	61	98	182
3	19	25	52	95	72	95	197	360
4	26	35	72	130	98	133	273	492
5	35	46	94	170	133	174	356	644
6	43	57	110	200	163	216	416	757
7	50	67	122	221	189	254	462	837
8	57	76	125	228	216	288	473	863

Colored Light

When the broiler house is lightproof, red light bulbs are often used to reduce cannibalism still further. Beak trimming can be eliminated. But if the house has open sides or windows, the use of red lights during the morning and evening hours to supplement the hours of natural daylight is a disadvantage as the birds will retain their agonistic habits developed during the hours of bright light. In lighttight houses never change from red light to white light as the birds will become cannibalistic within a few minutes.

Light Systems for Broilers

Light supplied by sunshine during the day, followed by normal darkness at night, is the most inferior of any lighting program for broilers. There are several methods used by growers to make improvements.

1. *Continuous light in the open-sided house.* Start with 48 hours of continuous light with an intensity of 3.5 fc at floor level; then supply dim artificial light during all dark hours except for 1 hour at midnight. This 1 hour trains the birds to darkness so that there will be no panic should there be an electrical power failure.

    The light intensity at floor level should be 0.50 fc. One 150-watt bulb for each 1,000 ft² (93 m²) of floor space will provide this light intensity. Do not use less light. The eye of the chicken will not adapt to lower intensities when there is bright light during daytime hours.

2. *Light in lighttight houses.* Provide 3.5 fc of continuous light at floor level for the first 5 days. This amount of time is necessary for the chicks to develop their eating habits and fully learn the location of the feeders and waterers.

    On the sixth day, reduce the light intensity to 0.35 fc. Supply about 125 watts for each 1,000 ft² (93 m²) of floor space. Place the bulbs directly over the feeders and waterers. Then use one of the following programs:

2A. *Continuous dim light.* After 5 days provide 23 hours of light with an intensity of 0.35 fc and 1 hour of darkness.

2B. *Intermittent dim light.* After 5 days, and during normal or cool temperatures, provide 1 hour of 0.35 fc light intensity (feeding time) followed by 3 hours of darkness (resting time); then repeat. This provides 6 hours of eating time each 24 hours.

During hot weather the feeding time must be increased. Provide about 1½ hours of dim light and 3 hours of darkness in each cycle.

Caution: Increase the feeder and waterer space when intermittent light systems are used. It may take up to 50% more equipment. Most birds must be able to eat at one time. The increased investment in feeders and waterers is a disadvantage of this lighting program.

Improvement with various lighting programs. Each of the three methods of lighting will be effective in producing better broiler growth, as shown by the following:

Lighting Program	Relative Growth Efficiency
1. Continuous light in open-sided house	100 (base)
2A. Continuous light in lighttight house	104–106%
2B. Intermittent light in lighttight house	106%

Why intermittent light produces better growth. Although the exact reasons for better growth on intermittent light programs are not known, it is thought that by giving chickens a meal (short feeding period), followed by a longer period of time for digesting the meal (no feed available), the efficiency of feed utilization is improved. This is a known fact with several species of mammals. The continuous eater excretes more protein than does the bird given a meal at regular intervals, thus wasting protein.

Do not alter the lighting program or change the intensity of light once one of the above two programs is started. Measure the light at floor level occasionally. Because the intensity is so low on these programs, dirty bulbs cut materially into the amount of light reaching the floor.

Light for Catching Birds

Most broilers and roasters ready for market are removed from the poultry house at night to reduce the incidence of bruising during the catching process. A small amount of red or blue light will enable the catchers to see, while the chickens will not be prompted to move about.

20-I. DISEASE PREVENTION

Management must provide a good program of disease prevention if successful broiler production is to result. With broilers, as opposed to birds raised for egg production, there is but a short growing period—too short for the birds to recover from most disease outbreaks prior to market time. Therefore, the disease-control procedure must be one of prevention rather than of treatment. Only when there is failure in the prevention program should medication be given.

Chicks Should Come from MG- and MS-Negative Breeders

If available, obtain chicks from *Mycoplasma gallisepticum* (MG)-free and *Mycoplasma synoviae* (MS)-free breeding flocks (see Chapter 37-H and 37-I).

Vaccination Program in the Broiler House

There is no common vaccination program. In fact, some broiler growers do not vaccinate for any disease, because they practice isolation and sanitation and feel that these programs will prevent most disease outbreaks.

When a vaccination program is to be followed, consult disease specialists in the area where the broilers are to be located. They are in an excellent position to know the prevalence of any disease and the control measures necessary to prevent it. As a suggestion and starting guide, see Table 39-1. In most instances, bronchitis, Newcastle disease, and Marek's vaccines are used for broilers. See discussions of these diseases in Chapter 37-L, 37-M, and 37-Q.

> *Remember:* Vaccination creates a stress. Be sure to use the right type of vaccine; ascertain that it is fresh; vaccinate according to directions and at the right age.

Coccidiosis Control

After a study of coccidiosis, it seems there would be a fool-proof method for eliminating this disease in broilers. But most methods are far from perfect; coccidiosis still is an important problem with many broiler flocks.

> *Complete suppression necessary.* Coccidiosis must be fully suppressed in broilers; there is not time to develop immunity (see Chapter 37-V).

Unabsorbed Egg Yolks

Up to 20% of marketed broilers have unabsorbed yolks varying in size from a speck to 1 in. (2.5 cm) in diameter, and up to 0.5% of the flock may be condemned if they rupture. The incidence is higher in males than in females.

Reducing early brooding temperatures will help eliminate the problem of unabsorbed yolk sacs found in broilers at the processing plant. The lower brooder temperatures induce the broiler chicks to utilize the yolk material for energy purposes, thereby decreasing yolk size (see Chapters 13-K and 20-E).

20-J. GROWTH AND FEED CONSUMPTION

Anyone producing broilers or roasters must make a critical study of the variations that occur in the growth and feed consumption of males and females. These variations are shown in Table 20-10 (pounds) and Table 20-11 (kilos) and involve

1. Live body weight
2. Weekly increases in body weight
3. Weekly feed consumption
4. Cumulative feed consumption
5. Weekly feed conversion
6. Cumulative feed conversion

The data given in the tables represent average annual figures for good flocks. There are seasonal variations in these figures; one could expect better results during the summer and poorer during the winter. The marketable weight of broilers varies, but most commercial growers produce a straight-run bird near 4.4 lb (2.0 kg) in about 45 days on an average year-round basis. When the average weight of the straight-run flock gets above 4.4 lb (2 kg), *some* birds are too heavy to be broilers and not heavy enough to be roasters. In this discussion, a flock with an average straight-run weight of 4.0 lb (1.8 kg) each is considered to be marketable broiler size. Those weighing 6.0 lb (2.7 kg) and over are in a roaster classification.

From an examination of the data in Tables 20-10 and 20-11 the following facts may be established:

1. *Chickens (and all animals and birds) do not grow at a uniform rate.* Growth starts slowly during a period of acceleration, then slows during a period of deceleration, giving rise to an "S-shaped" growth curve prior to sexual maturity.

Table 20-10. Growth, Feed Consumption, and Feed Conversion of Broilers and Roasters, in Pounds (Year-round Avg, 70°F)

	Males						Females						Straight-run					
	Live Weight		Feed Consumption		Feed Conversion		Live Weight		Feed Consumption		Feed Conversion		Live Weight		Feed Consumption		Feed Conversion	
Wk of Age	End of Wk	Wkly Gain	Wk	Cumulative	Wk	Cumulative	End of Wk	Wkly Gain	Wk	Cumulative	Wk	Cumulative	End of Wk	Wkly Gain	Wk	Cumulative	Wk	Cumulative
1	0.34	0.34	0.27	0.27	0.80	0.80	0.33	0.33	0.26	0.26	0.80	0.80	0.33	0.33	0.26	0.26	0.80	0.80
2	0.89	0.55	0.66	0.93	1.20	1.05	0.83	0.50	0.62	0.87	1.22	1.05	0.86	0.53	0.64	0.90	1.21	1.05
3	1.59	0.70	1.03	1.96	1.37	1.23	1.47	0.64	0.97	1.84	1.41	1.25	1.53	0.67	1.00	1.90	1.49	1.24
4	2.46	0.87	1.48	3.44	1.70	1.40	2.20	0.73	1.30	3.15	1.78	1.42	2.33	0.80	1.39	3.29	1.74	1.41
5	3.39	0.93	1.84	5.28	1.98	1.56	3.01	0.81	1.70	4.84	2.08	1.60	3.20	0.87	1.77	5.06	2.03	1.58
6	4.44	1.05	2.40	7.68	2.29	1.73	3.86	0.85	2.00	6.84	2.35	1.77	4.15	0.95	2.20	7.26	2.32	1.75
7	5.56	1.12	2.88	10.56	2.57	1.90	4.74	0.88	2.38	9.22	2.69	1.94	5.15	1.00	2.63	9.89	2.63	1.92
8	6.67	1.11	3.25	13.81	2.93	2.07	5.57	0.83	2.55	11.77	3.05	2.11	6.12	0.97	2.90	12.79	2.99	2.09
9	7.74	1.07	3.53	17.34	3.30	2.24	6.34	0.77	2.71	14.48	3.48	2.28	7.04	0.92	3.12	15.91	3.39	2.26
10	8.76	1.02	3.77	21.11	3.70	2.41	7.02	0.68	2.75	17.23	3.98	2.45	7.89	0.85	3.26	19.17	3.84	2.43
11	9.69	0.93	3.89	25.00	4.18	2.58	7.61	0.59	2.71	19.94	4.04	2.62	8.65	0.76	3.32	22.49	4.37	2.60
12	10.51	0.82	3.90	28.90	4.76	2.75	8.09	0.48	2.63	22.57	5.48	2.79	9.30	0.65	3.27	25.76	5.11	2.77

Table 20-11. Growth, Feed Consumption, and Feed Conversion of Broilers and Roasters, in Kilos (Year-round Avg, 21.1°C)

Wk of Age	Males Live Weight End of Wk	Wkly Gain Wk	Feed Consumption Wk	Feed Consumption Cumulative	Feed Conversion Wk	Feed Conversion Cumulative	Females Live Weight End of Wk	Wkly Gain Wk	Feed Consumption Wk	Feed Consumption Cumulative	Feed Conversion Wk	Feed Conversion Cumulative	Straight-run Live Weight End of Wk	Wkly Gain Wk	Feed Consumption Wk	Feed Consumption Cumulative	Feed Conversion Wk	Feed Conversion Cumulative
1	0.15	0.15	0.12	0.12	0.80	0.80	0.15	0.15	0.12	0.12	0.80	0.80	0.15	0.15	0.12	0.12	0.80	0.80
2	0.41	0.25	0.30	0.42	1.20	1.05	0.38	0.23	0.28	0.40	1.22	1.05	0.39	0.24	0.29	0.41	1.21	1.05
3	0.72	0.32	0.47	0.89	1.37	1.23	0.67	0.29	0.44	0.84	1.41	1.25	0.69	0.30	0.45	0.86	1.49	1.24
4	1.12	0.40	0.67	1.56	1.70	1.40	1.00	0.33	0.59	1.43	1.78	1.42	1.06	0.36	0.63	1.49	1.74	1.41
5	1.54	0.42	0.84	2.40	1.98	1.56	1.37	0.37	0.77	2.20	2.08	1.60	1.45	0.40	0.80	2.30	2.03	1.58
6	2.01	0.48	1.09	3.48	2.29	1.73	1.75	0.39	0.91	3.10	2.35	1.77	1.88	0.43	1.00	3.29	2.32	1.75
7	2.52	0.51	1.31	4.79	2.57	1.90	2.15	0.40	1.08	4.94	2.69	1.94	2.34	0.45	1.19	4.49	2.63	1.92
8	3.03	0.50	1.47	6.26	2.93	2.07	2.53	0.38	1.16	5.34	3.05	2.11	2.78	0.44	1.32	5.80	2.99	2.09
9	3.51	0.49	1.60	7.87	3.30	2.24	2.88	0.35	1.23	6.57	3.48	2.28	3.19	0.42	1.42	7.22	3.39	2.26
10	3.97	0.46	1.71	9.58	3.70	2.41	3.18	0.31	1.25	7.82	3.98	2.45	3.58	0.39	1.48	8.70	3.84	2.43
11	4.40	0.42	1.76	11.34	4.18	2.58	3.45	0.27	1.23	9.05	4.04	2.62	3.93	0.35	1.51	10.20	4.37	2.60
12	4.47	0.37	1.77	13.11	4.76	2.75	3.67	0.22	1.19	10.24	5.48	2.79	4.22	0.30	1.48	11.69	5.11	2.77

2. *Males grow faster than females.*   At near broiler market age, the males will attain approximately the same weight 4 days before the females (see Table 20-10).

3. *Weekly increases in weight are not uniform.*   Gains increase each week until reaching a maximum at about the seventh week for straight-run flocks, after which they decrease.

4. *Weekly feed consumption increases as weight increases.*   Each week the birds eat more feed than they did the week before.

5. *First gains require less feed.*   Feed conversion, or the units of feed necessary to produce a unit of gain in weight, is lowest the first week, then increases each week thereafter. For example, a straight-run flock produces a unit of gain the second week from 1.21 units of feed, but requires 2.32 units the sixth week.

6. *Males convert feed to meat more efficiently than females.*   A male weighing 4.44 lb (2.01 kg) requires about 7.68 lb (3.48 kg) of feed while a female of the same weight requires 8.70 lb (3.94 kg) of feed. The feed conversion in each case is 1.73 for the males and 1.89 for the females.

7. *The heavier the weight of the straight-run flock, the greater the difference in weight between the sexes.*   At 2 weeks of age the male weight is 107% of the female weight; at 6 weeks of age it is 115%.

20-K. BROILER GROWTH CURVES

Typical weekly growth curves for male, female, and straight-run broilers are shown in Figure 20-4, as taken from Table 20-10. Note that increases in growth are low the first 2 weeks, then gradually become larger until the straight-run flock averages about 5.00 lb (2.3 kg), after which weekly increments show progressive decreases. The data in Figure 20-4 are well past the age of the 4-lb (1.82-kg) marketable flock, but the extension gives a better picture of what is happening to growth over the longer period.

20-L. WEIGHT SPREAD BETWEEN THE SEXES

At hatching time the broiler males are about 1% heavier than the broiler females. However, as the birds grow this spread increases so that by the time the broilers are marketed the males are about 17% heavier than the females. The actual spread is more closely associated with weight than with age.

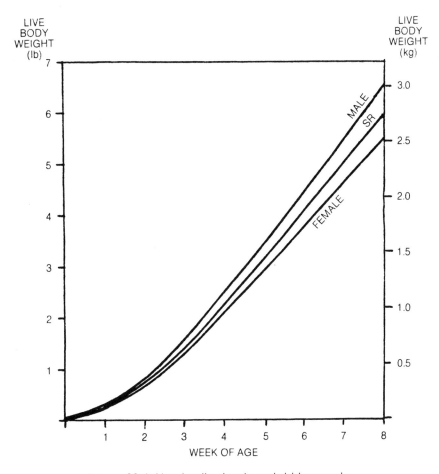

Figure 20-4. Live broiler body weight by weeks.

Figure 20-5 shows this relationship on a weekly basis. The system employed is calculating the male weight as a percentage of the female weight. From Figure 20-5 notice that the spread increases uniformly during the broiler growing period. At roaster weights, the males are about 25% heavier than the females.

20-M. WEEKLY VARIATIONS IN FEED CONVERSION

Weekly and cumulative variations in feed conversion of the straight-run broiler flock are given in Tables 20-10 and 20-11, and graphed in Figure 20-6. Feed conversion is correlated with growth, but the feed conversion curve does not parallel the growth curve (see Figures 20-4 and 20-6). The shorter the time period to produce a marketable broiler, the lower (better) the feed conversion.

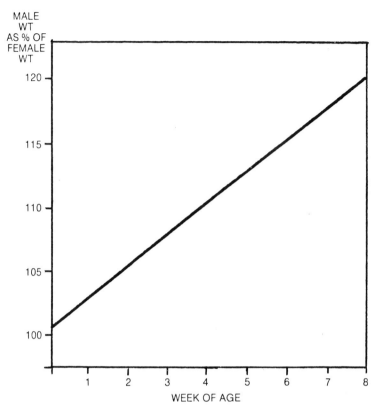

Figure 20-5. Broiler male weight as percent of female weight.

Table 20-12. Male Broiler Feed Consumption and Conversion by Growth Increments

| Male Live Body Weight lb | Feed Consumption | | | Feed Conversion | |
	To Add Last 0.1-lb Gain lb	Cumulative lb	Cumulative kg	To Add Last 0.1-lb Gain	Cumulative
3.1	0.18	4.68	2.13	1.92	1.51
3.2	0.19	4.90	2.23	1.94	1.53
3.3	0.19	5.08	2.31	1.96	1.54
3.4	0.20	5.28	2.40	1.98	1.56
3.5	0.20	5.50	2.50	2.00	1.58
3.6	0.20	5.70	2.59	2.02	1.59
3.7	0.21	5.91	2.69	2.05	1.61
3.8	0.21	6.15	2.80	2.06	1.63
3.9	0.21	6.38	2.90	2.09	1.64
4.0	0.21	6.62	3.01	2.12	1.66
4.1	0.22	6.86	3.12	2.15	1.68
4.2	0.22	7.12	3.24	2.17	1.70
4.3	0.22	7.40	3.36	2.20	1.71
4.4	0.23	7.68	3.49	2.25	1.73
4.5	0.23	7.94	3.61	2.27	1.74

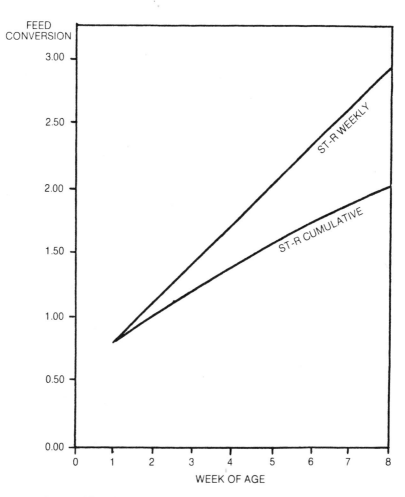

Figure 20-6. Weekly straight-run broiler feed conversion.

Table 20-12. (*continued*)

| Male Live Body Weight lb | Feed Consumption | | | Feed Conversion | |
| | To Add Last 0.1-lb Gain lb | Cumulative | | To Add Last 0.1-lb Gain | Cumulative |
		lb	kg		
4.6	0.23	8.20	3.73	2.32	1.76
4.7	0.24	8.46	3.85	2.35	1.78
4.8	0.24	8.72	3.96	2.37	1.79
4.9	0.24	8.97	4.08	2.38	1.81
5.0	0.24	9.22	4.19	2.40	1.83
5.1	0.25	9.49	4.31	2.45	1.84
5.2	0.25	9.77	4.44	2.47	1.86
5.3	0.25	10.04	4.56	2.50	1.87
5.4	0.25	10.30	4.68	2.54	1.89
5.5	0.26	10.56	4.80	2.57	1.90

20-N. FEED CONSUMPTION ACCORDING TO SEX OF BROILERS

Because feed represents the largest expense item in the production of broilers, and because feed conversion varies according to the age, sex, and weight of the birds, it is well that these variations—and their profit implications—be fully understood. Tables 20-12, 20-13, and 20-14 deal with these factors.

Table 20-13. Female Broiler Feed Consumption and Conversion by Growth Increments

Female Live Body Weight lb	To Add Last 0.1-lb Gain lb	Cumulative lb	Cumulative kg	To Add Last 0.1-lb Gain	Cumulative
		Feed Consumption		Feed Conversion	
3.1	0.21	5.02	2.28	2.12	1.62
3.2	0.21	5.25	2.39	2.14	1.64
3.3	0.22	5.48	2.49	2.17	1.66
3.4	0.22	5.71	2.60	2.20	1.68
3.5	0.22	5.95	2.71	2.23	1.70
3.6	0.23	6.19	2.81	2.26	1.72
3.7	0.23	6.44	2.93	2.30	1.74
3.8	0.23	6.69	3.04	2.33	1.76
3.9	0.24	6.84	3.11	2.36	1.78
4.0	0.24	7.22	3.28	2.40	1.80
4.1	0.24	7.46	3.39	2.44	1.82
4.2	0.25	7.73	3.51	2.48	1.84
4.3	0.25	8.00	3.64	2.52	1.86
4.4	0.26	8.23	3.74	2.56	1.87
4.5	0.26	8.51	3.87	2.60	1.89
4.6	0.26	8.79	4.00	2.64	1.91
4.7	0.27	9.07	4.12	2.68	1.93
4.8	0.27	9.36	4.26	2.72	1.95
4.9	0.28	9.65	4.39	2.76	1.97
5.0	0.28	9.95	4.52	2.80	1.99
5.1	0.28	10.25	4.66	2.84	2.01
5.2	0.29	10.56	4.80	2.88	2.03
5.3	0.29	10.87	4.94	2.92	2.05
5.4	0.30	11.18	5.08	2.97	2.07
5.5	0.30	11.50	5.23	3.02	2.09

Male and Female Feed Consumption and Conversion by Growth Increments

Feed consumption increases and feed conversion decreases as broilers age and get heavier. There are male and female variations. In Tables

Table 20-14. Straight-run Broiler Feed Consumption and Conversion
by Growth Increments

| Straight-run Live Body Weight lb | Feed Consumption | | | Feed Conversion | |
| | To Add Last 0.1-lb Gain lb | Cumulative | | To Add Last 0.1-lb Gain | Cumulative |
		lb	kg		
3.1	0.20	4.85	2.21	2.02	1.57
3.2	0.20	5.08	2.31	2.04	1.59
3.3	0.21	5.29	2.41	2.07	1.60
3.4	0.21	5.50	2.50	2.09	1.62
3.5	0.21	5.73	2.61	2.12	1.64
3.6	0.22	5.95	2.71	2.15	1.66
3.7	0.22	6.18	2.81	2.18	1.68
3.8	0.22	6.42	2.92	2.20	1.70
3.9	0.23	6.61	3.01	2.23	1.71
4.0	0.23	6.92	3.15	2.26	1.73
4.1	0.23	7.16	3.26	2.30	1.75
4.2	0.24	7.43	3.38	2.33	1.77
4.3	0.24	7.70	3.50	2.36	1.79
4.4	0.25	7.96	3.62	2.40	1.80
4.5	0.25	8.23	3.74	2.44	1.82
4.6	0.25	8.50	3.86	2.48	1.84
4.7	0.26	8.77	3.99	2.51	1.86
4.8	0.26	9.04	4.11	2.54	1.87
4.9	0.26	9.31	4.23	2.57	1.89
5.0	0.26	9.59	4.36	2.60	1.91
5.1	0.27	9.87	4.49	2.64	1.93
5.2	0.27	10.17	4.63	2.68	1.95
5.3	0.27	10.45	4.75	2.71	1.96
5.4	0.28	10.74	4.88	2.75	1.98
5.5	0.28	11.03	5.01	2.80	2.00

20-12, 20-13, and 20-14, body weight has been given in 0.1-lb intervals to show how feed consumption and feed conversion change as body weight increases. One table is for male broilers, one for female, and one for straight-run birds. It will be necessary to study these three tables carefully and relate the figures to the importance of feed costs.

Days Necessary to Produce Males and Females of Same Weight

Time is an element of expense in the production of broilers. As males will reach a desired weight several days before females, and the sexes are often raised separately, many cost calculations call for a table setting forth the number of days necessary to reach a desired body weight by sexes. Table 20-15 gives such figures.

Table 20-15. Days Necessary to Produce Desired Broiler Weight by Sexes (70°F, 21.1°C)

Desired Live Weight		Days Necessary to Reach Desired Live Body Weight		
lb	kg	Male	Female	Straight-run
2.6	1.18	29.2	31.9	30.6
2.7	1.23	30.0	32.6	31.3
2.8	1.27	30.7	33.4	32.1
2.9	1.32	31.5	34.2	32.9
3.0	1.36	32.2	35.0	33.6
3.1	1.41	33.0	35.8	34.4
3.2	1.45	33.7	36.6	35.2
3.3	1.50	34.5	37.4	36.0
3.4	1.54	35.2	38.2	36.7
3.5	1.59	36.0	39.0	37.5
3.6	1.63	36.7	39.9	38.3
3.7	1.68	37.5	40.7	39.1
3.8	1.72	38.2	41.6	39.9
3.9	1.77	38.7	42.4	40.6
4.0	1.81	39.3	43.3	41.3
4.1	1.86	39.9	44.2	42.1
4.2	1.91	40.5	45.0	42.8
4.3	1.95	41.1	45.8	43.5
4.4	2.00	41.7	46.7	44.2
4.5	2.04	42.3	47.6	45.0
4.6	2.09	42.9	48.4	45.7
4.7	2.13	43.5	49.3	46.4
4.8	2.18	44.1	50.2	47.2
4.9	2.22	44.8	51.0	47.9
5.0	2.27	45.4	51.9	48.7
5.1	2.31	46.0	52.7	49.4
5.2	2.36	46.6	53.6	50.1
5.3	2.40	47.2	54.4	50.8
5.4	2.45	47.8	55.3	51.6
5.5	2.49	48.4	56.1	52.3

20-O. INDIVIDUAL BIRD WEIGHT VARIATIONS

One of the great economic problems of broiler production is that all birds are not the same weight at market time. Not only are the males heavier than the females but neither sex is uniform; there are small, medium, and large cockerels and pullets.

Variability in broiler weight is often due to variability of hatching egg size (see Chapter 20-E). Males and females in the flock are more uniform the older they get. At market time, in the *normal healthy flock*, the extreme male weights are about ±31% of the mean flock weight; the extreme female weights are about ±27% of the mean flock weight. Male weights are more variable than female weights because the males are heavier. Even when each sex is the same average weight, the male weights are more variable than the female weights.

Normal Biological Flock Variation

If one were to weigh a group of mature broilers individually by weight intervals no greater than 0.1 lb (45.4 g), and then separate the data by sexes, a further computation could be made to arrive at the number in each weight group as a percentage of the total. Figures for a *normal, healthy flock* are given in Table 20-16. They show the variations for the males, females, and straight-run birds.

The Normal Curve

If the figures for the males and females in Table 20-16 are plotted on graph paper as a percentage of population against live body weight, and lines are drawn through the plotted points, two bell-shaped curves will result. These are shown in Figure 20-7.

When the figures for the straight-run flock are plotted, and the points are connected, a bimodal curve appears, as indicated in the same figure. These three curves are very typical when body weight is studied. However, every strain of birds will vary; some are more uniform than others. Furthermore, every group of broilers will vary.

The Abnormal Curve

Unfortunately, all broiler flocks are not normal, healthy flocks. Most have encountered some form of stress during the growing period. Temperatures other than optimum, mismanagement, medication, crowding, and other factors affect weight *variation* at market time. The bell-curves will no longer be uniform: The left-hand side extends farther because stresses produce a higher percentage of smaller birds; the right-hand side remains fairly uniform regardless of any reasonable stress. The extent of the left-hand malformation of the curve will depend on the severity of the stress. The malformation is highly variable among flocks. Because it is impossible to estimate the average effect of stress on the formation of the curves, the only reliable standard data must come from a normal curve, as shown in Figure 20-7.

Table 20-16. Distribution of Broiler Weights in Normal Healthy Flock (Approx. Avg. Wt.: Males 4.5 lb [2.04 kg]; Females 3.9 lb [1.77 kg]; St.-Run 4.2 lb [1.91 kg])

Weight of Birds		Percentage in Each Weight Group		
lb	kg	Male	Female	St.-run
2.6	1.18		0.03	0.01
2.7	1.23		0.07	0.03
2.8	1.27		0.19	0.10
2.9	1.32		0.51	0.25
3.0	1.36		1.1	0.55
3.1	1.41		2.4	1.2
3.2	1.45	0.01	4.0	2.1
3.3	1.49	0.02	6.8	3.4
3.4	1.54	0.06	9.4	4.7
3.5	1.59	0.13	12.2	6.2
3.6	1.63	0.27	13.3	6.8
3.7	1.68	0.51	13.3	6.9
3.8	1.72	1.1	12.2	6.6
3.9	1.77	1.8	9.4	5.6
4.0	1.81	2.5	6.8	4.6
4.1	1.86	3.4	4.0	3.7
4.2	1.91	5.2	2.4	3.8
4.3	1.95	6.9	1.1	4.0
4.4	2.00	8.4	0.51	4.5
4.5	2.04	9.3	0.19	4.7
4.6	2.09	10.3	0.07	5.2
4.7	2.13	10.3	0.03	5.2
4.8	2.18	9.3		4.7
4.9	2.22	8.5		4.3
5.0	2.27	6.9		3.5
5.1	2.31	5.2		2.6
5.2	2.36	3.5		1.8
5.3	2.40	2.5		1.3
5.4	2.45	1.8		0.90
5.5	2.49	1.1		0.55
5.6	2.54	0.51		0.25
5.7	2.59	0.27		0.13
5.8	2.63	0.13		0.07
5.9	2.68	0.06		0.03
6.0	2.72	0.01		0.01

What Percentage of Broilers Are Within Weight Requirements?

Many processors sell broilers that command a premium when within certain maximum and minimum live weights. Any birds between the two extremes are suitable for their use, but those outside the extreme

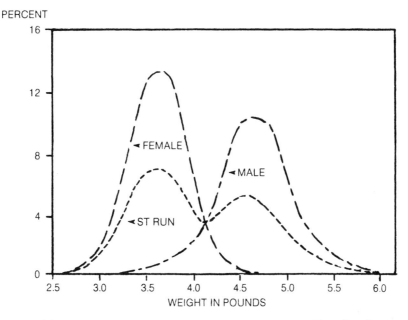

Figure 20-7. Distribution of broiler weights in the normal healthy flock.

must be considered as by-products and many times must be sold at a discount. The broiler grower is interested in this requirement too; he wants to know what percentage of his flock will fall within certain weight limitations. By referring to Table 20-16, he can determine this.

Problem: The processor's requirement is for live birds no less than 3.5 lb (1.59 kg) and no greater than 4.6 lb (2.1 kg). What percentage will fall between these weights when the average straight-run flock weight is 4.2 lb (1.9 kg)?

Solution: Refer to Table 20-16 and add the percentage figures between the above weight extremes. They will total as follows:

Males	49.8%
Females	75.6%
Straight-run	62.7%

What is a "Uniform" Flock?

Approximately 75% of the males and 78% of the females should be within 10% of the average weight of each sex for the flock to have satisfactory "uniformity" at maturity.

But many flocks do not attain this degree of uniformity, while others are better. Examples of the distribution of broilers by weight classifications for varying percentages of birds falling within the plus or minus 10% of the average weight of each sex are shown in Table 20-17.

Table 20-17.   Percentage of Male and Female Broilers
in Each Weight Classification

Classification	Percentage within 10% of Mean Sex Weight[1]	
	Male	Female
Excellent	86 and above	89 and above
Good	79–85	82–88
Average	72–78	75–81
Fair	65–71	68–74
Poor	58–64	61–67
Very poor	57 and below	60 and below

[1]With scale weight interval of 1 oz (28g), or less.

20-P. MEASURING BROILER GROWING EFFICIENCY

The efficiency of the broiler growing program, including the important feeding program, can be measured in the following three ways:

1. Mature live body weight
2. Feed conversion over the life of the bird
3. Age to reach a desired weight

Usually, as the programs become more efficient, feed consumption is reduced, feed conversion is improved, and the length of time necessary to reach a certain weight decreases. But growth is the most important. If you want to do a better job of raising broilers, speed up the growth rate.

Growth and Feed Conversion Not Uniform

The complex thing about growth is that it is not uniform week after week. As a basic example, if the average weight of the live, straight-run broiler flock is 4.0 lb (1.9 kg) at 41 days, it will be 2.0 lb (0.9 kg) at 25.4 days. In other words, one-half of the weight is put on the bird in 62% of the time; the last half requires only 38%.

Feed conversion is not uniform throughout the life of the broiler; it increases each week.

Feeding Efficiency Indices

There are two common methods of appraising feeding efficiency, as follows:

1. *Point Spread.* A measure of feed conversion efficiency used for a number of years is known as *point spread.* The formula is as follows:

(Live weight in pounds) − (Feed conversion) × 100
= Point Spread

Examples:

	Point Spread Reading			
	Excellent	Good	Fair	Poor
Average live weight (lb)	4.20	4.00	3.90	3.80
(Less) Feed conversion	1.80	1.80	1.90	2.00
Point spread	240	220	200	180

Point spread may be a poor indicator of efficiency. If one will study the point spread during different weeks of growth for the normal flock (Table 20-18), the figures will show that as birds age the point spread becomes greater. Therefore, only when weights of different flocks are similar can point spread be used as an indicator for comparing feed efficiency.

2. *Performance Index.* This is calculated as follows:

$$\frac{\text{Live weight in pounds}}{\text{Feed conversion}} \times 100 = \text{Performance Index}$$

But again, the figure changes according to the week of age of the bird (see Table 20-18).

Table 20-18. Variations in Point Spread and Performance Index According to Age of Straight-Run Broiler Flock

Age wk	Live Body Weight lb	Feed Conversion To Date	Point Spread	Performance Index
4	2.33	1.41	92	165
5	3.20	1.58	162	203
6	4.15	1.75	240	237

20-Q. BROILER PROBLEMS

Barebacks

Barebacks are usually the result of frequent picking among the birds and not to differences in feather development. Some coccidiostats induce a higher percentage of barebacks due to increased flightiness.

Oily Bird Syndrome

Oily or greasy bird syndrome is a condition in which excessive amounts of oil have accumulated as a type of fat under the skin of the broiler making it difficult to handle in the processing plant. Such birds appear healthy and normal when brought into the plant, but come from the picker with their fat cells destroyed and large amounts of liquid fat released under the skin. There is a slight weight loss, and the product is not attractive.

Evidently, there are two types of the syndrome: (1) associated with high body fat due to dietary problems, and (2) seen in birds not obese, but characterized by elevated skin hydroxyproline and hydroxylysine and a reduction in acid-insoluble collagen. Certain weed seeds have been associated with this type.

Leg and Bone Problems in Broilers

Broilers are subject to a variety of leg problems and abnormalities. Following is a list, along with probable causes:

1. *Redhocks.* Birds have pink or red hocks, even as young as newly hatched. Some young chicks recover, but in others the difficulty grows worse.
   Probable cause: Low level of vitamins in the breeder feed.
2. *Rickets.* This condition is evidenced by soft leg bones, bowed legs, and twisted keel. Bones have a low calcium content.
   Probable cause: Lack of vitamin D_3 and nutritional phosphorus. Improper calcium: phosphorus level in the feed or malabsorption of nutrients due to an infectious agent.
3. *Chondrodystrophy (Perosis).* This is a disorder of the growth plates with an enlargement and malfunction of the tibial metatarsal joint (hockjoint). Wire floors

and rations with a high mineral content increase the incidence.

> *Probable cause:* Inadequate manganese or choline in broiler diet.

4. *Tibial dychondroplasia.* This is a metabolic disorder in which the articular cartilages grow downward into the shaft of the tibia.

> *Probable cause:* This has genetic implications but is also of nutritional origin as the addition of 2.5% of dried brewers yeast to the breeder diet will almost completely eliminate it in the broilers. Lack of exercise is another cause. Low-calcium, high-phosphorus rations, another.

5. *Femoral head necrosis.* The head of the long bones is porous and discolored yellow or brown. The difficulty may extend to the tibia and metatarsus.

> *Probable cause:* Viral arthritis caused by a reovirus. There is a vaccine available.

6. *Staph arthritis.* This is seen most frequently in 2- to 4-week-old chicks.

> *Probable cause:* Secondary infection after viral arthritis.

7. *Twisted legs.* These are characterized by twisted or bowed legs (with no shortening of the leg bones) resulting in deformation of the hock joint.

> *Probable cause:* Lack of exercise, which weakens the leg muscles that hold the leg in a correct position, low feed consumption during hot weather, and new litter are possible causes.

8. *Osteoperosis.* A condition accentuated by a reduction in the mass or an increase in the airspace of the bone. In caged layers it results in caged layer fatigue.

> *Probable cause:* An improper mineral deposition in the bones.

9. *Synovitis.* This is the result of an inflammation of the synovial membranes of the joints of the leg and the adjacent tendon sheaths.

> *Probable cause:* Caused by a Staphylococcus organism, by a virus, or by the organism, *Mycoplasma synoviae.*

10. *Fragile bones.* This fragility is a form of osteoperosis, characterized by the withdrawal of calcium phosphate from the medullary and cortical portions of certain long bones.

> *Probable cause:* Caused by a lack of vitamin D, calcium, or phosphorus in the broiler feed.

11. *Osteomyelitis.* Leg weakness with no bone deformation.
   Probable cause: A bacterial infection.

Scabby-Hip

A dermatitis called scabby-hip is often found on the hip or thigh of broilers. Scabs develop on the skin between the feathers or near the feather follicles. The condition often first appears at 3 weeks of age, and females are more often affected than males.

Increasing the floor space allotment per chick and providing an even distribution of birds over the floor area will reduce the incidence of the dermatitis. Anything that will reduce feather pulling will be helpful. Broiler sexing by the feather method produces slow-feathering males and more scabby hip condition occurs in the males than in the females in sexed flocks.

20-R. RAISING SEX-SEPARATED BROILERS

It is common practice for some broiler producers to have the chicks sexed at the hatchery and raise each sex separately. The program has its advantages and disadvantages:

Advantages

1. It often makes it possible for the processor to better meet market demands. The males can be processed younger than the females and 60 to 70% of both sexes will fall within a 4-oz (114 g) variation.
2. Sex-separated flocks move through the processing plant better because the weights of the birds are more uniform when the males are slaughtered prior to the females.
3. It allows the grower to use different feeds or different feeding periods for the same ration for the males and females, sometimes reducing feed cost.

Disadvantages

1. Increased cost of sexing. Feather sexing will cost more than color sexing.
2. Larger breeding flocks are necessary in order to meet the demand for male and female broiler chicks. To over-

come this most hatcheries require that the customer take equal numbers of male and female chicks.

3. The integrated broiler growing contract is more complicated in order to satisfy the grower.

4. Some processors have a greater demand for small and large birds, and adequate numbers of these would not be available in sex-separated flocks marketed at the same average weight.

20-S. BROILER PRODUCTION COSTS

The cost of producing a pound (or kilo) of live broiler is highly variable from farm to farm, season to season, and country to country. The ability of the bird to convert feed to meat is important; yet there are many factors that affect its efficiency in doing this. Table 20-19 is one example of estimated broiler production costs. The figures would be representative in those areas with average feed costs and efficient production.

Table 20-19. Estimated Cost to Produce One Pound or Kilo of Live Straight-Run Broiler on Contract

Item	Cost per Lb of Live Broiler at 4.2 lb Avg Wt	Cost per Kilo of Live Broiler at 1.9 kg Avg Wt
Chick[1]	US$0.049	US$0.108
Feed[2]	0.160	0.352
Contract grower payment	0.055	0.121
Services	0.010	0.022
Vaccines & medications	0.010	0.022
Condemnations[3]	0.005	0.011
Total	0.289	0.636

[1]$0.20 + 3% mortality.
[2]$0.09/lb ($0.198/kg) × 1.76 conversion.
[3]1.3% condemnations.

20-T. BROILER/FEED RATIO

Inasmuch as feed represents the largest production cost of raising a broiler, this cost is often compared with the unit weight value of the bird at market time. The following formula is used to make the computation:

$$\frac{\text{Value of 1 lb (kg) of live broiler}}{\text{Value of 1 lb (kg) of feed}} = \text{Broiler/Feed Ratio}$$

The figures are often publicized and comparisons made on a monthly or annual basis.

20-U. MARKETING BROILERS

Feed and Water Withdrawal at Market Time

Just prior to marketing the feed must be withdrawn at the farm for a time ample to lower the contents of the gut so as to reduce Salmonella contamination during processing. But this feed withdrawal reduces the live weight (called *shrink*) and has a linear relationship with processing yield.

The shrink prior to processing may be divided into two time periods: (1) on the farm, and (2) from the farm to the processing line. The total of the two should be between 8 and 9 hours. Gorging of feed by the birds is to be avoided just prior to the feed withdrawal at the farm. In order to alleviate some of the shrink from feed withdrawal on the farm, the birds should have access to water until they are picked up in the house.

Live bird shrinkage from the time of feed withdrawal in the house, but with no water withdrawal, plus the time held in crates, is highly variable, depending on the temperature, and so forth, but the following is an approximation:

Hours Held in Pens Without Feed but with Unrestricted Water, Plus Hours Held in Crates	Live Weight Loss %
3	2
6	3
9	4
12	5
15	6

Loss in processed yield is about two-thirds the actual live weight loss from shrinkage, because during the course of handling and chilling the birds in the processing plant much of this weight loss is regained.

Preventing Bruises at Market Time

To raise good broilers is one thing; to retain their quality until they get on the processing line is another. Extreme caution should be taken to catch and handle the birds carefully. Employ the following suggestions:

1. Eliminate grit feeding at least 2 weeks prior to marketing time.
2. Catch and load the birds at night.

3. Allow the feeders to go empty about 2 hours before the catching crew arrives.
4. Prevent bruising of the birds.

Some have said that the processing loss from bruising and the consequent down-grading is as great as that due to condemnation from disease. Certainly bruises represent a severe economic loss to the broiler industry. The situation is deplorable when one realizes that most bruising is the result of improper handling of the birds just prior to and during catching and hauling.

Facts about bruises

1. Females bruise more easily than males. Usually there are about twice as many females as males with bruises.
2. Within each sex, the larger and older the bird, the greater the incidence of bruising.
3. There will be about 50% more bruises on birds raised during summer and fall than during winter and spring.
4. There will be fewer bruises when the broilers are raised under low light intensities.
5. Crowding during growing increases the percentage and severity of the bruises.
6. The use of dim light in the broiler house during catching will reduce the incidence of bruises.
7. Most bruises are the result of improper handling during the 24-hour period prior to slaughter.
8. The method of handling during marketing is more closely associated with bruises than any other factor.

Recommendations to prevent bruises

1. Keep the litter from getting wet.
2. Reduce ammonia fumes; increase the ventilation.
3. Provide adequate floor space.
4. Handle the birds carefully during catching, hauling, and removing from coops.
5. Do not excite the broilers during the day prior to catching. When excited, birds hit the feeders and waterers, causing bruises.
6. Trucks used to haul the birds should not come near the broiler house until after dark. Truck noises excite the birds during daylight hours.
7. Train the catching crews to handle the birds carefully.
8. Remove all floor equipment after dark, just before catching the birds.
9. Use dim blue or red lights in the house during catching and cooping.

10. Catchers should not carry too many birds at one time. The more birds carried, the higher the percentage of bruising.
11. Place the birds in the coops carefully.
12. Truck loaders should not throw the coops.
13. Personnel who unload the chickens should use caution in removing the birds from the coops.
14. Make a study of the causes of bruising, then detail a program to remove the causes.

Broiler Condemnations During Processing

Although most of the broilers produced are suitable for human consumption, some are not. From 1.0 to 1.5% of all broilers processed are condemned at processing time. The percentage varies greatly by processing plant and geographical location.

Condemnations during the winter are about twice as high as those during the summer; thus, there is a seasonal variation. Another important factor is that the entire bird may not always be judged unacceptable for human consumption; many times only a part of the bird is so classified. In these instances, the part is cut off or out. An example would be a severe wing bruise.

Broiler Downgrades During Processing

Any reduction in the processed grade of a broiler is considered a *downgrade*. Bruises are responsible for over half of these with about one-third occurring on the breast.

Age and Sex Associated with Processing

Table 20-20 shows the variations in dressing percentage, eviscerated yield, and chilling gains for broilers of varying ages and weights. The table clearly shows:

1. The heavier the bird, the lower the percentage of blood and feathers.
2. When the sexes are the same weight, the percentage loss from blood and feathers is greater in the males.
3. It is possible to restore a higher percentage of eviscerated weight loss in the females than in the males.

Table 20-20. Effect of Age and Sex on Processing Weight Loss

Age da	Live Weight				Loss in Blood and Feathers		Eviscerated Wt as % of Live Wt		Chilling Gains	
	Males		Females		Males %	Females %	Males %	Females %	Males %	Females %
	lb	kg	lb	kg						
31	2.7	1.23	2.2	1.00	12.6	13.7	66.1	64.7	8.7	9.9
34	3.3	1.50	2.7	1.23	11.9	12.0	66.6	65.7	8.4	9.3
37	3.9	1.77	3.1	1.41	11.9	12.7	67.4	65.9	8.5	9.0
41	4.6	2.09	3.6	1.63	11.1	11.9	68.2	68.0	7.6	9.0
46	5.4	2.45	4.2	1.91	11.0	12.1	69.7	69.7	7.4	7.9

Source: Orr, H. L., and E. T. Morgan, *Canadian Poultry Review*, July 1968 (adjusted).
¹Does not include giblets.

Processed Yield a Profit Criterion to Integration

The processed yield of a broiler is very important and of major importance to the producer as well as the processor. At present dressed broiler prices, 1% additional yield in the processing plant is equivalent to the following tradeoffs:

25 to 28 additional hatching eggs per breeder hen
13 to 15% better hatchability per breeder hen
20 to 25 lb (9.1 to 11.4 kg) less feed per breeder bird
5 points better broiler feed conversion

Yield of Parts at Processing

Sex has little effect on the yield of cut-up parts. These percentages are shown in Table 20-21.

Table 20-21. Yield of Cut-up Parts of Chilled, Ready-to-Cook, Male and Female Broilers (% of Carcass)

Item	Male	Female
Live weight (lb)	4.05	3.30
Live weight (kg)	1.84	1.50
Yield (%)		
Breast	26.6	28.3
Thigh	17.5	17.2
Drumsticks	16.3	15.5
Wings	11.7	12.1
Neck (without skin)	3.8	3.5
Tail back	9.6	9.4
Rib back	8.9	8.4
Heart	0.6	0.6
Liver	2.6	2.4
Gizzard	2.6	2.7

Source: Hayse, P. L., and W. W. Marion 1973. Poultry Sci. 52, 718.

20-V. BROILER RECORDS

The necessity of records cannot be overemphasized as a part of the broiler production program. Without careful records there is little indication of the economic progress of the flock. Certain records must be compared with a set of *standards* if there is to be any assurance that the flock is good as or better than average. There are three types of broiler records as follows:

1. Those involved with growing the flock
2. Those involved with making the contract settlement
3. Those involved with determining profit or loss

In many instances the above are computerized for detailed study.

20-W. PRODUCING SQUAB BROILERS

Squab broilers are very small broilers often merchandised as "Cornish Game Hens" or "Rock-Cornish Cross," both of which are misnomers. Squab broilers must not weigh more than 2 lb (0.9 kg) dressed, which means the live market weight should be between 2.25 and 2.50 lb (1.02 and 1.14 kg). The live weight is very important, and one or two days difference in growing time will mean a great economic loss to the processor. Normally, it will take 4 weeks to 4 weeks, 2 days of growing time for birds to reach the desired weight. Continuous floor operations are better than continuous cage or cage-and-floor. Usually squab broilers are not raised sex-separate. The processed broilers are sold as fresh or frozen whole-body birds, completely eviscerated.

20-X. ROASTER MANAGEMENT

When meat-type chickens are grown to live weights between 6 and 8 lb (2.7 and 3.6 kg), they are known as roasters in many countries. They are not to be confused with fowl (old hens or roosters) in this weight category. Roasters are young birds grown similarly to broilers, but to a heavier weight. In fact, the management program for roasters is quite similar to the program for broilers up until live body weight reaches 4.5 lb (2.04 kg), after which adjustments in the ration and some management practices must be made.

Age and Weight of Roasters

Maximum growth in a short period of time is a requisite to economical roaster production. Certain strains of meat-type birds are especially suited to rapid growth to roaster weight, while others are not. In many instances some strains grow rapidly to broiler weights, but weekly growth increments are poor at larger sizes. Be sure to select a strain of birds that has the genetic potential to attain roaster weight rapidly and economically. The birds should have white feathers or feathers with a white undercolor.

Tables 20-10 and 20-11 show the growth rate of meat-type birds up to roaster weights. Notice that the increments of weekly gain become smaller after birds reach 7 weeks of age. This is an important factor in roaster production.

Feed Conversion of Roasters

The *units* of feed necessary to produce *one unit* of gain in live weight (feed conversion) are given in Tables 20-10 and 20-11. These figures are of value when studying the economics of roaster production, for as birds increase in size, the amount of feed necessary to produce a unit of gain increases materially. Although a pound of live broiler may be produced on about 1.9 lb of feed, it will require approximately 2.4 lb of feed to produce 1 pound of live roaster.

Male and Female Roasters

As males grow more rapidly than females, the time consumed in getting a male to roaster weight is much less than for a female. It will take approximately 9 weeks to get a male to the required weight, but about 11 weeks for a female.

In many instances the day-old chicks are sex-separated and raised separately. One sex or the other may be sold at broiler weight, and the other kept for roaster production. Sometimes both sexes are sold as roasters. In these cases the males are marketed much sooner than the females. When the sexes are not separated at hatching, and the males and females are raised together, there will be a vast difference in the weights of the two sexes at market time; such a program is to be discouraged (see Tables 20-10 and 20-11).

Managing Roaster Production

Continue with the broiler management program. There are few changes except for the following:

1. *More floor space required.* Floor space requirements are given in Table 20-4. Optimal growth will not be possible unless the birds have plenty of room. Too often a quick decision is made to carry a group of broilers to roaster weight without providing additional floor space. In such cases, some of the birds should be marketed as broilers, thus providing more space for those kept as roasters.

2. *Additional feeder space required.* Roasters are larger; they require more room to eat. Provide 4 in. (10 cm) of

feeder space after 7 weeks of age. Fit the feeders with nonroosting devices.

3. *More waterer space necessary.* Provide more waterer space per bird after 7 weeks of age (see Table 20-8).

4. *More ventilation.* When the movement of air through a poultry house is based on the weight of birds in the house, the recommendations given previously are satisfactory. It is only when the roasters are crowded that difficulties with ventilation occur (see Table 20-2).

5. *Adequate feed consumption important.* Appetite begins to wane in heavier growing chickens. Adequate feed consumption is a must if roaster gains are to be produced economically. Many producers resort to 23 hours of light per day to induce birds to eat both night and day (see Chapter 20-G).

High Incidence of Breast Blisters in Roasters

One of the greatest difficulties with roaster production is associated with the high incidence of breast blisters. Heavy birds seem to sit more than lighter birds. Males sit more than females. The incidence of breast blisters, therefore, is related to body weight, sex, and the material with which the breastbone comes in contact.

The economic loss from breast blisters in roasters is tremendous. The blisters must be cut out at processing time and this downgrades the carcass. In the case of dressed broilers with breast blisters, the bird may be cut into parts and only the breast is downgraded. However, roasters are sold as whole-body birds; removing the breast blister at processing time downgrades the entire carcass.

Remedies for breast blisters. There is no method whereby all breast blisters may be eliminated; management must resort to a program of reducing their incidence to a minimum. Some ways of making the reduction are

1. Use of females instead of males. Even though the growing period of the females will be longer than in the case of males, reduction in the incidence of blisters usually will offset the costs of the longer growing period.

2. Keep the litter dry. Wet litter produces more and larger blisters.

3. Provide 4 in. (10 cm) of litter. When the litter becomes thin and hard, the difficulty from breast blisters increases.

4. Feed more often. When birds are eating, they are not sitting, and the breast is in contact with the litter for a shorter period of time.
5. Stir the birds occasionally. This keeps the birds moving rather than sitting.
6. Use a ration especially formulated for roaster production.
7. Remove objects or equipment on which the birds can roost.

20-Y. SURGICAL CAPONS

Caponizing, the surgical castration of male chickens, produces a unique type of poultry meat. Research results show that it is more tender, juicier, and flavorful.

Capon Weights

Caponized cockerels are raised to sizes greater than those attained by broilers and roasters. Most are grown to live weights of 12 to 14 lb (5.4 to 6.4 kg) in about 20 to 24 weeks; when processed, dressed weights are between 9 and 12 lb (4.1 and 5.4 kg). They are marketed whole-body, eviscerated. Producers feel that growing birds to these weights removes them from competition from roaster chickens and places them in a special category.

Caponizing Procedure

Cockerel chicks should be secured from breeders offering special strains capable of economical growth to heavy weights. They are caponized when they weigh about 1 lb and are 2½ weeks of age. A single incision is made on one side of the bird between the ribs and both testicles are removed through this one opening. If only one testicle is removed, the remaining one doubles in size, and the amount of the male hormone *testosterone* is identical to that produced by two normal testicles; the bird does not develop as a capon.

Growing Management

Young capons should be managed similarly to broilers until they are 8 weeks of age, after which they should be provided more space, as shown in Table 20-22.

Table 20-22. Floor, Waterer, and Feeder Space for Capons

Age wk	Space per Bird					
	Floor Space		Waterer Space		Feeder Space	
	ft²	m²	in.	cm	in.	cm
8–11	2	0.19	1.1	2.8	3	7.6
12–15	3	0.28	1.5	3.8	5	12.7
16+	4	0.37	2.5	6.3	7	17.8

Several problems occur with capons during the growing stage. Briefly, they are as follows:

Slips. Slips are improperly castrated cockerels. Some testicular material remains and can produce enough male hormone to remove the bird from the category of a capon, thus remaining a cockerel.

Wind-puffs. After caponization some birds will develop wind-puffs at the site of the operation, the result of a subcutaneous infection. Capons should be examined at intervals, and a small slit should be made on the puff to allow the air to escape.

Leg problems. Capons lose their male tendencies and become very docile. They want to sit, rather than stand, giving rise to leg problems. As they age, the birds should be made to move about by stirring the flock. Feeding some whole oats in the litter once a day will cause the birds to scratch to find the grain, a worthwhile exercise.

Breast blisters. Because of a tendency to sit, capons are prone to breast blisters where the breast comes in contact with the floor. Because they are processed whole-body, eviscerated, a breast blister is a major detraction to the processed capon, and usually such birds are downgraded. To alleviate some of the incidence, use deep dry litter. Remove or cover any equipment on which the birds can roost. Do not raise capons on wire or slat floors.

Amount of light during growing. Capons grown under continuous light (24 hours) will have fewer crooked keels, faster growth, and better feed conversion. The percentage of Grade A birds at market time will be higher.

Feeding Capons

Capons grow more slowly than normal uncastrated cockerels, yet will attain heavier mature weights. When growth is too rapid, capons produce too much carcass fat, which lowers their processed quality. See Chapter 33-M for a discussion of feeding programs.

20-Z. CHEMICAL CAPONS

Certain hormones will cause physiological and metabolic changes in the bird, often beneficial in the production of quality meat. There is an increased and more uniform deposition of fat. Cockerels have less activity; they are more docile.

One common hormone, *estradiol,* is often used. It is sold as a pastelike substance that is implanted under the skin in the nape of the neck. A small amount of the hormone is released to the bloodstream over a 4-week period. Birds must be at least 4 weeks of age when implanted, and should not be slaughtered until at least 6 weeks after implanting in order to void all traces of the hormone in the processed bird.

Caution: The use of hormones is illegal in some countries. Check with authorities before using such a program.

21

Genetics of the Chicken

Genetics is the science devoted to the study of inheritance. That "like begets like," or that offspring tend to resemble their parents, is the basis for the subject of this science. However, the theory is far from accurate. Although offspring do resemble their parents in some respects, in others they do not. The geneticist has worked out a great many rules governing the inheritance of certain characters pertaining to the chicken, and he is constantly using these to improve present strains of chickens and to develop new lines. In many cases the subject of genetics has been enlarged to include those principles associated with the behavior of birds and flocks. Such things as population distribution, physical control of the gene potential, hormones, management, physiology, and many others are now included in a study of poultry breeding.

21-A. THE CELL

The cell represents the basic constituent of all living material. Some living things are composed of but one cell, while others may be made up of millions. Not only does the number vary, but the cell may range from a tiny microscopic unit, as in most cases, to large cells as exemplified by an egg yolk.

Parts of the Cell

The cell has three basic parts:

1. *Cell wall.* This holds the interior contents of the cell together and prevents the contents of one cell from mixing with the contents of another.

2. *Cytoplasm.* When a cell is stained for microscopic examination, a portion of the interior takes on a light color. This is the cytoplasm, which lies outside the nucleus.

3. *Nucleus.* The area of the cell that stains dark is the nucleus. It is essential to many functions of the cell, and to the transmission of hereditary factors.

Types of Cells

Cells of chickens may be classified according to their function within the body.

1. *Somatic cells.* These are the cells that compose the tissues of the body and are to be differentiated from the sex cells.

2. *Sex cells.* These cells are specialized in that they are responsible for the continuation of the species. Somatic cells are not capable of creating new individual offspring; this is accomplished by the union of a female sex cell with a male sex cell. When the individual dies, the somatic cells die; the sex cells, if united, remain on to perpetuate the parents. Thus, it might be said that "sex cells never die," and the theory is correct at least for those sex cells involved with the creation of the new individual.

21-B. CHROMOSOMES AND GENES

Within the nucleus of the cell are rodlike structures known as chromosomes. Their number varies according to the species from two in certain lower forms of life to dozens in the higher forms. Although subject to some variation, it may be stated that the number of chromosomes remains the same, generation after generation. Chromosomes stain dark, and in most instances can be easily seen under a microscope.

Chromosome number in the chicken. Being small and of varying shapes and sizes, chromosomes in the nucleus are difficult to count. This is especially true in the case of the chicken. But evidently there are 5 or 6 pairs of macrochromosomes (large) and somewhere in the neighborhood of 33 pairs of microchromosomes (small). Whether all the latter carry gene material is not known.

Chromosomes in pairs. In the *resting body cell,* chromosomes are normally found in pairs. That is, although the shape and size of each set of chromosomes vary, they occur in pairs, the two members of which are alike in length, diameter, and shape. During certain phases of cellular division, members of each pair align themselves together in the center of the nucleus. In these stages the paired duplication may easily be seen through a microscope. At other stages, the pairs dissociate, and each member is intermingled at random with chromosomes of other pairs in the nucleus.

Types of chromosomes. There are two types of chromosomes, as follows, and at least a portion of their function is different:

1. *Autosomes.* All but one pair of the chromosomes in the body cells are known as autosomes.
2. *Sex chromosomes.* In the somatic cell, one pair of chromosomes is in part different from the other pairs. The chromosomes of this pair are known as the sex chromosomes, as they are associated with the determination of the sex of the individual.

The Genes

Within the chromosomes certain units of material are to be found that are partially responsible for the hereditary transmission of characters that differentiate one part of the body from another and one species from another. These units are known as *genes.* With proper cell staining, many genes may be seen under a high-powered microscope. They seem to be like discs or bands within the chromosomes. Genes are made up of chemicals that produce certain reactions and cause the cells to manifest themselves differently and to produce the various parts of the body and many of the body's different physiological potentialities.

Location of the genes within the chromosome. Identical genes are located in each of a pair of chromosomes. Furthermore, each specific gene is always at the same place in the chromosome. Thus, when two *allelomorphic* (like) chromosomes pair up in the cell, a specific gene in one chromosome will be exactly opposite the same gene in the other chromosome of the pair.

Gene characters. Each gene is responsible for a certain physical or physiological character. For example, one of the genes on one of the autosomes is responsible for the production of a rose comb in chickens. This gene is always in the same chromosome and at the same location within the chromosome. We might go on and on, for geneticists have established the existence of genes responsible for many characters in the chicken. A great many genes have been identified as to the chromosome in which they are located. Furthermore, the *locus* (location) of many genes within the chromosomes has been determined. Drawings have been made of these chromosomes showing the location of the genes. These are known as *gene maps.* As more information is brought to light, the gene maps are increased in scope.

DNA and RNA

Although the gene of the chromosome is the hereditary unit of inheritance, it is the molecular makeup of the gene that determines the gene action. This important genetic material is *deoxyribonucleic acid* (DNA), a chemical molecule of large molecular weight bound by a single strand. There are repeating units of four organic bases, *guanine, cytosine, thymine,* and *adenine,* all closely associated with phosphoric acid and sugar.

The special structure of the DNA molecule makes it a ''blueprint'' for another chemical compound of the cell, *ribonucleic acid* (RNA), and it in turn provides the design for what goes on thereby directing the protein and enzyme production of the cell.

DNA molecules also control cell function, thereby giving rise to the many variations in the types of animals or birds and the countless variations of each. It is the true hereditary material.

21-C. GENE BEHAVIOR

Maintenance of Chromosome Number in Somatic Cells

One of the accomplishments of nature is the ability of the somatic cells to divide and form two similar cells, thus giving rise to growth and differentiation of cell material. Furthermore, each new *daughter* cell has exactly the same number of chromosomes as its parent, thus continuing the basic relationship. How this is accomplished is interesting. Just before the parent cell divides to form two *daughter* cells, the number of chromosomes doubles, so there are two members of each of a pair where there was but one pair before. When the cell divides, each *daughter* cell gets one full set of chromosomes and the number remains intact, division after division.

Certain cells arise within the male and female that give rise to the sperm cell or the egg cell. In the case of these formations, the chromosomes do not double just before division, as in the case of somatic cellular multiplication; but rather, one member of each pair of chromosomes goes to each of the two new cells, the *gametes.* Thus, all egg or sperm cells have only half the number of chromosomes as the somatic cells, but there is always one member of each pair of chromosomes present. This phenomenon is known as the *reduction division* of the chromosomes of the cell. A new individual can only be the result of an egg cell uniting with a sperm cell, the *zygote.* Upon its formation, the single cell again has a full component of chromosomes, half coming from the dam and half from the sire, and the total number of chromosomes is restored. Furthermore, there are two members of each pair. There is continuity of the germ plasm.

Dominant and Recessive Genes

Rose comb has been designated above as being the result of a specific gene; this gene has a definite location on one of the *autosomal* chromosomes. A complete list of other known similar genes could be given. But genes have not always remained intact through the centuries, nor are they constant today. The substance responsible for their expression is subject to chemical reaction, even though such a reaction is exceptionally rare under natural conditions. It may occur only once in several centuries. When such a reaction occurs the gene material is changed; and the new material in the form of a new gene is passed on to coming generations. These changes are known as *mutations,* and by definition mean that the differences were sudden. Although nature has provided that all forms of life be subject to mutations in order to give rise to new and different features which could, in turn, produce new species, mutations may be easily produced in some forms of life by artificial means, e.g., chemicals, X-rays, and so forth.

Many natural mutations produce only a slight change in the individual, but others show such a strong effect that the entire structure associated with the gene is altered. Such a case is found with rose comb. In this case, the gene change through mutation produced a single comb. Thus, there is now one gene at a certain locus on the chromosome that produces rose comb, another that produces single comb. These are known as *allelomorphic genes* or *alleles.*

Centuries ago, and for centuries after the mutation took place, many chickens carried genes for rose comb and single comb, but each individual had either a rose comb or a single comb. It could not, and does not today, have both. Modern poultry breeders, through a process of genetic selection and testing, have been able to eliminate the gene not wanted and develop a pure line of either rose-comb or single-comb birds.

Dominant. When one bird has genes for both rose comb and single comb, the bird will have a rose comb. Thus, the gene for rose comb is *dominant* to the gene for single comb. Although the expression produces a rose comb, the single-comb gene is still there; only the rose comb is capable of manifesting itself.

Recessive. The single-comb gene is said to be *recessive* to the gene for rose comb. Or more simply, rose comb is dominant to single comb; single comb is recessive to rose comb.

Some dominant and their allelomorphic recessive genes. Many pairs of genes have been discovered by geneticists, and their dominance or recessiveness determined. Some common ones are as follows:

Character	Dominance
Rose comb	Dominant to single comb
Barred plumage	Dominant to nonbarred plumage
Silver plumage	Dominant to gold plumage
Slow feathering	Dominant to fast feathering
Side sprigs	Dominant to normal comb
Broodiness	Dominant to nonbroodiness
White skin	Dominant to yellow skin
Feathered shanks	Dominant to nonfeathered shanks

Phenotypic and Genotypic Relationship

From the above it may be seen that all rose-comb chickens may not be alike. Some could be pure for rose comb, having two genes for the character, one in each of the pair of chromosomes; yet another could be impure, having a gene for rose comb in one member of the pair, a gene for single comb in the other. When the bird has a rose comb, rose comb is its *phenotypic* expression; its gene makeup is its *genotypic* expression.

Homozygous genotype.   When the two genes in a pair of chromosomes are identical, either rose or single, the genotype is *homozygous* (like). Thus, there could be a bird homozygous for rose comb or one homozygous for single comb. When the genes are the dominant ones, the bird is *homozygous dominant*. When the genes are the recessive ones, the bird is a *homozygous recessive*.

Heterozygous genotype.   When one chromosome member of a pair carries a gene for rose comb and the other chromosome contains a gene for single comb, the genotype is said to be *heterozygous* (unlike).

Genetic Symbols

Rather than continue to write out a full description of each gene, like rose comb and single comb, geneticists have designated an abbreviation for each gene. For instance, the abbreviation or symbol for the dominant rose-comb gene is *R*. So as to correlate its allelomorph, the gene for recessive single comb is *r*. This procedure of abbreviations is similarly followed for all allelomorphic pairs of genes. Some symbols and their phenotypic and genotypic expressions for rose and single comb are shown below.

Type of Bird	Symbolic Expression	Genotype	Phenotype
Homozygous dominant for rose comb	*RR*	Homozygous rose	Rose comb
Homozygous recessive for single comb	*rr*	Homozygous single	Single comb
Heterozygous for rose comb	*Rr*	Heterozygous rose	Rose comb

21-D. INHERITANCE OF ONE PAIR OF CHARACTERS

It has been shown that there can be a pure rose-comb individual, *RR*, and a pure single comb individual, *rr*. Each will breed true because the germ material continues intact; there is no chance for the opposite allelomorphic gene, [R] or [r] to enter the picture. But what happens when a homozygous, rose-comb bird, male or female, is mated with a homozygous, single-comb bird, male or female? Figure 21-1 shows the gene makeup and the gene inheritance.

Explanation. A homozygous rose-comb parent, *RR*, produces two like gametes, [R] and [R]. A homozygous single-comb parent, *rr*, produces two like gametes, [r] and [r]. Each of the two male gametes has an equal chance of uniting with each of the two female gametes, initiating four combinations in the F_1 (first filial generation), *Rr*, *Rr*, *Rr*, and *Rr*. All are identical; genotypically they are heterozygous for rose comb; phenotypically, all have rose combs.

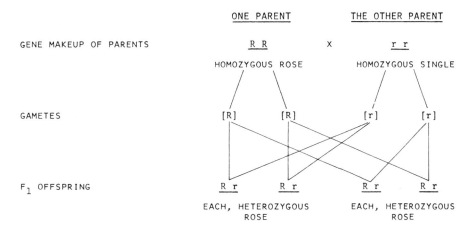

Figure 21-1. Inheritance of one pair of characters.

Note: The gene for rose comb, and its allele for single comb are located on one of the autosomes rather than on one of the sex chromosomes. Therefore, there is no involvement with the sex of the individual.

Dominance Complete

In the case of rose and single comb, the gene for rose is *completely dominant;* that is, the comb on a heterozygous rose-comb bird, *Rr,* looks exactly like the comb on a homozygous rose-comb bird, *RR.*

Crossing Two Heterozygous Parents

If two heterozygous rose-comb birds, *Rr* (as the F_1 individuals from Figure 21-1), are mated together the inheritance follows the pattern in Figure 21-2.

> *Explanation.* A heterozygous rose-comb bird produces two types of gametes, [R] and [r], in equal numbers. Each gamete from one sex has an equal chance of combining with each gamete from the other sex to give rise to four possibilities in the F_2 generation. (*Note:* rR combinations are always written with the dominant gene foremost, as *Rr.*)
>
> F_2 *ratio expression.* The F_2 offspring in Figure 21-2 may be expressed as the following two different ratios:
>
> > *Phenotypic ratio* = 3 rose: 1 single
> > *Genotypic ratio* = 1 homozygous rose, *RR:* 2 heterozygous rose, *Rr:* 1 homozygous single, *rr*

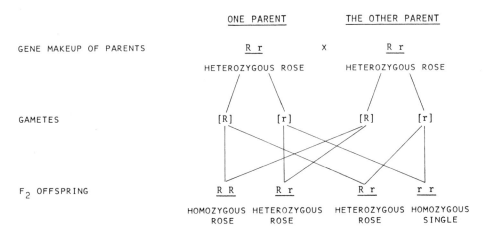

Figure 21-2. Inheritance when two heterozygous parents are crossed.

Another method of diagramming combinations. To show more clearly the various combinations of the genes when two parents are crossed, a gene-combination grid may be used, as shown in Figure 21-3.

Explanation. In Figure 21-2, the gametes from each sex are [R] and [r]. In Figure 21-3, the male gametes, [R] and [r], are placed at the top of the grid; the female gametes are on the left-hand side. (*Note:* The male and female gametes are always placed in these locations.) Using the grid to cross multiply, the various F_2 combinations appear in the squares.

Crossing Heterozygous and Homozygous Recessive Individuals

If a heterozygous rose-comb bird, *Rr,* is crossed with a homozygous recessive single-comb bird, *rr,* the genetic interpretation of the cross is shown in Figure 21-4.

Explanation. The heterozygous rose-comb bird produces two kinds of gametes, [R] and [r], but the homozygous single-comb bird produces only [r] gametes or sex cells. The gametic combinations in the offspring show the following ratios:

Phenotypic ratio = 1 rose: 1 single
Genotypic ratio = 1 heterozygous rose, *Rr:* 1 homozygous single, *rr*

Backcross. When an F_1 individual is crossed ''back'' to a homozygous recessive bird, the cross is known as a *backcross.* The procedure is used by geneticists to test the purity (homozygosity or heterozygosity) of a bird for a given character. An example is as follows:

Problem: A group of female chickens has rose combs, but the strain is not pure for rose comb inasmuch as single comb birds

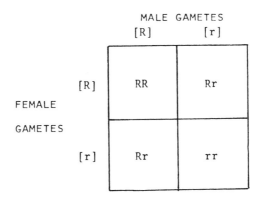

Figure 21-3. Grid showing F_1 gametic combinations.

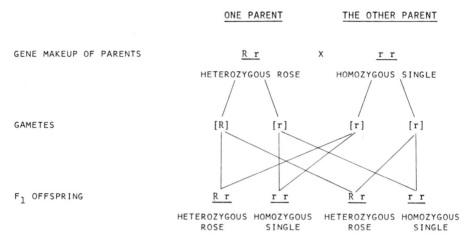

Figure 21-4. Inheritance when a heterozygous rose-comb bird is crossed with a homozygous single-comb bird.

appear occasionally. To determine which females are pure (homozygous *RR*) for rose comb they are mated with single-comb (homozygous recessive, *rr*) males, the females trap-nested, and the chicks from each dam hatched in separate containers.

Solution: In the above problem there are two types of rose-comb breeder females, *RR* and *Rr*, the former being pure, the latter, impure. When *RR* females are mated with *rr* males all the offspring will have rose combs, and will be heterozygous, *Rr*.

When *Rr* females are mated with *rr* males, half of the offspring chicks will have rose combs, *Rr*, and half will have single combs, *rr*. This group of breeder females is to be discarded because it is not pure for rose comb.

21-E. INHERITANCE OF TWO PAIRS OF CHARACTERS

Another somatic gene is known to produce black feathers; its symbol is *B*. Its recessive, *b*, is responsible for white feathers in one type of bird with white plumage, known as recessive white. The pattern of inheritance of these genes is exactly the same as in the case of rose and single comb. However, the genes *B* and *b* are on a pair of somatic chromosomes different from the pair containing the *R* and *r* genes.

Genes Considered Singly

We have seen what happens when *RR* birds are mated with *rr* birds. If we consider only the genes associated with *BB* (homozygous black) birds and *bb* (homozygous white), when a black chicken is mated with a white chicken all the offspring are heterozygous black, *Bb*. Crossing two F_1 individuals produces chicks in the phenotypic ratio of three blacks to one white. Genotypically, there is a ratio of one homozygous black, *BB*, to two heterozygous blacks, *Bb*, to one homozygous white, *bb*.

In other words, when we consider comb type and black color singly, the regular ratios for one set of characters appear in the F_2 generation. But what happens when both pairs of characters are considered together?

The Principle of Independent Assortment

When two pairs of characters are involved, each on different pairs of chromosomes, there are several more combinations of genes in the gametes. This increase in number is simply the result of mathematical chance, inasmuch as the chromosomes independent of one another align themselves on the spindle just before the reduction division.

How independent assortment works. For simplicity's sake, let us suppose there are but two pairs of autosomes in the chicken, "A" and "B." The two members of the pair of "A" chromosomes could be designated "A_1" and "A_2." Similarly, "B_1" and "B_2" for the "B" pair. When the cell division takes place to form the two sex cells, the chromosomes within each pair align themselves on the spindle at random. One possibility is:

$$\begin{bmatrix} A_1 \\ B_1 \end{bmatrix} \begin{bmatrix} A_2 \\ B_2 \end{bmatrix}$$

In the above case, chromosomes "A_1" and "B_1" go to one gamete; "A_2" and "B_2" would go to the other. But there is one more alternative for alignment on the spindle when two pairs of chromosomes are involved:

$$\begin{bmatrix} A_1 \\ B_2 \end{bmatrix} \begin{bmatrix} A_2 \\ B_1 \end{bmatrix}$$

When this alignment occurs, one gamete contains chromosomes "A_1" and "B_2," while the other contains "A_2" and "B_1." The alignment of the chromosomes in respect to each other is due to chance alone; therefore, in one-half of the cell reduction divisions the first example would hold true, and one-half would occur as in the second example. Male and female gametes (sex cells) would now contain four combinations of the "A" and "B" chromosomes as follows:

$$\begin{bmatrix} A_1 \\ B_1 \end{bmatrix} \begin{bmatrix} A_2 \\ B_2 \end{bmatrix} \begin{bmatrix} A_1 \\ B_2 \end{bmatrix} \begin{bmatrix} A_2 \\ B_1 \end{bmatrix}$$

Genes on the Chromosomes

For a discussion of the inheritance of two pairs of genes we can assume that the gene for rose comb, *R*, and its allelomorph, the gene for single comb, *r*, are located on the "A" chromosome. The gene for black plumage, *B*, and its allelomorph for white plumage, *b*, are on the "B" chromosome. A homozygous rose-comb and colored fowl would contain all dominant genes, *RR* and *BB*. A homozygous recessive single-comb and white bird would be described as *rr* and *bb*. More commonly these are written: *RR BB* and *rr bb*.

Crossing a Rose, Black and a Single, White Bird

The outcome of this cross is shown in Figure 21-5.

Explanation.   One parent produces only [RB] gametes, the other only [rb] gametes. Thus, the only offspring combination that could occur in the F_1 would be *Rr Bb*, or heterozygous rose and heterozygous black birds. Phenotypically, all offspring would have rose combs and black plumage.

Cross Involving Two Heterozygous Individuals

If two F_1 individuals from Figure 21-5 are crossed together, each parent produces four types of gametes, since there is independent assortment of the chromosomes in which the genes rest. The cross is shown in Figure 21-6.

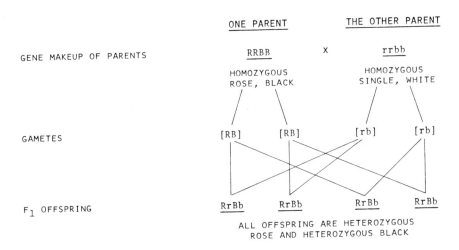

Figure 21-5. Inheritance of two pairs of characters.

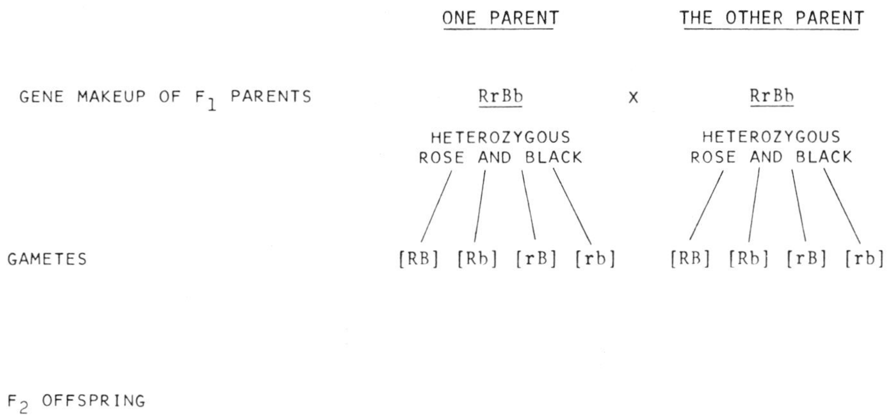

	ONE PARENT		THE OTHER PARENT
GENE MAKEUP OF F₁ PARENTS	RrBb	x	RrBb
	HETEROZYGOUS ROSE AND BLACK		HETEROZYGOUS ROSE AND BLACK
GAMETES	[RB] [Rb] [rB] [rb]		[RB] [Rb] [rB] [rb]

F₂ OFFSPRING

Figure 21-6. Inheritance of two pairs of characters when parents are heterozygous. (For F₂ offspring, see grid in Figure 21-7.)

Explanation. Since each F_1 parent in Figure 21-6 produces 4 types of gametes, and each produced by the male parent has an equal chance of uniting with each of the 4 different gametes produced by the female, there would be 16 different combinations as shown in the F_2 grid in Figure 21-7. The F_2 ratios expressed are as follows:

Phenotypic ratio = 9 rose, black: 3 rose, white: 3 single, black: 1 single, white

MALE GAMETES

FEMALE GAMETES	[RB]	[Rb]	[rB]	[rb]
[RB]	RRBB	RRBb	RrBB	RrBb
[Rb]	RRBb	RRbb	RrBb	Rrbb
[rB]	RrBB	RrBb	rrBB	rrBb
[rb]	RrBb	Rrbb	rrBb	rrbb

Figure 21-7. Gametic combinations when each parent is heterozygous for two characters.

Genotypic ratio = By checking against the genotypes in the grid in Figure 21-7, the combinations produce a ratio that is 1:2:1:2:4:2:1:2:1. Each of the groups in this ratio would be genetically different.

Of every 16 birds produced in the F_2 generation, on the average there will be 12 with rose combs and 4 with single combs, or a 3:1 ratio. There will be 12 blacks to 4 whites, or a 3:1 ratio. Considering the 12 with rose combs, 9 would be black and 3 white, or a 3:1 ratio. Of the 4 with single combs, 3 will be black and 1 white, or a 3:1 ratio. There will be 4 whites; 3 will have rose combs and 1 a single, or a 3:1 ratio.

21-F. SEX-LINKED INHERITANCE

The illustrations given for gene inheritance thus far have considered only those genes borne on the autosomes. There also are genes on the sex chromosomes, and the fact that the male has two sex chromosomes and the female but one complicates the diagrammatic picture of inheritance of these genes.

The Sex Chromosomes

The one pair of sex chromosomes in each cell of higher forms of life is different in one respect: the sex chromosomes are responsible for the sex of the individual. In the male chicken, each member of the pair of sex chromosomes is fully developed and functional inasmuch as it contains genes. In the female, however, only one member of the pair is functional; the other has atrophied and for all practical purposes contains no genes. When two sperm cells are formed from one complete cell, each sperm cell gets one member of the pair of sex chromosomes. But when an egg cell is formed by the female, only half such cells will receive a fully developed sex chromosome, the other half containing an atrophied one. When the sperm cell unites with an egg cell to form a new organism, the zygote will receive one fully developed sex chromosome from its male parent. However, half the zygotes will receive a fully developed sex chromosome from the female parent; the other half will receive an atrophied one. In this manner, those zygotes containing two fully developed sex chromosomes are destined to become males; those receiving one fully developed sex chromosome and one atrophied one will become females. This is the genetic interpretation of the inheritance of sex.

X and Y chromosomes. Geneticists have named the fully developed sex chromosome the X chromosome; the atrophied one, the Y chromosome. A simple example showing the inheritance of sex is given in Figure 21-8.

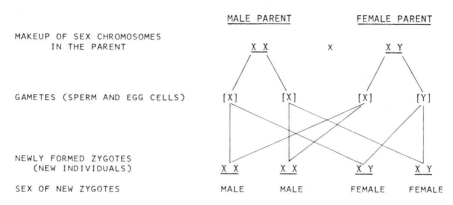

Figure 21-8. Inheritance of sex.

Interpretation. In Figure 21-8 it may be seen that each male gamete (sperm cell) has an equal chance of uniting with each of the two types of female gametes (egg cells). This relationship gives rise to the production of half males and half females in the next generation.

Inheritance of Genes on the Sex Chromosomes

One dominant gene found on the sex chromosomes produces silver (a certain type of white) feather color. Its symbol is *S*. The recessive allelomorph is *s*, the factor for gold feather color.

As the male nucleus carries two sex chromosomes, there will always be either an [S] gene or an [s] gene in each sex chromosome. If there were two [S] genes in the pair of chromosomes of the male, the bird would be *SS*, or homozygous silver. An *ss* makeup would represent homozygous gold. An *Ss* combination would be heterozygous silver. However, the female nucleus contains but one viable sex chromosome with its component of genes. The Y chromosome in the female does not carry genes and is involved with inheritance only as its presence produces a female zygote. Therefore, the female nucleus carries but one sex chromosome which could carry either one [S] or one [s] gene. It would be *hemizygous*.

Those genes located on the sex chromosomes are known as *sex-linked genes*; that is to say, such genes are linked with the sex chromosomes. The inheritance of one such pair of sex-linked genes is shown in Figure 21-9. It involves a cross between a gold male and a silver female.

Explanation. In Figure 21-9, the male has two functional sex chromosomes with a gene for gold plumage, [s], on each, while the female has but one functional chromosome with a gene for silver plumage, [S], on it. The female also has one nonfunctional chromosome carrying no genes, which for clarity has been designated [—]. In the F_1

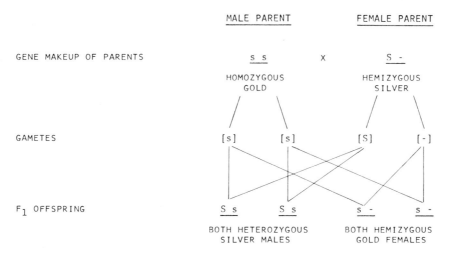

Figure 21-9. Inheritance when a gold male is crossed with a silver female.

generation, notice that when two genes appear (which would represent two X chromosomes), the sex is male. When only one gene appears, there is but one X chromosome and the sex is female.

Crisscross inheritance. In Figure 21-9, notice that a gold male and a silver female are used for the parents; yet the offspring chicks are silver males and gold females. This is sometimes known as *crisscross inheritance,* because the dominant and recessive characters crisscross in a diagram of this sort. In other words, male chicks have the same color as their female parent; female chicks have the same color as their male parent.

> *Practical example of this cross:*   The cross illustrated in Figure 21-9 is used by many breeders to provide a method of sexing the day-old chicks by down color. By mating a Rhode Island Red (or any other red, gold, or buff) male with a Barred Rock (or any other silver female), the male and female day-old, offspring chicks will be a different color, making it easy to sex them.

Reciprocal Cross Does Not Produce Sex Linkage

If a silver male is mated with a gold female, both the male and female offspring chicks receive a gene for silver, [S], from the sire, and all chicks will be silver. There is no possibility of sorting them by color to segregate the males and the females.

> *Important:*   To make a practical sex-linked mating in order to sex the day-old chicks, the male parent must be homozygous for the reces-

sive factor and the female parent must show the dominant factor. The reciprocal of this cross will not work.

21-G. INHERITANCE OF QUANTITATIVE CHARACTERS

The genetic characters described in the above paragraphs have been completely dominant or recessive. They may be spoken of as *qualitative characters,* and there are many. But all dominant characters do not show complete dominance. In some instances the recessive character bears as much weight as the dominant factor. In these cases there is a happy medium in the offspring. For example, when a large chicken is crossed with a small chicken, the offspring chicks will have a weight approximately midway between the weights of the two parents. Such genetic factors are known as *quantitative characters.* Probably the large proportion of genes in poultry are in this category. Egg size, egg production, skeleton size, egg shape, shell color, and breast width are but a few of the many quantitative characters.

Improvement in a Quantitative Character

There are often several quantitative characters working together to induce a change in the genetic makeup of a line of chickens. Egg production is in this group, for several sets of genes are responsible for the production of a large number of eggs.

How strain improvement is made. When the number of homozygous genes responsible for a given trait, such as egg production, increases, the number of eggs a bird lays also increases. Theoretically, at least, the birds within a flock that lay the best would be the most homozygous for the gene factors responsible for egg production. Thus, mating together only the best-producing birds would tend to increase the number of dominant genes within the line. And this is the procedure the geneticist uses to make improvements. He continues to breed from the best individuals, generation after generation, in order to increase the homozygosity of the gene material in his strain.

22
Genetic Management

Genetics is a science used mainly by the technician who develops new and better strains of chickens. But flock management plays an important part in the response to genetic development. Only the geneticist who has an insight into the behavior of genes under different environmental and other conditions will be able to make the greatest improvement in his lines of birds. For instance, a poultry breeder could develop a strain of broilers that would grow well on a litter floor but would do poorly on a wire floor. To be a practical genetic breeder he would have to develop two lines of birds especially suited to either litter or wire floors.

But the geneticist can do only a part of the job. Generally speaking, management has more to do with flock improvement than genetics. Although each line of birds has a bred-in potential to do a certain job, it is up to the poultryman to manage the flock so that the birds will have full expression of that potential.

This chapter is designed to give the practical poultry raiser a better insight into some of the reasoning behind genetic developments, why they have been instigated, and how he can best reap all the benefits of this science.

22-A. INHERITANCE OF SKIN COLOR

In some countries broilers with white skin are preferred to those with yellow skin. Special meat-type males with white skin are crossed with meat-type females having yellow skin, as white skin is dominant to yellow skin, and the broiler offspring have white skin. This procedure enables the producer of such chicks to use present-day, meat-type females with the special males, avoiding developing a white-skin female line as well as a white-skin male line.

Consumer demand for skin color. In the United States and some other countries there is a consumer demand for whole-body or cut-up broilers with a deep yellow skin color. But such a chicken, processed normally in water of relatively low temperature, is unsuited for the take-out restaurants. The batter used to coat the chicken pieces prior to cooking in deep fat does not adhere to the outer skin, and tends to flake off after cooking. If the water temperature is raised at processing time the batter tends to stick much better, but this bleaches out more of the yellow color in the skin. However, the whitish appearance of the parts is no detriment to the take-out restaurant trade; in fact, it is demanded.

22-B. INHERITANCE OF WHITE FEATHER COLOR

Dominant White Plumage Color

Unless the breeder male is involved with fathering chicks that can be color-sexed, most meat-type male lines of breeder parents are classified as *dominant white*. When such males are crossed with females that have colored feathers, the offspring chicks are white, or nearly white. (On occasion there may be some black or red showing.) This phenomenon is due to an autosomal gene that inhibits the production of color in the feathers. The White Leghorn is one pure breed with white feathers in this classification. Actually, the White Leghorn is a colored bird with an *inhibitor gene* that prevents expression of the genes for colored feathers. Dominant white meat-type male lines in use today are synthetics; most of them were developed by incorporating the inhibitor gene from the White Leghorn, then purifying the line.

The inhibitor gene is dominant and symbolized as I. Its recessive allele enables genes for colored feathers to produce their effects. Its symbol is i. The I gene is also only partially qualitative. That is, a homozygous bird for the inhibitor, II, will completely suppress colored feathers, but this is not always true in the case of a heterozygous bird, Ii.

If a dominant white male, $II\ CC$ (homozygous for the inhibitor and color), is crossed with a colored female, $ii\ CC$, all the offspring will show the genotypic makeup, $Ii\ CC$. Such offspring would be white or nearly white. Continual selection within the dominant white male line would eliminate practically all the slightly colored offspring.

The above cross is indicative of what happens when a dominant white, meat-type male line is crossed with certain colored female lines to produce broiler chicks.

Recessive White Plumage Color

A great many White Plymouth Rocks have been used for the female side of the mating in the production of broiler chicks. Such a bird is genetically known as a *recessive white*. It has neither an inhibitor gene, *I*, nor a gene for feather color, *C*. Its genotypic makeup is *ii cc*. When such a bird is crossed with a dominant white male, *II CC*, all the offspring broiler chicks are *Ii Cc*, or heterozygous for both characters. For all practical purposes they would be white, but on occasion some black or brown color would show in the outer feathers because of the interaction of other genetic factors.

In some dominant white male lines the gene for colored feathers, *C*, has been nearly eliminated by constant selection. When the *C* gene is removed, such birds have a genotypic combination of *II cc*. When crossed with a recessive white female, *ii cc*, all offspring broiler chicks are *Ii cc*, and completely white.

22-C. HERITABILITY OF QUANTITATIVE CHARACTERS

Quantitative genetic characters do not have a concise expression of their dominance or recessiveness, as in the case of qualitative characters such as rose comb and single comb. For instance, egg production per hen varies between 350 and 100 eggs or less per year. When a high-producing strain is crossed with a low-producing strain, the offspring produce eggs at some rate between two parent strains. Because quantitative characters are expressed as the result of several genes working together, there is no possibility of making simple genetic crosses to improve the occurrence of any desired trait.

Constant selection, generation after generation, and breeding from only the better birds is the most common method of improving the exemplification of characters that are quantitative in nature. Even then improvement in each character is not the same; in some instances there can be great accomplishments in the next generation, in other cases, they can be small. Another method of making genetic improvement is to cross two or more closely inbred lines of chickens, each line possessing great homozygosity for the desired characters. If the lines nick, improvement may be great for certain characters.

The ability of quantitative characters to be transmitted from parent to offspring is known as *heritability*. Each character, being quantitative, varies in its ability to be transmitted as compared to other characters. When heritability is high, progeny improvements are rapid; but when it is low, progress in improving future generations is slow.

Heritability percentages. It is common for the heritability of quantitative characters to be expressed in terms of percentage. Many geneticists have tried to determine these percentages, but the results have been highly variable because of the varying conditions under which the chickens involved have been housed and kept. A composite average of some of these results is given in Table 22-1.

Example: One would expect to be able to make a greater improvement in broiler weight than in egg production by continually breeding from the better birds, generation after generation. Hatchability, with only 10% heritability, could be improved very slowly under similar methods of selection.

Importance of management. The difference between the heritability percentages and 100% is attributable to management. For instance, only 5% of results of good chick livability are due to genetics; 95% are the responsibility of management. As with the other quantitative characters given in Table 22-1, there is a relationship between genetics and management.

Table 22-1. Some Estimated Heritability Percentages

Character	Heritability Percentage
Layers	
Chick livability	5
Adult livability	10
Age at sexual maturity	25
Keel length	20
Body depth	25
Adult body weight	55
Egg production	15
Egg weight	55
Shell texture	25
Egg shape	60
Albumen quality	25
Blood spots	15
Fertility	5
Hatchability of fertile eggs	10
Broilers	
6-week broiler weight	45
Total feed consumption	70
6-week feed conversion	35
Breast fleshing	10
Fat deposition	50
Dressing percentage	45

Nickability

As many genes are responsible for the expression of quantitative characters, it is possible to develop a line of chickens that is quite homozygous for several dominant genes. Another line might be homozygous for dominant factors other than those of the first line. When the two lines are crossed, new dominant genes are brought together in the offspring and the parent lines are said to *nick*. Actually, this means that the offspring will be better than either of the parent lines in respect to a certain character. However, other characters in the same lines may show little or no nickability. An example when two parent meat-type lines were crossed is shown in Table 22-2.

> *Explanation.* In Table 22-2, there is a definite nicking of characters responsible for egg production, as the offspring laid more eggs than either of the parent lines. However, this is not true of egg weight.

Table 22-2. Meat-type Offspring Compared with Their Parents

Item	Parent Male Line	Parent Female Line	Offspring
Total eggs, hen-day	148	165	166
Total eggs, hen-housed	139	155	156
Total hatching eggs	121	134	139
Average egg weight (oz/doz)	25.7	25.4	25.4
Average egg weight (g each)	60.7	60.0	60.0

Management Affects Expression of Gene Makeup

All quantitative genes do not express themselves equally under all systems of management. A typical example occurs in the case of egg production where a line of egg-type layers would lay an average of 260 eggs each when kept on a litter floor, but would produce only 248 eggs when kept in cages. Similarly, birds might produce well when given a good feed but poorly when given an inferior feed. With *qualitative* characters, the expression of the genes is complete regardless of the management. For instance, birds have either a rose or single comb; there is no variation.

> *Getting the genetic potential.* All strains of chickens have a bred-in inherent potential to produce a given number of eggs of a certain size, to attain a required weight, to be resistant to stresses, etc. Management of the flock will determine how well these genetic characters are expressed. As an example, data were collected on flocks containing 300,000 meat-type breeder females. Average figures for some characters involved with the entire 300,000 birds are shown in Table 22-3. The figures for the best flock in the group also are given; these are much higher than those for the average of all birds. But even though one flock was outstanding, there is no reason to believe that they produced according to the genetic potential; improved management might have increased their productivity. However, the

Table 22-3. Production Averages for All Flocks and Best Flock of Meat-Type Breeders

Item	Average of All Flocks (300,000 Pullets)	Average of the Best Flock (5,438 Pullets)
Pullets housed of those started (%)	95	97
Eggs per pullet housed	156	166
Hatching eggs per pullet housed	140	156
Hatching eggs (%)	91	94
Average monthly mortality (%)	1.4	0.8
Average feed/100 hens/day (males incl) (lb)	37.6	36.0
Average feed/100 hens/day (males incl) (kg)	17.1	16.4
Feed per dozen eggs (males incl) (lb)	7.5	6.7
Feed per dozen eggs (males incl) (kg)	3.4	3.04
Average total hatchability (%)	86	88

Source: Arbor Acres Farm, Inc., Glastonbury, Conn.

flock did come the closest to reaching the genetic potential. Failures in the management program produced less favorable results in the other flocks.

22-D. INHERITANCE OF BODY CHARACTERS

Although many factors such as depth of body, width of body, length of keel bone, length of leg bones, etc., have a bearing on body conformation, these characters must be left to the professional geneticist in his endeavor to develop strains of birds that best meet the demands of the poultryman. However, the inheritance of body weight is a factor that is the responsibility of both the geneticist *and* the poultryman. This is particularly true of meat-type strains involved with the production of commercial broilers and roasters.

Body weight in broilers is closely correlated with the weight of the parents at 6 weeks of age. The heritability is about 45%, and few quantitative genetic characters present a higher figure.

What 45% heritability means. Heritability is the relationship between genetic responsibility and the responsibility of the poultryman for improvements in the next generation. In this case 45% of growth is genetic, 55% is management.

Example: A flock of broiler breeder female parents has an average weight of 3.5 lb (1.6 kg) at 6 weeks of age. The geneticist wants to increase the weight of the offspring broilers, so he selects for breeding purposes only those females in the breeder parent flock that weigh 4.0 lb (1.8 kg), and breeds from them when they reach maturity. Therefore, those selected are 0.5 lb (227 g) heavier than the flock average. Broiler weight is 45%

heritable. Therefore, offspring broilers from the 4.0-lb (1.8-kg) breeders would inherit a weight increase of 0.225 lb (102 g) over the parent female flock (45% × 0.5 lb or 45% × 227 g). Assuming that no such selection was made in the male line parents, the genetic increase in broiler weight must be divided by 2, giving an increased broiler weight of 0.1125 lb (51 g). Had the same selection pressure been made in the male line as in the female line, the increase in average broiler weight would have been 0.225 lb (102 g).

Selection pressure. The genetic term denotes increased selection within a given flock of birds so that a smaller segment of the flock population will be used as breeders. Or, to say it another way: The more birds culled from the breeding flock, the higher the selection pressure. The term here is expressed in *percent*, denoting that percentage of the flock retained for breeders. The lower the percentage, the *higher* the selection pressure.

Selection pressure and body weight. Body weight in broilers is more closely correlated with parent body weight at 6 weeks of age than with parent body weight at sexual maturity. The more pressure exerted on the parents at 6 weeks of age, the greater the improvement in the offspring broiler weights.

Because commercial meat-type flockowners secure parent chicks from primary breeders, these flockowners have the opportunity of exerting selection pressure at the parent level in their flocks, i.e., they may use a lower percentage of the 6-week-old birds as breeders. The higher the pressure, the more the improvement. The exact relationships are shown in Table 22-4.

How to use Table 22-4. Select the percentage of the heaviest male birds to be kept after weighing them at 6 weeks of age. Find the figure at the top of the table. Extend the male percentage column downward until it reaches the line for the female percentage. The figure indicates the expected gain in pounds of broiler weight as a result of the selection at the parent level.

> *Example:* Male selection retention is 60%, female retention is 80%. The expected gain in weight in the broiler offspring is 0.081 lb.
>
> *Note:* The effect of exerting pressure on one sex is exactly the same as exerting pressure on the other sex, showing that no sex-linked genes are involved.

22-E. SELECTING MEAT-TYPE BREEDING BIRDS AT 6 WEEKS

Most breeding organizations require the exertion of *selection pressure* in meat-type lines when the birds are 6 weeks of age, and more for the

Table 22-4. Table for Calculating Expected Gain in Pounds as a Result of Selection Pressure at Various Levels in the Parent Population

Percentage Selection Pressure (Retained) of Females	Percentage Selection Pressure (Retained) of Males									
	100	90	80	70	60	50	40	30	20	10
	Expected Broiler Weight Gain in Pounds									
100	0.000	0.014	0.027	0.040	0.054	0.067	0.081	0.096	0.114	0.139
90	0.014	0.028	0.041	0.054	0.068	0.081	0.095	0.110	0.128	0.153
80	0.027	0.041	0.054	0.067	0.081	0.094	0.108	0.123	0.141	0.166
70	0.040	0.054	0.067	0.080	0.094	0.107	0.121	0.136	0.154	0.179
60	0.054	0.068	0.081	0.094	0.108	0.121	0.135	0.150	0.169	0.193
50	0.067	0.081	0.094	0.107	0.121	0.134	0.148	0.163	0.181	0.206
40	0.081	0.095	0.108	0.121	0.135	0.148	0.162	0.177	0.195	0.220
30	0.096	0.110	0.123	0.136	0.150	0.163	0.177	0.192	0.210	0.235
20	0.114	0.128	0.141	0.154	0.168	0.181	0.195	0.210	0.228	0.253
10	0.139	0.153	0.166	0.179	0.193	0.206	0.220	0.235	0.253	0.278

males than for the females. The program calls for starting more day-old chicks than will be necessary in the breeding pens at sexual maturity. Some of these extra chicks will be used as replacements for the birds that die during the growing period; others will be needed because a certain percentage of the birds—the smaller ones—will be culled at 6 weeks as a part of a genetic program of improving the growth of the broiler off-spring.

Weight Selection of Meat-Type Lines

There is a close correlation between the weight of meat-type parents at 6 weeks of age and the weight of their broiler offspring. Therefore, there must be a weight selection of the breeders at 6 weeks of age. As there is little correlation between body weight at sexual maturity and weight of the broiler offspring, the weight selection cannot be made when the birds go into the laying pens, or at any time after 6 weeks.

More pressure (culling) usually is placed on the males than on the females, although if the same *percentage* of birds are removed from each sex group there is no difference in the effects in the next generation. Because there are fewer males than females, it is more economical to cull the males than the females (see Chapter 22-D).

How to select males. First it is necessary to know the percentages of the day-old males to be retained at 6 weeks of age. This figure can be supplied by the genetic breeder. Next, use a poultry catching screen and catch and individually weigh a minimum of 15% of the males in *each pen* when the males are 6 weeks of age. The sample must be representative.

On a sheet of paper, record the weights in order from the heaviest to the lightest. Using the percentage figure, calculate the number of males to be kept, and count these off starting with the heaviest and working toward the lightest. The figure at which the percentage is reached will be the minimum weight for males in the pen to be kept.

Next, individually weigh all males and discard those below the minimum established weight.

Remember: There will be weight variations among the pens in the house; thus, sample weighings must be taken in every pen, and a minimum body weight established.

How to select females. Most primary breeders do not stipulate that se-lection pressure for body weight be exerted on the females of meat-type lines, but some do. If the latter is the case, a procedure identical with that used to process the males must be followed.

Selection of Egg-Type Lines

Inasmuch as 6-week selection is made to improve the weight of meat-type offspring, there is no reason for following the procedure with egg-type breeder or commercial lines; individual weight is not involved here. Only inferior birds should be removed at a young age, and it is better to wait until the birds are 10 to 14 weeks of age before removing any inferior ones.

Negative (Antagonistic) Correlation of Genetic Characters

From a genetic standpoint certain quantitative characters have a positive correlation; that is, when selection is made to improve livability within the line, egg production also increases. However, many other quantitative characters have a negative correlation. When selection is used to increase body weight in the next generation, for instance, egg production decreases. A detailed example is given in Table 22-5. When there is genetic selection pressure exerted in the parents so as to increase the broiler weight by 0.1 lb (45 g) notice the effect on some other characters.

Another way of stating negative correlation. If one were to weigh a group of broiler chicks, one would find that the dams of the heavier birds would produce relationships as outlined in Table 22-5. The heavier the broilers, the fewer the eggs produced by their female parents, the lower the hatchability, and so forth.

Table 22-5. Effect of Genetically Increasing the Broiler Weight by 0.1 lb (45.4 g) at Maturity

Item	Effect	
Egg production	Decreases	10.0 %
Pounds of feed per dozen eggs	Increases	0.9 %
Laying house livability	Decreases	2.3 %
Hatchability	Decreases	1.2 %
Chicks hatched per hen housed	Decreases	8.3 %
Broiler livability	Decreases	0.25%

22-F. SEX-LINKED DWARFISM

Although there are several types of dwarfism in chickens, the sex-linked type is most common. It is produced by the sex-linked, recessive gene, *dw.* Birds of normal size carry the sex-linked, dominant gene, *Dw* (see Chapter 1-D).

Dwarf Commercial Layers

Inasmuch as the economic value of dwarfism rests almost entirely with the commercial female, and because the responsible gene is sex-linked, there is one practical method of making a Leghorn mating.

Parents		Offspring	
Male	Female	Male	Female
Dwarf (*dw dw*)	Normal (*Dw*—)	Normal (*Dw dw*)	Dwarf (*dw*—)

Using the above procedure it is only necessary to develop a male line of dwarf breeding birds; the regular normal-size female line may be used for the female side of the mating. The offspring pullets are dwarfed, but not the males, which are normally destroyed when commercial Leghorn pullet chickens are produced.

Dwarf Broiler Breeders

With broiler breeders the economics rest with the production of a dwarf female *parent* so that hatching eggs may be produced at a lower cost. The same type of mating shown to produce commercial Leghorn pullets can be used to produce the broiler breeder parent females, which are then mated with normal males to produce normal commercial broiler chicks as follows:

Parents		Broiler Offspring	
Male	Female	Male	Female
Normal (*Dw Dw*)	Dwarf (*dw*—)	Normal (*Dw dw*)	Normal (*Dw*—)

22-G. INHERITANCE OF EGG PRODUCTION

The inheritance of egg production presents a complicated genetic picture, for egg production is the end product of several separate genetic characteristics working together to cause the bird to produce a given number of eggs in a given time. Of the several factors involved, the five most important are as follows:

1. *Early sexual maturity.* The younger the bird when she begins to produce eggs, the greater her egg production will be during her laying year. However, this takes on a negative value, because layers must be prevented from laying eggs at an early age, because eggs laid by young birds are smaller than those laid by older birds. Feed-control and light-control programs are used to prevent early egg production. Genetically, however, the gene relationship still stands.

2. *High intensity of lay.* This character is manifested by the ability of the bird to lay at a rapid rate. Since chickens lay their eggs in clutches—that is, they lay an egg on each of several consecutive days before they miss a day—clutch size (length) is an important genetic factor.

When a hen is producing at the rate of 80%, she must lay one egg on 4 out of every 5 days. Some birds have been known to lay eggs for 200 consecutive days before missing a day.

3. *Persistence of lay.* The longer the laying cycle before the hen enters her molting period, the better egg producer she is. Persistence is a definite genetic factor associated with egg production. In the early days of commercial egg production, a 12-month period of production was thought to be ample, but today the requirement is for a bird that can profitably produce eggs over a 13-month or longer period.

4. *Incidence of pauses.* Pauses longer than 2 or 3 days between clutches have a phenomenal bearing on the total number of eggs laid during the production period. These long pauses are usually the result of some type of stress, but may be due to inheritance.

5. *Number of broody periods.* Although broodiness has been reduced or removed in many strains of chickens, it still is a factor to be reckoned with, particularly in meat-type strains. Chickens do not normally produce eggs when broody.

Egg production has a low heritability percentage of 15; in other words, 85% of the egg-producing ability of a hen rests with management. There are so many management factors involved that it would be repetitious to mention them here. Indirectly, anything that furnishes a better environment for the hen, prevents stresses and excessive mortality, and provides for her physiological well-being will also cause the production of more eggs.

22-H. INHERITANCE OF EGG SIZE

What is ideal egg size? Modern geneticists have induced hens to lay large eggs through breed improvement. But egg size is a highly variable factor; therefore, what may be ideal in one case or one segment of the laying period may not be in another.

Large Egg Size May Be Important

Egg size is the result of a quantitative genetic character or characters with a high heritability. Therefore, it is relatively easy for the geneticist to increase the size of the eggs in a given strain of birds; it is less easy for the poultryman to manipulate egg size.

Regardless of whether eggs are sold by graded weight or by bulk weight, it is important to have large eggs. Extra-large eggs often command no higher price than large eggs, yet it costs more to produce them. When this is the case, strains of chickens producing excessively large eggs may have a disadvantage.

Variations in Egg Size

There are many reasons for variations in egg size. Some are the result of the natural pattern of egg production; others are the result of feeding and management; still others bear a genetic relationship.

Bird variations in egg weight. Individual birds within a given flock lay eggs of different sizes (weights). However, each bird tends to lay consecutive eggs that are similar in size. The only variation is when double-yolk eggs are produced by multiple ovulation. Furthermore, each bird lays eggs that are similar in shape. Although a high proportion of the hens will lay eggs that approach the average weight of all eggs laid by the flock, there will be many that will produce larger or smaller eggs.

Variations in egg size throughout the laying year. When a pullet begins egg production her eggs are relatively small. Gradually egg size increases as she becomes older, until maximum size is reached near the end of her first laying cycle. If she is molted, eggs laid during her second cycle of production will be larger than those laid during her first cycle, but no larger than had she not been molted. Egg size is correlated with the age of the bird (see Table 19-7).

Most of the first eggs laid are of relatively little value, either as commercial eggs or as hatching eggs, because of their small size. This has led poultrymen to delay the onset of egg production until later in the bird's life, when her larger egg size is commensurate with her greater age.

Commercial egg grades. The producer of commercial eggs is quite cognizant of the fact that egg size increases during the laying year, because he markets his eggs according to a grade weight. The greater the percentage of the heavier sizes, the greater his cash return. Egg weight grade sizes vary widely from country to country, but in the United States they are usually as follows:

Weight Classification	Minimum Net Weight per Dozen Eggs (oz)
Jumbo	30
Extra Large	27
Large	24
Medium	21
Small	18
Peewee	15

Egg size as it affects income.   Only when a premium is received for Jumbo and Extra Large eggs is there an advantage in increased egg size. However, if eggs are to be sold on a weight basis, added size does have an economic advantage. For further details regarding average egg sizes, see Chapter 16-S.

Early egg-size variation in the breeder hen.   From a flock standpoint, even though the first eggs laid are smaller than those laid later in the life of the hen, they are genetically similar. Such small eggs should impart identical gene material to the next generation as larger eggs produced later. In many instances hatcherymen use hatching eggs with a smaller minimum size at the start of production and raise the minimum requirement after about 10 weeks of egg production.

Chick-size variations.   Chick size is directly related to the size of the hatching egg from which the chick is hatched. Large eggs produce large chicks; small eggs produce small chicks. It is also known that small chicks do better if raised separately from large chicks. This fact has caused many hatcherymen to segregate their hatching eggs into two or three weight classifications, such as large, medium, and small, prior to setting. Chicks from each group are sent to individual customers, so that the chicks will be more uniform in size. From a brooder house management program, this is an excellent procedure, but genetically it does have a disadvantage. After the first few weeks of egg production by the breeding flock, small eggs produce small chicks, which in turn produce mature pullets that lay smaller eggs. Pullets coming from large eggs would produce larger eggs.

Recommendation.   Where there is an egg-size problem in the strain of egg-type birds being used, eggs should not be set according to size unless the minimum hatching weight is raised.

22-I. INHERITANCE OF EGG QUALITY

Egg quality includes a study of the quality of the eggshell and the quality of the interior contents. Although not as important in meat-type strains as in commercial egg-producing strains, egg quality has a greater value today than it did several years ago.

Eggshell Quality

One measure of eggshell quality is its thickness. The thicker the shell the more resistant it is to breakage. Eggshell thickness is a quantitative genetic factor with relatively low heritability. The thickness of the shell may be altered by various factors of management: temperature, stress, disease, feed, and others.

Some shell abnormalities are inherited. Chalky shells, ridges, and so forth, may be the result of gene variations. Other abnormalities are often due to imperfections in the area of the oviduct where the shell is deposited.

Interior Egg Quality

Albumen quality has a heritability of about 25%. Therefore, inferior quality is more often the result of improper flock management than of the manifestation of genetic action.

Blood spots have a very low heritability. However, they have become associated with some breeds of chickens. Eggs from those breeds laying brown-shelled eggs have a higher incidence of blood spots than eggs with white shells (see Table 16-35).

Interior egg quality may be improved within a strain of chickens, and commercial poultrymen are ever on the lookout for those strains that produce good eggs.

22-J. VARIABILITY OF INDIVIDUAL BIRDS WITHIN A FLOCK

The "Average" Bird

Unfortunately, practically all poultry management procedures are built around flock averages. Feed consumption is based on a bird of flock *average weight;* medication is similarly calculated; egg production figures are always for the average bird. If we take weight, only a very small percentage of the birds in the flock are average weight. Half are larger; half are smaller. These variations are normal even in the healthy flock because there is no strain of birds so uniform that there are no variations. But if it were not for this variability the geneticist would not be able to make improvement through breeding techniques. The variability makes it possible for him to select the best (largest, etc.) individuals and breed from them.

Body Weight Variability

The variability in weights of mature broilers is shown in Table 20-16. Here it may be seen that the variation is different in males than in females, and in the straight-run flock. Obviously, some of the variability is inherited, for strains differ in their uniformity.

Laying Strain Variability in Weight

Table 22-6 shows the approximate extremes in individual body weight for laying strains at varying weights. For variations in sexually mature weight see Table 16-8.

> *Small birds remain small.*   Variations in body weight at 20 to 22 weeks continue through the laying period. The lighter birds remain lighter; the heavier birds remain heavier.

Table 22-6.   Average Maximum and Minimum
Pullet Body Weights

| Average Flock Body Weight for Leghorn Pullets | | Approximate Extremes in Individual Body Weight in a Healthy Flock | | | |
| | | Maximum | | Minimum | |
lb	kg	lb	kg	lb	kg
1	0.45	1.24	0.56	0.76	0.34
2	0.91	2.50	1.13	1.50	0.68
3	1.36	3.78	1.71	2.22	1.01
4	1.81	5.08	2.30	2.92	1.32

Largest Pullets Lay First

At the end of the growing period, the largest pullets are the ones that lay first. This differential is accentuated because an individual pullet will gain 0.5 lb (227 g) or more just prior to laying her first egg. This sudden increase in weight is nature's way of preparing the pullet for her egg production period.

Those pullets laying first attain a certain weight when egg production begins. As the smaller birds reach sexual maturity later, they, too, will attain the same weight as the first birds to lay. Therefore, for practical purposes, all birds in the flock will weigh approximately the same when they lay their first eggs, but as all birds continue to gain weight, the first birds to lay always will be the heaviest, and there always will be large, medium, and small birds throughout the laying period.

Weight Gain on an Individual Egg-Type Bird Basis

For years, poultrymen have been accustomed to a body weight growth curve based on a *flock average* that shows a gradual gain during the growing period and continuing on during the laying period, although weekly weight increases are materially less during egg production.

On an *individual bird basis* the curve is vastly different than the one for flock averages. A large share of the increase in weight in the bird near sexual maturity occurs just prior to, and within a week after, the first egg is laid. For the next 10 weeks of egg production, the individual bird will show little or no gain in weight. In fact, many birds will lose weight.

Percentage of Production on Individual Hen Basis

Beginning with the start of egg production, the curve for *flock average* egg production shows a rapid and almost uniform weekly increase to 90%, or higher, hen-day egg production the eighth week, then generally decreases during the next 44 or more weeks.

But this is not the picture on an *individual bird* basis. Most pullets will lay their first egg, then skip the next day, then waver somewhat during the remainder of the week, laying about four eggs during the first 7 days. After that, the pullet is in high gear, and her second week of egg production is usually her best, with little or no reduction during the third week, after which there is a slow weekly decrease.

Age at First Egg on an Individual Hen Basis

All pullets in the flock do not begin to lay eggs at the same time (age). In fact, most Leghorn flocks show a period of at least 7 weeks between the time the first pullets lay and the time the last pullets lay. In some instances, flocks show a time period of up to 10 weeks. Furthermore, the same percentage of pullets do not begin to lay during each of the 7 to 10 weeks.

Flock Average and Individual Bird Lay Compared

Table 22-7 has been constructed to show the first weeks of egg production for an average flock of pullets starting to lay with production figures for the first week, second week, and so forth, through the tenth. If the egg production is averaged for these eight groups it is easily seen that the weighted flock average figures are common and those usually seen by the flock manager, while on an individual bird basis the figures are vastly different.

Table 22-7. Variation in Individual Percentage of Hen-day Egg Production and How It Affects Flock Average Production

% of Flock	Week of Egg Production									
	1	2	3	4	5	6	7	8	9	10
	Individual % Hen-day Egg Production									
0.5	60	94	94	93	92	91	90	89	88	87
3.0		60	94	94	93	92	91	90	89	88
11.5			60	94	94	93	92	91	90	89
35.0				60	94	94	93	92	91	90
35.0					60	94	94	93	92	91
11.5						60	94	94	93	92
3.0							60	94	94	93
0.5								60	94	94
Weighted Average % Hen-day Egg Production	0.3	2.3	10.2	35.1	68.0	86.6	91.8	92.2	91.5	90.5

Body Weight Best Criterion of Bird Uniformity

It is obvious that once a flock begins to lay, nothing can be done to improve its uniformity of sexual maturity. Improvement must be made during the growing period. But unfortunately we have no measure of how fast a flock is progressing toward sexual maturity; we must look to an associated factor, and that is body weight. If the flock is uniform for body weight during growing it will be uniform for sexual maturity.

Flock Uniformity and Egg Production Curves

Degree of flock weight uniformity at sexual maturity affects the egg production patterns and the resulting production curves. Three flocks of varying uniformity at sexual maturity have been selected to show these variations:

1. *Excellent uniformity:* 78% of the birds are within 10% of the mean flock body weight.
2. *Satisfactory uniformity:* 70% of the birds are within 10% of the mean flock body weight.
3. *Very poor uniformity:* 58% of the birds are within 10% of the mean body weight.

Production graphs for these three flocks are shown in Figure 22-1. Several points are of interest.

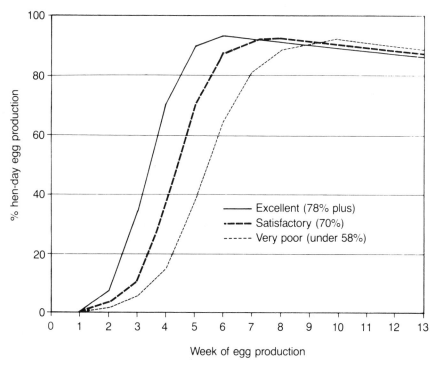

Figure 22-1. Flock production curves in relation to flock weight uniformity at sexual maturity.
(From M. O. North, Poultry Digest, Oct. 1980.)

1. The more uniform the body weight at sexual maturity, the earlier the flock peaks in egg production.
2. The flock average egg production curve just prior to peak production is more abrupt with more uniform flocks.
3. The lower the weight uniformity at sexual maturity, the fewer eggs produced during the laying period when all the pullets are liquidated (sold) at the same time.

Individual Versus Flock Administration of Medicaments

It is not possible, nor is it economical, to try to medicate birds on an individual basis. Feeding chickens is on a group basis, unlike the feeding of large animals such as the cow where feed allocations are based on age, body weight, and milk production of each individual. The best that can be done with a flock of chickens is to administer to the "average" bird. Obviously those that are smaller than average will get more than an adequate amount; those that are larger will receive less. But the poultry manager must be ever cognizant that these variations occur, and wherever possible, he must make adjustments to compensate for them.

22-K. FEATHER SEXING

Slow feathering in the young chicken is due to a *qualitative sex-linked, dominant gene* "K." Its allele, rapid feathering, is the response of the recessive gene, "k." Although the predominant feature of the recessive gene is to cause the feathers to grow more rapidly during the first 6 to 9 weeks of the chick's life, the difference between slow and fast feathering is obvious at the time the chick is hatched, but only in the relationship of the length of the primary wing feathers to the length of the primary wing coverts, which are small downy feathers covering the base of the primary feather shafts. A description of the fast-feathering and slow-feathering wing in the day-old chicks is as follows:

Rapid-feathering (k) female. When the chicks hatch, the primary wing feathers of the females are longer than those of the slow-feathering males, and the coverts are always shorter than the primaries.

Slow-feathering (K) male. At hatching time the primary wing feathers are short, and the coverts are always as long as, or longer than, the primaries.

Diagrammatic view. See Figure 22-2.

Day-Old Feather Sexing

Slow-feathering and rapid-feathering genes may be used in a breeding program to make it possible to determine the sex of day-old chicks by feather sexing, the term applied to this type of inheritance.

How to make the cross. When a rapid-feathering male, *kk*, is mated with a slow-feathering female, *K—*, sex-linked inheritance is involved, and the speed of feathering reverses itself in the chicks of the next generation. Thus, the offspring chicks from such a mating would show males that are slow-feathering, *Kk*, and females that are rapid-feathering, *k—* (see Chapter 21-F).

Important: The *reciprocal* of this cross will not provide sex linkage and feather-sexing possibilities in the offspring chicks. Remember that the male parent must be homozygous for recessive rapid feathering, *kk*.

Special lines involved. To make it possible to sex-segregate day-old chicks, it is necessary that special parent lines of birds be developed. The male line must be homozygous for rapid feathering; the female line must be hemizygous (pure) for slow feathering.

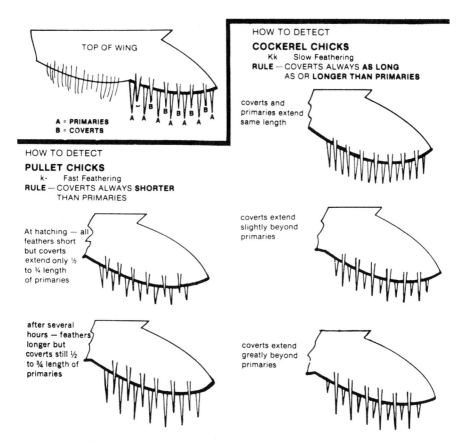

Figure 22-2. Feather sexing chicks at hatching. Sexing is accomplished by examination of primary (A) and covert (B) feathers on the tip of the outspread wings as illustrated above for a typical pullet. Note that coverts emerge from well up on the top surface of the wing and primaries emerge from the lower edge of the wing. The wing should be examined from the top surface. A good light is essential. As can be seen, the relative length of primaries and coverts is more important than the overall length of feathers, since overall length depends upon the length of time that the chick has been out of the shell.
(Source: Arbor Acres Farm, Glastonbury, Conn.)

Slow-Feathering Broiler Chicks

When matings are made so that it will be possible to feather-sex broiler chicks at hatching time, the cockerel chicks are slow-feathering *Kk*, while the pullet chicks are rapid feathering *k—*. Although most strains of such slow-feathering males produce fully feathered birds by the time they reach 5 or 6 weeks of age, often because of stress and hot weather some males may be poorly feathered with excessive pin feathers at market time, and the feather sexing feature of these crosses may represent a disadvantage.

22-L. COLOR SEXING

The genes for silver, *S*, and gold, *s*, are of great economic importance because they are used as one basis for sex determination in day-old chicks. The breed difference gives rise to two classifications.

1. *Color-sexing laying strains.* Such a cross is confined almost entirely to a cross between a gold male, *ss*, and a silver female, *S—*, to produce medium-size birds laying brown-shelled eggs (see Chapter 21-F).

2. *Color-sexing broiler strains.* Many prime broiler breeders have used the silver, *S*, and gold, *s*, genes to develop autosexing in day-old broiler chicks. Although most of the broiler breeder parents involved in these crosses are synthetics, the male line possesses a great deal of Cornish blood and is gold in color. The female line usually is recessive white, carrying the silver genes.

   Day-old female chicks from these crosses are gold or buff colored and sometimes the back is striped. The male chicks are light yellow or white.

   Mature broiler females are gold and white, with most of the gold in the outer portion of the feathers. The undercolor and quills usually are white so there is no disadvantage in the processing plant. The male broiler chicks are white, or almost white.

22-M. SOCIAL ORDER IN CHICKENS

Within a group of chickens there are certain individuals that dominate others of the same sex. This dominance is the result of agonistic conflict through fighting, pecking, and chasing until a social hierarchy (each individual subordinate to the one above) is established. Such a social order should not, and cannot, be prevented. It is nature's way! Aggressiveness and submissiveness are common to all vertebrates, and in the chicken reduce tension and conserve energy.

Establishing the Social Order

Origin of the social order begins when the birds are about 8 to 10 weeks of age, but is not fully consummated until they approach sexual maturity. Because social dominance in either sex is not evident prior to 8 to 10 weeks of age, it is possible to raise large numbers of broilers in a single pen without fear that the birds will develop social agonistic characteristics. There is no challenge among individuals for social dominance during broiler age.

The hierarchical type of social organization is the result of sexual hormone activity, and because chickens can recognize one another by observing a bird's head and its appendages. When the number of mature birds in a pen is large, suborders are set up to divide the group. These suborders involve between 200 and 300 ft² (19 and 28 m²) of floor space with no area containing more than 100 birds, the maximum number that can be identified by one individual.

Birds within one suborder are hesitant about going to another location for they know they will be chased back to their own territory. Therefore, equipment should be scattered throughout the house so that no birds will have to go more than 10 ft (3.05 m) to get to feed, water, and a nest.

Types of Social Hierarchy

The chicken uses agonistic conflict to establish its place in a social order. There are two types of social hierarchy, and the differences are great.

1. *Complete social dominance hierarchy.* This type of hierarchy is established when mature male and female chickens are kept on a litter or slat floor. Within the pens or suborders each sex establishes its own social order known as the *peck order.* For example, one male finally dominates all other males in the group and ranks number 1 in the social order. Then there is the next male in the order. He is number 2 and dominates all males but number 1, etc., until finally there is the male at the bottom of the order; he is dominated by all other males. Similar social relationships are established by the females in a group kept on litter or slats.

2. *Social dominance by despotism.* Despotism is a type of social ranking among females held in multiple-bird cages (never on litter or slats). The despot (a ruler with absolute authority over all others) dominates all other females in the cage, and no peck order is set up among the subordinate pullets.

T. R. O'Keefe et al. (1988, *Poultry Sci. 67(7)*, 1008–1014) compared Leghorn hens in 6-bird cages with similar hens on a litter floor. From their work, several statements may be made, as follows:

1. Only one hen in each cage (the despot) will show any agonistic activity with other hens, winning all encounters.

2. Hens in cages develop their social order by despotism, not by the peck order.

3. In cages, hens subordinate to the despot may develop some type of social order, but it is not by agonistic encounters. Because these subordinates do not establish a peck order, they cannot be visually classified according to any degree of hierarchy.

4. When hens within a despotic social order are removed from cages to a litter floor, they establish a peck order involving all the female individuals.

5. When hens with an established peck order are removed from a litter floor to multiple-bird cages, all in each cage become a part of the despotic social order, with one becoming the despot.

6. Although increasing the density of hens in multiple-bird cages reduces individual bird egg production, the reduction is not from agonistic changes in the social order from one of despotism to one of a peck order but because of reduced feeder, waterer, and floor space.

23

Record
Management

Flock production records are a necessary part of good flock management. Some records, such as those of mortality, egg production, and feed consumption, must be kept on a daily basis; others can be kept on a less frequent basis. In most cases, daily records should be summarized at the end of each week and placed in a permanent file. Certain calculations are necessary at this time too. The amount of feed required to produce a dozen eggs, feed consumption per 100 females, hatching eggs per hen housed, and others fall in this category.

Weekly summary records may be divided according to the following groups:

1. Growing records
2. Laying records
3. Hatchability records
4. Egg grade-out records

23-A. SUMMARY RECORDS

Growing Summary

This is a record of flock behavior from 1 day of age to sexual maturity. In this record certain standards should be inserted at weekly intervals. This makes it possible for the poultryman to make regular comparisons between the actual and the standard.

There should be columns for numbers of birds alive, depletion (mortality), culling, cumulative depletion, feed consumption, body weight and notations regarding vaccinations, problems, weather conditions, etc.

Flock Production Summary

Similar to the Growing Summary is a Flock Production Summary. An illustration is given in Figure 23-1 for a flock of meat-type breeder females. By deleting the columns having to do with production of hatching

	Page 1							ARBOR ACRES FARM, INC. **FLOCK PRODUCTION SUMMARY**				

Flock No._____ House No._____ Pen No._____ No. Housed:_____ Males_____

Standard Hatching Egg Min.:   22 Ozs./ dz.   52.0 Gms./each

(1)	(2)	(3)	(4)	(5)	(6)	(7)	(8)	(9)	(10)	(11)	(12)	(13)
Date End of Week	Wks. in Prod.	NO.BIRDS END WK.		FLOCK DEPLETION				TOTAL EGG PRODUCTION			% H-D PROD.	
		Female	Male	Female		Male		No. Eggs For Week	Cum. Eggs to Date	Cum. Eggs H-H to Date	This Week	Stand-ard
				% Wk.	% Cum.	% Wk.	% Cum.					
	1											5
	2											18
	3											44
	4											68
	5											82
	6											81
	7											80
	8											79
	9											77
	10											76
	11											75
	12											74
	13											72
	14											71
	15											70
	16											69
	17											67
	18											66
	19											65
	20											64
	21											63
	22											62
	23											61
	24											60
	25											58
	26											57

Figure 23-1. Flock production summary.

eggs and substituting data for the production of commercial eggs, the sheet can be used for commercial layers. Items included in Figure 23-1 are number of birds, flock depletion, total egg production, percentage of hen-day egg production, hatching egg production, feed consumption, and body weight. Egg producers should also include average egg

___Females   Date Chicks Started :_____ Branch_____

Your Hatching Egg Min.: _____ Ozs./dz. _____Gms./each   Age 5% Prod. _____ Wks.

(14)	(15)	(16)	(17)	(18)	(19)	(20)	(21)	(22)	(23)	(24)	(25)	(26)
HATCHING EGG PRODUCTION							FEED CONSUMPTION IN LBS./KGS.					
% Hatching Eggs		Hatching Eggs Produced			Cum. Hatching Eggs/H-H		This Week	Cumulative Feed	Av./100 Fem./Day for Wk.(Males Inc.)		Feed per Doz. Eggs to Date (Incl. Male)	Female Body Wt.
		For Week		Cumulative								
Act.	Std.	Total	Cases	Total	Act.	Std.			Actual	Est.		
	—					—						
	—					—						
	20					.6						
	50					3.0						
	66					7						
	77					11						
	83					16						
	87					20						
	91					25						
	93					30						
	94					35						
	95					40						
	95					44						
	96					49						
	96					53						
	96					58						
	97					62						
	96					67						
	96					71						
	96					75						
	96					79						
	96					83						
	96					87						
	95					90						
	95					94						
	95					97						

Figure 23-1. (*continued*)

weights. Notice how certain standards of production are inserted to make it possible to compare actual weekly figures with standard figures. Two special computations are needed each week:

1. Average feed consumed per 100 birds per day for the week
2. Feed per dozen eggs

These two calculations serve as excellent guidelines for the amount of feed being consumed. Most feed recommendations are given according to (1) above, thus exemplifying the importance of the calculation.

Hatchability Summary

A weekly summary of hatchability is a necessary part of flock records. An example of such a form is shown in Figure 23-2, to be used for meat-type breeder females. Included in the data are number of hatching eggs set, total chicks hatched, percentage of total hatch, percentage of grade-outs, percentage of salable chicks, and number of salable chicks. Weekly standard figures for percentage of total hatch are included in order to make a direct comparison with the actual.

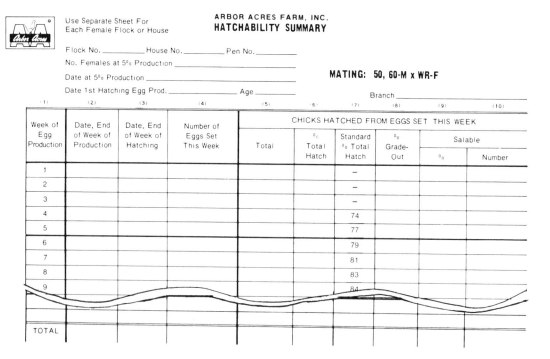

Figure 23-2. Hatchability summary.

How to record the weekly data. Weekly hatchability data should be inserted on the form according to the week the eggs were laid rather than the week the chicks were hatched. This gives a more direct correlation between hatchability and egg production.

Total hatch and salable hatch. Total hatch refers to the number of chicks "scooped" from the trays prior to any grading. Salable hatch includes only those chicks that are invoiced. Thus, the total hatch, minus the culls and extra chicks, equals the salable hatch. Both figures have been included on the form, because neither one alone is adequate for a complete analysis of the hatchability records.

Egg Grade-Out Records

Commercial egg producers must evaluate the size and quality of eggs produced. Egg weights should be taken at weekly intervals and compared with breeder standards. Egg grade-outs should be monitored both by hand candling and through the normal grading procedure in the processing plant.

Total egg mass, egg numbers times average egg weight, should be used in calculating nutritional requirements and the true productivity of a flock. Similarly, feed conversion should be calculated by dividing weight of feed by weight of eggs.

The routine monitoring of egg quality is necessary to pinpoint shell and interior quality problems. In many cases, shell problems can remove all the profits in the business.

23-B. SUMMARY GRAPHS

As the weekly data are inserted on the various Summary Records, visual analysis of results becomes more difficult. There are too many figures on too many forms. But the Summary Record form is important, for it is the sheet on which the figures are kept.

To get a better picture of flock performance, the figures on the Summary Record should be transferred to a graph. Furthermore, the graph should show the standards for the factors measured. Variations in actual production from the standards will be quickly evident. They are usually supplied by the prime breeder, and some become very detailed as Figure 23-3 shows.

23-C. HOUSE RECORD RECAP

An analysis of actual flock production should be made at the end of the laying period. Certain calculations are necessary in such a *House Record Recap;* they should include the growing period and the egg produc-

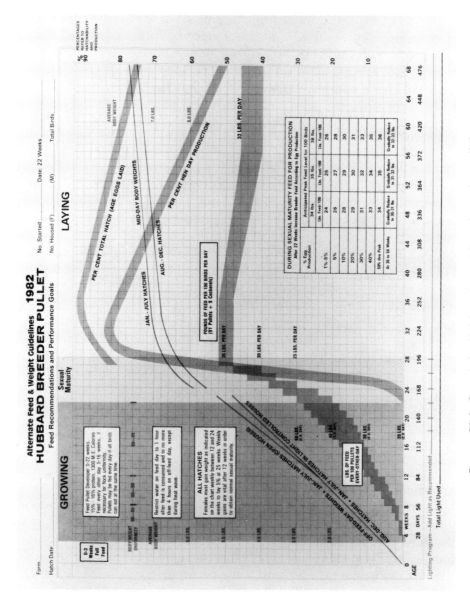

Figure 23-3. Graph for Hubbard meat-type breeder pullet.

tion cycle. An example is given in Figure 23-4 for breeder flocks and is self-explanatory. Notice that a column is available for inserting the standard figures when they are available.

23-D. PATTERNS OF EGG PRODUCTION

This subject deals with the relationship between weekly hen-day egg production and total egg production for the entire laying cycle. It is a

| FOR FLOCKS COMPLETING 1ST PERIOD OF EGG PRODUCTION | HOUSE RECORD RECAP |

AFFILIATE _____

MATING: MALE_____ X FEMALE _____   FLOCK_____ HOUSE_____

GROWING PERIOD

1	DATE CHICKS STARTED (Spell the month)	_____
2	NUMBER OF PULLET CHICKS STARTED, INCLUDING EXTRAS	_____
3	% CUMULATIVE PULLET DEPLETION TO 5% EGG PRODUCTION	_____ %
4	NUMBER OF COCKEREL CHICKS STARTED, INCLUDING EXTRAS.	_____
5	% CUMULATIVE COCKEREL DEPLETION TO 5% PULLET EGG PRODUCTION	_____ %
6	FEED CONSUMED PER PULLET TO 5% H-D PRODUCTION (MALES INCL.) (Total feed consumed by males and females in flock divided by number females alive at 5% production. Indicate lbs. or kilos).	_____

1ST LAYING PERIOD

		ACTUAL	STANDARD
7	AGE PULLETS REACHED 5% H-D EGG PRODUCTION (WEEKS)	wks.	
8	NUMBER PULLETS AT 5% H-D EGG PRODUCTION		
9	% PULLETS AT 5% H-D PRODUCTION OF THOSE STARTED	%	%
10	ACTUAL DATE FLOCK COMPLETED 1ST LAYING PERIOD (Spell the month)		
11	NUMBER WEEKS IN EGG PRODUCTION		
12	% CUMULATIVE FEMALE DEPLETION	%	%
13	% CUMULATIVE MALE DEPLETION	%	%
14	TOTAL NUMBER OF EGGS PRODUCED BY FLOCK		
15	TOTAL EGGS PRODUCED PER HEN HOUSED (Line 14 ÷ line 8)		
16	YOUR HATCHING EGG MINIMUM WEIGHT (OZS./DOZ. OR GMS./EACH)		
17	TOTAL NUMBER HATCHING EGGS PRODUCED BY FLOCK		
18	HATCHING EGGS PRODUCED PER HEN HOUSED (Line 17 ÷ line 8)		
19	AVERAGE % TOTAL HATCHABILITY FOR LAYING PERIOD	%	%
20	POUNDS OR KILOS FEED CONSUMED PER DOZEN TOTAL EGGS PRODUCED (MALES INCL.)		

COMMENTS:

DATE:_____ PREPARED BY: _____

Figure 23-4. House record recap.

well-known fact that when egg production begins in a flock, the number of eggs produced increases rapidly during the first 8 or 9 weeks, then drops off at a constant rate during the remainder of the laying period. Egg production on a weekly basis is usually expressed in terms of hen-day percentages. Accumulated egg production is kept on a hen-housed curve that is expressed in numbers of eggs.

How to Construct a Production Curve

In constructing a production curve, a uniform percentage figure of egg production must be used as the starting point. The week in which the flock attains 5% hen-day egg production is used in this text as the *first week of production*. To wait until the flock is in higher production before starting the curve gives an erroneous picture.

With a *normal* curve for any line of birds the following facts are evident.

1. *Rapid increase in egg production.* Egg production increases rapidly after 5% hen-day production until it reaches a peak in 8 to 9 weeks. This time interval may be altered by management. The more feed restriction during the growing period, the sooner the peak is reached. Out-of-season flocks peak later than in-season flocks. Underweight flocks, caused by any form of mismanagement, will be delayed in starting into production.

2. *The peak of egg production is abrupt.* Percentage of *flock* egg production is nothing more than a mathematical average of the percentage figure for each hen in the flock (see Table 22-8). If all birds could peak on the same day, the peak of the production curve would be very abrupt. But flocks are not uniform in sexual maturity. Thus, some birds start egg production sooner than others and reach their peak of production sooner. If the flock is "normal" in respect to the age at which 5% production is reached, the peak of the production curve is fairly abrupt. But if the flock is not uniform, the peak is not abrupt; rather, the curve will take on an arched form (see Figure 22-1). Associated with a decrease in uniformity at sexual maturity and peak of production are such factors as full feeding rather than controlled feeding during the growing period, out-of-season flocks, and lack of a light-control program during growing.

3. *Descending curve a straight line.* In the normal or stan-

dard production curve the percentages show an equal drop each week after the peak is reached, but the rate of drop is a genetic response varying with each breed and line. When management is good, the actual productivity of the flock will coincide with this line; but when it is poor, when there are stresses, or when the environmental temperature is high, the rate of weekly decrease will be greater than the genetic standard; flocks will do poorly at the end of the laying period. In practical situations, most flocks tend to depart from the theoretically straight line as the effects of various flock problems tend to accumulate.

Flocks Must Come into Production Fast

For a flock to attain its standard peak of egg production, the birds must reach the peak quickly. In all probability this is the most critical period of egg production. It is a time when management must be aware of the pitfalls that await annual egg production when peaks are late.

Peak Egg Production Associated with Annual Production

If the flock manager is to attain the bred-in ability of the strain of birds to produce a standard number of eggs, birds must peak high on the production curve. If the standard for a flock of Leghorn pullets is 266 eggs in 52 weeks of lay on a hen-day basis, and the peak production is 92%, the flock is unlikely to produce 266 eggs unless the peak percentage is reached. In fact the correlation between peak egg production and annual egg production is very close. For example, if the flock peaked at 83% instead of 92%, the chances are great that the total egg production will decrease by the same percentage.

Example: 83% is approximately 90% of 92%; 90% of 266 eggs (hen-day) is approximately 239 eggs.

University of California studies of commercial flocks hatched during 1981 to 1986 illustrate the egg production peaks of flocks during their first and second laying cycles. Table 23-1 summarizes these observations.

Poor Peak Production Cannot Be Made Up Later

The old adage that if the flock does not peak high it will make up the egg production later in the laying year is unfounded. Although there are

Table 23-1. Egg Production Peaks of First and Second Cycle Layers[1]

Strain	First Cycle			Second Cycle		
	Age (wk)	No.[2]	Peak (%)	Wk	No.	Peak (%)
C	30.7	317	90.9	13.1	277	82.5
I	30.3	266	88.6	11.9	166	77.5
J	29.2	239	87.1	10.9	101	77.8
G	30.7	102	89.2	12.6	90	78.9
D	30.3	77	90.0	12.8	79	82.0
F	30.7	71	90.4	12.1	67	79.5
H	31.2	64	89.4	12.0	43	79.4
B	30.3	41	85.8	11.9	17	75.1
A	28.6	40	89.0	11.7	31	78.6
E	32.0	13	89.9	12.0	16	81.6
Total/avg	30.2	1230	89.0	12.3	887	80.0

[1]One week peaks (1981 to 1986 data), White Leghorns.
[2]Number of flocks.

environmental and management conditions that affect the relative slope of the downcurve during this period, most flocks will show weekly decreases in egg production that are comparable to the decreases exemplified by the normal or standard flock. In most cases, if the flock lays at 10% below the standard production at the peak, it will continue to stay 10% below the standard rate throughout the rest of the laying period.

Break on the Upswing

When a flock shows a *break* in production due to a pronounced stress, disease, or other factor, egg production does not continue at the normal rate, but decreases rapidly, and it may be days or weeks before the flock returns to "normal." When these breaks come during the first 6 weeks of egg production, the break is on the "upswing" portion of the production curve. These are most disastrous; flocks never reach their standard peak and eggs are "lost," never to be regained. The flock will not be uniform after recovery, and the production curve will take on an "arched" appearance before it begins its downswing. When flock recovery is made, the best the flock can do is to return to an egg production percentage commensurate with the standard percentage egg production on the week of recovery. That is the bred-in ability of the flock to produce at a given rate for that particular week of egg production.

Break on the Downswing

If a stress causes a drop in egg production after the peak is reached, it is said to be a *break on the downswing* of the production curve. Breaks during this period usually are not as severe as those on the upswing period. If the egg production of the flock has been normal and at standard before a break on the downswing, the egg production after recovery from the break can only attain a percentage figure identical with the *standard* percentage figure. Production will not return to the same figure that it was before the break. Again, the genetic potentiality is for a flock to lay at a given percentage according to the specific week of egg production.

Table 23-2. Sample Computer Printout for a Table-Egg Flock Weekly Record

Weekly Individual Flock Summary (From 20 weeks of age)			
Flock: A Week: 30		Week Ending Date: 3/10/90	
Strain: XYZ Hens Housed at 20 Weeks:		10,000	

Flock Data	This Week	Last Week	To Date
Starting count	9,820	9,840	10,000
Number died	20	20	200
Number culled	0	0	0
Ending count	9,800	9,820	9,800
Mortality (% hen-day)	0.20	0.20	2.02
Mortality (% hen-housed)	0.20	0.20	2.00
Hen-days	68,670	68,810	693,000
Body weight (lb)	3.45	3.41	3.45
Egg Production			
Total eggs	63,516	62,784	452,160
Total dozens	5,293	5,232	37,680
Hen-day production (%)	92.5	91.2	65.2
Eggs per hen-housed	6.35	6.28	45.2
Egg Size			
Ounces per dozen	22.8	22.5	21.5
Pounds per case	42.8	42.2	40.4
Daily egg mass (g)[1]	49.9	48.5	33.2
Egg mass/hen-housed (kg)[1]	0.349	0.340	2.300
Feed			
Feed consumed (lb)	14,568	14,352	135,838
Per 100 hens/day (lb)	21.2	20.9	19.6
Per hen-housed (lb)	1.46	1.44	13.58
Per dozen (lb)	2.80	2.70	3.61
lb feed/lb eggs	2.07	2.10	2.68
Daily kilocalories/hen (ME)	270	266	250
Daily protein/hen (g)[1]	17.3	17.1	16.0
Summary			
Hen-day egg production (%)	92.5	91.2	65.2
Av. egg weight (oz/doz)	22.8	22.5	21.5
Mortality (% hen-day)	0.20	0.20	2.00
Daily feed intake (lb/100)	21.2	20.9	19.6
Feed/doz (lb)	2.80	2.70	3.61
Feed:egg ratio	2.07	2.10	2.68
Daily egg mass (g)[1]	49.9	48.5	33.2

[1]Metric system is suggested.

Manager Must Study the Curves

As the stock market analyst studies his charts, so must the poultry farm manager study his egg production graphs. In this way he will be able to prevent many production difficulties before they become an actuality. Graph your production figures. Graphs will give you a much better picture of laying house performance than do the data.

Computerized Records

The concentration of ownership and multiple flock firms have demanded more timely and comprehensive record systems. Managers and owners need to make daily decisions based on flock performance and environmental conditions. They cannot wait for weekly summaries.

Egg production, mortality, and feed consumption records are now analyzed daily with the use of in-house computer systems. Records are available within minutes of their collection and managers can make decisions regarding tomorrow's feed formulas. In many cases, flock performance, environmental measurements, egg processing data, and feed formulations are all integrated into one network of analysis. Some sophisticated systems can relate two or more measurements and even suggest possible causes for various problems.

Computers allow immediate comparisons with previous flocks and standards of previous periods for the same flock. Flock histories are easily prepared and records may be as detailed as required. Table 23-2 illustrates a typical weekly record for a table egg flock. Computers are excellent tools to speed up the analysis of records, but they do not replace the need for daily observation of flock comfort and conditions by well-trained personnel.

24

Digestion and Metabolism

The following several chapters are devoted to the feeding of chickens. It is not the intention to detail the many facets of the subjects of digestion, metabolism, and feed formulation, but rather to treat the subject of feeding from a practical standpoint. The technical aspects of nutrition may be pursued in many textbooks and scientific articles on the subject. Only a brief summary will be given here.

24-A. BASIC NUTRITIONAL COMPONENTS

All animals and birds require certain basic nutritional constituents to be able to live, grow, and reproduce. The list includes

carbohydrates	minerals
fats	vitamins
proteins	water

The digestion of these dietary components varies greatly, and each section of the digestive tract is responsible for its own processes (see Chapter 2-E).

24-B. WHY A CHICKEN EATS

The lack of satiety (fullness) in certain sections of the alimentary tract induces the primary need for feed. Chickens are continuous nibblers compared with most animals that resort to eating a meal, then resting while the meal digests. But even then chickens do not eat every minute the light is ample for them to find the feed trough. They fill their crop and gizzard to capacity, then wait until some feed leaves these organs before they eat again. The process will be repeated many times a day if feed is present.

Nutritionists have taken advantage of the above regulatory phenomenon by increasing the density of a feed so that birds will eat more before they feel "full." Rations high in energy (less bulk) and compressed feeds (pellets) are typical examples.

Satiety may be the explanation for why birds visit the feeder often during a day, or even on a short-time basis, but there is a more powerful regulatory mechanism operating over longer periods. Although several theories have been advanced, all involve the hypothalamus, the organ that turns feed eating on and off as the needs of the birds change. But the activity of the hypothalamus has regulators too, for its activity in this respect is ever changing. Set points determine the triggering mechanism that causes the hypothalamus to be activated or deactivated, and in turn, the amount of feed eaten is increased or decreased. For example, if the requirement for nutrients decreases, the proper metabolic set point is altered and the bird eats less because if feels "full" sooner.

Many things affect the set points that regulate feed consumption. A few of the important ones among the group are strain of birds, genetics, size, sex, age, degree of egg production, egg size, feather cover, activity, type of housing, feed palatability, energy content of the feed, quality of feed ingredients, water consumption, body temperature, body fat content, and degree of stress.

24-C. DIGESTION

A large percentage of the feed ingredients consumed by a chicken is in a form that necessitates chemical and other reactions before it can be utilized by the bird. The alimentary canal is a long tube through which the food passes while these reactions take place. Therefore, *digestion* refers to those changes that occur in the alimentary canal to make it possible for the feed to be absorbed through the intestinal wall and enter the bloodstream.

Within certain sections of the digestive tract, chemicals are produced to facilitate the digestive process. These are known as *enzymes*, and each of the several types has a specific function in producing the necessary chemical reaction. Enzymes are catalysts produced by living cells to aid certain chemical reactions without entering into them. All enzymes are conjugated proteins.

Other chemicals are secreted to alter the acidity or alkalinity of the tract so that the chemical reactions may be expedited. Bacteria play a certain role. All-in-all, the digestive process is quick, continuous, and constant. For the most part, the processes are the same in a young chick as they are in a laying hen (see Chapter 2-E).

Mouth

Secreted in the mouth of the chicken is a fluid known as *saliva*. It is very slightly alkaline and contains the enzyme *ptyalin*, which has the capacity to hydrolyze starch, converting it to sugars. However, food is held just a short time in the mouth of the chicken and the hydrolysis in this area is minor.

Crop

After leaving the mouth, food continues down the gullet to the crop, a reservoir for storage. The food material remains here for varying lengths of time depending on its particle size, on the amount consumed, and on the quantity of material in the gizzard. In the crop the feed particles are softened, and ptyalin from the mouth continues to hydrolyze the starches. No enzymes are produced in the crop.

Proventriculus

The *proventriculus* is a bulbous organ situated just before the gizzard, and is sometimes known as the *glandular stomach*. It is here that the gastric enzyme, *pepsin*, is produced, along with hydrochloric acid. The pepsin acts to break down the complex protein molecules; the hydrochloric acid changes the contents of the digestive tract from alkaline to acid and aids in protein digestion.

The proventriculus is small and holds little food material; food passes quickly through it to the gizzard. Because food is held in the proventriculus for such a short time, little or no actual digestion takes place here.

Gizzard

The *gizzard* is a highly muscular portion of the alimentary tract and is capable of exerting pressures of up to several hundred pounds per square inch. It is here that large particles of feed material undergo mechanical grinding, usually in the presence of "grit" in the form of sand, granite, or other abrasives to help facilitate the process. Although highly variable, the contents comprise about 50% water when in the gizzard. No enzymes are secreted in the gizzard, but digestion continues as the result of the secretions of the proventriculus.

Small Intestine

The foremost portion of the small intestine is known as the *duodenum.* It takes the form of a loop known as the *duodenal loop;* imbedded within the loop is the *pancreas,* a gland that empties its secretions into the intestine. The pancreas produces *pancreatic juice* that contains *amylase, lipase,* and *trypsin.* These, along with other enzymes, continue the process of digestion in the duodenum, although most of the absorption takes place in the next section of the small intestine, the *jejunum.* The third section is the *ileum* where enzymes are produced.

Bile is secreted by the liver and flows into the duodenum as a thick green material. It does not contain enzymes, but helps emulsify the fats and plays a part in other digestive processes. When the feed contents leave the gizzard they are slightly acid as the result of the hydrochloric acid secreted in the proventriculus, but the contents become alkaline as they pass through the *jejunum* and the *ileum.* Relatively speaking, little digestion takes place until the food reaches the small intestine. Here, most of it is completed.

Large Intestine

Some of the processes of digestion may continue in the large intestine, although no enzymes are secreted here; any digestion is merely a continuation of processes initiated in the small intestine.

Water moves in and out of the large intestine, but outward transfer predominates to bring the intestinal contents into a more solid state. This movement of water is related to conditions associated with dehydration and edema of the tissues. *Dehydration* is a condition produced as the result of a loss of sodium or potassium from the muscle cells. Retention of water produces edema, a condition arising when too much salt is consumed, and the body tries to dilute the salt in the cells of the tissues and in the space between the cells by osmosis. Both dehydration and edema of the tissues affect the transfer of water through the walls of the large intestine.

Ceca

At the juncture of the small and large intestines are two blind pouches, called the *ceca.* Fermentation and some digestion take place here. Fermentation is instrumental in digesting the very small quantity of crude fiber the chicken is able to utilize.

Endpoints of Digestion

The endpoints of digestion may be briefly described.

Carbohydrate. Carbohydrates are complex chemical structures composed of starches, celluloses, pentosans, some sugars, and other forms. The carbohydrates undergo hydrolysis during the course of digestion, reducing the complex structures to *maltose* and finally to *glucose.* The latter is easily absorbed from the intestine and is the main form in which simple sugars enter the bloodstream.

Crude fiber. In ruminants and some other animals, crude fiber forms an important part of the diet. However, the chicken is not endowed with processes necessary to digest any quantity; only a token amount of the crude fiber in the feed is digested and most of that by fermentation. Thus, for all practical purposes, crude fiber in feed is not utilized by the chicken.

Fat. Fats cannot be absorbed unless they are at least partially soluble in water. Digestion includes the formation of fatty acids and glycerols through the fat-splitting enzyme. The bile material is helpful in providing this reaction, and fats are absorbed to enter the lymphatic system, thence to the portal system by way of the liver.

Although fats are high in calories, there is no relationship between their original form and their available energy. The makeup of the fat and the age of the bird enter into the relationship. Beef tallow, for example, is a better fat source for older birds than for young chicks. The role of fat digestion is not the final answer either; it is but the first step in making the energy of consumed fat available for productive processes.

Protein. Proteins must be broken down into *amino acids* in order to pass the intestinal wall. Proteins, as generally known in the Plant and Animal Kingdoms, are composed of various combinations of the 22 amino acids. But each protein does not contain all 22, nor is the quantity of each acid constant in each. Therefore, there are many combinations of amino acids that comprise proteins; the list runs into the hundreds. Furthermore, proteins are often combined with carbohydrates, fats, and minerals, to add to the many combinations. Consumed proteins not only vary in their amino acid relationship but in their digestibility. For instance, fish protein is more digestible than protein from blood.

Vitamins. Many vitamins occur in combinations that prevent absorption through the intestinal wall; they must undergo a type of digestion, or at least change, to enable them to pass into the bloodstream. The diet is not the sole source of all vitamins. Vitamin D, for instance, is synthesized at the skin surface by the ultraviolet radiation in sunlight, but it may also be included in the diet.

Minerals. Minerals cannot be said to undergo digestion; they are absorbed from the intestinal tract in the same form as they are fed, but solubility is related to their absorption.

Drugs and antibiotics. Many drugs and antibiotics are administered to chickens, either through the feed they eat or the water they drink, to enter the digestive tract. Some of these are of nutritional value, as they produce an increase in growth or egg production, but most of them are used to suppress or regulate the growth of pathogenic organisms. In practically all cases, the drugs enter the bloodstream in their original form. However, there are variations between the amount absorbed and that consumed.

Total Digestible Nutrients

All the food material that is consumed is not digested. The percentage is quite consistent with each feed ingredient, although there are some variations according to the age of the bird. That portion of the food that does not pass the intestinal wall is excreted in the fecal material, though not necessarily in its original form. As an example of this variability, corn has a digestibility of about 80%, wheat middlings 48%, and alfalfa meal about 25%.

Time Required for Food to Pass Through the Alimentary Tract

Many factors affect the flow of food through the alimentary tract. The "call" of the gizzard for more food will determine the length of time that feed remains in the crop. Sometimes it may be only minutes; at others it may be several hours. If the feed is in fine form it can pass the gizzard in a very short time, but if it is coarse it must first be broken down into small particles before it can enter the intestines. Some feed may leave the gizzard after a few minutes; in other cases, as with whole grains, the grinding action may take hours.

Actually, the entire process of digestion is rapid. If the alimentary tract is empty, feed will pass through it in about 3½ hours. When feeding is more or less continuous, the entire process of transfer will take about 12 hours. Digestion is more rapid in a laying hen than in one that is not laying. The transfer is quicker during daylight hours than at night.

24-D. METABOLISM

Metabolism is a term used to denote those chemical changes in food components that occur after digestion and absorption. Since the various portions of feed (protein, carbohydrates, fats, vitamins, and minerals)

have been converted to structures capable of absorption during digestion, they must be reconverted to complex forms before they are of value
to the bird. For the tissues of the body to be able to utilize the simpler
compounds carried to them by the blood system, therefore, further
chemical reactions must take place. By these additional processes energy
is developed, fat is stored, heat is liberated, and many end products not
of value are eliminated through the kidneys.

How Food Material Is Utilized in the Body

The body has almost an hourly need for certain food materials in order
to carry out its physiological processes. These materials perform the following general functions:

1. Maintenance of life
2. Growth
3. Production of feathers
4. Egg production
5. Deposition of fat

To carry out these functions, food must be metabolized. As the subject
is detailed and scientific, only a general approach will be presented here
to familiarize the reader with the metabolic processes involved.

Carbohydrate Metabolism

A portion of the simple sugars entering the bloodstream is used to
produce energy. During the process, body heat is generated. The procedure is relatively quick; there is a close correlation between feed consumption and energy produced. The bulk of the glucose and a few other
simple sugars are first converted to *glycogen* by the liver. Glycogen has
the common name of *animal starch*. It is in the form of glycogen that
excesses of simple sugars are stored in the liver and on occasion in some
other parts of the body. But the storage capacity in this form is not great.
When there is a demand for additional glucose, the stored glycogen is
converted to glucose, in which form the sugar is released into the bloodstream. The bird has a governing mechanism to keep the level of glucose
in the bloodstream nearly constant so that the supply is uniform. When
the bird's storage capacity for glycogen reaches its maximum, additional
glucose in the bloodstream is quickly converted to fat to keep the blood
at its tolerance level; the fat is deposited in the fat cells at various locations in the bird.

Fat Metabolism

The metabolism of fats is a process by which the fatty acids are converted and used for energy, egg production, or stored as body fat. Stored fats are species-specific; that is, the consistency, as indicated by their texture, melting point, etc., varies according to the bird or the animal. The relationship between the fat consumed and the stored fat in the chicken can be altered only when large amounts of fat are consumed.

Unlike some other nutrients, fat is not excreted either in an original form or as a by-product. Excesses can only be deposited in the fat cells. If the carbohydrate or fat consumed is greater than that required by the bird, deposits of fat continue; seemingly there is no limit. If the energy portion of the diet is lowered below the amount necessary for body processes, the stored fat will be called upon to make up the difference, and the fat deposits decrease.

Protein Metabolism

Once the amino acids enter the bloodstream they are transferred to the various tissues of the body. Here the cells use them in many ways, such as for the repair of tissue structure, new tissues, egg production, etc., and for the rebuilding of various complex protein structures. However, all the amino acids entering the bloodstream may not be necessary to manufacture a type of protein for a particular bird at a particular time. Excesses of amino acids may be used for energy through a process of deaminization, which splits off the nitrogen from the molecule, after which the nitrogen is excreted by the kidneys most generally in the form of *uric acid*; this is found in small quantities in the urine of most mammals, but in high amounts in the excrement of chickens in the form of *urates*.

Proteins are essential for life; the actual need by the bird is the result of its demand for the amino acids. Most amino acids are formed in the body, but when their production is low, or they are not made at all, they are said to be *essential amino acids*. The inadequacy must be made up through dietary ingestion. When body production of an amino acid is sufficient for normal physiological processes, it is included in the group known as *nonessential amino acids*. Of the 22 amino acids, about 12 are classified as essential and must be incorporated in the ration either separately or as a component of a feed. Of these 12, there are 5 that are especially critical.

Mineral Metabolism

Many minerals are necessary for the physiological well-being of the individual. For example, calcium is required in relatively large quantities in bone formation and in the deposition of eggshells, while another, phosphorus, is especially needed for the production of bones. Other minerals fall in the classification of *trace minerals*; since their relative requirement is small, only trace amounts are required. In this group are copper, zinc, iron, manganese, selenium, etc. Another group comprised of sodium and potassium has a different relationship, but the requirement for this group is also low.

Minerals are not metabolized in the strict sense; rather, they are incorporated as a part of certain protein or enzyme molecules. In some instances, chemical reactions that produce these molecules cannot take place without the mineral. Therefore, many minerals are an important part of the metabolic process, although sometimes indirectly. In some cases a small amount of a trace mineral is absolutely essential, but an excess leads to difficulties, as with selenium.

Water Metabolism

Probably none of the items consumed by the bird is more important to its physiological processes than water. Not only does the bird derive its supply from normal water and the moisture in the feed but water is one of the end points of metabolic reactions (see Chapter 39-C).

Hormones

Hormones are body chemicals generated by specialized cells; when transported to other cells in the body, hormones influence their activity. Probably the most common are the hormones of the sex glands, which produce the differences associated with maleness and femaleness. Although both sexes produce a quantity of male and female hormone, one or the other predominates to produce the sex variation. When male chickens are castrated, the production of the male hormone is reduced and the shape of certain feathers associated with a male are altered to female-type feathers, as a result of the female hormone.

Some hormones are excreted by the pituitary gland, some by the adrenals, one by the thyroid, etc. Some have an effect on digestion, some on metabolism, and others on reproductive performance.

Estrogens. The female hormones are known as estrogens. Certain estrogens have a bearing on fat deposition. *Diethylstilbestrol* (DES) is one commonly used during the production of broilers and roasters to increase and alter the fat content of the males and females. A small pellet of DES is embedded under the skin of the neck where it is slowly absorbed.

Caution: The use of DES is illegal in the United States and some other countries. Check with authorities before using.

Thyroxine. Thyroxine is the chief secretion of the thyroid gland. Iodine in the ration undergoes digestion to form iodide. Once in the bloodstream it finds its seat in the thyroid gland, where it is changed to organic iodine. Thyroxine contains the iodine so necessary for certain body reactions. When there is a deficiency of iodine in the diet, the thyroid increases in size in its endeavor to increase the production of thyroxine. A goiter is produced.

Other hormones. Many other hormones play a part in keeping the body functioning properly. The hormones of the parathyroid are involved with maintaining the calcium level in the blood; prolactin is responsible for broodiness; secretions of the adrenals have a bearing on carbohydrate metabolism; and the islets of Langerhans secrete insulin so important in keeping the level of blood sugar constant. There are many more, some of which work together to produce their effect.

25
Major Feed Ingredients

Commercial poultry rations today are known as *complete rations;* that is, they contain all the essential ingredients for the bird to do a job well, whether it be in growth, feather renewal, egg production, or the production of meat. For the most part, the bird, being closely confined to its quarters, has no other source of food material. What it needs it must get from the feed it is given each day.

Certain parts of this feed come from the common and major feed ingredients such as cereal grains, protein and fat supplements, certain mill by-products, and the major minerals. But in most cases, a mixture of these ingredients would not satisfy the bird's nutritional requirement, nor would it be economical. Certain vitamins, minerals, by-products, and other ingredients must be added to "balance" the diet. This chapter deals with the major feed components; Chapter 26 includes the others.

25-A. CARBOHYDRATES

Weight per Bushel

Cereal grains are measured in 100 lb (cwt) or bushels. The major ones and their bushel weights are as follows:

Grain	Weight per bushel	
	lb	kilo
Barley	48	21.8
Buckwheat	50	22.7
Corn, shelled	56	25.4

	Weight per bushel	
Grain	lb	kilo
Sorghums	56	25.4
Oats	32	14.5
Rice, rough	45	20.4
Rye	56	25.4
Soybeans	60	27.2
Wheat	60	27.2

Barley

Barley is produced abundantly in some areas and is used in many poultry feeds as a fine-ground ingredient. Compared with corn, it contains about 75% as much energy and three times as much fiber. Therefore, its use is limited, especially in feed mixtures that must be high in energy and low in fiber. Although the fiber of barley is practically indigestible, the grain may be soaked at high temperatures or treated with enzymes to improve its qualities. The cost of energy in normal barley must be considered when it is substituted for a high-energy cereal such as corn. In many areas it would be uneconomical to use barley.

Buckwheat

Buckwheat is seldom used as a poultry feed because of its limited production and its unpalatability. Ground buckwheat may be used to replace up to 15% of the cereal grain portion of the ration.

Cassava

Cassava or cassava root is produced in abundance in many tropical areas under a variety of names: mandioca, manioca, tapioca, yucca, and manioc. By enzymic action the roots release a poisonous compound, prussic acid. Special washing is necessary to make the root edible. In ground form, cassava root may replace up to half the cereal grains in a ration if its inadequate content of methionine and protein is provided for.

Corn (Maize)

In most areas corn is the predominant source of energy in poultry feeds, mainly because of its abundance, economy, and high digestibility. Corn is, however, a variable cereal grain, and in many countries is sold by "grade," which gives an indication of its moisture content, weight, kernel composition, and the presence of foreign material. Corn also has a variable protein content, from 8 to over 11%. Most corn is now the result of hybrid breeding in an endeavor to produce plants adaptable to certain climates, rainfall, and soil composition. Corn is a good source of linoleic acid, an essential fatty acid.

Yellow corn. Yellow corn contains an abundant quantity of carotenoid pigments called xanthophylls, which impart yellow pigment to the fat deposits of chickens and to egg yolk. Yellow corn is a fair source of vitamin A activity, but storage tends to reduce its content by as much as 30%.

White corn. White corn is similar to yellow corn in most respects except that it contains little or no xanthophyll and has practically no vitamin A activity.

High-lysine corn. A special hybrid corn has been developed that is high in the amino acid, lysine, but costs more to produce. The hybrid is specifically known as *Opaque-2*, after the gene responsible. The corn contains about 11% total protein, about 30% more than normal dent corn. The amount of lysine is about 50% greater than the lysine content of normal corn. It is questionable whether the hybrid can be economically fed to chickens.

Millet (Proso)

Proso millet is grown abundantly in many areas. There are several varieties, all of which are low in linoleic acid. In general, millet has an energy content nearly that of corn, but is fed in combination with corn or milo in broiler rations. Millet is not often used in poultry feeds, but on occasion may be combined with corn, milo, oats, or barley in growing and laying rations. It should be coarsely ground.

Molasses

Usually, molasses is a by-product of the cane sugar and beet sugar industries. Beet molasses contains about 6% protein; cane molasses, about 3%. Although both are relatively high in energy, molasses is used more in poultry feeds to prevent dustiness, but care must be exercised in mixing to prevent balling of small molasses particles.

Oats

Although an excellent feed for chickens, oats are limited in their use. They contain a large amount of fiber because of their husk, and are therefore low in energy. With a fiber content of about 12% compared with 2% for corn, oats contain only about 75% as much energy as corn. In most instances the energy from corn is more economical than from oats. Because of this, oats cannot be used in quantity in a high-energy broiler ration; their value lies in growing, laying, and breeding feeds. Because oats vary in weight, their protein content is highly variable. When incorporated in a mash, oats should be finely ground in order to pulverize the hulls thoroughly.

Naked oats.   Naked oats, a hull-less variety, is becoming more prevalent because of its high protein content of 16 to 19%, compared with 11% for common oats and 9% for corn, and because of its well-balanced amino acid content.

Rice

Rice is second to wheat in worldwide production. However, only where it is produced in abundance is any incorporated in a poultry feed, and then the use of only inferior grades and broken kernels is common. Rice hulls and rice bran are often fed to poultry.

New varieties of rice have materially increased the yield of rice by several times. They are short-strawed, lodging-resistant, respond better to nitrogen as a fertilizer, and have a much shorter growing period.

Rye

Rye has a property that produces a laxative effect when fed to chickens; the droppings become sticky and adhere to the feet of the birds. Rye also contains factors that depress the retention of protein and fat in the digestive tract. Furthermore, chickens do not like the taste of rye. When they have a choice between whole-kernel rye and other whole-kernel cereals, they will eat practically no rye. But if rye is fed in abundance in a mash the intestinal flora changes and the birds better adapt to it. Rye is growth depressing at any level, and is not usually used in broiler rations, but may be ground and mixed in other feeds provided it does not replace more than 15% of the cereal grains in the ration for young chicks or more than 25% for older birds. It has a high energy content.

Sorghums

There are several sorghum grains, but kafir and milo are the two generally used in poultry rations. They are difficult to store because they tend to hold moisture and do not dry easily. Sorghums are grown extensively in many areas and make up an important part of many poultry feeds. Although somewhat unpalatable in ground form, they may be used effectively to replace two-thirds of the cereal grain portion of most rations. If the feed is pelleted, the percentage can be higher. Kafir and milo are quite comparable to yellow corn in feeding value except that they have no vitamin A activity or any pigmenting xanthophylls.

Bird-resistant sorghums. Special strains of sorghums, high in *tannins*, have been developed to prevent wild birds from eating the grain in the fields. In general, darker-colored sorghums contain more tannin than lighter-colored sorghums. However, tannins are known to cause growth depression in chickens, and egg mottling in the yolks of eggs produced by layers. Bird-resistant sorghums should not replace over 40% of the cereal grain portion of the ration.

High-lysine sorghums. These variants have been shown to produce results superior to normal sorghums because of their higher protein content.

Triticale

Triticale is a cereal developed by crossing durham wheat and rye, and will outyield either. At first, it showed great promise as a crop in arid regions, but production has waned because of the selling price. It contains about 16% protein, but is not equal to corn, wheat, or milo in growth-promoting or egg-promoting properties, even though containing more protein. At high levels, triticale should be supplemented with 0.5 to 1.0% lysine.

Wheat

Whole wheat has an energy relationship analogous to corn and contains a higher percentage of protein. The protein may vary between 10 and 17%, depending on the type of wheat and the area where grown. However, wheat is practical in poultry diets only when it is available in quantity and will provide an economical source of energy. Because of its great use in human diets, it generally carries a high price.

Wheat is gelatinous, and when ground and used in high percentages, it has a tendency to "paste" on the beaks of birds. The pasting may sometimes produce *beak necrosis*. If the wheat incorporated in a poultry mash is coarsely ground, or if the feed is pelleted, most of the difficulty is overcome. Wheat has no vitamin A activity or pigmenting properties. As far as metabolizable energy is concerned, 109 lb of wheat are equivalent to 100 lb of corn.

25-B. MILL BY-PRODUCTS

Hominy Feed

This is a by-product mixture, the result of producing pearl hominy. The product used for poultry feeding should not be the result of solvent extraction, as this removes most of the fat, and thus reduces the energy value to a low level. Good hominy feed should contain at least 1,350 kcal ME/lb (2,970 kcal ME/kg) and no less than 5% fat.

Rice Bran

Rice bran is composed mainly of the pericarp and germ of rice as a by-product of the milling of raw rice to produce an edible product. It contains about 13% protein, slightly less than wheat bran, and about half as much energy as corn. The high fat content of rice bran (13–15%) makes it a fairly good poultry feed.

Rice Hulls

Rice hulls have little nutritional value, but sometimes are used as a filler to build low-energy rations.

Wheat By-products

Wheat bran. Wheat bran is composed of the outer layer of the wheat kernel. It is one of the by-products of wheat milling and contains about 15.6% protein and 510 kcal ME/lb (1,322 kcal ME/kg).

Wheat middlings, shorts. These are mixtures of milling by-products including the finer particles of bran, germ, flour, etc. Wheat shorts contain about 16% protein and 890 kcal of ME/lb (1,958 kcal ME/kg).

25-C. FATS AND OILS ($C_{57}H_{105}O_6$)

Although the fat content of a feed is usually calculated as that percent that will dissolve in ether, known as lipids, fats are better identified with only pure fatty acid esters of glycerol, called triglycerides. Fats are solid, while oils are liquid.

Fatty acids contain carbon, oxygen, and hydrogen and are classified as saturated, monosaturated, or polyunsaturated. A saturated fatty acid contains all the hydrogen it can hold; a monosaturated fatty acid has room for two additional atoms per molecule; and polyunsaturated fatty acids have room for four or more hydrogen atoms.

It has long been known that when the vegetable oil content of the diet is increased, egg size is larger, even when the total calories in the ration remain the same. Most of the effect of the fat in the oil is due to the increases of readily absorbable fatty acids including linoleic and oleic acid. Even when large amounts of yellow corn are used in the diet with no added fat, some rations may be marginal for these fatty acids. The problem may become acute when milo, barley, or oats are substituted for corn.

Most commercial feed ingredients are low in linoleic acid. Consequently, fats and oils, such as certain stabilized vegetable oils, should be added to many rations in order to prevent a deficiency of linoleic acid, which has a requirement of about 1.5% of the ration.

Types of Fats and Oils

Because of their high content of energy, relatively large amounts of fat or oils are added to some poultry rations, particularly those in the high-energy category. As a side effect they reduce the dustiness of the mixed mash and improve its palatability. Up to 8% of the mash as added fat is practical, but chickens can tolerate over twice this amount. In many instances the practical use of fats or oils is determined by the price relationship between their energy and the energy derived from corn, milo, wheat, or rice. When fat energy is cheap compared with the energy of any of these four, it is economical to use more fat or oil. There are several feed grades of fats:

1. *Hard fats.*   Most of these are solid at room temperature and come from slaughtered cattle; they are known as *tallow* and *lard*. Their melting point is above 104°F (40°C).
2. *Soft fats.*   These are semisolid, and are termed *greases*. Their melting point is below 104°F (40°C).
3. *Hydrolyzed animal fats.*   These are by-products, mostly from the manufacture of soaps, and are sold as *hydrolyzed animal fat* or *hydrolyzed vegetable fat*. They must contain no less than 85% total fatty acid. Hydrolysis splits the glycerine from the fatty acids.
4. *Vegetable oils.*   Oils in this group come from plants such as coconut oil, and so forth, and are used as an energy source in poultry feeds.

The following shows the comparison between corn and several fats in regard to their metabolizable energy content and the utilization of the energy:

	Approximate Kcal of ME per		Energy Utilization
	lb	kg	%
Corn	1,530	3,366	70
Lard	4,000	8,800	80
Hydrolyzed animal and vegetable fat	3,400	7,480	72
Grease (yellow)	3,400	7,480	84
Tallow (beef, feed grade)	3,130	6,886	80

Polyunsaturated Fatty Acids in Egg Yolk

Recent experiments by D. B. Bragg, University of British Columbia, Canada, have shown that increasing the linoleic acid in a chicken's diet could raise the deposition of polyunsaturated fatty acids in an egg yolk from a basic level of about 5% to approximately 28%, but not affect the total fat.

Antioxidants for Fats

Fats, particularly the unsaturated fatty acids, are subject to oxidative rancidity. To prevent the oxidation, antioxidants are usually added, particularly if the fats are to be stored.

25-D. PROTEINS OF ANIMAL ORIGIN

Dried Blood

This protein supplement, composed of ground dried blood, contains about 80% crude protein and is an excellent source of the amino acid lysine, of which about two-thirds is available to the bird. But blood meal is deficient in isoleucine, therefore rendering it a protein of poor quality, and only token amounts should be included in the ration if maximum growth and egg response are to be realized.

Dried Poultry Waste (DPW)

DPW is a highly variable product produced mostly from dried cage layer manure. When dried, poultry waste contains about 15% fiber and 25% protein, although only 8% is true protein, the remainder being non-protein nitrogen. Under certain legal restrictions it may be fed to chickens, but most of the product goes to feed ruminants.

Liver Meal

Liver meal is an excellent animal protein supplement but is in short supply and expensive. Because of this, it is seldom used except in special laboratory diets.

Meat By-products

Two meat by-products are of value in poultry feed formulation, although their use has largely given way to vegetable protein supplements in the ration.

Meat scrap. This is a dry-rendered product made from animal flesh and tissues. It must be guaranteed low in phosphorus to show that little or no bone was incorporated. It contains about 50 to 55% protein. It is high in lysine, but low in methionine, cystine, and tryptophan. Meat scrap is used in moderate amounts in many poultry rations. Usually, 5% is a maximum when efficiency of the ration is needed. Otherwise, up to 10% may be included.

Meat and bone meal (scrap). This product is more readily available than meat scrap and is a good supplement with 47 to 50% protein. It contains a high percentage of ground bone, making it a carrier of calcium and phosphorus. It may be used up to 10% of the ration, but often it is restricted to about 5%.

Milk Products

Most milk products used for poultry feeding are in dried form and are used in the mash, although some milks have been condensed to a thick consistency (27% solids) and fed separately. The quality of milk protein is excellent, and for years some milk products were incorporated in great amounts in poultry rations. However, large amounts are laxative, milk protein is comparatively expensive, and the products are not generally available in abundance. Seldom does a poultry ration contain more than 2% today, but several times this amount could be included under unusual circumstances. Dried skim milk and dried buttermilk, for all practical purposes, are equal in feeding value.

Dried skim milk. When the fats (cream) have been removed from whole milk and the remaining liquid (skim milk) is dried, the product is known as dried skim milk. It has about 32% protein.

Dried buttermilk. Drying the liquid remaining after the production of butter results in a product known as dried buttermilk, containing 32% protein.

Dried whey. In the production of cheese, the liquid portion remaining (whey) is dried. Dried whey contains a minimum of 65% lactose (milk sugar) and about 12.5% protein, depending on the cheese-making process used.

Poultry By-product Meal

This product consists of ground dry-rendered poultry offal including the heads, feet, intestines, and so on, but excluding the feathers. It contains 55 to 60% protein, and unless extracted, about 12% fat. It is an excellent protein source, but its short supply limits its use to 1 to 2% of the ration.

Poultry Feather Meal (Hydrolyzed)

Hydrolyzed poultry feather meal contains 70% and over of protein, of which 75% is digestible. However, the protein is high in cystine and deficient in the amino acids methionine, tryptophan, histidine, and lysine. Feather meal must be used sparingly in the ration, with thought given to its deficiencies. It should not replace more than 10% of the soybean oil meal in the ration.

Poultry Hatchery By-product Meal

This meal is a cooked, dried, and ground conglomerate of eggshells, unhatched and infertile eggs, and culled chicks. In some cases, the fat is removed; in others, it is not. The product is highly variable, and will contain between 22 and 32% protein, 17 and 20% calcium, and 10 and 18% fat.

25-E. PROTEINS OF FISH ORIGIN

There are many types of protein supplements derived from fish, the variations arising from the many types of fish and the part of the fish used in producing the meals. The number also is increased because of the four different methods of processing: (1) sun-dried; (2) vacuum-dried; (3) steam-dried; and (4) flame-dried. Sun-dried fish meal is usually of low quality, and little flame-dried material is produced today.

Most fish meals get top priority as a source of good-quality protein for poultry feeding because of their well-balanced amino acid content. But all fish meals are not equal in the makeup of amino acids or in their digestibility. Broadly, fish meals may be grouped into two categories.

1. *White fish meals.*   These are processed from the nonedible portions of tuna, cod, halibut, and other fish. They are low in fat.
2. *Dark fish meals.*   These come from such fish as sardine, herring, menhaden, etc., and are usually high in fat.

Fish meals vary in their content of crude protein from 55 to 75%. Herring meal is high, for instance; menhaden and sardine meals, medium; while tuna meal is low.

Protein efficiencies.   When the protein efficiency of casein is given a base rating of 100, the following variations will be derived from various fish meals:

Casein (base)	100
Vacuum-dried white fish meals	104
Steam-dried white fish meals	104
Domestic sardine meals	94
Asiatic sardine meals	91
Flame-dried menhaden fish meals	80

Antioxidant in manufacture.   Antioxidants are added to many fish meal
products to prevent oxidation. This materially improves the value
of the meals and tends to smooth out the above variability in efficiencies.

Salt in fish meals.   Associated with the method of salting fish to preserve them is the salt (NaCl) content of the resulting fish meal. As
salt produces a laxative effect in the chicken, the salt content of the
various fish meals should be carefully determined. They should
contain less than 3% salt for best results, but legally may contain as
much as 7%.

Pricing fish meals.   Because of their variability in protein content, many
fish meals are priced on a point system, one point being equal to
1% crude protein on a ton basis.

   Example: If a point of protein costs US$8.00, 1 ton (2,000 lb) of
   fish meal containing 70% protein would have a value of US
   $560.00 ($8.00 × 70).

Amount of fish meal in the diet.   Because of their relatively high cost
plus a usual shortage of supply, fish meals are confined to about 5%
in broiler rations and about 2% in others, but levels up to 8% will
usually show productive improvement.

Fish flavor in meat and eggs.   The oil from fish carries a definite "fishy"
taste and odor that are noticeably imparted to the poultry meat and
eggs when the diet contains more than 6 to 10% fish meal or 1%
fish oil, depending on how much fat is in the meal or oil.

Fish solubles.   The wet processing procedure of producing fish meal
leaves a water by-product, known as *stick,* that may be condensed or
dried. The value of these products lies not in the fish protein, but
in vitamin B_{12} and certain unidentified growth factors (UGF). Fish
solubles are laxative, more so than dried skim milk or dried buttermilk.

Shrimp Meal

A by-product of shrimp processing, shrimp meal contains about 43 to
47% protein and is higher in calcium than fish meals. The salt content
should be less than 7%.

25-F. PROTEINS OF VEGETABLE ORIGIN

Except for the cereal grains, protein supplements of vegetable origin comprise the largest component of most poultry rations. Soybean oil meal is most commonly used because of its available supply, good nutritional value, and relative economy. The aim of most nutritionists is to build a diet composed of corn and soybean oil meal, adding other ingredients only to make up for the deficiencies of this mixture.

Raw seeds cannot be used efficiently.   Many of the vegetable protein supplements are derived from seeds produced by various plants. However, such seeds in their raw form cannot be used satisfactorily by chickens. The seeds must undergo heat or other treatment to eliminate certain toxic factors and increase the availability of the methionine-cystine fraction. Such treatment thereby increases the nutritional value of the seeds.

Corn Gluten

Corn gluten comes in two products.

1. *Corn gluten feed.*   This is the part of the corn remaining after extraction of most of the starch and germ when corn starch and syrup are made. It contains about 22% protein.
2. *Corn gluten meal.*   The meal is similar to corn gluten feed, except that the bran portion of the corn kernel has been removed and this meal is usually used for poultry feeding. Although a good vegetable protein, its main value lies in its ability to impart yellow coloring matter to the skin of chickens and to the egg yolks. Two grades, containing 50 and 60% protein, are available.

Coconut (Copra) Oil Meal

Coconut meal is the result of grinding the portion remaining after a part of the oil has been extracted from the coconut. Its average protein content (solvent product) is about 22%. Evidently there is great variability in coconuts and some may contain a material toxic to chickens. Light-colored meals are better than dark-colored. Ten percent of the diet seems the limit of coconut meal when certain amino acids are supplemented, although chickens will consume up to twice this amount.

Cottonseed Meal

This meal is generally available in many areas as the result of oil extraction of cottonseed. The expeller process was first used, but in many instances it has given way to solvent extraction, which removes more oil from the seed but leaves less in the meal.

Although cottonseed meal is a vegetable protein of good quality with about 41% protein, it is inferior to soybean oil meal. A dehulled cottonseed meal will carry 50% protein. Neither should be used as the only vegetable protein source in the ration.

> *Gossypol content.* Cottonseed oil contains gossypol in minute quantities, yet the amount left in cottonseed meal after oil extraction is adequate to cause the production of eggs with pink to dark mottled yolks. Free gossypol is toxic and reduces growth and egg production. These properties have led to the production of special cottonseed meals very low in gossypol and these are the products best used in quantity in laying rations. They are sold as *degossypolized cottonseed meals* containing less than 0.04% gossypol.

Guar Meal

Guar is an annual legume and its seed is used for the production of special gums. The meal contains a trypsin inhibitor that is destroyed by cooking. The gum is a growth depressor when used in excess of 2% of the diet. It also causes very sticky fecal material.

Linseed (Flax) Oil Meal

This product is unpalatable and is generally not suitable for poultry feeding, but in the absence of good vegetable protein supplements a modest amount could be incorporated in the ration.

Peanut (Groundnut) Meal

Peanut meal is a good vegetable protein supplement and, where available, large amounts may be used in the ration. It contains 24 to 47% protein depending on the type of processing. Although peanuts contain a trypsin inhibitor, it is destroyed in the heating process. Peanut meal is best used to replace no more than 10% of the soybean oil meal in the ration.

Rapeseed Oil Meal (Canola Meal)

Rapeseed oil meal should be fed cautiously as it tends to irritate the digestive system. But it is a good, well-balanced product, containing from 32 to 34% protein. A solvent-extracted product, containing 44% protein, is also available. Meal made from the older varieties of rapeseed should not make up more than 10% of the diet, and preferably only 5%. Such meals when feed in excess cause liver degeneration, thyroid hypertrophy, reduced feed efficiency, and egg production because they contain high levels of glucosinolate.

New rapeseed varieties (canolas) have been developed during the past few years that are low in glucosinolate and the meal from these may replace up to 75% of the soybean oil meal in the ration with only minor adversities. Egg weight and shell thickness seem to be the only factors affected.

Safflower Meal

Decorticated safflower meal has had long use in moderate amounts in poultry rations. It is low in lysine but can be fed up to 5% of the ration during the first 5 weeks of a chick's life, and 15% thereafter. Much more may be fed if adequately supplemented with lysine.

The following two products are generally available:

1. 22% protein product
2. 42% protein product (dehulled)

Sesame Meal

Sesame meal contains about 47% protein and except for its great deficiency of lysine it is a good vegetable protein supplement. But it also contains phytic acid, which appears to bind dietary calcium, making it necessary to proportionally increase this mineral in the diet. The phytic acid also forms a chelate with zinc in the intestines, making the zinc unavailable. Probably sesame meal should not make up more than half the vegetable protein of the ration, with a maximum of 15% of the total feed intake.

Soybean Meal

The abundance of soybean production and the high nutritional value of the processed bean have made it possible to use high percentages of the meal in most poultry rations. Soybean meal is best supplemented with some animal or fish protein to make up its deficiencies of certain amino acids. Synthetic amino acids also may be used. Raw soybeans should not be fed. They contain a trypsin inhibitor that must be destroyed by heat or other methods. Soybean meal is the byproduct of oil extraction and contains from 42 to 50% protein, depending on the method of processing.

1. *Expeller sobean meal.* This process does not remove as much valuable oil as solvent extraction, although the meal is nutritionally comparable. It has 43% protein.
2. *Solvent soybean meal.* Solvent extraction of the oil from soybeans is predominantly in use today. The resulting meal is of excellent quality, although lower in fat than those resulting from expeller processing. The protein content is 44%.
3. *Dehulled (solvent) soybean meal.* A meal higher in protein (50%), lower in fiber, and higher in energy can be produced through special processing methods to give a meal with less than 3.3% crude fiber. When high-energy diets such as broiler rations are required, dehulled meal is recommended.

Full-Fat Soybeans

The low availability of the oil and the presence of a toxic factor in *raw* soybean seeds have made them unsuitable for poultry feeding. In the case of the young growing chick, raw soybeans will produce only about two-thirds of the growth achieved with soybean oil meal.

Of late, however, renewed interest has been shown in trying to develop methods of heating raw, whole-fat beans in order to eliminate the toxic factor and make the fat more available. Several methods of heat treatment have been used. Any procedure of heating has close limitations in regard to the amount of heat and length of the heating period; too much heat is more detrimental than too little. But when conditions are optimum, the growth value of such treated beans will approximate about 90% of that of soybean oil meal.

Sunflower Seed Meal

This product contains about 44% protein, but is low in lysine. It may be substituted for 50% of the soybean oil meal in the ration, and up to 100% if lysine is added. But the meal is sticky and may cause necrosis of the beak at the higher levels. Pelleting a feed containing sunflower seed meal will prevent the stickiness in the beak. The product is becoming more available because of great increases of sunflower planting.

25-G. GREEN LEAFY PRODUCTS

The tops from many grasses and legumes may be dried and fed to chickens as a source of carotene, xanthophyll, and the unknown growth factor (UGF). Some are high in vitamin K. Most common are those from alfalfa.

Alfalfa Products

There are several alfalfa products as the result of different methods of curing the hay and that portion of the plant used to make the meal.

1. *Sun-cured alfalfa meal.* Originally, alfalfa hay was sun-cured and ground, but the product was highly variable.
2. *Dehydrated alfalfa meal.* Alfalfa hays are now properly and uniformly dried artificially by heat, and then ground.
3. *Dehydrated alfalfa leaf meal.* This is a product made from only the leaves of the alfalfa plant.

Analysis of alfalfa products. Most alfalfa products are known by their protein content; the higher, the better the quality, and the lower the crude fiber. Following are the recognized grades (guarantees):

Dehydrated Alfalfa Meals

Protein	Crude Fiber
%	%
13	33
15	30
17	27
18	25
20	22
22	20

Vitamin A activity. This activity in alfalfa meals is due to the precursor of vitamin A, which is carotene, mostly in the beta form. Dehydrated alfalfa products are much higher in carotene than sun-cured products. But the carotene in the meal is easily lost through oxidation. To prevent this loss, an antioxidant is added to the ground alfalfa, or it is kept under refrigeration during storage, or the meal is pelleted to reduce air exposure. The pellets are ground prior to feed mixing.

Alfalfa meals are measured by their vitamin A activity instead of their carotene content. A meal of high quality should contain no less than 100,000 units of vitamin A activity per lb (454 g).

25-H. MACROMINERALS

This section is devoted to sources of four major minerals, calcium, phosphorus, sodium, and chloride, commonly used in quantity.

Curaçao (Island) Rock Phosphate ($CaHPO_4$) ($CaHPO_4 \cdot H_2O$)

This is a special rock phosphate containing about 15% phosphorus and 34% calcium.

Dicalcium Phosphate ($CaHPO_4 \cdot 2H_2O$)

Dicalcium phosphate comes from rock phosphate or bone after special processing. Dicalcium phosphate derived from rock phosphate may contain an appreciable amount of fluorine, most of which must first be removed before the product is acceptable for poultry feeding. Dicalcium phosphate contains approximately 18% phosphorus and 23% calcium.

Rock Phosphate

Much ground phosphate rock is so high in fluorine that the raw rock must be defluorinated before it is fed. Such a product is sold as *defluorinated rock phosphate*, containing no more than one part fluorine to 100 parts of phosphorus. Raw rock phosphate contains about 18% phosphorus and 0.5% fluorine.

Steamed Bone Meal ($Ca_3 (PO_4)_2$)

A source of phosphorus that comes from bones of animals, steamed bone meal contains an appreciable amount of calcium. Most products contain about 31% calcium, 14.5% phosphorus, and 6.5% protein.

Aragonite ($CaCO_3$)

Aragonite is a mineral-like calcite but differing in its orthorhombic crystallization having greater density and less distinct cleavage. It is an excellent source of calcium for chickens.

Limestone (CaCO₃)

Used as a source of feed calcium, limestone contains 35 to 38% calcium. Care should be taken to use a limestone that is low in fluorine, sometimes known as *high-calcium limestone.*

Oystershell (CaCO₃)

In most areas this is the main source of supplemental calcium, containing about 94% calcium carbonate (38% calcium).

Gypsum (CaSO₄·2H₂O)

There is every indication that the calcium in gypsum (calcium sulphate) is as available to the chicken as the calcium in calcium carbonate. Gypsum contains about 22% calcium.

Salt (NaCl)

Salt is a source of sodium and chlorine. Although necessary in small quantities, either through other feed ingredients or free salt, large percentages in the diet increase water consumption and have a laxative effect. Generally, no more than 0.25% of free salt is added to the poultry ration.

    Iodized salt.   In most of the areas of the world iodine must be added to the chicken's diet. This is easily accomplished by adding iodine to the salt at a level of 0.007%, equivalent to 70 ppm.

26

Vitamins, Minerals, and Trace Ingredients

Besides the carbohydrates, fats, and macrominerals there are many nutrients in the chicken's diet that are necessary in much smaller quantities to enable the bird to live, produce meat and eggs economically, and to reproduce efficiently. The list includes the vitamins, microminerals, and certain other additives.

26-A. VITAMINS

Generally speaking, vitamins are organic chemical compounds that are usually not synthesized by the body cells, but are necessary for maintenance, growth, and egg production. They are used in small amounts, and when they are deficient or absent from the diet, characteristic manifestations result. Many of them are enzyme-associated. There are 13 vitamins usually listed as necessary for the chicken; they occur in feedstuffs in varying quantities and in different combinations. All feedstuffs do not include all vitamins, and some contain a greater quantity of certain vitamins than of others. Some vitamins are produced by microorganisms of the intestinal tract, one by irradiation at the area of the bird's skin, while others are manufactured synthetically. As vitamins are definite chemical compounds, commercially produced vitamins are as valuable as those found in natural feedstuffs.

Vitamins are often segregated into two groups: fat-soluble and water-soluble.

Fat-soluble vitamins. The fat-soluble vitamins are

        A (Retinol and its precursors)
        D
        E
        K

The fat-soluble vitamins contain only carbon, hydrogen, and oxygen, and require some body fat for their metabolism. They are often present in plants as provitamins which are quickly converted to the true vitamins in the body of the chicken. They are easily stored in the fat cells of the bird and excesses are excreted through the feces. Only one, Vitamin K, is synthesized in the body intestinal tract.

Water-soluble vitamins. The important vitamins in the water-soluble group are

C (ascorbic acid)	pyridoxine (B_6)
thiamin (B_1)	choline
riboflavin (B_2)	biotin
pantothenic acid	folacin (folic acid)
niacin	B_{12} (cobalamin)

Water-soluble vitamins contain carbon, hydrogen, and oxygen plus either sulfur, cobalt, or nitrogen. In the main they are needed for the transfer of body energy. Chickens require all the known water-soluble vitamins in their diet except vitamin C.

When the feed contains more water-soluble vitamins than the bird needs, excesses of all but one are excreted in the urine. Vitamin B_{12} has the capacity of being stored. Those not stored must be included in the daily diet; the bird has no reservoir on which to draw.

Vitamin A (Retinol)

True vitamin A exists only in the animal kingdom. Its precursor, carotene, is found in the vegetable kingdom, and it too is fat-soluble. Carotenes are consumed through their vegetable sources to undergo conversion to provitamin A, then to vitamin A, which can be stored in the body, mainly in the liver, but this process is poor in young chicks. Vitamin A is essential for normal vision, growth, egg production, and reproduction.

Unit of measurement. Vitamin A activity is expressed as IU (International Units). Usually the activity of carotene in plants is rated according to its vitamin A production in the body.

Deficiency symptoms

1. Retarded growth
2. Weakness, ruffled feathers
3. Absence of liquid from the tear glands. Xerophthalmia and blindness may result. There are cheesy exudates of the eyes in adult birds.
4. Impaired egg production and hatchability
5. Increased incidence of bloodspots in eggs
6. Lowered resistance of the bird to some poultry diseases

Sources of vitamin A. Precursors of vitamin A are to be found in green leafy plants, alfalfa meal, yellow corn (maize), and corn gluten meal. A good, dehydrated alfalfa meal (17% protein) should contain at least 100,000 IU of vitamin A activity per lb (454 g). Broiler rations are best prepared by using an alfalfa meal containing 20% protein and a minimum of 150,000 IU of vitamin A. Yellow corn contains about 2,200 IU per lb (454 g); corn gluten meal, about 12,000. Vitamin A is found in many fish and animal liver meals. It is produced commercially as a synthetic product of high and variable concentrations.

Oxidation of vitamin A. Carotenes and vitamin A are easily oxidized, reducing their potency in the feed ingredient in which they are present. To reduce this oxidation, alfalfa meals are pelleted, antioxidants are sometimes added to certain feedstuffs, and synthetic vitamin A products are manufactured as small particles, then coated with fat, oil, or wax to produce stabilized forms. In formulating feeds, care should be taken to use the actual present vitamin A activity of the ingredient rather than its original potency.

Vitamin D (Cholecalciferol) (Ergocalciferol)

This vitamin has several forms, but D_2 and D_3 are the most important. Vitamin D_3 (cholecalciferol) is utilized by birds, human beings, and four-footed animals, while vitamin D_2 (calciferol) is of value to human beings and four-footed animals. Thus, D_3 becomes essential for poultry. Vitamin D aids the absorption of calcium and phosphorus from the intestinal tract, thus increasing the amounts of these two minerals available for bone development, and the amount of calcium for eggshell deposition.

Under natural conditions, the ultraviolet rays of sunshine or fluorescent light act on 7-dehydrocholesterol, synthesized in the bird, to produce cholecalciferol, which in turn is absorbed to become the only source of vitamin D the bird has. With the advent of commercial poultry production, chickens are closely confined to houses with no irradiation from the sun. Even glass does not allow the ultraviolet rays to pass. Vitamin D_3 supplements must be added to the feed, either from fish liver oils or from synthetic products.

Unit of measurement. Vitamin D_3 is measured as IU (International Units).

Deficiency symptoms

1. Rickets: Calcium and phosphorus are not deposited in the bones in normal amounts. The hock joints are enlarged, the ribs are "beaded," and the beak and shanks in young chicks are soft and pliable.

2. General unthriftiness.
3. Soft-shelled eggs.
4. Calcium crystals on eggshells.
5. Lowered egg production.
6. Reduced hatchability.

Sources of vitamin D_3. Good natural sources of vitamin D_3 are the fish-liver oils. However, 7-dehydrocholesterol can be irradiated with ultraviolet light to produce cholecalciferol. This process is the basis for manufacturing commercial vitamin D products, some of which have potencies of 400,000 IU per gram.

Vitamin E (Alpha-tocopherol)

Vitamin E (tocopherol) is necessary for adequate productivity of the cells and for blood formation. When the diet is lacking in adequate amounts, there are several manifestations, but these vary because other dietary components affect the requirement for vitamin E. The tocopherol involved is alpha-tocopherol. The measure of alpha-tocopherol is in IU.

Deficiency symptoms

1. Nutritional encephalomalacia, evidenced by a twisted neck, prostration, curled toes, and ''crazy chick'' disease.
2. Exudative diathesis: There is some indication that selenium is involved, as additions of this mineral have been shown to reduce this difficulty when it results from a lack of adequate vitamin E.
3. Male sterility: With prolonged deficiencies, the sterility may be permanent as a result of a degeneration of the testes.
4. Nonproduction in the female: Birds stop laying on diets low in tocopherol, but cessation is not permanent, for additions of vitamin E to the diet restore the bird to normal egg production.
5. Embryonic mortality: There is a circulatory failure at about the fourth day of egg incubation.

Sources of vitamin E. Whole grains and alfalfa meal are the best natural sources of vitamin E. Synthetic tocopherols are available and these are usually added to chick starter and breeder rations.

Destruction of vitamin E. This vitamin is easily oxidized and the process is increased when minerals and unsaturated fatty acids are present in the feed. An antioxidant should be used in the ration when vitamin E is added.

Vitamin K

Vitamin K, the antihemorrhagic vitamin, is necessary for the synthesis of prothrombin, a chemical necessary for blood clotting. When vitamin K is low or lacking, the blood vessels rupture, causing excessive bleeding.

There are several types of vitamin K, all of which are active but of varying potency; three of these are

1. Vitamin K_1 (phylloquinone): Present in plant tissues.
2. Vitamin K_2 (menaquinone): Synthesized in small amounts in the intestinal tract, and found in fish meal.
3. Vitamin K_3 (menadione).

Unit of measurement. When supplementary vitamin K is added to the ration, it is usually in the form of menadione sodium bisulfate, or menadione dimethylpyrimidinol bisulfite, and measured in milligrams.

Deficiency symptoms. There is a hemorrhagic syndrome. Hemorrhages that are pinpoint in size at first and large later occur on the flesh. If the skin of the chicken is pulled off, bleeding may be seen on the breast, thighs, and ribs.

Sources of vitamin K. A good natural source of vitamin K is alfalfa meal. Meat scrap and fish meal are fair sources. A synthetic water-soluble form of vitamin K_3 is available.

Thiamin (B₁)

Thiamin is necessary to stimulate the appetite, to form certain enzymes necessary for digestion, and to prevent nervous disorders that culminate in polyneuritis. However, as many feed ingredients carry abundant supplies of this vitamin, the symptoms are seldom seen.

Sources. Thiamin is relatively abundant in cereal grains, mill by-products, vegetable oil meals, alfalfa meals, and synthetic thiamin.

Riboflavin (B₂)

This vitamin is of major importance not only because of its effect on body processes but because it generally is inadequate in rations composed of ordinary feedstuffs. Riboflavin is a part of an enzyme probably needed by all living cells. It may be isolated as yellow crystals. Commercially it is manufactured by fermentation processes to produce the same form. Most rations contain added riboflavin.

Unit of measurement. The riboflavin unit of measurement usually is milligrams of the pure product.

Deficiency symptoms

1. Curled-toe paralysis: The toes curl and sometimes the legs are affected to produce paralysis.
2. Poor hatchability: Embryos not hatching are dwarfed and have abnormal down termed ''clubbed down.''

Sources. Although fish meals, fish solubles, alfalfa meal, and milk products are relatively high in riboflavin, their limited use in most poultry feed formulas means that practically all poultry rations must include supplemental riboflavin.

Pantothenic Acid

This vitamin is associated with many protein molecules and is involved with protein, carbohydrate, and fat metabolism. It is relatively unstable. The requirements of young and growing chicks for pantothenic acid are high.

Unit of measurement.   Milligram.

Deficiency symptoms

1. Retarded growth in young chicks
2. Ruffled feathers
3. Granulated and stuck eyelids in young chicks
4. Scabs at the corner of the mouth
5. Dermatitis of the feet
6. Lowered egg production
7. Lowered hatchability

Under field conditions, diets are seldom so low in pantothenic acid as to cause major exemplification of many of the symptoms.

Sources.   Several feedstuffs are good sources of pantothenic acid, the list including liver meal, peanut meal, milk products, mill by-products, and alfalfa meal. Calcium pantothenate is manufactured commercially to serve as the supplementary source of this vitamin. It is added to most rations.

Niacin (Nicotinic Acid)

This vitamin is important to the metabolism of the carbohydrates, proteins, and lipids. Nicotinic acid is an important part of two enzymes, and is found as nicotinic acid in the plant kingdom and nicotinamide in the animal kingdom. The amino acid, tryptophan, is a body precursor of niacin, but the conversion is very poor (40 to 1), yet must be considered when adding niacin to a feed.

Cereal grains contain little niacin, which is in a bound form, unavailable to the chicken. High-energy diets, young chickens, and chickens under stress have a higher niacin requirement. In the presence of adequate tryptophan (0.215%), and diets of medium to high energy when less feed is consumed, the niacin requirement is about 50 to 60 mg/kg of feed for chicks and broilers, 20 to 25 mg/kg for layers, and 30 to 40 mg/kg for broiler breeders.

Unit of measurement. Milligram.

Deficiency symptoms

1. Swollen hocks, similar to perosis, but the tendon seldom slips from the condyle.
2. Reduced growth.
3. Inflamed tongue and mouth (blacktongue).
4. Scaly skin and feet, ruffled feathers.
5. Hysteria.
6. Increased deposits of liver and body fat in layers.

Sources. Yeast, fish meal, fish solubles, distillers solubles, and a synthetic product.

Pyridoxine (B₆)

This vitamin plays a part in protein, carbohydrate, and lipid metabolism. It forms a part of several enzymes and is a muscle conditioner.

Unit of measurement. Milligram.

Deficiency symptoms. Diets low in pyridoxine show reduced chick growth.

Sources. Most natural feed ingredients contain some pyridoxine so dietary deficiencies are very uncommon.

Choline

The chick's demand for choline is great. Choline forms a part of the phospholipid, lecithin, rather than an enzyme. Therefore, choline is seldom considered a true vitamin. At times it may be synthesized by the chick, but the amounts are small and usually inadequate. The older a bird gets, the better the synthesis.

The vitamin has a great many functions in the body: It helps in fat movement in the bloodstream; it has a sparing action on methionine; it aids in growth; it prevents a type of slipped tendon; and it helps to reduce excessive fat deposits in the liver.

Unit of measurement. Milligram.

Deficiency symptoms

1. Perosis
2. Fatty liver (syndrome)
3. Retarded growth

Sources.   Fish meal, fish solubles, yeast, liver meal, soybean oil meal, and distillers solubles are good sources, but a choline chloride product is the usual feed supplement and contains 25% choline chloride.

Biotin

Biotin seems adequate in the diet when the composition of normal feedstuffs is concerned, but only about half is available to the chicken. Thus, it is possible that deficiencies may exist when some diets are fed. At times biotin may be synthesized in the intestinal tract, but this process is highly variable.

Unit of measurement.   Milligram.

Deficiency symptoms

1. Scaly dermatitis
2. Mild perosis
3. Retarded growth
4. Reduced hatchability

Sources.   Good sources are alfalfa meal, yeast, liver meal, and soybean oil meal.

Folic Acid

Folic acid, an antianemia factor, is a complicated chemical compound necessary for many physiological functions: growth, muscle formation, blood formation, and feather growth. Diets are seldom low in this vitamin.

Unit of measurement.   Milligram.

Deficiency symptoms

1. Depressed growth
2. Poor feathering, with feathers lacking pigment
3. Anemia
4. Necrosis
5. Dermatitis
6. Perosis
7. Increased embryonic mortality

Sources.   Good sources are alfalfa meal, yeast, liver meal, and soybean oil meal.

Vitamin B₁₂ (Cobalamin)

This vitamin is associated almost entirely with feeds of animal and fish origin. Plant products contain little or no vitamin B_{12}. It is synthesized by microorganisms of the intestinal tract as a cobalt-containing compound. However, the process is inadequate to supply the amount needed by most chickens. Lack of a normal amount causes anemia (pernicious anemia in human beings). Most rations contain a supplemental supply, particularly chick-starting and breeder diets. The bird's own droppings are a source of vitamin B_{12}. Therefore, birds raised on wire are more likely to show a deficiency than those kept on a litter floor.

Unit of measurement.   Milligram.

Deficiency symptoms

1. Anemia
2. Reduced chicken growth
3. Poor hatchability
4. Fatty liver

Sources.   Meat scrap, fish meal, fish solubles, and poultry manure.

Ascorbic Acid (Vitamin C)

In all probability vitamin C is not required in the feed of chickens, for birds synthesize an adequate small amount. Ascorbic acid helps embryo growth, aids bone development in young chicks, and stabilizes body fat.

26-B. MINERALS

Besides proteins, carbohydrates, fats, and certain vitamins, many other elements form a part of the bird's nutritional requirements. In most cases, the necessary quantity of each is small, often infinitesimal. Many have interrelationships with other nutrients. In some instances, trace amounts are necessary, but excesses are toxic. Although most of these elements must be added to the diet in their inorganic form sometimes organic forms are important sources. High percentages of some are absorbed through the intestinal wall; in other instances, the amount is small.

Calcium

Calcium is primarily necessary for bone and eggshell formation but also has certain other functions. The mineral is deposited in bone mainly as calcium phosphate, but there is some calcium carbonate. Eggshells are almost entirely calcium carbonate.

Eggshell formation. At the approach of egg production (sexual maturity) estrogens are released from the ova in greater abundance, which in turn increase the level of blood calcium. The parathyroid secretes hormones to keep the blood level of calcium constant. Calcium is then deposited in the medullary bone to be released later for eggshell formation. The amount deposited is not related to the amount of calcium fed during the growing period. In fact, too much calcium during the growing period is a detriment to maximum egg production, probably because of injury to the developing parathyroid gland. An increase in dietary calcium is needed only about 1 to 2 weeks before the first egg the pullet lays. Feeding more oystershell or other form of calcium during this period is the accepted field practice. Once egg production begins, the source of calcium for eggshell formation is from both the dietary calcium and the medullary bone.

Phosphorus

Although a major constituent of the blood, phosphorus plays an important part in metabolic processes and is found in the cells, enzymes, and other body compounds. Not all the phosphorus in the feed is available to the chicken. Normally, the phosphorus content of the ration is represented by two measurements:

1. Total phosphorus
2. Available phosphorus

Usually, chicks can utilize about 30% of the total plant phosphorus, while adult birds can utilize an average of 75%. Phosphorus derived from other than plant sources is available to the chicken in varying percentages based on solubility.

Availability of phosphorus. There are many inorganic phosphates, but relatively few are used for feeding chickens because they must have a high biological value. Values for several phosphates are as follows:

Phosphate Source	Biological Value
Hydrated dicalcium phosphate	110
Dicalcium phosphate	96
Steamed bone meal	96
Defluorinated rock phosphate	90
Curaçao rock phosphate	75

The available phosphorus, as a percentage of the total, varies with the type of ration, as follows:

Ration	Available Phosphorus as % of Total Phosphorus
Chick starter	60–65
Broiler	60–65
Grower	55–60
Layer	80–90

Calcium:phosphorus ratio. Not only must the diet contain minimal amounts of calcium and phosphorus but there must also be a correct ratio between the two minerals, particularly for starting and growing rations. A ratio of 1.5 to 2.0 parts of calcium to one part of total phosphorus is optimum for starting and growing rations, with 2.5 to 3.5 being too great and generally producing rickets.

Calcium, Phosphorus, and Vitamin D

Not only are calcium and phosphorus essential dietary minerals for the production of bone, and calcium for the deposition of eggshell material but vitamin D plays an important part in the processes. Evidently, vitamin D helps to form a protein in the intestinal tract to keep the calcium in solution so that it can pass the intestinal wall and reach the cells. Vitamin D also aids in other ways to get the calcium to those areas of the body that need it.

Amount of calcium and phosphorus in the diet. The dietary amounts of these two minerals must be maintained within close limits according to the age and type of bird involved. The following are typical examples:

Type of Ration	% Calcium	% Phosphorus	
		Total	Available
Starting	0.9	0.6	0.40
Growing	0.9	0.6	0.35
Laying, egg-type	3.5–4.0	0.5	0.42
Laying, meat-type	2.9–3.1	0.5	0.42

Calcium:phosphorus (Ca:P) dietary ratio. In many instances this is expressed as the ratio of calcium to *total* phosphorus in the ration, but a better meaning is given when it is the *dietary* ratio of calcium to *available* phosphorus. Examples of the latter are given below:

Type of Ration	Calcium: Available Phosphorus Ratio
Starting	2.2 : 1
Growing	2.5 : 1
Laying	9.0 : 1

Note: The figures given above for the amounts of calcium and phosphorus and the ratios are generalized. More precise figures are given in later chapters.

Sodium, Chlorine, and Potassium

These three elements are involved with acid-base equilibria in the body. Natural feedstuffs usually require supplemental feeding of salt (NaCl) to satisfy the bird's requirement for sodium and chlorine. The amount of salt added to the ration should seldom be over 0.25 to 0.35%. Too much salt produces a laxative effect. Additions over 8% are lethal.

Potassium is a necessity, but ordinary poultry rations are seldom deficient in this element. No supplementation is given.

Sulfur

Sulfur is a part of two amino acids, cystine and methionine, which are often in short supply in natural feedstuff protein. The naturally occurring form, cysteine, is readily oxidized in the body to form cystine. Sulfur is important to certain enzymes and hormones. Cystine and methionine are often grouped together as total sulfur-containing amino acids (TSAA) because of their complementary action.

Iodine

Iodine has a relationship to the thyroid and its hormone, thyroxine. When the ration is low in iodine, the thyroid increases in size, producing a goiter. Besides being a part of many metabolic functions in growing and adult birds, iodine is needed by the developing embryo. When the iodine content of hatching eggs is low, hatchability is reduced. Iodine usually is added to the diet as potassium iodide in iodized salt (see Chapter 25-H).

Fluorine

Fluorine, a dietary necessity of the chicken, is associated with proper bone development. Small amounts are required. In many areas the requirement is supplied naturally by the fluorine in the water the chicken drinks, since many soils contain an abundance of fluorine which finds it way to the water supply. But in other areas the water supply is almost void of fluorides. In some localities fluorides are added to public drinking water (1 to 2 ppm) to aid in the formation of better teeth enamel in children. But in large amounts, fluorides accumulate in the tissues and are toxic to the chicken. Although fluorine is often added to poultry rations, it occurs in many minerals, such as limestones and phosphates. Before being fed to chickens, these must be processed to lower their fluorine content. The products are sold as *defluorinated rock phosphate* or *high-calcium limestone*. Most of these have a fluorine content of less than 0.5% and can be safely fed to chickens.

Iron and Copper

Nutritional anemia occurs when there are deficiencies of copper and iron. The red blood cells contain iron. The mineral also is needed to pigment the feathers of certain breeds of chickens. Copper is necessary for iron utilization when hemoglobin is formed; therefore, if absent from the diet, anemia results. The amount of iron and copper needed in the diet of the chicken is quite specific; excesses may be toxic. About 5 to 10 times as much iron as copper is required. Usually, only small amounts, if any, are ever added to the feed formula.

Manganese

The chief function of manganese is to prevent perosis (slipped tendon), a condition where the hock joint becomes enlarged and the gastrocnemius tendon at this location slips from the condyle, twisting the shank to one side. But manganese is also needed for normal growth, eggshell deposition, egg production, hatchability, and to prevent ataxia. Since all rations composed of normal feedstuffs are deficient, manganese is added to the feed as manganese sulphate or manganous oxide. A 70% feeding grade is commonly used. From 30 to 50 g of manganese are added to a ton (2,000 lb) of feed to prevent perosis and 50 to 75 g to increase eggshell strength.

Magnesium

Magnesium is one of the essential trace elements of nutrition. When it is absent from the diet, chicks grow slowly, exhibit convulsions, and may eventually die. Deficiencies in laying rations produce a rapid drop in egg production. Calcium is poorly utilized in the absence of magnesium. An excess of magnesium in the diet is about as detrimental as too little. One pronounced effect of an excess is wet droppings. Some limestones (the dolomites) contain a high percentage of magnesium and are to be avoided.

Selenium

This element is required in small amounts by the chicken. Not only is it essential in itself, but it reduces some of the symptoms of vitamin E deficiency. Exudative diathesis is one evidence of the lack of vitamin E, yet selenium has been shown to be capable of eliminating the condition. It does not, however, affect encephalomalacia, another condition often present when vitamin E is lacking. More vitamin E is needed only when selenium is deficient.

The selenium content of plants is closely related to the amount of selenium in the soil on which the plants grow. Where the soil is deficient, the plants and their seeds are deficient. Excesses of selenium are toxic, reducing growth, hatchability, and increasing the incidence of embryonic malformations. The optimal dietary level of selenium is 0.1 ppm for chickens up to 16 weeks of age. Diets low in selenium decrease egg production and hatchability and produce anemia. Sodium selenite is a compound that can supply selenium; 1 lb (454 g) to 2,250 lb (1,023 kg) of ration will supply 0.1 ppm.

> *Caution:*   In some countries it is illegal to add selenium to poultry feed because of its residual effect in poultry meat and eggs. In others, the quantity fed is rigidly controlled. In some it cannot be fed to birds producing eggs for human consumption. Check with authorities before making additions. It will take 4 weeks after the removal of a selenium supplement in the ration for selenium to disappear from the bird's body tissues and eggs.

Vanadium

Too little vanadium is of no consequence to chickens, but too much is harmful. Some types of dicalcium phosphate and defluorinated rock phosphate reduce the quality of egg albumen, but the difficulty has been traced to high amounts of vanadium in these minerals. Either should not contribute more than 4 ppm of vanadium to diets of laying hens.

Zinc

A small amount of zinc is needed by the chicken to foster good egg production, hatchability, proper feathering, and growth. As feedstuffs generally are low in zinc, this mineral is usually added to the ration as zinc carbonate (about 57% zinc) or zinc oxide (about 80.5% zinc). Normally, from 15 to 30 g of zinc are added to 1 ton (2,000 lb) of feed.

26-C. AMINO ACIDS

Of the 22 amino acids, five are deemed critical as the others are usually adequate from combinations of feedstuffs found in most poultry rations, or by internal synthesis. The five are methionine, cystine, lysine, tryptophan, and arginine.

When a poultry ration is low in one or more of these five amino acids, certain protein supplements carrying large amounts of pure amino acids must be added to the feed formula to make up the deficiencies. Since methionine is most often lacking, most formulas call for supplementation in pure form as DL-methionine. One cause of methionine deficiency is the fact that large amounts of vegetable protein supplements are now used in feeds, plus low levels of the animal and fish proteins. Lysine and cystine often are inadequate when normal feedstuffs are used in poultry feeds.

Proper amino acid formulation requires minimum requirements of all, with little excess of any. This is practically impossible; there is some waste. Usually the value of the protein portion of the ration is determined by the limiting amino acid. Large amounts of others are usually of no value to the limiting one.

Some amino acid characteristics of the major feed ingredients are as follows:

Barley	low in tryptophan and lysine
Corn	low in lysine and arginine
Milo	low in lysine
Soybean oil meal	low in methionine; high in lysine
Corn gluten meal	low in lysine

Amino Acid Relationship

As dietary protein is expensive, most laying rations have but a minimum in them. Anything that reduces the daily consumption of feed per bird reduces the daily intake of protein, and egg size suffers. If continued for some time, egg production will also be affected.

But more important than total protein is the daily intake of amino acids necessary to produce the protein of the egg yolk and albumen. If only one such essential amino acid is deficient, the production of egg protein will be reduced, as will egg size. The amino acids most often involved and lacking are lysine, cystine, methionine, isoleucine, and valine.

Normally, 100 laying Leghorn hens will consume about 22 to 24 lb (10 to 11 kg) of feed per day. At peak egg production of 90% or over for Leghorns, the bird average intake of the critical amino acids will be about 720 mg of lysine, 360 mg of methionine, and 650 mg of sulfur amino acids per day. Normally, a bird eating 17 g of quality protein per day would consume these levels of amino acids, yet 16 g or less are often shown to be adequate. But how?

On an individual bird basis, when the flock peaks in egg production some birds are just beginning to lay, and consequently are eating less feed and amino acids. The precocious individuals have a high demand for feed because of higher egg production, and thereby eat more, resulting in the consumption of more protein. On the average, however, it appears that all birds are consuming the average amount of protein.

26-D. WATER

Water makes up the basic medium in the body for the transportation of nutrients, many metabolic reactions, for the elimination of waste products, and to aid in maintaining body temperature.

Chickens consume from two to seven times as much water by weight as they do feed, the variance depending on bird age and environmental temperature. For a full discussion see Chapters 16-L (layers) and 20-G (broilers). Drinking water must be pure and clean (see Chapter 39-C).

26-E. OTHER FEED CONSTITUENTS

There are other items added to feed in relatively small amounts. Some are directly associated with metabolism; others are not.

Antibiotics

Low levels of certain antibiotics are added to the formula to promote growth. Their action is indirect; they alter the microbial environment of the intestines, thereby increasing the availability of certain other feed constituents.

Arsenicals

Supplements of 3-nitro-4-hydroxyphenylarsonic acid, arsanilic acid, or sodium arsanilate are made to broiler diets to promote growth and to improve the yellow color of the skin and shanks. Although the metal is deposited in the tissues, the amount is very small, and once the arsenical is withdrawn from the feed the tissues are depleted. Any dietary source should be withdrawn from the feed at least 5 days before broilers are slaughtered. Arsenicals in combination with certain other drugs should not be used in laying rations.

Caution: Be sure to check with authorities before using any arsenical. Its use is illegal in many countries.

Xanthophylls

Xanthophylls compose a group of chemicals within a larger group of plant pigments known as *carotenoids*. The xanthophylls impart yellow color to the fat deposits and skin of the birds and to egg yolks. See Chapters 31-I and 33-L for complete details.

Antioxidant

Fat rancidity in a feed tends to destroy the fat-soluble vitamins A, D, and E. Most of this oxidation of the fats may be prevented by adding an antioxidant to the mixture. The two usually added at 0.0125% level are ethoxyquin and butylated hydroxytoluene (BHT). Diphenyl-*p*-phenylenediamine (DPPD) has been barred because it affects reproduction in mammals.

Coccidiostat

As an aid in suppressing coccidiosis-producing organisms, certain chemicals known as coccidiostats are added to most chick starting, broiler, and some growing rations. Levels of these coccidiostats should be those designated by the manufacturer (see Chapter 37-R).

Electrolytes

Body water has substances called electrolytes dissolved in it. They may be divided into two groups:

1. Extracellular (sodium, chloride, and bicarbonate)
2. Intracellular (potassium and phosphate)

The electrolytes regulate enzyme activity, regulate the osmolarity of some body fluids, and help control the body pH.

In mammals, the loss of water from sweating will upset the electrolyte balance, but this is not a factor with chickens because they have no sweat glands. However, during certain diseases there will be excessive water loss from the body. Under these circumstances some remedial physiological improvement may be shown through the addition of electrolytes to the diet.

Pellet Binders

Normally, the pellets processed from most mash mixtures tend to crumble. To increase their hardness certain binders are commonly added (see Chapter 28-B).

Tranquilizers

On many occasions, tranquilizers are added to poultry feeds to quiet birds being moved from place to place, to reduce the incidence of cannibalism, or to calm flocks affected with hysteria. Although there are several tranquilizers, *reserpine* is used predominantly. Aspirin, ethylene glycol, and others are sometimes used.

Other Supplements

There are many other items added to feeds at times. Dozens are used for the treatment or prevention of disease. On occasion flavoring agents, enzymes, thyroactive compounds, and drugs to precipitate a molt, are used.

27

Analysis of Feedstuffs

In order to formulate poultry rations it is necessary to have tables showing the analyses of the various feedstuffs. These values are necessary to build formulas that are properly balanced for the type and age of bird involved, and for the environment under which it is kept. However, the ability to calculate a proper diet is the result of long experience and training in the field of poultry nutrition. There are many combinations of feedstuffs that would provide calculated requirements for growth and reproduction, yet would not be good diets. Many of the feedstuffs would have deleterious effects when given at percentages greater than the optimum—they would be unpalatable, toxic, or otherwise impractical.

But once the specified amounts of a feedstuff are included in a formula, the analysis tables do provide a basis for determining whether the minimum nutritive requirements for carbohydrates, fats, proteins, minerals, vitamins, and so forth have been met.

27-A. EXPRESSION OF NUTRITIVE REQUIREMENTS

There is no precise manner in which the component parts of a ration or the nutritional requirements are expressed. Some of these variations are due to the fact that all countries and all scientists do not use the same units of measure. Some of these variations are given below.

Major feed ingredients. Usually, these are expressed in percentages, by weight.

Minor feed ingredients. Vitamin A is most often given as IU per pound or kilo. Vitamin D_3 is expressed as IU per pound or kilo. In the case of vitamin E, IU or milligrams per pound or kilo are used. Most other vitamins are expressed as milligrams, while trace minerals and amino acids are given in terms of percentage.

Conversions

ICU (International Chick Units).   The measure of vitamin D_3 activity. Sometimes however, vitamin D_3 is measured in IU (International Units). For practical purposes the two are equal.

USPU (U.S. Pharmacopoeia Units).   The unit for measuring vitamin A activity. Often IU (International Unit) is used. The two are equal. One USPU of vitamin A is equivalent to 0.6 mcg of carotene.

IU (International Units).   One measurement for vitamin E is the IU. It is equivalent to 1 mg.

Gram conversions.

1,000 micrograms	= 1 milligram (mg)
10 millligrams	= 1 centigram (cg)
10 centigrams	= 1 decigram (dc)
10 decigrams	= 1 gram (g)
1,000 milligrams	= 1 gram (g)
1,000 grams	= 1 kilogram (kg)
1,000,000 milligrams	= 1 kilogram (kg)
1/1000 gram	= 1 milligram (mg)
28.349 grams	= 1 ounce (oz)
453.592 grams	= 1 pound (lb)
1 microgram per gram	= 1 part per million (ppm)
1 ppm	= 0.0001 percent (%)

Calories

Small calorie (cal).   A small calorie is the amount of heat required to raise the temperature of 1 g of water 1°C. The small calorie is not often used in nutritional work. It is designated by the small letter ''c.''

Large Calorie (Cal).   The amount of heat required to raise the temperature of 1,000 g of water 1°C. Thus, 1 (large) Calorie is equal to 1,000 small calories. The large Calorie is often spoken of as the kilocalorie (kcal), meaning 1,000 small calories. Often the energy value of a ration is given only as calories, meaning large calories. It is conventionally capitalized when a large Calorie is meant.

Therm.   1,000,000 small calories or 1,000 large calories equal 1 therm.

Joule.   In some European countries the use of joule is becoming more common as the measurement for the unit of energy, mainly because it was adopted by the International Union of Pure and Applied

Chemistry. One calorie is equal to 4.184 joules. One joule is equal to 0.239 calorie.

Expression of Requirements

Many feed elements are expressed in the formula as the quantity per pound; at other times, the amount per kilo is used. Still others show the amount necessary per 100 lb or per short ton (2,000 lb). The metric ton (2,204.6 lb) is sometimes used. When analyzing any feed requirements or feed formulas, care must be taken to determine which units are used.

Energy

The energy value of a feedstuff may be measured in several ways. First, there is the total or *gross energy* (GE), the energy released as heat by burning the feedstuff in a bomb calorimeter. But all of the GE consumed by the chickens is not used for productive purposes, as a considerable amount of energy is excreted in the feces. The amount of energy contained in the feces, urine, and gaseous products subtracted from the GE gives the *apparent metabolizable energy* (AME) of the feedstuff. These values are usually measured as the kcal of AME per pound (kilo) of the feedstuff, and are the ones usually used in poultry nutrition work.

The AME values of practically all feed ingredients have been determined (see Table 27-1). Notice that fats are the highest in AME with 3,130 to 3,720 kcal per pound. Alfalfa products are the lowest with 500 to 640 kcal of ME per pound.

As an example, the usual requirement of a chick starting ration is about 1,360 kcal of AME per pound, and feedstuffs varying in AME are mixed together to give this amount.

True Metabolizable Energy (TME)

In 1978, it was discovered that the AME caloric values were in error because up to that time it had been assumed that all feces consisted of only undigested food material, with no adjustment made for the urinary energy produced mostly by metabolic and endogenous reactions. When these adjustments were made, the *true metabolizable energy* (TME) could be calculated.

But the TME figures are still not errorless, inasmuch as feed spillage and different feed consumption by various birds still cause great variability in the computations. In this book only AME values are used for the

Table 27-1. Poultry Feed Ingredient Analyses

| Ingredient | ME kcal/lb | % | | | | | Vit. A IU/lb | Xanthophyll mg/lb |
		Protein	Fat	Fiber	Ca	P		
Alfalfa meal (20% protein)	640	20.0	2.6	17.0	1.90	0.22	150,000	150
Alfalfa meal (17% protein)	500	17.0	2.5	26.0	1.40	0.21	100,000	120
Barley, ground	1,190	11.0	2.1	6.0	0.07	0.36	—	—
Bone meal, steamed	450	6.5	0.6	2.6	31.30	14.50	—	—
Cereal tailings	1,460	9.0	2.0	2.5	0.02	0.27	—	—
Corn, yellow, ground	1,530	8.9	4.0	2.0	0.02	0.27	2,270	10
Corn DDG with solubles	1,050	28.0	8.0	7.0	0.19	0.35	1,100	8
Corn distillers dried solubles	1,250	27.0	5.0	2.0	0.35	1.55	—	—
Corn ferm. solubles, dried extract	980	23.4	0.25	0.2	0.07	1.55	—	—
Corn gluten feed (yellow)	650	21.0	1.0	10.0	0.30	0.70	1,000	30
Corn gluten meal (41% protein)	1,480	41.0	2.0	3.0	0.07	0.37	12,000	60
Corn gluten meal (60% protein)	1,691	60.0	2.7	2.0	0.18	0.39	20,600	160
Cottonseed meal (50% protein)	1,020	50.5	1.0	7.0	0.20	1.40	—	—
Defluorinated rock phosphate	—	—	—	—	32.00	18.00	—	—
Dicalcium phosphate	—	—	—	—	23.00	18.50	—	—
Dried bakery product	1,720	8.0	12.0	1.5	0.10	0.40	—	—
Fat, stabilized								
Animal tallow, feed grade	3,130	—	97.0	—	—	—	—	—
Fish oil	3,490	—	95.0	—	—	—	—	—
Hydrolyzed animal & vegetable fat	3,400	—	99.0	—	—	—	—	—
Poultry oil	3,720	—	98.0	—	—	—	—	—
Yellow grease	3,400	—	99.0	—	—	—	—	—

Feed							
Fish meal							
Canadian herring (72% protein)	1,450	72.0	9.0	0.5	2.40	2.20	—	—
Maine herring 65% (10% fat)	1,420	65.0	10.0	0.5	3.00	2.30	—	—
Atlantic menhaden (58–65% protein)	1,275	60.0	9.0	1.0	6.20	3.60	—	—
Atlantic whitefish	1,250	67.0	5.0	1.0	5.60	2.90	—	—
Peruvian anchovetta	1,300	65.0	4.0	1.0	3.80	2.50	—	—
Fish solubles, condensed	575	31.5	4.6	0.5	0.15	0.63	—	—
Grain sorghums, Milo	1,500	10.5	3.0	2.5	0.04	0.32	—	—
Hominy feed (yellow)	1,370	10.6	5.8	5.1	0.08	0.52	1,600	—
Hydrolyzed poultry feathers	1,160	87.0	5.0	1.4	0.22	0.77	—	—
Limestone, ground (38% calcium)	—	—	—	—	38.00	—	—	—
Limestone, ground (35% calcium)	—	—	—	—	35.00	—	—	—
Meat and bone meal (50% protein)	900	50.0	10.0	2.5	10.00	5.00	—	—
Meat and bone meal (47% protein)	870	47.0	11.0	2.0	12.00	6.00	—	—
Molasses, cane	890	3.0	—	—	0.65	0.08	—	—
Oats, ground	1,170	11.0	4.5	11.0	0.10	0.36	—	—
Oyster shells, ground	—	—	—	—	38.00	—	—	—
Peanut meal	1,200	44.0	6.1	11.8	0.17	0.56	—	—
Poultry by-product meal	1,260	55.0	12.0	2.5	3.00	1.35	—	—
Sesame meal	820	42.0	7.0	6.5	2.00	1.30	—	—
Soybean meal (dehulled)	1,120	49.0	0.5	2.5	0.20	0.60	—	—
Soybean meal (44% protein)	1,020	44.0	0.5	5.2	0.30	0.60	—	—
Wheat, ground	1,410	12.0	1.9	2.4	0.04	0.39	—	—
Wheat bran	510	15.6	4.2	9.0	0.11	1.21	—	—
Wheat middlings	890	16.0	4.5	7.5	0.08	0.93	—	—
Whey, dried	860	12.5	0.7	0.3	0.85	0.70	—	—
Yeast, dried	880	45.0	3.0	1.4	1.30	1.25	—	—

Source: New England College Conference Board, 1984.

Table 27-2. Poultry Feed Ingredient Analyses

| Ingredient | mg per pound | | | | | % | | | | | |
	Riboflavin	Panthothenic Acid	Choline	Niacin	Arginine	Lysine	Methionine	Cystine	Tryptophan
Alfalfa meal (20% protein)	8.40	17.00	500	24.00	1.10	0.98	0.34	0.38	0.27
Alfalfa meal (17% protein)	6.70	16.00	400	22.00	0.76	0.86	0.30	0.33	0.25
Barley, ground	0.59	3.00	500	31.10	0.54	0.34	0.19	0.22	0.14
Bone meal, steamed	—	—	—	—	1.70	0.90	0.20	0.10	0.05
Cereal tailings	0.50	2.30	200	9.60	0.35	0.22	0.13	0.17	0.12
Corn, yellow, ground	0.73	2.09	227	9.08	0.44	0.21	0.19	0.16	0.08
Corn DDG with solubles	3.45	5.30	1,800	37.00	1.10	0.65	0.44	0.38	0.22
Corn distillers dried solubles	7.00	11.00	2,500	68.00	0.94	0.84	0.50	0.37	0.18
Corn ferm. solubles, dried ext.	2.40	6.65	484	109.50	1.00	1.00	0.54	0.49	0.10
Corn gluten feed (yellow)	1.20	4.00	500	30.00	0.80	0.80	0.30	0.50	0.20
Corn gluten meal (41% protein)	1.00	6.00	750	13.60	1.34	0.74	0.98	0.64	0.22
Corn gluten meal (60% protein)	0.80	5.40	215	28.85	2.20	1.40	1.60	0.90	0.30
Cottonseed meal (50% protein)	2.10	8.10	1,300	20.70	4.20	1.80	0.62	0.85	0.56
Defluorinated rock phosphate	—	—	—	—	—	—	—	—	—
Dicalcium phosphate	—	—	—	—	—	—	—	—	—

Feed								
Dried bakery product	0.80	2.00	400	10.00	0.32	0.24	0.14	0.15	0.05
Fat, stabilized									
Animal tallow, feed grade	—	—	—	—	—	—	—	—	—
Fish oil	—	—	—	—	—	—	—	—	—
Hydrolyzed animal & veg. fat	—	—	—	—	—	—	—	—	—
Poultry oil	—	—	—	—	—	—	—	—	—
Yellow grease	—	—	—	—	—	—	—	—	—
Fish meal									
Candian herring, (72% protein)	4.50	8.00	2,200	35.00	5.30	6.50	2.10	1.00	0.90
Maine herring, 65% (10% fat)	4.00	4.00	1,800	40.00	4.80	5.70	1.90	0.80	0.80
Atlantic menhaden (58–65% protein)	3.10	4.00	1,230	38.59	3.97	5.50	1.80	0.79	0.73
Atlantic whitefish	3.50	3.75	1,500	25.00	4.90	5.75	1.90	0.85	0.82
Peruvian anchovetta	4.00	4.50	1,600	40.00	3.60	5.30	1.70	0.75	0.70
Fish solubles, condensed	7.70	17.25	1,250	117.50	1.55	1.65	0.56	0.21	0.16
Grain sorghums, Milo	0.50	3.90	200	33.00	0.36	0.23	0.16	0.17	0.10
Hominy feed (yellow)	1.06	3.98	501	20.68	0.50	0.25	0.15	0.13	0.12
Hydrolyzed poultry feathers	0.90	4.25	400	11.00	5.75	1.75	0.56	2.65	0.57
Limestone, ground (38% calcium)	—	—	—	—	—	—	—	—	—
Limestone, ground (35% calcium)	—	—	—	—	—	—	—	—	—

Table 27-2. (continued)

Ingredient	mg per pound				%				
	Riboflavin	Panthothenic Acid	Choline	Niacin	Arginine	Lysine	Methionine	Cystine	Tryptophan
Meat and bone meal (50% protein)	2.10	1.50	990	21.40	3.15	2.60	0.58	0.57	0.31
Meat and bone meal (47% protein)	2.10	1.50	990	21.40	2.77	2.30	0.50	0.50	0.27
Molasses, cane	1.23	15.58	366	14.40	—	—	—	—	—
Oats, ground	0.59	4.65	420	6.60	0.70	0.40	0.20	0.21	0.14
Oyster shells, ground	—	—	—	—	—	—	—	—	—
Peanut meal	2.40	25.05	795	76.20	4.40	1.33	0.47	0.71	0.44
Poultry by-product meal	4.50	4.00	2,720	18.00	3.75	3.70	0.99	0.98	0.45
Sesame meal	1.50	2.50	680	6.00	4.80	1.30	1.40	0.57	0.78
Soybean meal (de-hulled)	1.20	6.50	1,150	13.50	3.53	3.14	0.69	0.72	0.63
Soybean meal (44% protein)	1.49	6.22	1,050	9.08	3.16	2.82	0.63	0.66	0.57
Wheat, ground	0.55	5.00	330	27.50	0.53	0.38	0.20	0.22	0.16
Wheat bran	1.35	11.40	460	139.00	0.93	0.56	0.21	0.21	0.22
Wheat middlings	1.15	7.20	500	56.50	0.86	0.59	0.21	0.19	0.20
Whey, dried	10.90	21.00	700	8.50	0.38	0.97	0.28	0.34	0.21
Yeast, dried	16.00	50.00	1,500	200.00	2.10	3.23	0.78	0.48	0.56

Source: New England College Conference Board, 1984

many calculations shown. At the present time there are not enough data for TME values for the authors to do otherwise.

27-B. INGREDIENT ANALYSES

The analyses of some common poultry feedstuffs are given in Tables 27-1 and 27-2.

28
Feed Fundamentals

28-A. FEEDING SYSTEM

Most of the poultry rations first formulated were used to supplement locally produced cereal grains grown on the average small farm. But when commercialism entered the poultry business chicken farms increased in size, birds were closely confined to houses, and the knowledge of poultry feeding increased.

Complete Feed

Little by little it became possible to formulate poultry rations that would include all the *known* nutrients. They were *complete feeds*, requiring no supplementation, which was a decided advantage, since with commercial chicken production the units were so large that the grains had to be purchased.

However, feed formulation did not become stationary. New nutritional discoveries were made every year and the cost of ingredients changed, making formula substitutions necessary, and improvement in the management of chickens required that changes be made in the feed formula.

Self-Feeding Versus Controlled Feeding

Within limits, the chicken has the ability to control its feed intake according to its nutritional needs. In the early days chickens were self-fed, that is, a complete mash was kept before them at all times and they could consume all they needed.

Later, it was found that chickens did overeat and got too heavy or did not produce eggs or meat economically. Today, some rations are self-fed, particularly with Leghorn egg strains, while the heavier varieties of egg strains and broiler breeders must have their feed intake controlled (restricted) according to their body weight and condition (see Chapters 31-J and 32).

28-B. FORM OF FEED

Most poultry rations are available in mash, crumble, or pellet forms.

Mash Form

Many feed ingredients are in ground form; others, such as the whole grains, must be ground prior to mixing the ration. The essence of a "mash" is that, in theory at least, each *bite* of feed is a balanced diet containing all the known nutrients in finely ground form. But birds find the finely ground mashes unpalatable; they are too dry and sticky. Therefore, mashes composed of materials of medium particle size improve the bird's ability to eat them readily. But it is impossible to secure all feed ingredients with this texture; some are available only as a finely ground product. Therefore, it is of great importance that the cereal grains should not be ground to a fine consistency. In turn, this means that chickens will have a tendency to pick out the larger cereal grain particles from the mash first, leaving the finer material until last. A feeding problem exists, especially with automatic feeders that have a chain-and-drag method of propelling the feed in an open long trough, because the birds nearest the feed hopper tend to eat more of the palatable carbohydrate grain portion of the feed mixture, leaving the finer ingredients for the remainder of the birds in the house.

Usual form of feed.

> Leghorns and brown-egg varieties: mash
> Broilers and broiler breeders: mash or crumbles for 2 weeks, then pellets

Particle size and its effect.   Particle size of the mash mixture affects water consumption—the coarser the texture, the less water the birds drink. However, the particle size of the mash has no effect on the percentage of water in the feces.

Bulkiness affects water consumption.   The bulkier the diet (more fiber) the more water consumed, and, therefore, more water is voided in the feces. As high-energy diets are less bulky, the birds drink and excrete less water when they are consumed.

Pellet Form

The mash may be compressed by running it through specialized equipment to form pellets of varying sizes. The pelleting machine is composed of a die made up of dozens of holes of a specific diameter through which the feed is forced under pressure to form a pellet.

With pellets, the chicken cannot pick out certain parts of the feed, but must eat it all. This is particularly advantageous with young chicks because they are consuming such a small amount of feed. When medicated feeds are used, it becomes more important.

Producing firmer pellets. Steam is added to the mash at pelleting to produce a firmer pellet. This moisture, plus the heat generated during pelleting, increases gelatinization of the mixture, thereby forming a firmer pellet. When fat is added to the feed mix, the pellets tend to crumble because the fat acts as a lubricant rather than an adhesive. To overcome this, fat is injected into the die, or the pellets are cooled; then fat is sprayed on them.

Pellet Binders

In order to improve the hardness of pellets certain pellet binders are often added to the feed mix. There are four binders that are common:

1. Sodium bentonite (anhydrous silicate)
2. Cellulose products (from the wood pulp industry)
3. Lignin derivatives (Lignosol®)
4. Grain industry by-product (Nutri-Bond®)

Additions of up to 2.5% may be made, depending on the original feed formula. Rations containing large amounts of fat will require more of the binder.

Binders have other values. Sodium bentonite absorbs water from processed pellets, tends to reduce wet droppings, and improves growth of young chicks. Hemicellulose wood pulp products are good energy sources at levels necessary to form firm pellets, but lignin has no nutritive value.

Physical Makeup of Pellets

Size of pellets. The size of pellets is determined by their diameter and length. A knife cuts the material extruded from the die into pellets of varying lengths, but pellets are merchandised according to their diameter rather than their length. Small chicks are usually started on mash or crumbles; then some varieties, like broilers, are changed to pellets as they age.

Fine material not a disadvantage.   Although pellets look better if there is no fine material, experimental work has shown that small amounts of "fines" are not detrimental to feed consumption or feed conversion, but large amounts increase feed waste and reduce growth, particularly in broilers.

Pelleting alters nutritive value.   The heat generated during the pelleting process will destroy some of the carotene (provitamin A) in feedstuffs, and it is wise to increase the *minimum* allowances by 10 to 20% to compensate for the depletion. However, pelleting destroys some trypsin growth inhibitors found in soybean meal, which is an offsetting advantage.

Advantages and Disadvantages of Pellets

The production of pellets is an expensive procedure. If the costs are to be regained, the advantages of pelleting must outweigh the disadvantages.

Advantages of pellets

1. Wind loss is less with pellets than with mash.
2. Most feed dustiness is eliminated.
3. When handling feeds, there is no separation of ingredients when feed is pelleted.
4. Pelleting destroys some bacteria in the feed (e.g., *Salmonellae*).
5. Usually, it takes less labor to feed pellets than mash.
6. Pelleting increases feed density and birds can consume more low-energy (high-fiber) feeds.
7. Certain feed ingredients are unacceptable to chickens (e.g., rye, buckwheat, barley), but when feeds are pelleted, consumption is markedly increased.
8. The heat, moisture, and pressure from the pelleting process may increase the efficiency of the ration.
9. There is less feed waste from the feeders.

Disadvantages of pellets

1. There is the added cost of pelleting the mash.
2. Some pellets crumble when they are moved by automatic feeding systems, and the finer particles are wasted.
3. Pellets increase water consumption.
4. The droppings are wetter when pellets are fed.

5. Pellets increase the incidence and severity of cannibalism.

Crumble Form

When pellets are coarsely ground, or preferably run through special cracking rolls, a type of product midway between mash and pellets results. It has most of the advantages and disadvantages of pellets, but because of the smaller size, it may be fed to younger chicks. Often crumbles are used from 1 day of age.

Coarseness of crumbles. The texture of crumbles should be intermediate, neither too coarse nor too fine. In fact, crumbles are best prepared by leaving some finer material in them. This enables younger chicks to eat more rapidly, and prevents some of the cannibalism which results from compressing all the feed particles. With many feed mixtures it is necessary to sift out some of the fine material from the crumbles and repellet it, to start the crumbling process over again so as to secure a coarser material.

Dietary Fiber in Compressed Feeds

It has been found that the response to compressing a feed into pellets and crumbles is in part dependent on the fiber content of the original mash. Pelleting rations high in fiber (low in energy) shows more response than pelleting feeds low in fiber (high in energy). Undoubtedly this is because high-fiber diets are less palatable to begin with, which results in lowered feed intake.

Wet Mash Versus Dry Mash

A procedure of past years was to moisten the mash and feed it as a *wet mash,* supposedly to induce greater feed consumption. This may be true for the first day or two the wet mash is fed, but birds soon learn to adjust for the increased palatability, and feed consumption reverts to its normal level. Wet-mash feeding will *not*

1. Increase egg production
2. Increase egg weight
3. Increase growth
4. Increase feed conversion

28-C. CALORIE/PROTEIN RATIO (C/P RATIO)

There is a close association between the number of calories of ME in the ration and the percentage of protein necessary to balance the energy. The ratio varies with the age of the birds and the use to which they are put. The ratio is expressed as a figure calculated by dividing the number of kcal of ME per pound of feedstuff by the percentage of protein.

Example: The ration contains 1,200 kcal of ME per lb, and 20% protein. Thus, the C/P ratio is 60 (1,200 ÷ 20 = 60).

Some recommended ratios on a pound basis according to the type of bird are given below:

Type of Bird		C/P Ratio (ME)
Broilers	0–2 weeks	58
Broilers	3–7 weeks	75
Chicks	0–5 weeks	67
Growing	6–22 weeks	90
Laying and breeding 50% production		91
Laying and breeding 60% production		86
Laying and breeding 70% production		81
Laying and breeding 80% production		76
Laying and breeding 90% production		70

29
Poultry Rations

There are definite basic components of a poultry feed. Within each component several different feedstuffs may be used to satisfy the nutritive requirements. For example, the carbohydrate portion of the diet might be made up of one of the following:

corn
corn and milo
corn and oats
milo and barley

29-A. DIETARY REQUIREMENTS

Basic Components of a Ration

1. *Carbohydrates.* This includes the cereal grains and other high-carbohydrate ingredients. Carbohydrate sources make up the largest segment of the ration.
2. *Fats.* Supplementary fats are often added to build high-energy diets.
3. *Mill by-products.* By-products from the milling of wheat, rice, and corn are included in this list.
4. *Green, leafy material.* Dried products produced from alfalfa, green grasses, and other plants.
5. *Fish protein.* This group includes fish meals and related products.
6. *Animal protein.* Included are meat and bone meal, poultry by-product meal, milk by-products, etc.

7. *Vegetable protein.* Meals produced from soybeans, cottonseed, peanuts, etc. fall in this group to make it the second-largest portion of the formula.

8. *Amino acid supplements.* A great many mixtures of natural feedstuffs are deficient in one or more amino acids, and supplementation must be made.

9. *Macrominerals.* These are sources of calcium, phosphorus, salt (NaCl), potassium, magnesium, and sulfur.

10. *Trace minerals.* Manganese, iron, copper, zinc, selenium, etc. are included in this list.

11. *Antibiotic supplement.* Certain rations, such as those used for broiler feeding, include a small amount of an antibiotic as a growth stimulator. Such antibiotics are not to be confused with those used for medicinal purposes.

12. *Vitamins.* These are supplementary vitamin concentrates.

13. *Antioxidants.* To prevent rancidity and destruction of certain feed components by oxidation, antioxidants are added to many feed mixtures.

14. *Medicaments and drugs.* Coccidiostats and other drugs are incorporated in many feed formulas.

15. *Other.* These include supplementary sources of xanthophylls, hormones, enzymes, pellet binders, flavors, and several other items. They are used only under certain conditions.

The above list gives the major feed segments under which practically all feedstuffs may be classified. There are about 40 dietary nutrients in a ration, and most poultry feeds should contain 20 to 25 ingredients to supply the minimum nutritional needs of the chicken.

Basis for Formulation

The units of measure used in feed formulation have been given in Chapter 27-A. The requirement of each nutritional item is usually computed as a number of units per pound or per kilo or percentage of the feed mixture. To be more practical, however, rations are computed on the basis of the number of units or percentage per ton (2,000 lb) of feed, this being the amount in a "batch" of feed when produced in a feed manufacturing plant.

Nutrient Requirement Based on Feed Energy

The energy content of the ration will govern the chicken's daily feed consumption. The amount of each of the other nutrients in the ration must be related to the feed's caloric content. This recommendation is based on the premise that the bird has a daily requirement of each nutritional factor; when there is a variation in feed consumption as a result of dietary changes in the caloric value of the ration or environmental or other factors, causing the bird to eat more or less feed, an adjustment must be made in the nonenergy portion of the diet. To follow this rule explicitly would entail the devotion of countless hours to formula changes plus a high-speed computer. Normally, such changes in the ration are made only when there is a major need for alteration.

High-Energy Rations More Efficient

Normally, high-energy rations are more efficient and more economical than low-energy feeds. But the terms involving energy levels are not definite; a high-energy diet for broilers would have a higher ME value than a high-energy ration for young replacement egg-type strains of chicks.

Phase Feeding

The use of phase feeding during egg production, whereby the protein percentage is lowered during the last one-half to two-thirds of the laying year, has an economic significance. Such rations usually call for a reduction of about 3% in their protein content.

29-B. EXAMPLES OF FEED FORMULAS

Replacement Chicken Rations

Table 29-1 lists four rations and their calculated analyses.

1. *Chick starter.* Includes 20 and 18% starter to be used for egg-type and meat-type strains from 1 day of age through the fifth week (35 days).
2. *Chick grower.* These are 14 and 12% protein egg- and meat-type cockerel and pullet-growing rations to be used from the beginning of the sixth week (36 days) until they reach sexual maturity.

Table 29-1. Replacement Chicken Rations[1]

	Starter		Grower	
	20% Pro	18% Pro	14% Pro	12% Pro
Ingredient	lb	lb	lb	lb
Ground yellow corn (2, 3)	1,267	1,310	1,438	1,481
Wheat middlings	130	200	254	323
Alfalfa meal (17% protein)	25	25	25	25
Soybean meal (dehulled)	422	309	217	104.8
Fish meal, herring (65% protein) (4, 5)	50	50	—	—
Meat and bone meal (47% protein) (5)	50	50	—	—
Lysine	—	—	1	1.2
Dicalcium phosphate (6)	10	9	30	29
Ground limestone (7)	19	20	28	29
Stablized yellow grease, or equivalent	20	20	(16)	(16)
Iodized salt (4)	7	7	7	7
Antibiotic supplement	(8)	(8)	—	—
Antioxidant	(9)	(9)	(9)	(9)
Coccidiostat	(10)	(10)	(10)	(10)

Vitamin and mineral supplements (12)					
Vitamin A	(IU)	3,000,000	3,000,000	3,000,000	3,000,000
Vitamin D_3	(IU)	1,000,000	1,000,000	1,000,000	1,000,000
Vitamin K (20)		—	—	—	—
Vitamin B_{12}	(mg)	6	6	6	6
Riboflavin	(mg)	1,500	1,500	1,500	1,500
Niacin	(mg)	10,000	10,000	10,000	10,000
Calcium pantothenate	(mg)	4,000	4,000	3,000	3,000
Choline	(mg)	213,000	298,000	125,000	209,000
Manganese (11)	(g)	52	52	52	52
Selenium (25)	(mg)	90.8	90.8	90.8	90.8
Totals (21)	(lb)	2,000	2,000	2,000	2,000

Calculated Analysis

Metabolizable energy	(kcal/lb)	1,361	1,362	1,361	1,362
Protein	(%)	20.03	18.01	14.01	12.01
Lysine	(%)	1.04	0.89	0.63	0.49
Methionine	(%)	0.34	0.32	0.24	0.21
Methionine + cystine	(%)	0.64	0.59	0.46	0.41
Fat	(%)	4.48	4.70	3.54	3.74
Fiber	(%)	2.67	2.83	3.00	3.15
Calcium	(%)	0.90	0.90	0.90	0.90
Total phosphorus	(%)	0.66	0.66	0.66	0.65
Available phosphorus (13)	(%)	0.41	0.41	0.40	0.40
Vitamins					
Vitamin A activity	(IU/lb)	4,188	4,237	4,381	4,430
Vitamin D_3 (added)	(IU/lb)	500	500	500	500
Riboflavin	(mg/lb)	1.78	1.76	1.64	1.62
Niacin	(mg/lb)	19.08	20.49	20.53	21.84
Pantothenic acid	(mg/lb)	5.34	5.27	4.72	4.64
Choline	(mg/lb)	600.20	600.11	420.20	419.76

Source: New England College Conference Board, 1984.
[1]Numerals in parentheses refer to footnotes presented on pages 630–631.

Cage and Floor Egg-Type Layer Rations

Five rations for layers kept on the floor or in cages are given. These layer rations are formulated so as to include 15, 16, 17, 18, and 19% protein, and are found in Table 29-2.

No endeavor has been made to provide formulas for other types of layer rations. These would include high- and low-energy rations, those used for phase feeding, those used for environmental temperature changes, and those used for high-calcium diets. Many of these are detailed in Chapter 31.

Breeder Rations

Two rations are given in Table 29-3:

1. *Egg-type breeder.* For 3½- to 5-lb (1.6- to 2.3-kg) egg-type female birds to be used for producing hatching eggs including Leghorns and medium-size layers producing brown-shelled eggs.
2. *Meat-type breeder.* For 5- to 8-lb (2.3- to 3.6-kg) meat-type female birds (broiler breeders) to be used for producing hatching eggs.

Note: Rations 1 and 2 above are also fed to male breeders.

Broiler Rations

1. *Broiler starter.* This ration is to be used until male and female broiler chicks are 14 days of age.
2. *Broiler grower.* This ration is to be used beginning with the 15th day until the cockerel chicks are 37 days of age and the pullet chicks are 41 days of age.
3. *Broiler finisher.* This ration is to be fed to cockerels beginning with the 38th day until they are marketed, and to pullets beginning with the 42nd day until they are marketed. It may also be used as a drug-withdrawal feed by removing all medicinal drugs from the formula.

The three formulas are found in Table 29-4.

Table 29-2. Cage and Floor Egg-Type Layer Rations[1]

Ingredient	Protein Level of Ration				
	15% lb	16% lb	17% lb	18% lb	19% lb
Ground yellow corn (2, 3)	1,457	1,403	1,339	1,242	1,177
Alfalfa meal (17% protein)	25	25	25	25	25
Soybean meal (dehulled)	292.2	340.6	393.6	451.6	504.6
Meat and bone meal (47%) (5)	50	50	50	50	50
DL-Methionine or equivalent	1.0	1.0	1.0	1.0	1.0
Dicalcium phosphate (6)	9	8	8	7	7
Ground limestone (7)	159	159	159	174	174
Iodized salt (4)	7	7	7	7	7
Stablized yellow grease, or equivalent	—	7	18	43	55
Antioxidant	(9)	(9)	(9)	(9)	(9)
Vitamin and mineral supplements (12)					
Vitamin A (IU)	6,000,000	6,000,000	6,000,000	6,000,000	6,000,000
Vitamin D$_3$ (IU)	2,000,000	2,000,000	2,000,000	2,000,000	2,000,000
Vitamin K (20) (mg)	—	—	—	—	—
Vitamin B$_{12}$ (mg)	6	6	6	6	6
Riboflavin (mg)	2,000	2,000	2,000	2,000	2,000
Niacin (mg)	12,000	12,000	12,000	12,000	12,000
Calcium pantothenate (mg)	5,000	4,500	4,500	4,500	4,000
Choline (mg)	274,000	231,000	184,000	140,000	94,000

Table 29-2. (continued)

Ingredient		Protein Level of Ration				
		15% lb	16% lb	17% lb	18% lb	19% lb
Zinc (17)	(g)	16	16	16	16	16
Selenium (25)	(mg)	90.8	90.8	90.8	90.8	90.8
Manganese (11)	(g)	52	52	52	52	52
Totals (21)	(lb)	2,000	2,000	2,000	2,000	2,000
Calculated Analysis (27)						
Metabolizable energy	(kcal/lb)	1,306.2	1,303.9	1,303.4	1,304.1	1,304.5
Protein	(%)	15.08	16.02	17.03	18.01	19.03
Lysine	(%)	0.68	0.75	0.83	0.91	0.98
Methionine	(%)	0.31	0.32	0.33	0.34	0.35
Methionine + cystine	(%)	0.54	0.57	0.59	0.62	0.64
Fat	(%)	3.29	3.54	3.98	5.05	5.54
Fiber	(%)	2.20	2.20	2.21	2.18	2.18
Calcium	(%)	3.25	3.24	3.24	3.50	3.50
Total phosphorus	(%)	0.52	0.52	0.53	0.52	0.53
Available phosphorus (13)	(%)	0.45	0.45	0.45	0.45	0.45
Vitamins						
Vitamin A activity	(IU/lb)	5,904	5,842	5,770	5,660	5,586
Vitamin D_3 (added)	(IU/lb)	1,000	1,000	1,000	1,000	1,000
Riboflavin	(mg/lb)	1.84	1.85	1.86	1.86	1.87
Niacin	(mg/lb)	15.40	15.48	15.55	15.50	15.56
Pantothenic acid	(mg/lb)	5.01	4.88	4.99	5.07	4.95
Choline	(mg/lb)	500.13	500.34	500.05	500.39	500.48

Source: New England College Conference Board, 1984.
[1]Numerals in parentheses refer to footnotes given on pages 630–631.

Table 29-3.   Breeder Rations[1]

Ingredient		3½–5 lb Egg-Type lb	5–8 lb Meat-Type lb
		Female Body Weight	
Ground yellow corn (2, 3)		1,305	1,379
Wheat middlings		—	70
Alfalfa meal (17% protein)		50	50
Soybean meal (dehulled)		324	248
Fish meal, herring (65% protein) (4, 5)		50	50
Meat and bone meal (47% protein) (5)		50	50
Dicalcium phosphate (6)		6	4
Ground limestone (7)		157	142
DL-Methionine, or equivalent		0.5	0.4
Stablized yellow grease, or equivalent		51	(16)
Iodized salt (4)		7	7
Antioxidant		(9)	(9)
Vitamin and mineral supplements (12)			
Vitamin A	(IU)	4,000,000	4,000,000
Vitamin D₃	(IU)	2,000,000	2,000,000
Vitamin E	(IU)	2,000	2,000
Vitamin K (20)		—	—
Vitamin B₁₂	(mg)	6	6
Riboflavin	(mg)	3,000	3,000
Niacin	(mg)	10,000	10,000
Calcium pantothenate	(mg)	6,000	6,000
Choline	(mg)	172,000	200,000
Zinc (17)	(g)	16	16
Manganese (11)	(g)	52	52
Selenium (25)	(mg)	90.8	90.8
Totals (21)	(lb)	2,000	2,000
Calculated Analysis (27)			
Metabolizable energy	(kcal/lb)	1,337	1,295
Protein	(%)	17.01	16.03
Lysine	(%)	0.87	0.78
Methionine	(%)	0.33	0.31
Methionine + cystine	(%)	0.59	0.56
Fat	(%)	5.81	3.57
Fiber	(%)	2.42	2.67
Calcium	(%)	3.27	2.98
Total phosphorus	(%)	0.54	0.54
Available phosphorus (13)	(%)	0.46	0.46
Vitamins			
Vitamin A activity	(IU/lb)	5,983	6,067
Vitamin D₃ (added)	(IU/lb)	1,000	1,000
Riboflavin	(mg/lb)	2.49	2.51
Niacin	(mg/lb)	15.21	17.00
Pantothenic acid	(mg/lb)	5.67	5.74
Choline	(mg/lb)	500.40	500.57

Source: New England College Conference Board, 1984.

[1]Numerals in parentheses refer to footnotes given on pages 630–631.

Table 29-4.   Broiler Rations[1]

Ingredient		Starter lb	Grower lb	Finisher lb
Ground yellow corn (2, 3)		1,106	1,238	1,282
Alfalfa meal (17% protein)		—	25	25
Soybean meal (dehulled)		605	420	370
Corn gluten meal (60% protein)		50	75	75
Fish meal, herring (65% protein) (4, 5)		50	50	50
Meat and bone meal (47% protein) (5)		50	50	50
Dicalcium phosphate (6)		10	9	9
Ground limestone (7)		16	14	14
DL-Methionine, or equivalent		0.8	—	—
Stablized yellow grease, or equivalent		106	112	115
Iodized salt (4)		7	7	7
Antibiotic Supplement		(8)	(8)	(8)
Antioxidant		(9)	(9)	(9)
Coccidiostat		(10)	(10)	(10)
Vitamin and mineral supplements (12)				
Vitamin A	(IU)	4,000,000	4,000,000	4,000,000
Vitamin D₃	(IU)	1,000,000	1,000,000	1,000,000
Vitamin E	(IU)	2,000	2,000	2,000
Vitamin K (20)	(mg)	2,000	1,000	1,000
Vitamin B₁₂	(mg)	12	12	12
Riboflavin	(mg)	3,000	3,000	3,000
Niacin	(mg)	20,000	20,000	20,000
Calcium pantothenate	(mg)	5,000	5,000	5,000
Choline	(mg)	503,000	672,000	672,000
Zinc (24)	(g)	30	30	30
Manganese (15)	(g)	75	75	75
Selenium (25)	(mg)	90.8	90.8	90.8
Organic arsenical supplement (19)		0.1	0.1	—
Totals (21)	(lb)	2,000.9	2,000.1	2,000.2

Calculated Analysis (27)

Metabolizable energy	(kcal/lb)	1,399	1,451	1,500
Protein	(%)	24.08	21.00	18.50
Lysine	(%)	1.30	1.05	1.00
Methionine	(%)	0.45	0.38	0.31
Methionine + cystine	(%)	0.81	0.71	0.68
ME:Protein Ratio		58.1	69.1	81.1
Fat	(%)	8.20	8.92	8.92
Fiber	(%)	1.97	2.22	2.22
Calcium	(%)	0.84	0.80	0.80
Total phosphorus	(%)	0.64	0.60	0.60
Available phosphorus (13)	(%)	0.40	0.38	0.38
Vitamins				
Vitamin A activity	(IU/lb)	3,769	5,424	5,465
Vitamin D₃ (added)	(IU/lb)	500	500	500
Riboflavin	(mg/lb)	2.44	2.49	2.49
Niacin	(mg/lb)	21.36	21.33	21.33
Pantothenic acid	(mg/lb)	5.69	5.51	5.51
Choline	(mg/lb)	800.03	800.48	800.48
Xanthophyll (28)	(mg/lb)	9.50	14.05	16.00

Source: New England College Conference Board, 1984, adjusted.

[1]Numerals in parentheses refer to footnotes given on pages 630–631.

Footnotes for Tables 29-1, 29-2, 29-3, and 29-4

1. Whenever substitutions are made in the rations, the total nutrient content should be adjusted to meet established requirements.

2. Two hundred to four hundred pounds of coarsely ground wheat or yellow hominy may be used to replace an equal amount of corn. If wheat is used, add 200,000 IU of vitamin A for each 100 pounds of corn removed.

3. There is usually some loss of provitamin A activity in corn and alfalfa meal during storage. If stored ingredients are used, it may be advisable to increase the added vitamin A level of the ration by 1,000 or 2,000 IU per pound. This may be accomplished by increasing the recommended supplement by 2,000,000 or 4,000,000 IU per ton of feed.

4. The added salt level should be reduced by the amount supplied by the fish meal and other by-product ingredients.

5. Poultry by-product meal may be substituted for all the meat and bone scrap and up to 50% of the fish meal. Correct for calcium and phosphorus loss due to substitution of poultry by-product meal.

6. Based on an 18.5% phosphorus product. Steamed bone meal or defluorinated rock phosphate may replace the dicalcium phosphate on a phosphorus basis.

7. Based on 35% calcium, low-magnesium limestone.

8. An antibiotic may be used in these rations at the level recommended by the manufacturer.

9. 1,2-dihydro-6-ethoxy-2,2,4-trimethylquinoline (ethoxyquin) is recommended in the chick starter, broiler, and breeder rations at the 0.0125% level to help prevent the appearance of encephalomalacia (crazy chick disease). If desired, ethoxyquin or an equivalent antioxidant may be added to help prevent the oxidation of dietary components. Total ethoxyquin from all sources must not exceed 0.25 lb per ton.

10. A coccidiostat or antihistomonal drug may be used in these rations, as required, at levels recommended by the manufacturer.

11. This amount of manganese will be furnished by 0.5 lb of manganese sulfate or 0.21 lb of manganous oxide (70% feeding grade.) An equivalent amount of manganese may be added from other acceptable sources.

12. Caution should be used when high-potency vitamin mixes are involved. It is recommended that 10 lb be the minimum amount of any item added to a ton of feed to ensure proper mixing. Thus, high-potency vitamin, mineral, or drug mixes should be premixed with a carrier, such as corn meal, to such a dilution that 10 lb of final mix will be added for each ton of feed mixed. Minerals and vitamins should not be premixed together.

13. Available phosphorus has been taken as 30% of total phosphorus from plant sources for chicks and 75% of total phosphorus from plant sources for adult birds. Phosphorus from other than plant sources is considered to be 100% utilized.

15. This amount of manganese will be furnished by 0.7 lb manganese sulfate or 0.3 lb manganous oxide (70% feeding grade). An equivalent amount of manganese may be added from other acceptable sources.

16. Stabilized fats may replace an equal amount of cereal grains to provide a higher energy level, control dust, and aid pelleting. Where maintaining body weight in layers is a problem, increase the fat by 1 or 2% during winter by replacing an equal amount of cereal grains.

17. Approximately this amount of zinc will be furnished with 20 g of zinc carbonate or 20 g of zinc oxide. An equivalent amount of zinc may be used from other acceptable sources.

19. Based on 3-nitro-4-hydrophenylarsonic acid at level of 45 g (0.1 lb) per ton. Other compounds that may be used at a level recommended by the manufacturer are sodium arsanilate or arsanilic acid.

20. In the absence of alfalfa meal, or if the birds are raised on wire, 2 g of vitamin K activity should be added. Values in the broiler rations are based on menadione. Other compounds supplying equivalent levels of vitamin K may be used.

21. If an even 2,000 lb is desired, adjust by removing or adding ground yellow corn.

23. This amount of manganese will be furnished by approximately 0.3 lb of manganese sulfate or 0.13 lb of manganese oxide (70% feeding grade). An equivalent amount of manganese may be added from other acceptable sources.

24. This amount of zinc will be furnished by approximately 53 g of zinc oxide. An equivalent amount of zinc may be added from other acceptable sources.

25. Federal law, which strictly regulates the addition of selenium to poultry rations, should be consulted. Selenium, as sodium selenite or sodium selenate, may be added to complete feed for chickens at a level not to exceed 0.1 part per million. It shall be incorporated into each ton of complete feed for chickens by a premix containing no more than 90.8 mg of added selenium and weighing not less than 1 lb.

28. These diets do not produce high skin pigmentation. The xanthophyll of natural ingredients is variable, so if more pigment is desired use a high-potency source of xanthophyll.

[*Authors' Note:* Footnotes 14, 18, 22, 26, and 27 from the source publication for these tables have been omitted because they refer to data about turkeys.]

30

Feeding Egg-Type Growing Pullets

This chapter deals with egg-type pullets; the feeding of meat-type birds is covered in Chapter 32. The period involved is from the time the chickens are 1 day of age until they reach sexual maturity, which is approximately 20 weeks.

Growing and developing a good pullet is one of the most important items in the operation of a poultry farm. Undoubtedly, the quality of the bird at the time her production cycle begins will greatly determine how profitable she will be during her period of lay. Therefore, special emphasis must be placed on feeding the growing bird so that she may develop into a healthy productive individual and one that can fulfill her genetic potential. Mistakes made during the growing phase cannot be corrected during the laying cycle.

30-A. FACTORS AFFECTING PULLET DEVELOPMENT

There are several factors of importance in pullet development. But the major rule that applies is in two parts, and each part is extremely important. Each group of egg-type pullets must reach sexual maturity

1. At the correct weight for that particular strain.
2. At an age that is optimum to produce eggs economically during her laying year.

From a feeding standpoint, the following have a bearing on the above rules:

1. *Genetics.* The size (weight) of a strain of layers at sexual maturity is a derivation of genetics. Although there

is little evidence that the basic feed formula need be altered according to the strain involved, feed management during the growing period is important so the birds may reach their desired weight when they lay their first eggs.

2. *Season of hatch.*   Two problems that confront the poultryman when chicks are started at different months of the year, with only normal daylight and full feeding, are as follows, and are shown in Table 30-1.

a. Those pullets raised during decreasing light days reach sexual maturity at an older age.

b. The younger the bird at sexual maturity, the smaller the size of her first eggs.

These normal variations in flock behavior must be corrected in order to secure profitable laying house performance.

3. *Light stimulation.*   Table 30-1 clearly indicates the importance of having a light-control program during the growing period. Feed control (restriction) will also delay the onset of egg production, but its effects are most pronounced with flocks maturing in open-sided houses during days of increasing day length.

University of California research in 1980 to 1982 with 68 commercial flocks of White Leghorn chickens showed a somewhat different picture under commer-

Table 30-1.   Effect of Date of Hatch on Age at Sexual Maturity and Egg Size

Month Hatched	Days to 25% Egg Production	12 Months Production	
		Large Eggs %	Eggs under 22 oz per doz %
January	164	87.1	10.5
February	172	86.5	10.6
March	184	89.0	7.9
April	187	93.1	4.8
May	189	94.1	3.9
June	195	93.8	3.7
July	190	86.4	8.9
August	202	93.6	4.1
September	200	93.4	4.1
October	179	80.2	14.3
November	150	78.5	16.6
December	147	72.3	22.6

Source: Skoglund, Univ. of Delaware.

cial conditions. The season in which the birds were grown significantly affected body weight and subsequent egg weight throughout the life of the flocks. These flocks were raised under commercial lighting programs and California temperature extremes were probably greater than in the Delaware research (see Table 30-2).

4. *Stress.* Chickens are subjected to stresses during the growing stage; many of these are manmade. Most stresses cause a reduction in the daily feed consumption, and some method of increasing the nutritional intake becomes the responsibility of the poultry caretaker.

One of the most stressful events in the life of a growing pullet is beak trimming. Feed consumption is depressed following trimming and body weight is often reduced. Table 30-3 illustrates the effects of beak trimming at 12 weeks of age on daily feed consumption and body weight during the following week.

5. *Management practices.* Many management practices call for a change in the feed formula and feeding method. Whether birds are to be raised on a litter floor or in cages with wire floors is a typical example. As birds exercise less in cages than on the floor, their daily energy requirement is lower.

Table 30-2. Effect of Season of Hatch on Performance in Commercial Leghorn Flocks

| Month of Hatch | 18-wk Body Weight | | Egg Weight | | | | Egg Production | |
| | lb | kg | 24 wk | | 60 wk | | 24 wk | 60 wk |
			oz/doz	g/egg	oz/doz	g/egg	%	%
Jan through Mar	2.62	1.188	20.4	48.2	26.8	63.4	54.1	71.7
Apr through June	2.62	1.188	20.7	48.8	26.4	62.3	44.7	70.4
July through Sept	2.60	1.179	21.0	49.5	25.9	61.2	63.0	72.1
Oct through Dec	2.76	1.252	21.1	49.9	27.2	64.2	62.8	70.9
Avg	2.65	1.202	20.8	49.1	26.6	62.8	56.1	71.3

Source: University of California, 1982.

6. *Nutritional imbalance.* To delay the onset of egg production, a vast array of imbalanced feeding programs
have been suggested. Rations high in iodine, low in
lysine, low in protein, all-corn, etc., have been used.
Some can be self-fed; others must be subjected to controlled feeding.

7. *Feed management.* When full-fed, each strain of layers
has an inherent ability to grow at a certain rate and to
reach sexual maturity with a certain body size. But the
inherent size may not be optimum. Only by careful
feed control during the growing period will the best
mature weight be attained. Although the bird has some
control over her caloric intake to meet her demands,
this mechanism is far from perfect. She cannot compensate for all the environmental variations, stresses,
etc., with which she comes in contact.

Table 30-3. Effects of Beak Trimming at 12 Weeks of Age on Feed Consumption
and Body Weight

| Days Post Trim | Feed Consumption | | | | Body Weight | | | |
| | Controls | | Trimmed | | Controls | | Trimmed | |
	lb	g	lb	g	lb	g	lb	g
0					1.99	903	2.07	939
1	0.14	63	0.07	30	2.00	907	1.98	898
2	0.15	67	0.01	5	2.04	925	1.91	866
3	0.15	69	0.01	4	2.07	939	1.81	821
4	0.15	69	0.01	5	2.09	948	1.75	794
5	0.15	68	0.01	4	2.13	966	1.71	776
6	0.15	68	0.05	22	2.16	980	1.72	780
7	0.15	70	0.09	41	2.18	989	1.75	794

Source: University of California, 1988.

Growth During the First 6 Weeks

When based on the percentage increase over the weight at the end of
the previous week, the most rapid gains are made when the chick is
young. As the chick grows older, the weekly increments of weight increases become materially less, as shown in Table 30-4. Leghorn pullets
generally reach 1 lb (0.45 kg) of body weight in 6 weeks, 2 lb (0.90 kg)
by 12 weeks, and 3 lb (1.35 kg) by 20 weeks.

Table 30-4. Increases in Weekly Weight Through 6 Weeks

Age		Increase in Weight over Previous Week (%)	
wk	da	Leghorn Pullets	Medium-size Pullets[1]
1	7	90	90
2	14	60	68
3	21	41	48
4	28	33	32
5	35	25	25
6	42	20	20

[1]Producing brown-shelled eggs.

Feed Consumption

Weekly feed consumption during the first few weeks will vary according to the strain of birds, the energy content of the ration, temperature, and a vast number of other conditions. Guidelines for averages are shown in Table 30-5. Figures are for chicks on a self-feeding program.

Table 30-5. Daily Feed Consumption per 100 Pullets During the First 6 Weeks

	Feed Consumption per Day			
	Leghorn-type		Medium-size[1]	
Week	lb	kg	lb	kg
1	2.2	1.0	2.6	1.2
2	3.3	1.5	4.2	1.9
3	4.6	2.1	5.8	2.6
4	6.2	2.8	7.3	3.3
5	8.0	3.6	8.6	3.9
6	8.8	4.0	9.7	4.4

[1]Producing brown-shelled table eggs.

30-B. ENERGY REQUIREMENTS FOR GROWTH

Starting and growing feeds are formulated to contain a prescribed amount of energy, usually measured in terms of kilocalories per pound or per kilo of ration. Although the bird's *daily* need for energy would be a better criterion of its requirements, this is not practical because chickens are fed on a flock basis rather than individually. Only because the chicken can partially govern its feed intake according to its need for energy is it possible to feed one ration to a group of birds of assorted sizes and development. The smaller birds have a lower daily energy requirement and eat less feed, while larger birds with a greater need for energy consume more feed.

Reducing the energy in the growing diet by about 20% to 1,000 kcal ME per lb (2,200 kcal/kg) of ration has been shown to increase future egg production up to 12 eggs per pullet per year with some strains of Leghorns, while others must be full fed.

ME Requirement of Young Egg-Type Pullets

Chick starting diets used during the first 5 to 6 weeks of the growing period should contain 1,300 to 1,350 kcal of ME per lb (2,860 to 2,970 kcal/kg) of ration. Seldom is the energy in the chick starting ration altered much from these figures even during periods of higher or lower environmental temperature, but on occasion some nutritionists do increase or decrease the ME content of the diet for other reasons.

30-C. PROTEIN REQUIREMENT FOR STARTING DIETS

The protein requirement of the growing chick is based on its need for amino acids in correct proportion. Therefore, *quality protein* is a requisite of proper feed formulation as well as the correct percentage of protein in the diet. As most poultry rations are modifications of a corn-soybean meal base, certain amino acids are likely to be deficient, methionine being the one most often affected. Any amino acid deficiencies in natural feedstuffs can be overcome by adding other protein supplements or synthetic amino acids to the ration.

An analysis of feed mixture or feed formulation must include the amounts of certain essential amino acids as well as the total protein. These requirements are given in Table 30-6.

In practical rations, most nutritionists formulate at levels 10 to 20% higher than those listed to compensate for feed consumption problems and the variability of the ingredients. Most poultry producers would rather pay a little more for their feed than risk growth problems due to marginal formulations. Care must be taken, though, to not adjust levels of individual nutrients arbitrarily as the requirements of some are closely associated with others and imbalances may occur.

Table 30-6. Protein and Amino Acid Requirements of Young Leghorn-Type Chickens

Item	Starting Chickens 0 to 42 Days %
Protein	18.00
Arginine	1.00
Glycine + serine	0.70
Lysine	0.85
Methionine	0.30
Methionine + cystine	0.60
Tryptophan	0.17

Source: Nutrient Requirements of Poultry, 1984; © by the National Academy of Sciences, Washington, DC.

Protein–Energy Relationship

There must be a proper ratio between energy and the total protein in the diet to assure adequate intake levels of critical amino acids. The recommended relationship for starting diets is as follows:

Age of replacement egg-type pullets	0–42 days
ME kcal per lb	1,350
ME kcal per kg	2,970
Protein (%)	20
ME/P ratio (lb)	67.5

When adjustments are made in the energy content of a ration, the protein percentage must be adjusted accordingly to maintain a correct ME/P ratio. This same principle applies to all essential nutrients.

Protein percentage changes. As the energy of the diet increases, birds will consume less feed. Therefore, the protein in the diet must be increased in order to maintain a comparable daily intake. Table 30-7 shows the ME/P ratio for varying caloric and protein contents of the ration.

Example: When the ME/P ratio of a diet is to be 66, several combinations of ME and protein could be formulated, among which are

1,200 kcal ME and 18.25% protein
1,350 kcal ME and 20.50% protein
1,500 kcal ME and 22.66% protein

Note: Table 30-7 can be used to calculate growing rations too.

Table 30-7. ME/P Ratios for Varying Caloric and Protein Content of the Diet

kcal ME per lb	Protein (%)													
	12	13	14	15	16	17	18	19	20	21	22	23	24	25
1,200	100	92	86	80	75	71	67	63	60	57	55	52	50	48
1,250	104	96	89	83	78	74	69	66	63	60	57	54	52	50
1,300	108	100	93	87	81	76	72	68	65	62	59	56	54	52
1,350	113	104	96	90	84	79	75	71	68	64	61	59	56	54
1,400	117	108	100	93	88	82	78	74	70	67	64	61	58	56
1,450	121	112	104	97	91	85	81	76	73	69	66	63	60	58
1,500	125	115	107	100	94	88	83	79	75	71	68	65	63	60
1,550	129	119	111	103	97	91	86	82	78	74	71	67	65	62
1,600	133	123	114	107	100	94	89	84	80	76	73	69	67	64

30-D. MINERAL REQUIREMENTS OF YOUNG CHICKENS

The mineral requirements of young growing chickens are given in Table 30-8. The list contains macrominerals and some trace minerals adequate in natural feedstuffs.

Table 30-8. Mineral Requirements of Young Leghorn-Type Chickens

Mineral	Starting Chickens 0 to 42 days		
	%	per lb of diet	per kg of diet
Calcium	0.80		
Phosphorus (available)	0.40		
Sodium	0.15		
Potassium	0.40		
Manganese (mg)		28	60
Magnesium (mg)		273	600
Iron (mg)		36	80
Copper (mg)		3.6	8.0
Zinc (mg)		18	40
Selenium (mg)		0.08	0.15

Source: Nutrient Requirements of Poultry, 1984, © by the National Academy of Sciences, Washington, DC.

30-E. VITAMIN REQUIREMENTS OF YOUNG CHICKENS

Some vitamin requirements for young chickens are given in Table 30-9. These figures include no margin of safety, and feed formulators will

Table 30-9. Vitamin Requirements of Young Chickens

Vitamin	Starting Chickens 0 to 42 Days	
	per lb	per kg
Vitamin A activity (IU)	682	1500
Vitamin D (ICU)	91	200
Vitamin E (IU)	4.6	10
Vitamin K_1 (mg)	0.22	0.5
Thiamin (mg)	0.82	1.8
Riboflavin (mg)	1.6	3.6
Pantothenic acid (mg)	4.6	10
Niacin (mg)	12.3	27
Pyridoxine (mg)	1.4	3
Biotin (mg)	0.067	0.15
Choline (mg)	591	1300
Vitamin B_{12} (mg)	0.004	0.009

Source: *Nutrient Requirements of Poultry*, 1984, © by the National Academy of Sciences, Washington, DC.

often prescribe much larger amounts than those listed, particularly for vitamins that are subject to oxidation.

30-F. FEEDING DURING THE FIRST 6 WEEKS

During the first 6 weeks of the life of the chick, a well-balanced diet should be fed. In some instances, the starter is fed for more than 6 weeks, especially if body weights are below standard, but any period longer than 8 weeks is usually uneconomical.

30-G. FEEDING FROM 6 THROUGH 20 WEEKS

This is a critical period in the development of an egg-type pullet, for how well she is grown will have an important bearing on her productivity during her laying period. A pullet must develop at a rate appropriate for her strain and reach sexual maturity at an opportune and economical age.

The nutritional requirements during the growing phase are vastly different from those during the starting period. However, the primary difference involves the amount of protein in the growing ration. Compared with the starting ration, protein must be reduced materially not only to justify the bird's requirement but to produce a pullet at the lowest cost possible. Guidelines for feed consumption for 7 to 20 weeks of age are given in Table 30-10.

Table 30-10. Daily Feed Consumption per 100 Pullets Between 7 and 20 Weeks of Age

Week	Feed Consumption per Day			
	Leghorn-type		Medium-size[1]	
	lb	kg	lb	kg
7	9.5	4.3	10.6	4.8
8	10.0	4.5	11.3	5.1
9	10.6	4.8	11.9	5.4
10	11.0	5.0	12.5	5.7
11	11.5	5.2	13.3	6.0
12	12.0	5.5	13.8	6.3
13	12.6	5.7	14.4	6.5
14	13.0	5.9	15.0	6.8
15	13.6	6.2	15.8	7.2
16	14.2	6.5	16.3	7.4
17	14.8	6.7	17.0	7.7
18	15.3	7.0	17.8	8.1
19	15.7	7.1	18.5	8.4
20	16.4	7.5	19.1	8.7

[1]Producing brown-shelled table eggs.

30-H. DIETARY ENERGY FOR EGG-TYPE GROWING PULLETS

The amount of energy in the growing ration should be from 1250 to 1318 kcal ME per pound (2750 to 2900 kcal/kg) of ration. However, this type of growing diet will not prove optimum under all conditions. A problem often arises during hot weather because the birds will not eat enough feed and body weights will be low. During cold weather they will eat too much; birds will gain weight too rapidly.

From 8 through 16 weeks the body fat layer develops in the bird as the result of the energy consumed in the ration, but too much or too little body fat development is to be avoided. Although the energy in the growing ration should be about 1,318 kcal ME per pound (2,910 kcal/kg) from 6 to 14 weeks of age, it should be reduced to about 1,250 kcal ME per pound (2,750 kcal/kg) after 14 weeks to more closely regulate the body fat deposits. Each pullet grower will need to assess this problem for his own conditions. Exceedingly fat pullets usually suffer an increased rate of prolapse. Insufficient energy consumption, on the other hand, will result in poor laying house performance.

Temperature and Feed Consumption

Table 30-11 shows that there is a change in feed consumption as house temperatures increase or decrease, but the relationship is not constant at various house temperatures. The figures are much larger during hot weather than during cold weather.

Tables 30-12 and 30-13 put the figures in Table 30-11 on a different basis. Table 30-12 shows how feed consumption decreases as temperatures increase; Table 30-13 shows how feed consumption increases as temperatures decrease.

Table 30-11. Percentage Change in Feed Consumption for Each 1°F Change in House Temperature at Various Temperatures

Average Daytime House Temperature Between		Change in Feed Consumption for Each 1°F (0.6°C) Change in Temperature
°F	°C	%
90–100	32.2–37.8	3.14
80–90	26.7–32.2	1.99
70–80	21.1–26.7	1.32
60–70	15.6–21.1	0.87
50–60	10.0–15.6	0.55
40–50	4.4–10.0	0.30

Table 30-12. Percentage Decrease in Feed Consumption as Average Daytime House Temperatures Increase

		% Decrease in Feed Consumption Between Two Temperatures as Temperatures Increase					
From		To					
°F	°C	50°F 10.0°C	60°F 15.6°C	70°F 21.1°C	80°F 26.7°C	90°F 32.2°C	100°F 37.8°C
40	4.4	3	8	16	27	42	60
50	10.0		6	14	25	40	59
60	15.6			9	21	37	56
70	21.1				13	31	52
80	26.7					20	45
90	32.2						31

Table 30-13. Percentage Increase in Feed Consumption as Average Daytime House Temperatures Decrease

		% Increase in Feed Consumption Between Two Temperatures as Temperatures Decrease					
From		To					
°F	°C	90°F 32.2°C	80°F 26.7°C	70°F 21.1°C	60°F 15.6°C	50°F 10.0°C	40°F 4.4°C
100	37.8	46	82	110	130	143	151
90	32.2		25	44	58	67	72
80	26.7			10	26	34	38
70	21.1				10	16	20
60	15.6					6	9
50	10.0						3

30-I. DIETARY PROTEIN FOR EGG-TYPE GROWING PULLETS

The amount of protein in the ration for the growing pullet should be reduced as her body weight increases. This is necessarily brought about because the daily protein requirement of the growing bird is relatively constant. But because she is eating more feed each day, her daily protein intake would increase if the percentage in the ration were not reduced. Any protein that was consumed above her daily requirement would do the bird no good and would certainly add to the cost of maturing the pullet.

The growing bird's requirement for protein is better indicated by her body weight than by her age, as shown later. Normally, the total protein in the ration should be reduced by about 1% per week after the sixth week until it is 13%. This point is reached when the birds are about 14 weeks of age. However, weekly adjustments are frequently impractical. Changes are usually made but twice during the growing period: 6 to 14 weeks (first period) and 15 to 20 weeks (second period). On occasion, the growing period may be divided into three phases. But regardless of the breakdown, there also are additional necessary formula adjustments due to environmental temperatures and other factors.

Body weight and production changes. The protein of the growing diet affects future egg production. For each percent increase in dietary protein between 10 and 18 weeks, age of sexual maturity will be reduced by 1 day. Similarly, body weight at first egg will be increased by 1 oz (28 g) and annual egg production will be increased by one egg.

Amino Acid Requirement of Leghorn Growing Pullets

Feed formulations must include the minimum requirement of certain amino acids. These are given in Table 30-14.

Requirements of a Single Growing Ration

Where one growing ration, rather than two as shown in Table 30-14, is used from 6 through 20 weeks of age, the major requirements are

ME (kcal/lb)	1,300
ME (kcal/kg)	2,860
Protein (%)	14
ME/P ratio (lb)	93

Table 30-14. Protein and Amino Acid Requirements of Egg-Type Growing Pullets

Item	Amount in Ration	
	6 through 14 wk %	15 to 20 wk %
Protein	15.00	12.00
Arginine	0.83	0.67
Glycine + serine	0.58	0.47
Lysine	0.60	0.45
Methionine	0.25	0.20
Methionine + cystine	0.50	0.40
Tryptophan	0.14	0.11

Source: Nutrient Requirements of Poultry, 1984, © by the National Academy of Sciences, Washington, DC.

Energy and Protein as They Affect Growth

To show the variations encountered when the dietary energy and protein are altered and full-fed to growing Leghorn pullets, an experiment is presented in Table 30-15. It is obvious from a study of this table that

1. Increasing the protein in the diet increased the body weight at maturity, and also decreased the days to sexual maturity.
2. Increasing the energy and the protein reduced the feed necessary to raise a pullet.

Table 30-15. Protein and Energy Requirements of Leghorn Growing Pullets from 5 Through 20 Weeks[1]

Protein in Diet %	ME per lb of Ration kcal	Body Weight at 21 wk lb	Total Feed Consumed per Pullet lb	Age at First Egg da	Hen-Day Egg Production %	First Egg Weight g
9	959	2.55	18.5	174	65.4	47.8
12	959	2.70	18.2	169	66.9	47.6
16	959	2.76	16.4	167	65.6	48.0
9	1288	2.14	11.0	179	64.9	47.3
12	1288	2.73	12.6	168	67.4	47.7
16	1288	2.94	12.3	164	67.2	47.2
20	1288	3.01	12.1	161	66.1	47.8
25	1288	2.95	12.1	160	66.1	48.1
16	1616	2.93	10.9	166	66.4	47.6
20	1616	2.98	10.6	160	68.4	48.0

[1]Full fed.

3. Within certain limits, the growing pullet had the ability to adjust her feed intake according to her energy needs, regardless of the energy value of the ration.

4. With the exception of the 9%-protein diets, all diets performed similarly with respect to hen-day egg production.

5. Increasing the protein in the growing diet increased the weight of the first egg slightly. The data in Table 30-15 are quite indicative of results when rations varying in protein and energy are self-fed. Within certain confines of dietary energy, the bird has the ability to regulate her feed consumption. For example, there were three diets containing 16% protein with 959, 1,288, and 1,616 kcal of ME per pound of ration respectively. The total number of kilocalories of ME consumed by the three groups are shown as follows:

Protein in Diet (%)	kcal ME per Pound of Ration	kcal ME per Kilogram of Ration	Total Feed Consumed Pullet (lb)	Total Feed Consumed per Pullet (kg)	Total kcal ME Consumed per Pullet
16	959	2,116	16.4	7.5	15,727
16	1,288	2,834	12.3	5.6	15,842
16	1,616	3,555	10.9	5.0	17,614

Phase Feeding Egg-Type Growing Pullets (6–20 wk)

Although the minimum protein percentages in the growing ration have been established by the National Research Council as 15% between 6 and 14 weeks, and 12% thereafter, these figures are somewhat low under field conditions where birds are subjected to stresses of varying degrees. There is also some recent evidence that lowering the protein percentage too much at the end of the growing period reduces egg production later.

Most producers of egg-type pullets prefer not to drop below 13% protein in the diet during the last growing phase. At the present time the most practical diets under commercial conditions appear to be those given in Table 30-16.

Step-Up Protein Growing Rations

A few years ago scientific research at the University of Guelph, Canada, on self-selection of protein and energy showed that protein consumption was correlated with the age of the growing pullet. As the birds

Table 30-16. Energy and Protein Requirements During Growing for Egg-Type Pullets

Item	Two-Phase System		Three-Phase System		
	6–14 wk	15–20 wk	6–12 wk	13–16 wk	17–20 wk
ME (kcal/lb)	1,318	1,318	1,318	1,318	1,318
ME (kcal/kg)	2,900	2,900	2,900	2,900	2,900
Protein (%)	15	13	15	14	13
ME/P ratio (lb)	88	101	88	94	101
Kcal ME per gram of protein	19	22	19	21	22

aged, they consumed more protein rather than less as was the case with most feeding procedures.

Later, to test this discovery, a step-up (sometimes called *reverse feeding*) program was the basis for research. The rations and results are shown in Table 30-17.

The results of the experiment shown in Table 30-17 indicated the following:

1. The growing pullets consumed less protein on the step-up program.
2. Body weight was lower by 0.5 lb (227 g) at maturity with the step-up program and the difference continued throughout the laying period.
3. Egg production was not statistically different, but sexual maturity was delayed on the step-up program and

Table 30-17. Step-Up Protein for Growing Leghorn Pullets When Diets Are Self-Fed

Age Wk	Protein in Diet %	Protein Consumed per Bird per Day, 0–20 Wk Avg g	Body Weight at 20 Wk	Hen-Day Egg Production %	Avg Egg Weight	
					oz/doz	g/each
Step-up Program						
0–12	12					
12–16	16	10.6	2.97 lb	80.1	23.9	56.4
16–20	20		1347 g			
Step-down Control Program						
0–8	18					
8–12	15	11.3	3.42 lb	78.7	24.7	58.0
12–20	13		1552 g			

Source: Leeson, S., and J. D. Summers, *Poultry Sci. 58*, 681, 1979.

this program produced more eggs after 24 weeks of age.

4. Egg size was materially reduced on the step-up program and the program may be best suited to strains possessing inherent large egg size.

30-J. MINERAL REQUIREMENTS FROM 6 THROUGH 20 WEEKS

The mineral requirements of egg-type growing pullets are given in Table 30-18.

Table 30-18. Mineral Requirements of Egg-Type Pullets from 6 Through 20 Weeks

	Feed Requirement					
	6 to 14 wk			15 to 20 wk		
	%	per lb	per kg	%	per lb	per kg
Calcium	0.70			0.60		
Phosphorus (avail.)	0.35			0.30		
Sodium[1]	0.15			0.15		
Potassium	0.30			0.25		
Manganese (mg)		13.6	30		13.6	30
Magnesium (mg)		227	500		182	400
Iron (mg)		27.3	60		27.3	60
Copper (mg)		2.7	6		2.7	6
Zinc (mg)		15.9	35		15.9	35
Selenium (mg)		0.045	0.001		0.045	0.001

Source: Nutrient Requirements of Poultry, 1984, © by the National Academy of Sciences, Washington, DC.
[1]Equivalent to 0.37% sodium chloride (salt).

30-K. VITAMIN REQUIREMENTS FROM 6 THROUGH 20 WEEKS

The minimum vitamin requirements for egg-type growing pullets are given in Table 30-19.

30-L. ATTAINING OPTIMAL BODY WEIGHT AT SEXUAL MATURITY

The importance of correct body weight during the growing period and at sexual maturity cannot be stressed too highly. Changes must be made in the feeding program, and often in the ration, so that the pullet will mature not only at an optimal body weight, but at an optimal age.

Table 30-19. Vitamin Requirements for Egg-Type Pullets from 6 Through 20 Weeks

| Vitamin | Feed Requirement | | | |
| | 6 to 14 wk | | 15 to 20 wk | |
	Per lb	Per kg	Per lb	Per kg
Vitamin A activity (IU)	682	1,500	682	1,500
Vitamin D (ICU)	91	200	91	200
Vitamin E (IU)	2.3	5.0	2.3	5.0
Vitamin K_1 (mg)	0.22	0.50	0.22	0.5
Thiamin (mg)	0.59	1.30	0.59	1.30
Riboflavin (mg)	0.82	1.80	0.82	1.80
Pantothenic acid (mg)	4.55	10.00	4.55	10.00
Niacin (mg)	5	11	5	11
Pyridoxine (mg)	1.36	3.00	1.36	3.00
Biotin (mg)	0.045	0.100	0.045	0.100
Choline (mg)	454	900	227	500
Vitamin B_{12} (mg)	0.0014	0.0030	0.0014	0.0030

Source: Nutrient Requirements of Poultry, 1984, © by the National Academy of Sciences, Washington, DC.

Optimal Mature Body Weight

The weight of the sexually mature egg-type pullet varies with the strain of bird. Added to this is the variability of individual birds within the flock; some mature earlier than others and some attain heavier weights. As with all chickens, flock averages must be used in making feeding recommendations, and we may assume that most strains of Leghorns will reach sexual maturity when about 20 weeks of age and will weigh about 3 lb (1.4 kg). Medium-size pullets for the production of brown-shelled eggs will mature at the same age, with a weight of approximately 4 lb (1.8 kg). However, these optimums are not always reached. If the growing pullets are self-fed, some variations expected are shown in Table 30-20.

Some primary breeders recommend rations lower in energy during hot weather than those during cold, the reduction being from 25 to 50 kcal ME per pound (55 to 110 kcal/kg) of ration. Corresponding adjustments are made in the protein content of the diet when energy changes are made.

Most Egg-Type Strains Should Be Full Fed

If the energy and protein in the ration are corrected for age of the bird and house temperature during different seasons, most breeders feel their growing pullets should be full fed, with no restriction being practiced.

Table 30-20.   Expected Effect of Weather and Light Day on Body Weight at Sexual Maturity (Full-Fed)

| Weather | Light Day | Body Weight at Sexual Maturity | | | |
| | | Leghorns | | Medium-size | |
		lb	kg	lb	kg
Average	Average, all conditions	3.00	1.36	4.00	1.81
Cold	Average	3.25	1.47	4.25	1.93
Hot	Average	2.75	1.25	3.75	1.70
Average	Decreasing (in-season flocks)	3.25	1.47	4.25	1.93
Average	Increasing (out-of-season flocks)	2.75	1.25	3.75	1.70

How to Feed When Full-Feeding Is Practiced

When birds are hand-fed, feed should be kept before them at all times. To keep a low level of feed in the troughs, feed must be added to the feeders two or three times a day. But when automatic feeders are used for full-feeding, the feeders should be operated intermittently. Run them approximately 20 minutes; then allow them to remain idle for 20 minutes, then run again, etc. The exact on-and-off time will be determined by the length of the house and the amount of feed in the troughs or pans at the end of the period when the feeders are not operating. Some feed should be in the troughs or pans at all times. As automatic feeding equipment is operated from a time clock, settings may be made for any time interval.

30-M. FEED CONTROL AND OPTIMAL MATURE WEIGHT

Obtaining Correct Body Weight

Feed control during the growing period varies from full feeding to some degree of feed restriction to attain a given body weight and age at sexual maturity. Unfortunately, the growing pullet offers no index of how rapidly she is advancing toward the production of her first egg. Growing body weight seems to be the best criterion, and is the only one available to the poultryman.

To keep growing pullets from becoming too heavy when exposed to certain conditions, the daily feed allotment (consumption) must be reduced. For this program to be effective, it is necessary to maintain the body weights on a weekly schedule beginning at 7 or 8 weeks of age. One cannot wait until near the end of the growing period to begin to reduce the feed intake to compensate for high body weights. With meat-type birds, the feed intake must be on a drastically reduced basis throughout the growing period, but egg-type pullets can withstand only moderate feed reduction. The reasons are as follows:

1. Egg-type strains are small in comparison, and feed restriction cannot be as great as with meat-type lines.
2. Egg-type strains do not respond as well to a delay in the onset of egg production when the growing diet is restricted.
3. Egg-type pullets do not lend themselves as well to prolonged feed restriction during the growing period as do meat-type pullets.

Development of abdominal fat in the layer. Developing an adequate abdominal fat layer is important. Research at Cornell University and in Sweden and England shows that tissue for the abdominal fat pad of White Leghorns is developed between 8 and 16 weeks of age. Those pullets that had their energy intake restricted during this time period had an abdominal fat layer approximately one-half as thick as full-fed pullets. The difference was maintained throughout the laying year during which the flock was full-fed.

How much feed restriction? Although meat-type pullets need a feed reduction of at least 20%, egg-type birds seldom can tolerate more than 7 to 8%.

Weighing the growing birds. Average flock body weights must be established once each week during the growing period. These weights are of the utmost importance if a good pullet is to be raised. Weigh a sample of birds individually in each house (see Table 16-1). Then calculate the average weight of the birds and their uniformity (see Figure 30-1). If there are no pens, weigh a sample of birds from several areas within the house. Be sure to note the location of each sample so that you can return to the same location in subsequent weighings.

Feeding During Stress

Stresses created by vaccination, beak trimming, disease, high and low temperatures, and moving must be compensated for in the feeding program. When stresses occur, return the birds to full feeding and continue with this program until they recover; then gradually reduce the feed intake to the control recommendations. Body weight should be at or above standards prior to beak trimming. Because of the severity of this particular stress, body weight will usually drop well below the standard for several weeks, which will result in a slower growth rate.

Figure 30-1. Periodic weighing of laying hens in cages.

Stress and low-protein rations. Low-protein rations do not produce
quick recovery when the birds are subjected to severe and prolonged stresses. The difficulty is that under such conditions feed
consumption is drastically reduced, even when the pullets are self-
fed, thus reducing the daily intake of protein. In such circumstances, it is advisable to feed a ration higher in protein.

Anticipating stresses. Many times it will be obvious that certain management procedures, beak trimming, vaccinating, moving, etc., are
going to create a stress. Full-feed for about 2 or 3 days before such
procedures and for a like period afterward.

30-N. GROWING FEED FORMULA VARIATIONS

Although controlled feeding during the growing period makes it possible to maintain optimum growing body weight, at times this is not the
most economical program. Adjustments in the basic feed formula must
be made to supplement feed restriction.

Hot versus cold weather. It must be kept in mind that during hot
weather birds of all types require less energy to maintain their body
temperature than during cold weather. If the caloric content of the

diet remains the same, birds will eat less feed when the environmental temperature rises, which will result in slower growth rates.

Daily Protein Consumption During Growth

The growing bird's need for protein is low, and because she regulates her feed intake according to the temperature and the energy content of the feed, the daily protein intake becomes important. Regardless of how many calories of energy she consumes each day, her protein requirement increases only moderately.

Measurement of protein consumption. Protein is measured as the grams of total protein consumed per bird. It may be computed on a daily, per pound, or kilo, or other basis.

Protein requirement per day. The daily protein requirement of a growing egg-type pullet is as follows:

Mini-Leghorn	5 to 6 g
Standard Leghorn	7 to 8 g
Medium-size	9 to 10 g

Equalizing the protein intake. To maintain a near constant daily intake of protein per bird when the temperature increases, there are two recommendations.

1. *Decrease the energy content of the ration.* Decreasing the energy content will cause the birds to eat more feed, thus increasing the daily protein consumption.
2. *Increase the protein content of the ration.* To maintain daily protein consumption as the temperature rises, the percentage of protein in the ration may be increased, since feed consumption will drop.

Rule of thumb. The rule of thumb for making the above two adjustments is to decrease the energy content of the ration by 1% or increase the protein by 1% of the total protein for each 1°F rise in temperature. Reverse when the temperature drops below 70°F.

Floor versus cage birds. As birds in cages exercise less than those on a litter floor, their energy requirement is less, and they eat less feed. To adjust the feed formula in these cases, decrease the caloric value of the cage feed or increase the protein percentage. Usually the former is more economical unless a low-energy feed is already involved.

Caged birds are generally heavier than their sisters raised on the floor. University of California studies in 1980 to 1982 showed that 18-week weights for cage-reared pullets averaged 2.74 lb (1.24 kg) compared to 2.61 lb (1.18 kg) for floor-reared pullets. Interestingly, 73.1% of the floor pullets were within 10% of the average weight, compared to only 67.6% for the cage-reared pullets. The cage-reared pullets were 4% heavier than the primary breeder's standards, while the floor-reared pullets were 2% under the standards.

Out-of-season flocks. These are flocks that are raised during increasing lengths of light day. They mature early. If self-fed, the early maturity will cause them to increase their feed consumption during the latter part of the growing period. Little can be done to offset the situation by changing the feed formula. Feed restriction is the salvation here, because it delays the onset of egg production.

Started pullets. Most started pullets are sold several weeks prior to the onset of egg production. This is often an awkward age for chickens; they are not fully feathered and have not "combed up." Although no special formulas are used to produce started pullets, the need for well-developed pullets is obvious. Excessive feed restriction cannot be used during the growing period. Most buyers of started pullets prefer that their pullets are not restricted.

30-O. FEED SUPPLEMENTS

Several feed supplements are often fed to growing pullets. These are fed separately from the well-formulated starter and grower rations. The merits of these are as follows:

Grit

Although grit is not a feed, it is eaten by the bird. Grit is used by the chicken as a grinding material to help the gizzard break down large particles of feed into smaller ones. It should be fed if there are large particles of corn, wheat, milo, or rice in the mash, and when chickens have access to litter or feathers, since it helps to grind any of these that the birds eat. But when chickens are given a feed composed of fine material only, no grinding action of the gizzard is necessary, yet it has been shown that some grit is advantageous.

How much grit? Chickens will consume tremendous quantities of grit if it is self-fed, but only a small amount will be retained by the gizzard; the excess will be excreted with the feces. Never self-feed grit; rather, use the following recommendations for growing layer strains.

Birds on litter floor. Beginning the eighth week, feed 1 lb (454 g) of hen-size grit per 100 birds per week. Feed all the weekly allowance on 1 day.

Birds on wire or slat floors. Beginning the eighth week, feed 1 lb (454 g) of hen-size grit per 100 birds every 6 weeks. Feed all the 6-week allowance on 1 day.

Caution: When skip-a-day feeding programs are being used, feed the grit on the day that feed is given. Never feed the allotment of grit on a day when no feed is given.

How to feed grit. Grit is best fed from hanging feeders, but it may be scattered in the litter. Supply one tube feeder for each 250 birds. Do not feed grit in automatic feeders as grit tends to lodge in the corners, and being abrasive, it wears out the tubes, chains, and troughs.

Oystershell

The growing feed should be complete, including the necessary calcium. However, as the pullets approach egg production, additional calcium must be supplied. Directions for this procedure are given in Chapter 31-A.

Whole Grain

In the past, many skip-a-day feeding programs called for the feeding of a small amount of whole oats in the litter on the days when no feed was given, in an endeavor to give the birds something to do on these days. Nutritionally, the practice has no merit, and because of the labor, the procedure is seldom used today.

31

Feeding Egg-Type Layers

Feeding the laying bird is only a continuation of feeding the growing pullet plus supplying the necessary ingredients in the correct proportions so that the bird may produce an abundant number of eggs. The nutritive demand for egg production in modern strains of chickens is tremendous. The eggs produced by a pullet during a layer year weigh eight times as much as she weighs and she will increase her body size by one-third. To do this, she will have to eat nearly 20 times her body weight in feed.

Today, most commercial layers are kept in cages rather than on a litter floor, as was the custom several years ago. This method of management has necessitated some changes in laying feed formulas and in feeding methods. Except where otherwise noted, this chapter deals with the caged layer (see Figure 31-1).

31-A. FEED CHANGES AT SEXUAL MATURITY

Just prior to the time the laying flock begins to produce eggs, the first of several management and feed changes must be started:

1. The total light day must be lengthened.
2. The growing ration must be replaced with a laying ration.
3. Feed consumption must be increased.
4. Calcium consumption must be increased.

Figure 31-1. A modern fully automated caged layer plant.

Feed Consumption Pattern Changes

The amount of feed consumed by the *individual* pullet just prior to and after the production of her first eggs has an almost unbelievable pattern, greatly different than that shown by flock averages. During the month prior to her first egg, the individual Leghorn pullet consumes an almost constant daily amount of feed—about 16.5 lb (7.5 kg) per 100 pullets—until 4 days prior to her first egg when daily feed consumption decreases by about 20% and remains at this low level until the first egg is produced. This is followed by a rapid daily increase in feed intake during the first 4 days of egg production, and a moderate increase thereafter until 4 weeks later, after which it increases very slowly.

On the basis of the above, the flock should be self-fed from the time the first eggs are produced. Individual birds tend to eat what they require. Average flock feed consumption prior to peak production is not representative of the actual consumption of the higher-producing individual birds.

Body Weight Increases

On an individual bird basis, an increase in body weight occurs during the 2 or 3 weeks prior to, and 1 week after, the production of her first egg. The weight gain for a Leghorn pullet will be between 0.50 and 0.75 lb (227 and 340 g) and between 0.75 and 1.00 lb (341 and 454 g) for a medium-size layer. During the following 10 to 12 weeks, the young pullet gains weight very slowly. In fact, many birds lose weight. This is a much different pattern than that observed by flock averages.

Calcium Needs Increase

The pullet's requirement for calcium is relatively low during the growing period, but when the first eggs are produced, the need is at least four times as great, with practically all of the increase being used for the production of eggshells.

It is a natural procedure for large amounts of calcium to be deposited in the long bones of the body just prior to the time the bird lays her first egg. This process has been shown to take place during the 2 weeks prior to the day the pullet lays her first egg, and not before. During this period, and for a week after, it is essential that there be ample calcium consumed if high egg production is to be attained. However, there is no evidence that excessive calcium intake during this period, and the weeks prior, is a deterrent to high egg production.

The best recommendation that can be made is to not increase the calcium intake until 10 days before the flock is expected to produce its first eggs. Certainly the calcium stored in the bones of the first pullets to lay will be somewhat inadequate, but the last pullets to lay will not suffer as much from earlier calcium feeding.

31-B. BASIC NUTRITIONAL REQUIREMENTS OF LAYING HENS

Feed is necessary for four reasons.

1. *Body maintenance.* The amount of feed necessary for body maintenance varies with the weight of the bird and the type of environment.
2. *Body growth.* A Leghorn pullet should gain from 0.75 to 1.00 lb (350 to 454 g) during her laying year. A medium-size layer (producing brown-shelled eggs) should gain from 1.00 to 1.25 lb (454 to 570 g).
3. *Feather production.* This includes the growing of new feathers to replace those molted or pulled out.
4. *Egg production.* The feed requirement for the production of eggs is determined by the number and size of the eggs laid (egg mass).

31-C. ENERGY REQUIREMENT FOR MAINTENANCE

The weight of the layer will, in part, determine the daily energy requirement for maintenance. Some figures are shown in Table 31-1.

Ambient temperature also influences the energy requirement for maintenance. Table 31-2 shows the effect of temperature on the average daily maintenance energy requirement of the laying Leghorn.

Table 31-1. Feed Requirement of Laying Hens for Maintenance (Moderate Temperature)

Weight of Hen		Feed Required for Maintenance[1]			
			Feed per Hen per Day		
		Feed per Unit of			kcal of ME per Hen
lb	kg	Body Weight	lb	g	per Day
3	1.4	0.042	0.126	57.2	164
4	1.8	0.039	0.156	70.8	203
5	2.3	0.037	0.185	83.9	240
6	2.7	0.035	0.212	96.2	276
7	3.2	0.034	0.238	108.0	309

Source: Nutrient Requirements of Poultry, © 1984 by the National Academy of Sciences, Washington, DC.
[1]1,300 Kcal/lb (2,800 Kcal/kg).

Table 31-2. Effect of Ambient Temperature on the Maintenance Requirement for Energy

| Temperature | | Maintenance Requirement in Kcal ME per Laying Hen per Day | |
°F	°C	Leghorn	Medium-Size[1]
50	10.0	230	243
60	15.6	204	217
70	21.1	184	197
80	26.7	172	185
90	32.2	162	175
100	37.8	154	167

[1]Producing brown-shelled eggs.

Example: The data in Table 31-2 show that a Leghorn pullet would have to consume 42 kcal more of ME per day when the ambient temperature is 60°F (15.6°C) than when it is 90°F (32.2°C) to maintain the same rate of egg production. This amount would be equivalent to 3.2 lb (1.5 kg) of additional feed per 100 pullets per day.

31-D. ENERGY REQUIREMENTS FOR EGG PRODUCTION

Leghorn and medium-size layers producing brown-shelled eggs require a diet with about 1,300 kcal of ME per lb (2,860 kcal/kg) of feed. This optimum energy level of the diet, though, is based on computer calculations that provide least cost daily allowances of all essential nutrients including energy. Layers on the floor should receive a diet slightly higher in ME than those kept on wire floors (cages).

The daily energy requirement of the laying bird is highly variable. Some reasons for this are

1. Variation in body weight of pullets
2. Environmental temperature
3. Amount of bird activity
4. Variations in egg production
5. Differences in egg size
6. Prevalence of stress
7. Age of the bird
8. Amount of feather cover

The only compensating fact in overcoming the above variations is that each bird is able to govern her feed intake according to her energy needs. But whether the governing mechanism is efficient is a question. Good feeding requires changes in the feed formula or in the method of feeding to help the bird regulate her energy intake.

Birds Gain Weight

The number of calories of energy in a laying ration is highly variable, as will be pointed out. Furthermore, the daily feed consumption is far from consistent throughout the egg production period. Not only does the weight of the layer influence consumption but birds must also gain weight, and this gain in weight is not uniform. Detailed tests have shown that practically all individual birds have periods of weight gain followed by intervals when they gain no weight. From a flock standpoint, however, there should be some weekly increase in average body size.

Energy Requirement

The ME requirement of a 4-lb (1.8-kg) layer, kept at a moderate temperature and laying at the rate of 75% hen-day production, is about 300 to 310 kcal per day. The figure will increase in cold weather and decrease in hot weather.

The amount of energy in the diet will control feed consumption. The relationship is shown in Table 31-3, which gives feed consumption necessary per day as the caloric content of the ration varies.

Note: The figures in Table 31-3 are calculated. Actually under field conditions the rations higher in energy will be more efficient than those lower in energy.

Table 31-3. Dietary Energy in the Feed and Daily Feed Requirement for a 4-lb (1.8-kg) Hen (Moderate Temperature)

Kcal of ME per		Feed Required per Day per 100 Hens to Supply 306 Kcal ME per Hen		Feed per Dozen Eggs Produced[1]	
Lb of Ration	Kg of Ration	lb	kg	lb	kg
1,200	2,640	25.5	11.6	4.1	1.86
1,250	2,750	24.5	11.1	3.9	1.77
1,300	2,860	23.5	10.7	3.8	1.73
1,350	2,970	22.7	10.3	3.6	1.64
1,400	3,080	21.9	10.0	3.5	1.59
1,450	3,190	21.1	9.6	3.4	1.55

[1]75% hen-day egg production.

Environmental Temperature and Feed Consumption

As the hen's requirement for energy is higher in cold weather than in hot weather, there are differences in the amount of feed she consumes under these conditions. These variations in feed consumption are smaller for each degree Fahrenheit change in temperature when the weather is cool than when it is hot.

Example: Between 40° and 50°F each degree change alters the feed consumption about 0.3%, while between 90° and 100°F each degree change alters the consumption by 3.14%. The relationship between temperature and feed intake is shown in Table 31-4.

Factors Affecting Daily Feed Consumption

The following have major importance:

1. Caloric content of the diet
2. Ambient temperature

The following have minor importance:

1. Strain of birds
2. Body weight

Table 31-4. Relationship Between Ambient Temperature, Energy Content of the Diet, and Feed Consumed per 100 Caged Laying Leghorns per Day

Avg Daytime Ambient Temperature		kcal of ME per Pound (Kilo) of Ration					
		1,250	(2,750)	1,300	(2,860)	1,350	(2,970)
		Feed/100 Birds/Day					
°F	°C	lb	kg	lb	kg	lb	kg
40	4.4	27.0	12.8	26.0	11.8	25.0	11.0
50	10.0	26.3	12.5	25.3	11.5	24.4	10.7
60	15.6	25.1	11.9	24.1	11.0	23.2	10.2
70	21.1	23.4	11.1	22.5	10.2	21.7	9.5
80	26.7	21.4	10.1	20.6	9.4	19.8	8.7
90	32.2	18.0	8.5	17.3	7.9	16.7	7.3
100	37.8	14.6	6.9	14.0	6.4	13.5	5.9

3. Daily egg mass
4. Feathering
5. Degree of stress
6. Bird activity

Table 31-4 shows the feed consumption of 100 Leghorn layers per day when there are variations in the two factors of major importance. Changes in the minor factors are included in the figures but are not obvious variables.

Directions for Table 31-4. To use this table, first ascertain the ME of the ration being fed to the nearest of the following three: 1,250 lb (2,750 kg), 1,300 lb (2,860 kg), and 1,350 lb (2,970 kg); then, find the feed consumption of the flock of birds for the appropriate average daytime ambient temperature. Feed consumption per 100 pullets per day for other temperatures will be found in the same column.

Summer and Winter Caloric Requirement

The consumption of feed and calories changes with increasing age and with temperature changes. Table 31-5 illustrates these relationships in 100 commercial California Leghorn table-egg flocks.

Dietary Caloric Content and Feed Consumption

As the energy content of the feed increases, hens will eat less, and vice versa.

Rule of thumb: Feed consumption will be reduced by 4% for each increase of 50 kcal of ME for each 1 lb (454 g) of ration, or vice versa.

Table 31-5. Consumption of Feed and Calories by Leghorn Layers in Relationship to Their Age and the Season

	21/24	25/28	29/32	33/36	37/40	41/44	45/48	49/52	53/56	57/60	Avg
Daily Feed Consumption						lb					
Summer	.164	.199	.213	.226	.227	.232	.231	.231	.221	.217	.216
Winter	.179	.214	.233	.240	.242	.246	.245	.246	.251	.240	.233
Year	.171	.205	.229	.236	.236	.239	.238	.238	.239	.235	.227
						g					
Summer	74	90	97	103	103	105	105	105	100	98	98
Winter	81	97	106	109	110	112	111	112	114	109	106
Year	78	93	104	107	107	108	108	108	108	107	103
Daily Energy Consumption						kcal ME					
Summer	212	252	268	286	290	294	293	291	278	273	274
Winter	229	272	296	302	306	313	313	315	321	306	297
Year	220	260	291	300	300	304	303	302	304	299	288

Header spanning: *Age in Weeks*

Source: University of California.
Note: 1978 data.

Rate of Egg Production and Feed Intake

An egg contains between 65 and 100 kcal of energy, depending upon its size. As the energy efficiency during digestion and metabolism is about 70%, each egg of average size will require about 121 kcal of dietary energy. If the flock were laying at the rate of 70% hen-day egg production, with an ambient temperature of 70°F (21.1°C), 121 kcal of the total daily intake of 306 kcal of ME per bird would go to produce eggs.

Rule of thumb: If egg size remains the same, each 10% change in hen-day egg production will alter the feed requirement by 4%.

Egg Size and Feed Requirement

As large eggs contain more calories of energy than small eggs, the dietary energy required to produce them is greater.

Rule of thumb: A hen needs 1.2% more feed as egg size increases 1 oz/doz (2.4 g/each).

Body Weight Alters Feed Intake

The larger the hen, the greater the feed requirement for maintenance; therefore, feed consumption increases during the egg production year.

Rule of thumb: For each 0.1 lb (45 g) increase in body weight, a laying hen requires 1.3% more feed.

31-E. FAT IN THE LAYER RATION

Fats are primarily used in poultry feeds to supply concentrated sources of energy. Most feed grade fats have more than two times the ME of feed grains. In addition, the utilization of energy in the diet is enhanced when energy derived from fats is substituted for carbohydrate or protein sources of energy. The fatty acid composition of eggs can also be manipulated by incorporating high levels of corn or sunflower oils in the diet.

31-F. DAILY NUTRIENT INTAKE

Egg-type laying hens are fed to meet their daily requirements for all major nutrients relative to the stage of performance. This concept replaces the concept of "percentage of the diet," which is still quoted throughout the industry and is used in this book for reference purposes.

The essential elements of the daily intake feeding program is to know the amount of feed each flock is eating, the requirements for each production level, and the nutrient content of the diet. Its success is dependent on the accuracy with which feed consumption can be predicted, the quality of the standards used in establishing the requirements, and the precision achieved in feed formulation and manufacture.

Nutrient intake is calculated by multiplying the measured feed consumption per hen per day by the percentage of the nutrient in the diet. For example, if a flock eats 100 g of feed per hen per day, and it is formulated to contain 0.35% methionine, the average daily intake of methionine is 350 mg.

Projecting feed consumption for next week's usage requires knowledge of present consumption and reliable estimates of whether or not a change is likely to occur because of changes in ambient temperature or other factors. Present consumption is usually obtained by estimating feed disappearance from the feed tank during the present week or by actually weighing the amount of feed used. Several systems are available that will monitor the quantity of feed in a feed tank electronically.

31-G. PROTEIN REQUIREMENTS FOR EGG PRODUCTION

The protein requirement of laying birds is closely associated with the rate of egg production. Protein in the egg production diet is much lower than the 18 to 20% required for early growth. Just prior to egg production, only 13% of the pullet's diet should be protein; but when egg production reaches its peak, the requirement may be as high as 17 to 19%. At the end of the production cycle, it may drop to as low as 14%.

Amino Acids

To speak of the protein requirement for egg production is to speak of the amino acid requirement. Protein must be well balanced and of high quality for a hen to lay her maximum number of eggs, and to produce them economically. Of the amino acids often deficient in the laying ration, methionine is most commonly involved. The laying bird's amino acid requirements are given in Table 31-6.

The commercial application of poultry nutrient requirements requires knowledge of the quality of local feedstuffs, allowances for milling or feed separation problems, allowances for unexpected changes in feed consumption, and an assessment of the losses, which could result from insufficient nutrient intake. For these reasons, practical diets are always formulated to higher specifications.

Table 31-7 summarizes the more critical amino acid requirements for three stages of egg production.

Table 31-6. Amino Acid Requirements of Layers

Amino Acid	Amount in the Diet
Arginine (%)	0.68
Lysine (%)	0.64
Methionine (%)	0.32
Methionine + cystine (%)	0.55
Tryptophan (%)	0.14

Source: Nutrient Requirements of Poultry, 1984, © by the National Academy of Sciences, Washington, DC.

Table 31-7. Recommended Daily Amino Acid Intake for Leghorn Layers by Weeks of Egg Production

Amino Acid	Phase (Weeks of Egg Production)[1]		
	Phase 1 1–20 wk	Phase 2 21–40 wk (mg/day)	Phase 3 41+ wk
Arginine	950	875	825
Lysine	800	750	725
Methionine	385	360	350
Methionine + cystine (TSAA)	700	650	625
Tryptophan	190	175	170

[1]First cycle of lay.

Weather Affects ME/P Ratio

When one diet is used throughout the laying year, it should contain approximately 1,300 kcal ME per lb (2,680 kcal/kg) of feed and about 16% protein. These amounts would provide an ME/P ratio of 81 on a pound basis. Because this type of feeding program is not as economical as altering the ration as egg production changes, it has given way to phase feeding. The environmental temperatures may also necessitate variations in the energy and protein content of the ration. General recommendations for a phase-feeding program are given in Table 31-8.

Table 31-8. Energy and Protein Variations in Laying Rations

Egg Production (Hen-Day) %	Hot Weather				Cool Weather			
	ME		Protein %	ME/P (lb) Ratio	ME		Protein %	ME/P (lb) Ratio
	per lb	per kg			per lb	per kg		
80 and over	1,250	2,750	18	69	1,400	3,080	17	82
70–80	1,225	2,695	17	72	1,375	3,025	16	86
Under 70	1,200	2,640	16	75	1,350	2,970	15	90

Note: Rations used during moderate weather should be between the extremes shown above. Many commercial feed manufacturers make changes in their laying feed formulas as the weather changes.

Daily Protein Requirement of Leghorn-Type Chickens

One of the most controversial points about the nutrition of the laying Leghorn is the minimum daily requirement for protein. There have been countless experiments and much has been written, but there is still confusion. As protein is expensive, every feed formulator wants to be sure he has enough in his rations, yet doesn't want to waste any.

Flock behavior has great effect. To understand minimum protein requirement one must understand the variables involved. In the first place, nothing is constant during the laying year. Birds increase in body weight, egg production rises rapidly then falls gradually, and egg size increases. These are the important flock variables. To add to the confusion, individuals within the flock are not uniform. Some start to lay at an early age; some lay later. Body weight varies, as does egg size, all of which influence the amount of protein necessary.

When body weight increases, more protein is required. During the year, feather growth decreases, necessitating less protein. Egg production is highly variable, and the dietary protein necessary to produce the egg protein therefore is highly variable. As egg size increases, more protein is needed to deposit the additional protein in the larger eggs. To add to these variables, protein in itself is not well utilized in the production of eggs. Most agree that the efficiency is about 55% in young layers and less in older ones.

Determining the protein need.   The above relationships have been put together in Table 31-9 by making the calculations from a mathematical formula. The variables you see are body weight, egg production, and egg size. The others are not shown, but are included.

After casually glancing at the table you can see that there is a great variation in the daily protein requirement according to flock behavior. The protein requirement shows a high of 20.2 g per day to a low of 10.9 g. But all variations would not apply during the laying year. We would not expect large eggs to be produced at the peak of lay, nor would we expect body weight and egg weight to be low at the end of the laying cycle.

Most Leghorn flocks average about 3 lb (1.4 kg) in body weight at sexual maturity and lay a small number of small eggs. Body

Table 31-9. Body Weight, Egg Weight, and Egg Production as They Affect Daily Protein Requirement

| Body Weight | | Egg Weight | | Hen-Day Egg Production (%) | | | | | |
lb	kg	oz/doz	g each	50	60	70	80	90	100
				g of Protein Required per Hen per Day					
3.00	1.36	22	52.0	11.5	12.7	13.8	14.9	16.1	17.2
3.00	1.36	24	56.7	12.0	13.2	14.5	15.7	17.0	18.2
3.00	1.36	26	61.4	12.4	13.7	15.0	16.3	17.6	18.9
3.00	1.36	28	66.1	13.0	14.4	15.9	17.3	18.8	20.2
3.25	1.48	22	52.0	11.4	12.6	13.7	14.8	16.0	17.1
3.25	1.48	24	56.7	11.9	13.1	14.4	15.6	16.9	18.1
3.25	1.48	26	61.4	12.3	13.6	14.9	16.2	17.5	18.8
3.25	1.48	28	66.1	12.9	14.3	15.8	17.2	18.7	20.1
3.50	1.59	22	52.0	11.2	12.4	13.5	14.6	15.8	16.9
3.50	1.59	24	56.7	11.7	12.9	14.2	15.4	16.7	17.9
3.50	1.59	26	61.4	12.1	13.4	14.7	16.0	17.3	18.6
3.50	1.59	28	66.1	12.7	14.1	15.6	17.0	18.5	19.9
3.75	1.71	22	52.0	11.1	12.3	13.4	14.5	15.7	16.8
3.75	1.71	24	56.7	11.6	12.8	14.1	15.3	16.6	17.8
3.75	1.71	26	61.4	12.0	13.3	14.6	15.9	17.2	18.5
3.75	1.71	28	66.1	12.6	14.0	15.6	16.9	18.4	19.8
4.00	1.82	22	52.0	10.9	12.1	13.2	14.3	15.5	16.6
4.00	1.82	24	56.7	11.4	12.6	13.9	15.1	16.4	17.6
4.00	1.82	26	61.4	11.8	13.1	14.4	15.7	17.0	18.3
4.00	1.82	28	66.1	12.4	13.8	15.3	16.7	18.2	19.6

weight will steadily increase until adult weights of 3.6 lb (1.6 kg) to 4.0 lb (1.8 kg) are reached. Egg mass will reach its maximum level at 35 to 40 weeks of age.

Protein consumption varies with birds. The example in Table 31-10 shows that the peak requirement for protein is at the peak of egg production, with an average daily requirement of 16.6 g. But during the peak of lay, many hens in the flock are laying at a 100% rate—an egg a day—and this amount of protein is not adequate for 100% egg production. How then will this amount of dietary protein maintain this high rate with many birds in the flock? The solution is understandable when one visualizes that if the flock average production is 92%, and some hens are laying at a 100% rate, some have to be laying at an 84% rate or lower, and would eat less feed and therefore consume less dietary protein daily. This reasoning is more obvious when one considers that when the flock peaks at about 8 weeks after the first hens start to lay, some birds are just starting to produce eggs (see Chapter 22-J). These birds would be eating less because their energy requirement is less, thereby creating lower figures for protein on a flock average basis.

Table 31-10. Daily Protein Requirement per Hen During the Course of the Laying Year

Period of Egg Production wk	Avg Body Weight		Avg Egg Production % Hen-Day	Avg Egg Size		Protein Required per Hen per Day g
	lb	kg		oz/doz	g each	
0–6	3.2	1.45	32	22	52.0	11.3
7–9 (Peak)	3.5	1.59	92	23	54.3	16.6
10–22	3.6	1.64	88	24	56.7	16.3
23–35	3.7	1.68	79	25	59.1	15.4
36–48	3.8	1.73	70	26	61.4	14.5
49–60	3.9	1.77	64	27	63.8	13.6

Note: Data for Leghorn type.

Protein requirement less than thought previously. This approach to the protein problem seems to result in figures that closely agree with those of most recent research work, but Table 31-9 gives the necessary daily intake for many more combinations of variables than are found in any experimental data. The table adds evidence to other work to show that the minimum daily protein requirement for Leghorns is less than shown previously. But it must be remembered that the dietary protein must be thought of in terms of body weight, egg production, and egg size. The figures in Table 31-10 reflect a peak egg mass of 50 g of eggs per day; many flocks today peak at 54 or more g of eggs.

Daily Protein Requirement
of Other Layers

The daily protein requirement for medium-size layers producing brown-shelled eggs is higher than the needs shown for Leghorns. To make the necessary compensation, add 2 g to all figures in Table 31-9 and Table 31-10. To adjust for Mini-Leghorns, subtract 1 g from all figures in these two tables.

Older Birds Utilize Protein
Less Efficiently

This statement is important because the dietary protein level at the end of the laying year may drop to suboptimal levels particularly when eggs are large and production is good. Particular concern must be directed to recycled flocks because of their large egg mass. Daily egg mass often exceeds egg mass produced during the first laying cycle. This variable has been included in Table 31-9.

Larger Hens Get More Protein

Bird weight within the flock is not uniform, and as the larger birds consume more feed they also get more protein each day. Egg size is greater in the larger birds, but feed protein to egg conversion is poorer. Theoretically, at least, the more uniform the body weight, the easier it will be to formulate to the flock's needs.

Feed Necessary to Supply Protein

The amount of feed per day necessary to furnish the required daily protein is given in Tables 31-11 (lb) and 31-12 (kg) according to the percentage of protein in the ration. The amino acid composition in the protein is considered to be balanced as shown in Table 31-7. *Note:* In extremely hot weather, feed consumption may be insufficient to attain required levels of protein (amino acid) intake.

Feed Consumption as It Relates
to Protein Consumption

If the feed consumption per 100 hens per day and the percentage of protein in the diet are known, it is possible to calculate the daily intake of protein in grams. These figures are given in Table 31-13.

Example: If the birds were eating 25 lb (11.3 kg) of feed per 100 hens

Table 31-11. Pounds of Feed Necessary to Supply Daily Protein Requirements of Laying Hens

Protein in Ration %	Grams of Protein Required per Hen per Day						
	20	19	18	17	16	15	14
	lb Feed Required per 100 Hens per Day						
20	22.0	20.9	19.8	18.7	17.6	16.5	15.4
19	23.2	22.0	20.9	19.7	18.6	17.4	16.2
18	24.5	23.3	22.0	20.8	19.6	18.4	17.1
17	25.9	24.6	23.3	22.0	20.7	19.4	18.1
16	27.5	26.2	24.8	23.4	22.0	20.7	19.3
15	29.4	27.9	26.4	25.0	23.5	22.0	20.6
14	31.5	29.9	28.3	26.8	25.2	23.6	22.0

Table 31-12. Kilograms of Feed Necessary to Supply Daily Protein Requirements of Laying Hens

% Protein in Ration	Grams Protein Required per Hen per Day						
	20	19	18	17	16	15	14
	kg Feed Required per 100 Hens per Day						
20	10.0	9.5	9.0	8.5	8.0	7.5	7.0
19	10.5	10.0	9.5	9.0	8.5	7.9	7.4
18	11.1	10.6	10.0	9.4	8.9	8.3	7.8
17	11.8	11.2	10.6	10.0	9.4	8.8	8.2
16	12.5	11.9	11.3	10.6	10.0	9.4	8.8
15	13.4	12.7	12.0	11.3	10.7	10.0	9.3
14	14.3	13.6	12.9	12.1	11.4	10.7	10.0

Table 31-13. Grams of Protein Consumed per Bird per Day According to Feed Consumption and Percentage of Protein in the Ration

Feed per 100 Pullets per Day		Protein in Laying Ration (%)						
lb	kg	14	15	16	17	18	19	20
		Grams Protein Consumed per Bird per Day						
18	8.2				14	15	16	16
19	8.6			14	15	16	16	17
20	9.1		14	15	15	16	17	18
21	9.5		14	15	16	17	18	19
22	10.0	14	15	16	17	18	19	20
23	10.4	15	16	17	18	19	20	21
24	10.9	15	16	17	19	20	21	22
25	11.3	16	17	18	19	20	22	23
26	11.8	17	18	19	20	21	22	
27	12.2	17	18	20	21	22	23	
28	12.7	18	19	20	22	23		
29	13.2	18	20	21	22			
30	13.6	19	20	22	23			
31	14.1	20	21	23				

per day, and the ration contained 16% protein, the protein consumed per bird per day would be 18g.

Protein and Egg Size

Although the size of the egg yolk has a greater relationship with egg size than the amount of albumen, the amount of the latter is very important to the size of the egg. Increasing the protein content of the diet has a marked effect on increasing egg size, particularly when there are small eggs. Excessive protein consumption may increase egg size too much; too little may result in an excessive number of medium eggs. The size of the egg in relation to the period of egg production will determine the practicability of altering egg size.

Smaller egg size during the summer months is commonly the result of lower energy intake since egg producers usually adjust their feed formulas to maintain uniform protein intake throughout the year. The solids in egg albumen are almost entirely protein. Because the egg demand for protein is so great, any lack of dietary protein results in a decrease in the amount of albumen, and the egg size is smaller even though the quantity of yolk is adequate.

31-H. MINERAL REQUIREMENTS DURING EGG PRODUCTION

Mineral requirements for laying rations are shown in Table 31-14. These do not include any margin for safety.

Table 31-14. Average Mineral Requirements of Laying Rations

	Leghorn		Medium-size[1]	
	21–40 Wk of Age	40 and over Wk of Age	21–40 Wk of Age	40 and over Wk of Age
Calcium (%)	3.25	3.50	3.00	3.25
Phosphorus (total %)	0.5	0.5	0.5	0.5
Phosphorus (inorganic %)	0.15	0.15	0.15	0.15
Sodium (%)	0.15	0.15	0.15	0.15
Manganese (mg/lb)	50	50	50	50
Manganese (mg/kg)	110	110	110	110
Zinc (mg/lb)	23	23	23	23
Zinc (mg/kg)	50	50	50	50

[1]Producing brown-shelled eggs.

Daily Mineral Requirements

Table 31-15 lists the recommended daily intake for several of the major minerals. These are recommended levels for practical diets.

Table 31-15. Recommended Daily Mineral Intake for Leghorn Layers by Weeks of Egg Production

	Phase (Week of Egg Production)[1]		
	Phase 1 1–20 wk	Phase 2 21–40 wk	Phase 3 41+ wk
Mineral		(mg/day)	
Calcium	3,600	3,800	4,000
Available phosphorus	450	440	420
Sodium	180	180	180

[1]First cycle of lay.

Calcium Requirement

During the latter stages of the growing period, the diet should contain approximately 0.6% calcium and 0.3% available phosphorus. But once egg production begins, the need for calcium is much greater because of eggshell formation. The requirements are given in Table 31-15 and there should be little variance from the figures listed. Of particular importance is that too much calcium during egg production is detrimental; it depresses the appetite. It is also uneconomical since surpluses are excreted in the fecal material.

Only a portion of the calcium consumed by the laying hen is retained by the bird, the balance being excreted. The retention is about 60% for young laying hens, and 40% for older layers. Of equal importance is that the calcium increase from the growing need to the laying need must begin when the first egg is laid. The recommended procedure is to feed a pre-lay diet with 2% calcium for 2 weeks. Half of the calcium should be supplied in a coarse particle size so that the earlier maturing pullets may self-select their calcium needs.

Dietary Calcium Variations

The amount of calcium necessary in the laying ration is determined by several major factors, all of which alter the dietary formula:

1. Rate of lay. (The higher the rate, the more calcium needed.)
2. Size of bird. (The larger birds consume more feed.)
3. Age of birds. (Those past 40 weeks of lay require more dietary calcium.)
4. ME content of the ration. (The higher the figure, the less feed consumed.)

5. House temperature. (Birds eat less when temperatures are high; the ration should contain more calcium.)

Table 31-14 shows the average percentage of calcium needed in the ration, but because the above five factors enter into the figure, the actual percentages needed are subject to much variation. More accurate requirements are given in Table 31-16.

Table 31-16. Percentage of Calcium Needed in the Laying Ration with Varying Feed Consumption, Age, and Egg Production

Feed Consumed per Day		Age in Weeks							
		21–40				After 40			
		% Hen-Day Egg Production							
lb per 100 Hens	g per Hen	90	80	70	60	80	70	60	50
		% Calcium in Ration							
17.6	80	4.7	4.2	3.7	3.2	5.2	4.7	4.1	3.4
19.8	90	4.2	3.8	3.3	2.9	4.7	4.2	3.6	3.1
22.0	100	3.8	3.4	3.0	2.6	4.5	3.8	3.3	2.8
24.2	110	3.5	3.1	2.7	2.3	3.8	3.5	3.0	2.6
26.4	120	3.2	2.9	2.5	2.1	3.5	3.2	2.8	2.4
28.6	130	3.0	2.7	2.3	1.9	3.3	3.0	2.6	2.2

Decline in Eggshell Quality as Hen Ages

Although eggshells are poorer in quality at the end of the laying year than at the start, no one has been able to determine the exact cause. One hypothesis is that the hen is capable of generating a uniform daily quantity of eggshell material throughout her life, and as eggs get progressively larger, the shell material must be spread over a larger area, and thus is thinner. Experiments have shown that the shell weight remains almost constant throughout the egg production period. Some researchers have reported that efforts to reduce egg size during the latter stages of production by limiting protein consumption have resulted in improved eggshells. This method should be applied with caution, though, because egg numbers may be affected as well.

Larger eggs within a sample at a given age tend to have thicker shells and greater shell mass, reflecting the individual bird's ability to compensate for the size of her eggs. See Table 31-17.

Table 31-17. Effect of Egg Weight on Various Eggshell Characteristics (Three Leghorn Strains)

| Egg Weight g | Shell Characteristics | | | |
| | Weight | | Thickness microns | Specific Gravity |
	g	%		
44 wk of age				
<60	5.7	9.9	363	1.0870
60–64	6.1	9.8	368	1.0864
65–69	6.5	9.7	373	1.0860
>70	6.7	9.3	384	1.0854
Avg	6.1	9.8	368	1.0863
56 wk of age				
<60	5.4	9.4	356	1.0815
60–64	5.8	9.4	366	1.0816
65–69	6.1	9.2	363	1.0808
>70	6.6	9.1	371	1.0815
Avg	5.9	9.3	363	1.0813

University of California data, 1987.

Eggshell Quality Is Poorer During the Summer

During the hot weather, feed consumption is reduced and the daily intake of critical minerals may be less than optimum. Even though it is a common practice to offset lower feed consumption with higher concentrations of calcium and phosphorus, it is still common to see eggshell thickness decrease during the summer months. Table 31-18 illustrates this problem.

Table 31-18. Effects of Season on Eggshell Thickness

Season	Age wk	Shell Thickness micron	Shells Less Than 356 Microns in Thickness %
Winter	50	365	30
Summer	50	355	43
Winter	60	369	26
Summer	60	352	47

University of California data, 1982.

Feeding Coarse-Particle Oystershell or Limestone

The formation of eggshell material and its deposition usually occurs during the night hours when the hen is not eating. If the source of dietary calcium is finely ground the calcium passes through the gizzard quickly; little is available to the bird when the eggshell is being deposited.

To improve the situation, two-thirds of the dietary calcium supplement should be in the form of large-size flaked oystershell or coarse limestone. This material leaves the gizzard more slowly so a larger amount passes the intestinal wall during the dark hours when the eggshell is being formed. There is more benefit during the latter part of the laying year than during the first from such a procedure (see Chapter 26-B).

The *pulverized* source of calcium supplement is best supplied as ground aragonite or high-calcium limestone. The utilization of calcium is best in oystershell, next in aragonite, and poorest in limestone.

Phosphorus Requirement of Layers

Much of the phosphorus in plant ingredients is in the form of phytin phosphorus, an organic compound not well-utilized by the chicken; only about 30 to 40% is available. Phosphorus recommendations are now based on available phosphorus. Various inorganic forms are available (see Chapter 25-H).

The laying hen's need for phosphorus is low, mainly because there is little phosphorus in the eggshell. Too little will prevent proper shell calcification, as will too much. One of the chief causes of poor eggshell quality and strength is an excess of phosphorus in the diet, but rations low in total phosphorus increase hen-house mortality. The recommended daily intake of available phosphorus in the laying ration is controversial, but levels of 400 to 450 mg per hen per day are considered adequate.

Trace Minerals Requirement of Layers

The requirement of the laying bird for trace minerals is very indefinite. Except for manganese and zinc, natural feedstuffs seem to supply the necessary quantities. Most layer rations include supplementary manganese and zinc. Some include selenium (see Chapter 26-B).

31-H. VITAMIN REQUIREMENTS FOR EGG PRODUCTION

The dietary vitamin requirements of a laying ration are given in Table 31-19. The vitamins most often added to a laying ration include

A	riboflavin
B_{12}	pantothenic acid
D_3	choline
K	niacin

Table 31-19. Vitamin Requirements of Laying Hens

Vitamin	Amount per Unit of Feed	
	per lb	per kg
Vitamin A activity (IU)	1818	4000
Vitamin D (ICU)	227	500
Vitamin E (IU)	2.3	5.0
Vitamin K (mg)	0.23	0.5
Thiamin (mg)	0.36	0.8
Riboflavin (mg)	1.0	2.2
Pantothenic acid (mg)	1.0	2.2
Niacin (mg)	4.6	10.0
Pyridoxine (mg)	1.4	3.0
Biotin (mg)	0.05	0.1
Choline (mg)	?	?
Vitamin B_{12} (mg)	0.001	0.004

Source: Nutrient Requirements of Poultry, 1984, © by the National Academy of Sciences, Washington, DC.

31-I. XANTHOPHYLLS AND EGG-YOLK COLOR

The xanthophylls in the feed are the main contributors to yolk color. In the United States consumer preference is for yolks that are pale to deep yellow rather than darker shades, but this is not the preference in some other countries. Huge quantities of egg yolks are used in the preparation of noodles, cake mixes, and many other bakery products. A deep orange-colored yolk is necessary to prepare these products.

There are many xanthophylls and they represent a group of hydroxy-carotenoids. They are absorbed from the intestinal tract of the chicken and deposited in the egg yolks and fatty tissues in the same form as consumed.

Not only do the xanthophylls impart yellow color to egg yolk but they affect the color of the skin of yellow-skin chickens. However, these pigments are not deposited in the skin of white-skin chickens, yet are laid down in the yolks produced by such birds.

In the United States, diets of yellow corn rarely have yolk color problems. Diets that depend more on grain sorghum or wheat will usually have very pale yolks unless supplemented with products of the type listed in Table 31-20.

Table 31-20.   Mixed Xanthophyll Content of Various Feedstuffs

Feedstuff	Total Xanthophyll Content	
	mg per lb	mg per kg
Marigold petal meal	3,182	7,000
Algae (common, dried)	909	2,000
Alfalfa meal (20% protein)	127	280
Alfalfa meal (17% protein)	118	260
Coastal Bermuda grass	122	270
Corn gluten meal (60% protein)	132	290
Corn gluten meal (41% protein)	80	125
Yellow corn	8	17

Sources of Xanthophylls

Xanthophylls suitable for pigmenting egg yolk are usually found in two common poultry feeds: alfalfa and yellow corn. Alfalfa leaf products are a major source of several with five being abundant, but the one of greatest abundance is *lutein*. The higher the protein in alfalfa meals, the higher the potency. Although variable, egg yolk contains about 70% lutein and 30% zeaxanthin.

Yellow corn and its by-product, corn gluten meal, are excellent sources of certain xanthophylls, the main one being zeaxanthin. The quality of zeaxanthin contained in corn is highly variable, and gives rise to variability in corn gluten meal. A new corn gluten meal, containing 60% protein, contains twice the quantity of xanthophylls as the common 41%-protein product.

Other products have been found that contain large quantities of the xanthophylls. One of these is the petals of a marigold species, *Tegetes erecta*. Lutein is the chief xanthophyll in these petals and the potency is extremely high. The product is available in quantity as a stabilized dried petal meal or as a hexane extract.

A synthetic carotenoid, beta-apo-8'-carotenal, has value as a yolk pigmenter and produces a color similar to the xanthophylls, lutein and zeaxanthin. A synthetic product, canthaxanthin, has a higher color potency than lutein or zeaxanthin; however the yolk it produces is orange-red, which is often undesirable.

Feedstuffs are analyzed for their *total* content of xanthophylls as it is very difficult to segregate the individual ones. Recognized contents are found in Table 31-20.

Caution: There are xanthophyll-containing feedstuffs other than those included in Table 31-20, but the FDA in the United States has approved just a few for use in poultry rations. Before using any of these forms one should check with the FDA to see if they are approved.

Measuring Egg-Yolk Color

The first and simplest procedure to measure yolk color was by visual comparison, matching yolk color with various colors in color fans. The Roche Yolk Color Fan, a series of plastic paddles with reference numbers from 1 to 15, became the most popular.

A photometric method is based on the procedure of using graded solutions of potassium dichromate as the reference standard with a photometer and comparing these with the ether extract of egg yolk. The solutions are given NEPA (National Egg and Poultry Association) numbers from 1 (light color) to 10 (dark color) to express color in terms of dichromate percentage. Normally, a NEPA yolk-color value of 2 or 3 is suitable for eggs for human consumption, but values of 5 and above are necessary for the bakery trade.

Although NEPA numbers represent a measure of yolk-color density, they are not directly correlated with the quantity of xanthophylls necessary in the feed to produce the desired color. For example, it requires twice as much dietary xanthophylls to go from NEPA 6 to 7 as it does from NEPA 1 to 2. These relationships are shown in Table 31-21, columns 1, 4, and 5.

Table 31-21. Dietary Mixed Xanthophylls Necessary to Produce Egg-Yolk Densities

Yolk-Color Densities			Approximate Mg of Xanthophylls in Ration	
NEPA Number	Roche Color Fan Number	Micrograms of Beta-carotene per g of Yolk	per lb	per kg
(1)	(2)	(3)	(4)	(5)
1	5	25.0	6.2	13.6
2	7	43.0	12.5	27.5
3	9	57.0	19.0	41.8
4	11	70.0	25.7	56.5
5	13	82.5	32.9	72.4
6	15	96.0	42.6	93.7
7		108.5	55.8	122.8
8		120.0	73.0	160.6

Another system uses beta-carotene as the standard and identifies yolk color in terms of beta-carotene equivalents per gram of yolk and is known as the AOAC method. It is today's most accurate and consistent method of determining yolk color. The average relationship of the measurement to NEPA numbers is shown in Table 31-21, columns 1 and 3. With all this evidence, it is obvious that the figures in Table 31-21 are highly variable and they can only be considered as approximate.

Causes of Yolk-Color Variations

The quantity and type of dietary xanthophylls are not the only causes of variation in yolk color. Some others are as follows:

Strain difference.   These can cause as much as 14% variation in density of yolk color.

Individual bird variation.   The genetic capability to absorb and deposit xanthophylls in egg yolk varies between hens within a single strain.

Cages.   Hens kept in cages are able to make better use of yolk pigmenters than hens kept on a litter floor.

Morbidity.   Disease reduces the bird's ability to absorb xanthophylls from the intestinal tract. This is particularly true with certain coccidial strains of *Eimeria*.

Stress.   Any stress reduces the amount of xanthophylls getting to the ovary.

Fat in the diet.   There is an increase in the xanthophyll absorption as fat in the diet is increased.

Oxidation of the xanthophylls.   Xanthophylls are easily oxidized in pure state or in mixed feeds, thereby reducing their ability to color egg yolk. Where possible, an antioxidant should be used.

Certain ingredients.   On occasion meat scraps, soybean oil meal, charcoal, and sulfur have been shown to reduce egg-yolk color, probably because of lowered intestinal absorption of the xanthophylls.

Egg/feed ratio.   Rate of egg production is a cause of variability in yolk color. As flock egg production increases, the dietary xanthophylls are spread over more egg yolks with a corresponding decrease in yolk color, and vice versa. Rations for flocks laying at a high rate should contain more xanthophylls than those laying at a low rate.

31-J. FEED REQUIREMENT

As has been shown, the daily feed requirement for egg production is based on the energy and protein requirement. Furthermore, the bird varies her feed intake according to her caloric need, thus affecting the amount of protein consumed. To show the weekly feed intake for all conditions to which the flock is subjected, and for all strains, is an impossibility. Only average figures can be itemized; these are given in Tables 31-22 and 31-23, as feed consumption during moderate weather and when the ration contains 1,275 kcal of ME per lb (2,805 ME/kg).

Table 31-22. Feed Consumption of 100 Leghorn Layers per Day (Hen-Day Basis)

	Feed Consumed					Feed Consumed			
	Per 100 Hens per Day		Cumulative per Hen			Per 100 Hens per Day		Cumulative per Hen	
Age wk	lb	kg	lb	kg	Age wk	lb	kg	lb	kg
21	18.0	8.2	1.3	0.6	49	23.0	10.4	44.1	20.0
22	18.4	8.3	2.6	1.2	50	23.0	10.4	45.7	20.7
23	18.7	8.5	3.9	1.8					
24	19.1	8.7	5.2	2.4	51	23.0	10.4	47.3	21.5
25	19.4	8.8	6.6	3.0	52	23.0	10.4	48.9	22.2
					53	23.0	10.4	50.6	22.9
26	19.8	9.0	8.0	3.6	54	23.0	10.4	52.2	23.7
27	20.1	9.1	9.4	4.3	55	23.0	10.4	53.8	24.4
28	20.5	9.3	10.8	4.9					
29	20.9	9.5	12.3	5.6	56	23.0	10.4	55.4	25.1
30	21.2	9.6	13.8	6.2	57	23.0	10.4	57.0	25.9
					58	23.0	10.4	58.6	26.6
31	21.6	9.8	15.3	6.9	59	23.0	10.4	60.2	27.3
32	21.9	9.9	16.8	7.6	60	23.0	10.4	61.8	28.0
33	22.3	10.1	18.4	8.3					
34	22.6	10.3	20.0	9.1	61	23.0	10.4	63.4	28.8
35	23.0	10.4	21.6	9.8	62	23.0	10.4	65.0	29.5
					63	23.0	10.4	66.7	30.2
36	23.0	10.4	23.2	10.5	64	23.0	10.4	68.3	31.0
37	23.0	10.4	24.8	11.2	65	23.0	10.4	69.9	31.7
38	23.0	10.4	26.4	12.0					
39	23.0	10.4	28.0	12.7	66	23.0	10.4	71.5	32.4
40	23.0	10.4	29.6	13.4	67	23.0	10.4	73.1	33.2
					68	23.0	10.4	74.7	33.9
41	23.0	10.4	31.2	14.2	69	23.0	10.4	76.3	34.6
42	23.0	10.4	32.8	14.9	70	23.0	10.4	77.9	35.3
43	23.0	10.4	34.4	15.6					
44	23.0	10.4	36.1	16.4	71	23.0	10.4	79.5	36.1
45	23.0	10.4	37.7	17.1	72	23.0	10.4	81.1	36.8
					73	23.0	10.4	82.7	37.5
46	23.0	10.4	39.3	17.8	74	23.0	10.4	84.4	38.3
47	23.0	10.4	40.9	18.5	75	23.0	10.4	86.0	39.0
48	23.0	10.4	42.5	19.3					
					76	23.0	10.4	87.6	39.7

Table 31-23.   Feed Consumption of 100 Medium-Size Layers per Day (Hen-Day Basis)

Age wk	Feed Consumed				Age wk	Feed Consumed			
	Per 100 Hens per Day		Cumulative per Hen			Per 100 Hens per Day		Cumulative per Hen	
	lb	kg	lb	kg		lb	kg	lb	kg
21	20.0	9.1	1.4	0.6	49	25.0	11.4	48.1	21.9
22	20.4	9.3	2.8	1.3	50	25.0	11.4	49.9	22.7
23	20.7	9.4	4.3	1.9					
24	21.1	9.6	5.8	2.6	51	25.0	11.4	51.6	23.5
25	21.4	9.7	7.3	3.3	52	25.0	11.4	53.4	24.3
					53	25.0	11.4	55.1	25.1
26	21.8	9.9	8.8	4.0	54	25.0	11.4	56.9	25.9
27	22.1	10.0	10.3	4.7	55	25.0	11.4	58.6	26.6
28	22.5	10.2	11.9	5.4					
29	22.9	10.4	13.5	6.1	56	25.0	11.4	60.4	27.4
30	23.2	10.5	15.1	6.9	57	25.0	11.4	62.1	28.2
					58	25.0	11.4	63.9	29.0
31	23.6	10.7	16.8	7.6	59	25.0	11.4	65.6	29.8
32	23.9	10.9	18.5	8.4	60	25.0	11.4	67.4	30.6
33	24.3	11.0	20.2	9.2					
34	24.6	11.2	21.9	9.9	61	25.0	11.4	69.1	31.4
35	25.0	11.4	23.6	10.7	62	25.0	11.4	70.9	32.2
					63	25.0	11.4	72.6	33.0
36	25.0	11.4	25.4	11.5	64	25.0	11.4	74.4	33.8
37	25.0	11.4	27.1	12.3	65	25.0	11.4	76.1	34.6
38	25.0	11.4	28.9	13.1					
39	25.0	11.4	30.6	13.9	66	25.0	11.4	77.9	35.4
40	25.0	11.4	32.4	14.7	67	25.0	11.4	79.6	36.2
					68	25.0	11.4	81.4	37.0
41	25.0	11.4	34.1	15.5	69	25.0	11.4	83.1	37.8
42	25.0	11.4	35.9	16.3	70	25.0	11.4	84.9	38.6
43	25.0	11.4	37.6	17.1					
44	25.0	11.4	39.4	17.9	71	25.0	11.4	86.6	39.4
45	25.0	11.4	41.1	18.7	72	25.0	11.4	88.4	40.2
					73	25.0	11.4	90.1	41.0
46	25.0	11.4	42.9	19.5	74	25.0	11.4	91.9	41.8
47	25.0	11.4	44.6	20.3	75	25.0	11.4	93.6	42.6
48	25.0	11.4	46.4	21.1					
					76	25.0	11.4	95.4	43.4

Notes on Tables 31-22 and 31-23

1. Feed consumption is based on breeder standard body weights at the start of egg production.
2. It is assumed that there are no temperature or seasonal influences.
3. Figures are based on the number of birds present with zero mortality.
4. Statistics from the two tables are

	Leghorn-Type	Medium-Size
Eggs per hen	306	306
Feed consumed per hen (lb)	86.0	95.4
Feed consumed per hen (kg)	39.0	43.4
Feed consumed per doz eggs (lb)	3.37	3.74
Feed consumed per doz eggs (kg)	1.53	1.70

Note: Hen-day basis 21 to 76 weeks of age.

31-K. PHASE FEEDING OF EGG-TYPE LAYERS

Because of reduced rate of lay as the bird continues through her laying year, less protein is required per day by the commercial layer. To save protein and its expense during this period, it is practical to reduce the percentage of protein in the diet (see Table 31-10). Although the procedure, known as phase feeding, does not improve the rate of lay, it does prevent waste of protein and lowers the cost of producing a dozen eggs.

Number of Feeding Phases

To be practical, three levels of daily nutrient intake are usually employed during the laying period to satisfy the flock's basic nutrient requirements. The feeds are associated with three periods known as phases.

> Phase   I: From day of first egg through 20 weeks of egg production.
> Phase II: Beginning week 21 through week 40 of egg production.
> Phase III: After week 40 of egg production.

Importance of phase I.   The first 20 weeks of egg production are important from a nutritional standpoint. During this period the birds will peak in egg production and egg mass will reach its maximum level. With today's feeding programs, the production peak should occur between 26 and 32 weeks of age. The egg mass peak will usually occur about 10 weeks later. Flocks peak earlier and more abruptly when the body weight at sexual maturity is more uniform (see Chapter 22-J).

At the peak of egg production, better than 90% egg production will be achieved by most strains of birds. This requires about 18 g of dietary protein per Leghorn per day, including that necessary for growth and feather production.

Analysis of a Three-Phase Feeding Program

The nutritional implications of a three-phase feeding program for Leghorn layers are shown in Table 31-24.

Note that feed, energy, and protein consumption all appear to be low during the first 20 weeks, which is a reflection that all birds are not into full production during the first half of this period. When feed intake is calculated for the average hen, it actually reflects the feed consumed by layers and a large percentage of nonlaying pullets. For this reason, nutrient intake calculations for the first two months of lay are not too meaningful.

Table 31-24. Analysis of Phase Feeding Programs for Leghorn Layers by Weeks of Egg Production (Moderate Temperature)

Item	Phase (Weeks of Egg Production)		
	Phase 1 1–20 wk	Phase 2 21–40 wk	Phase 3 41+ wk
Protein in ration (%)	17.0	16.5	16.0
kcal of ME per lb of ration	1300	1300	1300
kcal of ME per kg of ration	2860	2860	2860
kcal of ME consumed per hen per day	269	294	294
Egg production at peak production (%)	90+	—	—
Avg % egg production (hen-day)	77.6	82.3	72.3
Avg feed consumed per hen per day (lb)	0.207	0.226	0.226
Avg feed consumed per hen per day (g)	94	103	103
Avg g of protein consumed per hen per day	16.0	17.0	16.5

Experimental Results of a Three-Phase Feeding Program

Table 31-25 shows the actual results of phase feeding birds on a litter floor and in cages compared with birds fed a diet of a constant protein percentage. Although the experimental diets used for phase feeding are different in protein than those shown in Table 31-24, the results are typical and bear study.

In actual practice, phase feeding involves separate requirements for each critical amino acid and mineral and these requirements are adjusted to compensate for fluctuations in feed consumption.

Table 31-25. Comparison of Three-Phase and Single-Phase Feeding of Standard Leghorn Layers on the Floor and in Cages (336 Days of Production)

Item	Litter Floor		Cages	
	Phase Fed[1]	Constant Protein[2]	Phase Fed[1]	Constant Protein[2]
Hen-day egg production (%)	74.9	74.4	73.2	73.7
Large-or-better egg size (%)	81.3	82.8	88.9	88.6
Avg feed consumption				
100 birds per day (lb)	24.0	24.1	23.7	23.7
Per lb of eggs	2.44	2.45	2.38	2.35
Per doz eggs	3.86	3.88	3.89	3.86
Avg egg weight (oz/doz)	25.3	25.4	26.2	26.2
Avg egg weight (g each)	59.8	60.0	61.9	61.9

Source: Florida Extension Service, October, 1977.
[1]3 phases, 19, 17, and 15% protein respectively.
[2]One feed, 17% protein.

Phase Feeding When the Temperature Changes

The data given in Table 31-24 pertain to moderate house temperatures; but when the temperature changes, the feed formula recommendations must be corrected. The main change has to do with the energy requirement of the flock for maintenance and the fact that birds eat less as the temperature rises. It becomes necessary to alter the nutrient content of the diet to prevent a nutritional deficiency. Examples for these alterations are given in Table 31-26.

These recommendations allow for normal variations of feed quality and feed consumption. Greater safety margins would be required if ingredient quality was poor and if house temperatures fluctuated widely from one week to the next.

Table 31-26. Phase Feeding for Temperature Variations

Temperature	Phase 1: 1–20 Wk		Phase 2: 21–40 Wk		Phase 3: 41+ Wk	
	Amount in Feed					
	Protein %	Calcium %	Protein %	Calcium %	Protein %	Calcium %
Moderate (65°–70°F) (19°–21°C)	17.0	3.6	16.5	3.8	16.0	4.0
Cold (50°–55°F) (10°–13°C)	15.6	3.3	15.2	3.5	14.7	3.7
Hot (85°–95°F) (30°–35°C)	18.6	3.9	18.1	4.2	17.5	4.4

Note: Phase feeding periods during egg production.

31-L. EGG MASS AS A MEASURE OF EGG PRODUCTION

The use of egg mass rather than egg numbers will lead to better comparisons of flocks or strains of birds, along with feeding and management programs. To calculate egg mass it is first necessary to determine the average egg weight of eggs laid by the flock. It is necessary to weigh only a sample of the eggs laid. Total the weight of the entire sample; then divide this weight by the number of eggs in the sample. After the mean egg weight has been determined in grams each, the following formula is used to compute egg mass on a daily basis:

$$P \times W = M$$

where P equals % hen-day egg production, W equals average egg weight in grams per egg, and M equals average egg mass per hen per day in grams.

> Note: Average daily egg mass is a measurement not commonly used in the United States. The metric system (grams) is used to conform to the system used in other countries.

Egg Mass Equivalents

Inasmuch as large eggs command a higher sales price than small eggs, not only is it important that the flock produce a goodly number of eggs but the eggs must be large. Some strains or flock of pullets produce large numbers of small eggs; others produce fewer eggs, but the eggs are large. Table 31-27 shows the comparative trade-off values of daily egg mass when measured in grams per hen per day over a laying period of 365 days, using various average egg weights in grams per egg and

Table 31-27.   Average Daily Egg Mass at Varying Egg Weights and Rates of Lay for Layers

Avg Annual Egg Wt		Avg Hen-Day Egg Production for Year (%)				
		60	65	70	75	80
		Avg Eggs per Hen for Year				
		291	237	256	274	292
lb/case	g/egg	Average Egg Mass per Hen per Day (g)				
44.0	55.4	33.2	36.0	38.8	41.6	44.3
46.0	58.0	34.8	37.7	40.6	43.5	46.4
48.0	60.5	36.3	39.3	42.4	45.4	48.4
50.0	63.0	37.8	41.0	44.1	47.3	50.4
52.0	65.5	39.3	42.6	45.9	49.1	52.4

percentages of hen-day egg production. In Table 31-27, any combination of egg weight and percentage production producing the same average egg mass per hen per day would be comparable.

Strains and Flocks Differ in Egg Mass

Generally, flocks are compared on the basis of hen-day or hen-housed egg production. These calculations are easy to make, but both disregard egg weight. A better procedure is to include egg weight as well as egg production and mortality. To bring the three measurements into focus as one index, the total egg mass produced per hen-housed basis should be used to compare various strains or management programs. It will show differences where others fail. Comparison to a standard age is required if the comparison is to be meaningful.

Production, Egg Weight, Egg Mass, and Feed Consumption

These figures have been projected in Table 31-28 for average Leghorn hens in the laying flock by weeks, through 52 weeks of production. Average and total figures accumulated from Table 31-28 for 21 to 76 weeks of age are as follows:

Total hen-housed egg production (number per hen)	289.2
Total hen-housed egg production (doz per hen)	24.1
Avg hen-day egg production (%)	78.0
Avg egg weight (oz/doz)	25.1
Avg egg weight (g each)	59.2
Avg feed consumed per 100 hens per day (lb)	22.3
Avg feed consumed per 100 hens per day (kg)	10.1
Avg feed consumed per dozen eggs (lb)	3.43
Avg feed consumed per dozen eggs (kg)	1.56
Avg feed consumed per unit of egg weight	2.19

One can see from the above summary figures that they represent a good flock, better than average, but not exceptional.

Feed Consumption Needs Adjusting

As a general rule, except in extremely hot weather, pullets will eat enough feed to balance their energy requirement, but they may not always be getting enough protein. On an egg-mass basis (Table 31-28, col. 7) the period of probable inadequacy would be between the weeks 28 and 34 of age when the ratio of feed consumed to egg mass produced is

Table 31-28. Hen-Day Egg Production, Egg Weight, Egg Mass, and Feed Consumption of White Leghorn Laying Hens by Week of Age

Age wk	Hen-Day Egg Production %	Average Egg Weight g	Avg Egg Mass per Hen per Day g	Feed Consumed per 100 Hens/day lb	Feed per Dozen Eggs lb	Feed per lb of Egg Mass lb
(1)	(2)	(3)	(4)	(5)	(6)	(7)
21	10.0	44.1	4.4	18.0	21.60	18.51
22	23.0	45.5	10.5	18.4	9.60	7.98
23	39.8	46.9	18.7	18.7	5.64	4.55
24	60.0	48.1	28.9	19.1	3.82	3.00
25	76.5	49.3	37.7	19.4	3.04	2.33
26	84.5	50.4	42.6	19.8	2.81	2.11
27	87.0	51.4	44.7	20.1	2.77	2.04
28	89.5	52.3	46.8	20.5	2.75	1.99
29	91.2	53.2	48.5	20.9	2.75	1.95
30	92.5	53.9	49.9	21.2	2.75	1.93
31	92.0	54.7	50.3	21.6	2.82	1.95
32	91.5	55.4	50.7	21.9	2.87	1.96
33	91.0	56.1	51.0	22.3	2.94	1.98
34	90.5	56.6	51.2	22.6	3.00	2.00
35	90.0	57.2	51.5	23.0	3.07	2.03
36	89.5	57.7	51.6	23.0	3.08	2.02
37	89.0	58.1	51.7	23.0	3.10	2.02
38	88.5	58.6	51.9	23.0	3.12	2.01
39	88.0	59.0	51.9	23.0	3.14	2.01
40	87.5	59.2	51.8	23.0	3.15	2.01
41	87.0	59.6	51.9	23.0	3.17	2.01
42	86.5	59.9	51.8	23.0	3.19	2.02
43	86.0	60.2	51.8	23.0	3.21	2.01
44	85.5	60.5	51.7	23.0	3.23	2.02
45	85.0	60.7	51.6	23.0	3.25	2.02
46	84.5	60.9	51.4	23.0	3.27	2.03
47	84.0	61.1	51.3	23.0	3.29	2.03
48	83.5	61.2	51.1	23.0	3.31	2.04
49	83.0	61.5	51.0	23.0	3.33	2.04
50	82.5	61.6	50.8	23.0	3.35	2.05
51	82.0	61.7	50.6	23.0	3.37	2.06
52	81.5	61.9	50.4	23.0	3.39	2.07
53	81.0	62.0	50.2	23.0	3.41	2.08
54	80.5	62.1	50.0	23.0	3.43	2.09
55	80.0	62.2	49.8	23.0	3.45	2.10
56	79.5	62.2	49.5	23.0	3.47	2.11
57	79.0	62.4	49.3	23.0	3.49	2.12
58	78.5	62.5	49.1	23.0	3.52	2.13
59	78.0	62.5	48.7	23.0	3.54	2.14
60	77.5	62.6	48.5	23.0	3.56	2.15

Table 31-28. *(continued)*

Age wk	Hen-Day Egg Production %	Average Egg Weight g	Avg Egg Mass per Hen per Day g	Feed Consumed per 100 Hens/day lb	Feed per Dozen Eggs lb	Feed per lb of Egg Mass lb
61	77.0	62.6	48.2	23.0	3.58	2.16
62	76.5	62.7	48.0	23.0	3.61	2.17
63	76.0	62.7	47.7	23.0	3.63	2.19
64	75.5	62.7	47.4	23.0	3.66	2.20
65	75.0	62.9	47.2	23.0	3.68	2.21
66	74.5	62.9	46.8	23.0	3.70	2.23
67	74.0	62.9	46.5	23.0	3.73	2.24
68	73.5	63.0	46.3	23.0	3.76	2.25
69	73.0	63.0	46.0	23.0	3.78	2.25
70	72.5	63.0	45.7	23.0	3.81	2.28
71	72.0	63.0	45.4	23.0	3.83	2.30
72	71.5	63.0	45.0	23.0	3.86	2.32
73	71.0	63.1	44.8	23.0	3.89	2.33
74	70.5	63.1	44.5	23.0	3.91	2.34
75	70.0	63.1	44.2	23.0	3.94	2.36
76	69.5	63.1	43.9	23.0	3.97	2.38

the lowest. A protein deficiency at this time would, no doubt, reduce egg size. The relationship of feed consumed to egg mass produced may be the basis for phase feeding. With a II-phase system of feeding, a ration higher in protein should be fed through about week 30 of production, with a reduction from weeks 31 to 52 to produce more economy in feeding.

Maintaining Body Weight During Laying

Caged layers must gain weight during their production year. As caging is conducive to heavier body weights than when pullets are kept on the floor, there is usually no problem in maintaining the proper weight increase when the weather is cool or normal. However, during periods of hot weather, birds do not consume as much feed, and there is always the possibility that body weight will suffer. Lowering the environmental temperature, giving fresh feed early in the morning and later in the afternoon to increase feed consumption during the cool hours of the day, along with plenty of cool, fresh water will be of help.

If there is a problem with excessive body weight, it may be best to initiate some form of feed control during the laying cycle. If body weight declines, you are not feeding enough.

More feed energy will not offset crowding. Feeds with higher energy values will not compensate for stress and lower body weights brought on when birds are crowded in cages.

31-M. WATER RESTRICTION OF LAYERS

As an aid to reducing the moisture content of the feces, it may be practical to restrict the drinking water intake. Usually this is accomplished by allowing the birds to drink for 15 minutes, followed by a period of from 2 to 4 hours without water. The procedure is then repeated through the light hours. The moisture content of the feces can be reduced up to 7%, depending on the amount of water consumed.

Be careful. Water restriction during hot weather may be disastrous. Birds drink two to three times as much water during hot weather, and restricting water consumption may result in mortality or production problems. Water restriction is usually not practical in nipple or cup watering systems due to the problem of leakage associated with turning the system on and off.

Many experiments have compared restricted watering with full watering programs. In most of these experiments, egg production is not significantly different between the two systems.

Hourly consumption of water. With any water restriction or medication program, keep in mind that the laying pullets normally drink 25% of their daily intake during the 2 hours prior to the time the lights go off.

31-N. FEEDING GRIT TO CAGED LAYERS

There is experimental evidence to support the recommendation for feeding grit to cage layers. Some reports show that egg production is improved, but the amount of grit necessary is small. Well-ground rations and large-particle limestone or oystershell will eliminate the need for supplemental grit.

How much grit to feed. Feed 0.5 lb (227 g) of hen-size grit per 100 laying pullets in cages per week or feed 2 lb (908 g) per 100 laying pullets every 4 weeks. Each allotment of grit should be fed on 1 day.

Caution: Do not feed more grit than recommended, and never self-feed.

32
Feeding Breeding Birds

Breeding chickens may be divided into two groups: egg-type and meat-type. Because meat-type breeders tend to become obese their feed and feeding programs are much different than those used for egg-type breeders, including the starting, growing, and laying phases.

32-A. FEEDING EGG-TYPE BREEDERS DURING STARTING PERIOD

From day old through 5 weeks, the starter rations should be approximately the same for egg-type male and female breeders as for commercial egg-type pullets; however, if the chicks are raised on a litter floor or slats, a coccidiostat should be added.

Use one of the feeding programs in Table 32-1, which also includes the standard weights when the birds are 5 weeks of age. If the average weights are below these figures, continue feeding the starter until the weights are attained.

Table 32-1. Starter Feeding Programs for Egg-type Breeders, Weeks 0 through 5

| | | | 5-Wk Weight[1] | | | |
| | | | Male | | Female | |
Breed Type	ME kcal/lb	Protein %	lb	kg	lb	kg
Leghorn	1,360	20	1.0	0.46	0.8	0.36
Medium-size, brown-egg	1,360	18	1.3	0.59	1.0	0.46

[1]Or those furnished by the primary breeder.

32-B. FEEDING EGG-TYPE BREEDERS DURING GROWING PERIOD

Most egg-type breeders are raised on a litter floor although some are kept on slats or wire, but the type of floor will not alter the starter and growing formulae except for the addition of vitamin K for birds on wire, and a coccidiostat for birds on litter or slats. Use Table 32-2 for a growing feed program for Leghorn breeders and medium-size brown-egg breeders.

Table 32-2. Egg-Type Growing Feed Programs

Breed	6–12 Wk			13–20 Wk			21 Wk to 5% HD Production		
	Type[1]	ME[2]	% Pro	Type	ME	% Pro	Type	ME	% Pro
Leghorn	Grower	1,340	15	Grower	1,340	15	Breeder	1,340	17
Medium-size,									
brown-egg	Grower	1,340	14	Grower	1,340	12	Breeder	1,340	17

[1]Type of ration.
[2]kcal/lb of ration.

Body Weight During Growing

Breeder pullets should become sexually mature (produce eggs) at a specified age. One of the measures of advancement toward this optimal age is body weight during growing. Weekly average weights for Leghorn pullets and cockerels are given in Table 17-2, and are given in Table 17-3 for medium-size brown-egg pullets and cockerels. Use these tables or similar ones prepared by the primary breeder.

A representative sample of pullets and cockerels should be weighed weekly beginning at 4 weeks of age and continuing until the pullets begin egg production. For directions see Chapter 16-G. If birds are underweight at any time during the growing period, increase the protein in the ration, lengthen the light day to increase feed consumption, or increase the chain speed in chain-and-trough automatic feeders. If they are overweight, the feed must be restricted by one of the following methods:

1. *Limited-every-day feeding program.* The birds are given a measured amount of feed each day, but less than they would eat if full-fed.
2. *Skip-every-other-day feeding program.* Birds are fed every other day, and the amount of feed given on feed days must be regulated to twice the amount of feed allocated on the limited-every-day program.

32-C. CHANGING EGG-TYPE BREEDERS
FROM GROWING TO BREEDER DIET

The program for changing egg-type breeders from the growing to the breeding ration is identical to the program for changing egg-type commercial pullets from a growing to a laying ration (see Chapter 31-A). Necessarily, a breeder ration capable of producing high hatchability of the eggs laid is to be used rather than a laying ration. Substitute the breeder ration when the flock is about 20 weeks (140 days) of age. This gives ample time for the pullets to build yolk reserves of certain vitamins and other feed components before production of the first hatching eggs.

32-D. FEEDING EGG-TYPE BREEDERS
DURING EGG PRODUCTION

Although the feed formula must be changed during the period of egg production to compensate for the production of eggs that will hatch into quality chicks, the feed management program is identical to the one used for the production of commercial eggs by egg-type strains (see Chapter 31).

32-E. FEEDING THE EGG-TYPE BREEDER MALE

Common procedure during the growing period is to keep the cockerels with the pullets. Thus, both sexes have access to the same feed and feeding program. Growing males of the egg-type group seldom get too heavy if some program of maintaining the correct female body weight is followed.

Although males will eat about 25% more feed than females during the growing period, during the laying (breeding) period a male will consume about the same amount of feed per day as a female. Average weights of egg-type males are given in Tables 17-2 and 17-3.

32-F. FEEDING GROWING MEAT-TYPE
BREEDER FEMALES

Meat-type breeder females (broiler breeders), producing broiler offspring, possess the inherent ability to grow rapidly. When full-fed during the growing period they gain excessive weight and deposit too much internal fat for maximum egg production.

When feeding broiler breeder females during the growing period the nutritional object is to restrict the caloric intake to produce pullets that are smaller and older when they lay their first eggs. The process of weight reduction must encompass the entire growing period; it cannot wait until just before egg production begins.

What Growing Feed Restriction Does

As early as 1937, it was found that restricting the feed intake of the growing, meat-type bird would delay sexual maturity and increase the size of the first eggs laid. From this early beginning the method of feed restriction has been improved; today, the reults from the program show

1. Restricting the growing feed intake will delay the onset of sexual maturity from a few days to 3 or 4 weeks, depending on the severity of restriction.
2. Feed restriction will reduce the body weight of the bird at sexual maturity, usually by reducing the amount of body fat.
3. Mortality during the growing period is not normally affected unless the feed restriction approaches starvation.
4. Restriction of an ordinary growing diet may lead to nutritional deficiencies because all feed components are restricted.
5. Restricting the feed intake usually means that the cost of raising a pullet is reduced, but this is not necessarily true under all conditions. The additional time necessary to reach sexual maturity may involve the use of more feed.
6. Restricting the feed intake during the growing period usually produces better livability during egg production.
7. Egg production is not greatly affected during an equal number of months of lay, regardless of any growing program of feed restriction.
8. Egg weight is regulated by the age of the bird. Therefore, birds grown on feed restriction will produce larger first eggs only because they are older.

Comparison of Restricted and Full-Fed Meat-Type Growing Programs

It has been established that for a meat-type flock of breeders to have top egg production the pullets must have an average body weight of 5.5 lb (2.50 kg) at 24 weeks (168 days) of age when the first eggs are produced and be laying at 5% hen-day egg production during the 25th week. Body weight will be about 0.2 lb (91 g) heavier if the chicks are hatched during the warmer months and raised during the colder months.

If present-day, meat-type pullets are *full fed* a ration moderate in calories and protein, the average female flock weight at 24 weeks of age will be about 8.5 lb (3.89 kg). A high-calorie, high-protein diet will produce an average weight of up to 1 lb (454 g) heavier at the same age.

The relationship between full feeding and restricted feeding is shown in Table 32-3 when both groups are fed the same ration. The last column in the table shows the percentage reduction in full feeding to get the recommended weekly average weights when controlled feed is practiced. After 6 weeks of age these are between 41 and 57%.

Table 32-3. Comparison of Restricted Feeding Versus Full Feeding of Growing Meat-Type Pullets

| Week of Age | Restricted Feeding | | | | Full Feeding | | | | Restricted Feed Reduction (Weekly Basis) % |
| | Feed Consumption per 100 Pullets per Day | | Desired Body Weight | | Feed Consumption per 100 Pullets per Day | | Approximate Body Weight | | |
	lb	kg	lb	kg	lb	kg	lb	kg	
4	9.5	4.3	1.1	0.50	11.1	5.0	1.3	0.59	14
6	11.0	5.0	1.5	0.64	15.9	7.2	2.2	1.00	31
8	12.1	5.5	1.9	0.86	21.2	9.6	3.3	1.50	41
10	13.3	6.0	2.3	1.05	27.9	12.7	4.3	1.95	49
12	14.7	6.7	2.7	1.23	34.3	15.6	5.2	2.36	57
14	16.1	7.3	3.1	1.41	36.6	16.6	6.0	2.72	56
16	17.5	8.0	3.5	1.59	37.9	17.2	6.7	3.04	54
18	18.9	8.6	3.9	1.77	38.7	17.6	7.3	3.31	51
20	20.3	9.2	4.3	1.96	39.6	18.0	7.8	3.54	49
22	21.7	9.8	4.8	2.18	40.4	18.3	8.2	3.72	46
24	23.1	10.5	5.5	2.50	41.2	18.7	8.5	3.86	44

Growing Feed Reduction Less Than Indicated

To evaluate this point, Table 32-3 shows that a flock of meat-type pullets restricted in feed intake will average 4.3 lb (1.96 kg) in body weight and consume 20.3 lb (9.2 kg) of feed per 100 birds per day on the 20th week, and its feed intake will be 49% lower than a full-fed flock of the same age.

The figures may be better compared with a full-fed flock of the same average weight rather than the same age. For example, in Table 32-3 note that birds in a full-fed flock weigh 4.3 lb on the 10th week and consume 22.9 lb (10.4 kg) of feed per 100 pullets per day. On the basis of body weight a restricted flock of the same average weight will eat only 11% less feed. This amount is the real criterion of feed restriction for it is doubtful if any flock could survive a restriction of 49% as calculated on an age basis.

Weekly Weight Gain During Growing

Table 32-4 shows the percentage gain in weight for the respective weeks when the pullets are on a restricted feeding program. Notice that to be practical feed restriction must be started early in the chick's life if the program is to be effective. It is much more difficult to control body weight of meat-type pullets after 12 weeks than before.

Table 32-4. Weekly Percentage Weight Gain for Meat-Type Growing Pullets (Restricted Feeding Program)

Week of Age	Gain in Weight for Week %
4	22.2
6	15.4
8	11.8
10	9.5
12	8.0
14	6.9
16	6.1
18	5.4
20	4.9
22	4.4
24	4.2

Meeting the Nutritional Requirements During Growing

Energy. The body weight of meat-type growing pullets must be reduced early in life, which calls for starter and grower diets moderately low in ME. To further reduce the growth rate, these rations must be restricted (controlled) as early as 2 weeks of age, depending on the feeding program. With the present rapid growth of broilers, it is no longer possible to wait until their dams are 10 to 12 weeks of age.

Protein in the starter diet. A starter ration containing 20% is recommended. This percentage is somewhat higher than formerly suggested, but as the starter is fed for such a short period, the additional protein seems warranted.

The sulfur amino acids, methionine and cystine, along with lysine, are as important as total protein. The daily requirements must be met.

Protein in the grower diet. The pullet's daily protein need is lower as she ages, starting with diets near 20% and lowering to diets as low as 10% near sexual maturity. Following is a method of feed formulation to meet these requirements:

Phase feeding of protein diets. Although it is possible to reduce the protein in these rations as often as every 2 weeks, a more practical solution is to use three grower diets with 18, 15, and 12% protein, respectively, as the pullets age.

Minerals in the grower diet. Calcium and phosphorus are the important minerals to consider when restricting feed during growing. Furthermore, it must be remembered that calcium must be increased during the 23rd and 24th weeks by feeding a breeder ration higher in calcium plus 3 lb per week per 100 birds of flaked oystershell between 23 and 27 weeks.

Formula Variations of Grower Feeds

The feed formula variations of major ingredients are given in Table 32-5.

Table 32-5. Feed Specifications for Meat-Type Growing Pullet Diets

Item	Starter	Grower I	Grower II	Grower III
Weeks fed	0–4	5–9	10–15	16–22
Weeks of feeding	4	5	6	7
Kcal of ME per lb	1,275	1,300	1,300	1,300
Kcal of ME per kg	2,805	2,860	2,860	2,860
Protein (%)	20	18	15	12
Methionine + cystine (%)	0.70	0.60	0.55	0.50
Lysine (%)	1.00	0.90	0.70	0.50
Calcium (%)	0.9	0.9	0.9	0.9
Phosphorus, total (%)	0.66	0.66	0.66	0.66
Phosphorus, available (%)	0.40	0.40	0.40	0.40

Programs for Feed Restriction

There are two general types of feed restriction.

1. Using skip-day feeding, a restricted (allotted) amount of feed is given on feed days and no feed is given on ''skip'' days.
2. Feed the birds every day, but restrict the daily amount of feed given them.

32-G. SKIP-EVERY-OTHER-DAY PULLET GROWING FEED PROGRAM

The breeder starter should be fed the first 4 weeks. It is to be self-fed the first 2 weeks, then restricted during the third and fourth weeks, but fed every day. On the average, meat-type pullets should be eating about 7 lb (3.18 kg) of feed per 100 pullets per day at 14 days of age.

If the flock is in good health and on the targeted weight, beginning with the 5th week (29 days) feed a specified allotment on one day, no feed the next, feed the next, then no feed, etc., so that feed is provided every other day. A guide for feed allotments on feed days is given in Table 32-6 (pounds) and 32-7 (kilos), along with the average desired live flock body weight.

It is imperative that the growing pullets gain about 0.2 lb (91 g) per week from the 3rd week through week 11, then 0.25 lb (1.2kg) per week through the 24th week. However, the suggested daily feed allowances will be slightly greater for flocks hatched between April and September, and slightly less for those hatched between October and March in the Northern Hemisphere (reverse for Southern Hemisphere). The in-season flocks will be about 0.2 lb (91 g) heavier at sexual maturity than the standards given, and out-of-season flocks will be about 0.2 lb (91 g) lighter.

Exact Feed Allotments Depend on Body Weights

The feed allotments given in Tables 32-6 and 32-7 are only a guide; many things affect the exact figure: strain of birds, date of hatch, caloric and protein content of the feed, season of the year, ambient temperature, hours of light per day, physical condition of the flock, age of the pullets, etc.

Representative samples of the flock must be weighed weekly on the afternoon of no-feed days, beginning at the end of the 3rd week (21 days) (see Chapters 16-E and 17-G). For each 1% the flock average weight is below the standard each week, increase the daily feed allotment by 1%. If the flock average weight is above the standard by more than 1% on any one week, the feed allotment given on a day the birds are weighed should be continued until the correct body weight is reached. It is best not to reduce the feed allotment of a flock; for best results the weight of the pullets should increase some each week, and they should never be forced to lose weight.

Important measurements of the skip-every-other-day feeding program include daily consumption of grams of protein and kilograms of ME per bird as shown in Table 32-8.

Table 32-6. Estimated Body Weight and Feed Allowance of 100 Meat-Type Growing Pullets by Weeks (in Pounds)

Week	Desired Avg Female Weight lb	Approximate Feed Amounts for 100 Birds on Feed Days		Cumulative Feed Consumed per 100 Birds lb
		Skip-every-other-day Program lb	Feed-every-day Program lb	
		Full-feed Starter Every Day	Full-feed Starter Every Day	
1	0.3	3.5	3.5	25
2	0.5	5.5	5.5	63
		Restrict Starter Every Day	Restrict Starter Every Day	
3	0.9	9.0	9.0	126
4	1.1	9.5	9.5	193
		Restrict Grower I Every Other Day	Restrict Grower I Every Day	
5	1.3	21.0	10.5	266
6	1.5	22.0	11.0	343
7	1.7	23.0	11.5	424
8	1.9	24.2	12.1	508
9	2.1	25.4	12.7	597
		Restrict Grower II Every Other Day	Restrict Grower II Every Day	
10	2.3	26.6	13.3	690
11	2.5	28.0	14.0	788
12	2.7	29.4	14.7	891
13	2.9	30.8	15.4	999
14	3.1	32.2	16.1	1112
15	3.3	33.6	16.8	1229
		Restrict Grower III Every Other Day	Restrict Grower III Every Day	
16	3.5	35.0	17.5	1352
17	3.7	36.4	18.2	1480
18	3.9	37.8	18.9	1612
19	4.1	39.2	19.6	1749
20	4.3	40.6	20.3	1892
21	4.5	42.0	21.0	2039
22	4.8	43.4	21.7	2190
		Restrict Breeder Every Other Day	Restrict Breeder Every Day	
23	5.1	44.8	22.4	2347
24	5.5	46.2	23.1	2509

Table 32-7. Estimated Body Weight and Feed Allowance of 100 Meat-Type Growing Pullets by Weeks (in Kilos)

Week	Desired Avg Female Weight kg	Approximate Feed Amounts for 100 Birds on Feed Days		Cumulative Feed Consumed per 100 Birds kg
		Skip-every-other-day Program kg	Feed-every-day Program kg	
		Full-feed Starter Every Day	Full-feed Starter Every Day	
1	0.14	1.59	1.59	11.4
2	0.23	2.50	2.50	28.6
		Restrict Starter	Restrict Starter	
3	0.41	4.09	4.09	57.3
4	0.50	4.32	4.32	87.7
		Restrict Grower I Every Other Day	Restrict Grower I Every Day	
5	0.59	9.55	4.77	120.9
6	0.68	10.00	5.00	155.9
7	0.77	10.46	5.23	192.3
8	0.86	11.00	5.50	230.9
9	0.96	11.55	5.77	271.4
		Restrict Grower II Every Other Day	Restrict Grower II Every Day	
10	1.05	12.09	6.05	313.6
11	1.14	12.73	6.36	358.2
12	1.23	13.36	6.68	405.0
13	1.32	14.00	7.00	454.1
14	1.41	14.64	7.32	505.5
15	1.50	15.27	7.64	558.6
		Restrict Grower III Every Other Day	Restrict Grower III Every Day	
16	1.59	15.91	7.96	614.6
17	1.68	16.55	8.27	672.7
18	1.77	17.18	8.59	732.7
19	1.86	17.82	8.91	795.0
20	1.96	18.46	9.23	860.0
21	2.05	19.09	9.55	926.8
22	2.18	19.73	9.86	995.5
		Restrict Breeder Every Other Day	Restrict Breeder Every Day	
23	2.32	20.36	10.18	1066.8
24	2.50	21.00	10.50	1140.5

Table 32-8. Estimated Protein and Metabolizable Energy Consumed by Meat-Type Growing Pullets (Moderate Temperature)

Week	Ration	Desired Body Weight End of Wk		Feed Consumed per Bird per Day	Protein Consumed per Bird per Day	ME Consumed per Bird per Day	Kcal ME Consumed per Unit of Body Weight per Day	
		lb	kg	g	g	kcal	lb	kg
1	**Starter**	0.3	0.14	15.9	3.2	45	149	67.7
2	20% protein	0.5	0.23	25.0	5.0	70	140	63.6
3	1275 kcal ME	0.9	0.41	40.9	8.2	115	128	58.2
4	per lb	1.1	0.50	43.2	8.6	121	110	50.0
5	**Grower I**	1.3	0.59	47.7	8.6	137	105	47.7
6	18% protein	1.5	0.68	49.9	8.1	143	95	43.2
7	1300 kcal ME	1.7	0.77	52.2	9.4	150	88	40.0
8	per lb	1.9	0.86	54.9	9.9	157	83	37.7
9		2.1	0.96	57.7	10.4	165	79	35.9
10	**Grower II**	2.3	1.05	60.4	9.1	173	75	34.1
11	15% protein	2.5	1.14	63.6	9.5	182	73	33.2
12	1300 kcal ME	2.7	1.23	66.7	10.0	191	71	32.3
13	per lb	2.9	1.32	69.9	10.5	200	69	31.4
14		3.1	1.41	73.1	10.9	209	68	30.9
15		3.3	1.50	76.3	11.4	218	66	30.0
16	**Grower III**	3.5	1.59	79.5	9.5	228	65	29.6
17	12% protein	3.7	1.68	82.6	9.9	237	64	29.1
18	1300 kcal ME	3.9	1.77	85.8	10.3	246	63	28.6
19	per lb	4.1	1.86	89.0	10.7	255	62	28.2
20		4.3	1.96	92.2	11.1	264	61	27.7
21		4.5	2.05	95.3	11.4	273	60	27.3
22		4.8	2.18	98.5	11.8	282	59	26.8
23	**Breeder**	5.1	2.32	101.7	16.2	291	57	25.9
24	16% protein 1300 kcal ME per lb	5.5	2.50	104.9	16.8	300	56	25.5

Note: Constructed from Tables 32-6 and 32-7.

32-H. IMPROVED SKIP-DAY MEAT-TYPE GROWING FEED PROGRAMS

Difficulties with Skip-Every-Other-Day Feeding Program

Although skip-every-other-day feeding has been the most popular program in recent years, there have been some problems with it. As an explanation, this program calls for feeding 2 day's supply of restricted feed allotment with no feed the next. Birds are very hungry following 1 day without anything to eat and are capable of consuming large quantities of feed in a short period of time, thus tending to gorge themselves when feed is dispersed. Because of this gorging, the crop and gizzard enlarge, and the birds not only can eat a larger amount of feed but can gorge themselves more, and the process is repeated over and over during the growing period.

Once the feed is consumed on feed days the birds spend their time drinking, and drink even more on the days no feed is given, in order to produce a degree of satiety to the digestive tract. This additional water results in loose droppings and wet litter. The only remedial measure is to restrict the *availability* of water on feed and no-feed days.

Be sure to read Chapter 17-F detailing the program for water restriction.

Another criticism of the program is that the amount of feed allocated on feed days after the birds are about 18 weeks of age is more than they need or will eat. Consequently, the growing pullets leave some feed in the troughs or pans, to be eaten the next day—a no-feed day. Evidence shows that when birds gorge themselves with feed an increased amount passes the digestive tract undigested, resulting in poor feed efficiency. Thus, there is more variation in body weights; there are more larger and more smaller birds.

Two Choices for Improving Skip-Day Feeding Programs

1. *Change to feed-every-day program.* When the birds begin to leave feed in the trough or pan at the end of a feeding day (about 18 weeks of age), change to a feed-every-day *restricted* feeding program. See Tables 32-6 and 32-7 for daily feed allocations starting at this age.

2. *Use decreased no-feed days per week program.* With this program the starter is self-fed for the first 3 weeks, then restricted for the next 3 weeks. A 15% protein grower is fed beginning with the 7th week (43 days) through the 22nd week (154 days) using a different skip-day program, first skipping feeding every other day, then gradually increasing the feeding days to 5 days per week, and no feed on 2 days by the end of the 22nd week (see Table 32-9). This program lowers feed gorging and reduces the daily feed intake so that birds clean up their *allotment* of feed each day.

32-I. DAILY RESTRICTED FEEDING MEAT-TYPE GROWING PROGRAM

This program is fast gaining in popularity, but it calls for specialized flock management. Use the same rations as shown in Table 32-8: Starter, Grower I, Grower II, Grower III, and Breeder, but feed and restrict them

Table 32-9. Meat-Type Growing Pullet Feeding Program: Decreased No-Feed Days per Week

Weeks of Feeding	Type of Feed			Feeding Program (Limited Feeding)
	Name	Protein %	ME per lb Kcal	
1–3	Starter	19	1,380	Self feed every day
4–6				Restrict, but feed daily
7–11	Grower	15	1,335	Skip every other day (e.g., 3½ days per wk)
12–19				Feed 2, skip 1 day (e.g., 4⅔ days per wk)
20				Feed 5, Skip 2 days (e.g., skip Sun and Wed)
21–22	Breeder	16	1,320	
23–24				Self feed every day

Source: Arbor Acres Farm, Glastonbury, Conn.

daily. During the 3rd week, sample weighing of the pullets must be instigated for bird weight determines the daily feed allocation. If the pullets meet their target weight (see Tables 32-6 and 32-7) or similar figures furnished by the primary breeder, continue with the recommended daily feed allocations in the table. When the weekly average weight is below the target weight, increase the daily feed allocations; when above, reduce the daily feed allocations.

Daily Feed Restriction When Feeding in Blackout Houses

Inasmuch as separate houses for pullets and cockerels are used with blackout houses, restricted feeding of males and females is easily accomplished (see Figure 17-1 and Chapter 32-K).

Weekly Feed Consumption Does Not Change

Regardless of how many feed days are skipped, or which feeding program is used, the *weekly* feed consumption and standard weekly body weights remain the same as those shown in Tables 32-6 and 32-7.

Feeding Grit During Growing

On a feed day, feed 1 lb (454 g) of large-size grit per week per 100 birds raised on litter. Do not feed grit on a no-feed day because the birds will gorge themselves with grit.

When the birds are on slats, plastic, or wire, feed 1 lb (454 g) of grit per 100 birds every 6 weeks, and do not feed on a no-feed day. Start the grit feeding the 7th week in the litter or from separate grit feeders. Never feed grit in automatic feeders.

32-J. IMPORTANCE OF PULLET BODY WEIGHT UNIFORMITY DURING GROWING

It is necessary to begin sample weighing of the growing breeder flock during the 3rd week and weigh every week thereafter. During the 3rd and 4th weeks weigh five pullets at a time. Weigh the birds in the afternoon on the same day each week. When a skip-a-day feeding program is used, weigh on a no-feed day.

Beginning the 5th week, individually weigh the sample of birds using a scale with no greater than 1 oz (28 g) graduations and determine the average weight; then compare this figure with the recommended weight. Not only should the flock average pullet weight meet the standard but flock uniformity must be high. It may be a better measure of a quality pullet flock than average weight.

Uniformity of the pullet flock is best measured by determining the percentage of pullets within 10% (plus or minus) of the average weight of the birds in the sample. Degree of flock uniformity may be measured according to the following weight variance:

Terminology	% of Pullets within 10% of Flock Average Weight
Superior	81 and above
Excellent	77–80
Good	73–76
Average	69–72
Fair	65–68
Poor	61–64
Very poor	60 and below

32-K. FEEDING MEAT-TYPE GROWING COCKERELS

Meat-type cockerels, to be used as broiler breeders have standard weekly weights, and it is just as important that male weights be maintained as the weights of the growing pullets. To get these weights, the cockerel-growing feed must be restricted. In past years restriction was impossible when the cockerels were raised with the pullets that were on a restricted feeding program, since the robust males pushed the females away from the feeders. But with the advent of blackout housing (see Chapter 17-H) the males are raised in separate houses and the feed is controlled, so it is relatively easy to maintain the correct cockerel weight during growing.

Weigh a sample of the cockerels at the same time (age) the pullets are weighed, using the same system. Table 32-10 gives the target weights for growing males when the feed is restricted. Weights are given for in-season and out-of-season hatches. At 24 weeks of age the males should average about 35% heavier than the females (see Table 17-4).

Table 32-10. Weights of Meat-Type Cockerels Fed on a Restricted Feeding Program (Moderate Temperature)

Week of Age	Guidelines for Approximate Male Body Weights			
	Aug–Jan Hatches		Feb–July Hatches	
	lb	kg	lb	kg
1	0.31	0.14	0.33	0.15
2	0.53	0.24	0.57	0.27
3	0.98	0.45	1.10	0.49
4	1.2	0.54	1.3	0.59
5	1.4	0.64	1.6	0.73
6	1.7	0.77	1.9	0.86
7	1.9	0.86	2.1	0.95
8	2.2	1.00	2.4	1.09
9	2.4	1.09	2.6	1.18
10	2.7	1.22	2.9	1.32
11	3.0	1.36	3.2	1.45
12	3.2	1.45	3.4	1.54
13	3.5	1.59	3.7	1.68
14	3.8	1.72	4.0	1.81
15	4.1	1.86	4.3	1.95
16	4.4	2.00	4.6	2.09
17	4.6	2.09	4.8	2.18
18	4.8	2.18	5.2	2.36
19	5.1	2.31	5.5	2.50
20	5.4	2.45	5.8	2.63
21	5.7	2.59	6.1	2.77
22	6.2	2.81	6.6	2.99
23	6.7	3.04	7.1	3.22
24	7.2	3.27	7.6	3.45

Note: Data are for the Northern Hemisphere. Reverse for Southern Hemisphere.

Growing Management Programs
for Meat-Type Breeders

There are four meat-type rearing programs, the fourth involving blackout houses.

1. *Cockerels separate to 7 days.*   Coming from small eggs, small meat-line cockerel chicks should be started within guards under separate brooders using the same feed and feeding program for the cockerels and pullets. The earliest that cockerels should be mixed with pullets is 7 days.

2. *Cockerels separate to 10 weeks.*   During the first 7 days, keep the cockerel chicks under separate brooders confined to one part of the house by a high fence. Both cockerels and pullets should get a ration with the same formula.

   At 10 weeks of age, mix the cockerels with the pullets. Feed the pullets and cockerels the same feed in the same room, using male exclusion grills on the automatic feeders so that the daily feed allotment of males and females can be independently controlled.

   Meat-type males should not be fed low-protein diets to reduce weight, particularly before 8 weeks of age, as such a practice reduces fertility later.

3. *Cockerels separate to 20 weeks.*   Formerly, this program was the best of all growing ones; growth rate of each sex could be accurately controlled by feed allocation.

   At 20 weeks of age, move the cockerels with the pullets after dark to reduce male fighting. Continue with the grower feed until the birds are 20 weeks of age or are up to standard weight; then feed a breeder feed.

4. *Feeding in blackout houses.*   With this program the sexes must be raised in *separate* houses that are environmentally controlled with forced-air ventilation, cooled, and capable of being fully blacked out. See Chapter 17-H for details.

   Full feed males and females a 20% protein starter ration for 5 days; then go to controlled feeding of the starter using *daily* feed restriction rather than a skip-day program. This procedure is easily accomplished because the males and females are in separate houses.

   At 4 weeks of age, change from the Starter to Grower Ration I, shown in Table 32-5, changing to Grower II at 9 weeks, and Grower III at 15 weeks. Continue to restrict the daily feed allotments according to standard weekly body weight.

This program will induce chicks hatched out of season to come into egg production earlier, resulting in smaller eggs at the start of production. To prevent this occurrence, care should be taken to see that pullets on this program do not become sexually mature (first eggs) too early. But cockerels should start mating when the first eggs are laid (during the 22nd week of age).

Change to controlled feeding of a *breeder* feed when the first egg is laid.

32-L. FEEDING BROILER BREEDERS DURING THE CHANGEOVER PERIOD

From the end of the growing period until the flock is well into egg production is the *changeover period.* It is now recognized as an exceptionally important period as there are many changes in management, lighting, and feeding that are most critical to the bird and her future egg production.

Body Weight Variability at Sexual Maturity

For simplicity, it has been stated that the female body weight should average 5.5 lb (2.5 kg) at 24 weeks of age. But on a seasonal basis this weight is variable, as shown in the following (reverse for Southern Hemisphere):

	Female Average Body Weight at Sexual Maturity (Northern Hemisphere)	
Month of Hatch	lb	kg
Aug	5.5	2.50
Sept–Jan	5.4	2.45
Feb	5.3	2.41
Mar	5.4	2.45
Apr	5.6	2.55
May, June	5.7	2.59
July	5.6	2.55

Although hatching date is used as the basis for the above variations in body weight at sexual maturity, it is the changing length of the light day and temperature during growing that are the cause. This variability, caused by differences in hatching date, necessitates changes in the feeding program during the changeover.

Normally, the largest birds in the flock are the first to produce eggs.

About a week before a pullet starts to lay her body weight begins to increase rapidly. Between this time and 1 week after she lays her first egg she should gain about 0.5 lb (227 g) or 10%. During the next 8 to 10 weeks she should gain a similar amount.

At the time an individual pullet lays her first egg she should weigh between 5.5 and 6.0 lb (2.5 and 2.7 kg), depending on the month she was hatched. As the smaller birds reach sexual maturity, they too will attain a weight that approaches the weight of the first birds to lay. But there are still large, medium, and small birds in the flock, and always will be. The largest birds remain large; the smallest birds remain small.

First Week of Flock Egg Production

Even though today's breeder flocks are best brought into 5% hen-day egg production at 24 weeks of age, there will be variability because of hatching date, season, strain, temperature, ration, feeding program, etc., so flocks may vary 2 or 3 weeks from this age. Because of this variability, further feeding recommendations during the changeover period must be geared to the time at which the flocks begin egg production rather than age.

The common base for early egg production is the day when the flock first averages 5% hen-day egg production. About 8% of the flock will be in production at this time.

Timetable for Feeding and Management Changes

The basic schedule is as follows, but this can only be a guide. Individual flocks still will need some adjustment in the schedule.

> −14 days: Flock should lay first eggs (about 1% production). Change from grower to breeder ration.
>
> −12 days: Increase the length of the light day. It should approximate 14 hours with this change.
>
> −7 days: Feed 3 lb (1.3 kg) of flaked oystershell per 100 birds per week by feeding the shell on top of the mash.
>
> Note: Do not use this procedure prior to this time as early calcium feeding may permanently impair the function of the parathyroid gland, necessary for calcium assimilation during egg production. Feed flaked oystershell for 4 weeks only.
>
> 0 day: The flock first reaches 5% hen-day egg production.
>
> +1 day: Some feed restriction must be used during the changeover period. Change to daily feeding. If pullets have been on a skip-day feeding program during grow-

ing select a day on which no feed is normally given, compute the average daily amount fed per 100 birds per day during the previous week and increase this figure by 3 lb (1.4 kg). Include both males and females in the count. Feed this daily amount for 1 week.

+ *8 days:* Increase the daily feed allotment by 4 lb (1.8 kg) per 100 birds when the temperature is moderate, 4.5 lb (2.0 kg) when the temperature is cold, and 3.5 lb (1.6 kg) when the temperature is hot.

+ *15 days:* Continue to increase the daily feed allotments *each week* by amounts similar to those made the previous week as the feed allotment is consumed by mid-afternoon. Birds should be eating between 34 and 39 lb (15.4 and 17.7 kg) of feed per 100 birds per day at peak of pullet egg production, depending on the strain, rate of lay, egg size, body weight, and ambient temperature. At peak of production the flock should be consuming about 10% less feed than if full fed.

+ *5 weeks:* Increase the length of the light day to 15 or 15½ hours about 2 weeks prior to the time the flock will peak in egg production, where it should remain throughout the laying period (see Chapter 18-H).

Important: If pullets are overweight at the time first eggs are produced by the flock do not make additional daily feeding reductions during the changeover period to try to reduce the weight. Birds in such flocks should always remain heavier than normal, including the entire laying period.

If pullets are underweight at start of lay, the feed allotment should be increased in order to bring the pullets up to their standard weight.

32-M. NUTRITIONAL REQUIREMENTS OF BREEDERS DURING EGG PRODUCTION

Energy Requirements During Egg Production

The energy requirement of the breeder diet for egg-type strains is slightly higher than diets for egg production alone. Meat-type strains tend to get too heavy if full fed during laying as well as during growing. Besides, meat-type breeders produce fewer eggs, particularly after they have been laying for several weeks. For these reasons the energy content of the meat-type breeder ration is usually lower than that of egg-type breeder formulas. The comparison is shown below:

	Kcal of ME in Ration	
Ration	per lb	per kg
Egg-type layer	1,300	2,860
Egg-type breeder	1,350	2,970
Meat-type breeder	1,300	2,860

Even then, feed should be restricted when meat-type layers are in egg production. The procedure is given in Chapter 32-N, and is very specific.

Protein Requirement

As with other types of rations, the percentage of protein in the egg-type breeder ration must be governed in part by its energy content (see Table 31-8). An average protein content for a breeder formula should be about 17% for egg-type layers and 16% for meat-type, but is subject to some variation according to the environmental temperature, caloric content of the diet, rate of egg production, size of birds, and so forth.

Amino acid requirements. Most of these are similar or slightly more than those required for egg production alone (see Table 31-7).

Mineral Requirements

As with the bird's need for large amounts of calcium for egg production, there is a similar requirement in its need for calcium for the production of quality hatching eggs. This, along with other mineral requirements, are given in Table 32-11.

Table 32-11. Mineral Requirements of Breeder Rations (Moderate Weather)

| | Leghorn | | Medium-size[1] | | Meat-type | |
Mineral	21–40 wk	Over 40 wk	21–40 wk	Over 40 wk	23–40 wk	Over 40 wk
Calcium (%)	3.50	4.00	3.25	3.75	3.00	3.50
Phosphorus (total %)	0.5	0.5	0.5	0.5	0.5	0.5
Sodium (%)	0.15	0.15	0.15	0.15	0.15	0.15
Potassium (%)	0.1	0.1	0.1	0.1	0.1	0.1
Manganese (mg/lb)	30	30	30	30	30	30
Manganese (mg/kg)	65	65	65	65	65	65
Zinc (mg/lb)	30	30	30	30	30	30
Zinc (mg/kg)	65	65	65	65	65	65
Selenium (mg/lb)	0.045	0.045	0.045	0.045	0.045	0.045
Selenium (mg/kg)	0.1	0.1	0.1	0.1	0.1	0.1

[1]Producing brown-shelled eggs.

Vitamin Requirements

Table 32-12 gives the vitamin requirements for hatching egg production. In order to secure good hatchability, the breeder feed requirement is greater for many vitamins than for laying diets.

Table 32-12. Vitamin Requirements of Breeder Rations

Vitamin	Amount per Unit of Feed[2]	
	per lb	per kg
Vitamin A activity (IU)	1818	4000
Vitamin D (IU)	227	500
Vitamin E (IU)	4.6	10.0
Vitamin K (mg)	0.2	0.5
Thiamin (mg)	0.36	0.8
Riboflavin (mg)	1.73	3.8
Pantothenic acid (mg)	4.55	10.0
Niacin (mg)[1]	4.55	10.0
Pyridoxine (mg)	2.05	4.5
Biotin (mg)	0.068	0.15
Choline (mg)	?	?
Vitamin B_{12} (mg)	0.0019	0.004

Source: Nutrient Requirements of Poultry, © 1984 by the National Academy of Sciences, Washington, DC.
[1]In diet that contains 0.15% trytophan.
[2]No margin of safety.

32-N. FEEDING STANDARD-SIZE MEAT-TYPE BREEDERS AFTER PEAK OF PRODUCTION

A program of continued feed restriction should be followed during the egg production period past the peak of production. One must be sure the flock is given ample amounts of feed necessary to produce the maximum number of eggs, but not amounts excessive to the extent that the birds gain too much weight. There are two segments to the program.

1. *More eggs.* Once the peak of production is reached, challenge the flock to produce more eggs by increasing feed consumption by 2 lb (0.9 kg) per 100 birds per day for 4 days. If production rises, repeat the procedure until production no longer increases. If production remains the same after a feed increase, return to the preexisting feed level. This procedure will work only if the flock is getting about 10% less than if full fed when it is first tried.

2. *Less feed.* When the flock has passed its peak and production has dropped about 5%, challenge the flock to lower feed cost by reducing feed intake by 0.5 lb (227 g) per 100 birds per day for 4 days. If, after this feed reduction the drop in egg production is not greater than normal, continue with the new feed allotment for about 10 days; then make another similar reduction. If production drops more than normal after a feed decrease, return the flock to the preexisting feed level.

Continue this program throughout the remainder of the laying period. Do not make feed reductions during stress, disease outbreaks, sudden drops in temperature, or if body weight shows a great decrease.

Ambient Temperature and Feed Intake

Even though maximum and minimum guides have been shown for feed consumption throughout the laying period, there are times when these limits will not suffice, and the variations in ambient temperature are usually the cause. There is nothing that disrupts a feeding schedule more than temperature change. Extremes can cause variations in feed consumption of up to 40%. Variations are smaller for each degree Fahrenheit change in temperature when the weather is cool than when it is hot as seen in Table 30-12.

Because meat-type layers should be fed less than if they were on full feed, the daily feed allotment should be consumed by midafternoon. If the weather turns cold, the hen's demand for more energy means the feed supply will be consumed earlier in the day, and to offset this consumption more feed must be given. When the weather turns warmer, the daily feed allotment will be consumed later in the day or may not be completely eaten. It is then that less feed should be supplied.

> *Example:* An example of variations in feed consumption as the ambient temperature changes is given below, using Table 30-11 as the basis for the computation.
>
> Suppose the temperature is 70°F (21.1°C) and the laying flock is allocated 36 lb (16.4 kg) of feed per 100 birds per day. Variations in feed supplied at 90°F (32.2°C) and 50°F (10.0°C) are given as follows:

90°F (32.2°C)		70°F (21.1°C)		50°F (10.0°C)	
lb	kg	lb	kg	lb	kg
24.1	10.9	36.0	16.4	41.2	18.7

32-O. ESTIMATED DAILY FEED ALLOWANCES FOR STANDARD MEAT-TYPE BREEDER FLOCKS

Estimated feed allowances for standard-size meat-type breeder flocks are given in Table 32-13. A further guide for feed consumption of standard-size, meat-type breeder flocks showing the ME and protein consumed per day during egg production is given in Table 32-14.

Table 32-13.   Guide for Feed Consumption When Standard-Size Meat-Type Pullets Are Control-Fed During Egg Production

Week of Egg Production	Hen-Day Egg Production %	Feed Consumed per 100 Birds per Day		Female Body Weight	
		lb	kg	lb	kg
1	5	24–28	10.9–12.7	5.2–5.7	2.4–2.6
2	20	28–32	12.7–14.6	5.4–5.9	2.5–2.7
3	38	30–34	13.6–15.4	5.6–6.1	2.6–2.8
4	56	32–36	14.5–16.4	5.7–6.2	2.6–2.8
5	73	33–37	15.0–16.8	5.8–6.3	2.6–2.9
6	84	34–38	15.5–17.3	5.3–6.3	2.6–2.9
7	86	34–38	15.5–17.3	5.9–6.4	2.7–2.9
8	85	34–38	15.5–17.3	5.9–6.4	2.7–2.9
9	84	34–38	15.5–17.3	6.0–6.5	2.7–3.0
10	84	34–38	15.5–17.3	6.0–6.5	2.7–3.0
11	83	33–37	15.0–16.8	6.1–6.6	2.8–3.0
12	82	33–37	15.0–16.8	6.1–6.6	2.8–3.0
13	81	33–37	15.0–16.8	6.2–6.7	2.8–3.1
14	81	33–37	15.0–16.8	6.2–6.7	2.8–3.1
15	80	33–37	15.0–16.8	6.3–6.8	2.8–3.1
16	79	32–36	14.6–16.4	6.3–6.8	2.8–3.1
17	78	32–36	14.6–16.4	6.3–6.8	2.8–3.1
18	77	32–36	14.6–16.4	6.4–6.9	2.9–3.1
19	77	32–36	14.6–16.4	6.4–6.9	2.9–2.1
20	76	32–36	14.6–16.4	6.5–7.0	3.0–3.2
21	75	31–35	14.1–15.9	6.5–7.0	3.0–3.2
22	74	31–35	14.1–15.9	6.5–7.0	3.0–3.2
23	74	31–35	14.1–15.9	6.6–7.1	3.0–3.2
24	73	31–35	14.1–15.9	6.6–7.1	3.0–3.2
25	72	31–35	14.1–15.9	6.6–7.1	3.0–3.2
26	71	30–34	13.6–15.4	6.6–7.1	3.0–3.2
27	70	30–34	13.6–15.4	6.6–7.1	3.0–3.2
28	70	30–34	13.6–15.4	6.7–7.2	3.1–3.3
29	69	30–34	13.6–15.4	6.7–7.2	3.1–3.3
30	68	30–34	13.6–15.4	6.7–7.2	3.1–3.3
31	67	29–33	13.2–15.0	6.7–7.2	3.1–3.3
32	66	29–33	13.2–15.0	6.7–7.2	3.1–3.3
33	66	29–33	13.2–15.0	6.8–7.3	3.1–3.3
34	65	29–33	13.2–15.0	6.8–7.3	3.1–3.3
35	64	29–33	13.2–15.0	6.8–7.3	3.1–3.3
36	63	28–32	12.7–14.6	6.8–7.3	3.1–3.3
37	62	28–32	12.7–14.6	6.8–7.3	3.1–3.3
38	61	28–32	12.7–14.6	6.9–7.4	3.1–3.4
39	60	28–32	12.7–14.6	6.9–7.4	3.1–3.4
40	60	28–32	12.7–14.6	6.9–7.4	3.1–3.4
41	59	27–31	12.3–14.1	6.9–7.4	3.1–3.4
42	58	27–31	12.3–14.1	6.9–7.4	3.1–3.4
43	57	27–31	12.3–14.1	7.0–7.5	3.2–3.4
44	56	27–31	12.3–14.1	7.0–7.5	3.2–3.4

Table 32-14. Metabolizable Energy and Protein Consumption of Meat-Type Breeders During Egg Production

Week of Egg Production	Hen-Day Egg Production %	ME Consumed per Bird per Day kcal	Protein Consumed per Bird per Day g
1	5	312–364	17.4–20.3
2	20	364–416	20.3–23.3
3	38	390–442	21.8–24.7
4	56	416–468	23.3–26.2
5	73	429–481	24.0–26.9
6	84	442–494	24.6–27.6
7	86	442–494	24.6–27.6
8	85	442–494	24.6–27.6
9	84	442–494	24.6–27.6
10	84	442–494	24.6–27.6
20	76	416–468	23.3–26.2
30	68	390–442	21.8–24.7
40	60	364–429	20.3–23.3

Feed per Dozen Eggs Produced

It is interesting to study the variations in the amount of feed necessary to produce a dozen eggs as the pullet progresses through her laying year. These figures are given in Table 32-15 and are for hens kept on a controlled feeding program. Notice that the feed allowances per dozen eggs produced are highly variable and increase as the hen goes through her

Table 32-15. Feed to Produce a Dozen Eggs by Standard Meat-Type Breeding Hens (Controlled Feeding Program)

Week of Egg Production	Egg Production		Feed Allocation per Bird[1] per dozen Eggs[2]	
	Hen-Day %	Dozens per Day per 100 Hens	lb	kg
5	73	6.08	5.8	2.64
10	84	7.00	5.1	2.32
15	80	6.67	5.3	2.41
20	76	6.63	5.4	2.46
25	72	6.00	5.5	2.50
30	68	5.67	5.6	2.55
35	64	5.33	5.8	2.64
40	60	5.00	6.0	2.72
44	56	4.60	6.2	2.82

[1]No male feed included. Add 10% if male feed is to be included.
[2]Moderate temperature.

production cycle, but not as rapidly as her rate of egg production decreases. A great deal of the variation is due to changes in body weight.

32-P. FEEDING SCHEDULE FOR MEAT-TYPE MINI-BREEDER FEMALE

Estimated feed allocations for Mini-meat-type breeders are given in Table 32-16. Although Mini-meat-type females are mated with Standard-size meat-type males, count both the males and the females in the flock when calculating bird number.

32-Q. FEEDING THE MEAT-TYPE MALE DURING THE BREEDING PERIOD

In the past, recommendations for feeding and managing the meat-type breeding flock have been established for the female only; the male has been neglected. While the pullet's feed has been restricted, the cockerels have had all the feed they could eat. They became obese, and after a few weeks developed foot and leg problems that decreased their reproductive performance, which got progressively worse.

Dual Feeding System Needed

Dr. G. R. McDaniel of Alabama Experiment Station has shown that the remedy to the problem of male overweight is to feed two separate rations during the breeding season, known as *dual feeding*. The pullets are fed their regular ration, but the cockerels are fed a ration low in protein, energy, and calcium. The main differences in the two rations are shown in Table 32-17. In order to be sure the pullets and cockerels get their respective rations, two automatic independent feeding systems must be installed in the same house.

Specifics for the Pullet Feeding Program

Either trough-and-chain or trough-and-pan automatic feeders may be used. To prevent cockerels from eating from the pullet troughs or pans, male exclusion grills must be placed on the pullet feeders. See Figure 32-1 for an exclusion grill on a trough.

If the house is slat-and-floor type, place the pullet feeder on the slats with the bottom of the feeder trough or pan 1 in. (2.5 cm) above the slats. Pullets do not readily eat through an exclusion grill when the feeder is higher. The male exclusion grill should have an opening $1\frac{5}{8}$ in. (4.1 cm) wide through which the pullets can eat, but excluding the cockerels. Pro-

Table 32-16. Estimated Daily Feed Allocations for 100 Mini-Meat-Type Breeder Pullets (Restricted Feeding Program)

| Week of Age | Desired Avg Female Weight | | Approximate Feed Amounts per 100 Birds on Feed Days | | Cumulative Feed per 100 Birds | |
| | | | Skip-every-other-day Program | | | |
	lb	kg	lb	kg	lb	kg
			Full-Feed Starter Every Day			
1	0.25	0.11	3.0	1.36	21	9.6
2	0.45	0.21	4.7	2.14	54	24.6
			Restrict Starter Every Day			
3	0.70	0.32	7.2	3.27	105	47.7
4	0.95	0.43	7.6	3.46	158	71.8
			Restrict Grower Every Other Day[1]			
5	1.15	0.52	16.8	7.64	217	98.6
6	1.30	0.59	17.6	8.00	278	126.4
7	1.45	0.65	18.4	8.36	343	155.9
8	1.60	0.72	19.4	8.82	411	186.8
9	1.75	0.80	20.3	9.23	482	219.1
10	1.90	0.86	21.3	9.68	557	253.2
11	2.05	0.93	22.4	10.18	653	296.8
12	2.20	1.00	23.5	10.68	717	325.8
13	2.35	1.07	24.6	11.18	803	365.0
14	2.50	1.14	25.8	11.73	894	406.4
15	2.65	1.21	26.9	12.23	987	448.6
16	2.80	1.27	28.0	12.73	1085	493.2
17	2.95	1.34	29.1	13.23	1188	540.0
18	3.10	1.41	30.2	13.73	1293	587.7
19	3.25	1.48	31.4	14.27	1403	637.5
20	3.40	1.55	32.5	14.77	1517	689.6
21	3.55	1.61	33.6	15.27	1635	743.2
			Restrict Breeder Every Other Day[1]			
22	3.70	1.68	34.7	15.77	1754	797.3
23	3.85	1.75	35.8	16.27	1879	854.1
			Restrict Breeder Every Day[2]			
24	4.00	1.82	19–21	8.6– 9.6		
25	4.20	1.91	22–25	10.0–11.4		
26	4.35	1.98	24–27	10.9–12.3		
27	4.50	2.05	26–29	11.8–13.2		
28	4.65	2.11	27–30	12.3–13.6		
29	4.80	2.18	27–30	12.3–13.6		
30	4.90	2.23	27–30	12.3–13.6		
40	5.10	2.31	26–29	11.8–13.2		
50	5.30	2.40	25–28	11.4–12.7		
60	5.40	2.45	24–27	10.9–12.3		

[1]When feed is restricted and fed every day, divide the following figures by two.
[2]Use a challenge feeding program similar to that given in Chapter 32–N.

Table 32-17. Pullet and Cockerel Breeder Rations for Dual Feeding (Major Feed Differences)

Item	Pullet Ration	Cockerel Ration
ME (kcal/lb)	1300	1275
ME (kcal/kg)	2860	2805
Protein (%)	16.0	12.0
Methionine (%)	0.31	0.22
Methionine + cystine (%)	0.56	0.41
Lysine (%)	0.78	0.50
Calcium (%)	3.00	0.90
Phosphorus, available (%)	0.46	0.40

Figure 32-1. Male exclusion grill on a trough feeder.
(Courtesy of Container Products, Inc., Chesterfield, Missouri, U.S.A.)

vide 6 in. (21.1 cm) of trough space per pullet or one 13-inch pan for every 11 pullets. Each day, start the pullet feeders about 15 minutes before the cockerel feeders.

Specifics for the Cockerel Feeding Program

A tube-and-pan automatic feeder must be used for the cockerel feeding system, and it may be placed on the slats or on the litter floor. It will take only 1 to 2 hours for the cockerels to consume their daily allotment of feed; therefore, it is very important that enough feeding space be available so all cockerels can eat at one time. Furthermore, when the feeder is started, feed must flow simultaneously to all pans to prevent cockerels from migrating to get feed.

When the cockerel feeders are first used, the lip of the feeders should be about 10 in. (25.4 cm) above the slats, then gradually increased to 18 in. (45 cm) by the time the pullets are added to the house. Provide one 13-inch pan for every 10 cockerels. Between feedings, some systems provide for raising the pan feeders by a winch.

Feeding and Management on the Dual System

The cockerels should be moved from the growing house to the dual-system breeding house when about 19 to 20 weeks of age, and the inferior cockerels should be culled. Change the ration to a low-protein cockerel breeder feed, which is fed in the elevated feeders at this time.

Move the pullets to the breeder house 7 to 10 days later. Continue to feed the pullets a growing feed in the pullet feeders with the male exclusion grills. When the pullets lay their first eggs (about 22 weeks of age) change them to a regular breeder feed. Cull the cockerels again, retaining about 10 males per 100 females.

Body Weight Governs Feed Allocation

It is imperative that the body weight of both pullets and cockerels be on target when in the breeding house. Adhere to the weights in Table 32-18 (or those of the primary breeder).

Both males and females should be on a program of daily feed restriction to maintain their target body weights. Do not use a skip-day pro-

Table 32-18. Target Body Weights on the Dual System of Feeding

| | Live Body Weight | | | |
| | Male | | Female | |
Age wk	lb	kg	lb	kg
20	5.6	2.54	4.3	1.96
24	7.4	3.36	5.5	2.50
30	8.5	3.86	6.0	2.73
40	9.1	4.14	6.5	2.96
50	9.5	4.32	6.8	3.09
60	9.7	4.40	7.0	3.18

gram. Sample weigh the two sexes when they are moved to the breeding house, and each week thereafter. Weigh the birds in the afternoon after they have consumed their daily supply of feed.

Increase or decrease the daily feed allotment of both males and females so as to maintain their weekly standard body weights. Do not allow males to lose weight; their weight should increase regularly, but slowly.

Improvement by Using the Dual Feeding System

According to laboratory and field tests, the dual system of feeding will accomplish the following:

1. From the first week on the program, the cockerel weights will be materially lower than when they are conventionally fed with the pullets, and the differential will be more than 2 pounds (0.91 kilos) at 48 weeks of age and more than 3 pounds (1.36 kilos) at 60 weeks of age.
2. Compared with the old system, fertility will be better from the start of egg production, the differential increasing to 8% by the end of the breeding season.
3. Hatchability will increase from the start of egg production, the differential gradually increasing to 8 to 10% at the end of the breeding period.
4. About eight more broiler chicks will be produced by each female during the breeding season.
5. The cockerels will eat less of a less expensive feed.
6. The cost of the additional equipment for the dual system will be paid back during one egg-production period.

33

Feeding Broilers, Roasters, and Capons

In all probability, more is known about the nutrition of the broiler than any other type of chicken. In the clamor for rapid growth and high feed conversion scientists have spent countless hours developing feed formulae that will produce rapid and economical gains in the broiler house.

Except in a few unusual cases, broilers are full fed from start to market. They should be encouraged to eat as much as possible, since the more they eat, the faster they grow, and the faster they grow, the better the feed conversion.

33-A. BROILER FEEDING

There are two main feeding programs: One involves two rations and is used more often when lightweight broilers are raised; the other involves three rations and is used for heavier birds. Average feeding programs for straight-run broilers are as follows:

	Average Time Period of Feeding	
Feed Name	Two-Feed Program da	Three-Feed Program da
Starter	1–21	1–14
Grower	22–market	15–39
Finisher		40–market
Drug withdrawal	last 5	last 5

Drug withdrawal period alters feeding program. In order that poultry meat will be void of drugs at the time the birds are slaughtered, many drugs used in broiler feeds have a withdrawal period of from

3 to 5 or more days prior to marketing. Often the finisher feed may also be used as the withdrawal feed by eliminating these drugs from the formula and altering the feeding period to coincide with the withdrawal period (see Chapter 36-E).

Form of Broiler Feed

Broiler feed comes in three forms.

1. *Mash.* Mash is usually used for at least 2 weeks if crumbles are not available.
2. *Crumbles.* Broilers may be started on crumbles and continued on them during the entire growing period.
3. *Pellets.* When the chicks are 2 or 3 weeks of age they will eat starter pellets in preference to mash or crumbles, and most broiler feeding programs call for pellets at this age. At about 4 weeks of age broiler finisher pellets, larger in size, should be fed.

Difficult to Measure Pelleting Response

Because of the many contributing factors, it is difficult to appraise the improvement from pelleting or crumbling a broiler feed. Variations may be itemized as follows:

1. Pelleting a high-fiber feed will show more improvement in broiler growth rate than pelleting a low-fiber feed.
2. Pelleting reduces the time the birds spend at the feeders. Although a lot depends on the density of the mash, if it takes 1 hour for broilers to eat a given weight of pellets, it will take 1.8 hours to eat the same weight of crumbles; 2.1 hours for ground pellets; and 2.4 hours for mash.
3. Pelleting per se does not increase the growth rate of broilers. It is only because they eat more that birds grow faster.
4. Up to 2 weeks, chicks eat more mash than crumbles or very small pellets, but after this age pellets are preferred.
5. When a mixture of three-fourths pellets and one-fourth mash is fed, some individual broilers show a preference for mash and some for pellets, and the average bird weight is greater than when pellets alone are fed. Although such a mixture is more costly, it does show that

some fines in pellets is not detrimental to broiler growth.

6. If pellets are too large, acceptance will be reduced. Broilers under 4 weeks of age should have pellets ⁵⁄₃₂ in. in diameter; over 4 weeks, ³⁄₁₆ in. in diameter.

Broiler Feeding Behavior

Broiler feeding behavior consists of a large number of short meals throughout the day, but age influences the interval as shown below.

Age of Broiler	Meal Interval
wk	min
2	21
4	30
6	37
8	40

Average meal length remains about the same during the growing period, ranging from 2.8 to 3.2 minutes. Females eat more frequently than males, while the male meal length is of shorter duration.

33-B. ENERGY IN BROILER RATIONS

The primary sources of energy in broiler feed are the carbohydrates and fats. However, when protein is fed in excess, it too may become a source of energy. But to feed protein for energy is uneconomical; the balance between carbohydrates, fats, and protein in the diet must be carefully constructed.

Metabolizable Energy (ME) Content of Broiler Rations

Following are recommended ME contents of broiler rations for straight-run birds that are provided a three-feed program, but these recommendations should vary with changes in the ambient temperature. They should be higher during cold weather and lower during hot weather.

	Males			Females			Straight-Run		
		kcal ME			kcal ME			kcal ME	
Feed	Days Fed	per lb	per kg	Days Fed	per lb	per kg	Days Fed	per lb	per kg
Starter	1–14	1,400	3,080	1–14	1,400	3,080	1–14	1,400	3,080
Grower	15–37	1,450	3,190	15–41	1,450	3,190	15–39	1,450	3,190
Finisher	38–mkt	1,500	3,300	42–mkt	1,500	3,300	40–mkt	1,500	3,300

Calorie Consumption of Broilers

Tables 20-10 and 21-11 show the weekly growth and feed consumption of broilers, by sexes. Calculations for the energy consumption as made from these figures are shown in Tables 33-1 (pounds) and 33-2 (kilos). Of factual interest are that

1. The daily energy requirement per bird increases as the birds age, mainly because they are larger.

Table 33-1. Energy Consumed by Growing Broilers per Day on the Three-Feed Program (in Pounds)

Avg for Week	Males[1]		Females[2]		Straight-Run[3]	
	kcal of ME Consumed		kcal of ME Consumed		kcal of ME Consumed	
	Per Bird per Day	Per lb of Body Weight per Day	Per Bird per Day	Per lb of Body Weight per Day	Per Bird per Day	Per lb of Body Weight per Day
1	54	159	52	158	52	158
2	132	148	124	149	128	149
3	213	134	201	137	207	135
4	307	125	269	122	288	124
5	381	116	352	114	367	115
6	505	114	414	110	464	112
7	617	112	510	108	564	110

[1]1–14 da: ME/lb 1,400; 15–37 da: ME/lb 1,450; 38 da–mkt: 1,500 ME lb.
[2]1–14 da: ME/lb 1,400; 15–41 da: ME/lb 1,450; 42 da–mkt: 1,500 ME/lb.
[3]1–14 da: ME/lb 1,400; 15–39 da: ME/lb 1,450; 40 da–mkt: 1,500 ME/lb.

Table 33-2. Energy Consumed by Growing Broilers per Day on the Three-Feed Program (in Kilos)

Avg for Week	Males[1]		Females[2]		Straight-Run[3]	
	kcal of ME Consumed		kcal of ME Consumed		kcal of ME Consumed	
	Per Bird per Day	Per kg of Body Weight per Day	Per Bird per Day	Per kg of Body Weight per Day	Per Bird per Day	Per kg of Body Weight per Day
1	54	350	52	347	52	348
2	132	326	124	327	128	328
3	213	295	201	301	207	297
4	307	275	269	268	288	273
5	381	255	352	251	367	253
6	505	251	414	242	464	246
7	617	246	510	238	564	242

[1]1–14 da: ME/kg 3,080; 15–37 da: ME/kg 3,190; 38 da–mkt: 3,300 ME kg.
[2]1–14 da: ME/kg 3,080; 15–41 da: ME/kg 3,190; 42 da–mkt: 3,300 ME/kg.
[3]1–14 da: ME/kg 3,080; 15–39 da: ME/kg 3,190; 40 da–mkt: 3,300 ME/kg.

2. The daily energy requirement per pound (kilo) of body weight decreases as the birds grow older.
3. When compared on an age basis, there is little difference between the males and females in the caloric requirement per pound (kilo) of bird.

Effect of Energy Value on Growth and Feed Conversion

Within limits, the energy value of a feed affects the amount eaten. The chicken has the ability to regulate its feed consumption so that it eats less of a high-energy feed and more of a low-energy feed.

From many research reports Table 33-3 has been constructed, which shows

1. Decreasing the feed energy reduces the 6-week weight.
2. Decreasing the feed energy increases the total feed consumed.
3. Total feed consumed decreases by about the same *percentage* as the caloric content of the ration increases.
4. Decreasing the feed energy results in poorer feed conversion.
5. When the energy in a feed is increased or decreased from 1,450 kcal ME per pound, the total ME consumed during the 6-week growing period is increased.
6. Lowering the feed energy below 1,450 kcal ME per pound of feed, or raising it above 1,450, increases the ME required per pound of gain.

Feed Density

The subject is partially discussed in Chapter 28-B. Results at Ohio State University showed that density of broiler rations, as measured by

Table 33-3. Response of Straight-Run Broilers to Diets Containing Different Energy Levels

ME in Ration[1]		6-wk Weight		Total Feed Consumed per Broiler		6-wk Feed Con- version	kcal ME per lb Gain	kcal ME per kg Gain	kcal ME Con- sumed per Broiler
kcal/lb	kcal/kg	lb	kg	lb	kg				
1,350	2,970	4.00	1.82	8.19	3.72	2.05	2,761	6,074	11,044
1,400	3,080	4.10	1.86	7.70	3.50	1.88	2,629	5,784	10,779
1,450	3,190	4.15	1.89	7.26	3.30	1.75	2,537	5,581	10,529
1,500	3,300	4.20	1.91	7.22	3.28	1.72	2,579	5,674	10,832
1,550	3,410	4.22	1.92	7.17	3.26	1.70	2,634	5,795	11,116
1,600	3,520	4.23	1.93	7.15	3.25	1.69	2,705	5,951	11,442

[1]Average during 6-week period.

weight per cubic foot of feed, had an effect on broiler growth. A comparison of results with mashes of many densities showed that broilers fed less dense feeds grew more slowly. The same trend was found with pelleted diets. Broilers fed pellets grew faster and reached a weight of 3.89 lb (1766 g) about 3 days earlier than those fed mash.

Ambient Temperature, Growth, and Feed Conversion

Bird growth and feed conversion are better during moderate temperatures than during hot or cold. Of special importance is that feed consumption drops off drastically as house temperatures increase. These weekly differences are fully covered in Table 20-9.

33-C. FAT IN BROILER RATIONS

Adequate fat deposition in market broilers is necessary to give a pleasing appearance to the dressed carcass and to improve the quality of the flesh, but too much fat is a detriment. Triglyceride is the major type of fat deposited in the tissues of the chicken. About 95% of the triglycerides come from the diet; 5% are synthesized. Dietary fats are delivered to the fat cells in the body as lipoproteins, and therefore they represent the limiting factor in fat deposition. Fats may leave the fat cells to reenter the blood system and be delivered to other sections of the body when the need arises. But excesses of fats are never eliminated from the body. If too much fat is consumed, the excess is deposited in the fat cells where it remains except for a small amount that may be needed by the bird when there is an increased demand for energy.

How Much Fat in Broiler Rations

The gross energy value of fat is approximately 2.25 times that of most carbohydrates (starch); therefore, fat is usually added to broiler rations in order to increase the ME value of the ration to the high levels necessary.

When fats are included in broiler rations the utilization of all consumed energy is also improved, so the value of added fat is twofold. Up to 8% of fat may be added to broiler feeds, with more being added to diets used after 4 weeks of age than prior to this age. The usual added fat percentage is 5 or 6.

The availability of fat in the diet is highly variable. Not only do fats themselves differ but age of the bird, strain, type of diet, level of fat in the diet, fat composition including free fatty acid content, and degree of saturation and fat purity produce variability in availability and the figures used herewith can only be considered averages.

Abdominal Fat in Broilers

As the level of fat in the diet increases so does body weight and percentage abdominal fat. About 60% of the total abdominal fat is in the fat pad. Most differences in abdominal fat amounts are the result of differences in growth rate. Decreasing the energy value of the ration, or increasing the percentage of protein, will increase growth rate and therefore increase the amount of abdominal fat and the size of the fat pad. But an increase in body fat is associated with poorer feed conversion because it requires more feed to produce a unit of fat than a unit of meat.

Carcass Fat and Water Relationship

When the percentage of fat in the broiler increases there is a similar decrease in body water content. The total weight of the two should approximate 76 to 79% of the live, mature body weight.

When broilers are raised at 45°F (7.2°C) about 5% of the live weight is fat (ether extract) and 72% is water. At 90°F (32.2°C), 12% will be fat and 66% will be water. The total of the two will be about 77 to 78%.

High-fat Diets During Hot Weather

Broilers consume less feed during hot weather than during cool, thereby reducing the consumption of the daily amount of protein and other feed constituents. The often-used procedure of removing fat from the dietary ration in order to cause the birds to consume more feed so as to meet the critical amino acid requirements has been shown to produce a negative growth effect.

Dietary fat has a lower heat increment, requiring less energy to be utilized by the bird compared with carbohydrates and protein. Research results have shown that fat should not be withdrawn from the ration during hot weather, but perhaps should be increased in order to dissipate more heat from the body.

Reducing Fat in Broilers

Throughout the years of commercial broiler production there has been a never-ending endeavor to grow a larger bird with better feed conversion. But with these increases the birds have carried more fat, and the fat was not evenly distributed throughout the bird.

The abdominal fat pad (leaf fat) has become so large that it has to be discarded, partially in the processing plant, but mainly by the person who cooks the chicken. Table 33-4 has been prepared to show approximations for the fat content of broilers of varying ages and sexes.

Table 33-4.   Age as It Affects Total and Abdominal Fat Percentages in Broilers

Age[1] da	Males			Females		
	Live Weight lb	Total Fat %	Abdominal Fat %	Live Weight lb	Total Fat %	Abdominal Fat %
10	0.71	7.7	—	0.65	7.6	—
23	1.98	8.7	—	1.70	8.4	—
36	3.65	9.5	2.62	3.01	9.2	3.05
47	5.45	9.9	2.64	4.36	9.6	3.30

[1]Adjusted.

Body Weight and Fat in Broilers

In Table 33-4, the data have been classified for male and female broilers according to age, and the only discerning fact is that males have a higher percentage of total fat but a lower percentage of abdominal fat than the females.

In Table 33-5, the same data have been expanded and reclassified according to weight of each sex, and the table shows that through 3 lb (1.36 kg) of weight the males and females have identical percentages of total fat. At all weights, the females have a higher percentage of abdominal fat than the males.

Table 33-5.   Body Weight as It Affects Total and Abdominal Fat in Broilers

Live Weight		Total Fat		Abdominal Fat	
lb	kg	Male %	Female %	Male %	Female %
1	0.45	7.8	7.8	2.4	2.6
2	0.91	8.6	8.6	2.5	2.8
3	1.36	9.2	9.2	2.6	3.0
4	1.82	9.6	9.7	2.7	3.2
5	2.27	9.9	10.2	2.8	3.4
6	2.73	10.1		2.9	

Light Intensity and Abdominal Fat Deposition in Broilers

Two experiments conducted by J. W. Deaton et al. (1988, *Poultry Science 67(9)*, 1239–1242) showed that light intensity of 0.2 fc (2 lx) and 4.8 fc (52 lx) had no influence on the amount of abdominal fat in either male or female broilers at 48 to 49 and 62 to 63 days of age; nor did the two light intensities have any effect on body weight, feed conversion, or mortality during the same time periods for broilers of the same sex.

Greasy Broilers

On occasion certain deposited fats and oils remain fluid in processed chilled broilers. These cause a condition known as *greasy broilers.* As such fluid fat encompasses most of the body of the broiler, the parts are difficult to coat with batter just prior to cooking. Reducing the fat content of the diet or changing the source of any added dietary fat will be of some benefit.

33-D. PROTEIN IN BROILER RATIONS

It is not the broiler's requirement for *total* protein that is important but the daily need for the *individual* amino acids. The age and sex of the broiler also alters the protein requirement. The kcal of ME per pound (kilo) of ration affects the protein requirement: The higher the ME, the greater the protein percentage required.

Age and Sex as They Affect Dietary Protein

Theoretically, the diet of the broiler should contain about 24% protein the first 2 weeks, and should decrease thereafter as follows:

Week	Protein in Ration %
1	24
2	24
3	23
4	22
5	21
6	20
7	19

However, it is not practical to make so many changes in the diet. A three-feed program pretty well equalizes the necessary protein requirement during the starting, growing, and finishing periods, and matches the feeding schedule involved with the ME program, as follows:

Feed	Males Days Fed	Males Protein %	Females Days Fed	Females Protein %	Straight-Run Days Fed	Straight-Run Protein %
Starter	1–14	24	1–14	24	1–14	24
Grower	15–37	21	15–41	20	15–39	21
Finisher	38–mkt	18.5	42–mkt	18	40–mkt	18.5

Protein Consumption of Broilers

When Tables 20-10 and 20-11 are used as the standard for feed consumption and growth, and the three-feed formulae above denote the protein percentages, the grams of protein consumed per day and per pound (kilo) of live broilers are shown in Tables 33-6 and 33-7. These two tables show that

1. The grams of protein consumed per male, female, and straight-run broiler increase markedly as the chicks age, and the increases are greater for the males than for the females because of the more rapid growth.

Table 33-6. Protein Consumed by Growing Broilers per Day (per Pound of Body Weight)

	Males		Females		Straight-Run	
	Grams of Protein Consumed		Grams of Protein Consumed		Grams of Protein Consumed	
Week	Per Bird per Day	Per lb of Body Weight per Day	Per Bird per Day	Per lb of Body Weight per Day	Per Bird per Day	Per lb of Body Weight per Day
1	4.5	13.2	4.1	12.4	4.3	12.8
2	10.3	11.6	9.7	11.2	10.0	11.4
3	14.0	8.8	13.2	9.0	13.6	8.9
4	20.2	8.2	17.7	8.1	18.9	8.2
5	25.1	7.4	22.1	7.3	23.6	7.4
6	31.2	7.0	24.7	6.4	28.0	6.8
7	35.5	6.4	29.3	6.2	32.9	6.3

Table 33-7. Protein Consumed by Growing Broilers per Day (per Kilo of Body Weight)

	Males		Females		Straight-Run	
	Grams of Protein Consumed		Grams of Protein Consumed		Grams of Protein Consumed	
Week	Per Bird per Day	Per kilo of Body Weight per Day	Per Bird per Day	Per kilo of Body Weight per Day	Per Bird per Day	Per kilo of Body Weight per Day
1	4.5	29.0	4.1	27.3	4.1	27.3
2	10.3	25.5	9.7	24.6	10.0	21.2
3	14.0	19.4	13.2	19.8	13.6	19.6
4	20.2	18.0	17.7	17.8	18.9	18.0
5	25.1	16.3	22.1	16.1	24.1	16.5
6	31.2	15.4	24.7	14.1	29.0	15.4
7	35.5	14.1	29.3	13.6	31.4	13.4

2. The grams of protein consumed per pound (kilo) of body weight decrease as the birds age. A comparison of protein consumption is different, however, when sexes of the same weight are compared. Grams of protein consumed per day are only slightly greater for males than for females of the same weight (see Table 33-8).

Table 33-8. Protein Consumed Per Day per Unit of Body Weight by Broilers

| Body Weight | | Grams of Protein Consumed per Day per Unit of Body Weight | | | |
| | | Males | | Females | |
lb	kg	Per lb Body Weight	Per kg Body Weight	Per lb Body Weight	Per kg Body Weight
1	0.46	12.0	26.4	11.6	25.5
2	0.91	9.1	20.0	8.7	19.1
3	1.36	8.5	18.7	8.2	18.0
4	1.82	7.9	17.4	7.3	16.1
5	2.27	6.2	13.6	5.6	12.3

Calorie:Protein Ratio

There is a definite relationship between the kcal of ME and the protein requirement of the growing broiler. This relationship is known as the calorie:protein ratio, calculated as follows:

1. *Per pound basis*

$$\frac{\text{kcal ME/lb ration}}{\% \text{ Protein}} = \text{Calorie:Protein Ratio}$$

Example: If the kcal ME/lb of ration is 1,400, and the protein is 22%, the formula shows

$$\frac{1400}{22} = 63.6$$

2. *Per kilo basis*

$$\frac{\text{kcal ME/kg ration}}{\% \text{ Protein}} = \text{Calorie:Protein Ratio}$$

Variations in the calorie:protein ratios. The calorie:protein ratio increases with the age of the broiler because older birds require more calories of energy and less protein in their diet. The ratios are shown in Table 33-9 for male, female, and straight-run broilers for the three different broiler feeds on a pound basis.

Important: If either the kcal of ME or the percentage of protein of the feed formula is altered, then the remaining one must be adjusted so the calorie:protein ratio remains the same. If this adjustment is not done then calories, protein, or both are wasted.

Table 33-9.   Calorie:Protein Ratios for Three Broiler Feeds (Pound Basis)

Ration	Male			Female			Straight-Run		
	Days Fed	ME/lb % Pro	Cal:Pro Ratio	Days Fed	ME/lb % Pro	Cal:Pro Ratio	Days Fed	ME/lb % Pro	Cal:Pro Ratio
Starter	1–14	1,400 24	58.3	1–14	1,400 24	58.3	1–14	1,400 24	58.3
Grower	15–37	1,450 21	69.1	15–41	1,450 20	72.5	15–39	1,450 21	69.1
Finisher	38–mkt	1,500 18.5	81.1	42–mkt	1,500 18	83.3	40–mkt	1,500 18.5	81.1

33-E.   AMINO ACID REQUIREMENTS OF BROILERS

The requirements for some critical amino acids are shown in Table 33-10. Most primary breeders and others publish more complete lists.

Table 33-10.   Amino Acid Requirements of Broiler Rations

Amino Acid	Ration		
	Starter	Grower	Finisher
Arginine (%)	1.44	1.20	1.00
Glycine + Serine (%)	1.50	1.00	0.70
Lysine (%)	1.20	1.00	0.85
Methionine (%)	0.50	0.38	0.32
Methionine + cystine (%)	0.93	0.72	0.60
Tryptophan (%)	0.23	0.20	0.17

Source: Adapted from *Nutrient Requirements of Poultry,* © 1984 by the National Academy of Sciences, Washington, DC.

33-F.   VITAMIN REQUIREMENTS OF BROILERS

The vitamin requirements of broiler rations are given in Table 33-11 (see also Chapter 26-A).

Table 33-11. Vitamin Requirements of Broiler Rations

| | Age of Broilers in Days | | | |
| | 0–21 | | 22–Mkt | |
Vitamin	Per lb	Per kg	Per lb	Per kg
Vitamin A (IU)	682.	1,500.	682.	1,500.
Vitamin D_3 (IU)	90.9	200.	90.9	200.
Vitamin E (IU)	4.6	10.0	4.6	10.0
Vitamin K (mg)	0.23	0.5	0.23	0.5
Thiamin (mg)	0.8	1.8	0.8	1.8
Riboflavin (mg)	1.6	3.6	1.6	3.6
Pantothenic acid (mg)	4.6	10.0	4.6	10.0
Niacin (mg)	12.3	27.0	12.3	27.0
Pyridoxine (mg)	1.4	3.0	1.4	3.0
Biotin (mg)	0.07	0.15	0.07	0.15
Choline (mg)	591.	1,300.	386.	850.
Vitamin B_{12} (mg)	0.004	0.009	0.004	0.009

Source: Adapted from *Nutrient Requirements of Poultry,* © 1984 by the National Academy of Sciences, Washington, DC.

33-G. MINERAL REQUIREMENTS OF BROILERS

The mineral requirements of broiler rations are given in Table 33-12.

Table 33-12. Mineral Requirements of Broiler Rations

| | Age of Broilers in Days | | | | | |
| | 0–21 | | | 22–Mkt | | |
Mineral	%	Per lb	Per kg	%	Per lb	Per kg
Calcium (%)	0.95			0.90		
Phosphorus, total (%)	0.75			0.67		
Phosphorus, available (%)	0.45			0.40		
Salt (%)	0.35			0.35		
Sodium (%)	0.15			0.15		
Potassium (%)	0.40			0.35		
Manganese (mg)		27.	59.		27.	59.
Magnesium (mg)		273.	600.		273.	600.
Selenium (mg)		0.07	0.15		0.07	0.15
Zinc (mg)		18.	40.		18.	40.

Source: Adapted from *Nutrient Requirements of Poultry,* © 1984 by the National Academy of Sciences, Washington, DC.

33-H. OTHER DIETARY SUPPLEMENTS

Following are some other dietary supplements used for broiler feeding (see also Chapter 26-F).

Antibiotics. Although most antibiotics are used for disease prevention and treatment, there are several that have growth-promoting properties when fed continuously at low levels. Feed levels should be those recommended by the manufacturer.

Coccidiostat. A coccidiostat is usually added to broiler rations (see Chapter 37-R).

Antioxidant. Ethoxyquin may be added to broiler diets to help prevent the appearance of encephalomalacia, and it or another antioxidant must be added according to the manufacturer's directions.

Xanthophylls. Although natural feedstuffs may contain xanthophylls in a quantity ample to properly color the skin and shanks of broilers, on occasion it may be necessary to add other sources of these pigmenters to the ration (see Chapter 33-J).

Flavor producer. En-hance®, a flavor producer composed of natural ingredients, has been shown to improve flavor, tenderness, and moisture content of chicken meat when it is added to broiler feed at the rate of 3 lb per ton (2,000 lb) and fed for 10 days prior to slaughter (C. J. Wabeck and J. L. Heath, *Poultry Sci. 61:* 719).

Sand in the broiler diet. Common builders sand may be added to broiler rations at a level up to 6% provided the starter diet contains 23% protein and 1,384 to 1,454 kcal of ME per pound (3,050 to 3,200 kcal/kg) of ration. Such additions will not adversely affect the body weight. Six percent may seem like a lot, but because of the heavy weight of sand, it is very little on a volume basis. The addition of sand to the broiler diet is said to improve feed efficiency.

> *Equipment wear:* Sand in the ration increases equipment wear and should be used cautiously. Some feed the sand in feeder lids prior to 3 weeks of age.

Grit Feeding Schedule for Broilers

The value of feeding grit to broilers is highly debatable, but some growers feel that there are advantages. If grit is to be fed, give the following allocations:

Age of Bird da	Grit Size	Amount per Week per 1,000 Birds	
		lb	kg
1–14	Fine	5	2.3
15–35	Coarse	10	4.5
After 35		None	None

The grit should be insoluble and should be fed in the feeder lids at first, then in the litter. Do not feed grit in automatic feeders as it is too abrasive. To void the gizzard of grit prior to processing do not feed grit after broilers are 35 days of age, or 2 weeks prior to slaughter.

33-I. BROILER RATIONS

Typical broiler starter, grower, and finisher rations are given in Table 29-4.

33-J. FEEDING FOR BROILER SKIN COLOR

The color of yellow-skin chickens is due almost entirely to a group of chemicals known as *xanthophylls,* substances closely related to the carotenoids (see Chapter 31-I). Not only are xanthophylls easily oxidized from natural feed ingredients but the fat and skin of the chicken also lose their yellow color in this manner. To maintain any skin color the quantity of xanthophylls consumed must approximate that lost from the bird.

There are many xanthophylls capable of imparting a yellow-orange color to the skin of chickens, all classified as hydroxycarotenoid pigmenting compounds. Alfalfa leaf meal is a source of *lutein.* Yellow corn is an excellent source of several xanthophylls, but corn contains mostly *zeaxanthin.* The petals of a marigold species, *Tagetes erecta,* are a very potent source of xanthophylls. Another is algea, mostly *Spongiococum excentricum.* A synthetic carotenoid, beta-apo-8'-carotenal, produces a skin color similar to the xanthophylls.

Measuring Skin Colors

There are several methods, as follows:

Roche color fan. The simplest procedure is to visually match the skin color with those on the Roche color fan.

IDL color-eye. This is a type of photometer used for determining skin colors.

NEPA standard (National Egg and Poultry Association). Graded solutions of potassium dichromate are used and compared to the ether extract of a small piece of skin. The scores range from 0 to 5, as follows:

NEPA Number	Approximate Skin Color
0	Very pale
1	Light yellow
2	Dark yellow
3	Normal orange
4	Dark orange
5	Very dark orange

Skin-color variations. Skin color is not the same in all sections of the bird as shown in Table 33-13, which represents a comparative guide, using the toe web as a base of 100, the lightest color.

Table 33-13. Variations in Surface Skin Color of Broilers

Area	Color Index
Web of toe	100 (base)
Foot pad	108
Shank	122
Back	132
Breast feather tract	163

Xanthophyll Content of Feedstuffs

The *total* xanthophyll content of some feedstuffs is shown in Table 33-14.

Table 33-14. Total Xanthophyll Content of Feedstuffs

	Approximate Total Xanthophyll	
Feedstuff	mg per lb	mg per kg
Marigold petal meal	3,182	7,000
Algae meal	909	2,000
Dehydrated alfalfa meal (20% protein)	127	280
Dehydrated alfalfa meal (17% protein)	118	260
Coastal Bermuda grass	123	270
Corn gluten meal (60% protein)	132	290
Corn gluten meal (41% protein)	80	175
Yellow corn	8	17

Source: Nutrient Requirements of Poultry, © 1984 by the National Academy of Sciences, Washington, D.C.

Color Transmission of Xanthophylls

Xanthophylls differ in color transmission. The ability of the xanthophylls in alfalfa leaves, yellow corn, and corn gluten meal to increase the density of yellow-orange color in the skin of the chicken is not equal. For example, per unit of xanthophyll, alfalfa leaves will impart only 75% the density of yellow corn.

Time necessary to color broilers. It takes about 3 weeks to produce the desired color in a broiler. The older the broiler, the higher the percentage of xanthophylls transferred from the feed to the skin, but oxidation of xanthophylls is also greater in older birds. Of the two xanthophylls, lutein and zeaxanthin, found in alfalfa and yellow corn, respectively, zeaxanthin is far superior in producing a darker pigment, and it will increase the density more in the breast than in the shank.

Xanthophylls necessary to produce various skin colors. Amounts of mixed xanthophylls to produce various NEPA skin numbers are found in Table 33-15.

Dark broiler houses intensify pigmentation. The red pigments in certain xanthophylls produce a deeper orange pigment when broilers are raised in environmentally dark houses rather than in open-sided buildings.

Table 33-15. Dietary Mixed Xanthophylls Necessary to Produce NEPA Scores in Broilers

NEPA Number	Approximate Xanthophyll Quantity Necessary per Unit of Ration	
	mg/lb	mg/kg
1	5	11.0
2	10	22.0
3	16	35.2
4	23	50.6
5	30	66.0

33-K. LEAST-COST FEEDING OF BROILERS

There are two types of least-cost feeding. One has to do with formula substitutions that do not alter the feed's nutritional value. In the large commercial feed mill these are generally calculated through the use of computers.

The other recognizes that formula changes to increase growth and/or feed conversion may cost more than is returned from the sale of the birds. Diets high in energy are expensive, and may lead to a higher feed cost per pound (kilo) of the live mature bird. An example is illustrated in Table 33-16.

The figures in Table 33-16 show that as the energy of the starter and grower rations increased, an improvement was made in broiler growth and feed conversion. A further analysis of the original data showed that there was less response to high-energy diets in summer than in winter, which is almost always the case. However, at no time was there adequate response to the higher-energy diets to produce a lower feed cost per pound (kilo) of live market broiler.

The results of this experiment showed that there must be a happy medium between caloric content of the broiler diet and the ingredient cost per pound (kilo) of live mature weight when economy is considered.

Table 33-16. Flock Performance and Ingredient Cost per Pound (Kilo) of Broiler Weight on Four Rations Different in Energy

Kcal of ME in Ration				Avg Body Weight		Feed Conversion	Ingredient Cost per Unit of Live Broiler (US¢)	
Starter		Grower						
Per lb	Per kg	Per lb	Per kg	lb	kg		Per lb	Per kg
1,400	3,080	1,425	3,135	4.26	1.94	2.12	15.52	34.23
1,450	3,190	1,475	3,245	4.31	1.96	2.08	16.21	35.56
1,475	3,245	1,500	3,300	4.40	2.00	2.02	16.78	36.92
1,500	3,300	1,525	3,355	4.48	2.04	1.98	17.34	38.15

Source: Ward, J. D., *Poultry Digest,* Dec. 1978.
Note: Data based on averages of five replicates over five seasons.

33-L. FEEDING SQUAB BROILERS

Squab broilers are marketed at a very young age of about 4½ to 5 weeks (see Chapter 20-W).

A broiler starter diet containing 23 to 24% protein and 1,450 kcal of ME per pound (3,190 kcal ME/kg) of ration is fed to squab broilers throughout the growing period. The remaining components of the diet are the same as for a good broiler starter ration.

33-M. FEEDING ROASTERS

When roasters are raised, broiler starter and broiler developer rations are fed, followed by a roaster finisher ration, comprised of high energy and low protein. Ultrarapid growth is not always advantageous when raising roasters. It is more important that the birds deposit large amounts of fat in the tissues. Furthermore, when growth is forced to roaster weights there is likely to be a high incidence of breast blisters (see Chapter 20-X).

Feeding Program for Roasters

A recommended feeding program for roasters is given in Table 33-17.

Cost of Producing Roasters

Feed is the main item of cost in producing a roaster. Sex is also a factor not only because females must be kept for a longer period than males to attain a desired live weight but because females are poor converters of

Table 33-17.   Requirements for Roaster Rations

Ration	Age of Birds wk	Kcal of ME in Ration		Protein in Ration %
		Per lb	Per kg	
Starter	0–4	1,400	3,080	21
Developer	5–8	1,425	3,135	20
Finisher	9–Market	1,460	3,212	17

feed to meat compared with males. Thus, 1 lb (or 1 kg) of live female roaster is more expensive to produce than 1 lb (or 1 kg) of live male roaster (see Chapter 20-X).

33-N. FEEDING SURGICAL CAPONS

Surgical capons are commonly sold at about 17 to 20 weeks of age (see Chapter 20-Y). They are usually fed a broiler starter diet for 4 or 5 weeks, then given a high-fiber diet until about 12 to 13 weeks, when they should weigh about 8.0 lb (3.6 kg), in order to reduce breast blisters. After this age they are given a high-energy diet. The usual marketable live weight is 10 lb (4.5 kg). Feed conversion for the entire growing period is from 3.8 to 4.0. A suggested feeding program is found in Table 33-18.

A deep yellow carcass color is necessary and diets are altered during the last 6 weeks of the growing period by increasing the xanthophyll-containing ingredients to induce the necessary yellow pigment in the skin and fat.

Table 33-18.   Feeding Program for Surgical Capons

Age of Capons wk	Kcal of ME in Ration		Fiber in Ration %	Protein in Ration %
	Per lb	Per kg		
0–4	1,450	3,190	3.5	23
5–13	1,200	2,640	7.0	18
14–Market	1,300	2,860	4.6	17

Restricted Feeding of Capons

The key to success with capons is to produce a high percentage of Grade A birds. According to N. D. Magruder, Cargill Research Farm, one method of attaining this is to restrict the feed intake of the capon between 8 and 14 weeks of age. Daily feed allocations are reduced 10 to 15% below the amount of feed consumed when the birds are on full feed. The ration should contain about 1,300 kcal of ME per pound (2,860 kcal/kg) between 8 and 14 weeks of age during such feed restriction. In a test, full-fed birds had 66.7% Grade A birds at slaughter time, while those restricted during this time period had 80.6% Grade A.

34

Bacteria, Viruses, Protozoa, and Fungi

34-A. DISEASE AND RELATED TERMS

active immunity: Immunity produced in the bird either by natural exposure or by vaccination.

acute disease: A disease of short and severe duration.

agglutination test: A test for the presence of antibodies performed by mixing blood or serum and an antigen.

agglutinin: A substance that causes bacteria or blood corpuscles to coalesce or clump together.

air sacculitis: Inflammation of the air sacs.

anamnestic response: The improved immune response exhibited by an individual with former exposure to the specific antigen.

anion: A negatively charged ion.

anionic (detergent): Any class of synthetic compounds whose anions are alkali salts, as soap, or whose ions are ammonium salts.

anthelmintic: Capable of expelling or destroying parasitic worms especially of the intestines.

antibiotic: A dilute substance produced by microorganisms that has the power to kill or inhibit the growth of other organisms.

antibody: A substance formed in the body as the result of infection or administration of suitable antigens.

antigen: A protein or carbohydrate that produces antibodies when injected into the bird.

antiseptic: A substance applied to animals that reduces microorganisms to a harmless state, either by killing them or preventing their growth.

antiserum: A serum containing antibodies specific to a certain disease.

antitoxin: A specific antibody capable of neutralizing a specific toxin.

aseptic: Free from pathogenic organisms.

ataxia: Uncoordinated muscular movements (as in avian encephalo-myelitis).

attenuated: A disease organism that has been weakened to reduce its virulence.

autogenous vaccine: A vaccine prepared from cultures derived from infected birds and used to immunize against further contagion.

avirulent: An organism that is not virulent or pathogenic.

avitaminosis: A disease or malfunction caused by a vitamin deficiency.

B-cells: Cells of the immune system that are transformed and matured in the bursa of Fabricius.

bacteria: Microscopic organisms that are composed of a single cell.

bactericide: A substance that kills bacteria, but not necessarily their spores.

bacterin: A suspension of killed or attenuated bacteria that brings about immunity when injected into the chicken (antigen).

bacteriostat: A substance that inhibits the growth of bacteria without killing them.

blood test: *See* agglutination test.

broad-spectrum antibiotic: An antibiotic that inhibits the growth of many kinds of microorganisms (the more numerous the kinds, the wider the spectrum).

bursa of Fabricius: A small gland adjacent to the upper part of the cloaca involved in the processing and maturation of cells of the immune system.

carrier: A chicken that shows no evidence of a disease yet harbors the organism and is capable of transmitting the disease to others.

caseous: Cheesy in appearance.

catarrhal: Capable of producing an inflammation of the mucous membranes.

cation: Positively charged atom.

cationic: Compounds whose cations are colloidal in solution, such as antiseptics, wetting agents, and emulsifiers.

cellulitis: Inflammation of cellular tissue.

chronic disease: One that has a long duration, usually evidenced by morbidity rather than mortality.

coccidiostat: A chemical compound added to the feed or drinking water to combat coccidiosis.

COFAL-free: (Complement fixation avian leukosis). Fertile eggs that are free of avian leukosis.

congestion: An overaccumulation of blood in the blood vessels causing excessive blood in the tissues.

contagious disease: An infectious disease that is readily transmitted to other birds.

culture (noun): A group of microorganisms grown on artificial media in a laboratory.

culture (verb): A procedure used to remove organisms from the bird and to isolate them.

cyanosis: Bluish in color as a result of lack of oxygen.

cytoplasm: All the contents of the cell, excluding the nucleus.

detergent: Usually a soapless, synthetic, water-soluble agent that reduces surface tension and thus emulsifies oils and has cleansing properties.

disease: An impairment of the normal function of any body organ or part of the bird.

disinfectant: A substance that kills pathogenic organisms but not necessarily spores and is usually applied to inanimate objects.

dropsy: An illness characterized by excessive water in the abdominal cavity.

edema: An excess of fluid in the tissues of the bird.

endemic: A disease confined to a small locality.

endocardium: Tissue lining the heart.

enteritis: An intestinal inflammation.

erythrocyte: A red cell or corpuscle of the blood used for transporting oxygen.

estrogens: Certain hormones secreted by the ovary that are capable of governing some of the secretions of the oviduct, and certain synthetic compounds having similar properties.

fomites: Inanimate objects such as clothing, feed bags, etc., that harbor disease-producing organisms.

germicide: Any agent that kills bacteria, especially those bacteria that are disease-producing.

gram-negative bacteria: Those that retain a violet color even in the presence of alcohol or acetone.

gram-positive bacteria: Those that lose their color in the presence of alcohol or acetone.

hemorrhage: A condition occurring when blood escapes from the circulatory system.

hepatitis: An inflammation of the liver.

histology: Study of the body tissues.

hormone: A substance produced by specialized body cells, which when transported by the blood system, has the power of effecting a change in other body cells.

host: An animal that supports a parasite or a pathogenic organism.

host cell: A cell invaded by a foreign infection.

host-specific: An organism confined to a single host species.

hydrogen-ion concentration: A value that indicates the acidity or alkalinity of a solution, running from 1 to 13 (exponentially). Seven is neutral; above seven is alkaline; below seven is acid. Also called pH.

immune: A bird is said to be immune when it has some degree of resistance to a particular disease.

immunity: The state of being resistant or immune.

implant (verb): To deposit beneath the skin of the bird, as when a caponizing pellet is implanted beneath the skin of the neck.

inclusion bodies: Bodies, thought to be virus particles, found in the cell contents when the bird is infected with certain diseases.

induced immunity: Immunity resulting from vaccination.

infection: The invasion of a pathogen into susceptible tissue resulting in disease.

infectious disease: A disease produced by the invasion of living microscopic organisms.

infectious organism: An organism that has the capability of producing disease.

isolation: Keeping poultry in areas separate from other poultry and other vectors.

lentogenic: Low virulence.

lesion: A variation in the normal appearance of tissue as the result of a pathogen or injury.

lymph: Circulating body fluid mainly concerned with transporting the components of the immune system.

lymphocyte: White blood cells produced by the lymphatic system.

lyophilized: Freeze dried.

macroscopic: Observable without magnification.

memory cells: T-cells that ''remember'' previous immune responses and accelerate repeated responses.

mesogenic: Medium virulence.

microscopic: Visible with a microscope.

morbidity: A sickness in a bird or flock caused by disease.

mortality: Death of birds in the flock.

mycosis: Any disease caused by a fungus.

necrosis: Death of living tissue usually caused by lack of blood supply.

neoplasm: Tissue that develops abnormally and usually has no physiologic function, such as a tumor.

parasite: An organism that lives in or on another organism, from which it derives its nourishment.

passive immunity: Usually that parental immunity passed from mother to offspring through the egg (by antibodies), or artificially by the administration of an antiserum.

pathogen: An organism capable of causing disease.

pathogenicity: The capability of an organism to produce a disease; a quantitative term.

pericarditis: Inflammation of the sac surrounding the heart.

peritonitis: Inflammation of the peritoneum, the thin membrane found in the abdomen.

perosis: Deformity of the leg bones.

pH: *See* hydrogen-ion concentration.

plasma: The clear solution remaining after the corpuscles have been removed from the blood.

polyvalent: An antigen or bacterin containing several strains of an organism or organisms.

protozoa: Minute protoplasmic accellular or unicellular animals with varied morphology and physiology.

renal: Relating to the kidneys.

sanitizer: A preparation capable of reducing the number of bacteria present, sometimes combined with a detergent.

septicemia: Invasion of the bloodstream by pathogenic microorganisms.

serological test: A test performed on blood serum to determine the presence or absence of specific antibodies.

serotype: A particular strain of a microorganism.

serum: The clear fluid remaining after the corpuscles and clotting properties have been removed from the blood.

sterilizer: Any chemical or agency (steam, heat, etc.) that destroys all forms of life (bacteria, mold, viruses, etc.)

stress: Anything that affects the bird's well-being and lowers its resistance to disease.

surfactant: Chemicals that lower the surface tension of the solvents in which they are dissolved, such as detergents.

syndrome: A group of symptoms common to a specific disease.

T-cells: Cells of the immune system that are transformed and matured in the thymus.

thymus: A gland located in the neck responsible for immune system development in the young chick.

titer: A value placed on the potency of a biological agent; when applied to the agglutination test, it is the weakest dilution at which clumping of the antigen occurs.

toxin: A poison produced by the metabolic processes of microorganisms.

tranquilizer: A drug that slows the metabolic rate of the bird as exemplified by reduced heartbeat, lowered blood pressure, reduced mental awareness, etc.

vaccine: A preparation of microorganisms (killed, living attenuated, or living totally virulent) that when placed in the body of the bird produces or increases immunity to a certain disease.

variant: In microorganisms, one that exemplifies a variation from the original form, often the result of a mutation.

vector: An animal that carries and transmits parasites to poultry, such as the earthworm, which carries the chicken tapeworm eggs.

velogenic: High virulence.

vermicide: A preparation that kills poultry worms, either within or outside the host.

virulence: The relative ability of a microorganism to produce disease, usually a quantitative term.

virus: An organism, ultramicroscopic in size, that multiplies only in living cells, some of which are capable of causing disease.

wetting agent: Any substance that, when added to water, reduces surface tension and improves the cleaning action. *See also* detergent; surfactant.

34-B. MICROORGANISMS AND DISEASE

Microorganisms constitute a large group of living cells that complete their life cycle in their minute state; they do not form conglomerates of cells as are found in the higher forms of life.

Most such cells remain in a single form, although they may align themselves with other similar cells to form long filaments or chains. Being living organisms, their function is similar to that of higher organisms; they digest and assimilate food, and excrete waste products.

As do higher forms of life, they necessarily must fight for survival. Being small, they have been provided by nature with a survival potential through their ability to reproduce in extremely large numbers. These large numbers mean that dissemination is great; most microorganisms are abundant everywhere.

Some microorganisms are necessary to complete various reactions required to sustain the complex forms of life. Others are pathogenic; they produce certain reactions that are detrimental to the higher forms, among which is the chicken.

The organisms producing diseases in chickens are too small to be seen with the naked eye; a microscope must be used. All such organisms are not similar in structure, size, chemical composition, mode of nutrition, or in the manner in which they attack their host. In the main such organisms may be divided into four classes: (1) bacteria, (2) viruses, (3) protozoa, and (4) fungi. Although the organisms within each of these groups are similar in many respects, most may be identified only by using intricate laboratory techniques.

34-C. BACTERIA

Disease-producing bacteria are abundant in the poultry world. These organisms are capable of invading the chicken, where they multiply rapidly through cellular division and produce a physiological change in the host bird that it cannot withstand. Sickness follows, and if the reaction is severe enough mortality may result.

How Bacteria Grow

Commonly, the bacteria divide by fission that separates each cell into two equal halves. Once the disease-producing bacteria gain entrance to the chicken, multiplication of the cells is rapid in susceptible birds. One cell divides into 2, 2 into 4, 4 into 8, 8 into 16, and so forth. But geometric multiplication (commonly called exponential growth) does not continue forever; soon there is competition for a food supply and oxygen and many cells may become incapable of division. But the numbers are soon so large that their ability to produce disease in the bird is far beyond the quantity necessary.

Organisms grow similarly in the laboratory. Bacteria may be made to grow and reproduce in the laboratory by using artificial means. The bacteria are removed from the host, or from other laboratory cultures, and placed on or in a medium known to furnish the correct food material and moisture. The medium is then warmed to a certain temperature to create optimum growth. When the bacteria are placed on a medium such as agar, cell division is rapid and *colonies* of the growing organisms may easily be observed with the naked eye. Each organism produces its own pattern colony of different shape, color, and makeup, and this pattern presents one means of identifying the type of bacteria.

Variant forms of bacteria. Many bacteria, particularly those in certain Salmonella groups (e.g., *S. pullorum*), have through the years developed variations in their makeup, giving rise to several types of the same organism. Some of these variations may be due to mutations, and others to differences in the environment in which the bacteria live. But these variations have caused many misinterpretations in the reading of agglutination tests and in the production of antibodies. Many antigens in use today include one or more of the variant strains of bacteria, in order that the antibodies being produced by the bird being tested will detect carriers of all forms of the *Salmonella pullorum.*

How Bacteria Produce a Contagious Disease

Poultry diseases may be classified as *contagious* or *noncontagious*. When contagious, they are capable of being transmitted from one bird to another. Some bacteria produce the actual condition of the disease; others lower the bird's resistance, thus increasing the possibility of outbreaks of other diseases.

In most instances a disease is the result of the entrance of the disease-producing bacteria into the bird and the ensuing multiplication. The organisms produce toxins that are in turn antagonistic to the host. Being quantitative, the greater the amount of the toxin, either by increased cellular production or because the quantity of bacteria is overwhelming, the greater the effect of the disease.

Toxins produced by bacteria are of varying sorts.

1. Those produced by living organisms (exotoxin, tetanus)

2. Those not liberated from the cell until the cells dies or disintegrates (endotoxin, staphylococcal food poisoning)
3. Those that liberate hemoglobin from the red blood cells
4. Those that break down barriers that prevent the invasion of disease-producing organisms

The *virulence* (pathogenicity) of an organism is a measure of its ability to produce disease. Each organism varies according to its power to invade the tissues and in production of the necessary toxins. The rate of multiplication of the cells within the host is also a contributing factor.

Virulence within a species of bacteria may vary. Many disease-producing organisms remain in a quiescent state within many birds, but when conditions are right, they take on a new life, multiplying and producing toxins in abundance. On farms where successive groups of chicks are started, the virulence of the organism will be increased in each lot once an outbreak occurs. Not only does this disease strike harder but it attacks the birds at a younger age.

Host Must Be Susceptible for a Disease to Develop

Conditions must be right, and the host susceptible, for the bacteria to invade, establish themselves, and produce the disease. Nature has endowed the bird with protective immune systems to prevent invasion, and with certain other factors to help reduce the incidence of disease.

1. *Those that prevent (or attempt to prevent) organisms from entering the body.* The three are secretions, skin, and mucous membrane. However, these protections are generally minor; invasion through skin abrasions, via the respiratory tract, and by other means (or methods), usually overshadow their benefits.
2. *Those that fight the organisms that have entered the body.* Each type of bacteria causes the production of an unassociated chemical that acts to destroy the organism that produced it. This self-destruction is, in most instances, complete to the extent that all the disease-producing bacteria in the body are destroyed and the disease in the flock subsides.
3. *Species resistance.* Certain bacteria may invade hosts other than the chicken. These bacteria will grow and multiply but will produce no discomfort or other evidence of the disease. Some other bacteria may invade and cause disease symptoms in several hosts. Some

diseases quite common in the chicken occur in several types of poultry (duck, quail, turkey, etc.).

4. *Age susceptibility.*   Some diseases attack chickens of a certain age and produce disastrous results, yet show few, if any, symptoms in birds of other ages.

5. *Climate and season.*   Some poultry diseases affect birds more during cold weather than during warm weather; others are more prevalent during periods of warm or hot weather.

6. *Freedom from stress.*   The physiological well-being of the individual has an effect on the incidence of disease outbreaks. Good nutrition, freedom from stress, adequate housing, and temperature control improve the condition of the bird, making bacterial outbreaks less likely (see Chapter 39-A).

34-D. VIRUSES

A second group of disease-producing organisms is represented by the viruses. They are living organisms, exceptionally small in size. Commonly known as *filterable viruses,* they are capable of passing specific filters that bacteria will not permeate. The common microscope is of no value in detecting these as they are so small, but the electron microscope is capable of photographing them.

Parts of a virus.   A mature virus is known as a *virion* and is made up of a central core of genetic material (nucleic acid) plus a protein coat surrounding the nucleic acid for protective purposes and to aid some viruses in penetrating the host cell. Other viruses have a coat of lipoprotein called an *envelope* that develops when the virus penetrates the host cell.

DNA and RNA Groups of Viruses

All viruses contain RNA (ribonucleic acid) or DNA (deoxyribonucleic acid) but not both. Thus, viruses may be classified as the RNA group or the DNA group, but there are several subclassifications in each, as follows:

RNA group

> Picornaviruses (example: avian encephalomyelitis)
> Reoviruses (examples: infectious bursal disease, viral arthritis)
> Arboviruses (example: turkey meningo-encephalitis)
> Myxoviruses (examples: avian influenza, Newcastle disease)

Oncornaviruses (example: lymphoid leukosis)

Coronaviruses (examples: infectious enteritis, infectious bronchitis)

DNA group

Adenoviruses (examples: hemorrhagic enteritis, aplastic anemia, egg-drop syndrome)

Herpesvirus (examples: laryngotracheitis, Marek's disease)

Poxvirus (example: fowl pox)

Properties of the Viruses

Although infinitely small, there is a great variation in the size of viruses, some being 25 times as large as the smallest. They invade the host cells, and it is this invasion that is thought to precipitate the respective disease, rather than any toxin produced.

Secondary invaders. When a virus attacks the cells of the lining of the respiratory tract and the air sacs, the cell walls are ruptured and become a point for each invasion of other viruses and bacteria. In many cases the disease developed by the primary invader virus may produce little damage to the bird, but the secondary invader may cause all kinds of trouble.

How Viruses Multiply

Surprisingly, little is known about the exact life cycle of most viruses. Their small size has confused experimental work. Some think that the virus particles do divide and multiply as do bacterial cells. Yet others feel that a virus is inanimate, and must rely on the cell it invades for its reproduction, using the enzymes from the cell to complete the process. The consensus of this group of workers is that viruses are specific in their need for certain enzymes and seek out a location in the body where they are produced. Thus, some viruses invade tissue, some the bursa of Fabricius, others the respiratory tract, and so forth.

Viruses multiply only in cells. Bacteria can multiply almost anywhere in the body, but a virus can live and reproduce only within a host cell.

Treatment of a viral disease is difficult. To attack the living virus it is necessary to get to the seat of infection, namely, the cell. This is most difficult, and for this reason any chemicals or antibiotics administered in the feed or drinking water are seldom effective in treating a viral infection. Sometimes injections are of little value even though they quickly get into the bloodstream. Generally, viruses show little or no effect from drug treatment.

Laboratory Propagation of Viruses

Evidence points to the fact that viruses multiply only in living cells, yet all viruses cannot be propagated in this manner in the laboratory. Although tissue culture is predominantly used for laboratory techniques, other viruses are propagated on the chorioallantoic membrane of the developing chick embryo, some in other ways. Plant viruses cannot be propagated on animal tissue.

Vaccines. The specific virus is used to produce a vaccine against the disease. Culture and multiplication of the virus is important in the manufacture of these vaccines, and the medium on which they can be cultivated is quite specific.

34-E. PROTOZOA

Third in the group of disease-producing organisms are the protozoa. They are animals and are similar to higher forms of life except for the fact that all the functions are carried out within a single cell. In poultry they are parasitic in nature; that is, they live on the contents of the cell, eventually destroying it.

Protozoa may have a complex life cycle. An example would be coccidiosis in chickens. In this case *oocysts* are expelled in the fecal material. Heat, oxygen, and moisture cause them to sporulate. They are then consumed and enter the intestinal tract. The cell wall of the sporulated oocyst ruptures; eight infectious organisms exude and soon penetrate the cells of the intestinal lining. In turn these grow until the intestinal cells rupture and hemorrhage. This is the *asexual* portion of the life cycle, and may be repeated several times; but eventually a *sexual* phase occurs, giving rise to both male and female organisms. After other cell invasions, more oocysts are produced and leave the body by way of the fecal material, and the life cycle is complete.

Number of oocysts extremely large. A teaspoonful of droppings from a bird with severe coccidiosis will contain several million unsporulated oocysts. Consumption of as few as 10,000 *sporulated* oocysts will produce evidence of coccidiosis; 200,000 will produce death.

Other Protozoal Diseases (Extracellular)

There are other protozoal diseases in chickens. The most important ones are

1. *Trichomoniasis.* Caused by a motile protozoan, which locates in the upper digestive tract (esophagus, crop

and proventriculus). The name of the organism is *Trichomonas gallinae,* and its life cycle is different from that of coccidia.

2. *Hexamitiasis.*   Caused by *Hexamita meleagridis,* a motile protozoan; although it seldom affects chickens it is prevalent in quail, ducks, and pheasants.

3. *Blackhead.* This disease is seldom found in the chicken, but is quite common in turkeys. It is caused by the protozoan *Histomonas meleagridis.* The organism can live but a short while outside the host; however, it enters the egg of the cecal worm and can live in it for a long period. Reinfection occurs when the cecal worm eggs are consumed by the chicken or turkey.

34-F. FUNGI

Fungi represent a group of organisms containing the *molds* and *yeasts.* Generally, they grow outside the chicken, producing toxins. When consumed they may continue to grow, the toxin causing a distinct setback in the bird's well-being. In other cases, continued ingestion of the mold and toxin produces the disease-like effect.

Important poultry diseases from fungi.   The four common poultry diseases caused by fungi are

1. *Aspergillosis.*   Caused by a mold, *Aspergillus fumigatus.*
2. *Thrush.*   Caused by a yeast, *Candida albicans,* and other fungi.
3. *Favus.*   Caused by a fungus, *Trichophyton megnini.*
4. *Dactylariosis.*   Caused by a fungus, *Dactlyaria gallopava.*

35
Developing Immunity

Once a bird undergoes infection from a natural means of transmission of a disease-producing organism, or from a vaccination, chemicals are produced in the body that tend to kill the organism and endow the bird with *immunity* so that it may withstand future invasions of similar organisms. The intricacies of the development of this immunity are varied, and seemingly each organism has its own program.

34-A. THE IMMUNE SYSTEMS

The operation of the development of immunity to an antigen, referred to as any foreign substance that causes the production of antibodies, is initiated in the very young chick. It is a highly specialized defense system, called the immune system, and is nature's way of resisting disease caused by the early invasion of the many infectious bacteria, viruses, fungi, and so forth.

The immune system owes its origin to specialized cells of which lymphocytes and other derivative cells are the most important. Actually, two immune systems develop in the body of the chick, each the responsibility of two types of lymphocytes. These two systems are

1. *T-system (Thymus system).* In the very young chick certain immature lymphocytes originating in the yolk sac and bone marrow pass through the thymus (in the neck) and are known as T-lymphocytes. Here they mature, then grow and accumulate in the lymphoid organs such as the spleen, cecal tonsils, Harderian gland, and so forth, for a few weeks after hatching. The T-

lymphocytes do not produce antibodies, but they do have the power to develop *lymphokines*, often called defector cells, that can destroy foreign cells by direct contact without the presence of an antibody. This is called *cell-mediated immunity* or *cellular immunity*.

2. *B-system (Bursal system).* T-cells and other lymphocytes in the young chick pass through the bursa of Fabricius, a small gland located dorsal to the cloaca, where maturation takes place. The T-cells locate in the region near the bursal duct opening and the bursa may have a secondary function similar to a lymph node.

Plasma cells develop in the B-system, including the bursa, spleen, and cecal tonsils, and are responsible for the production of most of the antibodies in the young chick. In fact, the B-system produces more than 700 times the number of antibodies produced by the T-system although they live for less than a week and must be replenished to keep up the bird's defenses.

The B-lymphocytes passing the bursa of Fabricius soon locate throughout the body, including the blood system where they no longer require the bursa for maturation. A similar procedure takes place with the T-lymphocytes. Any disease that affects the thymus or bursa in the very young chick disrupts the early development of both the B- and T-systems.

Anamnestic response. Next in the B-system is the development of lymphocytes that act as *memory cells*, referred to as *anamnestic response*. These cells have great longevity and cause the T-cells to "remember" former immune responses and accelerate repeated responses. An example is the greater response of the second vaccination against a specific disease. The body defenses are produced more quickly after the second response than after the first. Evidently the cells that produce the antibodies do not forget how, and when restimulated start production quickly and rapidly.

The Immune Systems and Vaccination

Although the immune systems are nature's way of protecting the bird against early disease, they are far from perfect. Massive numbers of invaders can produce an overwhelming effect, and as the foreign invader must first be present in the bird in order to initiate the immune systems, the lag in time is usually such that some morbidity or mortality occurs before the systems produce adequate antibodies. Vaccination against the disease is the most successful remedy.

Suppression of the Immune Systems

Certain diseases and other conditions are known to affect the developing thymus and bursa of Fabricius in the young chick, thereby causing varying amounts of gland destruction followed by disruption in the immune systems. Among those having been shown to have such an effect are the following:

Suppression of the T-system

1. Marek's disease
2. Heat and cold
3. Genetic origin
4. Incomplete vaccine reaction
5. Aflatoxins

Suppression of the B-system

1. Infectious bursal disease
2. Heat and cold
3. Lymphoid leukosis
4. Nutritional deficiency
5. Toxins
6. Low antibody production
7. Inclusion body hepatitis (temporary)
8. Aflatoxins

35-B. PATTERN OF A DISEASE OUTBREAK

The pattern of most poultry diseases is the same. There are three phases.

1. *Infection.* Disease-producing microorganisms invade the chicken, and if no immunity is present these organisms attack various parts of the body and produce a sickness in the birds, the type of sickness being specific for the particular disease involved. Morbidity is usually rapid; mortality may follow, depending on the severity of the outbreak.
2. *Development of resistance.* Once the microorganisms establish themselves in the host chicken the production of a protective chemical begins, the chemical being specific for the organism involved.
3. *Disease subsides.* Protection develops rapidly and the chemical destroys the causative organisms; in the case of most diseases the outbreak subsides or is reduced

to a very low level. The bird recovers except for any permanent damage developed during the course of the disease.

35-C. ANTIBODIES AND IMMUNITY

Physiology of Antibodies

Once a foreign substance enters the chicken, the body acts to eliminate it. Some such substances never are assimilated; the body eliminates them through the feces. Others, such as some of those generated by the bird, are eliminated through the urinary tract.

In essence, bacteria are proteins, composed of one or more protein molecules that are foreign to the bird. These foreign protein particles produce a toxic reaction, resulting in what we call disease. In trying to eliminate them, the body system generates a chemical that reacts with the organisms or inactivates them. This chemical is known as an *antibody*. Each antibody is specific for the bacterium or virus that initiated its production. As the antibody is especially specific for the protein of the invading organism, the fact that there may be one or more protein molecules involved means that there may be one or more antibodies formed.

Antibody Behavior

Independent antibody for each disease. The antibody is specific for each type and strain of virus or bacterium that caused its production. The antibody produced as the result of one organism will not protect the bird from other organisms; e.g., antibodies against infectious bronchitis will have no effect on the Newcastle disease virus.

Length of time for antibody production. Once the disease-producing organism establishes itself within the bird, antibody production varies according to

1. Number of organisms involved at invasion time
2. Virility of the organism
3. Condition of the bird (freedom from stress)
4. Type of organism

But for the most part the time required for immunity to develop (through the production of antibodies) is dependent on the organism involved; some need several hours whereas others require several days.

Immunity long-lasting. Comparatively speaking, the value of antibodies in protecting the bird against future invasions of the same disease organisms is great and the immunity lasts for a long period. In some instances it continues for a lifetime; in others, for several weeks or months. Immunity is relative, and absolute immunity probably is nonexistent.

Antibodies may not destroy all organisms. In the case of certain diseases, pullorum disease for example, the *Salmonella pullorum* bacteria in the bird may not be completely destroyed. In this instance, the atrophied ova have no blood supply, which normally would move in antibodies produced elsewhere. Consequently, any pullorum bacteria harbored in the degenerated tissue of the ova would continue to live on and multiply. But this is not a serious condition, for if any bacteria should happen to be liberated into the bloodstream, their antibodies there would destroy them.

Variability in Antibody Production

In the course of a normal outbreak of a disease, antibody production reaches its height, after which it decreases and eventually may be reduced to zero. The rate of this decrease, and the longevity of adequate immunity, is a function of the type of antibody involved. It will vary according to the disease.

Quantitative measure of antibodies present. It is possible to determine the number of antibodies present in the bird. This determination involves computing the *titer.* Titer is really the strength of a solution, or, in this case, a quantitative determination of the presence of antibodies. Titers are usually expressed as a reciprocal of the highest dilution of the solution showing specific activity. Thus readings of 1:16, 1:64, and 1:256 would be examples. The higher the titer, the greater the number of antibodies.

Titer determines resistance. Following certain disease outbreaks, when the antibodies slowly decrease, the bird gradually loses its ability to withstand future invasions of the bacterium or virus. If the antibodies are reduced to a very low level, reinfection may occur and the bird again may contract the disease. The severity of the second invasion will be determined by the number of antibodies in the bird at that time. If it is in the medium range, the bird may show only slight effects from the disease, but if the quantity of antibodies is low, the disease may be severe. The titer or antibody as measured in the laboratory is generally, but not always, the ability to resist infection.

Age of the bird and antibodies. In respect to many diseases, young chicks seem more susceptible than older birds. They are *immunologically incompetent.* Usually this is because a greater number of antibodies is necessary to fight the effects of the invading bacteria or viruses. As the bird grows older, fewer antibodies will overcome the challenge. In some cases, however, as with lymphoid leukosis, older birds show more evidence of the disease than younger ones.

35-D. PARENTAL IMMUNITY

The protective device for the prevention of many diseases in newly hatched chicks can be attributed to *parental immunity,* sometimes called *passive immunity.*

If the breeding hen has had a specific disease, she has produced a large number of antibodies. Many of these localize in the ova, and consequently are passed to the chick through the ovum of the hatching egg. In turn, these antibodies aid in protecting the chick from invasions of the specific disease-producing bacterium or virus when it is quite young. But, as in the case of the adult mother, these antibodies in the chick wear away and their numbers decrease. Furthermore, the decrease of the antibodies is much more rapid in the chick than in the adult bird.

Antibody number in chick correlated with number in mother. When the amount of antibodies in the hen producing the hatching eggs is high (high titer), the number in the day-old chick is high. Usually, the concentration of antibodies in the egg yolk is identical with that in the hen. However, there is a decrease by the time the chick is hatched, and the number of antibodies is then about half that in the yolk of the fresh-laid egg and remain at this concentration for about 3 days, or until the yolk is absorbed. Thus, to provide ample immunity in the day-old chick, the titer must be high in the chick's mother.

Chicks with no parental immunity are easily invaded. When the hen producing the hatching egg has not had a specific disease or been vaccinated, she has produced no antibodies against the infection; thus, a chick hatched from an egg laid by such a mother would contain no antibodies and would have no protective device to withstand an invasion of the specific bacterium or virus. The chick would be completely susceptible.

Passing live organisms through the egg. Not only are antibodies passed from the dam to the chick through the egg but live disease-producing organisms are similarly transferred. One example is pullorum disease, in which viable bacteria are passed from the dam to the chick, along with some antibodies. However, the production of antibodies is not great enough in the dam to kill all the bacteria, so some live organisms will be found in the chicks, inducing an outbreak of pullorum.

What may be passed through the egg? There are several possibilities, such as

1. No bacteria or viruses, where the dam has neither been vaccinated nor has had the disease.
2. Live viruses or bacteria, where the dam is in the early stages of a disease or where a live-virus vaccine has been used and antibody production has not yet had a chance to destroy the organisms.

3. Antibodies, where, from an outbreak of a disease or a vaccination against it, antibodies have been produced in the dam and are transmitted to the chick.

Parental Immunity Not Effective with Some Diseases

The antibodies present in the egg yolk find their way to the bloodstream of the chick after the yolk is absorbed into the intestines just before hatching. Normally these antibodies completely protect the young chick against early invasion of most bacteria and viruses. However, in diseases that involve the respiratory tract the organisms gain entrance as the bird breathes, thus bypassing the bloodstream and its content of antibodies. An example is Newcastle disease. Because the antibodies in the bloodstream cannot destroy the virus of Newcastle in the respiratory tract, early disease outbreaks are prevalent; immunity is overwhelmed.

With laryngotracheitis, parental immunity is of little or no consequence inasmuch as the disease seldom occurs before birds are 5 weeks of age, and vaccinations can be delayed until the birds are 4 weeks of age.

Length of Parental Immunity

Any parental immunity begins to wane after the chick hatches. Half is lost in 3 days. By the end of the second week of the chick's life it is minor; it is generally ineffective by the end of the third week and completely disappears by the end of the fourth. Thus, at best, parental immunity can be considered an effective means in preventing disease outbreaks before the chicks are 2 weeks of age, but even during this period it must be recognized as a highly variable device.

35-E. THE IMMUNE SYSTEMS OF PROTOZOA

After an outbreak of a protozoan disease, most birds develop self-limiting immunity. One challenge to the organism will not produce complete immunity, but repeated challenges generally cause it to occur. Although the immunity from one challenge is short-lived, most protozoan disease outbreaks reinfect the bird, and seemingly the immunity lasts a lifetime.

Although antibodies are produced after the bird has been exposed to protozoa, these do not represent the sole immune system operating to alleviate the effect of the disease. Evidently localized tissue immunity continues in the intestine for the host will show immunity long after the antibodies have disappeared.

35-F. VACCINATION

Immunity is the result of antibody production, and under natural outbreaks of a disease, is a normal consequence. But there is also an artificial method of inducing the production of antibodies; this is through a process known as vaccination, using a *vaccine*.

How Vaccines Work

In most instances vaccines are used to produce a mild infection and a low manifestation of a specific disease. Normally, the formation of antibodies by the use of a vaccine will duplicate that from a natural outbreak, but not always.

The virulence of the vaccine may determine the behavior of a vaccination and the number of vaccine particles per unit of volume.

1. *Virulent organisms*
   a. *Virulent agent by unnatural route.* These are virulent agents that take an unnatural route in the body. Examples: Vent laryngotracheitis and fowl pox at a feathered area.
   b. *When virulent agent is less virulent.* Example: Wing-web Newcastle after 4 weeks of age.
2. *Low virulence.* Examples: B_1 Newcastle vaccine, and attenuated as tissue culture Newcastle.
3. *Inactive vaccine.* Examples: Cholera bacterins and killed Newcastle vaccine.

How Vaccines Are Administered

Vaccines may be classified according to the following methods used to administer them to chickens:

1. Intramuscular: into the muscle.
2. Subcutaneous: under the skin.
3. Ocular: in the eye (solution flows through the lacrimal duct to the respiratory tract).
4. Nasal: in the nostril.
5. Oral: in the mouth.
6. Water: into the respiratory tract by way of the throat.
7. Dust: into the respiratory tract by way of the nostrils.
8. Cloacal: into the tissues in the upper portion of the cloaca.
9. Wing web: by puncturing the web of the wing.

10. Feather follicle: by removing several feathers and swabbing or spraying vaccine over the area.
11. Spray: sprayed in the air, on the bird, or in the mouth. The spray must be a very fine mist to be effective. An automatic beak trimming machine that also sprays the vaccine into the mouth is available.

Types of Vaccines

Vaccines may be classified according to their efficacy or method of manufacture. First of all, a vaccine is produced from a live organism specific for a certain disease. Each vaccine is the result of harvesting bacteria or viruses in the laboratory, then treating them in such a manner that they will not produce their full effects when administered to the chicken. This procedure gives rise to the following classification involving viruses:

1. *Live virus vaccine.* The organisms in the vaccine are alive and completely capable of producing the disease in birds not affected or vaccinated previously. Containing a live virus, the vaccine is also capable of transmitting the disease to any susceptible bird that comes in contact with it.
2. *Attenuated vaccine.* By various methods the active organisms used to prepare a vaccine may be weakened (attenuated) so that when administered to a bird a milder form of the disease will be produced. In many cases there is no evidence of the disease. Generally it is impossible for such a vaccine to produce the disease in other birds except through the means employed for vaccination.
3. *Killed vaccine.* The organisms used to produce these vaccines have been killed; thus, there is no chance they will infect birds. They do, however, have the capacity to produce antibodies when used through vaccination. In some instances, however, their ability to do this is impaired, and immunity will not reach as high a level as with live or attenuated vaccines.

Wing-web Vaccine for Newcastle Disease

Many methods have been employed to find a suitable vaccine for Newcastle disease. The first involves a live-virus, wing-web product. It must be administered by dipping a needle into the vaccine, then pushing the needle through the wing web. Some active organisms enter the bloodstream, and if there has been no previous evidence of Newcastle disease or parental immunity, they make their way to the respiratory tract where antibodies develop. However, if antibodies are present in the bloodstream, the Newcastle organism is killed and does not reach the respiratory tract, and no additional antibodies are produced. Wing-web vaccine is of little value to birds vaccinated previously with any other form of Newcastle vaccine, because such vaccination produces antibodies that flow in the bloodstream. One must wait until the antibodies from the previous vaccination have had a substantial decrease.

Intranasal Vaccines

Certain vaccines may be administered intranasally; that is, they are dropped into one of the nostrils, after which they run into the throat. An alternate method is to drop the vaccine in one of the eyes, after which it slips down the lacrimal duct. The latter may have the advantage of uniformity of intake, as some chicks exhale the vaccine from the nostril, reducing the amount actually reaching the respiratory tract.

Intranasal vaccines are avirulent and produce only a mild reaction in the bird. One intranasal type is used against Newcastle disease. By dropping this vaccine into the eye or nostril it goes directly to the respiratory tract. However, there still may be difficulties from parental immunity. The trouble lies in the fact that some antibodies are contained in the cells of the respiratory tract. The greater the parental immunity, the less the chance that the virus will establish itself and form antibodies. Intranasal vaccination is more effective after the chicks are several days old.

Other Vaccines

There are many other variations in methods of vaccination, each depending on the virility of the vaccine and the manner in which it is administered. These are discussed in Chapter 37, according to the particular disease involved.

Parental Immunity and Vaccination

The function of parental immunity is to prevent pathogenic (and other) organisms from producing the effects of disease in young chicks when the incidence of disease might overwhelm their ability to survive. It is evident from past experimental work that in most cases it is useless to vaccinate chickens until most of the parental immunity has worn off, which means not before the birds are 14 days of age and preferably, not until they are 21 days of age. The principal exceptions to this policy

involve the use of day-old Marek's disease and infectious bronchitis/
Newcastle disease vaccinations in the hatchery. Maternal antibody inter-
ference is not considered as critical as the necessity for uniform early
protection.

Revaccination

Unless there is some circumstance preventing full results of vaccination, the first vaccination should suffice to build a high quantity of antibodies to the specific disease. However, the number of antibodies is at
its highest level shortly after vaccination; they gradually wear away and
eventually are reduced to a low level.

Variability in parental immunity. Because of the reduction in antibodies
after vaccinating the mother, parental immunity in the chicks produced from a vaccinated flock is highly variable. The longer the time
since the vaccination of the mother, the lower the parental immunity. If hatching eggs are coming from several flocks of breeders,
each in a different period of egg production, some chicks would
have a high parental immunity; others would have a very low parental immunity.

Revaccination of breeders to keep a high titer. In the case of those diseases
where antibody population deteriorates during the course of a year
of egg production, thus involving the degree of parental immunity
in the chicks, breeder birds should be revaccinated to reestablish the
high titer and the parental immunity in the chicks. This procedure
is of no value to the breeding hens, as they have ample immunity
from one or two vaccinations during the growing period to withstand another disease challenge; but it does maintain uniformity of
immunity in newly hatched chicks. Not only is the latter of value to
the chicks, but it makes any program of vaccinating young chicks
more uniform, and this purpose is its prime one.

Note: Flocks to be recycled should be reevaluated at the end of
their first laying cycle. If revaccination is required, it should be
done at least 1 week before the flock is molted.

Stress and Vaccination

Most present-day vaccines produce a mild effect on the normal
healthy bird. However, stress can accentuate the effect, producing a
greater physiological change in the bird, sometimes with disastrous results (see Chapter 39-A).

What are the stresses? There are many periods of stress during which
birds should not be vaccinated such as

1. When the birds are "off feed"
2. During periods of extremely hot weather

3. When birds have some other disease, such as coccidiosis, etc.
4. When birds are to be moved before they recover from a vaccination
5. When birds are in a stage of recovery from another vaccination or have recently been moved
6. When birds are being medicated or are diseased
7. Following beak trimming
8. During the first few weeks of an induced molt

Contaminated Vaccines

Some time ago many vaccines on the market were contaminated with impurities, particularly other organisms. As many vaccines are prepared from material harvested from growing embryos, there was always the chance that other organisms growing in the embryos would find their way into the commercial vaccines. With modern techniques, however, this possibility is becoming very remote.

Purchase vaccines from a reliable company that will stand behind its products and supply the service necessary to accomplish the best of vaccination. In some countries all vaccines must be COFAL-negative and Mycoplasma-negative.

Lack of Immune Response

Some flocks respond poorly to a vaccine because of a lack of *immune competence.* The ability of the bird to respond may have been reduced because of bursal disease, when the bursa of Fabricius does not have full ability to provide the immune response.

Vaccination Failure

In spite of all the perfection in vaccine manufacture and the explicit directions for the use of vaccines, there are many failures. In all probability most of these can be traced to the handling and administration of the product. Some rules to follow are

1. Keep vaccines under refrigeration.
2. Keep frozen vaccines frozen.
3. Do not open the vials of vaccine until you are ready to use them.
4. Mix vaccines thoroughly.
5. Do not vaccinate more birds from a vial than the directions recommend.
6. Keep a record of the vaccine manufacturer and the serial number of the vaccine.

7. Follow the manufacturer's procedures for vaccination.
8. Do not rush the vaccination job.
9. When using water-type vaccines be sure there are no sanitizers in the water.
10. Certain vaccines may be mixed (e.g., bronchitis and Newcastle). With certain others, the vaccinations may be given at the same time, but each a separate operation.

35-G. TITER OF VACCINES

Just as titers may be established for blood serum, they also may be determined for a vaccine, where the titer means the concentration of the virus in the product when it is prepared for use.

Definition. Vaccine titer means the number of virus particles, either living or dead, in 1 milliliter (ml) of the product. How many times the original virus can be diluted and still infect 50% of the chicken embryos when injected into fertile eggs under incubation is the measure of the vaccine titer. The virus titer is generally given in logarithms to the base 10. For example, a titer of 6 or (10^6) indicates six zeroes after number one, or 1,000,000.

35-H. TITER FOR ANTIBODIES

Test for Antibody Production

Just to administer a vaccine is not enough. The vaccine must produce the effect of disease, and antibodies must be established in quantity. In the case of many diseases, titers should be established after vaccination to determine if antibody production is adequate.

Arithmetic and logarithmic dilution. Equivalents for two measures of dilution are

Arithmetic	Logarithmic
1/10	10^{-1}
1/100	10^{-2}
1/1,000	10^{-3}
1/10,000	10^{-4}
1/100,000	10^{-5}
1/1,000,000	10^{-6}
1/10,000,000	10^{-7}
1/100,000,000	10^{-8}
1/1,000,000,000	10^{-9}
1/10,000,000,000	10^{-10}

Determining the Titer

The laboratory techniques for determining the number of antibodies contained in blood serum are usually

1. Serum neutralization (SN) test
2. Hemagglutination inhibition (HI) test. This is used to detect antibodies of Newcastle disease and *Mycoplasma* infections

These tests determine how much virus is neutralized by mixing it with blood serum from vaccinated or infected birds and concurrently mixing normal blood serum with an equal amount of virus.

Titer Does Not Always Indicate Degree of Immunity

Proper evaluation of the SN and HI tests should be made by trained personnel to get a true reading of any degree of immunity. As an example, birds may continue to be immune to Newcastle disease even though the antibody level in the blood serum is very low, or even absent.

Drugs and Antibiotics
for Disease Control

Although vaccination offers a method of preventing many poultry diseases from establishing themselves, certain drugs and antibiotics are used often in the poultry industry to help alleviate the symptoms of a vast number of diseases that gain entrance to flocks. The drugs make up an unassociated list of chemicals, and a great many are specific for a certain disease or for a group of similar diseases. New ones come on the market regularly; many others are in the process of experimentation. Besides the drugs there are the antibiotics. They are of general use for three reasons:

1. To aid in promoting growth and better feed conversion
2. To help restore the diseased bird to normalcy
3. To help prevent certain diseases from becoming established in the bird

The latter two are discussed in this chapter.

36-A. DRUGS IN DISEASE CONTROL

How Drugs Are Administered

Drugs may be administered as follows:

1. Through the feed
2. Through the drinking water
3. Through injections into the body of the bird

Water Solutions and Suspensions

The preparations used to treat birds have varying properties, making some of them acceptable for administration in the drinking water or for injections, whereas others would not be.

1. *Some form solutions.* These are especially well suited for use in the drinking water. They may also be used as an additive to the feed. Usually the water-soluble forms are more expensive than the insoluble forms used as feed additives.
2. *Some form suspensions.* Drugs used for water application must have the ability to mix uniformly. When the drug goes into solution the mix is uniform. However, soluble forms of some drugs are not available. Some drugs go into suspension; that is, they do not dissolve but float in the water and do not settle out. These also may be used as feed additives.
3. *Some do not form solutions or suspensions.* These, for the most part, are the insoluble drugs, and their use is confined to administration in the feed.
4. *Some do not pass from the intestines to the bloodstream.* This property may or may not be an advantage. If the microorganism involved is one that localizes in the intestinal tract, it is advantageous that the drug not be absorbed, but remain in the intestines to act on the organism. However, if the drug must reach the bloodstream to be effective, its ability to pass the intestinal wall becomes important.

36-B. CLASSIFICATION OF DRUGS

Examples of drugs and some of their effects when used for disease control are shown below. The list is not complete by any means and some drugs are available in some countries yet have not been approved for general use in the United States. Many products are sold under other trade names in various countries.

Sulfonamides

In the main, sulfonamides represent a group of synthetic drugs that inhibit the use of para-amino-benzoic acid (PABA) by the bird, a chemical necessary for the synthesis of folic acid, which in turn reduces cell multiplication. Sulfonamides are bacteriostatic rather than bacteriocidal.

Most drugs in this group produce some degree of toxicity; thus, they must be used in prescribed low doses. Sulfonamides are especially active against *Salmonella* organisms. The sulfonamide group includes sulfachloropyrazine (Esb$_3$®), sulfadimethoxine (and ormetoprim), sulfaethoxypyridazine (SAE), sulmethazine (Sulmet®), sulfathiazole, and sulfaquinoxaline (SQ®).

Most sulfonamides depress egg production at high levels, particularly sulfaquinoxaline. They also produce secondary hemorrhagic anemia. Sulfadimethoxine is usually supplemented with ormetoprim to augment the action of the sulfonamide. Sulfathiazole is a weak sulfonamide and is used for treating infectious coryza.

Nitrofurans

The list is limited to furazolidone (nf-180®, Furox®) and nitrofurazone (nfz®, Amifur®). These drugs are weak control agents for *Salmonella* and have some coccidiostatic properties. They are also used to treat for vibrionic hepatitis.

Coccidial-Specific Drugs

In this group are the coccidiostats. However, some also are effective against poultry diseases other than coccidiosis. Furthermore, some drugs from other groups used to treat birds afflicted with certain diseases also have anticoccidial properties of some degree.

The list of coccidiostats is too long for this book. For additional drugs of this type consult the manufacturer or your pathology specialist. Some common coccidial-specific drugs are

Generic Name	Trade Name
Amprolium	Amprol®
Clopidol	Coyden®
Decoquinate	Deccox®
Lasalocid	Avatec®
Monensin sodium	Coban®
Nicarbazin	Nicarb®
Robenidine	Robenz®
Sulfaquinoxaline	Several brands
Zoalene	Zoamix®

Pasteurella-Specific Drugs

There is only one drug in this classification although others have limited action. It is racephenicol (SW-5063).

Anthelmintics

Anthelmintics have a purging action, and with some degree of variability, remove worms from the intestinal tract. A partial list is as follows:

Generic Name	Trade Name
Butynorate	Tinostat®
Coumaphos	Meldane®
Hygromycin B	Hygromix®
Phenothiazine	Several brands
Piperazine	Several brands

36-C. DRUG TREATMENT

Potency of Drug Treatment

For any drug to be effective, it must locate at the point of infection in a concentration high enough to combat the invasion of the microorganisms. The amount of the drug in the blood serum is one indication of concentration; another is the quantity in the urine; another is the amount in the intestinal contents.

Drugs differ in their ability to impart concentrations in the blood, urine, intestinal contents, and body fluids. The dosages recommended by the manufacturer have been worked out carefully to obtain body concentrations that will be adequate to reduce the infection in the bird. These recommendations should be followed carefully.

Once a dose of a drug is administered, the concentration in the body soon reaches a maximum level. This maximum will be reached in the blood in 3 or 4 hours. Then the elimination of the drug begins and concentration drops off rapidly. In many instances the drug is depleted in 24 hours. If the drug is effective only at maximum concentrations, the period during which it will affect microorganisms is short. To administer less of the drug than the recommended amount will result in body concentrations under those required to effect a kill; to overadminister is costly and of little value.

Overdoses sometimes detrimental.   With certain drugs an overdose may produce detrimental effects in the bird. These effects are twofold.

1. *Injurious to the bird.*   They may cause a toxic reaction or otherwise affect some physiological function of the bird.
2. *Form residues in the tissues.*   Some drugs are not broken down during the process of digestion and metabolism and are not eliminated, and as some drugs are toxic or poisonous, their accumulation in the tissues gradually

produces a more drastic reaction. Some are deposited in egg yolk and albumen.

Drug retention and humans. Those drugs that are not altered during the process of digestion and metabolism and are not expelled from the body of the chicken in the feces and urine after a chemical breakdown soon accumulate to such an extent that they may be potentially injurious to human beings when the poultry meat or the eggs are eaten.

In many countries governments have established levels of tolerance of the drugs in the tissues or eggs. Above these levels the safety of the meat or eggs for human food is questioned, and in many instances the government may condemn the produce as unfit for human consumption and destroy it. Drug manufacturers are aware of this problem and write their directions for administration very carefully.

Tolerance levels and administration levels. Because certain drugs may be retained in the tissues and accumulate to a detrimental level, feed regulatory agencies have set up standards showing the amount of each drug that may be used for all methods of administration (feed, water, injection, and so forth). Feed manufacturers and poultrymen must abide by these directives, and in most instances feed manufacturers must furnish a tag with the feed to show the name and amount of the drug added.

Combinations of some drugs cumulative. Some drugs similar in chemical makeup supplement the retention potential of other drugs. As a consequence, regulatory agencies will allow only certain drugs to be fed simultaneously; others may not be fed at the same time.

Length of Drug Treatment

To be effective any drug must remain at a high concentration in the body from 3 to 5 days. This is difficult, as the period of maximum concentration is of very short duration after a single administration of the drug. Administration in the feed or water on a continuing basis is the only acceptable method, but this may prove costly. Low-level feeding of a drug to alleviate the disease condition usually is of little benefit. Second administrations of some drugs and antibiotics usually are less effective than the first. The drugs are eliminated in the urine much more rapidly after the second administration than after the first.

Injecting a drug produces only temporary effects. Although the activity reaches a high level quickly, it dissipates more rapidly. Continuous injections every day or two require too much labor and are uneconomical. Some drugs used for injection are dissolved in water. These are absorbed very rapidly by the system, but their effective duration is short.

To improve this method some drugs are added to oil, glycol, or wax. These are absorbed more slowly and the drug duration is longer.

36-D. ANTIBIOTICS

An antibiotic is a substance produced by a microorganism that is able in dilute solutions to inhibit the growth of or kill another microorganism. There are hundreds of such antibiotics, but through biological testing, only a relatively few have been found to be of value against certain diseases or conditions in the chicken. Antibiotics are used for both intestinal and systemic medication.

How Antibiotics Work

Antibiotics are used for disease control in the chicken. Usually they are specific for those diseases caused by bacteria or related organisms. They are of little value against viral infections. The beneficial effects of antibiotics are due to their ability to disrupt various phases of cellular metabolism.

An antibiotic will prevent bacterial multiplication provided enough is present to attack all the bacteria present. If the amount of antibiotic is small and the number of bacteria is large, the antibiotic will not produce its full effect. Antibiotics also act by changing the intestinal flora thereby destroying many beneficial bacteria.

Most antibiotics are given in the feed. In some instances, however, they are added to the water so that they may reach the digestive tract and the bloodstream faster, for some birds will not eat during the course of a severe disease outbreak, but will drink. In other cases, certain antibiotics may be injected.

Facts about Some Antibiotics

Bacitracin (Zinc Bacitracin®)

> Form: Feed and water.
> Intestinal absorption, none.
> Used primarily for treatment of necrotic enteritis.

Chlortetracycline (Aureomycin®)

> Form: Feed and water.
> Never used for an injection.
> May be potentiated.
> Intestinal absorption, medium.
> Slightly anticoccidial.
> Levels over 100 g per ton (2,000 lb) are not to be fed to
>   laying chickens.

At high levels do not feed continuously to young chicks
for more than 5 days.

Erythromycin (Gallimycin 50®)

Form: Feed, water, injectable.
Intestinal absorption, low.

Gentamicin

Form: Injectable.
Absorption poor.

Lincomycin (Lincomix®)

Form: Feed, water.
Absorption poor.

Neomycin sulfate (Neomycin®)

Form: Feed, water.
Intestinal absorption, none.

Oxytetracycline (Terramycin®)

Form: Feed, water, injectable.
When injected, may cause skin irritation.
Do not feed to laying hens at levels higher than 200 g per
ton (2,000 lb).
In feed of low calcium content do not feed for more than
5 days.
Slightly anticoccidial.
Intestinal absorption, low to medium.

Penicillin

Form: Feed, water, injectable.
Generally used as an injection.
Intestinal absorption, low.

Streptomycin

Form: Feed, water, injectable.
Large injections somewhat toxic; birds go to sleep.
Intestinal absorption, none.

Tylosin (Tylan®)

Form: Feed, water, injectable.
Intestinal absorption, low.

Caution: The U.S. Food and Drug Administration has withdrawn several antibiotics from the list of those permissible for injection. Poultrymen should check with the FDA or other government agency before using any antibiotic as an injection.

Low-Level Antibiotic Feeding

Some antibiotics are added to feed continuously at a low level to improve growth and feed conversion. This supplementation is not to be confused with therapeutic feeding of high levels. Because of the possibility of resistance to antibiotics being transferrable to human beings this procedure has not been approved in many countries. But this probability is very low in the case of those antibiotics that are not absorbed from the intestinal tract.

Caution: Check with authorities before adding low-level antibiotics to a feed.

Resistance to Antibiotics

When antibiotics are administered to a bird over a long period of time, particularly at a low level, certain species of bacteria become resistant, and finally the resistance becomes so great that the antibiotic is ineffective. In most instances resistance develops only to those antibiotics that are absorbed from the intestinal tract. One antibiotic not absorbed is bacitracin. Therefore, bacteria causing a systemic infection will not become resistant to bacitracin.

Antibiotic Sensitivity Test

Some antibiotics used in the poultry industry produce major effects in treating specific diseases; others are less valuable, and some are ineffective. In some instances the birds have become resistant to the antibiotic, producing a change in the value of the drug. The laboratory employs a technique known as a *sensitivity test* to determine which antibiotics will be effective in treating a disease. The organisms are cultured and grown on media in which various antibiotics have been incorporated at prescribed concentrations. If an antibiotic is to be effective in treating a disease the antibiotic in the medium will prevent reproduction of the organisms. The test actually shows which antibiotics will be ineffective; other antibiotics may be of help.

Antibiotic Potentiation

Certain antibiotics are used to treat diseases localizing in the intestinal tract. In these cases the antibiotic is administered in the feed or the drinking water only, and soon reaches the portion of the digestive tract affected by the disease. The amount in the tract is identical with that consumed; there is no loss. However, many diseases are systemic, and for the antibiotic to be effective it must leave the digestive tract, be taken into the bloodstream, and be transported to the point of infection. During this process some of the capabilities of the antibiotic are lost. To prevent most of this loss certain antibiotics are given by injection. The loss of the antibiotic when given orally is not the same for all antibiotics, the variation probably being due to the fact that all are not equally absorbed from the intestinal tract.

Oxytetracycline (Terramycin®) and chlortetracycline (Aureomycin®) are examples of two commonly used antibiotics that show this difference. They are of equal importance in treating intestinal disorders, but more than twice as much chlortetracycline is absorbed from the intestinal tract as oxytetracycline. In order to get the same amount of the antibiotic into the blood, more than twice as much oxytetracycline as chlortetracycline must be fed.

Increasing activity of an antibiotic. Calcium from the feed forms an insoluble salt when combined with oxytetracycline and chlortetracycline in the intestinal tract. This salt is insoluble and cannot be absorbed into the bloodstream. If the calcium in that ration is reduced, the absorption is increased, as a smaller quantity of insoluble salts is produced. The amount of each antibiotic absorbed may be increased more than twofold by deleting the "added calcium" from the ration.

Caution: Low-calcium feeds should not be fed to laying hens.

Terephthalic acid and potentiation. This drug (TPA) produces its effect by reducing elimination of the antibiotic in the urine. It increases the response to chlortetracycline four times, and two times to oxytetracycline.

Caution: The use of terephthalic acid is illegal in some countries.

Potentiation cumulative. Potentiation from reducing the calcium in the feed and from TPA is additive; thus the value of chlortetracycline may be increased eight times when both methods of potentiation are used together. The cost of feeding these antibiotics is materially reduced, and it is possible to use high-potentiated levels economically.

How to potentiate a feed. There are three methods.

1. Remove the added calcium from the formula. This method is not always the most practical because the feed will be low in calcium. It should not be fed to young chicks for over 5 days. A rachitic condition may develop if fed longer.

2. Remove the added calcium from the formula and replace it with 39 lb (13.6 kg) of sodium sulfate per ton (2,000 lb) of feed. The sodium sulfate removes the soluble calcium from the intestinal tract by forming calcium sulfate rather than uniting with the oxytetracycline.

3. Add TPA at the rate of 0.4%. This means 8 lb (3.6 kg) per ton (2,000 lb) of feed.

36-E. WITHDRAWAL PERIOD FOR DRUGS AND ANTIBIOTICS

In order to protect the consumer of poultry and eggs, and still provide effective drugs in the United States, the Food and Drug Administration (FDA) has set very precise tolerances for drug residues. Failure to properly withdraw a drug before slaughter can result in illegal residues.

The preslaughter withdrawal time for many drugs is given in the number of days that must pass between the last treatment with the drug and the day on which the chicken may be shipped for slaughter. The following list is not complete, but shows the withdrawal periods for many common drugs. For a more complete list, and more information in the United States write: Bureau of Veterinary Medicine, Food and Drug Administration, U.S. Department of Health, Education, and Welfare, Rockville, Maryland 20857. Although the following list is in effect at this writing, the list and withdrawal periods change often. Secure the latest information from the Rockville, Maryland address.

Active Ingredient	Withdrawal Days for Chickens
Clopidol	5
Chlortetracycline	1
Dihydrostreptomycin sulfate and streptomycin sulfate (injectable)	30
Furazolidone	5
Monensin	5
Nicarbazin	4
Nitrofurazone	5
Novobiocin	4
Oxytetracycline	3 (feed)
Sodium sulfachloropyrazine monohydrate	4
Spectinomycin dihydrochloride pentahydrate	5 (water)
Streptomycin sulfate	4 (water)
Sulfadimethoxine	5 (water)
Sulfanitran and aklomide	5
Sulfaquinoxaline	10
Tylosin	5 (feed)

37
Diseases of the Chicken

There are many diseases affecting chickens, but only the most important ones will be discussed in this chapter. Since this text is on management, the disease section will be aimed at recommendations and directions for the prevention and treatment of each disease discussed. The material involving the cause, symptoms, transmission, and diagnosis has been reduced to include only information necessary to the important management programs.

37-A. PULLORUM

Cause

Pullorum is a highly contagious disease; the organism localizes in the ovary, liver, heart, testes, and other body organs. Although some bacteria are sluffed through the intestinal tract, it is not the seat of multiplication.

The disease is widespread, and unless precautionary measures are provided to control it, mortality will be high. The causative bacterium is *Salmonella pullorum*.

Symptoms

Pullorum disease occurs in the chicken, turkey, pheasant, quail, pigeon, and some wild birds. In chickens the age of the bird influences the symptoms; young chicks are affected differently than older birds.

External symptoms.   Chicks huddle together, and seem chilled. There is an acute, whitish diarrhea. Vent pasting is prevalent. The appetite drops, feathers are ruffled, and the chicks breathe with difficulty. Hock joints may become inflamed. If the infection comes through the hatching eggs, the disease has an early onset; death losses may begin as early as the second day. If other chicks are the source of the infection, most losses occur after 1 week of age. Death in most affected birds is rapid. Losses may run as high as 50%.

In older growing birds, and those in egg production, there are few, if any, external symptoms except for a greenish-brown diarrhea. Fertility and hatchability of the eggs laid by breeder hens may be affected.

Internal symptoms.   An acute septicemia may be involved, and blood infection is the cause of death in young chicks. However, a visual examination of the internal organs may show few changes from the normal, except for the highly mucous contents of the intestines, and unabsorbed yolk on occasion.

In adult birds the internal symptoms may or may not be observable. Localism of the organisms in the ovary sometimes causes some of the ova to atrophy. The sex organs of the male may be affected. Sometimes the heart and gallbladder show indications of a definite infection by the presence of grayish nodules, but generally a diagnosis cannot accurately be made from a visual observation.

Transmission from Bird to Bird

There are several avenues as follows:

1. *Through the droppings.*   Young chicks with the disease continue to sluff *S. pullorum* organisms through the fecal material, and this represents the major means of transmission. However, in adult birds the fecal material contains few *S. pullorum* and is not a major means of spread.
2. *Cannibalism.*   This is an important means of transmission, as blood is transferred from bird to bird.
3. *Birds eating eggs.*   Many of the *S. pullorum* bacteria localize in the ovary, and many of the ova may become infected. Some of the infected ones are ovulated. If a chicken eats a newly laid, infected egg, the organisms gain entrance to the bird.
4. *Equipment.*   Contaminated equipment may be the source of infection. Beak-trimming machines offer a means of transferring the bacteria from bird to bird.

Transmission Through the Egg

The most important route of transmission of infection is from the infected dam, through the hatching eggs, to the newly hatched chick. Although all ova from an infected layer are not involved, enough are infected to carry the causative organisms to the next generation. A few infected chicks will soon transmit *S. pullorum* bacteria to most chicks in the incubator (hatcher) or in the pen.

> *Important:* The hatcher section of the incubator is the seat of high contamination. Organisms from hatch debris—shells, down, and droppings—are easily blown throughout the cabinet by the electric ventilation fans.

Diagnosis

The *S. pullorum* organism is easily isolated and identified in the laboratory. Cultures taken from such organs as the ovary, testicles, heart, liver, and spleen are used to make the laboratory determinations.

Another method may be used in passing judgment on whether the bird has, or has not, had the disease. This test is for antibodies. It is not to be confused with the laboratory test for the presence or absence of active bacteria.

Testing for Antibodies

Once a bird has been infected with the *S. pullorum* organism, antibodies specific for *S. pullorum* bacteria make their appearance in the bloodstream. These antibodies clump with the *S. pullorum* bacterial cells, inactivating them. A similar artificial reaction is used outside the body through a procedure known as the *agglutination test.* This test is widely used to identify layers which have had the disease, but recovered, and are carrying antibodies in their blood.

> *What is needed for the test.* Among the items needed to complete the test are
>
> 1. Whole blood or blood serum.
> 2. An antigen, a specially prepared and standardized mixture of killed *S. pullorum* bacterial cells, dyes, and solvents. It is sold by many manufacturers.

Two types of tests are in general use, the preference being determined by the owner of the birds, or, in many cases, by state governmental regulatory bodies that establish requirements for the blood test in their region. These two tests are as follows:

1. *Rapid, whole-blood test.* A measured drop (0.5 ml) of antigen is placed on the testing plate. Next, the median vein on the underside of the wing near the "elbow" is punctured and a drop of blood is picked up by a loop of wire and the blood mixed with the antigen. The plate is rotated in a circular motion to facilitate mixing and clumping. If the blood taken from the bird contains *S. pullorum* antibodies, they will clump with the bacterial cells of the antigen; such a bird is known as a *reactor.* If no antibodies are present in the blood, the mixture remains clear, and the bird is known as a *nonreactor,* or is *negative* to the test.

Whole-blood antigen.   Most pullorum antigens in use today are of the *polyvalent type;* that is, they contain both the standard and variant strains of bacteria. They also may contain killed bacteria of *S. gallinarum* strains, along with other *Salmonellae.* Some antigens are known as "K"-type; they contain colloidal sulfur as an aid in producing a better reading.

Testing the antigen.   When antigens deteriorate they do not give an accurate reading. Freezing, heating, prolonged storage, and other factors reduce their suitability for completing the whole-blood test. All antigens should be tested prior to any blood tests. Secure vials of negative and positive sera. Complete the test, using a drop of each test serum instead of blood from the chicken. The reaction on the plate (clumping) to the positive serum should be rapid and defined. If not, the antigen should be destroyed, and a new supply procured and tested.

2. *Tube agglutination test.*   A large sample of blood is collected from the bird either by (a) slitting the median vein under the wing, or (b) withdrawing the blood from the vein with a hypodermic syringe. The blood is placed in a small tube until the serum separates. In the laboratory the blood serum is withdrawn from the tube and mixed with the antigen, using varying dilutions. If the bird is a nonreactor (not carrying antibodies) the mixture remains cloudy and uniform. If antibodies are present in the serum (from a reactor), a clumping appears, the clumps fall to the bottom of the test tube, and the solution above is clear. Another laboratory procedure involves plastic plates with tiny depressions. No test tubes are used.

The two tests compared

1. The whole-blood test is more rapid.
2. The whole-blood test may be conducted in the chicken house; the tube test can be completed only in the laboratory.
3. Birds must be banded when the blood is taken for the tube test. The band numbers must be written on the small tubes (or the tubes otherwise identified) as the blood is collected in the chicken house. The banding is necessary to identify and help locate any reactors in the flock.
4. Some feel the tube test is more reliable because specialized technicians run the test under controlled laboratory conditions. But if the whole-blood test is read by properly trained personnel, there is no reason to believe it is not as valid.
5. The cost of conducting the whole-blood test is lower than the cost of the tube test.
6. Certain states and countries recognize only the tube test, especially if hatching eggs or chicks are to travel interstate or intercountry.

Identifying Carriers

Carriers sluff *S. pullorum* bacteria through the egg to the newly hatched chick. When the breeders are blood-tested, these carriers should be identified and removed from the flock. To be adequately certain that all reactors (carriers) are identified, two blood tests, no less than 6 months apart, should be conducted. If, on a test, some reactors are found, the flock should be retested after 21 days. The procedure should be continued until there are no reactors on two consecutive tests.

Why the Test Is Not Infallible

The blood test has its limitations because

1. Cross-agglutination between organisms other than *S. pullorum*, particularly other *Salmonellae*, occurs, clouding the validity of the test. Also, there are variant strains of *S. pullorum*, and if the antigen does not contain bacterial cells of such strains, the test will be unreliable.
2. Often the test is not sensitive enough. Marginal reactors, where the agglutination in the rapid whole-blood

test is only partially complete after 2 minutes on the plate, cause erroneous readings.

3. Infected birds do not always react to the agglutination test. Some birds, particularly during the time they are in egg production, may not react to a blood test, but will react later when a pause in egg production occurs.

4. Antibodies may not have made their appearance. Once a bird becomes infected, antibodies begin to appear about a week later. These antibodies increase in number until they reach a maximum in about 3 weeks. During this 3-week period there may not be an adequate number of antibodies present in the blood of the bird to cause a reaction to either of the agglutination tests.

Should All Breeder Flocks Be Blood-tested?

As a general rule, all breeder flocks should be blood-tested for pullorum disease. However, the integrator, who uses his own chicks in his production program, may find it economically advantageous to dispense with the program. As all primary breeders blood test, the integrator is assured that his breeder chicks at least start their life free of disease. Under modern poultry sanitation programs, there is little chance that the birds will become infected as they grow. But there is no guarantee; some integrators will ''spot test'' a percentage of their breeder birds as added assurance. Furthermore, they take measures to identify the breeder source of all chicks produced. If there is reason to believe some breeding flocks are transmitting active organisms, the flocks can be located easily.

Confirming Agglutination Test When Reactors Are Found

If the agglutination test shows that certain birds are positive (reactors) these birds are to be taken to a diagnostic laboratory for confirmation of the test. This is necessary because the test is not completely accurate. In such circumstances, the bird is sacrificed, and the usual cultures are taken and tests made to determine if active *S. pullorum* organisms are present. Remember, this test is not for antibodies, but rather for the presence or absence of active bacteria.

Pelleting Feed Reduces Salmonella Organisms

Certain feed ingredients, particularly the animal and fish proteins, provide optimum conditions for the growth of many Salmonellae. However, in the course of pelleting feeds some are destroyed. Undoubtedly the heat and high pressure created in the pelleting die, plus steam, are the effective agents. However, penetration of the killing elements is not complete because the pelleting process takes only a few seconds. Therefore, the pelleting process cannot be considered completely bactericidal. Salmonellae must be heated to 180°F (82.2°C) to kill them instantaneously.

Treatment

Chicks. Death losses from pullorum disease in young chicks may be reduced when furazolidone is given to the birds. The drug is quite specific against certain *Salmonellae,* and is bactericidal.

Furazolidone dosage. Add 100 g of pure drug to each ton (2,000 lb) of feed (0.011%) until the disease subsides (usually 2 weeks); then reduce to 50 g per ton (0.0055%) for another 2 or 3 weeks.

Sulfonamides also may be given. They are water soluble and may be added to the drinking water.

Laying hens and replacement chickens. Furazolidone should not be fed to replacement pullets over 14 weeks of age or to laying hens.

Note: Although furazolidone will destroy S. *pullorum* organisms in the blood, it has no effect on those in the intestines. Thus, the bird is alleviated of the disease, and ceases to transmit S. *pullorum* through the egg, but transmission through the droppings continues.

Caution for using furazolidone when blood testing. Do not feed this drug (or most other drugs) the week before completing any agglutination test or submitting birds to a laboratory for a diagnosis of pullorum disease.

What to do when the disease strikes. Evidence of the disease in a group of chicks is the first indication that in all probability the breeders are sluffing viable bacteria through the hatching eggs. The breeders literally have infected the chicks. In such a case

1. If the breeder flock from which the hatching eggs came can be identified, do not use eggs from this flock until the diseased birds have been eliminated.
2. Remove from the incubators any eggs from this flock.
3. Disinfect and fumigate all hatchery equipment thoroughly.
4. Blood test all the birds in the suspicious breeding flock. Remove all reactors to the test; then after 30 days, test again. Repeat the procedure until there are no reactors on two successive tests.

Can infected chicks be saved for future breeders? Recovered birds or flocks should not be saved for breeding purposes unless several blood tests on adult birds show no reaction.

37-B. FOWL TYPHOID

Fowl typhoid is a septicemic disease similar to pullorum, except that mortality from typhoid may occur at any age; it is not confined to young chicks, as in the case of pullorum disease (see Chapter 37-A).

Cause

Fowl typhoid is caused by a bacterium, *Salmonella gallinarum.* In many respects the organism acts like that of *Salmonella pullorum.* Most species of poultry susceptible to pullorum are also susceptible to typhoid. The typhoid organism grows and develops inside the body in a manner similar to *S. pullorum.*

External Symptoms

In comparison with pullorum, typhoid is a slowly spreading disease. The first external symptoms are ruffled feathers, loss of appetite, and a greenish diarrhea. The comb and wattles may be pale, with an anemic-like appearance. At times death in adult birds may be sudden, without any prior indication of illness. In untreated flocks mortality may run as high as 50%.

Internal Symptoms

In the early stages of the disease few tissue changes occur, but in the acute stage the liver is enlarged, and may have a color from bronze to mahogany. In some cases it may be streaked. The spleen and kidneys may be enlarged. In young chicks the egg yolk is usually unabsorbed; the liver is white and friable. The digestive tract is usually empty. As in the case of pullorum, adult carrier pullets may show atrophied ova.

Diagnosis

The diagnosis is the same as for pullorum. Agglutination tests may be used to determine if adult birds are carriers. A laboratory diagnosis is the only accurate means of determining the presence or absence of *S. gallinarum* organisms. The organism grows readily on most laboratory media and may be identified either bacteriologically or serologically.

Since most antigens used for conducting the *S. pullorum* agglutination test are polyvalent (containing pullorum variants), and since *S. gallinarum* will cross-agglutinate with *S. pullorum,* it is unnecessary to run a specific agglutination test using only *S. gallinarum* bacteria. Thus, the so-called blood test for *S. pullorum* is really not specific for this organism; a reactor to the test might be affected with either or both diseases. In either case reactors should be removed from the flock.

Control

Eradication of the carrier birds in the breeder flock should be the first line of attack in handling an outbreak of typhoid in young chicks. Depopulation of the entire breeder house generally is not a practical means of control, but in severe cases such a procedure may be the only method; testing every 30 days to remove the reactors could become an uneconomical and lengthy process. Usually when the percentage of reactors on the first test after a disease outbreak is less than 5%, blood testing may be used; if 20% or more, the flock should be marketed.

Treatment

As with pullorum, furazolidone is the drug most often used in treating birds affected with fowl typhoid. Antibiotics seem ineffective. Do not feed furazolidone to replacement pullets older than 14 weeks of age or to laying hens.

> *Caution:* Fowl typhoid is peculiar in that the disease may reappear after administration of furazolidone. In chicks another treatment will then be necessary. In the case of adult breeder birds, complete eradication of infected birds from the laying flocks will be necessary.
>
> Furazolidone prevents passage of *S. gallinarum* organisms in the droppings. Blood testing commercial egg-laying flocks (and removing the reactors) does not seem economically feasible.

37-C. PARATYPHOID

Beside the Salmonella organisms, *S. pullorum* and *S. gallinarum,* and some Arizona types, over 40 other Salmonellae have been isolated from domestic fowl. These have been grouped as "paratyphoids." Important ones are *S. enteritidis, S. typhimurium, S. montevideo, S. derby, S. meleagridis, S. newport,* and *S. bredeney.* They are quite stable; some can survive moderate heat for long periods; most can live for weeks in media such as water, feed, human food, and soil. Many can infect human beings in a weakened condition as well as chickens, and therefore take on an aspect of great significance. Some illnesses in human beings have been traced to *Salmonella* infections, and when these same organisms are found in or on the tissues of dressed fowl, or in eggs, suspicion is aroused that they might be the contributing source of the disease in people. This has caused the poultry industry to take extreme measures of sanitation in poultry and egg-processing plants.

Cause

Paratyphoids have worldwide distribution. The cause of the specific infection is the result of infection by one of the *Salmonellae* in the paratyphoid group. The life cycle and method of dissemination are similar to *S. pullorum. S. typhimurium,* a serotype, is one of the most troublesome for the poultry producer. Especially involved is the turkey, although chickens are also affected. Most turkey breeders use an antigen specific for *S. typhimurium* in their blood-testing program for pullorum disease.

Symptoms

There seems to be no specific external symptoms, probably because of the wide variety of organisms involved. In some cases the young chicks are listless, and there may be a diarrhea, but some affected flocks show no visible evidence of the disease. In many outbreaks, the disease causes 2 definite periods of death losses: one at 4 to 5 days of age and another at 10 to 12 days of age.

Internally, few symptoms are noticeable, but none are specific; they easily may be confused with those produced by *S. pullorum.* The disease becomes pathogenic through endotoxins, a toxin produced within the organism that does not diffuse from the bacterial cell until the cell degenerates.

Transmission

In general, transmission is similar to that of *S. pullorum,* but certain modes are of more importance.

1. *Egg transmission.* Paratyphoid definitely is egg-transmitted, but eggshell penetration is more important than in the case of *S. pullorum* and *S. gallinarum.* Eggshells are abundantly covered with paratyphoid organisms as the egg passes through the cloaca; when the egg is laid these organisms are sucked through the shell pores, an avenue of embryo inoculation. Contaminated nesting material is a major source of the organisms on eggshells.
2. *Fecal contamination.* The paratyphoid organisms are found in great numbers in the intestinal tract; this ac-

counts for the greatest amount of transmission from bird to bird.

3. *Ovarian transmission.*   Paratyphoid organisms lodge in the ovary; this represents a possible method of transmission of the disease from dam to the newly hatched chick. In most cases, however, there is little ovarian transmission. Young chicks can be infected but will not show evidence of the disease until they are stressed.

4. *Personnel.*   People are capable of transmitting paratyphoid disease. Organisms are carried on clothing and footwear from one location to another. The bacteria are capable of surviving outside the host for many weeks.

5. *Feed.*   As with pullorum disease, the paratyphoid organisms may live in certain feed ingredients for long periods.

Diagnosis

A laboratory diagnosis is necessary to ascertain the paratyphoid involved. Bacteriological methods generally are used. Although the diagnosis is more or less definite in young chicks, trying to type the organisms taken from adult birds may not always lead to conclusions because no sickness or other evidence of the disease is present. Many times it is more practical to culture embryos after 10 days of incubation to determine any correlation with paratyphoid infection in the breeder flock.

Treatment

Most of the time paratyphoid is not a major disease of chickens, as the disease is usually limited to the region of the intestinal tract. Because of some cross-agglutination when the pullorum blood test is made, many adult carriers of paratyphoid are eliminated from the breeding flocks.

Where carefully diagnosed, cases do appear in young chicks, the use of furazolidone at 200 g per ton (2,000 lb) for 2 weeks, or until losses stop, followed by a 100-g level for 2 or 3 more weeks should suffice to prevent the spread of the disease. Tetracyclines have shown promise in control. Sulfadimethoxine plus an antibiotic has been proved to have merit. Do not feed sulfa drugs to birds over 14 weeks of age.

There is little evidence to show that treating the breeder flock with medicaments to prevent egg transmission of paratyphoid via infected ova is of any value, although some drugs may act indirectly to destroy some organisms in the intestinal tract, thus decreasing the number of bacteria deposited on the eggshell.

37-D. FOWL CHOLERA

Fowl cholera has been one of the most baffling of poultry diseases. Until bacterins and sulfa drugs entered the field, there was little that could be done to alleviate the condition in an infected flock. Although better sanitation programs, along with less range rearing of growing birds, have brought about a reduction in the incidence of the disease, there are still many outbreaks, and they still are difficult to control. The disease is very important with turkeys, probably because many are grown on the ground, and turkeys seem to be more susceptible.

Cause

The disease is due to a gram-negative bacterium, *Pasteurella multocida*. There are many immunologically different serotypes associated with different avian species.

External Symptoms

Acute form. Usually, the first observation of the disease is a high incidence of birds dead on the floor or roosts, or in the nests, many times without any obvious external or previous symptoms. Mortality is rapid; 50% or more of the birds may die. Birds between 12 and 18 weeks of age seem very susceptible. There may be a greenish-colored diarrhea.

Chronic form. Death losses are relatively light from the chronic form. The most obvious symptom is a swelling of the wattles, particularly male birds. Adult birds show more of this symptom than do young birds. One or both wattles appear swollen from a cheesy, hard deposit, often resembling a marble. The disease may affect the inner ear, causing a twisting of the head, and an unsteady gait. Some birds may appear lame.

Internal Symptoms

There may be very small hemorrhages on the liver, and sometimes in the heart and intestines. The liver may be enlarged. The disease takes on the form similar to that encountered with CRD. The air sacs may contain lesions, and the heart sac is enclosed in a yellowish-appearing film. Other organs can be affected.

Transmission

The organism responsible for fowl cholera may be passed easily from bird to bird. It will enter the body through either the respiratory or digestive tract. Transmission can be by people, clothing or footwear, and through contaminated feed and water. A healthy bird pecking at a contaminated bird may be a means of spreading the disease through the blood or nasal exudate. Artificial insemination is a means of spread. Ulcerated wattles are a source of infection. The disease is not known to be egg-transmitted. Neither are recovered breeders a source of transmission.

Diagnosis

Fowl cholera can be diagnosed accurately only in the laboratory. The method is bacteriological; the organism is isolated and identified. The test should take no longer than 24 to 48 hours.

Treatment

Treatment involves the use of the sulfonamides. The best seems to be sulfadimethoxine, preferably in water. Although the sulfonamides may bring the disease under control and reduce mortality relapses are frequent once the drug is withdrawn.

Control

Although bacterins have had a history of difficulty in the treatment of cholera, there is now on the market a live vaccine that is quite effective. Only one serotype is used, but it will cross-agglutinate with all others. It is a lyophilized culture of the live strain with a low virulence and must be handled carefully. It will produce immunity in 4 days. In chickens, the vaccine must be administered by wing-web or subcutaneous injection to both males and females at 6 to 12 weeks and again at 18 to 20 weeks of age. Some give injections to layers 34 weeks of age.

Complete Control

The only satisfactory method for the control of cholera is by eradication. Treatment of growing pullets or laying hens either with drugs or bacterins should be considered a temporary measure; the organism must be eliminated from the farm.

37-E. INFECTIOUS CORYZA

This was a serious disease of most farm flocks of chickens a few years ago, but with the practice of more rigid sanitation on poultry farms its incidence has waned. However, in highly concentrated poultry areas, particularly those involving caged layers housed on multiple-age farms, it is still a major problem.

Cause

The disease is due to a gram-negative, nonmotile bacterium, *Hemophilus paragallinarum*. Three serotypes have been isolated, but there may be as many as seven. This is a relatively weak type of organism, and although easily spread from bird to bird, it cannot live outside the body of the chicken longer than 5 to 6 hours.

Symptoms

External. The disease may affect birds of all ages. Usually the first sign is sneezing. This is followed by a watery condition of the eyes, then a discharge of the nasal and sinus passages. Mucus may be squeezed from the nostrils. As the disease continues these areas become filled with cheesy exudates, particularly the sinuses. Swelling occurs and lumps of material appear in the sinuses below the eyes. The mouth and nostrils have a peculiar odor.

Internal. Although the symptoms are usually external, there may be some distress in the air sacs.

Mortality. Death loss from coryza usually is low, but continued infection in the laying flock creates a loss of appetite, and egg production drops. The disease may persist for months.

Transmission

There are two major means of transmission of the disease.

1. *Drinking water.* Contamination of the drinking water from the infected discharges probably is the main method of dissemination. The organisms can remain viable in nonsanitized drinking water for several hours.

2. *Air.* Certain carrier birds transmit the disease to others during periods of stress—moving, vaccination, changes in temperature, etc. Stress triggers a general outbreak in the flock.

Diagnosis

Although visible evidence of the symptoms in the birds usually is enough to diagnose most cases of infectious coryza, laboratory techniques should be employed for positive proof. These techniques are

1. Identification of the organism.
2. Bird inoculation: Nasal exudate from infected birds dropped into the nostrils or eyes of young birds usually will reproduce the disease in 2 days.
3. Serological testing with antigens.

Treatment

The sulfonamides are quite specific in their coryza-control properties. Other drugs, such as oxytetracycline, erythromycin, and streptomycin, are less productive. The mycins should be administered by injection.

For chickens under 16 weeks of age, sulfadimethoxine may be used. The drug does not destroy the organism, but only suppresses its reproductive power. Consequently, after the drug has been withdrawn, coryza may reappear. Thus, treatment of the disease is troublesome and the disease is difficult to obliterate.

Control

Preventing outbreaks of infectious coryza is most difficult because of the nature and ease of transmission of the organism responsible. Several items are involved in such a program.

1. *Keep only birds of the same age on the farm.* Depopulation of the premises under such a program prevents older carrier birds from infecting younger birds.
2. *Use of a bacterin.* A killed, polyvalent bacterin, specific for several isolates of *H. gallinarum* has been prepared and used, mainly with growing birds. Two vaccinations are necessary: the first at 8 to 10 weeks; the second, 4 weeks later.

37-F. COLIBACILLOSIS (*E. coli* INFECTION)

The coli organisms are responsible for a variety of poultry diseases with a variation in manifestation. The *Escherichia coli (E. coli)* are bacteria that represent one of many of the Coliform group of organisms inhabiting the lower intestinal tract. Most are harmless and are called saprophytic; these aid in the process of digestion. Others are pathogenic and produce certain poultry diseases. Although most are harmless, a few pathogenic ones are capable of producing high mortality and morbidity with serious economic loss.

Specific Diseases

E. coli are responsible for several types of diseases, namely

air-sac infection	coli septicemia
Bumble foot	egg peritonitis
coli enteritis	synovitis
coligranuloma	yolk-sac infection

Only the important diseases produced by *E. coli* organisms will be discussed.

Coli Enteritis

The organisms, located in the upper portion of the intestinal tract, cause it to become congested with small blood vessels. Deadly toxins are produced. Vessels rupture, causing hemorrhages very similar to those occurring in coccidiosis.

Nodules also may appear in the cecal lining, but whether the coli organisms are primarily responsible for the intestinal and cecal disorders is not known. Most coli bacteria produce their effects by being secondary invaders. When coccidiosis is involved along with *E. coli* infection there is the question of which was there first. The fact that *Coliforms* are ever-present in the intestinal tract makes the differentiation most difficult.

Incorrect diagnosis a problem. Because the lesions of the intestinal tract produced by *E. coli* and coccidiosis are similar, care should be taken to arrive at the correct diagnosis. Continued treatment for coccidiosis when the disease is caused by *E. coli* can only aggravate *E. coli* infection and lead to more serious complexities.

Coli Septicemia

The next step in coli infection is coli septicemia, when toxins and bacteria enter the bloodstream after the toxin produced by the coli organisms ruptures the intestinal wall. Such lesions allow the organisms to gain entrance to the portal system. In turn, they find their way to the kidneys, a blood-filtering organ. As the filtering continues, the kidneys become congested and enlarged. The liver is next, and it too, becomes enlarged, its edges rounded, and its surface speckled. The discolored areas enlarge when the organisms kill sections of the liver tissue.

Air-Sac Infection

Eventually *E. coli* involve the air sacs, the organisms arriving by way of the bloodstream. Airsacculitis results, and the birds cough and wheeze. Morbidity, rather than mortality, becomes the economic problem, especially in broilers, where birds with infected air sacs will be condemned in the processing plant as unfit for human consumption.

But the *E. coli* organisms may find their way to the air sacs by a more direct path. They may enter the upper respiratory tract through the routine of breathing, and soon settle in the thoracic air sacs, then eventually find their way to the abdominal air sacs. When infection reaches its height, the air sacs become filled with a yellowish cheesy material. A similar material also surrounds the heart and lungs.

Transmission

There are several means of *E. coli* transmission.

1. *Fecal.* Organisms in the intestinal tract are continually being sluffed through the fecal material, and in turn these bacteria dry and float in the air and gain entrance to uninfected individuals by way of the respiratory tract. Of significance in this mode of spread is the fact that the intestinal organisms are almost immune to the production of antibodies; thus the *E. coli* organisms continue to reproduce in the intestines and birds may remain carriers for a long period of time.

2. *Eggshell contamination.* As the completed egg lies in the cloaca prior to being laid, it becomes contaminated with the excrement of the intestinal tract, including *E. coli.* More bacteria are added to the shell when the egg remains in the nest. Subsequently, some organisms enter the egg contents and reach the developing embryo with resultant loss in hatchability and chick quality.

3. *Respiratory.* As air-sac infection from *E. coli* can be the result of infection through the respiratory tract, contaminated dust in the poultry house can be a direct cause of transmission.

4. *Ovarian.* Transmission through the ovary is possible when birds are shedding the *E. coli* organisms through uterine infection. Infected breeder hens thus transmit the disease to the newly hatched chick.

5. *Feed.* Although not a primary route of infection, coliforms may gain entrance to the body through contaminated feed.

Diagnosis

A laboratory test is the only satisfactory method of accurate diagnosis. Coliforms are isolated and classified.

Treatment

Any treatment must begin with a cleanup campaign as most *E. coli* infections start with dirty surroundings.

Sulfadimethoxine plus Ormetoprim is the only feed treatment recommended for colibacillosis. Other drugs have been used including tetracyclines, sulfas, novobiocin, and gentamicin. An antibiotic sensitivity test may be helpful in determining effective medications.

37-G. OMPHALITIS (NAVEL INFECTION)

Omphalitis is a disease that affects chicks when they are hatching. It results in an infection of the umbilical opening.

Cause

Omphalitis is a disease of general bacterial infection due to several organisms. They may be *Coliform, Staphylococcus, Pseudomonas,* or other types. Bacteria invade the umbilical tissues as the result of improper conditions in the hatcher. The navel opening does not close and infection passes to the internal organs.

Symptoms

The chicks seem weak, huddle together, and may have a watery diarrhea. Upon close observation, an infected and open umbilical area will be noticed. It will be discolored to bluish black. There is a very noticeable pungent odor characteristic only of this disease. The abdomen feels soft, mushy, flabby, and enlarged. The infection may carry to the internal organs, particularly to a portion of the intestines. Peritonitis may be found when the yolk sac ruptures. Mortality may run as high as 10%.

Transmission

The disease is very infectious, and death occurs within 2 or 3 days after hatching. The seat of the infection is the incubator (hatcher) although the original source probably is bacterial contamination of the eggshells before the hatching eggs enter the hatchery. The ventilating fans quickly disseminate the organisms, which find the unhealed navel a likely seat for infection. Once the chicks are in the brooder house, the likelihood of transmission from chick to chick is very low, although possible.

Diagnosis

Suspect chicks should be submitted to the laboratory for diagnosis. A bacteriological examination of the yolk-sac contents will identify the disease and the causative organisms.

Treatment

If the chicks in the brooder house appear chilled, increase the brooding temperature. Although antibiotics or nitrofurans may check the disease in the brooder house, the possibility is very remote; little can be done. In fact, it is better that the infected chicks die.

Control

When an outbreak occurs in the hatchery, equipment and rooms must be fumigated with formaldehyde gas. Use 3× strength where possible. Lower concentrations are not effective in destroying all responsible organisms. Incubating eggs (1 to 19 days) should be fumigated at 2× (double) strength, never more, for this procedure (see Chapter 9-J).

Hatchery rooms and all equipment must be fumigated every second day until the infection is destroyed. Also, use a good liquid disinfectant where practical.

> *Remember:* An outbreak of omphalitis on the poultry farm is almost always the result of an infection in the hatchery. Extra measures must be employed to rid the hatchery of the organisms responsible and to keep it clean thereafter.

37-H. MYCOPLASMA GALLISEPTICUM (MG)

This disease is known everywhere and is extremely important to both the broiler grower and the table egg producer. Infection of the air sacs in broilers is cause for condemning the dressed birds as unsuitable for human consumption. Laying flocks positive for MG have been shown to produce as many as 20 fewer eggs per year than MG-negative flocks.

Cause

The *Mycoplasma gallisepticum* organism is very small and delicate and has no cell wall. Over 20 serotypes have been discovered. The one to which this disease is attributable is known as S-6 serotype. It has been found in the chicken, turkey, and duck. MG is a stress disease because the organism seems to remain dormant in many flocks but when birds are stressed it becomes active. More MG develops in cool or cold houses than in warm.

Symptoms in Young Birds

MG is a respiratory disease, affecting the entire respiratory tract, particularly the air sacs, where it localizes. All the air sacs may become involved, are cloudy in appearance, and filled with mucus. In the latter stages of the disease this mucus develops a yellow color and a cheesy consistency. Similar exudates may encircle the heart and heart sac.

MG in itself is not a killer; in fact, even morbidity is not great; but an outbreak is quickly followed by many secondary infections, and it is these that do the damage. *Coliform* organisms are particularly involved. Thus, visible identification of MG is often confused by symptoms of secondary invaders.

In young chicks there is a rattling, sneezing, and sniffling, all indicative of the respiratory difficulty. If complicated by other similar respiratory diseases these symptoms are accentuated. In severe cases mortality may run as high as 30%.

Symptoms in Mature Birds

Visible evidence of the disease in adult birds may go unnoticed. Occasionally the birds will appear depressed and inactive. There may be a definite diarrhea during the intestinal phase of the disease. Egg production may suffer. Mortality in adult birds is low.

Transmission

There are several methods whereby the *Mycoplasma gallisepticum,* S-6 serotype organism may be transmitted.

1. *Through the hatching egg.* This is the major means of spread; the organism passes from one generation to the next. Transmission is not uniform, for it is greater during the period of active infection in the adult birds, then subsides. But only a few organisms in a few infected chicks are necessary to cause the disease to spread to practically all young chicks in the flock.
2. *Through the air.* The organisms may move short distances by way of the air. This is enough to cause infection of birds within a pen, but probably has little consequence in the transfer from house to house.
3. *On clothing, feed bags, feed, egg filler flats, poultry equipment, and trucks.* Personnel represent the important carriers. Probably 60% of cross-contamination between houses is the result of the organisms transported by people.
4. *Through infected chickens.* Complete depopulation of poultry houses and farms is essential to prevent exposing the new flock to the disease.

Diagnosis

MG may be fairly accurately diagnosed by the coughing and sniffling involved. Laboratory tests used for diagnostic determination are

1. *Rapid serum plate or tube agglutination test.*
2. *Hemagglutination inhibition test.*
3. *Embryonic examination.* Embryos infected with MG have lesions in the air sacs. A few cull chicks or pipped embryos should be examined at hatching time. Positive evidence of infection is conclusive proof that the parents have MG infection.

Treatment

Tylosin is an antibiotic specific for the treatment of birds infected with MG.

Aureomycin, erythromycin, spectinomycin, doxycline, or streptomycin may also be used as drugs for the treatment of the disease in growing and adult birds.

Egg dipping using Tylosin or Gentamycin solutions is used for treating hatching eggs (see Chapter 9-L).

Control

Medication can only be considered a temporary solution and is usually quite expensive. Control can be by vaccination, controlled exposure, or eradication. Eradication is built around complete curtailment of embryo transmission. Thus, infected dams must be removed as a source of infection. Unlike pullorum disease, MG eradication is difficult. First, it is so contagious that one or two infected birds in the pen will infect practically all others in a very short time. Second, an infected bird continues to be a carrier, and sluffs virus through the egg. Thus, if any infected birds are found in the breeding pen, all the birds in the pen must be eliminated as a source of hatching eggs. There is no such thing as testing and retesting, as with pullorum disease. Furthermore, breeder flocks may remain clean (no infected birds) for long periods, but suddenly break with infection. But eradication at the breeder level is the best method of handling the disease.

Tests Involved with MG Elimination

There are several methods of testing involved in eliminating the infected breeder flocks and in testing the day-old chicks to determine their disease status:

1. *Testing the breeder flock.* The test is similar to the pullorum whole-blood test except that blood serum is used for conducting the MG test instead of whole blood.
    a. Bleed the bird from the wing and collect the blood in a small glass test tube. Place the tube in a horizontal position and allow the serum to separate. At 70°F (21°C) this will require 1.5 to 2 hours. Then refrigerate at 45°F (7.2°C). Do not freeze.
    b. When ready to make the test, remove the test tubes from the refrigerated compartment and warm them to room temperature. Then place a drop of MG antigen on the testing plate.
    c. After running positive-serum and negative-serum tests on the antigen to be used, place a drop of the bird's serum on the antigen, and mix. If the bird is

positive (infected), a clumping will appear within 2 minutes. If the mixture is clear, the bird does not carry antibodies, and is not an infected carrier.

There is also a laboratory tube test for detecting infected birds.

2. *Testing day-old chicks.* Since the breeder flock is blood-tested only every 3 weeks during the laying season, there is a possibility that it could "break" with the disease between tests. As assurance that the chicks are not affected, culls or dead embryos should be blood tested at hatching time. Test about 1% of each group of chicks, and test every hatch.

First test the antigen, using negative and positive sera. Then collect blood from the neck, and complete the plate test as outlined for breeding birds.

3. *Checking for lesions in day-old chicks.* MG lesions are to be found in the thoracic air sacs of infected day-old chicks. Pipped embryos should be examined at every hatch for any evidence of lesions. About 20 chicks from each group should be examined.

Vaccine to Control MG Infection

Primary breeders of broiler and egg-laying stock have gone the route of eliminating MG from their flocks. Most multiplier flocks are also now negative to MG but it is still a common disease in table-egg flocks because of the multiple-age flock practices in common use.

In order to protect layers in cages, a live MG vaccine has been produced and is approved for use in several states. This is an avirulent live vaccine administered in the drinking water when the growing pullets are about 12 weeks of age. The vaccine shows promise by reducing egg production drops when the MG organism invades a flock. Where the vaccine is used, farms should be known to be positive for MG.

There is also a killed bacterin vaccine for the prevention of MG. It prevents the clinical signs of the disease without the threat of spreading the organism in the flock. Administration of the vaccine is by intramuscular or subcutaneous methods. Breast injections are not recommended because of vaccine residue problems.

37-I. MYCOPLASMA SYNOVIAE (MS)

A prevalent disease for many years, the identification of the causative organism has brought new understanding of the intricacies behind MS.

Cause

The disease is due to an organism, *Mycoplasma synoviae*, similar to *Mycoplasma gallisepticum* (see Chapter 37-H). There is one serotype. It is small, delicate, and has no rigid cell wall.

Symptoms

Actually, MS is a respiratory disease but seldom is the respiratory tract involved with sickness or death, though air-sac infection is noticed when broilers are processed. However, the organisms soon locate in the synovial fluids of the hock and joints of the footpads. These two areas become swollen and inflamed. In some severe cases, the wing joints may become affected.

In most instances this is a disease of young and growing chickens between 6 and 14 weeks of age, but it can attack older birds. There is a loss of appetite and weight. Birds are lame when the joints become inflamed, and sit on their hocks. A persistent tenosynovitis may be evident in layers. Morbidity, rather than mortality, is a problem; seldom is there a very high death rate in older birds.

Transmission

There are several known means of transmission and probably others that are not clearly understood.

1. *Through the hatching egg.* The disease is definitely egg-transmitted, and although the percentage of infected eggs is very low, these will give rise to enough infected chicks so that eventually a high percentage of the flock becomes infected. After infection, breeders will sluff MS organisms for as long as 14 to 40 days.
2. *Through the air.* The organisms are easily transmitted from bird to bird within the pen by air, but probably not from house to house.
3. *On clothing, trucks, equipment, etc.* Mechanical means of transfer are of major importance in transporting the organisms over long distances.

Diagnosis

Swelling of the joints of the hock and footpad is a symptom of MS, but not necessarily a diagnosis; there are too many diseases that produce similar conditions. Birds should be submitted to the laboratory where two tests may be conducted.

1. *Plate agglutination test.* This is a test similar to the agglutination test for pullorum disease and MG. Blood serum is mixed with MS antigen. As it is an antibody test, serum drawn from birds that have had the disease will clump with the antigen. This also may be conducted in the poultry house. *Note:* There is some cross-agglutination between *M. gallisepticum* and *M. synoviae* organisms. *M. gallisepticum* antibodies may cross-agglutinate with MS antigen. The opposite is not true.

2. *Bird inoculation.* Material is drawn from suspected hocks, ground, strained, and injected into the joint of the footpads of 4-week-old chicks. If the donor is infected with *M. synoviae* or *M. gallisepticum*, there will be a definite swelling of the footpads of the recipient chicks in about a week. Differentiate MS from MG serologically.

Treatment

Some broad-spectrum antibiotics are of value.

1. *Chlortetracycline.* Add to the feed at the rate of 100 to 200 g per ton (2,000 lb) for about a week. Do not feed to laying birds.

2. *Oxytetracycline.* Add to the feed at the rate of 200 g per ton (2,000 lb) for about a week. Resume drug treatment if disease does not subside. The treatment may be continued longer if the disease is not brought under control by the first treatment.

3. *Individual injections.* Serious outbreaks in adult birds may be treated by injecting oxytetracycline or erythromycin. Individual treatment in this manner is not practical with broilers because of the labor involved and the stress of handling.

Control

Complete eradication is the solution to the control of MS, but to completely eliminate the disease from a farm is subject to some difficulty. The problem lies in the fact that some birds do not produce antibodies against the organism and therefore the presence of MS cannot always be detected by the blood test.

1. *Eradication.* Breeders that have had MS should be removed from the farm and never used to produce hatch-

ing eggs. All breeders should be blood-tested beginning at 7 weeks of age and every 4 weeks thereafter during growing and laying. Test about 2% of the flock each time. Market the entire flock if a reactor appears and destroy all eggs in the incubators that came from the flock.

2. *Heat treatment of hatching eggs.* A procedure has been established for heating hatching eggs before they are placed in the incubator. The eggs are heated to 115°F (46°C). At this temperature the Mycoplasma are destroyed, but hatchability is affected only slightly. Full details of this procedure are given in Chapter 9-K.

37-J. INFECTIOUS BRONCHITIS (IB)

Infectious bronchitis is a disease affecting chickens in every part of the world. It is a serious affliction of young chicks, causing high mortality. In laying birds it entails a great economic loss through reduced egg production and eggshell quality. The chicken is the only bird known to be susceptible.

Cause

The disease is caused by a filterable virus. There are 20 or more recognized serotypes. In instances some of the serotypes produce cross-immunity. The best known serotypes are

Massachusetts
Connecticut
Holland

Lesser-known serotypes are

Arkansas 99
Florida
JMK
SE-17
Australian T
Arkansas Type DPI Strain (3168)

Properties of important types

1. The Massachusetts strain produces the severest type of the disease.
2. The Massachusetts strain produces cross-immunity with the Connecticut strain.

3. The Connecticut strain produces a poor cross-immunity with the Massachusetts strain.
4. The Holland strain produces cross-immunity with Massachusetts, Connecticut, JMK, Florida, and SE-17 strains.
5. The JMK strain is the most prevalent in some areas; Arkansas-99 in others.
6. No other vaccine affects Arkansas Type DPI Strain (3168) isolate for which there is a specific vaccine.
7. Many strains produce uterine damage in young chicks that depresses egg production later.

Symptoms

The disease produces symptoms varying with age; chicks and adult chickens are affected differently.

Infectious bronchitis in chicks. In young chicks there is a noticeable wheezing and sneezing, particularly at night. There may be a nasal discharge, watery eyes, and swollen sinuses. The birds gasp for breath. The disease has rapid dissemination following almost instantaneous onset. Mortality may run as high as 50%; morbidity affects practically all the remaining birds. The disease affects the immature reproductive system of the young pullet chick, leading to a reduced egg performance in the laying house.

Secondary invaders follow. The severity of the disease is associated with the damage done by secondary diseases, particularly those produced by the Coliforms. The incubation period of infectious bronchitis is from 18 to 36 hours. Usually the disease will run its course in 5 to 20 days, but the effects of secondary invaders may linger for long periods of time.

Infectious bronchitis in adult birds. As in chicks, the disease starts fast, without notice. Infection from bird to bird is rapid. However, there are few noticeable symptoms involved with the bird itself; the disease is manifested by a severe drop in egg production. After the disease subsides, the return to normal egg production may take several weeks. Egg quality undergoes a drastic change. Eggs are soft-shelled, misshapen, and wrinkled; eggshells are porous, chalky, and light in color. Albumen quality is poor. Even though egg production eventually may return to normal, egg quality seldom does.

Many birds become internal layers, making visits to the nests without laying an egg. Mortality is generally negligible. The disease lasts from 4 to 10 days.

Internal symptoms. Posting shows a mucus in the trachea, nasal passages, and sinuses. The air sacs of young chicks may contain cheesy deposits. In older birds there may be a low incidence of lesions, and in many instances there are none.

Transmission

IB is not egg transmitted; birds contract the disease through the transfer of the virus. Bird-to-bird transmission is by means of organisms being carried

1. *By the air.* It takes just a few of the virus organisms to infect a bird. As the virus is easily airborne, inhaling infected air is the most important means of spread.
2. *By people, birds, and animals.* This represents a major means of spread from house to house and farm to farm.
3. *By equipment, etc.*
4. *In or on the feed.*
5. *By carrier birds.* Birds may shed the virus for as long as 4 weeks after recovery.

Diagnosis

Diagnosis of infectious bronchitis is difficult; it is often made by eliminating the incidence of other similar diseases as causative agents. Newcastle disease and laryngotracheitis are two such. The serotypes of the bronchitis virus must be tested separately in the laboratory for an accurate diagnosis. This is very time-consuming, and in most instances the disease will have run its course before the tests can be made. The procedures usually followed in the laboratory are

1. Serum neutralization (SN) test
2. Virus isolation test
3. Hemagglutination test
4. Fluorescent antibody test
5. Enzyme-linked immunosorbent assay (ELISA) test (see Figure 37-1)

Treatment

There is no known treatment for infectious bronchitis. However, when secondary infections are in evidence, treatment for these may alleviate the damage to the bird. Laboratories making post-mortem and laboratory examinations usually will analyze the situation thoroughly for the presence of accompanying diseases.

Figure 37-1. Enzyme-linked immunosorbent assay (ELISA) test.

Control

Control of infectious bronchitis is by vaccines of established type and quality. Live attenuated types are recommended. The vaccine used will depend on the age and type of bird, and the area. As so many variant strains of IB virus have been found, the vaccination program must be what will afford the greatest protection in a particular location.

General recommendations for type of vaccine. As most serotypes will produce cross-immunization with the Massachusetts strain, it should be considered as the main component of most bronchitis vaccines. But it must be remembered that several serotypes will not be affected by either the Massachusetts or Connecticut strains, including the Arkansas-99, Florida, and JMK viruses, but the Holland type will. However, many mixed vaccines, containing both the Massachusetts and Connecticut strains, are on the market, and are predominantly used under certain conditions.

There are two types of Holland strain vaccine: (1) mild, and (2) virulent. Do not use the virulent strain as the first vaccination.

Important: Polyvalent vaccines (containing more than one strain) are more likely to produce a greater stress following vaccination than those that are monovalent. This may be a disadvantage in some instances. Vaccines composed of some serotypes other than the Massachusetts and Connecticut strains may produce kidney damage, and therefore are not generally suitable for vaccine preparation.

How vaccines are administered. Vaccines may be classified according to their means of administration as nasal or ocular, dust, water, and spray.

Method of vaccinating broilers. Broilers should be vaccinated only where the disease has taken on acute proportions. The reasoning behind this is that the stress and side diseases produced by bronchitis vaccination may cause greater flock morbidity than the disease itself.

Most vaccines should be administered when the broilers are from 14 to 21 days of age, after most parent immunity has dissipated, but there are other programs, as follows:

> *Spray at day old.* The vaccine is sprayed into the mouth and trachea by a special machine used in conjunction with day-old beak trimming. Immunity is produced even in the presence of parental immunity through the manufacture of lymphocytes by the Harderian gland located in front of the eyeball. Vaccination at 6 to 10 days does not produce as much immunity as 1-day vaccination.
>
> *Intraocular at day old.* An intraocular vaccine may be used at day old with results similar to the spray method.
>
> *Note:* Either of the above two methods is satisfactory for broilers as the immunity established seems adequate for the short time period necessary to produce a broiler, but the day-old vaccination should not be used for birds to be kept for egg production or breeding purposes.

Method of vaccinating egg-type replacement pullets. Replacement pullet growers usually vaccinate two or three times for IB between 2 and 14 weeks of age. The vaccine is usually given in the drinking water in combination with Newcastle disease vaccine. Cage-reared pullets are commonly sprayed with IB vaccines using large droplet techniques.

Method of vaccinating birds for breeding purposes. During the growing period use IB vaccinations similar to those shown in Table 39-4. Breeders producing hatching eggs must be vaccinated for IB during the egg-production period. The longer the period after vaccination, the more the immunity decreases; and as the presence of antibodies in the blood of the laying, breeder hen passes on parental immunity to the chick, it becomes necessary to revaccinate the breeders to produce uniformity of parental immunity in the offspring. This program calls for administering a bronchitis vaccine once every 10 to 20 weeks during the laying cycle.

Combination vaccinations. Bronchitis vaccine is sometimes combined with Newcastle disease vaccine and birds are vaccinated for both diseases at the same time.

Bronchitis–Newcastle vaccine combination comments are as follows:

1. Bronchitis virus vaccine multiplies more rapidly than Newcastle vaccine and the bronchitis growth may interfere with the growth of the Newcastle virus.
2. Newcastle virus vaccine is more stable than bronchitis vaccine. Deteriorated mixed vaccine is more likely to produce immunity to Newcastle disease and not to bronchitis.
3. There is more variability in the potency of bronchitis vaccine than of Newcastle virus vaccine.

Establish the titer. There may be failures of bronchitis disease vaccinations due to improper vaccinating methods, deteriorated vaccine, parental immunity, and other causes. In order to ascertain the potency of the immunity, titers should be established by a competent laboratory technician. If the titers are low, the birds should be revaccinated.

Employ professional help with IB vaccination programs. Because of the complexities of bronchitis variants, and the available vaccines, along with area-differentiating types of the disease, a competent virologist should be contacted to detail your best IB vaccination procedure. Some states require permits to use vaccines made from other than the three common strains of the virus.

37-K. NEWCASTLE DISEASE (ND)

Newcastle disease was named for the town in which it was first diagnosed—Newcastle, England. A highly infectious disease, in some areas of the world it takes on increasing virulence, adding to its importance. In certain localities ND is known as Ranikhet disease.

Cause

The disease affects chickens, turkeys, pheasants and many other birds. It is due to a filterable virus, and although there is one serotype, there are four general forms classified according to their pathogenicity. The virus may be neurotropic (nervous), pneumotropic (respiratory), or viscerotropic (internal organs).

Death in the subacute form is usually the result of paralysis. In some cases the organisms invade the respiratory tract; in others, they may locate in the intestines and proventriculus.

Symptoms

Symptoms vary according to the age of the bird and the form of New-castle disease virus involved. Three symptoms usually are major:

1. Respiratory difficulty
2. Nervous disorders
3. Egg production and eggshell quality reduced

The virus localizes in the respiratory tract, and all affected birds show evidence of the respiratory type. If nervous symptoms are involved, they arise later. Older birds seldom show any manifestations of nervous disorders. Egg production and eggshell quality are affected quickly in laying birds.

There are four forms of the disease, and the symptoms for each are

1. *Viscerotropic velogenic* (VVND) (Exotic ND) (very high pathogenicity)
   Sometimes known as the Asiatic type.
   Highly virulent; high mortality.
   Respiratory and nervous signs less evident.
   Spasms and twisted necks in young chicks.
2. *Velogenic or neurotropic* (field-type) (high pathogenicity)
   Sudden onset, acute, often fatal.
   High morbidity.
   Evidence of nervousness (twisted neck) and respiratory difficulty.
   Lesions usually found in the respiratory tract only.
   Most ND outside the United States is of this type; therefore, vaccination programs for the United States may not be effective elsewhere.
3. *Mesogenic* (intermediate pathogenicity)
   Disease acute in young chicks.
   Respiratory and nervous symptoms in young chicks but not in older birds.
   Most common type of ND in the United States.
4. *Lentogenic* (mild pathogenicity)
   All ages of birds may have unnoticed infections.
   Mild respiratory difficulty.
   Egg production declines.
   Eggshell quality deteriorates rapidly

Vaccines and Vaccination Procedure

Vaccines against Newcastle disease are usually made from the mesogenic and lentogenic forms of the virus, and in some degree give protection to all four forms of the disease. However, each vaccine is highly variable in the response it produces. A summary of these effects is as follows:

Lentogenic strains (B₁-type). In the United States the lentogenic vaccines will prevent drops in egg production, but this is not always true outside the United States.

1. *F Strain.* The vaccines from F strain have the lowest virulence of the common lentogenic strains. They are most effective when a flock is vaccinated individually.

2. *B₁ strain* (Hitchner). This is slightly more effective than the F strain. It is usually given in the drinking water or by the spray method. There is little spread from bird to bird. It may be given at day old but then must be followed by a La Sota-type vaccine at 10 to 14 days of age.

3. *La Sota strain.* Of the lentogenic vaccines, this is the one most often used. The spray method is the usual route of early administration. The strain is particularly adaptable to the first vaccination and to booster vaccinations. But care should be taken because these vaccines vary in virulence. After La Sota-based vaccines are given, there is mild bird to bird spread. It will not prevent the drop in egg production when adult birds are severely challenged.

Mesogenic strains

1. *Mukteswar strain.* This strain is particularly pathogenic and its use for vaccines should be confined to those birds previously vaccinated with one of the lentogenic vaccines. It has had wide acceptance in tropical climates and in southeast Asia.

2. *Hartfordshire (H) and Komarov (K) strains.* Vaccines prepared from these strains are less pathogenic than those from the Mukteswar strain. The H strain may be administered by subcutaneous or intramuscular methods.

3. *Roakin strain.* Vaccines prepared from this strain are often called *wing-web vaccines.* They are isolates that are attenuated, but still very virulent. They are administered by dipping a needle in the vaccine, then forcing the needle through the web of the wing.

   Roakin-strain vaccine cannot be given to young chicks that carry any degree of parental immunity, which means before 3 weeks of age, but because of its virulent nature, vaccination is best delayed until after 8 weeks. However, most outbreaks of Newcastle occur before this age, so the vaccine has definite shortcomings.

   If given, Roakin-type vaccine should be administered

after one or two vaccinations with a lentogenic-type vaccine. It then has merit when there is a natural invasion of the Newcastle virus as it will help prevent mortality and will help prevent a drop in egg production and in eggshell quality.

Transmission

Newcastle disease virus particles are easily spread; the disease is highly contagious. Methods of transmission are

1. *Through the air.* Coughing dislodges the virus from the respiratory tract whence it is easily airborne. It travels quickly from bird to bird and from house to house over short distances.
2. *On clothing, unsanitized filler flats, equipment, feed trucks, and so on.* This category probably represents the major means of transfer of the virus to uninfected flocks and farms.
3. *No cleanup period on farm.* Poultry operations that start chicks on a regular basis, resulting in several ages of birds on the farm at one time, have continuous problems; the older birds reinfect the younger. Farms with an "all-in, all-out" program of management break any cycle of infection when the premises are depopulated.
4. *Feed.*
5. *Wild birds, neighboring poultry.*
6. *Exotic birds.* Particularly those shipped from other areas or countries.
7. *Predators.*

Diagnosis

Many times a diagnosis may be made on the basis of physical observation. Nervous disorders inflicting twisted necks and causing some young birds to be "tumblers" (head between legs causing the chicks to roll or tumble) are enough to warrant a fairly accurate diagnosis. When the diagnosis becomes more difficult, certain laboratory techniques must be employed.

1. Hemagglutination test.
2. Virus isolation test. Inasmuch as the virus disappears quickly from the bird, only birds in the stages of the disease should be submitted to the laboratory for this test.

3. Hemagglutination-inhibition (HI) test. This test, being quantitative as well as qualitative, enables the technician to establish a titer.
4. Fluorescent antibody test.
5. Enzyme-linked immunosorbent assay (ELISA) test.

Treatment

There is no known treatment, although broad-spectrum antibiotic medication for secondary diseases will probably reduce some of the flock morbidity.

Vaccination Programs

If Newcastle disease is in the area, all birds on the flockowner's premises should be vaccinated. Transmission from farm to farm is too easy to justify not vaccinating.

Vaccinating young chicks. Young chicks, both broilers and breeder replacements, should be vaccinated after parental immunity has subsided in order to activate the immune system. This is never before the chicks are 14 days of age; preferably they should be 21 days old.

Alternate vaccination method for broilers. Vaccinate at day old with a mild B_1-type vaccine, but this should be followed with a La Sota-type at 10 to 14 days. A B_1-type vaccine is the only one that should be given at such a young age.

According to J. J. Giambrone, Auburn University, the coarse-spray system of vaccinating broiler chicks with high levels of maternal antibodies at day old produces better immunity than other methods, but any method does not give complete immunity to market age. There should be a revaccination between 14 and 21 days, especially during the cooler months of the year when the incidence of viral respiratory disease is increased.

Important: Parental immunity is the result of antibodies in the chick. But, as these are confined to the bloodstream, they do not give protection when the Newcastle virus enters the bird by way of the respiratory tract. Thus, parental immunity does not protect the chick from natural outbreaks of the disease by this route.

La Sota strain of virus best used for chick vaccination. La Sota is a mild strain of vaccine and can be used on chicks at 7 to 10 days of age without undesirable effects. There are no nervous symptoms, and only mild spread of the virus from chick to chick. Water, ocular, or intranasal methods of vaccination may be used.

Give a second vaccination. Complete "take" on the first vaccination at 7 to 10 days of age may not develop immunity in all chicks, because some may have parental immunity. A second vaccination should be given when layer or breeder replacement birds are 6 to 7 weeks of age in order to vaccinate birds that are still susceptible or have low antibody titers. Broilers usually are given the early vaccination only. Replacement pullet flocks should be monitored to determine the antibody level and the uniformity of vaccination response. In most cases, a third and final vaccination is given between 14 and 18 weeks of age.

Killed virus vaccines.   These lose some of their efficacy during the manufacturing process and therefore lack some ability to produce antibodies; postvaccination titers are lower. On occasion they will have a place in the vaccination program. A new one is often used in combination with a bursal disease vaccine. A freeze-dried Newcastle-bronchitis vaccine for spray use is also available.

Changing from killed-virus to live-virus vaccine.   On occasion some poultry producers may wish to change from a killed-virus vaccine to a live-virus vaccine. Certain practices must be followed.

1. Do not administer the live-virus vaccine for at least 3 to 4 weeks after the last killed-virus vaccine was used.
2. It is best not to wait longer than 3 months after the last killed-virus vaccine was given before giving the live-virus vaccine.
3. Birds in egg production may be vaccinated with the live-virus vaccine whether they have been previously vaccinated with killed-virus vaccine or not. There may be some respiratory disturbance and a slight drop in egg production depending on their level of susceptibility and the vaccine used.

Vaccinating breeder birds.   As in the case of infectious bronchitis, parental immunity in the chicks should be uniform. This means that the titers of the breeder females should be kept at a high level, and uniform, in order to have uniform parental immunity in the chicks. Breeders should be vaccinated every 10 to 12 weeks while they are producing eggs.

Velogenic Newcastle disease.   There have been outbreaks of a highly virulent Newcastle disease in the United States and Canada. Such outbreaks of a "hot" strain are characterized by respiratory symptoms, hemorrhagic conditions of the intestinal organs, high mortality, and a drastic drop in egg production.

Although vaccination programs commonly in use by most of the poultry industry are adequate for the usual mild variety of Newcastle disease, these programs will not give complete protection against velogenic strains of the virus. Individually applied vaccines (intranasal, ocular, or intramuscular) may provide better immunity than mass vaccination. Intramuscular-type vaccines may be given at 2 weeks of age. If there is evidence that velogenic Newcastle disease is in your area consult the nearest diagnostic laboratory for advice and type of vaccine and vaccination program recommended. Usually, infected flocks will have to be destroyed under governmental supervision.

Inefficiencies of Vaccination

Effective control measures can be accomplished only by a rigid program of vaccination. However, vaccines and vaccination procedures are so highly variable that extreme care must be followed in carrying out the programs. Even then there are some failures. Variations occur because

1. Strains of the virus differ with various vaccines.
2. Potency of the vaccine differs at time of manufacture.
3. Killed and live viruses are used to make vaccines.
4. Deterioration of the vaccine causes wide variation in its efficacy.
5. Mode of administration of vaccine varies (spray, water, wing web, etc.).
6. Administration may not be uniform with each bird as the result of the method of administration.
7. Parental immunity in the bird influences the degree of response.
8. Season of the year influences the bird's acceptance of the vaccine.
9. Various vaccines differ in developmental stress of the birds.
10. ND vaccines are often mixed with vaccines used for other diseases, e.g., infectious bronchitis.

Necessary Titer

Titer is a measure of antibody production. Following a Newcastle vaccination, a titer of 100 is considered immune, but a reading of 500 to 600 shows excellent immunity. After a natural outbreak of the virulent form of ND, some titers have been as high as 20,000, but 2,500 is more common. A good vaccine should produce a titer of 2,000 soon after vaccination with a titer of 500 after 6 months.

Various antibody monitoring methods utilize different numbering systems. Check with your laboratory for interpretive advice.

Vaccination Programs Require Professional Help

Because of the many variations in the types of the Newcastle virus, the form of the disease in various areas of the world, along with many vaccines, any program for control should be tailormade to fit the conditions. Seek professional help before you start a ND vaccination program.

37-L. FOWL POX

Fowl pox may be found in most poultry regions, but due to the use of vaccines it is not often seen in commercial flocks. However, in certain areas of the world it still remains a disease of major importance and is commonly seen in numerous small unvaccinated flocks. It is caused by a virus, *Borreliota avium*. There is one serotype, but there are three strains with some cross-immunity.

Symptoms

Pox may occur at any age; parental immunity has no effect.

There are two forms of the disease, and the symptoms of each are different.

1. *Cutaneous type (dry pox).* Wartlike scabs are found on the facial appendages—comb, wattles, eyes, and earlobes. There may be a loss of appetite and general unthriftiness. Egg production drops and fertility is impaired. In most instances little mortality occurs from the skin type of fowl pox.
2. *Diphtheritic type (wet pox).* Yellowish, cankerous, and cheesy lesions appear on the internal wet surfaces of the mouth, tongue, esophagus, nasal passages, and sometimes the crop. When such lesions are removed, profuse bleeding occurs.

   Breathing is hampered by the exudates, and the birds may suffocate. Egg production is retarded in laying birds; fertility is lowered. Mortality is much higher with the diphtheritic type than with the skin type.

Transmission

The skin acts as a barrier to the entrance of the pox virus; the skin must be broken for the organism to gain entrance. Transfer of the virus from bird to bird is relatively easy. Spread may be categorized:

1. *Bird to bird.* Both forms spread slowly from one bird to another in the pen. Most of the transmission is the result of birds picking, fighting, or scratching one another.
2. *Mosquito.* The mosquito definitely is a means of spread. It punctures the lesions of affected birds, then transfers the virus to other birds when it "bites" them. Consequently, fowl pox is more prevalent during the mosquito season. In some instances the virus may enter the body of the mosquito and it will remain a carrier for several months.

Diagnosis

Pox lesions on the facial tissues of the bird are a definite indication of dry fowl pox; no other poultry disease produces these. But the diphtheritic type of the disease is difficult to diagnose by visual means. Several types of laboratory techniques are used to make a diagnosis.

1. *Bird transfer.* A small amount of scab material from an infected bird is scratched into the surface of the comb of an uninfected bird. About 5 days later typical pox scabs will arise when dry pox is involved.
2. *Virus isolation.*
3. *Inclusion bodies.* Bollinger bodies are easily demonstrated in material collected from pox lesions during the early period of the disease.
4. *Fluorescent antibody test.*

Treatment

Treatment generally is ineffective, although some drugs may be used to reduce the morbidity, particularly if secondary.

Control

Control is readily accomplished by vaccination. However, there are several types of vaccines and several methods of application.

1. *Fowl pox vaccine.* This is a virulent type of vaccine, being a live virus; it is capable of spreading the disease and therefore is used infrequently.
   a. Birds must be 5 weeks of age or older, because of its virulence.

    b. Birds must not be under stress when it is adminis-
      tered.

    c. Fowl pox vaccine may be used at the same time as
      the following vaccinations are given, but the pox
      vaccination must be conducted separately: laryngo-
      tracheitis vaccination and second Newcastle disease
      vaccination.

How to apply the fowl pox vaccine. Three methods are
used to apply fowl pox vaccine:

    a. Wing-web puncture method.

    b. Feather-follicle method: A few feathers should be
      pulled from the thigh of the bird, and the vaccine
      brushed on the bleeding follicle openings.

    c. Spray method: A variation in the feather-follicle
      method is to spray the vaccine on the area from
      which the feathers have been pulled. A specialized
      syringe, incorporating a spray, is used.

Examination for "takes." About 10 days after one of the
above vaccinations, the birds should be examined. If
the vaccination has "taken" a definite pox scab will be
obvious. If there is no scab the birds should be revaccinated.

2. *Attenuated fowl pox vaccine.* This is a mild form of fowl
   pox vaccine and gives good immunity without many of
   the side effects of the fowl pox vaccine.

3. *Pigeon pox vaccine.* This vaccine is milder than fowl pox
   vaccine. Consequently, it may be used in cases where
   fowl pox vaccine produces a severe reaction, as in

    a. Day-old or very young chicks: Parental immunity
      does not prevent immunity from vaccination be-
      cause the area (feather follicles) where the vaccine
      is applied is not involved with a rapid flow of blood.
      Few antibodies are at the site of vaccination.

    b. Birds undergoing stress.

    c. Birds in egg production.

Pigeon pox vaccine applied by feather-follicle method. This
method is the only acceptable method of application.
Do not use the wing-web method. When day-old
chicks are vaccinated, pull some down from the back
of the chick and brush the vaccine on the exposed area.

4. *Modified pigeon pox vaccine.* Some vaccine manufacturers produce a live pox vaccine made from a modified
   strain of pigeon pox. Being milder, it can be used for
   birds of any age, or for birds in egg production. This

vaccine does not produce the diphtheritic form of the disease, and it does not produce a viremia. The vaccine is applied by puncturing the wing web, or by the eye-drop method. Aerosol and water methods are less successful.

Vaccinating broilers. If it becomes necessary to follow a vaccination program, the modified type of pigeon pox vaccine should be used. In hot climates, broilers may be vaccinated at 1 day of age by the wing-web method. Use one needle only. Or they may be vaccinated at 7 to 10 days of age. A mosquito-control program should be made a part of the preventive procedure.

Caution: Do not mix other vaccines with pox vaccine.

37-M. LARYNGOTRACHEITIS (LT)

Although the disease is found in every area of the globe, often it is sporadic in nature; some areas do not seem to be affected for long periods, then the disease appears. It is found mainly in chickens, but the virus has been isolated from pheasants and turkeys.

Cause

Laryngotracheitis is the result of infection by a specific herpesvirus, *Tarpeia avium.* There is one serotype.

This respiratory disease is one that spreads rapidly. The virus produces only a limited evidence in chicks under 1 month of age. It may vary in virulence, thus producing severe attacks in some outbreaks and mild in others. The virus has an incubating period of from 6 to 10 days. Usually the disease runs its course in about 14 days, but in some cases it may linger on for a month. The LT virus does not enter the blood of the infected chicken, but forms lesions only where it contacts the tissue.

External Symptoms

Laryngotracheitis produces a very severe respiratory disease. The birds cough and have difficulty breathing. The trachea becomes clogged and the birds gasp trying to get air to their lungs. This gasping is so severe and common that it becomes a major criterion of the presence of the disease.

In young chickens there is an infection of the eye, causing pain. Tears flow freely; the eyes are watery.

Internal Symptoms

The trachea is filled with an exudate, loosely attached to the tracheal lining. In most cases it may be removed easily. This makes a differentiating factor when it is necessary to determine the type of disease infecting the flock. There is severe hemorrhaging, and blood is coughed up as the birds endeavor to breathe.

Mortality

As the disease is of varying severity, death losses fluctuate from flock to flock, and from year to year. In some cases as many as 30% of the birds will succumb; in others, the mortality may be light. Age of the bird does not seem to be a contributing factor.

Transmission

The disease spreads rapidly from bird to bird. Modes of spread are

1. *By air.* Although airborne for short distances, as in a pen of birds, seemingly the organism cannot be carried in the air over long distances. Thus, this is not a means of transferral from farm to farm, and it is doubtful if the virus will be carried by the air to adjacent poultry buildings on the same farm.
2. *By people, trucks, birds, rodents, and so forth.* Mechanical means of dissemination of the virus is, no doubt, the major means of spread.

Diagnosis

For accuracy of diagnosis the services of the laboratory must be employed. Tests used in establishing identity of the virus are

1. *Challenge.* Inoculate a suspicious material into the sinuses. If the disease is present there will be a discharge and swelling in 3 to 5 days.
2. *Virus isolation (and identification).* The virus is grown on egg embryos. Only birds in the early stages of the disease should be sent to the laboratory for this test.
3. *Serum neutralization (SN) test.* This test is of limited value, it takes 3 weeks.

Control

Control is by vaccination. There are two main types of vaccines.

1. *Modified LT vaccine.* This vaccine is not capable of producing the disease in vaccinated birds; consequently it will not cause any spread. The immunity is as good as that of any other vaccine. It can be administered to young birds and to birds in egg production with only slightly noticeable effects. Parental immunity may affect the ''take'' in birds under 21 days of age.

    How to vaccinate. The vaccine should be dropped onto the open eyeball. Some immunity will be developed in 2 days; maximum immunity will be reached in 6 days. Vaccinate at 6 to 8 weeks of age. Repeat at 14 to 16 weeks.

2. *Cloacal-type vaccine.* For many years, and until the modified type of vaccine appeared on the market, this was the recognized and only vaccine available. It is a virulent type, and the vaccine is potent to the extent of being capable of reproducing the disease. It is administered by brushing the vaccine inside the upper lip of the vent. In approximately 4 days a cherry-red color develops, indicative of a ''take.''

    Vaccine declared illegal in some areas. Because of the virulence of this vaccine, and because it is capable of spreading LT to uninfected farms and areas, it is not legal in some regions.

Management and Eradication

Although it is probably possible to eliminate LT from the premises by the use of the attenuated vaccine, management should become a part of any eradication program. Isolation of affected flocks after vaccination, or even complete isolation of the farm, should be stressed as important.

Some Birds Remain as Carriers

After a natural outbreak, some birds remain as carriers for life. These few make up a reservoir of infection and are capable of spreading the disease to uninfected flocks.

Vaccinating on Farms Having LT

If the LT outbreak has been in existence for some time, vaccination is of no value. But if it is in the early stages, the affected pen should be vaccinated along with other birds on the farm. Start the vaccination with those birds farthest from the infected pen or house.

37-N. AVIAN ENCEPHALOMYELITIS (EPIDEMIC TREMOR) (AE)

This disease is found in every area, but some isolated flocks may escape contamination. Although there is only slight economic loss in adult birds, AE may take on great proportions in young chicks. It is a disease of the intestinal tract, caused by a virus.

Symptoms in Young Birds

The disease is one that affects young chicks between 6 and 21 days of age. They show nervous symptoms of varying proportions, including paralysis. The body quivers, and this symptom is especially noticable when the chick is held in the palm of the hand. Many chicks lie on their sides and cannot motivate themselves. Mortality is high; death results not from the disease itself but because the chicks cannot get to feed or water. Birds over 4 weeks of age seldom show evidence of AE.

Symptoms in Adult Birds

In most outbreaks in adult flocks there will be no noticeable symptoms in the birds. However, egg production often is affected, as exemplified by a drop in the number of eggs produced by the flock each day.

Transmission

The disease is transmitted by two methods.

1. *Fecal transfer of the virus.* As the virus multiplies in the intestinal tract, the fecal material offers one direct method of transfer. Consequently, complete eradication is impractical. Contaminated water and feed implement the transmission from bird to bird in the pen. Similarly, transfer of fecal material to uninfected pens will cause an outbreak. Such a virus will live outside the body of the bird for several days. Virus particles

begin to occur about 3 weeks after the birds have been infected.

2. *Transmission through the hatching egg.* The virus is shed by the infected breeder hen through the hatching egg to the newly hatched chick. Unfortunately, because infection in the breeder flock may go unnoticed, the first indication of trouble comes from the customer who purchased the chicks. As the infected eggs were laid at least 3 weeks prior to hatching, the disease has almost always run its course in the breeders by the time chick infection is noticed in the field.

   Eggs in incubators must be destroyed. It takes just a few days for the infected breeder hen to recover, but because all birds in the flock do not have AE at the same time, most flocks will lay eggs containing the virus for a period of about 3 weeks. Therefore, under most conditions all eggs in the incubators from the infected flock must be destroyed. There may be other infected chicks in the field at the time the first customer notifies the hatchery of the infection. Some program of handling these customer cases should be established.

   Hatching eggs from a recovered flock. If a flock has had a natural outbreak of AE, and has recovered, hatching eggs will contain antibodies and may be used to produce chicks.

Diagnosis

Most cases of AE can be diagnosed in the poultry house, but not all. Suspect chicks should be submitted to a laboratory. The following tests are used for the identification of AE:

1. Histopathology: Brain tissues are fixed, sliced thin, stained, and examined under a microscope.
2. Inoculation of uninfected chicks.
3. Fluorescent antibody test.
4. Enzyme-linked immunosorbent assay (ELISA) test.

Control

No drug will alleviate the condition once there has been an active outbreak in the flock.

Prevention

Prevention of the disease is by using hatching eggs from breeder flocks with an immune status. A natural outbreak in young flocks produces this, but vaccination is the accepted practice.

When to vaccinate. Growing birds to be used for breeders should be vaccinated after 8 weeks of age and 3 weeks before the first eggs are laid by the pullets. However, there have been many failures with AE vaccination. In most instances, failures have been caused by vaccine of poor quality due to improper storage, or by inadequate methods of vaccination. AE parental immunity is stronger than is the case with most diseases, and lasts for a longer period.

Type of vaccine. AE standard vaccines have been used for some time, and they are of two types.

1. *Live virus vaccine.* Freeze-dried (lyophilized) vaccines are used. Usually they are administered in the drinking water. Not only will the vaccination come through the water but there will be a spread from bird to bird.

2. *Killed-virus vaccine.* The live virus will usually cause a drop in egg production if it is administered to laying birds. A killed-virus vaccine may be used for such flocks as it affects egg production very little.

Vaccinate cockerels the same as the pullets.

Determine status of vaccinated birds. About 4 weeks after vaccination, birds should be submitted to the laboratory for an evaluation of their immunity by checking for virus-neutralizing antibodies. A neutralizing index of less than 1.1 indicates that the birds have not been fully immunized. The flock should be revaccinated.

Check the immunity prior to using hatching eggs. Several of the first fertile eggs laid by the flock should be submitted to a laboratory for an immunity test. This will show whether the vaccination was adequate. If inadequate, the flock must be revaccinated immediately. Three weeks must elapse after the revaccination before hatching eggs are saved. Any eggs set during this 3-week period will be AE-contaminated and will infect the resultant chicks.

37-O. MAREK'S DISEASE (MD)

One phase of the leukosis complex is known as Marek's disease. It is an acute neural leukosis, causing tumors. The chicken is the only known susceptible host.

Cause

Marek's disease is caused by a DNA-cell-associated herpesvirus. Several serotype have been identified. A nonpathologic herpesvirus is common in the turkey. Seemingly the virus is present in every chicken in the world over, but there are many causes for the variations in the severity of the disease. Birds may become infected early in life, and undoubtedly remain infected until death. Not all infected birds show the presence of the disease, but the lack of visible evidence does not mean the bird does not harbor the virus. Tumor incidence is not the only indication of the presence of the virus. In fact, some other areas, as cells lining the feather follicles, may be a more important area of infection. Neither is the incidence of the disease a reliable index of the amount of infection in the bird; the virus does not seem to localize in the tumors.

Antibodies appear along with the herpesvirus and may have great longevity. The preponderance of one or the other probably is responsible for a part of the variation in the degree of morbidity and mortality. But eventually some lesions enlarge and cause death.

Transmission

Bird-to-bird transmission is the main method, but many of the exact processes are not known. Some established variations are as follows:

Airborne. MD has a unique method of air transmission, much different from that of other airborne respiratory infections. With MD the virus localizes in the feather follicles and is sluffed through the dander and feather particles. In fine form it floats in the air and is inhaled by the birds.

Mechanical. When feather dust and dander settle on clothing, feed bags, equipment, etc., the virus is easily transmitted to other poultry houses and to poultry-producing areas.

Other means of spread. Undoubtedly there are means of spread other than through material from the feather follicles. Certain beetles have been identified as carriers.

MD virus not embryo-transmitted. From a practical standpoint it may be assumed that this method is of very little importance in the transmission of the disease.

Marek's affects bursa. Marek's disease affects the bursa of Fabricius, a gland necessary in the production of antibodies, thereby reducing the bird's ability to resist other infectious diseases. See Chapter 35-D.

Diagnosis

The presence of tumors may easily be observed, and those in birds under 16 weeks of age are usually caused by Marek's disease. Marek's disease is RIF-negative, and the test is used to differentiate between MD and lymphoid leukosis, which is RIF-positive.

There are several tests used in the laboratory to establish the identity of MD. However, the cost of performing them is still beyond the realm of commercial practicability, and very few laboratories are properly equipped to complete the tests.

Control

There are now several vaccines used in the control of Marek's disease, each with the same origin.

Herpesvirus, turkey origin, vaccine. The turkey is susceptible to Marek's disease, but the causative virus is not the same strain as that causing the disease in chickens. But both strains are herpes-like, and bear a close similarity. Vaccines have been developed using the turkey virus as their origin. When injected into young chicks they have the ability to stimulate the bird's resistance to the production of tumors or other lesions of Marek's disease. The herpes turkey virus (HVT) will produce a viremia in the chicken, and such birds will not develop the symptoms of Marek's disease common to the chicken.

Another point of interest in this complex disease is that the chicken virus will not produce Marek's in the turkey. The fact that the chicken virus may be isolated from chickens vaccinated with the turkey-type virus would seem to indicate that the turkey virus in such vaccinated birds is not producing immunity to the chicken type of the disease, but is only suppressing the growth of the tumors produced by the chicken virus.

The strain of turkey virus (FC 126) showing the above characteristics was first isolated in 1969 at the USDA Regional Laboratory at East Lansing, Michigan, and is used predominantly to develop the vaccines being produced today by several manufacturers.

The HVT vaccines have no adverse effect in the chicken. They have not been shown to be tumor-forming, nor will they spread from bird to bird. These vaccines require special handling, and must be kept frozen or refrigerated, as follows:

1. *Frozen in nitrogen.* The turkey virus is propagated in embryonic duck or chicken cell cultures; the vaccine is standardized, then preserved at a very low temperature of $-120°$ to $-135°F$ in liquid nitrogen or nitrogen gas. It must be thawed when needed, mixed with the diluent, and used to vaccinate chicks when they are 1 day of age. The thawed and mixed vaccine must be used quickly; 2 hours is the maximum time allowance after thawing.

2. *Lyophilized (dry).* This cell-free vaccine must be refrigerated, but not frozen. The diluent, packaged separately, need not be refrigerated. The efficacy of this vac-

cine has been shown to compare fully with the vaccine frozen in nitrogen (1, above). It has the definite advantage of saving time and labor.

Quantitative vaccine measure.   The amount of the HVT vaccine administered is very important, as is the potency of the vaccine. The quantitative measure is in plaque-forming units (PFUs). Proper and adequate immunization per layer-type chick requires at least 5,000 PFUs. When the dosage is 0.1 cc, each dose should contain the above number of PFUs, or more. The quantity may be reduced for broilers. When the maternal antibodies are high, the dosage should be greater.

Method of vaccine administration.   There are four methods of vaccinating day-old chicks in the hatchery.

1. *Subcutaneously:* The vaccine is injected subcutaneously in the nape of the neck. This is the most common method.
2. *Intramuscularly:* This is not a common practice, but is used on occasion.
3. *Intra-abdominally:* This procedure is not too practical and is seldom used.
4. *Spray:* The vaccine is sprayed on the chicks. The recommendation is that the chicks should be kept in a cabinet during the process in order to prevent air movement. The procedure has many inefficiencies. Antibody response is slower to develop. Special refined sprayers must be used.

Combination Vaccines

One combination HVT and Gumboro vaccine has been licensed for sale.

Failures with HVT Vaccines

For several years following its introduction, the HVT vaccine gave almost complete suppression of tumors in the vaccinated chicks, but lately there have been some failures. The following causes have been suggested:

1. Certain strains of chickens are not susceptible.
2. There is failure of HVT to precipitate infection before exposure to the MD virus.

3. Variant strains of MD virus are capable of masking HVT immunity.
4. There is a deficiency of immune response in vaccinated chicks.
5. Improper diluents are used with the HVT vaccine.
6. Improper combination vaccines are being used.

These failures may be of such magnitude that professional help should be secured before entering into any change in the conventional HVT vaccination program.

37-P. INFECTIOUS BURSAL DISEASE (IBD)

This disease was formerly popularly called Gumboro disease because it was first found near Gumboro, Delaware, U.S.A.

Cause

IBD is a highly infectious disease. It develops rapidly but has a short duration. It is seldom seen in chicks under 3 weeks of age. The disease is probably caused by a specific virus with one serotype.

The bursa of Fabricius plays an important part in the disease. The gland is the productive area of the B-cells (see Chapter 35-A), and as IBD localizes in the bursa the immune system of the chicken is disrupted. Often the bursa of IBD-infected chicks triples in size but will return to normal once the disease subsides. However, in the course of infection the tissues of the bursa may be partially or permanently destroyed.

Because the bursa of Fabricius is the origin of antibodies to many diseases, IBD may reduce the flock's immune response to vaccination. The morbidity and mortality resulting from these other diseases may be greater than that from IBD itself. A poorly timed outbreak of IBD can disrupt the entire disease-control program.

Symptoms

The disease seems confined to growing birds. Inception is rapid between 20 and 60 days of age, and recovery will take from 1 to 3 weeks. Birds are listless, nervous, sleepy, dehydrated, and have a whitish diarrhea. The vents seem irritated, and birds continue to pick at their own vents.

Mortality is variable. Although most flocks of chicks harbor the virus or its antibody, most cases of the disease are so slight as to go unnoticed. Only on occasion are there severe outbreaks, with mortality running as high as 30%. Mortality increases with age up to 10 weeks.

Transmission

The virus is very stable, and is capable of remaining viable outside the host for several months. Unclean poultry houses and poultry equipment are definite sources of infection. The disease may be spread to other areas by people and equipment. It is not egg transmitted.

Diagnosis

Visible symptoms are not such as to warrant a positive diagnosis by observation alone. Birds must be sent to the laboratory. Several tests are used to arrive at a decision.

1. *Virus isolation (and identification).* The identification is made by infecting egg embryos. IBD-negative eggs must be used, for the virus will not survive in the presence of antibodies in the egg.
2. *Serological.* Agar gel precipitation test and the ELISA test.
3. *Chick innoculation test.* The bursa of noninfected birds will enlarge appreciatively a few days after inoculation with infective material.

Treatment

Treatment with drugs and antibiotics has been of little value except for some possible reduction of morbidity from secondary infections.

Control

Control is by vaccination, either by

1. *Vaccination of chicks to prevent infection by IBD.* In some cases the vaccine can be so severe as to damage the bursa and impair the bird's immune system. Attenuated or killed vaccines have been developed to alleviate this condition while some companies produce a special vaccine for chicks older than 3 weeks.
2. *Vaccination of growing and adult breeders* to produce a high degree of parental immunity in the resultant chicks.

Vaccination program. Confirmation that IBD is present on the farm should first be secured. Some states require a permit to use live virus vaccines, but not killed virus vaccines. There is a combination IBD and Marek's vaccine on the market. Consult professionals before entering into any vaccination program.

37-Q. VIRAL ARTHRITIS

Viral arthritis (VA) (infectious tenosynovitis) is a relatively new, but important, disease of chickens. It is especially significant with broilers because it affects growth, increases mortality, decreases feed efficiency, and increases condemnations in the processing plant.

Cause

Viral arthritis is caused by a RNA reovirus that initially infects the synovial membranes and the tendon sheaths. The chicken is the only known host.

Symptoms

Symptoms are especially apparent in broiler chicks because there is a severe infection with lameness and stunted growth. In severe cases the hock joint may be immobilized and the gastrocnemius tendon enlarged. The hock joint usually contains an exudate at the site of the swelling.

Transmission

Horizontal transmission is the prime means of spread of the disease, but various strains of the organism differ in this respect. Transmission is greater in chicks under 2 weeks of age than later in life. Egg transmission of the virus is low, though possible. The ELISA test can also be used for VA.

Diagnosis

A laboratory viral diagnosis is difficult because of similarity of the test to that for bacterial and mycoplasmal synovitis. Such a laboratory viral examination cannot be made unless the bird is negative for *Mycoplasma synoviae*. A fluorescent antibody or agar gel technique is then used to identify the organism.

Treatment

No known treatment for an infected flock is available.

Control

There is a VA vaccine with three periods of vaccination:

1. *Six days:* The vaccination is administered by injection. It is used to protect chicks to be used later as breeders. On occasion it is administered at day old, but the hazards are greater.
2. *Ten to 12 weeks:* Wing-web vaccination is used to provide parental immunity to the next generation. Do not give in the same wing as other wing-web vaccinations given at the same time.
3. *Seventeen weeks:* Vaccinate via the wing web or drinking water to boost parental immunity from a former similar vaccination.

37-R. COCCIDIOSIS

Coccidiosis is a term used to identify the disease produced by a group of protozoan (parasitic) organisms in the class "Coccidia." There are hundreds of types, but only nine are important to the raiser of chickens. It is one of the most devastating of poultry diseases.

Cause

Coccidia have specific hosts, and each species produces its own type of coccidiosis. The coccidia that inhabit the chicken are all in the genus *Eimeria*. Coccidiosis is spread by unicellular bodies known as *oocysts*. These are shed in the droppings, but are not infectious. They first must be *sporulated*, a process that takes place when conditions of air, temperature, and moisture are opportune. This sporulation, therefore, occurs outside the body, requiring 2 to 4 days. The sporulated oocyst is eaten by the chicken and finds its way to the intestinal tract, where a series of divisions and multiplications take place. Eventually more oocysts are produced, most of which again are expelled from the body in the fecal material, and the life cycle is complete. The time from ingestion to expulsion varies between 4 and 7 days, depending on the species.

Coccidia, being parasitic, live in the epithelial tissues of the intestinal tract, and there they inflict their damage. Usually one oocyst in the intestinal tract can destroy only a few cells. Therefore, the extent of destruction of the intestinal wall is closely related to the number of oocysts present. But when the disease is at its height there will be millions of oocysts, although not all will be sporulated. Continued ingestion of the sporulated oocysts means that more cell tissue is destroyed.

External Symptoms

Symptoms include bloody droppings, ruffled feathers, paleness, loss of appetite, lowered growth, poor feed conversion, drop in egg production, diarrhea, and many other manifestations. Most of the external symptoms are the result of general disability due to the destruction of the lining of the intestinal tract; this in turn prevents the transfer of food material from the intestines to the bloodstream. In many cases hemorrhaging of the intestinal lining occurs, and blood in many forms is deposited with the droppings.

Internal Symptoms

Internal symptoms usually are confined to the intestinal tract, including the ceca. Inflammation, hemorrhages, lesions, mucus, and exudates are all indicative of coccidiosis, but most species produce specific internal symptoms. These are outlined in Table 37-1.

Transmission

The only method of transmission is for the bird to consume *sporulated oocysts*. Within the pen this goes on constantly because birds have access to the droppings. But the transfer from house to house and farm to farm is by mechanical means. In the active stage of the disease, millions of oocysts are contained in each teaspoonful of fecal material. These are easily transferred to a new location by shoes, feed trucks, crates, pets, rodents, and moving equipment. Once sporulated, the oocysts soon cause an outbreak of coccidiosis in the new premises.

Diagnosis

Appearance of the bird, along with intestinal lesions, may suffice for a diagnosis of most outbreaks. However, many symptoms are similar to those produced by other diseases. A laboratory diagnosis then is necessary. Scrapings are made of the infected area of the digestive tract, and a microscopic examination is made for the presence of coccidia.

Table 37-1. Characteristics of Coccidia Species

Species Common Name	External Symptoms	Intestinal Area Most Affected	Mortality	Morbidity
E. necatrix Intestinal coccidiosis	Diarrhea Bloody droppings Ruffled feathers Weight loss	Whitish lesions on upper third of small intestine	Heavy	Heavy
E. tenella Cecal coccidiosis	Bloody droppings Drop in feed Droopy Fewer eggs	Hemorrhagic ceca	Heavy	Heavy
E. acervulina Layer coccidiosis	Diarrhea Fewer eggs Drop in feed Weight loss	Upper half of small intestine	Light	Medium
E. brunetti Intestinal coccidiosis	Diarrhea Emaciation	Lower half of small intestine, ceca, and cloaca	Light	Light
E. maxima Intestinal coccidiosis	Diarrhea Bloody droppings Drop in feed Pale color	Middle and lower sections of small intestines	Light	Medium
E. mivati Intestinal coccidiosis	Fewer eggs Ruffled feathers	Lower half of small intestine	Light	Heavy
E. hagani	Diarrhea Drop in feed	Upper half of small intestine	Light	Light
E. praecox	Diarrhea Weight loss	Upper third of small intestine	Light	Light
E. mitis	Diarrhea	Entire small intestine	Light	Light

Control

Coccidiosis is far more easily prevented than treated. Certain chemicals, known as coccidiostats, suppress or upset the life cycle of the protozoan. The coccidiostats usually are added to the feed at a designated percentage. However, all such chemicals do not have equal ability to suppress all of the Eimeria species. Such drugs reduce or eliminate the shedding of oocysts in the droppings, thus reducing or preventing oocyst contamination of the poultry house floor. Some drugs are so specific for certain Eimeria that they completely suppress them, yet have little effect on others. Because there are nine types of coccidia, a person might be using a drug against one type while an outbreak is arising from another type. This has made the use of coccidiostats difficult. Furthermore, some species of Eimeria have become resistant to certain coccidiostatic drugs, mainly those that have been used consistently for treating several generations of birds.

Most outbreaks of coccidiosis are produced by three Eimeria:

1. *E. tenella*
2. *E. necatrix*
3. *E. acervulina*

The others account for less than 15% of the attacks, and usually involve only slight economic loss. Any good coccidiostat should at least be specific for the above three Eimeria, plus perhaps *E. brunetti* and *E. maxima*.

Properties of good coccidiostats

1. Prevent infection from as many species of Eimeria as possible.
2. Make it possible to dilute the dosages so as to develop natural immunity in breeder replacement chicks.
3. Do not interfere with reproduction (egg production and fertility).
4. Are not electrostatic or hygroscopic.
5. Should be nontoxic, palatable, and stable.
6. Should be economically acceptable.

Names of Coccidiostats

Names of some coccidiostats are given in Chapter 36-B.

Coccidiosis Control Program for Broilers

Growing broilers should be fed a coccidiostat that will completely suppress coccidiosis. This is difficult inasmuch as coccidia eventually develop some degree of resistance to a specific drug. When this happens it becomes necessary to change the coccidiostat. Some broiler producers make this change every 8 to 12 months; others change only when they see a failure with the drug being fed; others resort to a "shuttle" program in which one coccidiostat is used the first 10 to 15 days of the growing period, then another for the remainder. Some use three coccidiostats.

Caution: Get professional advice before mixing coccidiostats. Remember that in the United States the use of all medications, including coccidiostats, are regulated by the Food and Drug Administration (FDA).

Side effects. Some coccidiostats reduce feed consumption and feed conversion. Others reduce the absorption of methionine from the intestine. Skin pigmentation may be reduced by some. Consult a specialist in this field before selecting a drug best suited to your coccidiosis-control program.

Withdrawal period. Most coccidiostats must be withdrawn from broiler feed for 5 days prior to slaughter to allow time for the drug to be eliminated from the tissues.

Immunity is not to be produced in broilers. There is not enough time prior to the age at which broilers are marketed to develop immunity to coccidiosis and allow the birds to recover. All drugs should be fed at full strength (that necessary for full suppression of coccidiosis) from 1 day of age to near broiler marketing age.

Coccidiosis Control Program for Replacement Birds

Under natural outbreaks, chickens with coccidiosis develop an immunity to the Eimeria species that caused the disease. This immunity does not last a lifetime; but as the bird is continually consuming more sporulated oocysts, they produce continued immunity throughout the life of the bird.

Immunity may be developed artificially in the bird without the stress of its enduring an attack of acute coccidiosis. This process is made possible by the fact that if the number of oocysts in the intestinal tract is kept at low level there will be no danger of serious coccidiosis, yet the number will be adequate to enable the bird to build an immunity.

Most coccidiostats suppress oocyst reproduction completely when they are fed at a designated level in the feed. No immunity can build up during such feeding. Before the bird can acquire immunity, the coccidiostat must be reduced in order that some sporulated oocysts, consumed from the litter, may be allowed to complete their life cycle. A small reduction builds a little immunity and the coccidiostat may then be further reduced, producing more immunity. Gradually, all the coccidiostat is removed, and immunity is complete. The trick is to reduce the coccidiostat fast enough to produce just a very little coccidiosis. There must be a little, or no immunity will result.

How much moisture in the litter? The litter should contain 20 to 30% moisture for optimum oocyst sporulation. Most types of litter should have approximately the same consistency and feeling as shavings cut from green lumber. A little too much moisture is better than too little.

Wetting the litter. During extremely dry periods, when the litter gets very dry, it may be advantageous to sprinkle it with water to raise the moisture content and to better control the sporulation of oocysts. *But be careful.* If the litter has been dry for some period, it will contain billions of unsporulated oocysts. Wetting the litter will cause a great number of the oocysts to sporulate, and if there is a limited amount of a coccidiostat in the feed, a severe attack of coccidiosis may be initiated. It is better to wet the litter a little each day during dry weather than to wait and try to restore its moisture content in one application. In any event, whenever water is added to the litter, be ready to treat for coccidiosis should an outbreak occur.

Starting the coccidiostat withdrawal period. The coccidiostat should be fed at its full level during the first 5 or 6 weeks of the chick's life; then the withdrawal should start. The amount of drug added to the feed should be reduced gradually over a period of 10 to 12 weeks. Never withdraw the drug suddenly. Remember that the drug with the greatest anticoccidial potential will require the longest withdrawal period. Only a careful study of the birds will determine whether withdrawal is proceeding correctly.

Coccidiosis Control Program with No Coccidiostat

Some producers of egg-type replacement pullets do not use continuous feeding and withdrawal of a coccidiostat, but resort to quick medication when an outbreak of coccidiosis appears. This program is subject to inconsistent results.

Coccidiosis Inoculation

Inoculation of birds for coccidiosis is a practical procedure. A product known as CocciVac,® composed of live oocysts, is given in the water or fed to chicks. Eventually, oocysts are deposited in the litter, they sporulate, the bird consumes relatively small numbers, and immunity begins. The program gives assurance that inoculation can start at an early age, under controlled conditions. Immunity may develop in 5 or 6 weeks.

The CocciVac program. Be sure to follow directions as printed on the container. Briefly the procedure is as follows:

1. Vaccinate only healthy birds.
2. A program of litter management must be followed.
3. Birds should be 10 to 12 days of age when CocciVac is given.
4. Birds must be on a floor with litter so that they may have access to their droppings.
5. Select a CocciVac containing the species of Eimeria involved on the farm. *Note:* There are several CocciVac products, each being a different combination of oocysts from various Eimeria species. Some contain only three or four; others may contain as many as eight.
6. Do not feed a coccidiostat when the CocciVac program is being followed. A coccidiostat given at this time will only suppress oocyst formation and the development of immunity.

7. Be ready to treat for a coccidiosis outbreak. The Cocci-Vac program is a quantitative one, in which immunity will be built up when the population of oocysts in the intestinal tract is small. If, however, too many sporulated oocysts are consumed when the bird eats the litter, coccidiosis of varying proportions may arise. Some small degree of coccidiosis is necessary to be sure that immunity is developing, but if coccidiosis is excessive, the flock must be treated.

8. Add vitamins A and K to the feed. Vitamin A helps repair epithelial tissue; vitamin K helps prevent intestinal bleeding.

Treatment

In spite of many excellent programs and chemicals for the control of coccidiosis, there are many outbreaks of the disease. Treatment, therefore, becomes a necessary part of any control program. Several drugs are used; some are to be given in the feed and some in the water.

1. Sulfaquinoxaline (do not feed to layers nor to birds within 5 days of slaughter).
2. Sulfamethazine.
3. Esb$_3$ (for use in the drinking water).
4. Agribon.
5. SEZ.
6. Amprolium (soluble type for drinking water).

Cause of Coccidiosis Outbreaks

Why is there still so much difficulty from coccidiosis? Why are there so many outbreaks? Some of the reasons are

Broilers

1. Coccidiostat too weak for Eimeria species involved.
2. Incorrect percentages of the coccidiostat in the feed.
3. Eimeria species has become resistant to the drug.
4. Litter too damp, causing increased sporulation of the oocysts.

Breeder replacements (or for laying purposes)

1. Coccidiostat too weak for control of specific Eimeria.
2. Eimeria have become resistant to the drug.
3. Litter too dry or too wet.
4. Coccidiostat fed at too high a level, thus making it impossible for the bird to develop immunity properly.

5. Coccidiostat removed too fast.
6. Other drugs are being fed, potentiating the effects of the coccidiostat.

How to handle outbreaks. Outbreaks of coccidiosis, regardless of the cause, should be handled quickly. Time is most important. As feed consumption is reduced during coccidiosis outbreaks, the medicaments should be given in the water. A chicken will drink even though it will not eat. Most applications are made through the water for 3 days, then withdrawn for 2 days, then repeated for 3 days. After this, the flock is returned to the normal coccidiosis-control program. If there are reasons for failures in the first place, make corrections at this time.

37-S. ASPERGILLOSIS

Although not a serious disease, aspergillosis is common in many instances where poor poultry husbandry is practiced. It is mainly a disease of young chicks.

Cause

The disease is caused by a fungus, *Aspergillus fumigatus.* Normally this fungus grows on decaying organic material in the chicken house and in the hatchery, but it also has the ability to reproduce itself in certain tissues of the bird.

Symptoms

The lungs are the major area of internal infection. A close examination of the lung tissue will show small nodules that are hard and yellow. There may be but a few in some cases; in others there may be hundreds. In some instances the fungus gains entrance to the air sacs, and a respiratory condition evolves. However, there is no coughing or sneezing.

Chicks have few external symptoms, except for an occasional involvement of the eye, causing semiblindness. As the fungus continues to grow in the lungs and air sacs, flock mortality increases. It may be quite high in young chicks. Older birds are able to withstand the fungal growth, and few birds are affected.

Transmission

Spores from the fungus dry and are transported easily from chick to chick by way of the air. This may occur in the hatchery or in the brooding house. The incubator may become a source of contamination.

Diagnosis

In many instances a visual examination of the sliced lung will show the nodules, and form a basis for diagnosis. In others, it will be necessary to submit chicks to the laboratory. Here, the suspected tissue is cultured and examined microscopically or the fungus isolated.

Treatment

Poultry house. If the disease can be diagnosed in its early stages, a thorough cleaning of the brooding premises will eliminate the source of infection, and the disease will subside. Any moldy feed should be removed, bulk feed bins cleaned, old litter removed from the house and replenished with new, and waterers and feeders cleaned and disinfected.

Hatchery. A general cleanup of the hatchery and incubators will remove this source of infection.

Important: Formaldehyde fumigation is not particularly effective against *A. fumigatus.* Good liquid disinfectants should be used during the cleanup procedure. Use a good fungistat in the feed, or use copper sulfate or Racon® in the drinking water.

37-T. MYCOTOXOCOSIS

Although mycosis has been known for years, its importance as a poultry disease was not recognized until rather recently. An increasing incidence has been found responsible for great economic losses in some poultry flocks.

Cause

Mycotoxocosis in its various forms is caused by the toxins produced by molds. There are many types of toxins produced. Mycotoxocosis may complicate the recovery from other diseases, such as coccidiosis and viral infections. On occasion, the prolonged feeding of some antibiotics in the treatment of other diseases may alter the intestinal flora to such extent as to allow the fungi to increase in number.

Symptoms

Although morbidity and mortality are usual, the symptoms of the four most toxic mycotoxins in the chicken are as follows:

Aflatoxin

Decreased growth and feed efficiency
Impaired immunity

Increased bruising
Decrease in blood clotting
Altered protein and fat metabolism
Lowered resistance to other diseases

Ochratoxin

Reduced feed consumption
Dehydration and emaciation
Deposits of urates throughout the body cavity
Impaired kidney function

Fusariotoxin

Sores in the mouth
Scabby lesions on the feet and shanks
Reduced egg production and shell thickness
Reduced feed consumption
Reduced weight
Poor feathering in growing birds

Citrinin

Decreased feed consumption
Reduced growth
Increased water consumption
Diarrhea
Pale and swollen kidneys

Treatment

The best treatment is to find and remove the source of the mycotoxin. Use a new source of feed if feed is the suspect. Treat with fat-soluble vitamins via the drinking water.

Control

There are several suggestions, all being a way of preventing mycotoxin consumption by the bird.

1. Use mature cereal grains, high in quality, in the feed.
2. Use corn of low moisture content.
3. When the feed is not of high quality, add a mold inhibitor.
4. Do not allow feed ingredients or mixed feed to have a temperature buildup.
5. Disinfect all feed milling and handling equipment.
6. Do not allow moldy feed to build up in feed troughs or storage bins.

37-U. FATTY LIVER SYNDROME

Fatty liver is a metabolic disorder in laying hens, causing excess fat in the liver. Factors that will cause an increased deposition of fat in the cells of the liver are: toxins, high egg production, nutritional disturbances, and endocrine imbalances. Fatty liver syndrome represents one specific type of fatty liver, and is sometimes called *fatty liver hemorrhagic syndrome* (FLHS).

Cause

The exact cause of increased fat deposits in the liver of the chicken affected with FLS is not known. It has been impossible to produce the conditions experimentally. However, there is now adequate evidence to state that FLS is of nutritional origin, and that it is not pathogenic. There appears to be a breakdown of the metabolic processes involved in the synthesis and mobilization of lipids, particularly during the stress created by heavy egg production.

Symptoms

FLS appears only in good-laying flocks. Most birds appear in good physical condition. For this reason there is little indication of the disease until egg production gradually drops between 10 and 40% or the flock fails to have a high peak of egg production. Body weight usually increases from 20 to 25%. The problem is more acute with birds housed in cages than with those housed on the floor.

Postmortem examination will reveal an enlarged, fat, and friable liver, tan in color. Obesity, as indicated by internal fatty deposits, will be evident. The normal liver will contain about 36% fat, while one with FLS will have about 55%.

Treatment

Various nutritional supplements have been tried with mixed results. The addition of the following to the feed has been shown to alleviate the condition in some laying flocks, but not all.

Add to ton (2,000 lb) of regular feed

10,000 IU vitamin E	12 mg vitamin B_{12}
1,000 g choline chloride	908 g inositol

The main component of the above is inositol, an aid to the transport of fat in the liver.

The success and failure of different treatments is an indication that FLS is not a single-source problem but one that involves complex nutritional and possibly environmental interactions that are still unknown.

Control

Reducing the energy intake, either by feed restriction or by lowering the ME content of the diet, has merit. The addition of 5% alfalfa meal and 20% wheat bran will help control the problem if it has not been recurrent. These ingredients may contain a factor that is essential for lipid metabolism in caged layers.

37-V. CAGE LAYER FATIGUE

It is known that the breaking strength of bones from layers held in cages is less than that of bones of birds kept on a litter floor. This difference is great enough at times to cause difficulties with caged layers. Furthermore, the brittleness of the bones of caged birds at the end of their laying year may be so great as to make the spent hens unacceptable for poultry processing; their bones disintegrate, causing fine splinters in the meat removed in the process of making soup and other food products.

Cause

There is some indication that the difficulty may be due to an inadequate amount of inorganic phosphorus in the ration. However, this thought is not shared by all scientists. Feeding experimental diets with little or no inorganic phosphorus will not produce the symptoms; nor will adding inorganic phosphorus to the ration increase the strength of the bones. There is some indication that inadequate calcium during the period of peak egg production may be a precursor to the difficulty.

Symptoms

After long periods of egg production caged layers have difficulty in standing, and their body is held in a vertical position. They may lose control of their legs and lie on their sides, indicative of a type of paralysis. Usually there is no loss of egg production, shell quality, or interior egg quality. Some of the bones may be fractured; some will break when the bird is handled. The ribs may be beaded at their cartilaginous junctures. There may be fractures of the fourth and fifth thoracic vertebrae. There is little mortality; the birds appear healthy and pert.

Treatment and Control

There is no known method of treatment. There is some evidence that the smaller the number of layers in a cage, the lower the incidence of cage layer fatigue. One experiment showed that feeding a ration containing 6% calcium just before the birds were slaughtered increased bone strength. Other tests have indicated some alleviation when the phosphorus content of the ration was increased by 0.3%.

37-W. AVIAN HYSTERIA

There is no doubt that avian hysteria may express itself to the extent that it becomes a major disaster. Hysterical flocks are not common, but once the birds are inclined there seems to be little that can be done to control the problem. Hysteria must not be confused with general nervousness found in some of the Leghorn strains. Also, there is no indication that nervous flocks will show a greater incidence of hysteria.

Symptoms

The first indication of hysteria is evident when a few of the birds seem to become almost "wild." Anything unusual seems to excite them and they fly over the tops of other birds until a piece of equipment or fence stops them. Gradually more birds become affected. Many birds are injured during their wildness. Finally, the whole flock will be affected. Birds pile up at the end of their flight, and many suffocate. More and more things seem to excite them, and the difficulty reaches major proportions.

Causes in Field Cases

Although there seems to be no pattern to the cause of hysteria, most cases may be traced to a mismanagement or stress factor. Each case will be the result of a different factor. Some of the causes are thought to be

1. Crowding
2. Poor ventilation
3. Picking (poor or no beak trimming)
4. Insufficient protein consumption
5. Hot weather
6. Decreased feed consumption
7. High levels of dust in the poultry house
8. Excessive ammonia in the poultry house
9. Lights flashing in the poultry house
10. Sudden noise
11. Insufficient space in floor houses
12. A large number of birds in an individual cage

Hysteria affects birds of all ages, but the damage to the flock is greater in the case of laying birds. Many times the hysteria will begin when the birds are young, then take on greater proportions when they begin to lay.

Drop in feed consumption first indication.   Evidently hysterical birds lose their appetites and do not consume a normal amount of feed. A drop in flock feed consumption, along with the flightiness should present a positive diagnosis.

Control

Permanent control can be accomplished only when the cause of the difficulty is removed. Every effort must be made to determine the management and other factors that are responsible. Any treatment to alleviate the condition must be considered as only temporary. Once the treatment is withdrawn the hysteria may return if the cause is still there.

Removal of toenails.   Apparently pain is a factor in producing hysteria and hysterical birds scratch and claw other birds. At the Washington Experiment Station it was shown that removing the toenails of day-old chicks prevented hysteria for as long as 8 weeks, after which the nails had grown back and some hysteria was indicated.

Fewer birds in large cages.   In a supplementary trial at the same experiment station hysteria developed when 40 pullets were kept in a cage, but seldom when 20 birds were in a unit. There was no hysteria when the pullets were in six-hen and individual cages.

Color of artificial light.   Further research work showed that red light did not reduce the incidence of hysteria, although it did reduce cannibalism. The onset of hysteria was delayed when blue lights were used, but once hysteria began the incidence was greater than with white light.

Chicken sounds.   Some contend that the sounds from a radio playing continuously in the chicken house will subdue the birds and help prevent hysteria. Others have recorded the voices of "happy" hens in the course of eating or laying and have played these recordings back in other chicken houses to help reduce hysteria. In some instances these broadcasts have been of little help, in others, they have reduced hysteria in affected flocks, but the cessation was not permanent unless the cause was eliminated.

Treatment

Treatment may have some effect, but by no means is any treatment consistent in reducing hysteria. Usually, a treatment may help return the birds to a normal condition only as long as it is continued. Once it is discontinued, the birds again become and remain hysterical until the original cause is eliminated. Some of the following recommendations may prove of value:

Beak trimming. If birds have not been beak trimmed or were improperly trimmed, a good trimming may be advisable. If hysteria has been a major problem on a poultry farm, extreme care should be used in removing the beak at trimming time, particularly when young chicks are trimmed at 6 to 8 days of age.

Tranquilizers. There are tranquilizers on the market that are practical for chickens. A small dose in the water may aid in quieting the flock. Administer the tranquilizer every fourth day for three times. One such tranquilizer is metoserpate hydrochlorate.

Niacin. There is some indication that additional amounts of niacin in the diet are of value in quieting the birds.

Dosage: Add 0.4 lb (182 g) of niacin to each ton (2,000 lb) of the ration. Feed for 9 days. If hysteria is not reduced, repeat the feeding for another 9 days.

Vitamin-electrolyte mixture. Commonly used to prevent stress, this mixture has been shown to remove the hysteria under field conditions.

37-X. INCLUSION BODY HEPATITIS (IBH)

Inclusion body hepatitis is a disease of young chickens characterized by the sudden onset of increased mortality, anemia, and hepatitis.

Cause

IBH is caused by an adenovirus, producing a very common infection in poultry. This specific disease is caused by at least 3 serotypes of DNA adenovirus.

Symptoms

The disease is commonly seen in chickens 3 to 18 weeks of age and manifests itself by a sharp increase of mortality during a 2-week period. Total mortality may reach 10%. General depression and listlessness is observed. The skin is pale and may contain hemorrhages over the legs and breast.

The liver is commonly swollen, the kidneys are frequently swollen, the bone marrow is yellow, the blood is thin and watery, and hemorrhages are often seen in skeletal muscles.

Transmission

Adenoviruses are quite widespread and transmission from farm to farm has been observed. The virus is eliminated in the feces and can be mechanically spread by human beings. It is suspected that the organism is also transmitted through the egg.

Diagnosis

The agar gel precipitin test is commonly used to demonstrate antibodies to IBH, which in conjunction with mortality patterns and gross lesions are used to identify the problem.

Treatment

No treatment for the disease is available.

Control

Normal sanitation recommendations should be followed. Breeding flocks with IBH problems should not be used for producing hatching eggs. Wild birds should be excluded from the poultry house. If infectious bursal disease (IBD) predisposes the flock to IBH, a vaccination program for IBD should be considered.

37-Y. AVIAN INFLUENZA (AI)

AI is a viral disease of many different species of poultry affecting the respiratory, enteric, and nervous systems. In its most virulent form, it results in extremely high mortality.

Cause

The AI virus is classified as an orthomyxovirus–type A. There are also numerous subtypes that are identified by their hemagglutinin (H) and neuraminidase (N) typing. Each subtype differs in its pathogenicity, ability to infect different species, and transmissability.

Symptoms

Outbreaks of low-virulent AI experience mostly respiratory symptoms including coughing, sneezing, and sinusitis. Egg production may suffer and there may be diarrhea, edema of the face and head, and/or nervous symptoms. Highly pathogenic outbreaks may exhibit 90% or higher mortality rates. Eggshell pigmentation and quality may also suffer.

Transmission

Waterfowl are the natural hosts of AI. The virus has been isolated from both wild and domestic species and from the water in lakes and ponds used by waterfowl. The disease is easily transported from farm to farm on contaminated clothing, crates, filler flats, and equipment.

Diagnosis

Serologic evidence may be useful in determining the exposure to AI virus, but positive confirmation requires virus isolation and identification.

Treatment

There is no known treatment for this disease. Treatment with broad-spectrum antibiotics may reduce complications from other infections. Depopulation of infected premises is considered to be the only remedy.

Control

Exclusion of waterfowl is essential. Because the disease is so easily transmitted, flock security is critical (see Chapter 39-P).

37-Z. EGG-DROP SYNDROME (EDS76)

Egg-drop syndrome is a disease of laying hens that causes sudden losses of egg production, depigmentation of brown-shelled eggs, and increased incidence of thin-shelled and shell-less eggs.

Cause

EDS76 is caused by an adenovirus, first described in 1976 in Holland. It has been seen in numerous countries, but at the present time has not occurred in the United States. The organism is commonly recovered from waterfowl but the disease does not cause reproductive problems in these species.

Symptoms

The symptoms are almost exclusively associated with the reproductive performance of the flock. Egg production may drop by 40 to 50% for periods of up to 10 weeks. Brown-shelled eggs are laid without pigment. Many soft-shelled and shell-less eggs are laid.

Transmission

While waterfowl are the natural host of the disease, their involvement in the transmission of EDS76 is extremely rare. Contaminated vaccines and needles were responsible for earlier cases of transmission.

Diagnosis

Along with the specific clinical signs observed, the hemagglutination-inhibition (HI) and serum neutralization (SN) tests are the procedures most commonly used.

Treatment

There are no known treatment procedures available to combat the disease.

Control

An inactivated vaccine is available and is used in replacement pullet flocks at 14 to 16 weeks of age. Visitors should not be permitted to enter poultry houses. Waterfowl should be excluded from all poultry premises.

37-AA. LYMPHOID LEUKOSIS (LL)

Lymphoid leukosis is a tumor-producing virus disease usually of chickens 16 weeks of age and older. It is commonly a problem of breeders and laying hens.

Cause

LL is caused by a family of retroviruses. The subgroup A virus is most common in the United States.

Symptoms

Birds suffering from LL are commonly pale and emaciated. The abdomen may be very large because of the bird's massive liver. The bursa of Fabricius is commonly enlarged and lymphomas are seen in the liver, kidney, ovary, and bursa.

Transmission

Egg transmission is considered to be the primary means of spreading the virus.

Diagnosis

LL is usually diagnosed by clinical signs associated with the age of the affected flock. Diagnosis is made more difficult because of the similarity of lesions to Marek's disease. Serological and virological methods are not considered to be useful.

Treatment

There is no known treatment for lymphoid leukosis.

Control

Eradication of individual infected breeding hens is considered the best means of eliminating or greatly reducing the incidence of LL. Since egg transmission is the number one method of spreading the disease, elimination of carrier birds from the breeding flock has been highly successful in reducing the infection in their progeny. Mortality due to tumors has been reduced and egg production has increased.

38

Parasites, Insects, Mites, and Rodents

Each year parasites and predators cost the poultry industry millions of dollars. No one knows the actual figure. Some costs are direct; some are indirect.

There are many poultry parasites, but relatively few are of major importance to the poultryman. Some live inside the bird, others on the bird's surface. Certain mites are parasites of chickens, and when their population is large, they affect growth and egg production. Some insects are a continual problem, and although they do not greatly affect chickens, they inhabit poultry farms where they breed and grow, constituting a general nuisance to people and animals in the vicinity. Rodents seldom attack any but young chickens, but they do devour and destroy huge quantities of poultry feed, damage house insulation, create a general uncleanliness about the premises, and spread many poultry diseases.

38-A. INTERNAL PARASITES

Large Roundworm

The large roundworm, *Ascaridia galli,* is an intestinal parasite of the chicken. Of the several similar parasites invading the intestines, the large roundworm probably inflicts the most damage.

Female worms produce a form of egg which is expelled from the chicken in the fecal material. The shell of the egg consists of three layers of tough material and protects the developing larva during its growth. Many acids and other chemicals will not affect the shell. The larva matures in 10 to 14 days, and the egg loses its outer shell. The remaining egg is ingested by the chicken and within a few hours it hatches in the crop, proventriculus, gizzard, or small intestine. At this time the larva imbeds itself in the tract lining. After another 21 days the larva develops into an adult in the small intestines, where it floats in their contents. It is capable of producing eggs 7 days later. Mature worms may reach a length of 3 in. (7.6 cm), are gray in color, and easily visible to the naked eye.

A small infestation of roundworms does little damage to the host bird but when the worms become numerous the birds become unthrifty, and growth and feed conversion are impaired. Young birds show more damage than older birds. Mortality is seldom increased by the worms themselves, but when other conditions are present, such as some diseases, death losses will become important.

Transmission. The embryonated egg is the infective stage of the life cycle. Each worm is capable of producing thousands of eggs, many of which soon reach other birds. The eggs are especially hardy; some will live outside the host for several years. Eggs may get to other poultry houses by way of shoes, equipment, truck tires, etc.

Treatment. Treatment involves upsetting the life cycle of the worms, either by causing the mature worms to be expelled from the intestinal tract, or preventing the production of eggs. The following two anthelmintics are used:

1. *Piperazine.* Piperazine represents a purge-type wormer for adult birds only. It removes the worms from the intestinal tract. It should be given in the water for a short period. Usually it will not affect the birds, but if layers are in a stress period piperazine may produce some drop in egg production.

    The drug does not prevent reinfestations as the purged birds soon pick up living worms from the litter, and the life cycle starts again. Removing the litter after drug use will remove the source of supply of purged worms.

2. *Hygromycin B* (Hygromix®). An antibiotic, hygromycin has been found to possess the property of disrupting the life cycle of roundworms by preventing them from producing eggs. It does not create a stress. It is given in the feed, and fed over a period of 6 weeks to growing and laying birds.

    Pullets: Feed 12 g of hygromycin B per ton (2,000 lb) of feed starting at 12 to 14 weeks of age.

    Layers: Feed 12 g of hygromycin B per ton (2,000 lb) of feed continually.

    Note: When Leghorns are consuming feed at the rate of 20 lb (9 kg) or over per 100 birds per day, and heavy breeds at 30 lb (13.6 kg) or over, reduce the feed level of the antibiotic to 8 g per ton.

Control. Clean litter is an effective method of reducing the number of worm eggs in the floor covering, but not an absolute method of control. Birds on wire floors are not affected with worms.

Cecal Worm

Small roundworms, *Heterakis gallinarum,* are sometimes known as *cecal worms.* The life cycle is similar to that of the large roundworm except that the worms end up in the ceca instead of the small intestine. They attain a length of ½ to ¾ in. (1.3 to 1.9 cm). In the chicken this worm is of little economic importance, but as it is associated with blackhead it takes on importance in the turkey.

Transmission. Same as for the large roundworm.

Treatment. Little attention is paid to the incidence of the small round-worm in chickens. The cecal location of the worms makes it difficult for any drug to reach the seat of infestation, as most of the intestinal contents bypass the ceca in the course of digestion.

Hygromycin B. This antibiotic has some value in the treatment for cecal worms. Feeding recommendations are the same as for treating birds with large roundworms.

Capillaria Worm

The capillaria worm, *Capillaria obsignata,* is a very small parasite in-habiting the small intestine. Its life cycle is similar to that of the large roundworm except that the capillaria worm locates in the upper two-thirds of the small intestine, usually the duodenal loop. It imbeds itself in the mucosa, where it spends its entire life when in the bird. The wall of the gut shows hemorrhages, and becomes thickened. The greater the number of worms, the more pronounced the difficulty. Growth and feed conversion are affected in growing birds, and egg production may suffer in layers. Utilization of vitamin A is impaired, producing the poor feed efficiency mentioned above. Capillaria worms are very small. Their presence may be observed if the mucus is scraped from the intestinal wall, water added, and the mixture passed through a special screen having a very fine mesh.

Treatment. Treatment is by the following:

1. *Hygromycin B* (Hygromix®). This antibiotic is quite specific for capillaria worms. The amount to add to the feed is the same as for treatment of large roundworms.

2. *Coumaphos* (Meldane®). Of the drugs used for the treatment of the capillaria worm, only one, couma-

phos, is legal in the United States. Special precautions
in its use are necessary.

Tapeworm

Although there are eight tapeworms involved with chickens, only one
is of importance, *Railleitina cesticullus*. It is a segmented flat worm, from
very short to several inches in length. The head imbeds in the intestinal
wall, and new segments continually grow behind it, each segment being
a separate entity capable of its own digestion and reproduction. Segments at the end of the worm contain eggs; these segments continue to
break off and pass from the bird through the feces.

Transmission.   Transmission is by an intermediate host. Eggs expelled
from the birds are consumed by several types of insects. Transmission is consummated when a chicken eats any insect containing the
ova of the worm. With the advent of better methods of producing
poultry, the tapeworm now has little importance. Even when their
incidence in the intestinal tract is great they do not seem to harm
the bird.

Treatment.   *Dibutyltin dilaurate* is sold in combination with piperazine
and is effective in the removal of tapeworms. Only when the head
of the worm is removed will control be effective; otherwise new
segments will appear. Dibutyltin dilaurate is effective in this respect, but caution should be taken in its use. Treat only highly infected flocks. Do not feed to laying flocks.

38-B. EXTERNAL PARASITES

Caution

Many chemicals used for the control of external parasites of poultry
may leave residues in poultry meat and eggs. Some of the residues are
injurious to humans, particularly when concentrations are high. Thus in
some countries the use of certain drugs is prohibited by law. Furthermore, the list of approved chemicals is constantly changing. Those given
in this chapter are on the U.S.-approved list today, but may not be tomorrow. Some that are not on the approved list today may be added in
the days ahead.

Beware

Before using any chemical for the control of external parasites, be sure
to confirm its legality. Then follow instructions exactly. Dispose of unused drugs according to method outlined on the container.

Governmental agencies have set levels on most drugs that are allowed in poultry meat or eggs. These are known as *tolerance levels*. When either poultry meat or eggs contain more of the drug than the tolerance level, the product will be condemned as unfit for human consumption. If the flock that laid condemned eggs can be located, it too will be condemned, and must be destroyed. The birds cannot be processed and used for food, as they too will contain residues.

Insect Control Programs

The control of insects, particularly flies, on poultry farms has become very complicated. Most College Experiment Stations update their recommendations annually, and publish complete programs. These colleges should be contacted early in the spring for the latest information. Insecticides used generally fall in the categories of space sprays, baits, larvicides, and residual sprays.

Program to Control Flies

The following has been extracted from a bulletin, "Integrated Management of Pest Flies on Poultry Ranches," published by the University of California:

Management of pest flies can best be achieved by a program of farm sanitation practices that reduces natural fly attractants and eliminates conditions favorable to their breeding and development. For best results, insecticide use should be combined with sanitation and manure management practices, the preservation of natural enemies of flies, and the use of various kinds of fly traps.

A number of chemicals are available for fly control on the poultry farm. They are sold under many brand names and may be formulated as baits, wettable powders, or emulsifiable concentrates. Insecticides are grouped into four principal categories: residual sprays, baits, larvicides, and space sprays.

1. *Residual sprays.* Apply as a coarse, low-pressure spray, wetting all surfaces but stop just before the point of runoff. Apply when newly emerged flies first become evident in the spring, and throughout the remainder of the season whenever fly populations start to increase. Special care must be taken to avoid spraying the birds, feed, drinking water, and eggs.
   a. Outside uses only: Use dimethoate (Cygon®), or permethrin (Atroban®, Ectiban®).
   b. Inside or outside: Use naled (Dibrom®, Fly Killer D) or stirofos (Rabon®).

2. *Baits.* Baits are generally most effective when used to supplement other fly control efforts. Baits should be placed

where birds and other animals cannot reach them. Baits can be purchased ready to use or can be prepared by the user by adding sugar to approved insecticides. Bait stations are commonly used to concentrate the collection of dead flies.

Use commercially prepared methomyl with synthetic attractant (Improved Golden Malrin®), or operator-prepared dichlorvos (Vapona®) or naled (Dibrom®, Fly Killer D®).

3. *Larvicides.* These chemicals are best used as spot treatments only to control fly larvae that develop in wet areas. Overall treatment with larvicides on a regular schedule is not recommended because of questionable effectiveness, chances of increasing chemical resistance, high cost of application, and most important, the indiscriminate kill of many natural enemies. Use dimethoate (Cygon®) or stirofos (Rabon®).

4. *Space sprays.* These materials are most valuable in situations where an immediate kill of large numbers of flies inside the poultry house is desired. They are also effective for a quick kill of flies congregated in an enclosed area such as an egg room. Space sprays have no residual effect. Use dichlorvos (Vapona®) or pyrethrins plus piperonyl butoxide.

Other Fly-Control Programs

Frequent manure removal. Daily or weekly cleanout is a popular practice where rapid undercage drying is impractical. For effective fly control, manure should be cleaned out frequently to prevent fly larvae from reaching maturity. Manure should be properly managed at deposition sites to favor solar drying and the activity of natural enemies of flies. The schedule of operations depends on final manure disposition and regional climatic variations. New designs for lay houses, cleanout machines, and methods of manure processing make frequent cleanout economically feasible for many poultry producers.

Feed-through larvicides. A larvicide, Larvadex®, is approved as a feed additive in some countries and in most states in the United States. It comes as a premix for use at the rate of 3.3 lb of premix to 2,000 lb of feed (1.65 kg per 1,000 kg of feed). It is an insect growth regulator and acts by disrupting the life cycle of the fly. *Caution:* Check with authorities to determine its legality.

Biological control. Measures that maintain fly predators and parasitoids already present in manure can help to control flies on poultry farms. General recommendations to enhance the effectiveness of beneficial insects include concrete floors beneath cages, avoidance of water leaks, alternate row manure cleanout, leaving a dry pad of manure following cleanout, and avoidance of overall use of insecticides.

These practices preserve the habitat of the existing natural enemies such as ants, mites, beetles, and parasitoids, so they can quickly invade adjacent fresh droppings. This is important, because the natural-enemy complex could take from 6 to 12 months to reestablish itself on a farm if all manure were removed in one cleanout.

Special Precautions When Using Pesticides

All pesticides are hazardous and must be used strictly according to the instructions on the label. Workers must be thoroughly trained regarding the safe handling of these materials. Check with local agricultural authorities before use.

38-C. OTHER INSECTS

Lice

Only biting lice attack chickens and the biting aggravates the birds. The more lice present, the more aggravation. When there are thousands, egg production, growth, and feed conversion may suffer.

Lice spend their entire lives on the birds. They do not migrate, for off the chicken they will die in 5 or 6 hours. Eggs are usually laid on the feathers where they are held by a type of glue. The eggs hatch in a few days to 2 weeks. Lice are chewing insects and live on scales of the skin and feathers.

Types of lice

> Body louse (two types and most important group)
> Head louse
> Shaft louse
> Wing louse
> Fluff louse
> Brown chicken louse
> Large chicken louse

Darkling Beetle (*Alphitobius diaperinus*) (Lesser mealworm)

This insect lives in the soil and is abundant where dirt floors are used in poultry houses. The adults are about $\frac{1}{4}$ in. long and dark brown in color. Some scientists have associated the lesser mealworm with leukosis, having found that the leukosis virus remains viable in both the larva and adult worm. When chickens eat either the larva or adult worm they may develop a form of leukosis. The Darkling Beetle also burrows in insulation materials, thereby reducing their effectiveness.

38-D. MITES

Mites are parasitic pests (not insects) attacking chickens. They differ from lice in that they suck the blood of the host to survive. Furthermore, in the adult stage mites have eight legs; lice have six. There are two mites attacking chickens.

> 1. *Chicken red mite:* This is a large mite that lives in the cracks and crevices of the poultry house rather than in the litter, and is very prolific. It migrates onto the surface of the chicken and there does its damage; thousands may be found on a bird.
> 2. *Northern fowl mite:* This is a very small mite. It seldom migrates from the bird and is the most important external parasite of chickens. Some Northern fowl mites may be found at all times on affected birds. They consume large amounts of blood and congregate near the moist area of the vent. Here the skin becomes reddened and scaly.

38-E. EXTERNAL PARASITE CONTROL

Several insecticides are available to control mites and lice on chickens. Because of the chicken's feathers, penetration to the site of parasite infestation is often extremely difficult. High-pressure sprayers are required and it is recommended that sprays be applied at the rate of 1 gal per 75 birds. Spot treatments may be satisfactory.

Chemicals that are effective include permethrin (Atroban®, Ectiban®, Permectrin®), carbaryl (Sevin®), malathion, and stirofos plus dichlorvos (Ravap®).

Sevin should not be applied more often than every 4 weeks or pesticide tolerances in the eggs will be exceeded. It is advisable to alternate chemicals to lessen the development of resistance in the parasites. Consult local authorities for specific regulations.

38-F. RODENTS

Predators on poultry farms are usually confined to rats and mice. The annual economic loss to the poultry industry because of these runs into the millions of dollars. They eat and destroy feed, transmit diseases, and create a general nuisance. Rats will produce three to six litters of seven to eight young each during a year. A few migrating rats will soon produce a large population.

Rodents are best controlled with a cleanup campaign first, followed by an effective baiting program. Four of the most effective baits are warfarin, zinc phosphide, alpha-chlorhydrin, and brodifacoum.

Warfarin

One killer is a chemical known as warfarin. Combined with sulfaquinoxaline it is sold under the name Prolin®. When Prolin is consumed for several days, the rats and mice die suddenly as a result of internal hemorrhage. Warfarin reduces the clotting powers of the blood, causing the animals to hemorrhage. Sulfaquinoxaline inhibits the growth of those intestinal bacteria producing vitamin K. When the sulfaquinoxaline is not added, the vitamin K tends to increase blood clotting, working against the warfarin. Sulfaquinoxaline reduces the amount of vitamin K available in the rat or mouse.

Other anticoagulants. There are other chemicals that reduce the clotting properties of the blood. Consult your chemical supply house for those available in your area.

Prolin is added to a bait such as corn, corn oil, sugar and so forth. However, rats and mice are very suspicious at times and may refuse the bait material unless an appetizer, specific for rats and mice, is added.

The number of bait stations is governed by the area involved and the density of the rat or mouse population. But stations should be abundant. As rats or mice seldom cross wide barren areas or expanses, but rather move about the margins of these places, bait stations should be near walls, next to burrows and paths.

How to bait. Place the bait in a pan under a long board attached to the wall. A long box may be used with openings at each end for the rats and mice to come and go. Place fresh water near the bait stations to increase the bait consumption.

Bait should be fresh. Rats and mice will not consume the bait once it becomes old, dusty, or stale. Bait stations must be replenished every day or two. Do not put out more bait than the animals will consume in 1 or 2 days.

The process of creating the hemorrhagic condition takes time. Rats and mice must eat the bait for several days before they die. Be sure they are eating the bait. Many times there will be other sources of feed in the poultry house: in the feeders, in feed sacks, or in the litter. Place feed sacks on horses with tin covering their legs. Allow the poultry feeders to become empty at night.

Zinc Phosphide

Another bait is zinc phosphide. It is mixed with fresh chicken feed, but always mix it in the open and wear a dust mask and rubber gloves.

Be sure to use according to directions. Keep the bait away from the chickens; it is best used in houses with no birds in them. A license for its use is usually required.

Alpha-Chlorhydrin (EPIBLOC®)

A chemical, sold under the trade name EPIBLOC®, is both a raticide and a rodenticide. It is unique in that sublethal doses produce sterility in male rats, while large doses will effect a kill in both sexes. It will cause from 85 to 95% of sexually mature male rats to become permanently sterile. Some other species of animals will show temporary sterility, but the numbers are few.

The male sex drive in rats is not impaired after *sublethal* drug consumption, and the males continue to mate with the females. When large doses of EPIBLOC are eaten by rats the death rate in both sexes is between 70 and 95% within a few hours.

EPIBLOC is a biodegradable rodenticide, and is destroyed by the rat eating the chemical. Thus, if a nontarget animal ate a poisoned rat it would not suffer toxic effects. EPIBLOC is a restricted-use rodenticide and must be used by professionals according to the correct governmental regulation.

Brodifacoum (HAVOC®, TALON®)

This is a modern anticoagulant used for rodent control. It differs from other anticoagulants in that the product can kill in a single feeding. First dead rodents appear in 4 or 5 days. Rats having a resistance to warfarin will be killed by brodifacoum products.

38-G. DRUG AND PESTICIDE PRECAUTIONS

Drugs should be used carefully. Most of them are poisons, and precautions necessary to the handling of such chemicals must be taken. The following rules should be followed:

1. Properly identify the problem.
2. Select the correct drug or insecticide. When in doubt, consult someone who knows which is the right one, and which one may be used legally.
3. Read the label carefully. Many drugs and insecticides are sold under a trade name, but the label will show the list of ingredients, including the correct name of the active ingredient.

4. Follow directions. Do not dilute the product and do not overadminister.
5. Store only in original containers.
6. Keep containers out of reach of children, pets, and livestock.
7. Know the antidote before opening the container.
8. After any container is empty, rinse it at least twice.
9. When burning any empty containers, stay out of the smoke.
10. Clean up after the use of any drug or insecticide.

39
Disease Prevention and Animal Welfare

Undoubtedly prevention is better than cure. Today's poultry programs incorporate many procedures necessary to keep flocks in a healthy condition. Necessarily these are not just vaccinations, but constitute a great number of management and other practices. Many are on a continuing basis; cleanliness, disposal of refuse, stress prevention, flock security, and pollution control are examples. All possible endeavors to keep the flock at high productive capacity are requisite to economical meat, chick, and egg production. Treating a flock that has lost some of its efficiency because of disease, stress, or abuse represents a stopgap approach; prevention is a much better and more practical means of dealing with disease.

A detailed analysis of continuing practices necessary to a prevention program follow. Even then there may be pitfalls; adhering to all such practices will not give complete assurance that there will be no difficulties with the flock.

The practices that must be followed regularly on a poultry farm may be grouped as follows:

1. Stress prevention
2. Management
3. Supply of good water
4. Blood testing
5. Sanitation
6. Vaccination
7. Coccidiosis control
8. Parasite, lice, mite, and rodent control
9. Dead bird and refuse disposal
10. Pollution control

39-A. STRESS

All stresses are due to hundreds of tensions which upset the bird. Stress is the bird's response to what it perceives as a threat or challenge.

The Cause of Stress

A stressor induces the bird to succumb to specific stress through hormonal secretions. The hypothalamus is the initiation point of a response from a stressor. It tells the pituitary gland to increase production of the adenocorticotropic hormone (ACTH), which finds its way to the bloodstream where it is transported to the adrenal gland. Here the increased ACTH causes hyperactivity of the adrenal cortical cells which are capable of synthesizing the glyco-steroid hormone, corticosterone, which in turn is carried by the blood to all cells in the body. Here the RNA messenger of the nucleus regulates the production of enzymes and proteins each cell produces.

Many changes take place in the bird with a high corticosterone level. The rate of heartbeat increases, the blood pressure rises, feed consumption is reduced, sexual activity is lowered, fewer antibodies are produced, there are changes in the cardiovascular system, ulcerative enteritis is more common, growth rate is lowered, and plasma glycogen is reduced, but still the bird is better able to withstand a harsh environment. Resistance to bacterial disease is increased as the corticosterone gains a high level in the blood, but resistance to viral diseases is decreased.

In reverse of the above, birds with low corticosterone levels produce and live better, they are more docile, grow faster to large weights, have better feed efficiency, have less carcass fat, have a lower heartbeat, produce more and larger eggs, and show a lower resistance to bacterial disease through decreased antibody production, but a higher resistance to many viral infections, and respond better to certain vaccinations.

Indication of Stress

Associated are high levels of nervousness and rapid heartbeat and these are probably the first indications of stress in a flock. Sudden changes in the environment are contributors to both. The appearance of a strange caretaker is an applicable example. Sudden changes in surroundings, equipment, and daily routine are others.

The recreation of the social order brought about by moving birds to new pens is a stressor of magnitude at times. Birds in a constant environment are harder to stress than those in a variable one. Sudden hot or cold weather, equipment changes, electrical failures, changes in lighting schedule, and absence of drinking water are examples of short-term stressors. Conventional housing, inadequate light in the laying pen, high concentrations of ammonia, improper or too little equipment are long-term stressors. Short-term stressors produce high corticosterone levels quickly, but they return to normal in about a minute as there is no body storage. But stressors of a continuing nature cause prolonged stress because the production of the hormone is retriggered and high levels continue to exist.

Birds under stress from almost any cause have a lymphatic involution with an atrophy of the bursa of Fabricius, thymus, and spleen, the lymphoid associates, and are less able to withstand bacterial invasion because antibody production is lower. The number of lymphoid cells is indirectly proportional to the increase in the corticosterone level in the blood, and with a lower lymphoid defense, the bird is more severely stressed.

Stress from two or more causes is more difficult to cope with than the problems from one. The flock manager must be ever alert to any flock deviation from normal if he is able to correct the ever-present numerous stressors.

39-B. MANAGEMENT

The poultryman must be cognizant of the many factors that aid in keeping his flock healthy, and his products high in quality. Many of these factors are on a day-to-day basis; others require attention less frequently.

1. *Isolation.* Modern disease control programs call for isolation of the poultry house, flock, or hatchery. The *all-in, all-out* system is favored; it constitutes starting a single group of chickens in one house and leaving them there until the end of their productive or growth period without the addition of other flocks. Many times the all-in, all-out system includes the entire poultry farm.

2. *Housing and equipment.* Good housing and good equipment are necessary in today's poultry production. Poor housing and equipment may cause unnecessary stress on the birds with resulting greater morbidity, mortality, and inefficiency. Profit will be lowered. Housing and equipment must be sufficient for the number of birds housed.

3. *Hatching.* The details of hatchery operation are many, the improvement in hatchability through each detail is small, yet a little improvement here and there adds up to profit increases that are meaningful.

4. *Quality chick.* Many things go into the production of a chick of quality. These have been detailed throughout this text. Most of them call for a regular vigil; there can be no laxity on anyone's part. The objective of a breeder and hatchery operation is to produce a good chick, the foundation on which the commercial poultryman builds his business.

5. *Hatchery chores.* Day-old beak trimming, dubbing, sexing, vaccination, and so forth are special services that are to be completed in the hatchery. Each operation should be done with infinite care. Each is a part of the program of producing a quality chick.

6. *Proper nutrition.* Feeding the breeding flock correctly is a daily necessity. Without adequate nutrition the birds do not produce gains in weight, feed conversion, egg production, or hatchability adequately or economically. Faulty diets will lead to increased stress, which in turn increases the incidence of many disorders.

39-C. WATER SUPPLY

A supply of good water is essential. There is no such thing as pure water. All water contains many substances in solution or suspension, many of which affect its palatability and value.

Water Analysis

Before being used, a sample of water should be submitted to a laboratory for analysis of chemicals and purity. Some of the determinations made to evaluate its condition are as follows:

Color. Any color is due to certain substances in solution such as tannin, iron salts, etc.

Turbidity. Particles in suspension rather than in solution cause the water to be turbid.

Hardness. Salts of calcium and magnesium form scale and sludge and cause the water to be "hard." Hardness affects the taste of water.

Iron. Although iron in water seldom affects chickens, it stains almost everything with which it comes in contact.

pH. The pH of a solution is a measure of its acidity or alkalinity. When above 7, it is alkaline; below 7, it is acid. Water is normally about 7 to 7.2. A water pH of 8.0 is the upper limit for poultry.

Total solids. Total solids represent the total amount of solid material in a suspension or solution.

Nitrogen. Two determinations may be made for the presence of nitrogen, but each is indicative of decaying organic material, and is a measure of contamination.

Poisonous metals. When in excess of 0.5 ppm certain metals in the drinking water may accumulate in the bird to produce pronounced difficulties and cause illness.

Bacteria. Type of bacteria, rather than number, is important to a water analysis. Some bacteria may be detrimental to human beings and chickens.

Maximum Levels in Water

Table 39-1 shows the recognized threshold levels in a safe water supply for chickens.

Table 39-1. Threshold Value of Microelements in Water for Chickens

Factor	Threshold Concentration mg/L
Total dissolved solids	2,500
Total alkalinity	500
Calcium	500
Magnesium	125
Sodium	500
Bicarbonate	500
Chloride	1,500
Fluoride	1
Nitrate	25
Nitrite	4
Sulfate	250
Copper	1
Cadmium	0
Salt, growing birds	590
Salt, laying birds	1,000
Iron	0

Chlorinating Water

When water is microbiologically impure it should be chlorinated. There are several suitable chlorinators on the market, most of which operate by superchlorinating the water at the farm source. This guarantees a satisfactory level of chlorine in the water in the poultry waterers. The addition of chlorine to water also reduces oxidation of any iron, thus eliminating some of the rust developing in pipes and valves.

Water Sanitizers and Vaccination

Many vaccines are administered in the drinking water. The presence of any sanitizer in the water will affect the viability of the vaccine, often making it worthless.

> *Warning:* Do not add vaccine to any drinking water containing a sanitizer. First flush the water system several times until it is free of any sanitizer; then provide clean water to which the vaccine has been added.

> *Skim milk to neutralize sanitizers.* Where there is no supply of unsanitized water, the sanitizer may be neutralized by adding *dried skim milk* to the drinking water. Add 3.2 oz (90.7 g) of dried skim milk to each 10 U.S. gal (37.9 liters) of water. This is approximately 1 part of dried skim milk to 400 parts of water. Add the vaccine to the milk-water solution and mix thoroughly.

39-D. BLOOD TESTING AND SANITATION

Blood Testing

Blood testing the breeding flock becomes a regular chore for those poultry operators following certain disease-control programs. Pullorum, typhoid, and mycoplasma are diseases in this group. Control programs are detailed in Chapter 37. Testing the birds must be completed on schedule; many times the testing is done by representatives of government agencies rather than by the flockowner. Prearranging a schedule will be necessary.

Sanitation Program

All poultry farms and hatcheries must be kept clean. Not only does this mean that debris should be removed on a continuing basis, but certain disinfectants must be used regularly to lower the incidence of pathogenic microorganisms.

Hatcheries must be cleaned and disinfected, either by chemicals or fumigation, or both. Drinking fountains must be cleaned daily. Nests should be cleaned and new nest litter added on a weekly basis. Chick trucks must be fumigated prior to each chick delivery. Hatching eggs must be fumigated immediately after they are laid. Employees must shower and change to clean clothing before entering the poultry premises. These are but a few of the programs which must be regularly followed to keep the operation clean and sanitary. There are many more.

39-E. VACCINATION PROGRAM

Not only are there many vaccines, but there are numerous vaccination programs. These programs involve

1. Type of vaccine to use
2. Vaccines to be mixed together
3. Vaccinations to be administered simultaneously
4. Age of the bird when vaccinations should be given
5. Type of bird (breeders, layers, broilers) involved
6. Route of vaccination

Important: There are many types of vaccines and vaccination programs. Those given in Tables 39-2, 39-3, and 39-4 are only examples. They must not be construed as being practical for all areas of the world nor under all conditions. Diseases in the area, availability of vaccines, periods of stress, climatic conditions, and many other factors are involved to make up a program for a specific area. Consult your vaccine supplier or specialist before initiating any vaccination program.

Broiler vaccination program. The fact that broilers are marketed at about 7 weeks of age and are not subject to poultry diseases that affect older birds makes it necessary that the vaccination program for broilers differ from that used for birds to be used for laying or breeding (see Table 39-2).

Layer vaccination program. Growing birds to be used later for the production of commercial eggs require a special vaccination program. Although the diseases of commercial laying strains are identical with those of breeder replacement strains, vaccinations given during the period when eggs are being produced are different or are not needed (see Table 39-3).

Breeder replacement vaccination program. This program is similar to the vaccination program for commercial laying strains except that

Table 39-2. Example of Broiler Vaccination Program

Age of Vaccination wk	Vaccine		Method of Vaccination
	Disease	Strain	
Day-old	Marek's	HVT	Subcutaneously
2-3	Bronchitis	Massachusetts	Ocular or water
2-3	Newcastle[1]	La Sota	Ocular or water

[1]See Chapter 37-M for alternate procedure.

Table 39-3. Example of Egg-Type Grower Vaccination Program

Age of Vaccination wk	Vaccine Disease[1]	Strain	Method of Vaccination
Day-old	Marek's	HVT	Subcutaneously
2–3	Bronchitis	Massachusetts	Ocular or water
	Newcastle	La Sota	Ocular or water
6	Bronchitis	Massachusetts and Connecticut	Ocular or water
	Newcastle	La Sota	Ocular or water
10	Laryngotracheitis	Modified	Ocular
	Fowl pox	Modified pigeon	Wing-web
6–16	Avian encephalo-myelitis		Wing-web or water

[1]Laryngotracheitis: Secure regulatory requirements.

Table 39-4. Example of Meat-Type Breeder Replacement Vaccination Program

Age of Vaccination wk	Vaccine Disease[1]	Strain	Method of Vaccination
Day-old	Marek's	HVT	Subcutaneously
1	IBD	Modified	Water
	VA	Lukert	Subcutaneously
7–12 da	Newcastle	La Sota	Ocular or water
	Bronchitis	Massachusetts	Ocular or water
6	Newcastle	La Sota	Water
	Bronchitis	Holland	Ocular, water
10	IBD	Modified	Ocular, water
	AE/Pox		Wing web
	VA		In other web
14	Newcastle	La Sota	Water
	Bronchitis	Holland	Water
17	VA		Wing web or water
21–22	IBD/Newcastle	Oil emulsion	Subcutaneously
30–34	Bronchitis	Holland	Ocular or water
	IBD	Modified	Ocular, water

Note: Seek professional advice in your area before starting this program.
[1]Laryngotracheitis: Secure regulatory requirements.

breeders are vaccinated for certain diseases during the laying period to maintain a constant level of parental immunity in the chicks hatched from the eggs produced by the breeders. Usually these vaccinations are made every 12 weeks during the period the breeder flock is producing hatching eggs.

Often there are other vaccinations used for breeder replacement birds such as day-old pox vaccine; viral arthritis at 6 to 8 days, 10 to 12 weeks, and at 17 weeks (Chapter 37-G); and infectious bursal disease at 3 to 4 weeks and 19 to 23 weeks (Chapter 37-P).

Check with college specialists or vaccine manufacturers before entering on a vaccination program in your area. For a preliminary and basic program see Table 39-4.

Water Proportioners

Many vaccines or medicaments are added to the drinking water. At other times disinfectants are added to sanitize the water. When automatic waterers are used it is often difficult to add anything to the water. To facilitate the procedure a water proportioner may be used. Usually this is a pump that operates when water from the water supply is forced through a cylinder. When the proportioner is installed in the incoming water line the flow of water to the fountains in the poultry house operates the pump. The pump has the ability to draw liquid from a container and inject it into the water line at a rate proportionate to the flow of water to the watering troughs. Thus, the proportion of the medicated or treated liquid always is constant. The proportion may be adjusted by altering controls on the proportioner. There are several proportioners on the market.

Temperature Affects Amount of Medicament Consumed

Most dosage of any medicament for use in either feed or water are based on their consumption at average daytime house temperature of 70°F (21.1°C).

Water consumption. When house temperatures increase more water is drunk. Unless there is a reduction in the amount of the medicament in the water, the daily dosage would be materially increased. For example, at 100°F (38°C) chickens will drink about twice as much water as at 70°F (21.1°C). Therefore, they would be consuming twice as much of the medicament; vice versa when temperatures drop.

Feed consumption. When house temperatures increase less feed is consumed, and the daily medicament consumption decreases; vice versa when temperatures decrease.

Remedies for correcting medicament consumed. The following will remedy the situation:

1. *House temperatures increase.* Decrease the medicament in the water or increase it in the feed, or both.
2. *House temperatures decrease.* Increase the medicament in the water or decrease it in the feed, or both.

Water consumption when temperature changes.   Refer to the following tables:

Standard Leghorn pullets, 1–14 days, Table 13-3.
Standard Leghorn pullets, 1–22 weeks, Table 14-3.
Standard Leghorn layers in cages, Tables 16-6 and 16-7.
Broilers, Table 20-9.

Feed consumption when temperature changes.   Refer to the following tables:

Standard Leghorn pullets, 1–22 weeks, Table 14-3.
Standard Leghorn layers in cages, Table 16-16.
Broilers, Tables 20-9 and 20-10.

Water evaporation from waterers.   Although highly variable because of type of waterer and humidity of the air, evaporation from jug-type waterers is about 7 to 10% of the bird's water consumption the first day, with daily decreases of about 0.5% each day thereafter until it reaches 3 or 4% where it remains during the brooding period. The evaporation increases the potency of any medicament in the water by the same percentage. Percentage evaporation of water is not significant when the birds are older and consume more water.

39-F. OTHER CONTROL PROGRAMS

Coccidiosis Control

The control of coccidiosis is a never-ending program when raising young chicks on litter. Either a coccidiostat must be used in the feed or water or a coccidiosis-vaccination program must be followed. Complete suppression of the organism is desirable when feeding broilers; but when the chicks are to be used for laying or breeding purposes, the coccidiosis-control program must be one that establishes immunity in the birds. These are explained fully in Chapter 37-R.

Parasite, Lice, Mite, and Rodent Control

There must be a planned program for the control of all in this category. Each has the ability to cause increasing damage to the flock, and when there is neglect, the birds cease to function at full capacity and feed efficiency is decreased.

Dead Bird and Debris Disposal

Dead birds and debris from the poultry farm and hatchery should not be allowed to accumulate. This refuse must be disposed of daily and in a manner that prevents transmission of disease. There are two methods.

1. *Incinerate.* Special burners are on the market which will consume large quantities of both wet or dry material. Most operate with either gas or oil. A blast-type furnace soon incinerates all the material. Check with local authorities for equipment specifications.
2. *Use plastic bags.* Bags of plastic or similar material may be used where incinerators are impractical. All material is to be placed in the bags, sealed, and removed from the premises.

39-G. VACCINATION SCHEDULE

There are two things involved in a vaccinating schedule.

1. Schedule of the vaccination, including
   a. Diseases for which vaccination is required
   b. Types of vaccines to be used
   c. Age of the birds when vaccinated
   d. Date of each vaccination

2. Report of the vaccination, including
   a. Dates the vaccinations were given
   b. Manufacturer and serial number of the vaccine
   c. Name of the person doing the vaccinating

Such a schedule is necessary in order that the exact dates of vaccinations may be set up before the chicks arrive on the premises. The more flocks and ages of birds on the farm the more important the schedule becomes.

The second part of the schedule is a report of the vaccination procedure. This is a record showing the actual dates of vaccination, the vaccine used, and name of the vaccinator. This report is of great importance if difficulties arise following vaccination. A suggested form for such a schedule and report is given in Figure 39-1.

Flock No. _____ Breed _____ No. Males _____ No. Females _____ Date Started _____

Age To Be Vaccinated	Date To Be Vaccinated	Date Actually Vaccinated	Person Who Vaccinated	Disease	Type of Virus	Method of Vaccinating	Vaccine Manufacturer	Vaccine Serial Number	Remarks

Figure 39-1. Vaccination schedule and report.

Flock No. _____ Breed _____ No. Males _____ No. Females _____ Date Started _____

Medication									
Date from	Date to	Age of Birds	Type of Medication	Medication Dosage	Drug Manufacturer	How Drug was Administered	Feed Manufacturer if Drug in Feed	Disease Diagnosed as	Response to Treatment

Figure 39-2. Medication report.

39-H. MEDICATION REPORT

Another important record is the Medication Report. This is to include all medicaments given to the flock, along with other pertinent data. An example of such a report is found in Figure 39-2.

39-I. DISEASE DIAGNOSTIC REPORT

On many occasions sick chickens will be taken to a laboratory for a diagnosis of the presence and identification of any diseases. A history of the flock will be of great help to the laboratory technician in making the diagnosis.

Care should also be taken to select for diagnosis birds that are representative of the sick birds in the flock. Often more than one disease will be affecting the flock. One disease could be producing high morbidity; another could be causing high mortality. Usually six to eight birds will be adequate. Include some sick, fresh-dead, and healthy birds.

Figure 39-3 shows an example of an outline that should be completed and taken to the laboratory with the birds. It gives the past and present history of the flock.

39-J. SUBMITTING BLOOD AND SERUM SAMPLES

Many times chicken blood or blood serum must be submitted to the laboratory for making tests. Certain techniques are necessary to draw the blood correctly, and to submit the sample without deterioration (see Figure 39-4).

Although not necessary, a plastic, disposable syringe is usually used. This not only provides a method for collecting the blood, but the tube of the syringe also acts as a container for the sample. The main points for using the syringe are:

1. Use a new syringe and needle for each bird.
2. Blood is drawn from the underside of the wing near the ''elbow'' or by heart puncture. Draw approximately 10 cm^3 of blood, filling the syringe.
3. Remove the needle, separate the syringe, and seal the opening (where the needle was attached) with a hot iron or cauterizing blade of an electric beak-trimming machine.
4. Set the tubes in an upright position.
5. Get the samples to the laboratory as soon as possible.
6. Do not allow samples to freeze or heat.

FLOCK HISTORY

Submitted by _____ Date _____

Flockowner _____ Serviceman _____

No. birds: On farm _____ In flock _____ Age _____

No. birds submitted _____ % of flock showing symptoms _____

Symptoms Notice (Check):

Coughing _____ Trembling _____

Sneezing _____ Diarrhea _____

Swollen heads_____ Lame _____

Eye discharge_____ Swollen hocks_____

Paralysis _____ Dark combs _____

Additional description of the disease:_____

_____

_____

Mortality pattern (Number of Birds)

Week Beginning	Sun	Mon	Tues	Wed	Thurs	Fri	Sat

Previous medication: _____

_____

_____

(Circle the day in above chart when medication was given)

Report of egg production: _____

_____

_____

Figure 39-3. Flock history for laboratory diagnosis.

39-K. SUBMITTING TISSUES FOR HISTOLOGIC EXAMINATION

On occasion it becomes necessary to submit body tissues to a laboratory for examination. Not only must these tissues be removed from the bird in a prescribed manner, but they must be placed in a preservative to prevent deterioration before the histologic examination is started.

How to take the tissues from the bird

> *Brain.* Remove the entire brain. Do not mutilate it.
> *Heart.* Extract the entire heart.

Figure 39-4. Drawing a blood sample from a caged laying hen.

Liver. Take several sections about ¼ in. (0.64 cm) thick, cutting with a sharp scalpel.

Other organs except the liver. Remove the entire organ from young chicks. In older birds dissect a section approximately ¼ in. (0.64 cm) thick with a sharp scalpel.

How to preserve the tissues. Fixation of the tissues is accomplished by placing them in the following solution:

10% Formalin (37 to 40% formaldehyde)
90% Distilled water

Use about 10 times as much of the above solution as the size of the tissue. Keep each tissue section in a separate vial. Leave in the solution for 48 hours, then pour off most of the solution, leaving only enough to keep the tissue moist. Seal the container, and transfer to the laboratory. Include a complete flock history when submitting samples.

39-L. SUBMITTING CHICK DOWN SAMPLES
FOR EXAMINATION

Chick down in the hatcher often shows a higher degree of pathologic contamination than that of other areas of the hatchery building or incubators. Therefore, if it is necessary to examine for disease-producing organisms, the hatcher is the first place to look.

How to collect down samples. Samples should be collected in the hatcher when the chicks are hatching or as soon as they are removed. Do not disturb the piles of down. With a sterile wooden tongue depressor, collect about one tablespoon of down and place it in a new clean paper bag, and identify the sample. Take the bags to a competent pathologist for the necessary laboratory examination.

39-M. EVERYONE SHOULD UNDERSTAND
THE PROGRAM

Many people become involved with poultry production today: the poultryman, the serviceman, the integrator, and others. One of these will have to initiate the program: what is to be done each day, what vaccines are to be administered and when, what medications will be necessary, what are the quality standards, and what is to be done when trouble begins. Confusion arises because some people will not know what the program is, why it is done, what to do, or who is to do it. There must be complete understanding among the parties.

Do you have an identifiable program? Has it been written in detail and in such a manner that no questions will arise? Has the program been discussed with those who will carry it out? Has the program been reviewed periodically? Has it been brought up to date? Are there changes? Why?

To neglect a definite written program is to ask for trouble. Prevention is much better than treatment. But prevention comes only through a complete understanding of the things that must be done to ward off trouble, and to know what to do when there are indications that things are not ''just right.''

39-N. FLOCK HEALTH MONITORING

The current causes of mortality and suboptimal flock performance must be routinely examined. Examination requires a systematic approach to determine the causes of death and a routine immune status evaluation. A representative sample of dead birds from each flock should be examined at least weekly. Serologic tests should be performed on each flock just prior to housing, every two or three months throughout their laying cycle, and just prior to molting. The objective of a vaccination program must be to stimulate a high level of disease resistance to prevalent diseases in a high percentage of the individual birds in each flock.

Accurate records of vaccinations, treatments, flock performance, and disease symptoms are essential to facilitate meaningful diagnoses. Better health maintenance programs can only be built on the correct evaluation of existing programs.

39-O. FLOCK HEALTH SECURITY

Endemic and epidemic diseases cost the poultry industry millions of dollars annually in preventative costs and the losses due to poor performance and mortality. Since 1970, the U.S. poultry industry has been plagued with at least three major epidemics that have involved millions of birds and hundreds of flocks in different regions of the country. Velogenic viscerotropic Newcastle disease was discovered in California flocks in 1971, Avian influenza appeared in chicken and turkey flocks in the mid-Atlantic area in 1983, and *Salmonella enteritidis* was isolated in the mid-Atlantic and New England regions in 1986.

The lack of adequate farm security was a common problem in each of the above disease epidemics. Numerous examples of disease transmission from farm to farm were observed. In practically all cases, the disease was being carried by personnel, birds, and equipment.

Disease security requires that farms be capable of disinfection once they become contaminated with a particular organism. It requires that healthy birds be biologically secure from infected birds by separation. And it requires that all traffic to the farm be minimized and under the control of management. Single-age flocks with complete depopulation at the end of each production cycle, complete cleanup including disinfection, and thorough control of all traffic to the farm are the basic elements of a sound security management program.

39-P. ANIMAL WELFARE

Through the years better technological information has enabled the raiser of animals, including chickens, to use production programs that have intensified the manner in which they are bred, held, raised, fed, and conditioned. There is no doubt that these new procedures have enabled producers to reduce labor, feed, and production costs to bring down the expense of producing poultry meat and eggs, thereby reducing the price the customer pays for these products.

But there are those people who are interested in animal welfare and are asking that in spite of more economical production, what has it done to chickens? Are they being subjected to excessive stress and physical discomfort?

"Welfare" is a wide term that embraces both the physical and mental well-being of the chicken. Unfortunately, there is practically no scientific method of measuring an animal's well-being. Some scientists reason that the total of all reproductive responses may be our best means of assessing an animal's "satisfaction" of the care it is receiving.

The interest in animal welfare has taken on vast proportions in the last few years, first in some European countries, then Great Britain, and now in the United States and other countries. At issue are a long list of practices that are interpreted by the "welfarists" as detrimental to the bird. These include cages, beak trimming, recycling (induced molting), transportation and handling practices, and slaughtering methods. Many countries now have extensive management guidelines to describe the conditions in which animals may be kept. In some countries, legislatures have gone so far as to outlaw the cage method of housing chickens or to require unrealistic space standards.

Compared with the barnyard methods of raising chickens years ago the industry has come a long way. Chickens are better fed (probably no other bird or animal—including human beings—has a daily diet so well balanced); they are better housed and protected from storms and adverse weather, often in environmentally controlled buildings; there is better control of disease and parasites; problems with predators have practically been eliminated; artificial lighting programs have taken away nature's variability in the length of the light day; genetics has improved the bird's ability to resist stress.

Management systems in use by the poultry industry are not the result of arbitrary decisions. They are the result of carefully evaluated research by noted scientific institutions. Cages are not equivalent to human housing and should not be compared as such. Cages were developed to enable the chicken to live in a clean, disease-free environment. Recycling results in a rejuvenated flock that will live a longer productive life. Beak trimming prevents chickens pecking each other to death, thereby increasing the longevity of the flock. Misinterpretation of these systems gives rise to unfounded claims.

The best example of an "animal welfarist," though, is the farmer that recognizes the benefits he receives from giving his flocks the best care available. No one knows better than he what results from good husbandry practices. The welfare of his flock is his most important concern.

40
Waste Management

Wastes associated with poultry farming have an increased significance in this day when we become aware of the effects of polluting man's environment. New words and terms have come into use to explain the problem and offer means of control. Some of these are as follows:

aerobic bacteria: Bacteria that require free oxygen (in air or water) for their metabolic processes.

aerobic digestion: Decomposition of organic material by aerobic bacteria, thereby decreasing the energy level of the material while producing little if any odor.

anaerobic bacteria: Bacteria that do not require free oxygen for metabolism and in most instances do not thrive when oxygen is present.

anaerobic digestion: Reduction in the energy level of organic material by anaerobic bacteria, usually accompanied by odors.

biological oxygen demand (BOD): A measure of the amount of dissolved oxygen absorbed by the material in question, expressed in milligrams per liter, thus indicating the amount of biochemically oxidizable organic matter.

biological stabilization: The tendency of organic matter to putrefy due to activity of bacteria.

composting: The biological stabilization of organic solids, usually through aerobic bacteria.

effluent: The discharge of liquid waste material.

faculative bacteria: Bacteria that can live and reproduce under aerobic or anaerobic conditions.

lagoon: A shallow body of water containing waste materials to be subjected to bacterial action.

oxidation ditch: A long continuous circular ditch or trough
through which liquid organic material is circulated and
aerated to provide aerobic action.

40-A. IMPORTANCE OF POULTRY FARM POLLUTION

All poultry farms have a problem with pollution as we define the term
today. Pressures will be made on farm owners to reduce their pollutants
more and more each year.

What constitutes pollution.   Most poultry farm pollution comes under
the following headings:

1. Poultry manure
2. Odors
3. Noise
4. Feathers
5. Air (dust and chemicals)
6. Water runoff
7. Insects
8. Dead birds
9. Hatchery debris
10. Dust from feed manufacturing plants
11. Processing plant wastes
12. Exhaust from internal combustion engines
13. Unsightliness
14. Toxic chemical residues in tissues and eggs
15. Lights

40-B. VALUE OF POULTRY MANURE

Although poultry manure is a waste product, it is one that has considerable value as a fertilizer and feed nutrient.

For Fertilizing the Soil

In the past, tons of poultry manure have been applied to the soil either
when birds were allowed to range over the land, or the material was
spread. It was the accepted method of "getting rid" of it.

Successful marketing programs to utilize poultry manure must include
a quality control program to assure that the product has consistent chemical composition, is easy to apply, and is free of problems. If these condi-

tions are met, manure disposal would result in an income rather than a loss.

Undoubtedly, poultry manure is a valuable soil fertilizer as Table 40-1 shows, but the day may come when soil application will not be allowed; disposal by other means will be mandatory.

Table 40-1. Average Plant Nutrients per Ton (2,000 lb) of Cage Layer Poultry Manure

Manure Condition	Moisture %	Nitrogen		Phosphorus		Potash	
		lb N	%	lb P_2O_5	%	lb K_2O	%
Wet, sticky	75	25	1.3	23	1.1	12	0.6
Moist, crumbly	50	40	2.0	46	2.3	24	1.2
Crumbly, no dust	25	60	3.0	66	3.3	36	1.8
Dry, dusty	15	70	3.5	70	3.5	46	2.3
Completely dry	0	80	4.0	90	4.5	56	2.8

Source: W. Va. Univ. Ext. B.496T and Penn. State Univ. Leaflet 255.

As Poultry and Animal Feed

The fact that poultry manure contains many feed components that pass the digestive tract without digestion and numerous byproducts from metabolism suggests that it would have nutritional value if recycled through the chicken and other birds, and even domesticated farm animals.

Manure is dried. The manure is dried to reduce its moisture content and to make storage possible.

Chemical analysis. Although analyses of dried manure vary with the age of the bird, the age of the sample, the condition of storage and handling, and the type of chicken involved, an average would be represented by the following figures:

Ash	26%
Crude fiber	10%
Crude protein	33.5%
N-free extract	22.5%

Many experimental stations have worked on the nutritional value of dried poultry manure. Because of its relatively high crude protein values, dried poultry waste has become an alternative feedstuff for ruminants. Experiments with monogastric animals, including chickens, have generally proven marginal because of the relatively low percentage of true protein and its high ash content.

40-C. VALUE OF HATCHERY WASTE

Dehydrated hatchery by-products (waste) support satisfactory performance when fed to chickens, particularly laying hens, because of its high calcium content. It has a protein percentage of between 22 and 32 (see Chapter 9-N).

> *Caution:* The use of dried poultry waste and dried poultry manure has not been cleared for use in all poultry feeds in the United States. Check with authorities before using these in poultry feeds.

40-D. MANURE DISPOSAL

The manure load from large poultry farms creates a problem of magnitude. In many instances, small farms can be well taken care of, but when thousands of birds are on the premises, manure disposal systems must dissipate tremendous volumes. Some data are given in Table 40-2 to show the production per 1,000 birds.

The manure production figures in Table 40-2 are based on the daily production of 0.25 lb (113 g) of manure per hen. As manure accumulates, depending on drying conditions, these weights and volumes will be lessened as moisture is lost and as the characteristics of the manure change. While 1.8 cubic feet of manure is produced for each hen during the year, natural drying and decomposition will reduce this to less than 1 cubic foot. Of the 0.25 lb produced per day, only 0.05 lb is dry matter.

There are several methods of disposing of poultry waste, and many innovations of each method.

Table 40-2. Manure Production of a 4-lb (1.8-kg) Bird[1]

	Per Bird lb	Per 1,000 Birds	
Time Period		Weight	Volume
Daily	0.25	250.0 lb	4.9 ft^3
Weekly	1.75	0.88 ton	34.0 ft^3
Monthly	7.6	3.8 ton	5.5 yd^3
Yearly	91.3	45.6 ton	66.0 yd^3

Note: Ton = 2,000 lb.
[1]Fresh daily basis.

Spreading

In many areas it is still practical to dispose of poultry manure by spreading it on cropland or grassland. But in many instances the amount of available land is not great enough. About 4 tons (8,000 lb) (3.6 metric ton) of fresh manure may be spread on an acre (0.4 hectare) of land devoted to the production of corn. Thus, 100,000 hens would require about 1,140 acres (450 hectares) of available land. To add to the problem is the fact that manure may be spread on the land only during certain periods of the year. Until spreading is practical, manure must be stored, and storage may produce obnoxious odors.

Obviously, wet manure is expensive to handle and has less value, on a weight basis, to the user. Table 40-1 illustrates the importance of drying to reduce manure weight and to enhance its value.

Dehydration

Some poultry producers use artificial dehydration to produce a higher-quality product, to reduce the volume of manure, and to prevent bacterial action that results in odor production. There are several types of dehydrators on the market, the temperature created in these varying from 700 to 1800°F (371 to 982°C). The length of the drying time is governed by the drying temperature, the moisture content of the incoming manure, the rate of flow, and the moisture content of the finished product. Most dehydrators will reduce the moisture content of manure from 70 to 10% in less than 10 minutes. The capacity of any dehydrator is rated by the number of pounds of moisture it will remove in 1 hour.

Natural drying (solar) is utilized in parts of the country where rainfall levels are low and drying conditions are suitable. In California, cage manure is commonly collected on a daily basis, spread thinly in a drying yard and windrowed into piles when the drying process is complete. It is common to take 75% moisture to less than 20% in one or two days.

New cage systems are being used that have manure belts between the different decks. Air is channeled to the droppings area to facilitate drying on the belts.

Composting

Collecting poultry manure in pits under cages or slat or wire floors is gaining favor as a practical and economical way to handle poultry waste. The manure may be allowed to accumulate for several years through the process of composting. Aerobic bacterial action is involved. Many composting pits have been in operation for several years without manure removal. After 6 years of use, the debris in the pit should be about 2 or 4 feet deep, depending on the number of hens above it. The top foot is composed of fresh manure, the bottom foot is in an anaerobic condition, and the central portion is undergoing composting (see Chapter 11-S).

The essential requirement in managing the deep pit is that the fresh, wet material be adequately aerated to remove the moisture. To further the composting process and to prevent odors, the pit must be so tight that seepage water cannot enter. Care must be taken to prevent waterers from leaking or overflowing into the pit, for such overflow prevents proper bacterial action in the accumulated wet manure. When the procedure operates correctly, there is little or no odor arising from the pits, and manure removal may be delayed for years. There is practically no problem with flies.

Multiple-deck cage systems commonly employ scraping devices or belts to remove manure on a daily basis. Some egg producers have combined these removal systems with a manure storage barn that incorporates hot air circulation systems and stirring to compost the manure.

Bressler system. Dr. G. O. Bressler has designed a two-stage drying system commonly known as the Bressler system. The stages involved are

1. Stirring with rakes, and drying the manure in a pit with high-velocity fans to reduce the moisture content to 35 to 40%.
2. Dehydrating the partially dried material so that it has a moisture content of 10 to 12%.

Lagoons

Fresh poultry manure may be flushed into an open shallow pond known as a lagoon. Bacterial action reduces the waste material to a smaller volume. As bacterial growth occurs only during the warm months, the use of lagoons is more common in warmer climate areas. When aerobic action takes place, the lagoon produces very little odor; but as the sludge builds up, anaerobic activity may take place and odors may be pronounced. Modern installations commonly recycle the water through the poultry houses on a daily basis.

Liquid Manure Holding Tank

Many farms use the liquid holding tank method of handling their manure. This system incorporates an underground tank for temporary storage and tanker trucks or tank trailers for delivery of the liquid manure to neighboring farms. Long-distance transportation is uneconomical because of the large amount of water added to the manure.

40-E. OTHER POLLUTION PROBLEMS

Most poultry farms are plagued with other pollution problems and a management program must be worked out to care for these. Dead birds are best incinerated, but care should be taken that little smoke and no obnoxious odors arise from the burning process. Many communities, though, do not allow this method of dead bird disposal. Hatchery debris may be handled similarly; there are numerous commercial burners on the market that will handle this type of material. Processing plant waste consisting of poultry offal and poultry feathers is usually dried as two separate products to be used for poultry or animal feed. Dust and chemicals in the air, not only from burning but from the chicken house, create a health hazard at times. Odors, noise, insects, and unsightliness create pollution factors of one type or another. Certain chemicals and drugs when fed to chickens may leave residues in poultry meat and eggs, and these are under close scrutiny by health experts.

Not long ago the subjects discussed in this chapter had little importance in poultry production. But times have changed, and the commercial poultryman will be held responsible for more and more factors that tend to produce an undesirable environment for human beings.

Appendix A
Tables of Equivalents

A-1. WEIGHT OF WATER

1 cubic foot of water weighs	62.4 lb
1 U.S. gallon of water weighs	8.33 lb
1 Imperial gallon of water weighs	10.00 lb
1 Imperial gallon of water weighs	4.54 kilos

A-2. LIQUID MEASURE

1 teaspoon	5 cc
3 teaspoons	1 tablespoon
1 tablespoon	15 cc
2 tablespoons	1 ounce
8 ounces	1 cup
2 cups	1 pint (U.S.)
16 fluid ounces	1 pint (U.S.)
128 fluid ounces	1 gallon (U.S.)

A-3. LINEAR, AREA, WEIGHT, AND VOLUME EQUIVALENTS

Table 41-1.   Linear Measure Equivalents

	Inches	Feet	Yards	Miles	Centimeters	Meters	Kilometers
1 Inch	1	0.0833	0.0278	0.000016	2.54	0.0254	—
1 Foot	12	1	0.3333	0.00019	30.48	0.3048	0.00031
1 Yard	36	3	1	0.00057	91.44	0.9144	0.0009
1 Mile	63,360	5,280	1,760	1	160,934	1,609.3	1.609
1 Centimeter	0.3937	0.0328	0.0109	—	1	0.01	0.00001
1 Meter	39.37	3.2808	1.0936	0.00062	100	1	0.001
1 Kilometer	39,370	3,280.8	1,093.6	0.621	100,000	1,000	1

Table 41-2.   Area Equivalents

	Inches²	Feet²	Yards²	Acres	Miles²	cm²	Meters²	Hectares
1 Sq Inch	1	0.0069	0.00077	—	—	6.4516	0.00065	—
1 Sq Foot	144	1	0.1111	—	—	929.03	0.0929	—
1 Sq Yard	1,296	9	1	—	—	8,361.3	0.8361	—
1 Acre	—	43,560	4,840	1	0.00156	—	4,046.9	0.4047
1 Sq Mile	—	—	3,097,600	640	1	—	2,589,988	259
1 Sq Centimeter	0.155	0.00108	.00012	—	—	1	0.0001	—
1 Sq Meter	1,550	10.764	1.196	0.00025	—	10,000	1	0.0001
1 Hectare	—	107,639	11,959.9	2.471	0.00386	—	10,000	1

Table 41-3.   Weight Equivalents

	Ounces	Pounds	Short Tons	Metric Tons	Long Tons	Grams	Kilos
1 Ounce	1	0.0625	—	—	—	28.349	0.0284
1 Pound	16	1	0.0005	0.00045	0.00045	453.592	0.4536
1 Short ton	32,000	2,000.0	1	0.9072	0.8929	907,184	907.2
1 Metric ton	35,274	2,204.6	1.1023	1	0.9842	1,000,000	1,000
1 Long ton	35,840	2,240.0	1.12	1.016	1	1,016,046	1,016
1 Gram	0.035	0.0022	—	—	—	1	0.001
1 Kilo	35.274	2.2046	0.0011	0.001	0.00098	1,000	1

Table 41-4.   Liquid Volume Equivalents

	Fluid Ounces (U.S.)	Pints (U.S.)	Quarts (U.S.)	Quarts (Imp)	Gallons (U.S.)	Gallons (Imp)	Cubic Inches	Cubic Feet	Liters	cc
1 Fluid Ounce (U.S.)	1	0.0625	0.0313	0.0261	0.0078	0.0065	1.805	0.001	0.0296	29.6
1 Pint (U.S.)	16	1	0.5	0.4163	0.125	0.104	28.875	0.0167	0.4732	473.2
1 Quart (U.S.)	32	2	1	0.8326	0.25	0.208	57.75	0.0334	0.9463	946.3
1 Quart (Imp)	38.43	2.4	1.201	1	0.3002	0.25	69.354	0.0413	1.1365	1,136.5
1 Gallon (U.S.)	128	8	4	3.33	1	0.833	231	0.1337	3.785	3,785
1 Gallon (Imp)	153.7	9.608	4.804	4	1.201	1	277.42	0.1605	4.546	4,546
1 Cubic Inch	0.554	0.0346	0.0173	0.0144	0.0043	0.0036	1	0.00058	0.0164	16.387
1 Cubic Foot	957.51	59.844	29.922	24.883	7.481	6.221	1,728	1	28.316	28,316
1 Liter	33.815	2.1134	1.0567	0.880	0.2642	0.220	61.026	0.0353	1	1,000
1 Cubic Centimeter	0.0338	—	—	—	—	—	0.061	—	0.001	1

Table 41-5.   Dry Volume Equivalents

	Dry Pints	Dry Quarts	U.S. Bushels	In.³	Ft³	Yards³	Liters
1 Dry Pint	1	0.5	0.0156	33.6	0.0195	0.00072	0.5506
1 Dry Quart	2	1	0.0313	67.2	0.0389	0.0014	1.1012
1 Bushel U.S.	64	32	1	2,150.4	1.2445	0.0461	35.238
1 Cu Inch	0.0298	0.0149	—	1	0.00058	—	0.0164
1 Cu Foot	51.428	25.714	0.8036	1,728	1	0.037	28.3161
1 Cu Yard	1,388.56	694.28	21.697	46,656	27	1	764.53
1 Liter	1.816	0.908	0.0284	61.026	0.0353	0.0013	1

Appendix B
Selected Sources of Poultry Information

Books

Curtis, S. E. 1983. *Environmental Management in Animal Agriculture.* Iowa State University Press, Ames, IA.

Ensminger, M. W. 1980. *Poultry Science.* The Interstate Printers and Publishers, Danville, IL.

Gordon, R. F., and F. T. Jordan. 1982. *Poultry Diseases.* Bailliere Tindall, London.

Hofstad, M. S. 1984. *Diseases of Poultry.* Iowa State University Press, Ames, IA.

Moreng, R. W., and J. S. Avens, 1985. *Poultry Science and Production.* Reston Publishing Co., Reston, VA.

Mountney, G. J. 1981. *Poultry Products Technology.* Van Nostrand Reinhold Co., New York, NY.

National Research Council. 1984. *Nutrient Requirements of Poultry.* National Academy Press, Washington, D.C.

Nesheim, M., R. E. Austic, and L. E. Card. 1979. *Poultry Production.* Lea and Febiger, Philadelphia, PA.

Parkhurst, C. R., and G. J. Mountney. 1988. *Poultry Meat and Egg Production.* Van Nostrand Reinhold Co., New York, NY.

Patrick, H., and P. J. Schaible. 1980. *Poultry: Feeds and Nutrition.* Van Nostrand Reinhold Co., New York, NY.

Peterson, E. H. 1975. *Serviceman's Poultry Health Handbook.* Better Poultry Health Co., Fayetteville, AR.

Scott, M. L., M. C. Nesheim, and R. J. Young. 1982. *Nutrition of the Chicken.* M. L. Scott and Associates, Ithaca, NY.

Stadelman, W. J., and O. J. Cottrell. 1977. *Egg Science and Technology.* Van Nostrand Reinhold Co., New York, NY.

Sturkie, P. C. 1986. *Avian Physiology.* Springer-Verlag, Inc., New York, NY.

Swartz, L. D. 1980. *Poultry Health Handbook.* College of Agriculture, Penn. State University, University Park, PA.

Whiteman, C. E., and A. A. Bickford. 1979. *Avian Disease Manual.* Colorado State University, Fort Collins, CO.

Trade Journals

American Poultry, 521 East 63rd St., Kansas City, MO 64110.

Avicultura Profesional, P.O. Box 84, Athens, GA 30603-0084 (in Spanish).

Broiler Industry, Watt Publishing Co., Mount Morris, IL 61504.

Canada Poultryman, Farm Papers Ltd, Suite 105A, 9547 152nd St., Surrey, B.C. Canada, V3R 5Y5.

Egg Industry, Watt Publishing Co., Mount Morris, IL 61054.

Feed Bag Magazine, 152 W. Wisconsin Ave., Milwaukee, WI 53203.

Feed Industry, Communications Marketing, 5100 Edina Industrial Blvd., Edina, MN 55435.

Feedstuffs, P.O. Box 2400, Minnetonka, MN 55343.

Industria Avicola, Watt Publishing Co., Mount Morris, IL 61054 (in Spanish).

Misset International Poultry, P.O. Box 4, 7000 BA Doetinchem, The Netherlands.

Poultry and Egg Marketing, P.O. Box 1338, Gainesville, GA 30503.

Poultry Digest, Watt Publishing Co., Mount Morris, IL 61054.

Poultry International, Watt Publishing Co., Mount Morris, IL 61054.

Poultry Times, P.O. Box 1338, Gainesville, GA 30503.

The Virginia Poultryman, Virginia Poultry Federation, P.O. Box 552, Harrisonburg, VA 22801.

World Poultry, Reed Business Publishing, Carew House, Wallington, Surrey, SM6 ODX, England.

Scientific Journals

Avian Diseases, University of Massachusetts, Amherst, MA 01002.

British Poultry Science, Oliver and Boyd Ltd., Tweeddale Ct.; 14 High St., Edinburgh, Scotland.

Japanese Poultry Science, Japan Publications Trading Co., P.O. Box 469, Rutland, VT 05701.

Journal of Heredity, American Genetic Assn., 1507 Main St., N.W., Washington, DC 20005.

Journal of Nutrition, Wistar Institute Press, 3631 Spruce St., Philadelphia, PA 19104.

Poultry Science, Poultry Science Assn., 309 W. Clark St., Champaign, IL 61820.

World's Poultry Science, World's Poultry Science Assn., Agricultural House, Knightsbridge, London, S.W.1, England.

Index